Signal Processing Technology and Applications

Signal Processing Technology and Applications

Dr. John G. Ackenhusen

EDITOR

IEEE Technical Activities Board

A Selected Reprint Volume under the sponsorship of the Products Council of the IEEE Technical Activities Board

The Institute of Electrical and Electronics Engineers, Inc.
New York City, New York

Library of Congress Cataloging-in-Publication Data

Signal processing technology and applications / John G. Ackenhusen,
  editor.
       p.   cm. -- (IEEE technology update series)
   Includes bibliographical references and indexes.
   ISBN 0-7803-2469-2
   1. Signal processing.   I. Ackenhusen, John G., 1953-
II. Series.
TK5102.9.S547 1995
621.382'2--dc20   94-46444
                                                      CIP

IEEE Technical Activities Board
445 Hoes Lane
Piscataway, New Jersey 08855-1331

Dr. John G. Ackenhusen
Editor-in-Chief

Associate Editors

Elias Manolakos   Robert M. Owens   Earl E. Swartzlander, Jr.

Robert T. Wangemann
Executive Editor

Tania Skrinnikov, Managing Editor
Harry Strickholm, Technical Editor
Jayne Cerone, Administrative Editor
Lois Pannella, Administrative Editor
Patricia Thompson, Administrative Editor
Mark A. Vasquez-Jorge, Administrative Editor
Ann H. Burgmeyer, Production Editor
Susan K. Tatiner, Production Editor

FOREWORD

Electrotechnology is constantly changing as new fields are discovered and established fields evolve because of new and emerging theories and applications. A leading example of a field that is evolving is signal processing. Due to ongoing changes to the theory and applications of this field, signal processing was chosen as the topic for the third volume in the IEEE Technology Update Series. This Series was developed, by the IEEE Technical Activities Board (TAB) and the TAB Book Broker Committee, to furnish readers with up-to-date research and practical applications information in specific fields of interest.

I would like to recognize the dedicated efforts of John Ackenhusen, Editor-in-Chief of *Signal Processing Technology and Applications,* who brought together the talents of three editors and more than 200 authors. In addition to serving as Editor-in-Chief of this volume, Dr. Ackenhusen is former President of the IEEE Signal Processing Society and was elected to the IEEE Board of Directors for 1994-95, where he leads its Signals and Applications Division. My thanks to him and his colleagues for their wisdom and achievement.

I would also like to publicly thank Donald M. Bolle, IEEE Vice President Technical Activities, and Dr. Jan Brown, Chair of the TAB Book Broker Committee, for their insight and guidance with this book.

Finally, my thanks to the Technical Activities Department staff who compiled the information and administered the production of this book. These staff include Tania Skrinnikov, Harry Strickholm, Mark A. Vasquez-Jorge, Lois Pannella, Patricia Thompson and Jayne F. Cerone. Thanks also to production editors Ann H. Burgmeyer and Susan K. Tatiner for their contributions to the book's production.

Robert T. Wangemann
Staff Director - IEEE Technical Activities

CONTENTS

Part I: Technology

Chapter 1: Signal Modeling   3

Chapter 2: Adaptive and Learning Systems   57

Part II: Applications

EDITOR'S INTRODUCTION

This book is for the signal processing engineer who is returning to the study of new forms of signal processing and seeks an update in the field, perhaps after being immersed in a long project. It is also for the specialist in an area within signal processing who may wish to learn about current trends in technology and applications within other areas of signal processing.

Here are contained 76 of the approximately 2400 papers pertaining to signal processing that have appeared across 47 of the conferences sponsored in 1993 by the Institute of Electrical and Electronics Engineers (IEEE). These were selected by a panel of editors based on timeliness of the topic, quality of the technical presentation, and attractiveness of the format. Each editor reviewed every paper included here, so the results reflect the integration of the views of diverse specialists in the field. Because it is based upon conference papers rather than journal publications and has proceeded through a rapid publication process, this compilation is over one year more current upon release than typical books in signal processing. Also included are two excellent tutorial articles from the *IEEE Signal Processing Magazine* to introduce two active areas of signal processing that underlie many of the papers in this book.

The editors chose to emphasize several emerging trends for special consideration in this signal processing technology update. The basis for these choices is described in the section titled "New Directions in Signal Processing." The topics receiving emphasis are:

o Higher-order and non-linear signal processing;

o Wavelets and time-frequency signal analysis;

o Vector quantization, including means to accelerate vector codebook searching;

o Neural networks and their applications and computational architectures;

o Image coding and motion detection and estimation;

o High Definition Television (HDTV);

o Large-vocabulary speech recognition and language modeling;

o Signal processing system design methods;

o Biomedical applications of signal processing.

Other areas were excluded due to space limitations, despite the availability of excellent papers, either because they had been active long enough to allow the introduction of textbooks, or because they are the topic of other books in the IEEE Technology Update Series. For example, fuzzy systems are only slightly discussed here, because the IEEE Technology Update Series book entitled *Fuzzy Logic: Technology and Applications* provides a more complete treatment. An emphasis on comparing several techniques for solving the same problem is made -- 10% of the papers present careful analyses of the relative performance of two or more methods on the same data.

As its title suggests, this book is organized into two major sections. Technology, the domain of the first section, provides the building blocks from which a solution to a signal processing application is drawn. This section is further divided into a portion on algorithms (Chapters 1-4) and a portion on architectures (Chapters 5-7). Algorithms are regarded as theoretically-grounded computational prescriptions that are presented as independent of the structure of hardware or software elements that may be chosen to execute them. Architectures include methods, tools, and hardware or software structures that can be used across a variety of applications to efficiently implement these algorithms.

Within the second major section, Applications, topics are further divided into speech processing, image processing, and other applications. Speech processing emphasizes the problem of large-vocabulary word recognition, relating the state of the art and describing new underlying techniques that promise performance enhancements. Image processing, which extends over three chapters and contains the greatest number of papers, reflects current interest in image coding, image recognition, and system issues. The final section contains papers on applying signal processing to acoustics and biomedical applications.

NEW DIRECTIONS IN SIGNAL PROCESSING

One may regard signal processing as the conversion of large amounts of *data* into concentrated forms of *information*. Several trends have emerged as the practice of signal processing matures -- it is useful to review these as a context for introducing this book.

Models are key to extracting information from data. Some assumption (or model) of the underlying process that produced the information, combined with the identification of a modest number of parameters that govern the model at a particular moment in time, forms the basis of estimation techniques that seek to minimize the error between the actual signal and a signal generated at a particular setting of model parameters. In the past, linear models and Gaussian statistics were the standard tools, as exemplified by linear predictive coding of speech signals and the underlying all-pole filter model of the vocal tract. Linear models continue to form the mainstay of signal processing, but in some areas, their application has been exhaustively explored and new tools are sought. This paves the way for higher-order models, which can provide better performance by accommodating deviations from linearity. For example, the estimate of the power spectrum of a signal has been applied across most aspects of signal processing, including speech, communication, radar, and sonar. The power spectrum, based on the autocorrelation sequence of a signal, provides a complete description for a Gaussian signal. However, processes need not behave in a Gaussian manner, and so more recent thrusts that apply higher-order spectral estimation are emerging, generalizing from the second-order representation of the power spectrum. Similarly, the use of algorithms that are based in non-linear dynamics, including chaotic systems and self-similar fractal models, are becoming tools to aid information extraction.

Another concept pertaining to the extraction of information from data is the effect of *information reduction*. Signal processing algorithms typically begin with a high data rate source as input (e.g., speech signal or image sequence), and then apply successive operations to highlight the information-bearing parts while suppressing the redundant or noisy parts, resulting in a reduction in the rate of information transfer. As the data pass through these stages of reduction, the operations change from regular, repetitive functions applied uniformly to all parts to irregular, data-dependent operations applied to selected portions of the data. Vector quantization reduces the information rate by quantizing a group of samples, or their spectral representation, by replacing it with a nearest neighbor chosen from a pre-computed dictionary and transmitting only the codebook index of the choice. Symbolic processing, through which signal information is represented by symbols, becomes necessary to allow reasoning in the presence of uncertainty and making classification decisions in the face of often-contradictory multiple sources of evidence.

As suggested by the consideration of models and information reduction, a third trend in this area is the emphasis on *replicating certain human cognitive functions*. This is most apparent in the signal processing applications of speech recognition and character recognition, but it is also important in detecting signals in noise, for which models are used that are tuned to human processes. Perceptually-based similarity (or distance) measures are used in coding algorithms to allocate bits in a manner that minimizes distortion as perceived by humans. Several efforts have emerged that seek to provide objective correlates to perceptual criteria, allowing a computational prescription to substitute for a human panel of information evaluators. Models of the human languages, both spoken and written, are important contributors to accurate automatic recognition performance.

The evolution toward human-like functions of signal processing has taken place in the context of substantial *advances in processor technology*. The amount of circuitry that can be placed on a unit of integrated circuit area continues to increase significantly each year, and portable systems that include signal processing functions have become affordable to consumers. The Fast Fourier Transform, which resulted in a simplification in the circuitry required to compute a desired spectral representation within a set time duration, began the trend toward computationally-affordable signal processing. This trend has continued as hardware multiplier circuits have become common in standard microprocessors and as signal processors become available within personal computers. As symbolic processing and large databases become important, the problems of rapid computation of spectral content will be replaced by challenges in rapidly accessing large amounts of memory in random order.

With the growth in ability to perform real-time, on-the-spot signal processing comes a related set of new trends. *Whole-system design* considers not only the algorithm but its ability to keep up with real time data while meeting portability constraints on size, cost, and weight of the system that executes the algorithm. As the ability increases to deploy signal processing systems to solve real-world problems, new problems arise that extend beyond the performance of the algorithm to such total-system issues as the user interface, the coupling of the signal processor to sensors, and the communication of the information extracted from the signal to influence the world around it. For example, the problem of optical character recognition expands from reading handwritten addresses on an envelope to include finding the address block, finding the envelope, and moving the next envelope into the field of view. The concept of "approximate signal processing" [1] becomes important, in which as accurate an answer as possible is provided within the real-time deadline imposed on the processor, with that answer being computed in a manner that will allow successive refinement using later time slots. In like manner, progressive transmission, which provides a low-resolution rendition of an entire image in the first time interval, then refines the resolution as time progresses, accelerates the communication of useful information when compared to techniques that provide higher-resolution partial images at each time slot. Multiresolution techniques, particularly the use of wavelet theory, find application for these purposes. The incremental acquisition of knowledge permeates many of the current topics in signal processing, including neural networks, adaptive language acquisition, and other adaptive systems that do not have the luxury of having all the data in hand before making a decision.

Incremental development concepts also emerge in system design, where a premium is placed on gaining early access to a capability, even if not in final form. A "spiral" development process replaces the traditional waterfall, and thus a system moves through a series of prototypes, many of which are used in trial applications, as the final design is developed. The ever-increasing penetration of simulation, once reliable only for integrated circuit designs operating in isolation, increases the certainty of "what you simulate is what you get" in the board and multi-module system domains. Reusable elements in hardware and software libraries become more capable, and many of the papers here describe complete functional building blocks instead of the smaller component algorithms or computing elements to which earlier libraries had been restricted.

Finally, the emerging *trends in personal lifestyle and technology* influence the signal processing topics treated here. The increasing importance of image information, and the resulting need to move, manipulate, recognize, and access image information with the same facility as has long been possible for text information, place emphasis on encoding images for transmission within available bandwidth and extracting information from images. Medical uses of signal processing, particularly in the image domain, continue to grow. Image manipulation and information enhancement aid medical diagnostics, image transmission becomes a means to provide high-quality, low-cost care to remote areas, and digital techniques begin to replace film-based techniques of medical image acquisition and storage. Within the consumer domain, high definition television and multimedia pace some areas of signal processing, and the ever-increasing proliferation of portable personal computers begins to include signal processing functionality. Signal processing now appears in items that are purchased from personal paychecks as well as company budgets -- automobiles and audio equipment use signal processors, and some home appliances include fuzzy logic and signal processing.

These trends in technology, society, and signal processing determine the topics to which signal processing researchers devote their skills, which in turn define the content of the conference publications from which this book is drawn. These areas of activity become the areas of progress, for which a technology update is needed.

Reference:

[1]A. V. Oppenheim, S. H. Nawab, G. C. Verghese, and G. W. Wornell, "Algorithms for Signal Processing," *Proceedings of the First Annual RASSP Conference* (Arlington, VA, USA) Aug. 1994, pp. 146-153.

SUGGESTED ADDITIONAL READINGS

The Signal Processing Society of the Institute of Electrical and Electronics Engineers (IEEE) showcases much of the world's accomplishments in signal processing technology and applications. It publishes the following journals:

° *IEEE Transactions on Signal Processing*

° *IEEE Transactions on Image Processing*

° *IEEE Transactions on Speech and Audio Processing*

° *IEEE Signal Processing Letters*

In addition, it sponsors the following annual major conferences, with locations around the world and attendance in excess of 1000 participants:

° *IEEE International Conference on Acoustics, Speech, and Signal Processing (ICASSP)*

° *IEEE International Conference on Image Processing (ICIP)*

The IEEE Signal Processing Society consists of several Technical Committees, each of which hold workshops of about 100-150 participants every two years:

° Digital Signal Processing

° Image and Multidimensional Signal Processing

° Neural Networks for Signal Processing

° Speech Processing

° Statistical Signal and Array Processing

° Underwater Acoustics Signal Processing

° Very Large Scale Integration (VLSI) for Signal Processing (Design and Implementation of Signal Processing Systems)

Many of the papers in this book were taken from the Proceedings of these conferences and workshops. Continued attention to the publications and conferences of the IEEE Signal Processing Society provides the reader with a continualupdate on achievements in the field of signal processing. Other IEEE Societies also include some aspects of signal processing within their activities. These include:

° Aerospace and Electronic Systems

° Circuits and Systems

° Communications

° Computers

- ° Engineering in Medicine and Biology

- ° Geoscience and Remote Sensing

- ° Information Theory

- ° Instrumentation and Measurement

- ° Oceanic Engineering

- ° Ultrasonics, Ferroelectrics, and Frequency Control

- ° Vehicular Technology

Each chapter of this book begins with a list of suggested additional readings that pertain to the topic of the chapter. A few publications update the state of the art across the breadth of many chapters, and therefore they are listed here:

[1] K. G. Shin and P. Ramanathan, "Real-Time Computing: A New Discipline of Computer Science and Engineering," *Proc. IEEE*, vol. 82, no. 1, pp. 6-24, Jan. 1994.

[2] J. W. S. Liu, W. K. Shih, K. J. Lin, R. Bettati, and J. Y. Chang, "Imprecise Computations," Proc. IEEE, vol. 82, no. 1, pp. 83-94, Jan. 1994.

[3] R. O. Duda and P. E. Hart, *Pattern Classification and Scene Analysis*. New York: Wiley, 1973.

[4] J. Serra, *Image Analysis and Mathematical Morphology*. London: Academic, 1982.

[5] L. M. Garth and H. V. Poor, "Detection of Non-Gaussian Signals," *Proc. IEEE*, vol. 82, no. 7, pp. 1061-1095, July 1994.

[6] United States Advanced Research Projects Agency, *Proceedings: 1st Annual RASSP Conference* (Arlington, VA, USA) August 1994.

[7] D. H. Johnson and P. N. Shami, "The Signal Processing Information Base," *IEEE Signal Processing Magazine,* vol. 10, no. 4, pp. 36-42, October 1993.

IEEE Technology Update Series

Signal Processing Technology and Applications

Part I: Technology

Chapter 1: Signal Modeling

The detection and analysis of signals is greatly aided by the assumption of an underlying model. Limited signal data can then be focused to provide a more robust estimation of a limited set of model parameters, rather than being required to provide an unreliable estimate of an unconstrained system. The chapter begins with a tutorial article from *Signal Processing Magazine* entitled "Signal Processing with Higher-Order Spectra" (paper 1.1), which distinguishes higher-order techniques from traditional linear approaches. The paper reviews several motivations for applying higher-order spectral analysis -- to suppress colored additive noise from an unknown power spectrum, to reconstruct non-minimum phase signals, and to accommodate deviations from Gaussian or linear behavior.

Despite the power of higher-order signal processing techniques, caution must be exercised in their application. "Pitfalls in Polyspectra" (1.2) sounds the warning that "...the newcomer to the field still has to learn the hard way, from his own mistakes...," and, "...all the pitfalls known from ordinary spectral analysis occur here, too, some of them with new twists..." A comparison of the performance of non-linear signal processing to Kalman filtering provides an example of the advantages that accrue to nonlinear signal processing techniques that can justify their added mathematical complexity. "Nonlinear Signal Processing vs. Kalman Filtering" (1.3) provides this analysis.

In contrast to the paradigm of modeling an irregular signal as if it were produced by a linear system excited by a purely stochastic process, e.g., white noise, one may apply the theory of chaos to model such signals as arising from a deterministic non-linear system. Higher-order spectra are important to such models, as described in "Higher Order Within Chaos" (1.4). The discussion of chaos continues in "Chaotic Signals and Systems for Communications" (1.5), which demonstrates approaches to communications based on synchronized chaotic signals and systems.

Suggested Additional Readings:

[1] L. M. Garth and H. V. Poor, "Detection of Non-Gaussian Signals," *Proc. IEEE,* vol. 82, no. 7, pp. 1061-1095, July 1994.

[2] P. A. Delaney and D. O. Walsh, "A Bibliography of Higher-Order Spectra and Cumulants," *IEEE Signal Processing Magazine,* vol. 11, no. 3, pp. 61-70, July 1994.

[3] L. Marple, T. Brotherton, R. Barton, K. Lugo, and D. Jones, "Travels Through the Time-Frequency Zone: Advanced Processing Techniques," *The Twenty-Seventh Asilomar Conference on Signals, Systems, and Computers* (Pacific Grove, CA, USA) Nov. 1994, pp. 1469-1473.

[4] B. X. Li and S. Haykin, "Chaotic Detection of Small Target in Sea Clutter," *Proc. IEEE International Conference Acoustics, Speech, and Signal Processing 93* (Minneapolis, MN, USA) April 1993, vol. 1 pp. 237-240.

[5] J. R. Boston, "Comparison of Techniques for Modeling Uncertainty in a Signal Detection Task," *Proc. of ISUMA '93 The Second International Symposium on Uncertainty Modeling and Analysis* (College Park, MD, USA) Apr. 1993, pp. 302-309.

[6] G. Andria, E. D'ambrosio, M. Savino, and A. Trotta, "Applications of Wigner-Ville Distribution to Measurements on Transient Signals," *1993 International Symposium on Circuits and Systems* (Chicago, IL, USA) May 1993, pp. 612-617.

[7] D. M. Honea and S. D. Stearns, "Lossless Waveform Compression: A Case Study," *The Twenty-Seventh Asilomar Conference on Signals, Systems, and Computers* (Pacific Grove, CA, USA) Nov. 1994, pp. 1514-1518.

[8] S. Mallat and Z. Zhang, "Adaptive Time-Frequency Transform," *Proc. IEEE International Conference on Acoustics, Speech, and Signal Processing 93* (Minneapolis, MN, USA) April 1993, vol. 3 pp. 241-244.

[9]   N. T. Thao and M. Vetterli, "A Deterministic Analysis of Oversampled A/D Conversion and Sigma-Delta Modulation," *Proc. IEEE International Conference on Acoustics, Speech, and Signal Processing 93* (Minneapolis, MN, USA) April 1993, vol. 3 pp. 468-471.

Signal Processing with Higher-Order Spectra

The estimation of the power spectral density or simply the power spectrum of discrete-time deterministic or stochastic signals has been a useful tool in digital signal processing for more than thirty years. Power spectrum estimation techniques have proved essential to the creation of advanced radar, sonar, communication, speech, biomedical, geophysical, and other data processing systems. The available power spectrum estimation techniques may be considered in a number of separate classes, namely, conventional (or "Fourier type") methods, maximum-likelihood method of Capon with its modifications, maximum-entropy, minimum energy, and minimum-cross-entropy methods, as well as methods based on autoregressive (AR), moving-average (MA) and ARMA models; and harmonic decomposition methods such as Prony, Pisarenko, MUSIC, ESPRIT, and Singular Value Decomposition. Developments in this area have also led to signal modeling, and to extensions to multi-dimensional, multi-channel, and array processing problems. Each one of the aforementioned techniques has certain advantages and limitations not only in terms of estimation performance but also in terms of computational complexity and, therefore, depending on the signal environment, one has to choose the most appropriate [Marple, 1987; Kay, 1988; Haykin, 1983].

In power spectrum estimation, the signal under consideration is processed in such a way that the distribution of power among its frequency components is estimated. As such, phase relations between frequency components are suppressed. The information contained in the power spectrum is essentially that which is present in the autocorrelation sequence; this would suffice for the complete statistical description of a Gaussian signal. However, there are practical situations where we have to look beyond the power spectrum (autocorrelation) of a signal to extract information regarding deviations from Gaussianity and presence of phase relations [Nikias and Raghuveer, 1987; Mendel 1991].

Higher order spectra (also known as polyspectra), defined in terms of higher order statistics ("cumulants") of a signal, do contain such information. Particular cases of higher order spectra are the third-order spectrum also called the bispectrum which is, by definition, the Fourier transform of the third-order statistics, and the trispectrum (fourth-order spectrum) which is the Fourier transform of the fourth-order statistics of a stationary signal. The power spectrum is, in fact, a member of the class of higher order spectra, i.e., it is a second-order spectrum [Rosenblatt, 1985]. Figure 1 illustrates the higher-order spectra classification map of a given discrete-time signal. Although higher-order statistics and spectra of a signal can be defined in terms

Chrysostomos L. Nikias and Jerry M. Mendel

1053-5888/93/$3.00©1993 IEEE

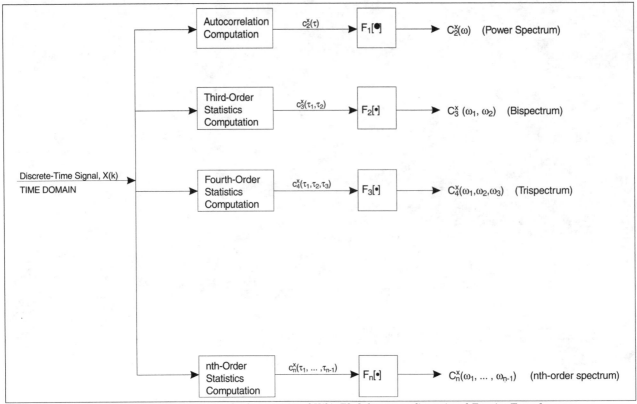

1. The higher-order spectra classification map of a discrete signal X(k). F[•] denotes n-dimensional Fourier Transform.

of moments and cumulants, moment and moment spectra can be very useful in the analysis of deterministic signals (transient and periodic) whereas cumulants and cumulant spectra are of great importance in the analysis of stochastic signals [Nikias and Petropulu, 1993].

There are several general motivations behind the use of higher-order spectra in signal processing. These include techniques to: (1) suppress additive colored Gaussian noise of unknown power spectrum; (2) identify non-minimum phase systems or reconstruct nonminimum phase signals; (3) extract information due to deviations from Gaussianity; and (4) detect and characterize nonlinear properties in signals as well as identify nonlinear systems [Nikias and Raghuveer, 1987].

The first motivation is based on the property that for Gaussian signals only, all cumulant spectra of order greater than two are identically zero. If a non-Gaussian signal is received along with additive Gaussian noise, a transform to a higher-order cumulant domain will eliminate the noise. Hence, in these signal processing settings, there will be certain advantages to detecting or/and estimating signal parameters from cumulant spectra of the observed data. In particular, cumulant spectra can become high signal-to-noise ratio (SNR) domains in which one may perform detection, parameter estimation or even entire signal reconstruction. Figure 2 illustrates time-delay parameter estimation results obtained by a cross-correlation (i.e., 2nd-order statistics) method and a technique based on cross third-order cumulants. The signal of interest is assumed to be non-Gaussian whereas the additive noise is Gaussian and spatially correlated. From Fig. 2, it is apparent that the third-order cumulants do sup-

press the effect of Gaussian noise and thus provide better time-delay estimates (TDE), especially in low SNR [Nikias and Pan, 1988].

The second motivation is based on the fact that polyspectra (cumulant and moment) preserve the true phase character of signals. For modeling time series data in signal processing problems, second-order statistics are almost exclusively used because they are usually the result of least-squares optimization criteria. However, the autocorrelation domain suppresses phase information. An accurate phase reconstruction in the autocorrelation (or power-spectrum) domain can only be achieved if the signal is minimum phase. On the other hand, non-minimum phase signal reconstruction or system identification can be achieved in higher-order spectrum domains due to the ability of polyspectra to preserve both magnitude and non-minimum phase information. Figure 3 illustrates two different signals which have identical autocorrelation but different third-order statistics. Consequently, these two signals have identical power spectra and different bispectra.

The third motivation is based on the observation that most "real world" signals are non-Gaussian and thus have nonezero higher-order spectra. As Fig. 1 demonstrates, a non-Gaussian signal can be decomposed into its higher-order spectral functions where each one of them may contain different information about the signals. This can be very useful in signal classification problems where distinct classification features can be extracted from higher-order spectrum domains.

Finally, introduction of higher-order spectra is quite natural when we try to analyze the nonlinearity of a system

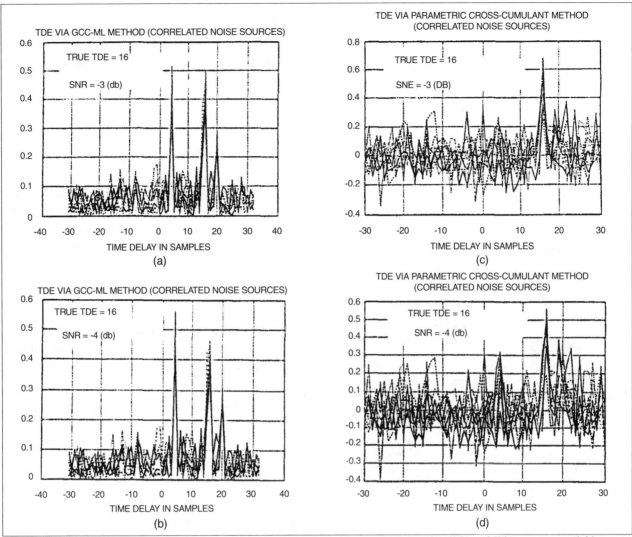

2. *Time delay estimation in the presence of spatially correlated Gaussian noise. Assume that X(k) and Y(k) are two available sensor measurements satisfying X(k) = S(k) + W₁(k) and Y(k) = A S(k − D) + W₂(k), where S(k) is an unknown non-Gaussian signal, D is the time delay and W₁(k), W₂(k) are spatially correlated Gaussian noises. The problem is to estimate D. A generalized cross-correlation method based on maximum likelihood window (ML) estimates the autocorrelation of S(k) which peaks at D (D = 16) and the spatial correlation between W₁(k), W₂(k) as shown in (a) and (b). On the other hand, a cross-cumulant method suppresses the effect of noise and thus makes the estimation of D more reliable, as shown in (c) and (d).*

operating under a random input. General relations for arbitrary stationary random data passing through arbitrary linear systems have been studied quite extensively for many years. In principle, most of these relations are based on power spectrum (autocorrelation) matching criteria. On the other hand, general relations are not available for arbitrary stationary random data passing through arbitrary nonlinear systems. Instead, each type of nonlinearity has to be investigated as a special case [Brillinger, 1977]. Polyspectra can play a key role in detecting and characterizing the type of nonlinearity in a system from its output data [Schetzen, 1989]. Several signal processing methods for the detection and characterization of nonlinearities in time series using higher-order spectra have been developed [Rao and Gabr, 1984; Nikias and Petropulu, 1993].

The organization of the article is as follows. First we discuss the strengths and limitations of correlation-based signal processing methods. Then we introduce the definitions, properties and computation of higher-order statistics and spectra with emphasis on the bispectrum and trispectrum. Following that, we describe parametric and non-parametric expressions of polyspectra of linear processes. The following section addresses polyspectra of nonlinear processes. We conclude the article with a discussion of the applications of higher-order spectra in signal processing.

Correlation-Based Signal Processing: Strengths and Limitations

In this section, we review important results from correlation-based signal processing in order to point out its strengths and limitations. All of our results are for a discrete-time random signal, because we are interested in digital signal processing applications and the data we work with is assumed to be sampled.

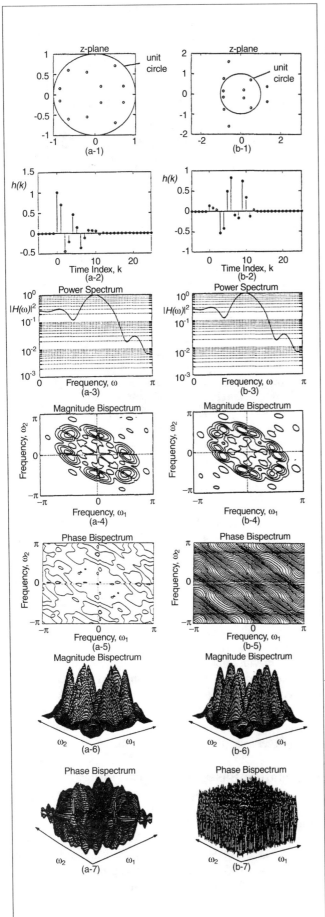

Given a real, stationary, zero-mean random signal $\{X(k)\}$, its **autocorrelation function,** $c_2^x(\tau)$, provides a measure of how the sequence is correlated with itself at different time points:

$$c_2^x(\tau) = E\{X(k)\,X(k+\tau)\} \qquad (1)$$

Note that $c_2^x(\tau)$ is a symmetric function about $\tau = 0$, i.e., $c_2^x(-\tau) = c_2^x(\tau)$; hence, $c_2^x(\tau)$ is a zero-phase function, which means that all phase information about $X(k)$ is lost in $c_2^x(\tau)$.

Given two real, stationary, zero-mean random signals $\{X(k)\}$ and $\{Y(k)\}$, their **cross-correlation function,** which is defined below, provides a measure of how the two signals are correlated with each other at different time points:

$$c_{xy}(\tau) = E\{X(k)\,Y(k+\tau)\} \qquad (2)$$

Note that $c_{xy}(\tau)$ is not a symmetric function about $\tau = 0$, but that $c_{yx}(-\tau) = c_{xy}(\tau)$.

The discrete-time Fourier transforms of $c_2^x(\tau)$ and $c_{xy}(\tau)$ are known as the **power spectrum** of signal $\{X(k)\}$ and the **cross-power spectrum** between signals $\{X(k)\}$ and $\{Y(k)\}$, respectively, i.e.,

$$C_2^x(\omega) = F\{c_2^x(\tau)\} \text{ so that } c_2^x(\tau) = F^{-1}\{C_2^x(\omega)\} \qquad (3)$$

and

$$C_{xy}(\omega) = F\{c_{xy}(\tau)\} \text{ so that } c_{xy}(\tau) = F^{-1}\{C_{xy}(\omega)\} \qquad (4)$$

where $F\{\bullet\}$ and $F^{-1}\{\bullet\}$ denote Fourier and inverse Fourier transforms. Note that $C_2^x(\omega)$ is a real function of ω whereas $C_{xy}(\omega)$ is a complex function of ω.

If a discrete random signal is uncorrelated from one time point to the next, i.e., $E\{X(i)\,X(j)\} = 0$ for all $i \neq j$, then we refer to that signal as (discrete-time) **white noise**. In general, white noise does not have to be stationary, i.e., its variance may vary with time. If it is stationary, so that $E\{X^2(k)\} = \gamma_2^x$, then $C_2^x(\omega) = \gamma_2^x$ for all values of ω; in this case, we associate white noise with a signal whose power spectrum is "flat" for all frequencies. Because this commonly-used description of

3. Signal representation: (a-1) zeros of minimum phase system, (a-2) minimum phase sequence, (a-3) power spectrum of the minimum phase sequence, (a-4) contour of the magnitude of the bispectrum of the minimum phase sequence; (a-5) contour of the phase of the bispectrum of the minimum phase sequence, (a-6) 3-D plot of the magnitude of the bispectrum of the minimum phase sequence, (a-7) 3-D plot of the phase of the bispectrum of the minimum phase sequence; (b-1) zeros of spectrally-equivalent nonminimum phase system, (b-2) nonminimum phase sequence, (b-3) power spectrum of the nonminimum phase sequence, (b-4) contour of the magnitude of the bispectrum of the nonminimum phase sequence. Signal representation: (b-5) contour of the phase of the bispectrum of the nonminimum phase sequence, (b-6) 3-D plot of the magnitude of the bispectrum of the nonminimum phase sequence, (b-7) 3-D plot of the phase of the bispectrum of the nonminimum phase sequence.

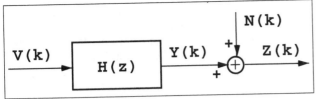

4. Single-channel system.

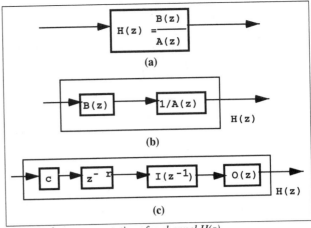

5. Equivalent representations for channel H(z).

white noise only involves second-order statistics, we could refer to such white noise as "second-order" white noise. We have not done so in the past, because we have not needed the notion of a white noise that is characterized by its higher-than second-order statistics. We will return to a discussion of such higher-order white noise.

A single channel model is depicted in Fig. 4. In this model, $V(k)$ is stationary white noise with finite variance γ_2^v ($V(k)$ may be Gaussian or non-Gaussian); $H(z)$ [$h(k)$] is causal and stable; $N(k)$ is Gaussian (white or colored); and, $V(k)$ and $N(k)$ are statistically independent. The convolutional equation describing the noise-free output of this system is

$$Y(k) = h(k) * V(k) = \sum_{i=0}^{k} h(i) \, V(k-i) \qquad (5)$$

which can also be expressed in the complex z-domain or frequency-domain, as

$$Y(z) = H(z)V(z) \text{ or } Y(\omega) = H(\omega)V(\omega) \qquad (6)$$

where $z = \exp(j\omega T)$, and we have assumed that sampling time T equals unity.

Three very popular parametric channel models are: (1) **Moving Average** (MA), in which $H(z) = B(z)$ and $B(z)$ is a rational polynomial in z, one with q zeros, so that $H(z)$ is an all-zero model; (2) **Autoregressive** (AR), in which $H(z) = 1/A(z)$ and $A(z)$ is a rational polynomial in z, one with p zeros, so that $H(z)$ is an all-pole model; and, (3) **Autoregressive-Moving Average** (ARMA), in which $H(z) = B(z)/A(z)$ where $A(z)$ and $B(z)$ are as described for the AR and MA models, respectively; hence, an ARMA model has q zeros and p poles. An MA model has a finite impulse response (FIR), whereas AR and ARMA models have infinite impulse responses (IIR).

When all the zeros of $H(z)$ lie inside the unit circle in the complex z-plane, then $H(z)$ is said to be **minimum phase**. When all of the zeros of $H(z)$ lie outside the unit circle, then $H(z)$ is said to be **maximum phase**. When some of $H(z)$'s zeros lie inside or outside of the unit circle, then $H(z)$ is said to be mixed phase or **nonminimum phase**. Many real-world systems (channels) are nonminimum phase.

Because of the linearity of the system in Fig. 4, different representations of the ARMA model, $H(z) = B(z)/A(z)$, are possible, including those depicted in Fig. 5. The representation in 5a considers the ARMA model in one shot; 5b considers the ARMA model as a cascade of an MA model followed by an AR model; and, 5c is based on expressing a nonminimum phase system in terms of its poles and zeros as follows:

$$H(z) = B(z)/A(z) = c \, z^{-r} I(z^{-1}) \, O(z) \qquad (7)$$

where c is a constant, r is an integer,

$$I(z^{-1}) = \frac{\left[\prod_{i=1}^{L_1} (1 - a_i z^{-1}) \right]}{\left[\prod_{i=1}^{L_3} (1 - c_i z^{-1}) \right]} \qquad (8)$$

is the minimum phase component of $H(z)$, with poles $\{c_i\}$ and zeros $\{a_i\}$ inside the unit circle, i.e., $|c_i| < 1$ and $|a_i| < 1$ for all $\{i\}$, and

$$O(z) = \prod_{i=1}^{L_2} (1 - b_i z) \qquad (9)$$

is the maximum phase component of $H(z)$, with zeros outside the unit circle at $1/|b_i|$, where $|b_i| < 1$ for all $\{i\}$. The different representations for $H(z)$ are each useful in their own right.

The **complex cepstrum** is widely known in digital signal processing circles (e.g., Oppenheim and Schafer, 1989, ch. 12). One starts with the transfer function $H(z)$, takes its logarithm, $\hat{H}(z) = \ln H(z)$, and then takes the inverse z-transform of $\hat{H}(z)$ to obtain the complex cepstrum $\hat{h}(k)$.

When $H(z)$ is decomposed as in Eq. 7 into the product of its minimum-phase and maximum-phase components, $i(k)$ and $o(k)$, respectively, then

$$h(k) = c \, i(k) * o(k) * \delta(k-r). \qquad (10)$$

It is well known that $i(k)$ and $o(k)$ can be computed recursively in terms of $\hat{h}(k)$ using the following formulas [Oppenheim and Schafer (1989), Pan and Nikias (1988)]:

$$i(k) = \begin{cases} 0, & k < 0 \\ \exp(\hat{h}(0)), & k = 0 \\ -\frac{1}{k}\left[\sum_{j=1}^{k} A^{(j)} i(k-j) \right] & k > 0 \end{cases} \qquad (11)$$

Table 1. *Minimum-, maximum- and mixed-phase systems with identical power spectra (or autocorrelations): $0 < a < 1$, $0 < b < 1$ (Nikias and Raghuveer, 1987).*

System	Minimum Phase	Maximum Phase	Mixed Phase
$H(z)$	$(1 - az^{-1})(1 - bz^{-1})$	$(1 - az)(1 - bz)$	$(1 - az)(1 - bz^{-1})$
Autocorrelations			
$c_2(0)$	$1 + a^2b^2 + (a + b)^2$		
$c_2(1)$	$-(a + b)(1 + ab)$		
$c_2(2)$	ab		

and

$$o(k) = \begin{cases} 0, & k > 0 \\ 1, & k = 0 \\ \dfrac{1}{k}\left[\displaystyle\sum_{j=k}^{k} B^{(-j)} o(k-j)\right] & k < 0 \end{cases} \quad (12)$$

in which the "A" and "B" cepstral coefficients are related to the minimum-delay and maximum-delay zeros, respectively, i.e.,

$$A^{(i)} = \sum_{j=1}^{L_1} a_j^i - \sum_{j=1}^{L_3} c_j^i \quad (13)$$

and

$$B^{(i)} = \sum_{j=1}^{L_2} b_j^i. \quad (14)$$

If the cepstral coefficients can be computed, then $i(k)$ and $o(k)$ can be reconstructed from Eqs. 11 and 12, respectively, after which h(k) can be reconstructed (to within a scale factor and a pure delay) from Eq. 10.

It is also well known (e.g., Papoulis, 1991), that

$$c_2^z(\tau) = c_2^y(\tau) + c_y^N(\tau) = \gamma_2^y \sum_{i=0}^{\infty} h(i)h(i + \tau) + c_2^N(\tau) \quad (15)$$

$$C_2^z(z) = \gamma_2^y H(z)H(z^{-1}) + C_2^N(z) \quad (16a)$$

or

$$C_2^z(\omega) = \gamma_2^y |H(\omega)|^2 + C_2^N(\omega), \quad (16b)$$

and,

$$c_{yz}(\tau) = \gamma_2^y h(\tau). \quad (17)$$

From Eqs. 16, we see that all phase information about $H(\omega)$ has been lost in the spectrum (or in the autocorrelation); hence, we say that **correlation or spectra are phase blind**. Observe, from Eq. 17, that if we have access to both the input and output of the system, then we can reconstruct the correct

phase IR, $h(k)$. In many important signal processing applications, we only have access to the output signal of the system; hence, we cannot use Eq. 17 in those applications.

Suppose that $H(z)$ has zeros both inside and outside the unit circle. When these zeros are reflected to their complementary locations [i.e., (some or all of) those inside the unit circle are reflected outside the unit circle, and (some or all of) those outside the unit circle are reflected inside the unit circle], we see that the power spectrum remains unchanged, e.g., $C_2(z^{-1}) = C_2(z)$. When all of $H(z)$'s zeros that are outside the unit circle are reflected inside the unit circle, so that the resulting transfer function is minimum phase, i.e., $H(z) \rightarrow H_{MP}(z)$, we again find that

$$\gamma_2^y H_{MP}(z)H_{MP}(z^{-1}) + C_2^N(z) = \gamma_2^y H(z)H(z^{-1}) + C_2^N(z).$$

The minimum phase transfer function $H_{MP}(z)$ is said to be **"spectrally equivalent minimum phase"** (SEMP) equivalent to $H(z)$. Table 1 provides a simple example of three MA systems that have exactly the same autocorrelations and spectra. It illustrates the point that it is impossible to reconstruct the correct phase model just from autocorrelation or power spectrum information.

Figure 6 depicts two time series that have exactly the same power spectrum. One is Gaussian and the other is non-Gaussian. The top time series was obtained by exciting the nonminimum phase MA system $H(z) = (1 - \frac{1}{2}z^{-1})(1 - 2z^{-1})$ with a ± 1 binary pulse train (unity variance, and $- 2$ fourth-order cumulant), whereas the bottom time series was obtained by exciting the same system with a Gaussian sequence (unity variance). Observe that these time series look quite different. If we only use correlation information, with its associated SEMP model, then we will be extracting the wrong information from the data when the actual model is nonminimum phase. In the next section we demonstrate that we can extract correct information from the data when we work with higher-order statistics.

Higher-Order Statistics and Spectra

a. Definitions and Properties

In this section, we introduce the definitions, properties and computation of higher-order statistics, i.e., moments and cumulants, and their corresponding higher-order spectra. The description is given for both stochastic and deterministic signals. However, the emphasis of the discussion is placed on the 2nd-, 3rd-, and 4th-order statistics and their respective Fourier transforms: power spectrum, bispectrum, and trispectrum.

If $\{X(k)\}$, $k = 0, \pm1, \pm2, \pm3, \dots$ is a real stationary discrete-time signal and its moments up to order n exist, then

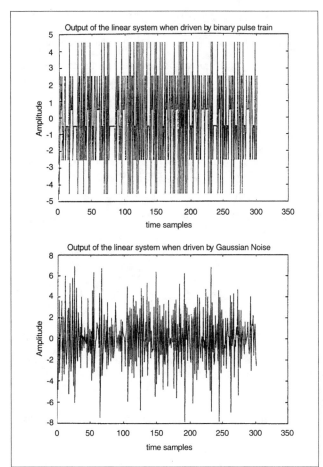

6. Time series with identical power spectra. One is Gaussian; the other is non-Gaussian.

$$m_n^x(\tau_1, \tau_2, ..., \tau_{n-1}) \triangleq$$
$$E\left\{X(k)\,X(k+\tau_1)\,...\,X(k+\tau_{n-1})\right\} \qquad (18)$$

represents the nth-order moment function of the stationary signal, which depends only on the time differences $\tau_1, \tau_2, ..., \tau_{n-1}$, $\tau_i = 0, \pm1, ...$ for all i. Clearly, the 2nd-order moment function, $m_2^x(\tau_1)$, is the autocorrelation of $\{X(k)\}$ whereas $m_3^x(\tau_1, \tau_2)$ and $m_4^x(\tau_1, \tau_2, \tau_3)$ are the 3rd- and 4th-order moments, respectively. $E\{\bullet\}$ denotes statistical expectation.

The nth-order cumulant function of a non-Gaussian stationary random signal $X(k)$ can be written as (for n = 3, 4 only):

$$c_n^x(\tau_1, \tau_2, ..., \tau_{n-1}) =$$
$$m_n^x(\tau_1, \tau_2, ..., \tau_{n-1}) - m_n^G(\tau_1, \tau_2, ..., \tau_{n-1}) \qquad (19)$$

where $m_n^x(\tau_1, ..., \tau_{n-1})$ is the nth-order moment function of $X(k)$ and $m_n^G(\tau_1, \tau_2, ..., \tau_{n-1})$ is the nth-order moment function of an equivalent Gaussian signal that has the same mean value and autocorrelation sequence as $X(k)$. Clearly, if is Gaussian, $m_n^x(\tau_1, ..., \tau_{n-1}) = m_n^G(\tau_1, ..., \tau_{n-1})$ and thus $c_n^x(\tau_1, \tau_2, ..., \tau_{n-1}) = 0$. Note, however, that although Eq. 19

is only true for orders n = 3 and 4, $c_n^x(\tau_1, \tau_2, ..., \tau_{n-1}) = 0$ for all n if $X(k)$ is Gaussian. The properties of cumulants are summarized in Table 2.

The following relationships between moment and cumulant sequences of $\{X(k)\}$ exist for orders n = 1, 2, 3, 4.

1st-order cumulants:

$$c_1^x = m_1^x = E\{X(k)\} \qquad \text{(mean value)} \qquad (20)$$

2nd-order cumulants:

$$c_2^x(\tau_1) = m_2^x(\tau_1) - (m_1^x)^2 \qquad \text{(covariance sequence)}$$
$$= m_2^x(-\tau_1) - (m_1^x)^2 = c_2^x(-\tau_1) \qquad (21)$$

where $m_2^x(-\tau_1)$ is the autocorrelation sequence. Thus, we see that the 2nd-order cumulant sequence is the *covariance* while the 2nd-order moment sequence is the *autocorrelation*.

3rd-order cumulants:

$$c_3^x(\tau_1, \tau_2) = m_3^x(\tau_1, \tau_2)$$
$$- m_1^x\left[m_2^x(\tau_1) + m_2^x(\tau_2) + m_2^x(\tau_1 - \tau_2)\right] + 2\left(m_1^x\right)^3 \qquad (22)$$

where $m_3^x(\tau_1, \tau_2)$ is the 3rd-order moment sequence. This follows if we combine Eqs. 18 and 19.

4th-order cumulants:

Combining Eqs. 18 and 19, we get

$$c_4^x(\tau_1, \tau_2, \tau_3) = m_4^x(\tau_1, \tau_2, \tau_3)$$
$$- m_2^x(\tau_1) \cdot m_2^x(\tau_3 - \tau_2) - m_2^x(\tau_2) \cdot m_2^x(\tau_3 - \tau_1)$$
$$- m_2^x(\tau_3) \cdot m_2^x(\tau_2 - \tau_1) - m_1^x[m_3^x(\tau_2 - \tau_1, \tau_3 - \tau_1)$$
$$+ m_3^x(\tau_2, \tau_3) + m_3^x(\tau_2, \tau_4) + m_3^x(\tau_1, \tau_2)]$$
$$+ (m_2^x)^2[m_2^x(\tau_1) + m_2^x(\tau_2) + m_2^x(\tau_3) + m_2^x(\tau_3 - \tau_1)$$
$$+ m_2^x(\tau_3 - \tau_2) + m_2^x(\tau_2 - \tau_1)] - 6\,(m_1^x)^4. \qquad (23)$$

If the signal $\{X(k)\}$ is zero-mean $m_1^x = 0$, it follows from Eqs. 21 and 22 that the second- and third-order cumulants are identical to the second- and third-order moments, respectively; however, to generate the fourth-order cumulants, we need knowledge of the fourth-order and second-order moments in Eq. 23, i.e.,

$$c_4^x(\tau_1, \tau_2, \tau_3) = m_4^x(\tau_1, \tau_2, \tau_3) - m_2^x(\tau_1) \cdot m_2^x(\tau_3 - \tau_2)$$
$$- m_2^x(\tau_2) \cdot m_2^x(\tau_3 - \tau_1) - m_2^x(\tau_3) \cdot m_2^x(\tau_2 - \tau_1). \qquad (24)$$

By putting $\tau_1 = \tau_2 = \tau_3 = 0$ in Eqs. 21–23 and assuming $m_1^x = 0$, we obtain

$$\gamma_2^x = E\{x^2(k)\} = c_2^x(0) \qquad \text{(variance)}$$
$$\gamma_3^x = E\{X^3(k)\} = c_3^x(0,0) \qquad \text{(skewness)} \qquad (25)$$
$$\gamma_4^x = E\{X^4(k)\} - 3\left[\gamma_2^x\right]^2 = c_4^x(0, 0, 0) \quad \text{(kurtosis)}.$$

Normalized kurtosis is defined as $\gamma_4^x/[\gamma_2^x]^2$. Equation 25 gives the variance, skewness and kurtosis measures in terms of cumulants at zero lags.

A 1-D slice of the n-th order cumulant is obtained by freezing $(n - 2)$ of its $n - 1$ indexes. Many types of 1-D slices are possible, including radial, vertical, horizontal, diagonal, and offset-diagonal. A diagonal slice is obtained by setting $\tau_1 = \tau$, $i = 1, 2, ..., n - 1$. All these 1-D slices are very useful in applications of cumulants in signal processing.

A logical question to ask is "Why do we need fourth-order cumulants, i.e., aren't third-order cumulants good enough?" If a random process is symmetrically distributed, then its third-order cumulant equals zero; hence, for such a process we must use fourth-order cumulants. For example, Laplace, uniform, Gaussian, and Bernoulli-Gaussian distributions are symmetric, whereas exponential, Rayleigh and K-distributions are nonsymmetric. Additionally, some processes have extremely small third-order cumulants and much larger fourth-order cumulants; hence, for such processes we would also use the latter. Finally, in some specific applications (e.g., retrieval of harmonics and cubic phase coupling) third-order cumulants equal zero whereas fourth-order cumulants are nonzero.

Higher-order spectra are defined in terms of either cumulants (e.g., cumulant spectra) or moments (e.g., moment spectra). As explained later, in the case of stochastic signals there are certain advantages to using cumulants; while for deterministic signals, it is better to use moments. Simply stated, higher-order spectra are multi-dimensional Fourier transforms of higher-order statistics. Thus, the power spectrum, bispectrum and trispectrum are defined in terms of cumulants as follows.

Power Spectrum:

$$C_2^x(\omega) = \sum_{\tau=-\infty}^{+\infty} c_2^x(\tau) \exp\{-j(\omega\tau)\}, \tag{26}$$

14

$|\omega| \leq \pi$ where $c_2^x(\tau)$ is the covariance sequence of $\{X(k)\}$. Eq. 26 is also known as the Wiener-Khintchine theorem. From Eqs. 21 and 26, we have

$$c_2^x(\tau) = c_2^x(-\tau)$$

$$C_2^x(\omega) = C_2^x(-\omega) \qquad (27)$$

$$C_2^x(\omega) \geq 0 \quad \text{(real, nonnegative function)}$$

Although the power spectrum was previously discussed, its definition and properties are repeated here, so they can be easily compared with the bispectrum and trispectrum.

Bispectrum: n = 3

$$C_3^x(\omega_1, \omega_2) = \sum_{\tau_1=-\infty}^{+\infty} \sum_{\tau_1=-\infty}^{+\infty} c_3^x(\tau_1, \tau_2) \exp\{-j(\omega_1\tau_1 + \omega_2\tau_2)\} \qquad (28)$$

$$|\omega_1| \leq \pi, \ |\omega_2| \leq \pi, \ |\omega_1 + \omega_2| \leq \pi$$

where $c_3^x(\tau_1, \tau_2)$ is the third-order cumulant sequence of $\{X(k)\}$ described by Eq. 22.
Important symmetry conditions follow from the properties of moments and Eq. 22:

$$c_3^x(\tau_1, \tau_2) = c_3^x(\tau_2, \tau_1) = c_3^x(-\tau_2, \tau_1 - \tau_2)$$

$$= c_3^x(\tau_2 - \tau_1, -\tau_1) = c_3^x(\tau_1 - \tau_2, -\tau_2) \qquad (29)$$

$$= c_3^x(\tau_1, \tau_2 - \tau_1)$$

As a consequence, knowing the third-order cumulants in any of the six sectors, I through VI, shown in Fig. 7a, would enable us to find the entire third-order cumulant sequence. These sectors include their boundaries so that, for example, *sector I* is an infinite wedge bounded by the lines $\tau_2 = 0$ and $\tau_1 = \tau_2, \tau_1 \geq 0$.

The definition of the bispectrum in Eq. 28 and the properties of third-order cumulants in Eq. 29 give

$$C_3^x(\omega_1, \omega_2) = C_3^x(\omega_2, \omega_1) \qquad (30)$$

$$= C_3^{x*}(-\omega_2, -\omega_1) = C_3^x(-\omega_1 - \omega_2, \omega_2)$$

$$= C_3^x(\omega_1, -\omega_1 - \omega_2) = C_3^x(-\omega_1 - \omega_2, \omega_1)$$

$$= C_3^x(\omega_2, -\omega_1 - \omega_2).$$

Thus knowledge of the bispectrum in the triangular region $\omega_2 \geq 0, \omega_1 \geq \omega_2, \omega_1 + \omega_2 \leq \pi$ shown in Fig. 7b is enough for a complete description of the bispectrum. For real processes, the bispectrum has 12 symmetry regions.

Trispectrum: n = 4

$$C_4^x(\omega_1, \omega_2, \omega_3) = \qquad (31)$$

$$\sum_{\tau_1=-\infty}^{+\infty} \sum_{\tau_2=-\infty}^{+\infty} \sum_{\tau_3=-\infty}^{+\infty} c_4^x(\tau_1, \tau_2, \tau_3) \exp\{-j(\omega_1\tau_1 + \omega_2\tau_2 + \omega_3\tau_3)\}$$

$$|\omega_1| \leq \pi, \ |\omega_2| \leq \pi, \ |\omega_3| \leq \pi, \ |\omega_1 + \omega_2 + \omega_3| \leq \pi$$

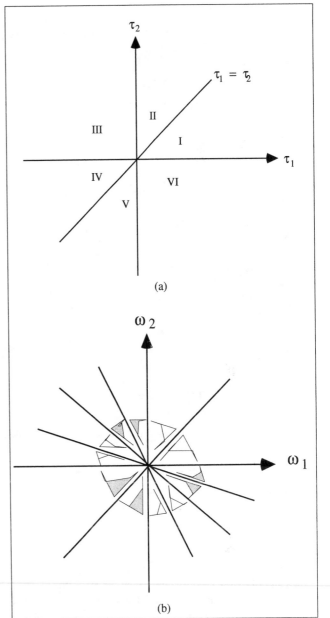

7. (a) Symmetry regions of third-order cumulants. (b) Symmetry regions of the bispectrum (Nikias and Petropulu, 1993).

where $c_4^x(\tau_1, \tau_2, \tau_3)$ is the fourth-order cumulant sequence given by Eq. 23. From the Eq. 31's definition of fourth-order cumulants, a lot of symmetry properties can be derived for the trispectrum, similar to those of the bispectrum. Pflug et. al. [1992] point out that the trispectrum of a real signal has 96 symmetry regions.

The bispectrum and trispectrum are generally complex functions, i.e., they have magnitude and phase:

$$C_n^x(\omega_1, \omega_2, ..., \omega_{n-1}) =$$

$$\left| C_n^x(\omega_1, \omega_2, ..., \omega_{n-1}) \right| \cdot \exp j\psi_n^x(\omega_1, \omega_2, ..., \omega_{n-1})$$

$$n = 3, 4, ... \qquad (32)$$

Cumulant spectra are more useful in the processing of random signals than are moment spectra, because: (a) cumulant spectra of order $n > 2$ are zero if the signal is Gaussian and thus non-zero cumulant spectra provide a measure of extent of non-Gaussianity; (b) cumulants provide a suitable measure of extent of statistical dependence in time series; (c) just as the covariance function of white noise is an impulse function and its spectrum is flat, the cumulants of (higher-order) white noise are multidimensional impulse functions and the polyspectra of this noise is multidimensionally flat; and (d) the cumulant of two statistically independent random processes equals the sum of the cumulants of the individual random processes, whereas the same is not true for higher-order moments. This last property lets us work with the cumulant very easily as an operator.

A normalized higher-order spectrum or nth-order coherency index is a function that combines the cumulant spectrum of order n with the power spectrum ($n = 2$) of a signal. The 3rd- and 4th-order coherency indexes are respectively defined by

$$P_3^x(\omega_1, \omega_2) \tag{33a}$$

$$= \frac{C_3^x(\omega_1, \omega_2)}{\sqrt{C_2^x(\omega_1) C_2^x(\omega_2) C_2^x(\omega_1 + \omega_2)}}, \quad \text{(biocoherency)}$$

$$P_4^x(\omega_1, \omega_2, \omega_3) \tag{33b}$$

$$= \frac{C_4^x(\omega_1, \omega_2, \omega_3)}{\sqrt{C_2^x(\omega_1) C_2^x(\omega_2) C_2^x(\omega_3) C_2^x(\omega_1 + \omega_2 + \omega_3)}}, \quad \text{(tricoherency)}$$

These functions are very useful in the detection and characterization of nonlinearities in time series and in *discriminating linear processes from nonlinear ones*. In fact, a signal is said to be a linear non-Gaussian process of order n if the magnitude of the nth-order coherency, $\left| P_n^x(\omega_1, \ldots, \omega_{n-1}) \right|$, is constant over all frequencies; otherwise, the signal is said to be a non-linear process [Hinich, 1982; Raghuveer and Nikias, 1985].

The higher-order spectrum of a linear non-Gaussian signal, $C_n^x(\omega_1, \omega_2, \ldots, \omega_{n-1})$, can always be written in the form

$$C_n^x(\omega_1, \ldots, \omega_{n-1}) = \gamma_n^v H(\omega_1) \ldots H(\omega_{n-1}) H^*(\omega_1 + \ldots + \omega_{n-1}) \tag{34}$$

where γ_n^v is a scalar constant and $H(\omega)$ is the frequency transfer function of a linear time-invariant (LTI) system. If a non-Gaussian signal, $X(k)$, is generated by exciting a LTI system, $H(\omega)$, with non-Gaussian white noise, $V(k)$, as shown later, then Eq. 34 is true for all orders. Special cases include

$$C_2^x(\omega) = \gamma_2^y \, |H(\omega)|^2 \quad \text{(Power Spectrum)} \tag{35a}$$

$$C_3^x(\omega_1, \omega_2) \tag{35b}$$
$$= \gamma_3^y H(\omega_1) \cdot H(\omega_2) \cdot H^*(\omega_1 + \omega_2) \quad \text{(Bispectrum)}$$

$$C_4^x(\omega_1, \omega_2, \omega_3) = \tag{35c}$$
$$\gamma_4^y H(\omega_1) \cdot H(\omega_2) \cdot H(\omega_3) \cdot H^*(\omega_1 + \omega_2 + \omega_3). \quad \text{(Trispectrum)}$$

However, there are non-Gaussian signals that can be linear in the bispectrum domain, i.e.,

$$C_3^x(\omega_1, \omega_2) = T(\omega_1) \, T(\omega_2) \, T^*(\omega_1 + \omega_2)$$

but non-linear in the trispectrum or higher-order spectrum domains.

Although the power spectrum of either a linear or non-linear process can always be written in the form of Eq. 35a, higher-order spectra can take the form of Eq. 34 only if the process is linear (has flat magnitude coherency). Combining Eq. 35a with Eq. 35b and Eq. 35c, we see that the power spectrum of a linear process can be reconstructed from its bispectrum or trispectrum up to a constant term (provided $H(0) \neq 0$), i.e.,

$$C_3^x(\omega, 0) = C_2^x(\omega) H(0) \cdot \left(\frac{\gamma_3^y}{\gamma_2^y} \right) \tag{36a}$$

$$C_4^x(\omega_1, \omega_2, 0) = C_3^x(\omega_1, \omega_2) H(0) \cdot \left(\frac{\gamma_4^y}{\gamma_3^y} \right) \tag{36b}$$

as well as the bispectrum from its trispectrum using

$$C_4^x(\omega_1, \omega_2, 0) = C_3^x(\omega_1, \omega_2) H(0) \cdot \left(\frac{\gamma_4^y}{\gamma_3^y} \right) \tag{37}$$

Equations 36a and 36b can be very useful in obtaining better estimates of the power spectrum of the signal from its bispectrum/trispectrum when the additive noise is Gaussian.

If $W(k)$ is a stationary non-Gaussian process with $E\{W(k)\} = 0$ and nth-order cumulant sequence

$$C_n^w(\tau_1, ..., \tau_{n-1}) = \gamma_n^w \cdot \delta(\tau_1, ..., \tau_{n-1}) \tag{38a}$$

where $\delta(\tau_1, ..., \tau_{n-1})$ is the delta function, then W(k) is said to be nth-order white. By taking $(n-1)$-dimensional Fourier transform of Eq. 38a, we obtain

$$C_n^w(\omega_1, ..., \omega_{n-1}) = \gamma_n^w \tag{38b}$$

which is a flat spectrum for all frequencies. Important special cases of higher-order spectra of a white process include

$$C_2^w(\omega) = \gamma_2^w \quad \text{(Power Spectrum)}$$
$$C_3^w(\omega_1, \omega_2) = \gamma_3^w \quad \text{(Bispectrum)} \tag{39}$$
$$C_4^w(\omega_1, \omega_2, \omega_3) = \gamma_4^w \quad \text{(Trispectrum)}$$

where γ_2^w, γ_3^w, and γ_4^w are the variance, skewness and kurtosis of $\{W(k)\}$, respectively.

The definitions of higher-order moment spectra of deterministic signals are summarized in Table 3.

b. Higher-Order Spectra Computation from Data

The problem met in practice is one of estimating the higher-order spectra of a process when a finite set of measurements is given. Two of the most popular conventional approaches are the *indirect* and *direct* methods which may be seen as direct approximations of the definition of higher-order spectra. While these approximations are straightforward, sometimes the required computations may be expensive despite the use of fast Fourier Transform (FFT) algorithms. We describe both conventional methods here as bispectrum estimators. Their extension to trispectrum estimation is described in [Nikias and Petropulu, 1993].

Indirect Method

Let $\{X(1), X(2), ..., X(N)\}$ be the given data set. Then we have the following:

1. Segment the data into K records of M samples each, i.e., $N = KM$.

2. Subtract the average value of each record.

3. Assuming that $\{x^{(i)}(k), k = 0, 1, ..., M-1\}$ is the data set per segment $i = 1, 2, ..., K$, obtain an estimate of the third-moment sequence

$$r^{(i)}(m, n) = \frac{1}{M} \sum_{\ell=s_1}^{s_2} x^{(i)}(\ell) \, x^{(i)}(\ell + m) \, x^{(i)}(\ell + n)$$

where

$$i = 1, 2, ..., K$$
$$s_1 = \max(0, -m, -n)$$
$$s_2 = \min(M-1, M-1-m, M-1-n).$$

4. Average $r^{(i)}(m, n)$ over all segments

$$\hat{c}_3^x(m, n) = \frac{1}{K} \sum_{i=1}^{K} r^{(i)}(m, n)$$

5. Generate the bispectrum estimate

$$\hat{C}_3^x(\omega_1, \omega_2) =$$
$$\sum_{m=-L}^{L} \sum_{n=-L}^{L} \hat{c}_3^x(m, n) \, W(m, n) \, \exp\{-j(\omega_1 m + \omega_2 n)\}$$

where $L < M-1$ and $W(m, n)$ is a two-dimensional window function. Let us note that the computations of the bispectrum estimate may be substantially reduced if the symmetry properties of third-order cumulants (Eq. 29) are taken into account for the calculations of $r^{(i)}(m, n)$ and if the symmetry properties of the bispectrum shown in Eq. 30 are also incorporated in the computations.

As in the case of conventional power spectrum estimation, to get better estimates, suitable windows should be used. The window functions should satisfy the following constraints:

17

a) $W(m, n) = W(n, m) = W(-m, n-m) = W(m-n,-n)$ (symmetry properties of third order cumulants);

b) $W(m, n) = 0$ outside the region of support of $\hat{c}_3^x(m, n)$;

c) $W(0, 0) = 1$ (normalizing condition);

d) $W(\omega_1, \omega_2) \geq 0$, for all (ω_1, ω_2).

A class of functions which satisfies the constraints for $W(m, n)$ is:

$$W(m, n) = d(m)d(n)d(n - m) \qquad (40)$$

where

$$d(m) = d(-m)$$

$$d(m) = 0, \; m > L$$

$$d(0) = 1$$

$$D(\omega) \geq 0, \; \text{for all } \omega$$

Equation 40 allows a reconstruction of two-dimensional window functions for bispectrum estimation using standard one-dimensional lag windows. Two good choices of 1-d windows [Nikias and Raghuveer, 1987] are:

a) Optimum window (minimum bispectrum bias supremum):

$$d_0(m) = \begin{cases} \frac{1}{\pi}\left|\sin\frac{\pi m}{L}\right| + (1 - \frac{|m|}{L})\,(\cos\frac{\pi m}{L}), & |m| \leq L \\ 0, & |m| > L. \end{cases} \qquad (41a)$$

b) Parzen window:

$$d_p(m) = \begin{cases} 1 - 6\left(\frac{|m|}{L}\right)^2 + 6\left(\frac{|m|}{L}\right)^3, & |m| \leq \frac{L}{2} \\ 2(1 - \frac{|m|}{L})^3, & \frac{L}{2} \leq |m| \leq L \\ 0, & |m| > L. \end{cases} \qquad (41b)$$

More windows and their properties for higher-order spectrum estimation can be found in [Nikias and Petropulu, 1993].

Direct Method

Let $\{X(1), X(2), ..., X(N)\}$ be the available set of observations for bispectrum estimation. Let us assume that f_s is the sampling frequency and $\Delta_0 = f_s/N_0$ is the required spacing between frequency samples in the bispectrum domain along horizontal or vertical directions; thus, N_0 is the total number of frequency samples. Then we have the following:

a) Segment the data into K segments of M samples each, i.e., $N = KM$, and subtract the average value of each segment. If necessary, add zeros at the end of each segment to obtain a convenient length M for the FFT.

b) Assuming that $\{x^{(i)}(k), k = 0, 1, 2, ..., M - 1\}$ are the data of segment $\{i\}$, generate the DFT coefficients

$$Y^{(i)}(\lambda) = \frac{1}{M}\sum_{k=0}^{M-1} x^{(i)}(k)\exp(-j2\pi k\lambda/M), \; \lambda = 0, 1, ..., \frac{M}{2}$$

$$i = 1, 2, ..., K.$$

c) In general, $M = M_1 \times N_0$, where M_1 is a positive integer (assumed odd number), i.e., $M_1 = 2L_1 + 1$. Since M is even and M_1 is odd, we compromise on the value of N_0 (closest integer). Form the bispectrum estimate based on the DFT coefficients

$$\hat{b}_i(\lambda_1, \lambda_2) =$$
$$\frac{1}{\Delta_0^2}\sum_{k_1=-L_1}^{L_1}\sum_{k_2=-L_1}^{L_1} Y^{(i)}(\lambda_1+k_1)Y^{(i)}(\lambda_2+k_2)Y^{(i)*}(\lambda_1+\lambda_2+k_1+k_2)$$

over the triangular region $0 \leq \lambda_2 \leq \lambda_1, \lambda_1 + \lambda_2 = f_s/2$. For the special case where no averaging is performed in the bispectrum domain, $M_1 = 1$, $L_1 = 0$, and therefore:

$$\hat{b}_i(\lambda_1, \lambda_2) = \frac{1}{\Delta_0^2}Y^{(i)}(\lambda_1)Y^{(i)}(\lambda_2)Y^{(i)*}(\lambda_1 + \lambda_2)$$

d) The bispectrum estimate of the given data is the average over the K pieces

$$\hat{C}_3^x(\omega_1, \omega_2) = \frac{1}{K}\sum_{i=1}^{K}\hat{b}_i(\omega_1, \omega_2)$$

where

$$\omega_1 = \left(\frac{2\pi f_s}{N_0}\right)(\lambda_1) \text{ and } \omega_2 = \left(\frac{2\pi f_s}{N_0}\right)(\lambda_2).$$

c. Properties of Conventional Estimators and Asymptotic Behavior

The statistical properties of the indirect and direct conventional methods for higher-order spectrum estimation have been studied by Rosenblatt and Van Ness [1965], Van Ness [1968], Brillinger and Rosenblatt [1967], Rao and Gabr [1984]; their implications in signal processing have been studied by Nikias and Petropulu [1993].

Let us consider the case of the bispectrum. Assume that $C_2^x(\omega)$ and $C_3^x(\omega_1, \omega_2)$ are respectively the true power spectrum and bispectrum of a stationary zero-mean signal. Let $\hat{C}_3^x(\omega_1, \omega_2)$ be a consistent bispectrum estimate computed by indirect or direct conventional methods using a single realization of the signal of length N. The key result associated with these methods is that for sufficiently large record size M and total length N, both direct and indirect approaches provide approximately unbiased estimates; viz.:

$$E\left\{\hat{C}_3^x(\omega_1, \omega_2)\right\} \cong C_3^x(\omega_1, \omega_2) \qquad (42)$$

with asymptotic variances

$$\text{var}[\text{Re}[\hat{C}_3^x(\omega_1, \omega_2)]] \cong \text{var}[\text{Im}[\hat{C}_3^x(\omega_1, \omega_2)]]$$
$$\cong \frac{1}{2}\sigma_2^3(\omega_1, \omega_2), \qquad (43a)$$

where

$$\sigma_3^2(\omega_1, \omega_2) = \begin{cases} \dfrac{VL^2}{MK}C_2^x(\omega_1)C_2^x(\omega_2)C_2^x(\omega_1+\omega_2) & \text{(indirect)} \\[3mm] \dfrac{N_0^2}{MK}C_2^x(\omega_1)C_2^x(\omega_2)C_2^x(\omega_1+\omega_2) & \text{(direct)} \end{cases} \quad (43b)$$

for $0 < \omega_2 < \omega_1$, where K is the number of records, M the number of samples per record and V is the total energy of the bispectrum window, which is unity for a rectangular window; L is defined in step (5) of the indirect method and $N_0 = (M/M_1)$ is defined in the description of the direct method. From Eqs. 43a and 43b, it is apparent that if a rectangular window is used with the indirect method and $L = N_0$, the two conventional methods give approximately the same estimates.

Brillinger and Rosenblatt [1967] and Rosenblatt [1985] showed that for large M and N, the error bicoherence

$$\frac{\hat{C}_3^x(\omega_1, \omega_2) - C_3^x(\omega_1, \omega_2)}{\sigma_3(\omega_1, \omega_2)} \sim N_c(0, 1) \quad (44)$$

is approximately complex Gaussian variant with mean zero and variance equal to unity. Another equally important large sample result that follows from the asymptotic results developed by Brillinger and Rosenblatt [1967] is that these statistics can be treated as independent random variables over the grid in the principal domain if the grid width is larger than or equal to the bispectrum bandwidth; i.e., $\hat{C}_3^x(\omega_j, \omega_k)$ and $\hat{C}_3^x(\omega_r, \omega_s)$ are independent for $j \neq r$ or $k \neq s$ if $|\omega_{j+1} - \omega_j| \geq 2\pi\Delta_0$ or $|\omega_{r+1} - \omega_r| \geq 2\pi\Delta_0$, where

$$\Delta_0 \triangleq \begin{cases} \dfrac{1}{L} & \text{(indirect)} \\[3mm] \dfrac{1}{N_0} & \text{(direct).} \end{cases} \quad (45)$$

The asymptotic independence and Gaussianity imply that the magnitude squared bicoherence statistic [Hinich, 1982]

$$ch_s(\omega_1, \omega_2) = \frac{\left| \hat{C}_3^x(\omega_1, \omega_2) - C_3^x(\omega_1, \omega_2) \right|^2}{\sigma_3^2(\omega_1, \omega_2)} \quad (46)$$

is approximately a noncentral chi-square statistic with 2 degrees of freedom. Hinich [1985] reported that this approximation holds for samples as small as $N = 256$ if $\Delta_0 \cong \sqrt{N}$. If the process has zero bispectrum $C_3^x(\omega_1, \omega_2) = 0$, then

$$ch_s(\omega_1, \omega_2)$$

is central chi-square variant with 2 degrees of freedom.

8. (a) Zeros of a nonminimum phase system, (b) magnitude response of the nonminimum phase system, (c) input random sequence (zero-mean one-sided exponential distributed with variance 1), (d) output sequence, (e) 2nd-order cumulants of the output, (f) power spectrum of the output. (g) Contour of the 3rd-order cumulants of the output, (h) 3-D plot of the 3rd-order cumulants of the output, (i) contour of the real part of the bispectrum of the output, (j) contour of the imaginary part of the bispectrum of the output. (k) 3-D plot of the real part of the bispectrum of the output, (l) 3-D plot of the imaginary part of the bispectrum of the output.

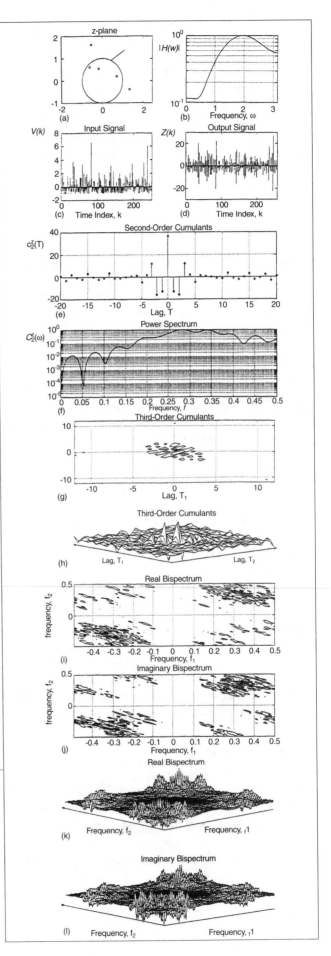

19

From Eqs. 43a and 43b, it is apparent that the bispectrum variance associated with conventional estimators can be reduced by: (i) increasing the number of records, (ii) reducing the size of the region of support of the window in the cumulant domain (L) or increasing the size of the frequency smoothing window (M_1), and (iii) increasing the record size M. However, increasing the number of records (K) is demanding on the computer time and may introduce potential nonstationarities. On the other hand, frequency domain averaging over large rectangles of size (M_1) or use of cumulant domain windows with small values of L_0 reduce frequency resolution and may increase bias. In the case of "short length" data, K could be increased by using overlapping records [Nikias and Raghuveer, 1987].

The conventional methods have the advantage of ease of implementation (FFT algorithms can be used) and provide good estimates for sufficiently long data. However, because of the "uncertainty principle" of the Fourier transform, the ability of the conventional methods to resolve harmonic components in polyspectra domains is limited. Raghuveer and Nikias [1985] point out that this could pose a problem in detecting quadratic phase coupling at closely-spaced frequency pairs.

One of the key advantages of conventional bispectrum (or higher-order spectrum) estimates is their asymptotic Gaussian properties illustrated by Eqs. 44–46. These asymptotic results serve as the bridge between Likelihood Ratio Test (LRT) detectors and Maximum Likelihood (ML) theory on the one hand, and higher-order spectra (HOS) on the other hand. Detection, parameter estimation and classification schemes can be developed for important signal processing applications using the asymptotic statistical properties of HOS estimates and LRT or ML theory. See, for example, Hinich [1985], Forster and Nikias [1991], and Giannakis and Tsatsanis [1992].

Linear Processes

a. Cumulants and Polyspectra
A major generalization to Eqs. 15 and 16 was established by Bartlett (1955) and Brillinger and Rosenblatt (1967). In this case the system in Fig. 4 is assumed to be causal and exponentially stable, and $\{V(k)\}$ is assumed to be independent, identically distributed (i.i.d.) and non-Gaussian, i.e.,

$$c_n^v(\tau_1, ..., \tau_{n-1}) = \begin{cases} \gamma_n^v, & \tau_1 = \tau_2 = ... \tau_{n-1} = 0 \\ 0, & \text{otherwise} \end{cases} \quad (47)$$

where γ_n^v denotes the nth-order cumulant of $V(k)$. Additive noise $N(k)$ is assumed to be Gaussian, but need not be white. Their generalizations to Eqs. 15 and 16 are

$$c_n^z(\tau_1, ..., \tau_{n-1}) = \gamma_n^v \sum_{k=0}^{\infty} h(k)h(k + \tau_1) ... h(k + \tau_{n-1}) \quad (48)$$

and

$$C_n^z(\omega_1, ..., \omega_{n-1}) = $$
$$\gamma_n^v H(\omega_1)H(\omega_2) ... H(\omega_{n-1})H(-\sum_{i=1}^{n-1}\omega_i) \quad (49a)$$

which can also be expressed in terms of multidimensional z-transforms, as

$$C_n^z(z_1, ..., z_{n-1}) = $$
$$\gamma_n^v H(z_1)H(z_2) ... H(z_{n-1})H(-\Pi z_i^{-1}) . \quad (49b)$$

Observe that when $n = 2$, Eqs. 48 and 49 reduce to Eq. 15 [subject to the addition of $c_2^N(\tau)$] and Eq. 16 [subject to the addition of $C_2^N(z)$]. Equations 49a and 49b have been the starting points for many nonparametric and parametric polyspectral techniques that have been developed during the past few years [e.g., Nikias and Raghuveer (1987) and Mendel (1988, 1991)]. One very important use for Eq. 48 is to compute cumulants for models of systems. The procedure for doing this is: (1) determine the model's IR, $h(k)$ $k = 0, 1, 2, ..., K$; (2) fix the τ_j values, and evaluate Eq. 48; and, (3) repeat step (2) for all τ_j values in the domain of support of interest (be sure to use the symmetry properties of cumulants to reduce computations). This is how the results shown in Fig. 8g and h were obtained. An important use of Eq. 49a is to compute polyspectra for models of systems. The procedure for doing this is: (1) determine the model's IR, $h(k)$ $k = 0, 1, 2, ..., K$; (2) compute the DFT of $h(k)$, $H(\omega)$; (3) fix the ω_j values, and evaluate Eq. 49a; and, (4) repeat Step (3) for all ω_j values in the domain of support of interest (be sure to use the symmetry properties of polyspectra to reduce computations). Of course, another way to compute the polyspectra is to first compute the cumulants, as just described, and then compute their multi-dimensional DFT.

The generalization of Eqs. 48 and 49 to the case of colored non-Gaussian input is more easily visualized in the polyspectral domain [Bartlett (1955) only considers the $n = 2, 3, 4$ cases; Brillinger and Rosenblatt (1967) provide results for all n], and is:

$$C_n^z(\omega_1, ..., \omega_{n-1}) = \quad (50)$$
$$C_n^v(\omega_1, ..., \omega_{n-1}) H(\omega_1) H(\omega_2) ... H(\omega_{n-1}) H\left(-\sum_{i=1}^{n-1}\omega_i\right)$$

Derivations of Eqs. 48–50 can also be found in Section C of the Appendix in Mendel (1991). The generalization of 48 to the colored noise case is also given in Mendel (1991); it is a $(n-1)$-fold convolution between $c_n^v(\tau_1^{2x}, ..., \tau_{n-1}^{2x})$ and $c_n^h(\tau_1^{2x}, ..., \tau_{n-1}^{2x}) = \sum h(k) h(k + \tau_1) ... h(k + \tau_{n-1})$.

Example 1. (Mendel, 1991): Suppose that $h(k)$ is the impulse response (IR) of a causal MA system. Such a system has a finite IR, and is described by the following model:

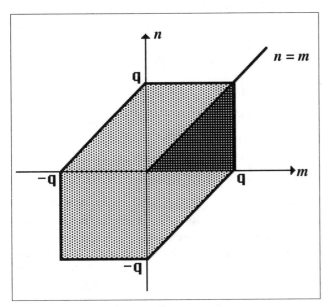

9. The domain of support for $c_3^y(m, n)$ for an MA(q) system. The dark shaded c_3^y region is the principal region defined in Eq. 54.

$$Y(k) = \sum_{i=0}^{q} b(i) V(k - i) \tag{51}$$

The MA parameters are $b(0)$, $b(1)$, ..., $b(q)$, where q is the order of the MA model, and $b(0)$ is usually assumed equal to unity [the scaling is absorbed into the statistics of $V(k)$]. It is well known that for this model $h(i) = b(i)$, $i = 0, 1, ..., q$; hence, when $\{V(k)\}$ is i.i.d., we find from Eq. 48 that

$$c_3^y(\tau_1, \tau_2) = \gamma_3^v \sum_{l=0}^{q} b(l) b(l + \tau_1) b(l + \tau_2) \tag{52}$$

and

$$c_4^y(\tau_1, \tau_2, \tau_3) = \gamma_4^v \sum_{l=0}^{q} b(l) b(l + \tau_1) b(l + \tau_2) b(l + \tau_3). \tag{53}$$

An interesting question is "for which values of τ_1 and τ_2 is $c_3^y(\tau_1, \tau_2)$ nonzero?" Of course, a comparable question can

be asked for $c_4^y(\tau_1, \tau_2, \tau_3)$. The answer to this question is depicted in Fig. 9. The domain of support for $c_3^y(\tau_1 = m, \tau_2 = n)$ is the six-sided shaded region. This is due to the FIR nature of the MA system. The dark shaded triangular region in the first quadrant is the principal region. In the stationary case, we only have to determine third-order cumulant values in this region, R, where

$$R = \{m, n: 0 \le n \le m \le q\} \tag{54}$$

Observe, from Fig. 9, that the third-order cumulant equals zero for either one or both of its arguments equal to $q + 1$. This suggests that it should be possible to determine the order of the MA model, q, by testing, in a statistical sense, the smallness of a third-order cumulant such as $c_3^y(q + 1, 0)$. See Giannakis and Mendel (1990) to determine how to do this.

Table 1 provided a simple example of three MA systems that have exactly the same autocorrelations and spectra; hence, they cannot be resolved using second-order statistics. Table 4 continues this example. It gives the third-order cumulants for each MA(2) system for $R = \{m, n: 0 \le n \le m \le 2\}$. Observe that the third-order cumulants for the three systems are sufficiently different so that we suspect that it should be possible to resolve them using these cumulants. We shall prove that this is indeed the case below.

b. Polycepstra
Beginning with the bispectrum

$$C_3^y(z_1, z_2) = \gamma_3^y H(z_1) H(z_2) H(z_1^{-1}, z_2^{-1}), \tag{55}$$

Nikias and Pan (1987) and Pan and Nikias (1988) take the logarithm of $C_3^y(z_1, z_2)$, $\hat{C}_3^y(z_1, z_2) = \ln C_3^y(z_1, z_2)$, and take the inverse 2-D z-transform of $\hat{C}_3^y(z_1, z_2)$ to obtain the **complex bicepstrum** $b_y(m, n)$. Note that these steps parallel the steps which previously led to the complex cepstrum of $h(k)$, $\hat{h}(k)$. Note, also, that comparable results are also obtained by them for the **complex tricepstrum**, $t_y(m, n, l)$. In general, $b_y(m, n)$, $t_y(m, n, l)$, and the even higher-order complex cepstra are known as "polycepstra."

When $H(z)$ is represented as in Eqs. 7–9 (see, also Fig. 5), then Pan and Nikias have shown that: (1) $b_y(m, n)$ has nonzero values only at: $m = n = 0$, integer values along the m and n

Table 4. Minimum-, maximum- and mixed-phase systems with identical power spectra (or autocorrelations), but with different third-order statistics: $0 < a < 1$, $0 < b < 1$ (Nikias and Raghuveer, 1987).

Third-Order Cumulants			
$c_3^y(0, 0)$	$1 - (a + b)^3 + a^3 b^3$	$1 - (a + b)^3 + a^3 b^3$	$(1 + ab)^3 - a^3 - b^3$
$c_3^y(1, 0)$	$-(a + b) + ab(a + b)^2$	$(a + b)^2 - (a + b)a^2 b^2$	$a^2(1 + ab) - (1 + ab)^2 b$
$c_3^y(1, 1)$	$(a + b)^2 - (a + b)a^2 b^2$	$-(a + b) + ab(a + b)^2$	$-a(1 + ab)^2 + (1 + ab)b^2$
$c_3^y(2, 0)$	ab	$a^2 b^2$	$-a^2 b$
$c_3^y(2, 1)$	$-(a + b)ab$	$-(a + b)ab$	$ab(1 + ab)$
$c_3^y(2, 2)$	$a^2 b^2$	ab	$-ab^2$

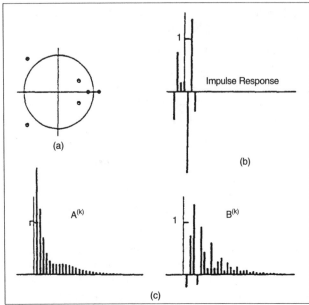

10. Mixed phase MA system with pronounced resonances: (a) zero location, (b) impulse response, and (c) cepstral coefficients [defined in Eqs. 13 and 14, respectively].

axes, and at the intersection of these values along the 45-degree line $m = n$, i.e.,

$$
b_y(m, n) = \begin{cases} \ln|\gamma_3^y c| & m = 0, n = 0 \\ -1/nA^{(n)} & m = 0, n > 0 \\ -1/mA^{(m)} & n = 0, m > 0 \\ 1/mB^{(-m)} & n = 0, m < 0 \\ 1/nB^{(-n)} & m = 0, n < 0 \\ -1/nB^n & m = n > 0 \\ 1/nA^{(-n)} & m = n < 0 \\ 0 & otherwise \end{cases} \tag{56}
$$

where the "A" and "B" cepstral coefficients are defined in Eqs. 13 and 14, respectively; and, (2) $t_y(m, n, l)$ has nonzero values only at: $m = n = l = 0$, integer values along the m, n and l axes, and at the intersection of these values along the 45-degree line $m = n = l$; for the specific values of $t_y (m, n, l)$, which are comparable to Eq. 56, see Appendix C of Pan and Nikias (1988).

Example 2. (Pan and Nikias, 1988): A noncausal MA(3,3) model with pronounced resonances is depicted in Fig. 10. Its transfer function is

$$
H(z) = \frac{(1 - 0.869z)(1 + 1.1z + 0.617z^2)}{(1 - 0.85z^{-1})(1 - 1.2z^{-1} + 0.45z^{-2})} \tag{57}
$$

Depicted, also in Fig. 10, are the "A" and "B" cepstral coefficients from which it is straightforward, using Eq. 56, to compute the bicepstral coefficients $b_y(m, n)$.

Table 5. Properties of Cepstral Coefficients
Here we give the properties of cepstrum coefficients as discussed by Nikias and Petropulu [1993, Ch. 5].

Let $h(k)$ be a deterministic energy signal with Fourier transform

$$
H(\omega) = |H(\omega)| \exp\{j\varphi(\omega)\}
$$

and complex cepstrum

$$
c_h(m) = F^{-1}[\ln H(\omega)] = \begin{cases} \dfrac{A^{(m)}}{m}, & m > 0 \\ \dfrac{-B^{(-m)}}{m}, & m < 0 \end{cases}
$$

where $F^{-1}[\bullet]$ denotes inverse Fourier transform.

The cepstrum of $|H(\omega)|$ is a function of $\{A^{(m)} + B^{(m)}\}$ i.e.,

$$
p_h^{(m)} = F^{-1}[\ln|H(\omega)|] = \begin{cases} -\dfrac{1}{m}[A^{(m)} + B^{(m)}], & m > 0 \\ \dfrac{1}{m}[A^{(-m)} + B^{(-m)}], & m < 0 \end{cases}
$$

whereas the cepstrum of $e^{j\varphi(\omega)}$ is a function of $\{A^{(m)} - B^{(m)}\}$, i.e.,

$$
f_h(m) = F^{-1}[\ln \exp\{j\varphi(\omega)\}] = \begin{cases} -\dfrac{1}{m}[A^{(-m)} - B^{(-m)}], & m < 0 \\ -\dfrac{1}{m}[A^{(m)} - B^{(-m)}], & m > 0 \end{cases}
$$

The cepstral coefficients $A^{(m)}$, $B^{(m)}$ can be computed directly from higher-order cumulants, without the use of phase unwrapping algorithms by solving a linear system of equations as shown by Pan and Nikias [1988].

The "A" and "B" cepstral coefficients have some very interesting and important properties, which are summarized in Table 5.

c. Identification of Nonminimum Phase Systems

Equation 17 demonstrates that, if we have access to both a system's input and output, then we can reconstruct its IR, $h(k)$, using the cross-correlation function, $r_{VZ}(\tau)$. In some signal processing applications (e.g., equalization, deconvolution in reflection seismology) we only have access to a system's output. Is it possible to determine $h(k)$ (or a scaled version of $h(k)$) just from output measurements? If signals are non-Gaussian, then Lii and Rosenblatt (1982) proved that it is indeed possible to reconstruct the correct phase IR just from output data using higher-order statistics. Their seminal paper has spawned a multitude of new methods for identifying nonminimum phase systems from just noisy output measurements. A very comprehensive survey of these methods is given in Mendel (1991).

Referring to Fig. 4, the problem is: given time-limited noisy measurements $Z(k)$, $k = 1, 2, ..., N$, estimate $H(z)$'s parameters, when $H(z)$ is represented as in Fig. 5, i.e.,

$$H(z) = B(z)/A(z) = \left[\sum_{j=0}^{q} b(j)z^{-j} \Big/ \sum_{i=0}^{p} a(i)z^{-i} \right] \quad (58)$$

The parameters to be identified are $b(1), ..., b(q), a(1), ..., a(p)$; $a(0) = b(0) = 1$ for scaling purposes, and orders p and q are assumed known. For a discussion of the more realistic case when orders p and q must also be determined, see Giannakis and Mendel (1990).

This output measurement identification problem occurs in many fields, including communications and reflection seismology. In the former, $V(k)$ is a "message," $h(k)$ is a "channel," and $Z(k)$ is a "distorted message" (distorted due to intersymbol interference). An accurate model of the channel is needed in order to restore the message at the receiver. This model is used in many equalization schemes. In reflection seismology $V(k)$ is the earth's "reflectivity sequence" (i.e., the earth's "message") $h(k)$ is the "seismic source wavelet," and $Y(k)$ is the "seismogram." An accurate model of the source wavelet is needed in order to estimate the earth's reflectivity sequence via deconvolution.

When the numerator parameters in Eq. 58 all equal zero except for $b(0)$, we have an all-pole model, in which case we are in the realm of AR parameter estimation. When all of the denominator parameters equal zero except for $a(0)$, we are in the realm of MA parameter estimation. These two special cases have been widely studied not only for their own interest, but also because some methods for estimating ARMA parameters proceed in two steps, by first estimating the AR parameters of the ARMA model and then estimating the MA parameters of the ARMA model, making use of the just-estimated AR parameters.

As a reminder, we use higher-order statistics to solve these identification problems because second-order statistics are phase blind [hence, they can only give rise to minimum phase

or maximum phase models, i.e., to SEMP models], and higher-order statistics are blind to additive Gaussian noise.

Because so many new and interesting methods have been developed for identifying the coefficients of an AR, MA or ARMA model, and these methods are carefully elaborated upon in Mendel (1991), we leave their details for the reader to explore. Here we demonstrate the rather remarkable result that it is possible to determine the parameters of an MA model in closed form. No result of this nature was available before the introduction of higher-order statistics into signal processing.

Example 3. This is a continuation of Example 1. Earlier, we showed that if we have access to both the input and output of a system, then we can reconstruct the correct phase IR, $h(k)$, using Eq. 17. Is there a comparable result, with a simple closed-form formula, for reconstructing the correct phase IR, $h(k)$, using just output measurements? Giannakis (1987) was the first to show that the IR of a qth-order MA system can be calculated just from the system's output cumulants using the following simple closed-form formula (stated here in terms of third-order cumulants; generalizations to arbitrary order cumulants can be found in Swami and Mendel [1988, 1990 (Eq.13)].

$$h(k) = c_3^z(q, k)/c_3^z(q, 0) \quad k = 0, 1, ..., q \quad (59)$$

A proof of this interesting result is given in Table 6.

Equation 59 is often referred to as the "$C(q, k)$ formula." Lohmann, et al (1983) and Lohmann and Wirnitzer (1984) provide a non-rigorous derivation of the $C(q, k)$ formula for 1-D and 2-D continuous-time processes. Note that Eq. 59 only uses the 1-D slice of the output cumulant along the right-hand side of the darkly shaded right triangle in Fig. 8. Note, also, that it requires exact knowledge of MA order q. It is interesting and important from a theoretical point of view, but is impractical from an actual computation point of view. This is because, in practice, the output cumulant must be estimated, and Eq. 59 does not provide any filtering to reduce the effects of cumulant estimation errors. Fortunately,

---

Table 6. Derivation of the $C(q, k)$ formula
Here, as in Giannakis (1987), we derive the $C(q, k)$ formula for third-order cumulants. We begin with (Eq. 48) for $k = 3$, i.e.,

$$c_3^z(\tau_1, \tau_2) = \gamma_3^v \sum_{k=0}^{\infty} h(k)h(k + \tau_1)h(k + \tau_2) \quad (1)$$

in which $h(0) = 1$, for normalization purposes. Set $\tau_1 = q$ and $\tau_2 = k$ in (1), and use the fact that for an MA(q) system $h(j) = 0$ $j > q$, to see that

$$c_3^z(q, k) = \gamma_3^v h(q)h(k) \quad (2)$$

Next, set $\tau_1 = q$ and $\tau_2 = 0$ in (1), to see that

$$c_3^z(q, 0) = \gamma_3^v h(q) \quad (3)$$

Dividing (2) by (3) we obtain the $C(q, k)$ formula (Eq.59).

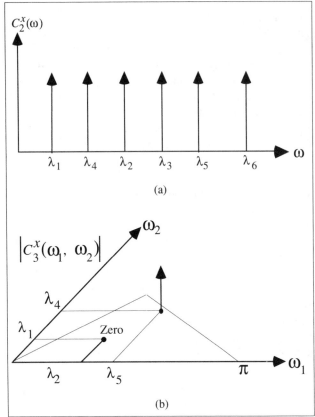

$$C_2^x(\omega)$$

$$\lambda_1 \quad \lambda_4 \quad \lambda_2 \quad \lambda_3 \quad \lambda_5 \quad \lambda_6 \qquad \omega$$

(a)

$$\left|C_3^x(\omega_1, \omega_2)\right|$$

$$\omega_2$$

$$\lambda_4$$

$$\lambda_1$$

Zero

$$\lambda_2 \quad \lambda_5 \quad \pi \qquad \omega_1$$

(b)

11. Quadratic phase coupling. (a) Power spectrum of the process described by Eq. 60. (b) Its magnitude bispectrum.

other cumulant-based methods have been developed that do provide such filtering.

Nonlinear Processes

Higher-order moment and cumulant spectra or polyspectra provide a means of detecting and quantifying nonlinearities in stochastic signals. These stochastic signals usually arise when a nonlinear system operates under a random input. General relations for arbitrary stationary random data passing through arbitrary linear systems have been studied quite extensively for many years. In principle, most of these relations are based on autocorrelation, power spectrum, or cross-correlation matching criteria. On the other hand, general relations are not available for arbitrary stationary random data passing through arbitrary nonlinear systems. Instead, each type of nonlinearity has been investigated as a special case. Polyspectra can play a key role in detecting and characterizing the type of nonlinearity in a system from its output sequence. In addition, cross-polyspectra may be used for nonlinear system identification from observations of input and output data.

There are situations in practice in which the interaction between two harmonic components causes contribution to the power at their sum and/or difference frequencies. For example, suppose the signal

$$X(k) = A_1 \cos(\lambda_1 k + \theta_1) + A_2 \cos(\lambda_2 k + \theta_2)$$

is passed through the following simple quadratic nonlinear system:

$$Z(k) = X(k) + \varepsilon X^2(k) \qquad (60)$$

where ε is a non-zero constant. The signal $Z(k)$ contains cosinusoidal terms in (λ_1, θ_1), (λ_2, θ_2), $(2\lambda_1, 2\theta_1)$, $(2\lambda_2, 2\theta_2)$, $(\lambda_1 + \lambda_2, \theta_1 + \theta_2)$ and $(\lambda_1 - \lambda_2, \theta_1 - \theta_2)$. Such a phenomenon, which gives rise to some or all of these phase relations, that are exactly the same as the frequency relations, is called quadratic phase coupling [Kim and Powers, 1978; Raghuveer and Nikias, 1985]. In certain applications it is necessary to determine if peaks at harmonically related positions in the power spectrum are, in fact, phase coupled. Since the power spectrum suppresses all phase relations, it cannot provide the answer. The third-order cumulants (the bispectrum), however, are capable of detecting and characterizing quadratic phase coupling.

Consider the process

$$X(k) = \sum_{i=1}^{6} \cos(\lambda_i k + \varphi_i) \qquad (61)$$

where $\lambda_2 > \lambda_1 > 0$, $\lambda_5 > \lambda_4 > 0$, $\lambda_3 = \lambda_1 + \lambda_2$, $\lambda_6 = \lambda_4 + \lambda_5$, $\varphi_1, \varphi_2, \dots, \varphi_5$ are all independent, uniformly distributed r.v.'s over $(0, 2\pi)$, and $\varphi_6 = \varphi_4 + \varphi_5$. In Eq. 61 whereas $\lambda_1, \lambda_2, \lambda_3$ and $\lambda_4, \lambda_5, \lambda_6$ are at harmonically related positions, only the component at λ_6 is a result of phase coupling between the components at λ_4 and λ_5; additionally, λ_3 is an independent harmonic component. The power spectrum of the process consists of impulses at λ_i, $i = 1, 2, \dots, 6$, as illustrated in Fig. 11. Looking at the spectrum one cannot say if the harmonically related components are, in fact, involved in quadratic phase-coupling relationships. The third-order cumulant sequence $c_3^x(n, \ell)$ of $X(k)$ can be easily obtained as, [Raghuveer and Nikias, 1985]

$$c_3^x(n, \ell) = \frac{1}{4}\{\cos(\lambda_5 n + \lambda_4 \ell) + \cos(\lambda_6 n + \lambda_4 \ell)$$
$$+ \cos(\lambda_4 n + \lambda_5 \ell) + \cos(\lambda_6 n - \lambda_5 \ell) \qquad (62)$$
$$+ \cos(\lambda_4 n - \lambda_6 \ell) + \cos(\lambda_5 n - \lambda_6 \ell)\}$$

It is important to observe that in Eq. 62, *only the phase coupled components appear.* Consequently, the bispectrum evaluated in one of the first quadrant triangular regions of Fig. 7 shows an impulse only at (λ_5, λ_4) indicating that only this pair is phase-coupled (see Fig. 11). In the total absence of phase coupling, the third-order cumulant sequence and hence the bispectrum are both zero. Thus the fact that only phase coupled components contribute to the third-order cumulant sequence of a process is what makes the bispectrum a useful tool for detecting quadratic phase coupling and discriminating phase-coupling components from those that are not.

Any of the existing bispectrum estimation techniques can be used for the analysis of quadratic phase-coupling phenomena; however, each of those techniques will exhibit certain

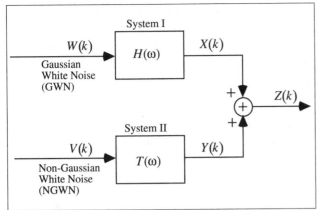

12. A signal $Z(k)$ whose power spectrum and bispectrum are modeled by different linear systems. The noise processes $W(k)$, $V(k)$ are assumed independent.

advantages and a number of limitations. For example, the conventional techniques can serve as better quantifiers of phase coupling (i.e., degree of coupling) whereas the parametric methods (AR and ARMA) are better as detectors rather than quantifiers [Raghuveer and Nikias, 1985; 1986].

The use of conventional methods for bispectrum estimation in conjunction with the magnitude bicoherence index $|P_3^x(\omega_1, \omega_2)|$ has been used extensively for the detection and quantification of quadratic phase coupling. When the magnitude bicoherence index takes on a value close to unity at a frequency pair where phase coupling has occurred this indicates an almost 100-percent degree of phase coupling. On the other hand, a near-zero value of $|P_3^x(\omega_1, \omega_2)|$ at harmonically related frequency pairs will suggest an absence of phase coupling (an almost zero-percent degree of phase coupling). Certainly, one of the advantages of using the conventional approach to bicoherence index estimation is its ability to serve as a good quantifier by providing good estimates of the degree of phase coupling at harmonically related frequency pairs. For example, the conventional bicoherence estimate of the signal.

$$X(k) = \sum_{i=1}^{4} \cos(\lambda_i k + \varphi_i) \qquad (63)$$

where $\lambda_3 = \lambda_1 + \lambda_2$, $\varphi_3 = \varphi_1 + \varphi_2$, $\lambda_4 = \lambda_1 + \lambda_2$ and φ_1, φ_2, φ_4 are i.i.d., uniformly distributed over $[0, 2\pi]$, will show a peak at (λ_1, λ_2) of magnitude approximately 0.5 indicating only a 50-percent degree of phase coherence. Possible limitations for using conventional methods for the analysis of quadratic phase coupling are the high variance of bispectrum estimates and the low resolution when harmonically related frequency pairs are close to each other [Nikias and Raghuveer, 1986]. Let us note that the effect of these limitations is reduced as the number and/or length of the data segments increase.

Quadratic phase coupling was studied by Raghuveer and Nikias, [1985, 1986], using AR models for bispectrum estimation. The motivation to use AR techniques was to take advantage of the high-resolution capability and low variance estimates associated with AR modeling. The justification for the use of AR methods for detection of quadratic phase coupling can be found in [Raghuveer and Nikias, 1985]. They have also shown that we cannot have just a single segment of data (i.e., just one set of fixed values for φ_1, φ_2, φ_3, φ_4 in Eq. 63) for detecting quadratic phase coupling between pairs of sinusoids. Segmentation of data into records is necessary to obtain consistent estimates of the third-order cumulants. The advantages of AR techniques arise in situations where the conventional estimators completely fail to resolve closely spaced frequency pairs. On the other hand, the limitation of the AR techniques lies on their inability to provide an accurate estimate of the degree of phase coherence when phase coupling does occur at harmonically related frequency pairs.

Swami and Mendel [1988] have shown that the trispectrum can be used to resolve cubic phase coupling, and, that if a signal contains components due to both quadratic and cubic phase coupling, the bispectrum of that signal is blind to the cubically-coupled components and can resolve the quadratically-coupled components, whereas the trispectrum of that signal is blind to the quadratically-coupled components and can resolve the cubically-coupled components. Hence, higher-order spectra can resolve different types of nonlinearities.

The linear model that can describe the bispectrum of a signal is generally different from the one that describes its power spectrum [Raghuveer and Nikias, 1985]. Consider the process, $Z(k)$, in Fig. 12. As we can see, $Z(k)$ is the sum of two signals, one which is the output of a linear system driven by white Gaussian noise $\{W(k)\}$ and the other, the output of a linear system driven by non-Gaussian white noise, $\{V(k)\}$. Further, $\{V(k)\}$ and $\{W(k)\}$ are independent, which implies that $X(k)$ and $Y(k)$ are also statistically independent, hence the bispectrum of $Z(k)$ is the sum of the bispectra of $X(k)$ and $Y(k)$. Since $X(k)$ is Gaussian, its bispectrum is zero and the bispectrum of $Z(k)$ is the bispectrum $Y(k)$, which means that the linear system, System II, of Fig. 12 models the bispectrum of $Z(k)$. In other words,

$$C_3^z(\omega_1, \omega_2) = \\ C_3^y(\omega_1, \omega_2) = \gamma_3^v T(\omega_1) T(\omega_2) T^*(\omega_1 + \omega_2). \qquad (64)$$

The power spectrum of $Z(k)$ is the sum of the power spectra of $X(k)$ and $Y(k)$. Since generally the contribution of $X(k)$ to the power spectrum of $Z(k)$ is significant, System II (alone does not model the power spectrum. In this case,

$$C_2^z(\omega) = C_2^x(\omega) + C_2^y(\omega) = \gamma_2^w |H(\omega)|^2 + \gamma_2^v |T(\omega)|^2. \qquad (65)$$

Clearly, if $T(\omega)$ and $H(\omega)$ represent two different AR models, the power spectrum of $Z(k)$ is truly an ARMA whereas its bispectrum is AR-type.

Nonlinear processes can also be represented by general Volterra systems [Schetzen, 1989]. We only examine here the popular case of a second-order Volterra model.

Suppose a signal is represented by the 2nd-order Volterra model

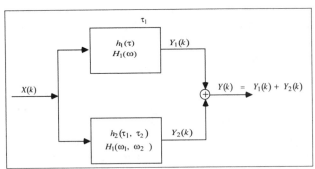

13. Second-order Volterra system: linear and quadratic parts in a parallel structure [Nikias and Petropulu, 1993].

$$Y(k) = \sum_{\tau_1} h_1(\tau_1)X(k - \tau_1) +$$
$$\sum_{\tau_1}\sum_{\tau_2} h_2(\tau_1, \tau_2)X(k - \tau_1)X(k - \tau_2) \qquad (66)$$

where $X(k)$ is a stationary random process with zero mean. The identification problem is to determine the impulse response, $h_1(\tau)$ and the kernel $h_2(\tau_1, \tau_2)$. Note that Eq. 61 can be viewed as a parallel connection of a linear system $\{h_1(\tau)\}$ and a quadratic system $\{h_2(\tau_1, \tau_2)\}$, as illustrated in Fig. 13. Assuming that the input signal, $X(k)$, is a stationary, zero-mean Gaussian process, Tick [1961] has established the following fundamental relationships:

$$H_1(\omega) = \frac{C_{xy}(\omega)}{C_2^x(\omega)} \qquad (67)$$

and

$$H_2(-\omega_1, -\omega_2) = \frac{C_{xxy}(\omega_1, \omega_2)}{2C_2^x(\omega_1)C_2^x(\omega_2)} \qquad (68)$$

where $C_{xy}(\omega)$, $C_{xxy}(\omega_1, \omega_2)$ are the cross-spectrum and cross-bispectrum, respectively. Consequently, when we have access to the input and output of the system illustrated in Fig. 13, Eqs. 67, and 68 represent the system identification formulas. It is important to remember that Eqs. 67 and 68 are valid only for a Gaussian input signal. More general results, assuming a non-Gaussian input, have been obtained by Hinich [1985] and Powers, et. al. [1989]. Additional results on particular simple nonlinear systems have been reported by Brillinger [1977] and Rozario and Papoulis [1989].

Applications

There have been numerous applications of cumulants and polyspectra during the past 25 years. See [USS Comprehensive Bibliography] for more than 500 references. The applications can be grouped into three major categories: physics-related, 1D, and 2D/3D. The **physics-related applications** deal with oceans (internal waves, noise, shoaling gravity waves, wave coupling, wave interaction, wave phenomena, ship resistance to waves, and sea-surface temperature anomolies), earth (free oscillations), atmosphere (pressure, turbulence), interplanetary (scintillation), wind (turbulence, currents), plasmas (wave interactions, nonlinear phenomena), electromagnetics (low frequency data), and crystallograpy (structures). The **1D applications** deal with: diagnosis (of surface roughness, machine faults, noisy mechanical systems), vibration analysis (signal pattern recognition, measurements, knock detection), speech (pitch detection, voiced/unvoiced decisions), noise (cancellation, ship radiated, from gears, bioelectric), system identification (nonminimum phase channels, input/output), detection, phase locking, FM signals, seismic deconvolution, range and doppler extraction, blind equalization, economic time series, brain potentials (eeg wave coupling) and biological rhythms. They also deal with a wide range of problems associated with: harmonic retrieval, nonlinear systems (Volterra, bilinear, phase coupling), array processing, sonar, and radar. The **2D/3D applications** deal with images (modeling, reconstruction, restoration, coding, motion estimation, sequence analysis), textures (model validation, discrimination), tomography (flow velocity field, 3D velocity field), speckle masking in astronomy, inverse filtering of ultrasonic images, and imaging photon-limited data.

In this section we give brief examples of how higher-order spectra have been applied to the following: array processing, classification, harmonic retrieval, time-delay estimation, blind equalization and interference cancellation. These examples are in no way exhaustive, but, instead, are meant to demonstrate the usefulness of higher-order spectra.

a. Array Processing

Array processing techniques play a very important role in the enhancement of signals in the presence of interference. Array processing problems include: direction of arrival (DOA) determination, determination of number of sources, beam forming, estimation of the source signal, source classification, sensor calibration, etc. See Van Veen and Buckley (1988) for an excellent introduction to array processing and its associated models.

So many novel and interesting array processing algorithms have appeared during the past ten years, why would one want to apply higher-order statistics to array processing problems? There are a number of answers to this question, including: (1) one of the most popular and important beamformers, namely, Capon's minimum-variance distortionless response (MVDR) beamformer, that has been the starting point for both signal enhancement and high-resolution DOA estimation, requires very specific and detailed information about the so-called array steering vector (e.g., source steering angles, array geometry, receiver responses), information that is often not available, or if available is not given to a high degree of accuracy; a technique recently developed by Dogan and Mendel (1992a), that is described below, shows how cumulants can be used to estimate the unknown steering vector, after which the MVDR beamformer can be used; (2) When additive noise is colored and Gaussian, a second-order statistics based high-resolution DOA algorithm, such as MUSIC, does not perform well; however a cumulant-based MUSIC algorithm does perform well; (3) most second-order statistics based beamformers assume that the received signals are not coherent, which rules out the important case of multipath propagation; cumulant-based beamformers can work in the presence of multipath (e.g., Dogan and Mendel, 1992a). The following references apply higher-order statistics to array processing: Cardosa (1989, 1990, 1991a, 1991b), Chiang and Nikias (1989), Comon (1989), Dogan and Mendel (1992a,b, 1993a,b), Duvaut (1990), Forster and Nikias (1990, 1991), Gaeta and Lacoume (1991), Giannakis and Shamsunder (1991), Jutten, et al. (1991), Lacoume and Ruiz (1988), Lagunas and Vazquez (1991), Mohler and Bugnon (1989), Moulines and Cardosa (1991), Porat and Friedlander (1990), Ruiz (1991), Ruiz and Lacoume (1989), Scarano and Jacovitti (1991), and Shamsunder and Giannakis (1991a, 1991b).

Dogan and Mendel (1992a) use cumulants of received signals to estimate the steering vector of a narrowband non-Gaussian desired signal in the presence of directional Gaussian interferers with unknown covariance structure. They assume no knowledge of the array manifold or DOA information about the desired signal. The desired signal could be voiced speech, sonar, radar return or a communication signal. The example below is for a communications scenario and requires the use of fourth-order cumulants, because the third-order cumulants for the communication signals are identically zero.

Consider an array of M elements, with arbitrary sensor response characteristics and locations. Assume there are J Gaussian interference signals $\{i_j(t), j = 1, 2, \ldots, J\}$, and a non-Gaussian desired signal $d(t)$, centered at frequency ω_0. Sources are assumed to be far away from the array so that a planar wavefront approximation is possible. The additive noise is assumed to be Gaussian with unknown covariance. Consequently, the M measurements can be collected together to give the following model:

$$\underline{r}(t) = \underline{a}(\theta_d)d(t) + A_I(\underline{\theta})\underline{i}(t) + \underline{n}(t) \tag{69}$$

where θ_d is the DOA of the desired signal; $\underline{a}(\theta_d)$ is the array steering vector of the desired signal; $A_I(\underline{\theta})$ is the array steering matrix for the J interference sources $\underline{i}(t)$; $\underline{\theta}$ is a $J \times 1$ vector of DOA's for the interferers; $\underline{r}(t)$ is the $M \times 1$ vector of received signals; and $\underline{n}(t)$ is the $M \times 1$ vector of Gaussian noises. We let R denote the covariance matrix of $\underline{r}(t)$.

The output of an MVDR beamformer can be expressed as (Capon, 1969)

$$y(t) = \underline{w}^H\underline{r}(t) = \left[\beta_1 R^{-1}\underline{a}(\theta_d) \right]^H \underline{r}(t) \tag{70}$$

where constant β_1 maintains a specified response for the desired signal and $\underline{w}$ denotes the weight vector of the processor. Clearly, MVDR beamforming requires knowledge of $\underline{a}(\theta_d)$. Without knowledge of the array manifold, it is not possible to determine $\underline{a}(\theta_d)$ even in the case of known θ_d. Consider the following vector of fourth-order cumulants $\underline{c} = \text{col}[c_1, c_2, \ldots, c_M]$, where $c_j = \text{cum}\left\{ r_1(t), r_1^H(t), r_1^H(t), r_j(t) \right\}, j = 1,2,\ldots,M$. Using the properties of cumulants stated in Table 2 and the receiver model in Eq.69, Dogan and Mendel (1992a) show that

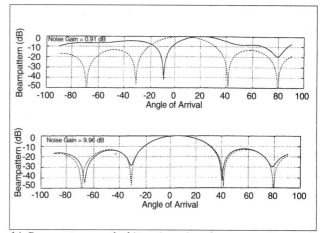

14. Beam patterns and white-noise gains of processors in a single realization for SNR = 20 dB: (a) MVDR and (b) Cumulant-based MVDR.

$$\underline{c} = \beta_2 \underline{a}(\theta_d) \tag{71}$$

where β_2 is another scale factor. Using this result, it is now possible to estimate $\underline{a}(\theta_d)$ directly from the given data using estimated fourth-order cumulants. Observe that $\underline{c}$ is a replica of the steering vector of the desired signal up to a scale factor. Substituting Eq. 71 into Eq. 70 we arrive at the following cumulant-based MVDR beamformer:

$$y(t) = \underline{w}_{\text{cum}}^H\underline{r}(t) = \left[\beta_3 R^{-1}\underline{c} \right]^H \underline{r}(t) \tag{72}$$

where $\beta_3 = \left(\underline{c}^H R^{-1}\underline{c} \right)^{-1}$

Many simulation results are given in Dogan and Mendel (1992a), not only for the incoherent source case but also for the coherent case (which requires a slight modification for the model in Eq. 69). Here we present results for the simplest scenario. The array is linear with 10 isotropic uniformly-spaced elements with uniform half-wavelength spacing. The record length is 1000 snapshots. This array is to be used for optimum reception of a BPSK signal which illuminates the array from 5 degrees broadside in the presence of temporally and spatially white, equal power, circularly symmetric sensor noise whose SNR is 20 dB. Fig. 14 compares the MVDR beamformer (assuming a priori knowledge of the DOA) and the cumulant-based MVDR beamformer. Plotted are the beam pattern responses $P(\theta) = \left| \underline{w}^H\underline{a}(\theta) \right|^2$. Also shown on the plots are the white-noise gains of the beamformers. All responses are normalized to have a maximum value of 0 dB. For comparison purposes, the optimum beamformer response, calculated by using the true statistics and true steering vector information in Eq.70, is presented as the dashed curves. Observe that even when the MVDR beamformer is given true steering vector information it does not perform as well as the cumulant-based MVDR beamformer. The cumulant-based beamformer yields excellent performance without any knowledge of source DOA. Note, also, that for 100 Monte-Carlo runs the mean values and standard deviations of the white-noise gains for the MVDR beamformer are 0.179

±1.360, whereas the comparable results for the cumulant-based MVDR beamformer are 9.954 ±0.015. The theoretical white noise gain for this scenario equals 10.

Recently, Dogan and Mendel (1993a, b) have shown that there are now two additional reasons to use higher-order spectra in array signal processing: (1) HOS can increase the effective aperture of an array, and (2) HOS can not only eliminate the effects of additive Gaussian noise, but it can also eliminate the effects of additive non-Gaussian noise. This is accomplished by computing the cross-correlations that are needed by all high-resolution direction-of-arrival techniques (e.g., ESPRIT) using fourth-order cumulants.

b. Classification

Pattern or signal classification can be done working directly with the pattern or signal samples, or with attributes related to them. It is difficult to handle additive colored Gaussian noise with traditional approaches. A new approach (Giannakis and Tsatsanis, 1992), that is blind to additive colored or white Gaussian noise, works with a vector of cumulants or polyspectra, and extends correlation-based classification to HOS-based classification. It is based on the important fact that estimates of cumulants or polyspectra are asymptotically Gaussian. Consequently, one is able to begin with an equation like

Estimate of HOS = HOS + estimation error   (73)

in which the "estimation error" is asymptotically Gaussian, and extend traditional classification or detection procedures to this formulation. Working with a vector of higher-order statistics is in the spirit of using attributes which are derived from the original data. Consequently, higher-order statistics now provide new attributes to be used in pattern and signal classification (including detection) problems. Additional work on using higher-order statistics for detection and classification problems can be found in: Giannakis and Dandawate (1991), Giannakis and Tsatsanis (1990), Hinich (1990), Jouny, et al. (1991a,b), Kletter and Messer (1989, 1990a, 1990b), Sadler (1991), Sadler and Giannakis (1990), Shamsunder and Giannakis (1991), Swindelhurst and Kailath (1989), and Tsatsanis and Giannakis (1989, 1992).

c. Harmonic Retrieval

The estimation of the number of harmonics and the frequencies and amplitudes of harmonics from noisy measurements is frequently encountered in several signal processing applications, such as in estimating the direction of arrival of narrow-band source signals with linear arrays, and in the harmonic retrieval problem. For the latter problem and real signals, we begin with the model

$$Y(k) = \sum_{j=1}^{p} a_j \cos(k\omega_j + \phi_j) + N(k) = X(k) + N(k) \quad (74)$$

where the ϕ_j's denote random phases which are i.i.d. and uniformly distributed over $[0, 2\pi]$, the ω_j's are unknown deterministic frequencies, and the a_j's are unknown deterministic amplitudes. The additive noise $N(k)$ is assumed to be white or colored Gaussian noise with unknown spectral density. The problem is to estimate the number of signals p, the angular frequencies ω_j's, and the amplitudes a_j's.

This problem has been very widely studied. When the additive noise is white then second-order statistics-based high-resolution methods, like MUSIC or Minimum Norm, combined with SVD (to determine p), give excellent results. When the noise is colored these methods break down They tend to overestimate the number of sinusiods by treating the colored noise as additional sinusoids. Higher-order statistics have no problem with any kind of Gaussian noise; hence, they have been successfully applied to this problem.

Third-order cumulants for $Y(k)$ equal zero; hence, this is an application where one must use fourth-order cumulants. The fourth-order cumulant of $Y(k)$ is, in general, a function of three lags (see Swami and Mendel (1991) or Mendel (1991) for the general expression); however, the diagonal slice of this cumulant is given as

$$c_4^y(\tau, \tau, \tau) \triangleq c_4^y(\tau) = -\frac{3}{8} \sum_{j=1}^{p} a_j^4 \cos(\omega_j \tau) \quad (75)$$

It is well-known (Marple, 1987) that the autocorrelation of $Y(k)$ is given by

$$c_2^y(\tau) = \frac{1}{2} \sum_{j=1}^{p} a_j^2 \cos(\omega_j \tau) \quad (76)$$

Comparing Eqs. 75 and 76 we see that (except for a difference in scale factors) $c_4^y(\tau)$ can be treated as an autocorrelation function of the following signal which is obviously related to $Y(k)$:

$$Y_1(k) = \sum_{j=1}^{p} a_j^2 \cos(k\omega_j + \phi_j) + N(k) \quad (77)$$

This means that already existing high-resolution methods, such as MUSIC and Minimum Norm, can be applied, just about as is, by replacing correlation quantitites with $c_4^y(\tau)$. For details on exactly how to do this, see Swami and Mendel (1991). For an extensive comparison of correlation-based and cumulant-based methods for determining the number of harmonics when the amplitude of one harmonic decreases, when the frequency of one harmonic approaches the other, and when different lengths of data are used, for the case of two harmonics in colored Gaussian noise, see Shin and Mendel (1991, 1992). Additional work on applying higher-order statistics to harmonic retrieval problems can be found in: Anderson and Giannakis (1991), Ferrari and Alengrin (1991), Moulines and Cardosa (1991), Pan and Nikias (1988), Shi and Fairman (1991), Swami (1988), and Swami and Mendel (1988).

d. Time-Delay Estimation

One important application of time delay estimation methods is for source bearing and range calculation.

Let us assume that $\{X(k)\}$ and $\{Y(k)\}$ are two spatially separated sensor measurements that satisfy the equations

$$X(k) = S(k) + W_1(k) \tag{78}$$

$$Y(k) = AS(k - D) + W_2(k) \tag{78b}$$

where $\{S(k)\}$ is an unknown signal, $\{S(k - D)\}$ is a shifted and probably scaled version of $\{S(k)\}$, and $\{W_1(k)\}$ and $\{W_2(k)\}$ are unknown noise sources. The problem is to estimate the time delay D from finite length measurements of $X(k)$ and $Y(k)$. This situation arises in such application areas as sonar, radar, biomedicine, geophysics, etc. The basic approach to solve the time delay estimation problem is to shift the measurement sequence $\{X(k)\}$ with respect to $\{Y(k)\}$, and look for similarities between them. The best match will occur at a shift equal to D. In signal processing, "look for similarities" is translated into "taking the cross-correlation" between $\{X(k)\}$ and $\{Y(k)\}$. That is

$$c_{xy}(\tau) = E\left\{X(k)Y(k + \tau)\right\}$$

$$= Ac_2^s(\tau - D), \qquad -\infty < \tau < \infty \tag{79}$$

provided that $\{W_1(k)\}$ and $\{W_2(k)\}$ are zero-mean stationary signals, independent with each other and with $\{S(k)\}$. Note that

$$c_2^s(\tau) = E\left\{S(k)S(k + \tau)\right\} \tag{80}$$

is the covariance sequence of $\{S(k)\}$.

The $c_{xy}(\tau)$ in Eq. 79 peaks at $\tau = D$. However, in practical situations, due to finite length data records and noise sources that are not exactly independent, the $c_{xy}(\tau)$ does not necessarily show a peak at the time delay D. Various window functions have been suggested to smooth the cross-correlation function in order to improve the quality of time delay estimates. These are ROTH, SCOT, PHAT, Eckart, and Hannan-Thompson (or maximum likelihood) just to name a few [Carter, 1987].

In practical application problems where the signal $\{S(k)\}$ can be regarded as a non-Gaussian stationary random process, and the noise sources are independent stationary Gaussian, the similarities between $\{X(k)\}$ and $\{Y(k)\}$ could also be "compared" in higher-order spectrum domains such as the bispectrum. Let us note that self-emitting signals from complicated mechanical systems contain strong quasi-periodic components and therefore can be regarded as non-Gaussian signals. The main motivation behind the use of higher-order spectra for time-delay estimation under the aforementioned assumptions is the fact that they are free of Gaussian noise. Assuming that $\{S(k)\}$ has also a nonzero measure of skewness, the following identities hold:

$$c_3^x(\tau, \rho) = E\{X(k)X(k + \tau)X(k + \rho)\} = c_3^s(\tau, \rho) \tag{81a}$$

$$c_{xyx}(\tau, \rho) = E\left\{X(k)Y(k + \tau)X(k + \rho)\right\} = c_3^s(\tau - D, \rho) \tag{81b}$$

where

$$c_3^s(\tau, \rho) = E\left\{S(k)S(k + \tau)S(k + \rho)\right\}$$

because the third-order cumulants of a Gaussian process are identically zero. Obtaining the bispectra of the third-order cumulants in Eqs. 81a and 81b we have

$$C_3^x(\omega_1, \omega_2) = C_3^s(\omega_1, \omega_2) \tag{82a}$$

$$C_{xyx}(\omega_1, \omega_2) = C_3^2(\omega_1, \omega_2) \exp\left[j\omega_1 D\right]. \tag{82b}$$

Assuming $C_3^s(\omega_1, \omega_2)$ is nonzero, the following ratio can be formed.

$$I(\omega_1, \omega_2) = \frac{C_{xyx}(\omega_1, \omega_2)}{C_3^x(\omega_1, \omega_2)} = \exp\left[j\omega_1 D\right]. \tag{83}$$

One way of computing the time delay D is to form the function

$$T(\tau) = \int_{-\pi}^{+\pi}\int_{-\pi}^{+\pi} \exp\left[-j(\tau - D)\omega_1\right] d\omega_1 d\omega_2 \tag{84}$$

which peaks at $\tau = D$. There are, of course, several other ways that have been developed based on parametric modeling of the third-order cumulants. It is important to note that Eq. 83 is "free" of Gaussian noise and thus better time delay estimates can be expected using $I(\omega_1, \omega_2)$. Figure 2 illustrates time delay estimation results in the presence of Gaussian noises obtained by 2nd-order and 3rd-order statistics-based methods. From this figure, it is apparent that the bispectrum-based method exhibits better noise reduction capability.

Time delay estimation methods based on higher-order statistics have been developed by Nikias and Pan [1988], Nikias and Liu [1990], Tugnait [1989], Zhang and Raghuveer [1991] and Oh et al. [1990].

e. Blind Deconvolution and Equalization

The blind deconvolution, or equalization problem, deals with the reconstruction of the input sequence given the output of a linear system and statistical information about the input. Blind deconvolution algorithms are essentially adaptive filtering algorithms designed in such a way that they do not need the external supply of a desired response to generate the error signal in the output of the adaptive equalization filter. In other words, the algorithm is blind to the desired response. However, the algorithm itself generates an estimate of the desired response by applying a non-linear transformation on sequences involved in the adaptation process. There are three important families of blind equalization algorithms depend-

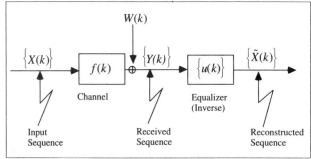

15. Block diagram of a baseband communication system subject to additive noise.

ing on where the nonlinear transformation is being applied on the data. These are:

(i) The Bussgang algorithms, where the nonlinearity is in the output of the adaptive equalization filter,

(ii) The Polyspectra algorithms, where the nonlinearity is in the input of the adaptive equalization filter, and

(iii) The algorithms where the nonlinearity is inside the equalization filter, i.e., nonlinear filters (e.g. Volterra) or neural networks.

Detailed discussion on blind equalization can be found in Haykin [1991] and Nikias and Petropulu [1993].

Let us consider a discrete-time linear transmission channel with impulse response, $f(k)$, which is unknown and possibly time-varying. The input data, $X(k)$, are assumed to be independent and identically distributed (i.i.d.) random variables with a non-Gaussian probability density function, with zero mean and variance, $E\{X^2(k)\} = \gamma_2^x$. Initially the noise will not be taken into account in the output of the channel. Then, the received sequence, $Y(k)$, (see Fig. 15) is

$$Y(k) = f(k)*X(k) = \sum_i X(k-i)f(i). \qquad (85)$$

The problem is to restore $X(k)$ from the received sequence $Y(k)$, or equivalently, to identify the inverse filter (equalizer), $u(k)$, of the channel.

From Fig. 15 we see that the output sequence $\widetilde{X}(k)$ of the equalizer is given by

$$\widetilde{X}(k) = u(k)*Y(k) = u(k)*f(k)*X(k). \qquad (86)$$

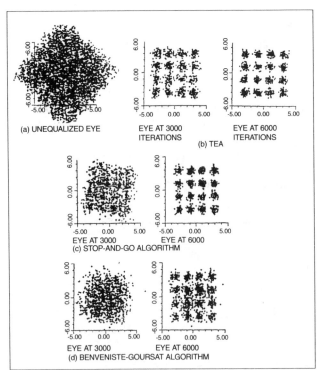

16. (a) Discrete eye pattern of a 16-QAM constellation distorted by the channel example, (b) Equalized using TEA, (c) Equalized using Stop-and-Go algorithm, (d) Equalized using the Benveniste-Goursat algorithm (Hatzinakos and Nikias, 1991).

To achieve

$$\widetilde{X}(k) = X(k-D)e^{j\theta} \qquad (87)$$

where D is a constant delay and θ is a constant phase shift, it is required that

$$u(k)*f(k) = \delta(k-D)\,e^{j\theta}, \qquad (88)$$

where $\delta(k)$ is the Kronecker delta function. Taking the Fourier transform of Eq. 88 we obtain

$$U(\omega)F(\omega) = e^{j(\theta-\omega D)}.$$

Hence, the objective of the equalizer is to achieve a transfer function

$$U(\omega) = \frac{1}{F(\omega)}e^{j(\theta-\omega D)} \qquad (89)$$

In general, D and θ are unknown. However, the constant delay D does not affect the reconstruction of the original input sequence $X(k)$. The constant phase θ can be removed by a decision device.

Blind equalization algorithms based on higher-order statistics (HOS) perform a nonlinear transformation on the input of the equalizer filter. This nonlinear transformation is a memory nonlinearity and it is identical to the generation of higher-order cumulants of the received channel data. One of the first blind equalizers based on HOS introduced in the literature is the Tricepstrum Equalization Algorithm (TEA) [Hatzinakos and Nikias, 1991] that estimates the equalizer impulse response by using the complex cepstrum of the fourth-order cumulants (tricepstrum) of the synchronously sampled received signal. Two extensions of the TEA have also been reported in the literature. The first one is the Power Cepstrum and Tricoherence Equalization Algorithm (POTEA) which recovers the Fourier magnitude of the equalizer using autocorrelations and its Fourier phase using fourth-order cumulants and the cepstrum of the tricoherence [Bessios and Nikias, 1991]. The second approach is an extension of TEA to the multichannel case using the cross-cumulants of the observed signals. It was thus designated as the Cross-Trispectrum Equalization Algorithm (CTEA) [Brooks and Nikias, 1991].

It has been well documented in the literature that the polyspectra-based blind equalizers achieve much faster convergence than the Bussgang-type algorithms at the expense of more computations per iteration. Fig.16 illustrates the eye-diagrams of a 16-QAM constellation distorted by a channel and then equalized by TEA and two other algorithms that belong to the Bussgang family, namely, the Stop-and-Go and Benveniste-Goursat algorithms (Hatzinakos and Nikias, 1991]. Clearly, the TEA algorithm opens the eye much earlier than the other two equalizers.

Blind equalizers based on HOS and parametric models have been developed by Porat and Friedlander [1991] and Tugnait [1991].

f. Interference Cancellation

When a signal of interest (SOI) is corrupted by an additive interference, and an auxiliary reference signal, which is

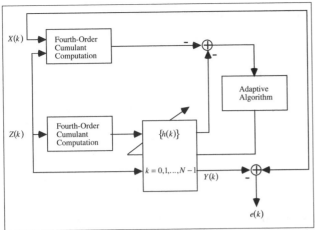

17. The configuration of the adaptive noise canceler using fourth-order statistics (ANC-FOS). The adaptive cancellation filter is {h(k)}.

A conventional transversal ANC, which is denoted in this paper as ANC-SOS algorithm, utilizes the LMS algorithm and second-order statistics (SOS) [Haykin, 1991]. Applying the ANC-SOS algorithm in practice, we usually encounter two major difficulties. The first is that the ANC-SOS filter is affected directly by uncorrelated noises at the primary and reference inputs. The second is that ANC-SOS algorithm is problem-dependent; i.e., it is very sensitive to both the reference signal statistics and the choice of step size.

Let $\{X(k)\}$ and $\{Z(k)\}$ denote measurements of the primary and reference sensors, respectively, satisfying

$$X(k) = S(k) + I(k) + N_p(k) \tag{90}$$

$$Z(k) = W(K) + N_\tau(k) \tag{91}$$

where $\{S(k)\}$ is the signal of interest (SOI), $\{I(k)\}$ is the interference (narrowband or wideband), $\{W(k)\}$ is a reference signal highly correlated with the interference, and $\{N_p(k)\}$ and $\{N_\tau(k)\}$ are uncorrelated sensor noises. We assume that the SOI is zero-mean and any kind of a signal, i.e., deterministic or random, or combination of both, and uncorrelated with the interference and the reference signal. The reference signal is a stationary, zero-mean, non-Gaussian random process. The noises $\{N_p(k)\}$ and $\{N_\tau(k)\}$ are zero-mean, white or colored Gaussian, uncorrelated with each other and independent of the SOI, interference, and reference signal.

Shin and Nikias [1992] have developed a new ANC based on fourth-order statistics (ANC-FOS) and have shown that the ANC-FOS filter is independent of white or colored Gaussian uncorrelated noises and insensitive to both the reference

highly correlated with only the interference, is also available, the elimination of the interference is accomplished by an adaptive noise canceling procedure. The reference signal is processed by an adaptive filter to match the undesired interference as closely as possible, and the filter output is subtracted from the primary input, which consists of the SOI and interference, to produce a system output. The objective of an adaptive noise canceller (ANC) is to produce a system output that best fits the SOI. Applications of ANC include the canceling of several kinds of interference in communications, speech, antenna sidelobe interference, and telephone circuits, just to name a few.

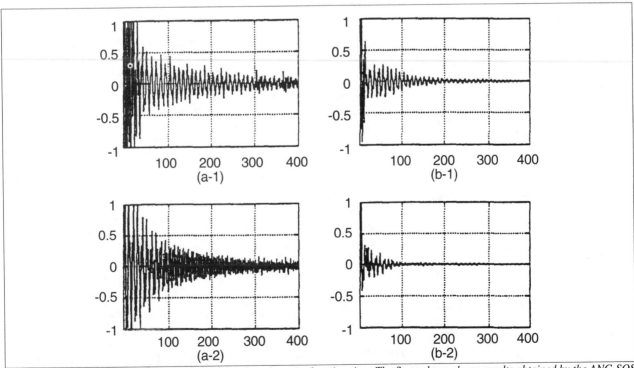

18. The error between the SOI and its reconstructed version as a function time. The first column shows results obtained by the ANC-SOS algorithm (N = 32) when $(a - 1)A_1 = \sqrt{2}$ and $A_2 = 0.5$ and $(a - 2)A_1 = 0.5$ and $A_2 = 1$; the second column shows results obtained by the ANC-FOS algorithm (N = 16) when $(b - 1)A_1 = \sqrt{2}$ and $A_2 = 0.5$ and $(b - 2)A_1 = 0.5$ and $A_2 = 1$.

signal statistics and the step size. Figure 17 illustrates the block diagram of the ANC-FOS algorithm.

We consider a typical example to compare the performance of the ANC-FOS algorithm with that of the ANC-SOS algorithm for eliminating interferences. Comparisons are presented in terms of the *error between the SOI and its reconstructed version by each ANC algorithm.* We assume that the SOI is deterministic BPSK having two states, and , satisfying

$$S(k) = \begin{cases} \cos(2\pi f_s Tk), & \text{for } s_1 \\ -\cos(2\pi f_s Tk), & \text{for } s_0 \end{cases} \qquad (92)$$

where $f_s T = 0.43$ and the duration of one state is 20 samples. The reference signal is assumed to be a sum of real-valued sine waves ($W(k) = W_1(k) + W_2(k)$)

$$W_i(k) = A_i \sin(2\pi f_i Tk + \phi_i), \quad i = 1, 2 \qquad (93)$$

where A_i's and f_i's denote amplitudes and frequencies, respectively and ϕ_i·s are independent random variables uniformly distributed over $[-\pi, \pi]$. Note that $f_1 T = 0.1$ and $f_2 T = 0.25$. Each interference signal $\{I_i(k), i = 1, 2\}$ is generated through three MA(2) systems excited by a reference signal $\{W_i(k), i = 1, 2\}$. The corresponding MA coefficients equal $[1, 0.1, -0.3], [1, 0.5, -0.1]$, and $[1, -0.2, 0.2]$, respectively. We assume that $I(k) = I_1(k) + I_2(k)$ with $A_1 = \sqrt{2}, A_2 = 0.5$ or $A_1 = 0.5$ and $A_2 = 0.1$. Figure 18 illustrates the results obtained by both ANC-SOS and ANC-FOS. From this figure, it is apparent that much faster convergence can be achieved with the ANC-FOS algorithm.

Adaptive filter schemes based on higher-order statistics have also been published by Chiang and Nikias [1988, 1990] and Dandawate and Giannakis [1989].

Conclusion

During the past two decades spectrum estimation techniques have proved essential to the creation of advanced communication, sonar, radar, speech, biomedical, geophysical and imaging systems. These techniques only use second-order statistical information, which means that we have been assuming that the signals are inherently Gaussian. Most real-world signals are not Gaussian. It is no wonder, therefore, that spectral techniques often have serious difficulties in practice.

There is much more information in a stochastic non-Gaussian or deterministic signal than conveyed by its autocorrelation or spectrum. Higher-order spectra (i.e., polyspectra), which are defined in terms of the higher-order statistics of a signal, contain this additional information. In this tutorial, an overview of higher-order spectral analysis and its applications in signal processing has been presented.

Signal processing algorithms based on higher-order spectra are now available for use in commercial and military applications. The emergence of low cost very high speed hardware chips and the ever growing availability of fast computers now demand that we extract more information than we have been doing in the past from signals, so that better decisions can be made. All of the new algorithms that have been developed using higher-order spectra are application driven.

Acknowledgment

Most of the material presented in this tutorial paper is based on the textbook by C. L. Nikias and A. P. Petropulu, *Higher-Order Spectral Analysis,* Oppenheim Series of Signal Processing, Prentice-Hall, Inc. 1993, as well as the tutorials by C. L. Nikias and M. R. Raghuveer, "Bispectrum Estimation: A Digital Signal Processing Framework," *Proc. IEEE,* Vol. 75, July 1987 and J. M. Mendel, "Tutorial on Higher-Order Statistics (Spectra) in Signal Processing and System Theory: Theoretical Results and Some Applications," *Proc. IEEE,* Vol. 79, March 1991. The plots in most of the figures were obtained using Hi-Spec™ which is a software package for signal processing with higher-order spectra. Hi-Spec™ is a trademark of United Signals & Systems, Inc. We thank Dae Shin and John Dogan for producing some of the figures in this paper.

Chrysostomos L. Nikias is Professor of Electrical Engineering and Director of the Center for Research on Applied Signal Processing (CRASP) at the University of Southern California (USC) at Los Angeles; *Jerry M. Mendel* is Professor of Electrical Engineering and the Director of the Signal & Image Processing Institute, also at USC.

References

Anderson, J. and G. B. Giannakis, "Two-Dimensional Harmonic Retrieval Using Cumulants," *Proc. of 7th Workshop on Multidimensional Signal Processing,* p. 210, Lake Placid, NY, September 1991.

Bartlett, M. S., *An Introduction to Stochastic Processes,* Cambridge University Press, UK, 1955.

Bessios, A. G. and C. L. Nikias, "FFT-Based Bispectrum Computation on Polar Rasters," *IEEE Transactions on Signal Processing,* November 1991.

Brillinger, D. R., *An Introduction to Polyspectra,* Ann. Math. Statist., **36**, pp. 1351–1374, 1965.

Brillinger, D. R. and M. Rosenblatt, *Asymptotic Theory of Estimates of kth Order Spectra,* in Spectral Analysis of Time Series, B. Harris, ed., Wiley, New York, NY, pp. 153–188, 1967.

Brillinger, D. R. and M. Rosenblatt, "Computation and Interpretation of kth - Order Spectra," in *Spectral Analysis of Time Series,* B. Harris, ed., John Wiley and Sons, New York, NY, pp. 189–232, 1967.

Brillinger, D. R., "The Identification of a Particular Nonlinear Time Series System," *Biometrika,* **64**, 3, pp. 509–515, 1977.

Brooks, D. H. and C. L. Nikias, "The Cross Bicepstrum: Properties and Applications for Signal Reconstruction and System Identification," *Proc. 1991 ICASSP,* pp. 3433–3436, Toronto, Canada, May 1991.

Capon, J., "High-Resolution Frequency-Wavenumber Spectral Analysis," *Proc. of IEEE,* **57**, no. 8, pp. 1408–1418, August 1969.

Cardosa, J.-F., "Source Separation Using Higher-Order Moments," *Proc. ICASSP,* Glasgow, Scotland, May 1989.

Cardosa, J.-F., "Eigen-Structure of the Fourth-Order Cumulant Tensor with Applications to the Blind Source Separation Problem," *Proc. 1990 ICASSP,* pp. 2655–2658, Albuquerque, NM, 1990.

Cardosa, J.-F., "Super-Symmetric Decomposition of the Fourth-Order Cumulant Tensor, Blind Identification of More Sources than Sensors," *Proc. 1991 ICASSP,* pp. 3109–3112, Toronto, Canada, May 1991a.

Cardosa, J.-F., "Higher-Order Narrow-Band Array Processing," *Proc. of Int'l. Workshop on Higher-Order Statistics,* pp. 121–130, Chamrousse, France, July 1991b.

Chiang, H. and C. L. Nikias, "Cumulant-Based Adaptive Time Delay Estimation," *Proc. 1988 Asilomar Conference,* Pacific Grove, CA, pp. 15–19, October 1988.

Chiang, H. and C. L. Nikias, "The ESPRIT Algorithm with Higher-Order Statistics," *Proc. Workshop on Higher-Order Spectral Analysis*, pp. 163–168, Vail, CO, 1989.

Chiang, H. and C. L. Nikias, "Adaptive Deconvolution and Identification of Nonminimum Phase FIR Systems Based on Cumulants," IEEE Trans. Automatic Control, AC **35**, pp. 36–47, January 1990a.

Chiang, H. and C. L. Nikias, "A New Method for Adaptive Time Delay Estimation for Non-Gaussian Signals," *IEEE Trans. Acoustics, Speech and Signal Processing,* ASSP **38**, pp. 209–219, February 1990b.

Comon, P., "Separation of Sources Using High-Order Cumulants," *SPIE Conf. on Advanced Algorithms and Architectures for Signal Processing, Real- Time Signal Processing Vol..XII*, pp. 170–181, San Diego, CA, August 1989.

Dandawate, A. V. and G. B. Giannakis, "A Triple Cross-Correlation Approach for Enhancing Noisy Signals," *Proc. Workshop on Higher-Order Spectral Analysis,* Vail, CO, pp. 212–216, June 1989.

Dogan, M. C. and J. M. Mendel, "Cumulant-Based Optimum Beamforming," USC-SIPI Report # 195, January 1992a; also accepted for publication in *IEEE Trans. on Aerospace and Electronic Systems* (to appear in 1994).

Dogan, M. C. and J. M. Mendel, "Single Sensor Detection and Classification of Multiple Sources by Higher-Order Spectra," *Proc. of IEEE Statistical Signal & Array Processing Workshop*, Victoria, BC, Canada, October 7–9, 1992b.

Dogan, M. C., and J. M. Mendel, "Joint Array Calibration and Direction-Finding With Virtual-ESPRIT Algorithm," *Proc. 1993 IEEE Signal Processing Workshop on Higher-Order Statistics*, Lake Tahoe, CA, June 7–9, 1993a.

Dogan, M. C., and J. M. Mendel, "Antenna Array Noise Reconditioning by Cumulants," *Proc. 1993 IEEE Signal Processing Workshop on Higher-Order Statistics*, Lake Tahoe, CA, June 7–9, 1993b.

Duvaut, P., "Principles of Source Separation Methods Based on Higher-Order Statistics," *Traitment du Signal*, **7**, pp. 407–418, December 1990.

Ferrari, A. and L. Alengrin, "Estimation of the Frequencies of a Complex Sinusoidal Noisy Signal Using Fourth Order Statistics," *Proc. 1991 ICASSP*, pp 3457–3460, Toronto, Canada, May 1991.

Forster, P. and C.L. Nikias, "Bearing Estimation in the Bispectrum Domain," *Proc. ASSP Workshop on Spectrum Estimation and Modeling*, pp. 5–9, Rochester, NY, October 1990; also, in *IEEE Transactions on Acoustics, Speech and Signal Processing,* **39**, pp. 1994–2006, September 1991.

Gaeta, M. and J. L. Lacoume, "Source Separation versus Hypothesis Testing," *Proc. of Int'l. Workshop on Higher-Order Statistics*, pp. 1269–1272, Chamrousse, France, July 1991.

Giannakis, G. B., "Cumulants: a Powerful Tool in Signal Processing," *Proc. IEEE*, **75**, pp. 1333–1334, 1987.

Giannakis, G. B. and J. M. Mendel, "Cumulant-Based Order Determination of Non-Gaussian ARMA Models," *IEEE Trans. on Acoustics, Speech and Signal Processing*, **38**, pp. 1411–1423, 1990.

Giannakis, G. B. and M. K. Tsatsanis, "Signal Detection and Classification Using Matched Filtering and Higher-Order Statistics," *IEEE Trans. on Acoustics, Speech and Signal Processing*, **38**, pp. 1284–1296, July 1990.

Giannakis, G. B. and S. Shamsunder, "Modeling of Non-Gaussian Array Data Using Cumulants: DOA Estimation with Less Sensors than Sources," *Proc. of 25th Conf. on Info. Sciences and Systems*, pp. 600–606, The Johns Hopkins University, Baltimore, MD, March 1991.

Giannakis, G. B. and A. V. Dandawate, "Detection and Classification of Non-Stationary Underwater Acoustic Signals Using Cyclic Cumulants," *Proc. Underwater Signal Proc. Workshop*, University of Rhode Island, October 1991.

Giannakis, G. B. and M. K. Tsatsanis, "A Unifying Maximum-Likelihood View of Cumulant and Polyspectral Measures for Non-Gaussian Signal Classification and Estimation," *IEEE Trans. on Information Theory*, March 1992.

Gratsteyn, I. S. and I. B. Ryzhik, *Table of Integrals, Series and Products,* New York Academic Press, 1980.

Hassab, J. C., *Underwater Signal and Data Processing,* CRC Press, Inc., Boca Raton, FL, 1989.

Hatzinakos, D. and C. L. Nikias, "Estimation of Multipath Channel Response in Frequency Selective Channels," *IEEE Journal Selected Areas in Communications*, **7**(1), pp. 12–19, January 1989.

Hatzinakos, D. and C. L. Nikias, "Blind Equalization Using a Tricepstrum Based Algorithm," *IEEE Trans. Communications,* **39**, pp. 669–682, May 1991.

Haykin, S., *Nonlinear Methods of Spectral Analysis,* 2nd edition, Berlin, Germany, Springer-Verlag, 1983.

Hinich, M. J., "Testing for Gaussianity and Linearity of a Stationary Time Series," *J. Time Series Analysis*, **3**(3), pp. 169–176, 1982.

Hinich, M. J., "Identification of the Coefficients in a Non-Linear Time Series of the Quadratic Type," *J. of Economics,* **30**, pp. 269–288, 1985.

Hinich, M.J., "Detecting a Transient Signal by Bispectral Analysis," *IEEE Trans. Acoustics, Speech and Signal Processing,* ASSP **38**, pp. 1277–1283, July 1990.

Huber, P. J., B. Kleiner, T. Gasser and G. Dumermuth, "Statistical Methods for Investigating Phase Relations in Stationary Stochastic Processes," *IEEE Transactions Audio Electroacoustics,* AU **19**, pp. 78–86, 1971.

Jouny, I., R. Moses and F. Garber, "Classification of Radar Signals Using the Bispectrum," *Proc. 1991 ICASSP*, pp. 3429–3432, Toronto, Canada, May 1991a.

Jouny, I. and E. K. Walton, "Applications of the Bispectrum in Radar Signatures Analysis and Target Identification," *Proc. of Int'l. Workshop on Higher-Order Statistics*, pp. 171–174, Chamrousse, France, July 1991b.

Jutten, C., L. N. Thi, E. Dijkstra, E. Vittoz and J. Caelen, "Blind Separation of Sources: an Algorithm for Separation of Convolutive Mixtures," *Proc. of Int'l. Workshop on Higher-Order Statistics*, pp. 273–276, Chamrousse, France, July 1991.

Kay, S. M., *Modern Spectral Estimation,* Prentice-Hall, Inc., Englewood Cliffs, NJ, 1988.

Kim, Y. C. and E. J. Powers, "Digital Bispectral Analysis of Self-Excited Fluctuation Spectra, " *Phys. Fluids*, **21**(8), pp. 1452–1453, August 1978.

Kletter, D. and H. Messer, "Detection of a Non-Gaussian Signal in Gausssian Noise Using High-Order Spectral Analysis," *Proc. Workshop on Higher-Order Spectral Analysis*, pp. 95–99, Vail, CO, 1989.

Kletter, D. and H. Messer, "Suboptimal Detection of Non-Gaussian Signals by Third-Order Spectral Analysis," *IEEE Trans. on Acoustics, Speech and Signal Processing*, **36**, pp. 901–909, 1990a.

Kletter, D. and H. Messer, "Optimal Detection of a Random Multitone Signal and its Relation to Bispectral Analysis," *Proc. 1990 ICASSP*, pp. 2391–2394, Albuquerque, NM, April 1990b.

Lacoume, J. L. and P. Ruiz, "Source Identification: a Solution Based on the Cumulants," *Proc. 4th ASSP Workshop on Spectral Estimation and Modeling*, pp. 199–203, August 1988.

Lagunas, M. A. and G. Vazquez, "Array Processing from Third Order Functions," *Proc. of Int'l. Workshop on Higher-Order Statistics*, p. 21720, Chamrousse, France, July 1991.

Lii, K.-S. and M. Rosenblatt, "Deconvolution and Estimation of Transfer Function Phase and Coefficients for Non-Gaussian Linear Processes," *The Annals of Statistics*, **10**, pp. 1195–1208, 1982.

Lii, K. S. and M. Rosenblatt, "Asymptotic Normality of Cumulant Spectral Estimates," *Theoretical Probability*, 1989.

Lohmann, A. W., G. Weigelt and B. Wirnitzer, "Speckle Masking in Astronomy: Triple Correlation Theory and Applications," *Applied Optics*, **22**, pp. 4028–4037, 1983.

Lohmann, A. W. and B. Wirnitzer, "Triple Correlations," *Proc. IEEE*, **72**, pp. 889–901, 1984.

Marple, S. L., Jr., *Digital Spectral Analysis with Applications,* Prentice-Hall, Inc., Englewood Cliffs, NJ, 1987.

Mendel, J. M., "Use of Higher-Order Statistics in Signal Processing and System Theory: an Update," *Proc. SPIE Conf. on Advanced Algorithms and Architectures for Signal Processing III*, pp. 126–144, San Diego, CA, 1988.

Mendel, J. M., "Tutorial on Higher-Order Statistics (Spectra) in Signal Processing and System Theory: Theoretical Results and Some Applications," *IEEE Proc.*, **79**, pp. 278–305, March 1991.

Mohler, R. R. and F. J. Bugnon, "A Second-Order Eigenstructure Array Processor," *Proc. Workshop on Higher-Order Spectral Analysis*, pp. 152–156, Vail, CO, 1989.

Moulines, E. and J.-F. Cardosa, "Second-Order versus Fourth-Order MUSIC Algorithms: an Asymptotical Statistical Analysis," *Proc. of Int'l. Workshop on Higher-Order Statistics*, pp. 221–224, Chamrousse, France, July 1991.

Nikias, C. L. and M. R. Raghuveer, "Bispectrum Estimation: A Digital Signal Processing Framework," *Proceedings IEEE*, **75**(7), pp. 869–891, July 1987.

Nikias, C. L. and R. Pan, "Non-Minimum Phase System Identification via Cepstrum Modeling of Higher-Order Moments," *Proc. ICASSP-87*, pp. 980–983, Dallas, TX, 1987.

Nikias, C. L. and R. Pan, "Time Delay Estimation in Unknown Gaussian Spatially Correlated Noise," *IEEE Transactions on Acoustics, Speech and Signal Processing*, Vol. **7**(3), pp. 291–325, 1988.

Nikias, C. L. and F. Liu, "Bicepstrum Computation Based on Second-and Third-Order Statistics with Applications," *Proc. ICASSP'90*, pp. 2381–2386, April. 1990.

Nikias, C. L. and A. P. Petropulu, *Higher-Order Spectral Analysis: A Nonlinear Signal Processing Framework,* Prentice-Hall, Inc., 1993.

Oh, W. T., S. B. Kim and E. J. Powers, "The Squared Skewness Processor for Time Delay Estimation in the Bispectrum Domain," *Signal Processing V: Theory and Applications*, L. Torres, E. Masgrau and M. A. Lagunas (eds.), Elsevier Science Publishers, pp. 111–114, 1990.

Oppenheim, A. V. and R. W. Schafer, *Discrete-Time Signal Processing,* Prentice-Hall, Englewood Cliffs, NJ, 1989.

Pan, R. and C. L. Nikias, "The Complex Cepstrum of Higher-Order Cumulants and Nonminimum Phase System Identification," *IEEE Transactions Acoust., Speech, and Signal Processing,* ASSP **36**(2), pp. 186–205, February 1988a.

Pan, R. and C.L. Nikias, "Harmonic Decomposition Methods in Cumulants Domains," *Proc. ICASSP'88*, pp. 2356–2359, New York, NY, April 1988b.

Papoulis, A., *Probability, Random Variables and Stochastic Processes*, 3rd ed., McGraw-Hill, New York.

Petropulu, A. P. and C. L. Nikias, "The Complex Cepstrum and Bicepstrum: Analytic Performance Evaluation in the Presence of Gaussian Noise," *IEEE Trans. Acoust., Speech and Signal Processing,* **38**(7), pp. 1246–1256, July 1990.

Petropulu, A. P. and C. L. Nikias, "Signal Reconstruction from the Phase of the Bicepstrum," *IEEE Transactions on Acoust., Speech and Signal Processing* **40**(3), pp. 601–610, March 1992.

Pflug, A. L., G. E. Ioup and R. L. Field, "Properties of Higher-Order Correlation and Spectra for Band Limited, Deterministic Transients," *J. Acoustics, Soc. Am.*, **91**(2), pp. 975–988, February 1992.

Porat, B. and B. Friedlander, "Direction Finding Algorithm Based on High-Order Statistics," *Proc. 1990 ICASSP*, pp. 2675–2678, Albuquerque, NM, 1990.

Powers, E. J., C. P. Ritz, C. K. An, S. B. Kim, R. W. Miksad and S. W. Nam, "Applications of Digital Polyspectral Analysis to Nonlinear Systems Modeling and Nonlinear Wave Phenomena," *Workshop on Higher-Order Spectral Analysis*, pp. 73–77, Vail, CO, June 1989.

Raghuveer, M. R. and C. L. Nikias, "Bispectrum Estimation: A Parametric Approach," *IEEE Trans. on Acous., Speech and Signal Processing*, ASSP **33**(5), pp. 1213–1230, October 1985.

Raghuveer, M. R. and C. L. Nikias, "Bispectrum Estimation via AR Modeling," *Signal Processing, 10*, pp. 35–48, 1986.

Rao, T. S. and M. M. Gabr, *An Introduction to Bispectral Analysis and Bilinear Time Series Models,* Lecture Notes in Statistics, **24**, Springer-Verlag, New York, NY, 1984.

Rao, S. S. and C. Vaidyanathan, "Estimating the Number of Sinusoids in Non-Gaussian Noise Using Cumulants," *Proc. 1991 ICASSP*, pp 3469–3472, Toronto, Canada, May 1991.

Rosenblatt, M. and J. W. Van Ness, "Estimation of the Bispectrum," *Ann. Math. Statist.*, **36**, pp. 1120–1136, 1965.

Rosenblatt, M., *Stationary Sequences and Random Fields,* Birkhauser, Boston, MA, 1985.

Rozario, N. and A. Papoulis, "The Identification of Certain Nonlinear Systems by Only Observing the Output," *Workshop on Higher-Order Spectral Analysis,* pp. 78–82, Vail, CO, June 1989.

Ruiz, P. and J. L. Lacoume, "Extraction of Independent Sources from Correlated Inputs: a Solution Based on Cumulants," *Proc. Workshop on Higher-Order Spectral Analysis*, pp. 146–151, Vail, CO, 1989.

Ruiz, P., "Sources Identification Using Cumulants: Limits and Precautions of Use," *Proc. of Int'l. Workshop on Higher-Order Statistics*, pp. 257–264, Chamrousse, France, July 1991.

Sadler, B. and G. B. Giannakis, "On Detection with a Class of Matched Filters and Higher-Order Statistics," *Proc. of 5th ASSP Workshop on Spectrum Estimation and Modeling*, pp. 222–226, Rochester, NY, October 1990.

Sadler, B., "Sequential Detection Using Higher-Order Statistics," *Proc. of Int'l. Conf. on Acoustics, Speech and Signal Processing*, pp. 3525–3528, Toronto, Canada, May 1991.

Scarano, G. R. and G. Jacovitti," Sources Identification in Unknown Coloured Noise with Composite HNL Statistics," *Proc. 1991 ICASSP*, pp 3465-3468, Toronto, Canada, May 1991.

Schetzen, M., *The Volterra and Wiener Theories on Nonlinear Systems,* updated edition, Krieger Publishing Company, Malabar, FL, 1989.

Shamsunder, S. and G. B. Giannakis, "Detection and Parameter Estimation of Multilple Sources via HOS," *Proc. of Int'l. Workshop on Higher-Order Statistics*, pp. 265–268, Chamrousse, France, July 1991a.

Shamsunder, S. and G. B. Giannakis, "Wideband Source Modeling and Localization: a HOS-Based Approach," *Proc. 25th Asilomar Conf. on Signals, Systems and Computers*, Pacific Grove, CA, November 1991b.

Shi, Z. and F. W. Fairman, "Cumulant Based Approach to Harmonic Retrieval Problem Using a State Space Approach," *Proc. 1991 ICASSP*, pp. 3505–3508, Toronto, Canada, May 1991.

Shin, D. C. and J. M. Mendel, "Comparison Between Correlation-Based and Cumulant-Based Approaches to the Harmonic Retrieval and Related Problems," USC-SIPI Report #177, University of Southern California, May 1991.

Shin, D. C. and J. M. Mendel, "Assessment of Cumulant-Based Approaches to Harmonic Retrieval," *Proc. 1992 IEEE ICASSP*, San Francisco, CA, March 1992a.

Shin, D. and C. L. Nikias, "Adaptive Noise Canceler for Narrowband and Wideband Interferences Using Higher-Order Statistics," USC-SIPI Report #220, September 1992b.

Swami, A. and J. M. Mendel, "Cumulant-Based Approach to the Harmonic Retrieval Problem," *Proc. of IEEE Conf. on ASSP,* pp. 2264–2267, New York, NY, 1988a.

Swami, A., "System Identification Using Cumulants," Ph. D. Dissertation, USC-SIPI Report #140, University of Southern California, Department of Electrical Engineering-Systems, Los Angeles, CA, 1988b.

Swami, A. and J. M. Mendel, "ARMA Parameter Estimation Using Only Output Cumulants," *Proc. IV IEEE ASSP Workshop on Spectrum Estimation and Modeling*, pp. 193–198, Minn., MN, 1988; also, *IEEE Trans. on Acoustics, Speech and Signal Processing*, **38**, pp. 1257–1265, July 1990.

Swami, A. and J. M. Mendel, "Cumulant-Based Approach to the Harmonic Retrieval and Related Problems," *IEEE Trans. on Signal Processing*, **39**, pp. 1099–1109, 1991.

Swindelhurst, A. L. and T. Kailath, "Detection and Estimation Using the Third Moment Matrix," *Proc. ICASSP 1989*, pp. 2325–2328, Glascow, Scotland, May 1989.

Tick, L. J., "The Estimation of Transfer Functions of Quadratic Systems," *Technometrics*, **3**(4), pp. 562–567, November 1961.

Tsatsanis, M. K. and G. B. Giannakis, "Object and Texture Detection and Classification Using Matched Filtering and Higher-Order Statistics," *Proc. of 6th Workshop on Multidimensional Signal Processing*, pp. 32–33, Monterey, CA, September 1989.

Tsatsanis, M. K. and G. B. Giannakis, "Object and Texture Detection and Classification Using Higher-Order Statistics," *IEEE Trans. on Pattern Analysis and Machine Intelligence*, 1992.

Tugnait, J. K., "Time Delay Estimation in Unknown Spatially Correlated Gaussian Noise Using Higher–Order Statistics," *Proc. 23rd Asilomar Conf. Signals, Systems, Computers*, pp. 211–215, Pacific Grove, CA 1989.

United Signals & Systems, Inc., "Comprehensive Bibliography on Higher-Order Statistics (Spectra)," Culver City, CA, 1992.

Van Ness, J. W., "Asymptotic Normality of Bispectral Estimates," *Ann. Math. Statist.*, **37**, pp. 1257–1272, 1986.

Van Veen, B. and K. Buckley, "Beamforming: a Versatile Approach to Spatial Filtering, *IEEE ASSP Mag.*, pp. 4–24, 1988.

Zhang, W. and M. R. Raghuveer, "Non-Parametric Bispectrum-Based Time-Delay Estimators for Multiple Sensor Data," *IEEE Trans. Signal Processing*, **39**(13), pp. 770–774, March 1991.

PITFALLS IN POLYSPECTRA

Ananthram Swami

CS^3, Los Angeles

Abstract. The area of higher-order statistics (HOS) has witnessed a tremendous resurgence over the last several years. However, Huber, Kleiner, Gasser and Dumermuth's 1971 observations [8], "**... the newcomer to the field still has to learn the hard way, from his own mistakes ...**", and, "**... all the pitfalls known from ordinary spectral analysis occur here too, some of them with new twists .....**", continue to be true today, despite at least two tutorials, and several special issues of journals. We will examine some of these pitfalls, particularly in the context of deterministic signals observed in noise.

Bispectrum: Non-Redundant Region

Consider a continuous-time zero-mean signal, $x(t)$, band-limited to $[-B, B]$. Its bispectrum, $B_x(w, v) := E\{X(w)X(v)\bar{X}(w + v)\}$ is non-zero only within the hexagon with vertices,
$(0, B), (B, 0), (B, -B), (0, -B), (-B, 0), (-B, B)$.
The non-redundant region is the triangle with vertices, $(0, 0), (B/2, B/2), (B, 0)$; denote this triangle by T_1.

For a sampled signal, the FT $X(w)$ is periodic with period F_s, the sampling frequency. Because of the replications of the term $X(w + v)$, the bispectrum of a sampled signal will not, in general, be zero outside the hexagon. It has been shown [7, 10] that the non-redundant region of the bispectrum is the triangle with vertices $(0, 0), (2B/3, 2B/3), (B, 0)$, which is the union of triangle T_1 and the 'outer' triangle, T_2 with vertices $(B/2, B/2), (2B/3, 2B/3), (B, 0)$.

Consider a sampled signal band-limited to $[B_1, B_2]$, with sampling rate $F_s \geq 2B_2$. If $2B_1 \geq B_2$ (signal has less than one octave bandwidth), the bispectrum will be identically zero over T_1. If $3B_1 \leq F_s \leq 3B_2$, the bispectrum will be non-zero over T_2. Note that if $2B_1 \geq B_2$ and $3B_1 \leq F_s \leq 3B_2$, then, the bispectrum is non-zero only over T_2. Hence, the claim that 'the bispectrum of a properly sampled (i.e., $F_s \geq 2B_2$) signal is identically zero inside T_2' is incorrect. Hence, tests for stationarity or aliasing based on checking whether or not the bispectrum is zero inside T_2 are also incorrect.

Finally, note that the bispectrum of a continuous-time signal is identically zero if $2B_1 \geq B_2$. For a sampled signal, the additional condition, $F_s > 3B_2$ (signal is oversampled at a rate greater than 1.5 times the Nyquist), or $F_s < 3B_1$, is required; [10, 12].

Since $C_{4x}(w, w, -w) = C_{2x}^2(w)$, the fourth-order spectrum cannot be identically zero.

The above results also hold for random processes. Note that, if $x(t) = \sum_k h(k)u(t-k)$, where $u(t)$ is zero-mean, i.i.d. non-Gaussian, $\gamma_{3u} := Eu^3(t) \neq 0$, does not guarantee a non-zero bispectrum.

Bicoherence, and alternatives

The bicoherence, defined by,
$$Bic(w, v) = B_y(w, v)/\sqrt{S_y(w)S_y(v)S_y(w + v)}$$
has been used as a basis for tests of linearity and non-Gaussianity [7]. However, the model assumes that signal is noise-free; for example, if $y(n) = x(n) + g(n)$, where $x(n)$ is a linear non-Gaussian process, and $g(n)$ is zero-mean stationary (colored) Gaussian (zmscg) noise, then the bicoherence of the process $y(n)$ is not a constant.

A time-domain test for Gaussianity is given in [6], where one uses the asymptotic normality of $\hat{c}_{3y}(m, n)$ to test whether or not $\hat{c}_{3y}(m, n)$ is zero over a set of lags. If $x(n)$ is a linear non-Gaussian process, with non-zero bispectrum, then appropriate weighted projections of $B_y(w, v)(= B_x(w, v))$ yield $S_x(w)$; the projection will not yield $S_x(w)$ if the process is non-linear. This estimate of $S_x(w)$ may be used in the definition of bicoherence.

An alternative test for linearity is given in [14]. The bicepstrum of the signal $y(n)$ is defined as the inverse FT of $\log B_{3y}(w, v)$ and exists if $B_{3y}(w, v)$ is non-zero for all w and v. For a linear process, the bicepstrum is non-zero only along the lines $m = 0$, $n = 0$, $m = n$. An index of linearity defined by the ratio of the energy along these three lines to the total energy in the bicepstrum, is used to quantify the 'amount of linearity' in [14].

Record Segmentation

A commonly used procedure in estimating cumulants consists of segmenting the given time-series into several records, estimating cumulants from each, and then averaging the estimates. Segmentation does not improve the sample estimates (reduce the variance). The asymptotic variance of the sample estimate is proportional to $1/N$ where N is the *total* number of 'independent' samples used. Segmentation causes a slight decrease in the asymptotic efficiency of the estimator.

Segmentation was introduced for estimating the bispectrum via the direct (FFT) approach, where the segment averaging (or frequency-domain smoothing) is required to ensure consistency of the estimates.

Cumulants and deterministic signals

If we use the conventional definition, then the cumulants of deterministic signals are identically zero. However, alternative useful definitions exist. We will differentiate between *transient signals*, i.e., deterministic signals with finite energy, and hence, limited time-duration, and *sustained signals*, i.e., signals with finite power; examples of sustained signals are harmonics.

Consider the mixed signal, $y(t) = s(t) + g(t)$, where $s(t)$ is deterministic, and $g(t)$ is zmscg. The process $y(t)$ is non-stationary, since it has a time-varying mean; however, under weak conditions, the process $y(t)$ is *quasi-stationary* and the usual time-averaging of moment products leads to useful results [3].

In [3], it is shown that under weak mixing conditions (summability conditions on cumulants, and existence of limits below)

$$M_{(k+1)y}(\tau_1, \cdots, \tau_k)$$

$$:= \lim_N \frac{1}{N} \sum_{t=1}^{N} y(t) y(t + \tau_1) \cdots y(t + \tau_k) \quad (1)$$

$$= \lim_N \frac{1}{N} \sum_{t=1}^{N} E\{y(t) y(t + \tau_1) \cdots y(t + \tau_k)\} \quad (2)$$

where it is assumed that $M_{1y} = 0$. Equation (1) is the definition, and (2) states the asymptotic mss equality. This is an important result, and states that sample estimates obtained by temporal averaging are asymptotically equal (in m.s.s.) with the time average of the time-dependent expected values.

For $k = 2, 3, 4$, and assuming that $g(t)$ is zmscg noise, we obtain immediately,

$$M_{2y}(\tau) = M_{2s}(\tau) + M_{2g}(\tau) \quad (3)$$

$$M_{3y}(\tau, \rho) = M_{3s}(\tau, \rho) \quad (4)$$

$$C_{4y}(\tau_1, \tau_2, \tau_3) = C_{4s}(\tau_1, \tau_2, \tau_3) \quad (5)$$

where

$$C_{4y}(i, j, k) := M_{4y}(i, j, k) - [M_{2y}(i) M_{2y}(j - k)]_3 \quad (6)$$

In general, the recoverability of $s(t)$ from M_{3y} or C_{4y} (defined in (6)) depends upon the nature of the signal $s(t)$.

For transient signals, the $1/N$ factor causes M_{ks} to be zero for all k. Consequently the above definition of M_{ks} is useful only for *sustained signals*, such as chirps and harmonics (observed in noise).

For transient signals, one must assume multiple realizations; the observed signal in the i-th realization is $y_i(t) = s(t + d_i) + g_i(t)$, where d_i is a delay term. The delay term may be arbitrary provided that the entire transient signal is contained within the observation window, $1 \le n \le N$.

We may define the moments via,

$$M_{(k+1)y}(\tau_1, \cdots, \tau_k) = \lim_K \frac{1}{K} \sum_{m=1}^{K} \sum_{n=1}^{N} \prod_{i=0}^{k} y_m(n + \tau_i) \quad (7)$$

where $\tau_0 := 0$. If $y(n) = s(n) + g(n)$, where $g(n)$ is zmscg noise, and $s(n)$ is a deterministic signal, we have

$$M_{3y}(\tau, \rho) = M_{1s}[R_g(\tau) + R_g(\rho) + R_g(\tau - \rho)]$$
$$+ M_{3s}(\tau, \rho) \quad (8)$$

$$C_{4y}(\tau_1, \tau_2, \tau_3) = C_{4s}(\tau_1, \tau_2, \tau_3) \quad (9)$$

If $M_{1s} \ne 0$, and $R_g(\tau)$ is not known, we cannot use $C_{3y}(\tau, \rho)$ to estimate $s(t)$. However, $M_{1s} \ne 0$ may be useful for signal *detection*.

Recovering $s(t)$ from M_{ks} is equivalent to estimating $h(t)$ from C_{ky}, where $y(n) = h(t) * u(t)$, and $u(t)$ is i.i.d. non-Gaussian, with $\gamma_{ku} = 1$.

Bicepstra and Transient Signals

Bicepstral methods assume that the system has no zeros on the unit circle (the signal should be full-band: it cannot be over-sampled). However, if bicepstra are to be used to estimate/model transient signals in additive noise, then one must assume that the transient signal has no DC component (to suppress the signal-noise cross-terms), violating the assumption of no zeros on the unit circle; see equation (8).

The cross-cumulant of three signals, $x(n) = s_1(n) + g_1(n)$, $y(n) = s_2(n) + g_2(n)$, and $z(n) = s_3(n) + g_3(n)$, where the g_i's are zero-mean, and mutually independent, is given by [1]

$$M_{xyz}(\tau, \rho)$$

$$:= \lim_K \frac{1}{K} \sum_{k=1}^{K} \sum_{n=1}^{N} x_k(n) y_k(n + \tau) z_k(n + \rho) \quad (10)$$

$$= M_{s_1, s_2, s_3}(\tau, \rho) \quad (11)$$

Hence, in order to avoid the zero-DC assumption,

36

one may estimate cumulants via,

$$\hat{C}_{3y}(\tau, \rho)$$

$$:= \lim_K \alpha(K) \sum_A \sum_{n=1}^{N} y_i(n) y_j(n+\tau) y_k(n+\rho)$$

which will be asymptotically equal to $C_{3s}(\tau, \rho)$, *provided that the signal is aligned across realizations*. Here, $A := \{(i,j,k) : i \neq j \neq k\}$; $\alpha(K) = K(K-1)(K-2)/6$.

For sustained signals, we use (1); from (4), we note that additive noise does not pose a problem; but, this does not guarantee that $s(t)$ can be recovered from M_{3y} or C_{4y}.

Transients: Trispectra

Since the third-order moment of $s(t)$ may be identically equal to zero, we will consider the fourth-order cumulants of $y(t) = s(t) + g(t)$, where $g(t)$ is zmscg noise. We will estimate C_{4y} from (7) and (6). In the frequency domain, we have,

$$C_{4y}(u,v,w) = S(u)S(v)S(w)\bar{S}(u+v+w)$$
$$- |S(u)|^2 |S(v)|^2 \delta(v+w) - |S(v)|^2 |S(w)|^2 \delta(w+u)$$
$$- |S(w)|^2 |S(u)|^2 \delta(u+v)$$

where $S(u)$ is the FT of $s(n)$. Hence, $C_{4y}(w,w,-w) = -[S(w)\bar{S}(w)]^2(1 + \delta(w))$, from which $M_{2s}(\tau)$ can be estimated. By (9), $C_{4y} = C_{4s}$; hence, we can now estimate M_{4s}. Known techniques may then be used to estimate $s(t)$ from M_{4s}.

Transients: MLE vs Cumulants

The model is, again, $y(t) = s(t) + g(t)$, where $g(t)$ is zmscg, and $s(t)$ is a transient signal. We will assume that multiple realizations are available, and that the signal is aligned across realizations.

Let $\mathbf{y}_i = [y_i(1), ..., y_i(N)] = \mathbf{s} + \mathbf{g}_i$, where N is the number of samples, subscript i refers to the realization number, $\mathbf{s}$ is the signal vector, and $\mathbf{g}$ is the Gaussian noise vector. We therefore have the following situation: we want to estimate the mean, $\mathbf{s}$, of a Gaussian r.v., $\mathbf{y}$, given K independent realizations. The ML solution to this classical problem is, [15]

$$\hat{\mathbf{s}} = \frac{1}{K} \sum_{k=1}^{K} \mathbf{y}_k .$$

Note that the r.v. $\mathbf{g}$ may have a non-diagonal covariance matrix, i.e., the process $g(n)$ may be colored.

Having estimated the signal (the mean of $\mathbf{y}$), we can now estimate the covariance of $\mathbf{y}$, i.e., $R_g(\tau)$. The sample estimate of the covariance is also the ML estimate. If desired, we can fit an ARMA model to $R_g(\tau)$.

If $\mathbf{g}$ is not Gaussian, the sample estimates of the mean and covariance are no longer MLE, but still have several attractive properties (consistent, easily computed, etc.).

We note that the signal alignment assumption is also required in the bicepstral approach. Given this assumption, the MLE must be preferred.

If the signals are not aligned, we first estimate $C_{3s}(\tau, \rho)$; next, we obtain a preliminary estimate of $s(t)$; then, we can estimate the color of the noise from $C_{2y}(\tau)$; since the color of the noise is now known (estimated), we can use classical techniques to estimate the signal delay of each realization wrt some reference realization, align the signals, and then obtain an improved estimate of $s(t)$ by using the MLE.

Harmonic Retrieval

Harmonics signals are sustained signals, and constitute one class of signals for which parameters may be estimated from M_{3y} or C_{4y} estimated from single realizations. Our signal model is

$$y(t) = \sum_{k=1}^{p} \alpha_k \cos(\omega_k t + \phi_k) + g(t) ,$$

where $g(t)$ is zmscg, and the ϕ_k's are assumed to be i.i.d. and uniformly distributed over $[0, 2\pi]$. In [13], it was shown that p can be determined and the α_k's and ω_k's estimated from the fourth-order cumulant of $y(t)$. Since the ϕ_k's are assumed to be random, the natural setting for this problem involves multiple realizations. Multiple realizations *may* be created artificially by proper segmentation.

Given only a single realization, the phases are no longer random. In this case, we estimate C_{4y} using the definition in (6). Using the results in (2) [3], it can be shown that

$$C_{4y}(\tau_1, \tau_2, \tau_3)$$
$$= -\frac{1}{8} \sum_{k=1}^{p} \alpha_k^4 [\cos(\omega_k(\tau_1 + \tau_2 - \tau_3) +$$
$$\cos(\omega_k(\tau_2 + \tau_3 - \tau_1)) + \cos(\omega_k(\tau_3 + \tau_1 - \tau_2))] \quad (12)$$

assuming that: (a) There is no cubic frequency coupling of the form $f_i + f_j + f_k = f_l$; and, (b) there is no bi-frequency coupling of the form, $f_i + f_j = f_k + f_l$, for i, j, k, l all distinct. The expression in (12) is identical to that obtained for the random phase case, under the assumption that there is no cubic phase coupling, i.e., $\phi_i + \phi_j + \phi_k \neq \phi_l$. The random-phase case does not involve an explicit counterpart for assumption (b).

In the presence of cubic frequency coupling, we get

additional terms involving the coupled frequencies,

$$\frac{1}{8}\alpha_i\alpha_j\alpha_k\alpha_l[\cos(\omega_i\tau_a + \omega_j\tau_b + \omega_k\tau_c$$
$$-\omega_l\tau_d + \phi_i + \phi_j + \phi_k - \phi_l)]_{24}$$

where the subscript 24 denotes that 24 terms are obtained by permuting $\{(a,b,c,d)\} = \{(1,2,3,4)\}$, with $\tau_4 := 0$. This is the same expression obtained in the random-phase case, with cubic phase coupling (the ϕ_m's sum to zero in this case), [13]

In the absence of quadratic frequency coupling, i.e., $f_i + f_j \neq f_k$, the third order cumulant is zero. If QFC is present, we get terms of the form

$$\frac{1}{4}\alpha_i\alpha_j\alpha_k\cos(\omega_i\tau_a + \omega_j\tau_b - \omega_k\tau_c + \phi_i + \phi_j - \phi_k)_6$$

where the subscript 6 denotes the six terms obtained by permuting $\{(a,b,c)\} = \{(1,2,3)\}$, with $\tau_3 := 0$. This is the same expression obtained in the random phase case, with QPC (the ϕ_m's sum to zero in this case). For related discussions between FC and PC, see [12].

GAR Models - MLE

In [2], the process $y(t) = s(t) + g(t)$ is considered. Here, $s(t)$ is a deterministic process, a sum of harmonics, and $g(t)$ is a Gaussian AR process. Simultaneous (ML) estimates of the AR parameters, and the amplitudes, phases and frequencies of the harmonics are obtained, based on single realizations. The approach is generalized in [9], where the process $s(t)$ is allowed to be the impulse response of an ARMA model. These approaches involve non-linear optimization, and can be computationally demanding especially when the number of harmonics (or poles) is greater than two. In this case, cumulant-based estimators may provide good initial estimates of the parameters for the harmonics problem. The transient problem, as noted above, demands multiple realizations, and cannot provide good cumulant-based estimates if only a single realization is available; in the latter case, the GAR model is the only recourse.

Wigner bispectra and "lag-centering"

A generalized Wigner bispectrum, parameterized by α was studied in [11]; lag-centering corresponds to the case $\alpha = -1/3$ of [5]. However, the choice $\alpha = -1$ is extremely useful, since it confines the effects of the additive noise to the bi-frequency axes. The choice of α should reflect the user's signal analysis objectives (signal estimation/detection; or only IF or GD).

Signals in Non-Gaussian Noise

Given a deterministic signal in linear non-Gaussian noise (i.e., the noise is modeled as the output of a linear system excited by an i.i.d. non-Gaussian process), we can use third-order cumulants, estimated via (1) to estimate $C_{3g}(\rho,\tau)$. (The terms due to $s(t)$ will be zero, if $s(t)$ is a transient, and for some classes of sustained signals, such as harmonics without QFC.) Since $g(t)$ is assumed to be linear, we can now estimate $R_g(\tau)$; then, we can use classical techniques for signal estimation/detection.

HOS versus SOS for AR(2) Models

Assume a second order AR process is observed in white-Gaussian noise. What is the relative efficiency of SOS-based methods to HOS-based methods (ratio of asymptotic variances) in terms of SNR and the pole location? Results will be presented at the conference.

Acknowledgements: It is a pleasure to acknowledge several stimulating discussions with Amod Dandawate and Georgios Giannakis, particularly on the issues of mixed signals and single realizations.

References

[1] D.H. Brooks and C.L. Nikias, *Proc ICASSP-91*, 3433-6, Toronto, Canada, May 1991.

[2] C. Chatterjee *et al*, *IEEE Trans ASSP*, 328-337, 1987.

[3] A.V. Dandawate and G.B. Giannakis, *Proc. 25th CISS*, 976-983, 1991.

[4] J.R. Fonollosa and C.L. Nikias, *Proc ICASSP-92*, San Francisco, March 1992.

[5] N.L. Gerr, *Proc. IEEE*, 290-292, March 1988.

[6] G.B. Giannakis and M.K. Tsatsanis, *Proc. ICASSP-91*, 3097-3100, Toronto, Canada, May 1991.

[7] M. Hinich, *J. Time Ser. Anal.*, 169-176, 1982.

[8] P.J. Huber, B. Kleiner, T. Gasser and G. Dumermuth, *IEEE Trans. AE*, 78-86, 1971.

[9] S.M. Kay and V. Nagesha, *Proc ICASSP-91*, 3137-40, Toronto, Canada, May 1991.

[10] L.A. Pflug *et al*, *J. Acoust. Soc. Am.*, 91 (2), 975-988, Feb 1992.

[11] A. Swami, *Proc SPIE-92*, Vol 1770, pp 290-301, San Diego, CA, July 1992;

[12] A. Swami, *Proc. Intl. Signal Processing Workshop on HOS*, 135-8, Chamrousse, France, July 1991.

[13] A. Swami and J.M. Mendel, *IEEE Trans. ASSP*, 1099-1109, 1991.

[14] A.M. Tekalp and A.T. Erdem, *Proc. Workshop on HOSA*, 186-190, Vail, CO, June 1989.

[15] H.L. Van Trees, *Detection, Estimation and Modulation Theory*, Part I, New York: Wiley, 1968.

Nonlinear Signal Processing Vs. Kalman Filtering

P. A. Ramamoorthy, Aleksandar Zavaljevski
Department of Electrical and Computer Engineering
University of Cincinnati, M. L. 30
Cincinnati, Ohio 45221 - 0030
FAX: (513) 556 - 7326; Tel: (513) 556 - 4757
Email: pramamoo@babbage.ece.uc.edu

Abstract - Most of the research activities in circuits and signal processing, whether analog or digital, are confined to linear signal processing (LSP). The interest in LSP is due largely to its mathematical tractability and ease of implementation, characteristics that were necessary for the technological climate of the earlier decades. However, LSP is limited in its usefulness and the technology for implementing algorithms and analyzing, modeling and design of complex systems has progressed tremendously. Further, we need to understand and utilize nonlinear signal processing (NLSP) before we can move to large-scale self-organizing or self-learning systems. In our work we have developed a unique and powerful paradigm or methodology for the design of nonlinear dynamical systems and signal processing algorithms. In this paper, we will discuss some aspects of this approach and show results of application of this approach in signal estimation. We also compare the obtained results with that of Kalman filtering, which is essentially filtering using a filter with a time-varying coefficient (Kalman gain). The results indicate that the new NLSP approach is superior and more robust as compared to Kalman filtering.

I. INTRODUCTION

Signal processing, which can be considered as a subset of intelligent information processing is stuck primarily at the simple level of linear processing, [1], [2]. However intelligent information processing is by nature nonlinear, and time varying (in terms of memory, application of rules and learning capability), and there are clear indications that the nonlinear/time varying processing account for most of the "intelligent" results outcome, [3], [4]. Thus, the need arises for systematic approaches for the design of nonlinear and time-varying signal processors. This paper is concerned with the first, that of designing stable nonlinear signal processing systems.

The analytical approach, that is defining proper models for nonlinear differential equations (NLDE) and analyzing the resulting models is the most commonly used technique in many areas, and NLSP is no exception to this trend. However, this approach has not been very successful in NLSP since it is known that even simple first order NLDE can lead to chaotic signals - a situation good to produce nice plots etc, but of not much use in design of stable systems. In our research, which is in its early stages, we have adopted what one may term as an engineering approach or building block approach. Under this approach, we have used the concept of passivity to define a number of elements as building blocks for passive nonlinear electrical circuits. Complex passive nonlinear networks can then be formed by proper interconnection of these various nonlinear elements. When energy storing elements are present in such a network (which themselves can be linear or nonlinear), we can obtain a set of input/output relationships as nonlinear differential equations. The basic property that the network is lossy (consumes energy) ensures that the nonlinear differential equations obtained from the networks would represent absolutely

stable systems, and this property holds as long as the individual element values are maintained in their permissible range of values. Thus, to design complex nonlinear systems (a nonlinear signal processor for tracking, for example) and self-organizing systems, one simply has to force the dynamics of those systems to mimic the dynamics of a properly constructed passive nonlinear network, a process akin to reverse engineering.

In our research we have developed the basis for the above approach and applied it with relative ease to a number of problems leading to encouraging results. The fruits of such an approach seems to be endless. For example, the approach can be applied to NLSP, linear and nonlinear controller design (for linear and nonlinear plants), self tuning controllers, model reference adaptive controllers, self-organizing networks, adaptive IIR filter design, adaptive beam-forming, two-dimensional systems, fuzzy systems etc. In this paper we provide some details of this approach and show results of application of this approach in signal estimation. We also show results comparing our approach to Kalman filtering, [5], and its superiority.

II. PROPOSED METHOD

Our approach for nonlinear signal processor design (which is equally applicable to a number of problem domains in the nonlinear systems arena) is based on an entirely new and interesting paradigm. It may be called an "engineering" or a "building block" approach for NLSP design as opposed to the analytical/mathematical point of view adopted by earlier researchers. An engineering or physically motivated approach tries to take into consideration physical properties and constraints based on physical properties at every stage of the design. Passivity formulation (to be defined shortly) is used here to obtain the necessary building blocks for a nonlinear system design. These building blocks can then be interconnected in a proper manner to obtain the general structure for a nonlinear system. To design any system, we can use this general structure and vary the parameters so as to obtain the desired properties. To obtain a digital nonlinear system, we can extract the nonlinear differential equations from the general structure and use a forward difference operator.

Passivity is a term commonly used in electrical network theory to indicate consumption of energy. A passive electrical element (linear or nonlinear) is one which always consumes power/energy (lossy) or at most, consumes no power/energy (lossless). They can be non-dynamic (no memory/can't store energy) or dynamic (stores energy and gives it back at some other time). They can be two-terminal (one-port) elements or multi-terminal (multi-port) devices. A passive linear/nonlinear network is simply an electrical network formed by proper interconnection of various passive linear/nonlinear elements. The interconnections must be such that the basic circuit laws are obeyed. An important property of such networks is that they are stable and remain so as long as the values of individual elements remain in the permissive range for passivity. Thus, if we have proper passive nonlinear elements, we can form stable nonlinear networks, obtain dynamical equations describing such networks in terms of the element parameters and use them as target equations for the proposed nonlinear system.

The above approach assumes that proper circuit elements exist. Only few such elements are available in the open literature. One such element is a passive resistor with a current-voltage relationship given by:

$$i_R(t) = G\tan^{-1}(v_R(t)); \qquad G > 0 \qquad (1)$$

The power p(t) consumed by this element is given by:

$$p(t) = i_R(t)\,v_R(t) \qquad (2)$$

and is always non-negative. In general, the v-i characteristic of a general nonlinear passive resistor has to be confined to the first and third quadrants in the v-i plane and has to pass through the origin.

Nonlinear dynamical elements (capacitors and inductors) and their characteristics have been defined. However, the use of nonlinearity in dynamical elements in passive networks may lead to chaos under external excitation. Hence, we use only linear dynamical elements in our design.

Other nonlinear elements do not exist in the literature. For this purpose, we have defined/invented a number of other passive nonlinear devices[1]. One such device is a nonlinear transformer, a two-port element and has the transfer characteristics given by:

$$\begin{vmatrix} v_2(t) \\ i_2(t) \end{vmatrix} = \begin{vmatrix} N(\ ) & 0 \\ 0 & -\dfrac{1}{N(\ )} \end{vmatrix} \begin{vmatrix} v_1(t) \\ i_1(t) \end{vmatrix} \qquad (3)$$

where N( ) is a nonlinear function of the current(s) and voltage(s) in an electrical network in which the transformer is embedded. It should be noted that:

$$v_1(t)\,i_1(t) + v_2(t)\,i_2(t) = 0 \qquad (4)$$

regardless of what ever form N( ) takes[2]. Thus, an ideal nonlinear transformer is a lossless, non-dynamic (or memoryless) two-port device. Though an analog implementation of this device (and other devices invented) is feasible, we will be using digital implementation for the NLSP application. The digital implementation would allow us to realize such elements/devices with ideal characteristic as defined.

Another device that we have invented, and that is highly useful, is the multiport nonlinear gyrator. If we denote V and I as:

$$V = [v_1, v_2, \ldots, v_N]^T \qquad (5)$$

$$I = [i_1, i_2, \ldots, i_N]^T \qquad (6)$$

where $v_1, \ldots, v_N$ are the voltages across the gyrator ports, and $i_1, \ldots, i_N$ the currents into the ports of the gyrator, we have the relationship between the voltages and currents as follows:

$$I = Y V \qquad (7)$$

[1] Patent applications are being submitted for some elements referred in this paper

[2] We can place constraints based on the nature of each element. For the transformer, we may require N( ) to be positive at any time instant and a continuous function of time.

Here,

$$Y = [\,y_{ij}(\ )\,], \qquad i,j = 1,\ldots,N \qquad (8)$$

is the admittance matrix of the nonlinear gyrator and satisfies the constraint:

$$Y + Y^T = 0 \qquad (9)$$

The elements $y_{ij}(\ )$ can be complex functions of the current(s) and voltage(s) in an electrical network. As an example, for a two port nonlinear gyrator, we may have:

$$\begin{vmatrix} i_1(t) \\ i_2(t) \end{vmatrix} = \begin{vmatrix} 0 & v_1^2 - v_2 \\ v_2 - v_1^2 & 0 \end{vmatrix} \begin{vmatrix} v_1(t) \\ i_1(t) \end{vmatrix} \qquad (10)$$

We can show that:

$$I^T V = 0 \qquad (11)$$

for any functions $y_{ij}(\ )$. That is, a nonlinear gyrator is a lossless and non-dynamic multiport device.

We can connect these devices to form a passive nonlinear electrical network. The number and type of elements chosen will depend upon the system to be realized. For example, a network with 4 state variables can be realized as shown in Fig. 1.

We have used this particular architecture (gyrator, nonlinear resistors and linear capacitors) in many of our applications as it is known in linear passive network theory that any kind of linear systems with complex poles can be realized using linear gyrator/capacitor combinations.

The dynamic equations of this network can be written in the state-space form as:

$$C\dot{V} = YV - F(V) + I \qquad (12)$$

where C is a diagonal matrix with the capacitor values, V is the vector of voltages, $\dot{V}$ is the vector of derivatives of voltages, F(V) the

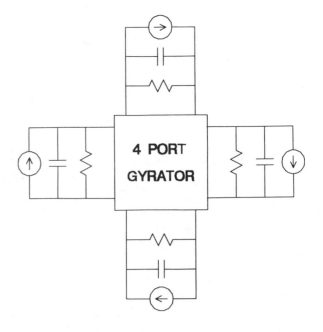

Fig. 1. A nonlinear network with a 4-port nonlinear gyrator

vector of currents through the nonlinear resistors, Y the admittance matrix of the nonlinear gyrator, and I vector of source currents. The set of equations in (12) represent a stable network or system as long as the element values are in the permissible range so as to retain the lossy or lossless property. The stability property holds good even if the terms in the various nonlinear elements are defined by complex functions and the network is highly complex.

Now, to design any stable system, we simply force the system dynamics to take the form given in equation (12). This concept can be applied with relative ease to design stable controllers, stable estimators, etc.

III. APPLICATION TO SIGNAL ESTIMATION

We can describe our approach for signal estimation and the difference from Kalman filtering using, block diagrams of both processes, shown in Fig. 2 and 3. As can be noted from Fig. 2, a Kalman filter is more of a time-varying filter in which K(t), the Kalman gain, is updated continuously and used in estimating the signal. However, the available estimated signal is not used in turn in the updating of K(t). On the other hand, the new approach leads to a truly nonlinear filter where there is high degree of coupling between the gain update block and the signal estimator, as shown in Fig. 3. The selection of the degree of coupling, various parameters/terms in the nonlinear filter etc., forms a part of the design procedure for the new approach and will be described in another paper. Though the circuit is highly nonlinear, since the dynamics of the complete system would be chosen so as to mimic the dynamics of a passive nonlinear network, we are guaranteed of the stability. In our simulations, we found the nonlinear filter to be more effective and highly stable as compared to the Kalman filter.

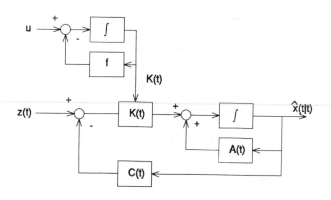

Fig. 2. A block diagram of signal estimator
using Kalman filter

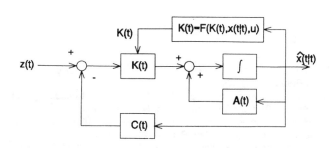

Fig. 3. A block diagram of signal estimator
using passive network approach

Let us consider the details for Fig. 3, the nonlinear filter architecture. For the sake of clarity, we will use a very simple model to illustrate the concept. Considering the problem of signal estimation, let the message and the observation models be as follows, [6]:

$$\dot{x}(t) = -x(t) + w(t) \tag{13}$$

$$z(t) = \frac{1}{2}x(t) + v(t) \tag{14}$$

$$x(0) = 1 \tag{15}$$

$$E\{w(t)w(\tau)\} = 6\delta(t - \tau) \tag{16}$$

$$E\{v(t)v(\tau)\} = \frac{1}{2}\delta(t - \tau) \tag{17}$$

$$E\{w(t)\} = E\{v(t)\} = E\{w(t)v(\tau)\} = 0 \tag{18}$$

The classical Kalman filter estimator equations for this problem can be represented in the form of the matrix equation as follows:

$$\begin{vmatrix} \dot{\hat{x}}(t|t) \\ \dot{V}_x \end{vmatrix} = \begin{vmatrix} 0 & (z - \frac{1}{2}\hat{x}) \\ 0 & 0 \end{vmatrix} \begin{vmatrix} \hat{x} \\ V_x \end{vmatrix} - \begin{vmatrix} \hat{x} \\ f(V_x) \end{vmatrix} + \begin{vmatrix} 0 \\ u \end{vmatrix} \tag{19}$$

where u equals 6, and:

$$f(V_x) = 2V_x + \frac{1}{2}V_x^2 \tag{20}$$

Now, let us show how our approach can be applied to the same problem. We choose the expression for the estimator to comply with equation (12). Then, we add the term $-(z-x/2)$ to the estimator (19) to obtain antisymmetric matrix Y and choose

$$f(V_x) = 2V_x + \frac{1}{2}V_x^3 \tag{21}$$

as a nonlinear function of V_x and representing the current through a passive resistor. It has to be noted here that at this point, we have not used any optimization procedure or an exhaustive search for the selection of the various terms. These and other issues are under investigation now. Finally, the source u is selected to force V_x to 2 as x approaches 2z, leading the estimation/Kalman gain update expressions:

$$\begin{vmatrix} \dot{\hat{x}}(t|t) \\ \dot{V}_x \end{vmatrix} = \begin{vmatrix} 0 & (z - \frac{1}{2}\hat{x}) \\ -(z - \frac{1}{2}\hat{x}) & 0 \end{vmatrix} \begin{vmatrix} \hat{x} \\ V_x \end{vmatrix} - \begin{vmatrix} \hat{x} \\ 2V_x + \frac{1}{2}V_x^3 \end{vmatrix} + \begin{vmatrix} 0 \\ u \end{vmatrix} \tag{22}$$

The corresponding differential equation for this problem is:

$$\frac{dx}{dt} = -x \tag{23}$$

and the solution for the initial condition x(0) = 1, is:

$$x = e^{-t} \tag{24}$$

Estimators described by equations (19) and (22) were simulated, and tested with several input signals.

We first tried the estimators with signals of the form:

$$x(t) = e^{-t} + N \qquad (25)$$

where N is Gaussian noise. In all the examples, we show the first 500 points of estimation with a sampling time of 0.01. In that case, both estimators, described with equations (19) and (22) showed identical results, and that is illustrated in Fig. 4 and 5. In Fig. 4, we have the input signal of the form (25), with standard deviation of the noise 0.01, and in Fig. 5, the absolute errors of estimations with both estimators. It can be noted that both estimators provide the same results.

An important characteristic of estimators is their robustness to initial conditions. Insignificant dependance on initial conditions is very desirable since the initial conditions in many cases represent no more than an educated guess on the part of the designer. For that reason, we tested estimators with the signals of type:

$$x(t) = Ae^{-t} + N \qquad (26)$$

In this case, for values A < 1 , both estimators performed identically, but for values A > 1, estimator described by equation (22) performed much better. This is illustrated in Fig. 6 and 7. In Fig. 6 we have input signal of the form (26), with A equal 100, and with standard deviation of the corrupting noise equal to 2. In Fig. 7 we have absolute errors of estimations with estimator (19) (curve 1), and with the estimator from (22) (curve 2). From these figures it is clear that

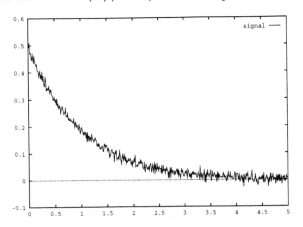

Fig. 4. Input signal of the form (25)

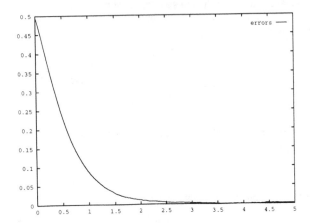

Fig. 5. Absolute errors of estimations with estimators (19) and (22)

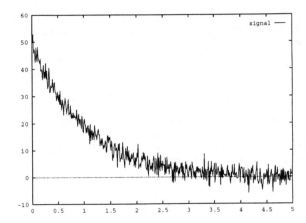

Fig. 6. Input signal of the form (26) with A=100

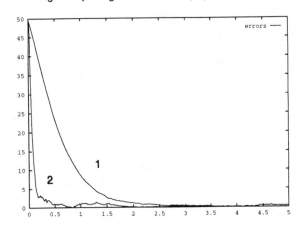

Fig. 7. Absolute errors of estimations
with estimator (19), curve 1, and (22), curve 2

the estimator (22) performs superiorly, ie. it converges much faster, and the error approaches very small values much quicker than with the classical estimator of the form (19). The estimator of the form (22) performed better or equal for all values of A, and for all reasonable signal to noise ratios.

IV. CONCLUSION

A new procedure for designing nonlinear signal processors is presented and applied to signal estimation problems. From the results, we can conclude that the new estimator has a superior performance as compared with Kalman filter. The new concept can be applied to other estimation problems, control theory and related problems.

REFERENCES

[1] Chi-Tsong Chen, "Linear System Theory and Design", 1984. Holt, Reinhart and Winston,
[2] David F. Delchamps, "State Space and Input-Output Linear Systems", 1988. by Springer-Verlag, New York, Inc.,
[3] Rajko Tomovic, "Introduction to Nonlinear Automatic Control systems", translated by Paul Pignon, Wiley, 1966.,,
[4] Jean-Jaccques Slotine and Weipeng Li, "Applied Nonlinear Control", Prentice Hall, 1990.,
[5] Simon S. Haykin, "Adaptive Filter Theory", Prentice Hall, 1991.,
[6] M. D. Srinath, P. K. Rajasekaran, "An Introduction to Statistical Signal Processing With Applications", John Wiley & Sons, 1979.

HIGHER-ORDER WITHIN CHAOS

Patrick Flandrin and Olivier Michel

Ecole Normale Supérieure de Lyon
Laboratoire de Physique (URA 1325 CNRS)
46 allée d'Italie 69364 Lyon Cedex 07 France
e-mail: flandrin@ens-lyon.fr, omichel@ens-lyon.fr

ABSTRACT

Although deterministic in nature, some non-linear systems may give rise to random-like signals. This situation, which is referred to as *chaos*, offers a new perspective in signal modeling and new challenges in signal analysis. The purpose of this paper is to briefly introduce key concepts related to chaos and to point out how higher-order based approaches may prove useful in the context of chaotic signal analysis.

1. MOTIVATION

Given an irregular signal, a classical approach is to model it as the realization of a stochastic process. Until a recent past, this point of view was even the only one, irregularity being implicitly associated with randomness within a generation mechanism. In this respect, one of the most common paradigms of signal processing consists in describing an irregular signal as the output of a linear system driven by a purely stochastic process, namely white noise.

However, although such an approach may prove extremely useful in numerous situations, it is nowadays well-recognized that, in some cases, irregularity can stem as well from some purely deterministic (i.e., non-random) non-linear systems, a situation referred to as *chaos*. Chaos offers therefore a new paradigm and it allows to describe the corresponding signals from a completely new perspective, thus requiring the development of specific analysis tools. (General references concerning chaos and its analysis are, e.g., [1, 4, 10, 22, 24, 27].)

Because of the nature of chaos (non-linear systems, non-Gaussian statistics, ...), higher-order techniques are likely to play an important role in algorithms aimed at chaotic signals. It is the purpose of this paper to provide a brief introduction to chaotic signal analysis and its main algorithms (some classical, some new), emphasizing on the usefulness and the relevance of higher-order concepts in this context.

2. CHAOS

2.1. DEFINITIONS

Instead of considering an observed signal $x(t)$ as a partial information on a stochastic process (a realization), we will view it as a partial information related to the deterministic evolution of a dynamical system. By definition, a *dynamical system* is characterized by a *state* $\mathbf{X} \in \mathbf{R}^p$, whose time evolution is governed by a *vector field* $f : \mathbf{R}^p \rightarrow \mathbf{R}^p$ according to the differential system

$$\frac{d\mathbf{X}}{dt} = f(\mathbf{X}). \tag{1}$$

The dimensionality p of the state vector $\mathbf{X}$ (i.e., the number of its coordinates) characterizes the number of *degrees of freedom* involved in the system and the whole space to which $\mathbf{X}$ is allowed to belong is called the *phase space*.

If we assume that f does not depend upon time (what will be done in the sequel), the system is said to be *autonomous*. In such a case, any solution of (1) can be written $\mathbf{X}(t) = \varphi_t(\mathbf{X}_0)$, where $\mathbf{X}_0$ stands for some initial condition. This solution is referred to as a *trajectory* in phase space, the mapping φ_t (such that $\varphi_t \circ \varphi_s = \varphi_{t+s}$ and $\varphi_0(\mathbf{X}) = \mathbf{X}$) being called the *flow*.

Asymptotically (i.e., as $t \rightarrow \infty$), a solution of Eq. (1) is characterized by a *steady-state* behavior which has to be bounded for making sense. The trajectory tends therefore to be restricted to a subset of the phase space (called the *attractor*), whose nature heavily depends on f. In the case of *linear* f's, the only (bounded) attracting sets are *points* or *cycles*, which means that the steady-state trajectories can only be fixed points or quasi-periodic orbits (countable sum of periodic solutions). Moreover, the steady-state behavior is unique and is attained whatever the initial condition is.

In the case of non-linear dynamics, the situation is quite different and much richer. Different steady-state behaviors can be observed, depending on the initial conditions, and the solutions themselves are not

0-7803-1238-4/93 $3.00 © 1993 IEEE

restricted to quasi-periodicity : chaotic motion is one among the many possibilities.

There is no unique and well-accepted definition of chaos. A signal will be here considered as chaotic if

1. it results from a (non-linear) autonomous deterministic system, and

2. the behavior of this system is highly dependent on initial conditions in the sense that trajectories initiated from neighbouring points in phase space diverge exponentially as functions of time.

Furthermore, we will restrict ourselves to situations which only involve a *low number of degrees of freedom*, i.e. a low dimensionality p.

2.2. CHARACTERIZATIONS

According to the above definition, different characterizations of chaotic signals exist, each of them putting emphasis on some specific property.

Attractor. Trajectories of a chaotic system with few degrees of freedom converge towards an attractor which only fills a low-dimensional subset of phase space. Because of the assumptions of (i) boundedness and (ii) non-periodicity, this attractor has necessary a very peculiar and intricate structure with possibly *fractal* properties, a situation referred to as a *strange attractor*. The existence of a fractal (i.e., non-integer) dimension for an attractor can therefore be used as a hint for a possibly chaotic behavior, the dimension itself being a lower bound for the number of degrees of freedom governing the dynamics of the system. (Let us remark that the fractal structure of the attractor associated to a given signal has not to be confused with fractal properties pertaining to the signal itself.)

Spectrum. Another consequence of no periodicity is the *broadband* nature of the spectrum for chaotic signals, which is therefore a necessary (but, of course, not sufficient) condition for the assessment of chaos.

Unpredictability. One key point in the definition of chaos is the high dependence on initial conditions, which drastically limits any possibility of *prediction*. Sensitivity to initial conditions can be expressed by the fact that trajectories diverge exponentially with respect to time, a measure of this divergence being provided by the so-called *Lyapunov exponents*. Suppose that we consider, at some initial time $t = 0$, an infinitesimal hypersphere of radius $\epsilon(0)$ centered on any point of the attractor. After some (non-zero) time

t, this hypersphere will be deformed by the action of the flow into some hyperellipsoid whose principal axes $\epsilon_i(t), i = 1 \ldots p$ characterize contracting (resp. dilating) directions if $\epsilon_i(t) < \epsilon(0)$ (resp. $\epsilon_i(t) > \epsilon(0)$). In this picture, Lyapunov exponents are defined as

$$\lambda_i = \lim_{t \to \infty} \lim_{\epsilon(0) \to 0} \frac{1}{t} \log \frac{\epsilon_i(t)}{\epsilon(0}.$$

Therefore, a necessary condition for a possible situation of chaos is the existence of at least one *positive* Lyapunov exponent.

2.3. EXAMPLES

A classical example of a chaotic behavior is observed in the so-called *Rössler system*, whose dynamics is given by

$$(\dot{x}, \dot{y}, \dot{z}) = (-y - z, x + 0.15y, 0.2 + xz - 10z). \quad (2)$$

Different representations of this system are plotted in figure 1. Although the sampled time series associated with the first coordinate $x(t)$ looks quasi-sinusoidal, its spectrum reveals a broadband structure. A time-frequency diagram (Choi-Williams) evidences that this frequency spreading is mostly due to phase irregularities occurring each time the amplitude is rapidly decreased, thus limiting the predictability. At last, a projected phase diagram ($x(t)$ vs $y(t)$) displays the characteristic structure of a fractal attractor.

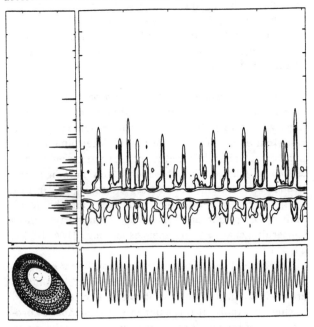

Figure 1. Rössler system: time signal, frequency spectrum, time-frequency distribution and attractor.

The chaotic situation of figure 1 contrasts with that of figure 2, in which different realizations of a stochastic signal are analyzed. The (discrete-time) signals obey the linear equation

$$x_n = 1.95x_{n-1} + 0.98x_{n-2} + \epsilon_n, \qquad (3)$$

where ϵ_n is white noise. In such a case, the resulting stochastic time series share some features with the deterministic ones of Eq. (2), in the time domain. However, other representations (such as the time-frequency diagram or a hypothetical "phase diagram" built on two independent realizations of x_n) show drastic differences, as compared to figure 1.

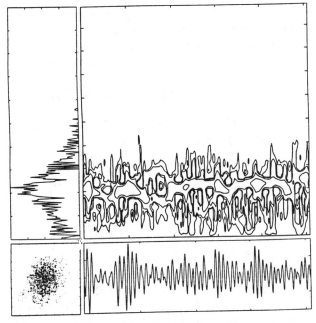

Figure 2. Filtered white noise: time signal, frequency spectrum, time-frequency distribution and attractor.

These simple examples have been chosen to exemplify (i) that, given an observed signal, a chaotic modeling may be appealing and (ii) that the discrimination "chaotic vs stochastic" is not necessarily phrased in terms of "regular vs noisy-like".

3. EMBEDDING

The characterizations of chaotic signals considered so far are based on properties of state vectors within phase space, which requires at least p independent measurements if p is the expected dimensionality of phase space. Unfortunately, in most cases, no sufficient information is available for properly composing a state vector and studying its evolution, the most critical situation corresponding to only one measurement. (In such a case, the

problem is somewhat similar to the one faced within a stochastic framework, when ensemble quantities are to be inferred from the observation of only one realization.)

Given an observed signal $x(t)$, considered as only one component of an unknown state vector, the problem is therefore to *reconstruct* an approximate phase space, with the requirement it be topologically equivalent to the true one.

3.1. METHOD OF DELAYS

A solution to this problem relies upon a theorem of Takens [30] which states the following : given one (noiseless) infinitely long observation $x(t)$ and any arbitrary (non-zero) delay τ, the collection

$$x(t), x(t+\tau), \ldots, x(t+(p-1)\tau)$$

defines a reconstructed attractor which is guaranteed to be equivalent (up to a diffeomorphism) to the actual one, provided that $p \geq 2D+1$ (where D stands for the dimension of the attractor). Geometrical properties of the attractor are therefore unchanged, which makes of this *method of delays* a tool of considerable interest.

When only one discrete-time signal x_n is available over N data points, the technique consists in building the vectors

$$\mathbf{X}_i = (x_i, x_{i+d}, \ldots, x_{i+(p-1)d}); \quad i = 1, 2, \ldots N_{pd}, \quad (4)$$

where $N_{pd} = N - (p-1)d$, d is the elementary delay and p the *embedding* dimension chosen (as previously) such that $p \geq 2D+1$, where D is the dimension of the attractor.

The embedding dimension p is a priori a free parameter, which can be used as a variable when looking for a chaotic dynamics, given a signal. Indeed, if we suppose that a dimension estimation of the (reconstructed) attractor is performed with increasing embedding dimensions p's, it is expected that the estimated dimension will increase significantly until the effective dimension D of the attractor is attained. Chaotic signals should therefore be associated with a saturation of the estimated dimension (at a low, and generally non-integer, value), as a function of the embedding dimension. In contrast, stochastic signals (with many degrees of freedom) are expected to always explore as many directions within phase space as offered, thus presenting no saturation effect for increasing embedding dimensions.

3.2. OPTIMUM CHOICE OF THE DELAY

Although the choice of the delay is theoretically of no importance in the case of noiseless observations of infinite length, it may become an important issue from a practical point of view : for too small delays, all the coordinates of the reconstruction are strongly correlated and the estimated dimension tends towards 1; conversely, for too large delays, the coordinates are almost independent so that the dimension is generally close to the embedding dimension p, with no significant relationship with the number of degrees of freedom involved in the dynamics.

A "good" delay should therefore correspond to the smallest value for which the different coordinates of the reconstructed state vector are almost "unrelated" in a sense which has to be made precise. A first proposal was to make use of the first zero of the cross-correlation function between the observation and its delayed version, but this is clearly a gross indicator which only concerns second-order independence. A more global criterion, relying on the concept of *general* (and not only second-order) independence, has therefore been proposed in terms of *mutual information* between the two signals (original and delayed), the optimum delay being chosen as the first minimum of the function [13]. Efficient algorithms have been developed at this end, thus permitting an improved phase space reconstruction, but at the expense of huge computations (in the general case) [13, 14].

This problem of optimum selection of a delay is a first instance in which higher-order techniques may prove useful. The idea is that they allow some improvement as compared to second-order based criteria, while remaining of a reduced complexity, as compared to "all-order" methods.

The general framework for *independent component analysis* of vectors can be described as follows [8, 7]. Let $p_{\mathbf{x}}$ and p_{x_i} be the probability density functions of a set of vectors $\mathbf{x} = (x_1, ..., x_i, ..., x_p)^T$ and of its components x_i, respectively. If all the components are statistically independent, then we have

$$p_{\mathbf{x}}(\mathbf{u}) = \prod_{i=1}^{p} p_{x_i}(u_i). \qquad (5)$$

The purpose of independent component analysis is therefore to search for a linear transformation that minimizes the statistical dependence between the vector components. At this end, a useful distance measure is the *Kullback divergence* which, in this case, is given by

$$d\left(p_{\mathbf{x}}, \prod_{i=1}^{p} p_{x_i}\right) = \int p_{\mathbf{x}}(\mathbf{u}) \log \frac{p_{\mathbf{x}}(\mathbf{u})}{\prod_{i=1}^{p} p_{x_i}(u_i)} d\mathbf{u}, \qquad (6)$$

and which turns out to be as well the average *mutual information*, hereafter noted $I_m\left(p_{\mathbf{x}}, \prod_{i=1}^{p} p_{x_i}\right)$. (It is recalled that $I_m(p_{\mathbf{x}}, p_{\mathbf{y}}) = S(p_{\mathbf{x}}) - S(p_{\mathbf{x}|\mathbf{y}})$, where

$S(p_{\mathbf{x}})$ stands for the differential entropy of $\mathbf{x}$.) It may then be proved that

$$\begin{aligned} I_m\left(p_{\mathbf{x}}, \prod_{i=1}^{p} p_{x_i}\right) &= -S(\phi_{\mathbf{x}}) + \sum_{i=1}^{p} S(\phi_{x_i}) \\ &\quad + J(p_{\mathbf{x}}) - \sum_{i=1}^{p} J(p_{x_i}), \quad (7) \end{aligned}$$

where $\phi_{\mathbf{x}}$ is the Gaussian probability density function having the same mean and variance than $p_{\mathbf{x}}$, and $J(p_{\mathbf{x}})$ is the negentropy of $\mathbf{x}$, defined by

$$J(p_{\mathbf{x}}) = S(\phi_{\mathbf{x}}) - S(p_{\mathbf{x}}). \qquad (8)$$

In our special case, negentropy is equivalent to Kullback divergence between $p_{\mathbf{x}}$ and $\phi_{\mathbf{x}}$.

Making use of the fact that $J(p_{\mathbf{x}})$ is invariant with respect to orthogonal changes of coordinates and that

$$S(\phi_{\mathbf{x}}) = \frac{1}{2}\left(p + p\log 2\pi + \log \det \mathbf{V}\right), \qquad (9)$$

where $\mathbf{V}$ is the covariance matrix of $\mathbf{x}$, the mutual information (7) can be expressed as

$$\begin{aligned} I_m\left(p_{\mathbf{x}}, \prod_{i=1}^{p} p_{x_i}\right) &= J(p_{\tilde{\mathbf{x}}}) - \sum_{i=1}^{p} J(p_{\tilde{x}_i}) \\ &\quad - S(\phi_{\mathbf{x}}) + \sum_{i=1}^{p} S(\phi_{x_i}), \quad (10) \end{aligned}$$

where $\tilde{\mathbf{x}}$ is the standardized vector obtained from a Choleski factorization of $\mathbf{x}$.

Instead of explicitly evaluating the mutual information, it is therefore possible to approximate it via a fourth-order expansion of $J(p_{\mathbf{x}})$ based on the *Edgeworth expansion* of probability density functions [17]. In the case of standardized data, and using the Einstein summation convention [7], this is given by

$$\begin{aligned} J(p_{\tilde{\mathbf{x}}}) &= \frac{1}{12} K^{ijk} K_{ijk} + \frac{1}{48} K^{ijkl} K_{ijkl} \\ &\quad + \frac{3}{48} K^{ijk} K_{ijn} K_{kqr} K^{qrn} \\ &\quad + \frac{1}{12} K^{ijk} K_{imn} K_{jr}^{m} K_{k}^{nr} \\ &\quad - \frac{1}{8} K^{ijk} K_{ilm} K_{jk}^{lm} + O(p^{-2}), \quad (11) \end{aligned}$$

where the $K_{ij...q}$'s stand for the cumulants of the standardized variables $\tilde{x}_i, \tilde{x}_j, ... \tilde{x}_q$ and p is the number of independent variables in $\mathbf{x}$. It is important to notice that the convergence of such an expansion is related to the fact that cumulants of order i behave asymptotically as $p^{\frac{2-i}{i}}$.

As an illustration, figure 3 compares an estimate of mutual information (between the time series and its delayed versions) based on Fraser's algorithm with its fourth-order approximation using Edgeworth expansion. The signal under study consists in 2048 data points from the first coordinate of the Rössler system (2). All the cumulants involved in the Edgeworth expansion have been estimated using k-statistics [17].

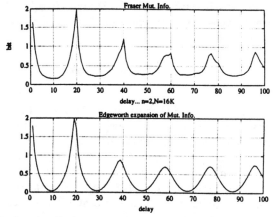

Figure 3. Estimation of mutual information between a Rössler time series and its delayed versions. Top: Fraser's algorithm. Bottom: fourth-order approximation via Edgeworth expansion.

This figure clearly shows a good agreement between the location of local extrema, and especially the first one, which is the quantity of major interest. Further advantages offered by the approach based on the fourth-order expansion are the following : (i) it allows to compute the variance of the estimated mutual information, (ii) it does not need any empirical threshold (as needed to test homogeneity of sub-classes in Fraser's algorithm), (iii) it has an improved efficiency for larger p's (cf. Eq. (11)) and (iv) it requires generally less data points than Fraser's algorithm in the case of well structured signals.

One can finally note that the global approach reported here generalizes recent studies and support the claim that, from an empirical point of view, the "embedding window" $(p-1)d$ may be chosen as the characteristic time for which some suitably chosen cumulants (up to order four) vanish simultaneously [3].

4. DIMENSION ESTIMATION

It has been said previously that a chaotic system is associated with a low-dimensional attractor whose complex geometry makes of it a *fractal* object. Estimating the fractal dimension of an attractor (true or reconstructed) is therefore a clue for evidencing a possibly

chaotic behavior within a given system.

Different definitions of dimensions can be adopted. The simplest one (*capacity* dimension) consists in covering the attractor by the minimum number $N(\epsilon)$ of hypercubes of size ϵ, and evaluating the quantity

$$D_c = \lim_{\epsilon \to 0} \frac{\log N(\epsilon)}{\log(1/\epsilon)}. \qquad (12)$$

A refinement of this definition takes into account the probability P_i with which each of the $N(\epsilon)$ hypercubes is visited, leading to the *Renyi's (generalized) information dimensions*

$$D_q = \frac{1}{q-1} \lim_{\epsilon \to 0} \frac{\log\left(\sum_{i=1}^{N(\epsilon)} P_i^q\right)}{\log(1/\epsilon)}. \qquad (13)$$

Estimating any of these quantities is an important issue but it must be emphasized that, even in the case of a reliable estimation, the significance of a result based on purely geometrical properties of the attractor is to be questioned. It is, e.g., not clear whether any dynamical information (such as the number of degrees of freedom governing the system) can be gained from such a static perspective. This problem is another instance in which higher-order statistics may prove useful.

4.1. CORRELATION DIMENSION

The most popular fractal dimension is the so-called *correlation dimension* (which identifies to Renyi's D_2). Its estimation can be achieved by the *Grassberger-Procaccia Algorithm* (GPA) [16] which measures, for each p, the number of pairs $(\mathbf{X}_i, \mathbf{X}_j)$ whose distance is less than a given radius r. More precisely, the algorithm computes the *correlation integral*

$$C_N(r;p) = \frac{1}{M(N_{pd}-1)} \sum_{i=1}^{M} \sum_{j=1, j\neq i}^{N_{pd}} U(r - \|\mathbf{X}_i - \mathbf{X}_j\|), \qquad (14)$$

where M is a number of test points $\mathbf{X}_i$ selected at random on the attractor and U the unit step function. The effectiveness of this approach is due to the fundamental property

$$\lim_{N \to \infty} C_N(r;p) \sim r^D, \quad r \to 0, \quad p \geq 2D+1, \qquad (15)$$

according to which the *correlation dimension* D can be estimated from a slope measurement in a log-log plot of $C_N(r;p)$.

Figure 4 presents some results obtained from GPA. When applied to a chaotic signal (Rössler) or white noise, it behaves as expected in the sense that the estimated dimension saturates in the first case (and keeps on growing in the second case) when the embedding dimension is increased. However, a misleading behavior can be observed as well in the case of some purely stochastic processes with a "1/f" spectrum. This is especially true for the Wiener-Lévy process

$$x_n = x_{n-1} + \epsilon_n, \qquad (16)$$

for which a thorough theoretical [23, 31, 32] and experimental [19] analysis justifies a saturation at the value $D = 2$. This is due to the fact that, although stochastic, "1/f" processes are fractal *signals*, thus leading to fractal reconstructed attractors, without any relationship to a chaotic dynamics. This counter-example evidences that, in the general case, GPA is much more an *estimation* algorithm (given a chaotic dynamics, what is the dimension of the attractor?) than a *detection* one (is there any chaotic dynamics?).

Another drastic limitation of GPA concerns its prohibitive computational load and its reduced effectiveness on small data sets. Roughly speaking, a reliable estimation of a dimension D requires approximately 10^D data points, which makes the method unapplicable as soon as D exceeds some units [28].

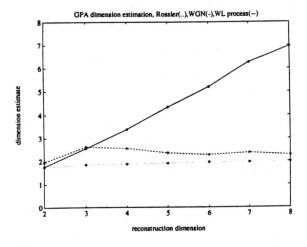

Figure 4. Results of the Grassberger-Procaccia algorithm on a chaotic signal (Rössler, dotted line) and two stochastic processes (white noise, full line, and Wiener-Lévy, dashed line).

4.2. RANK DIMENSIONS

The limitations of GPA cited above have motivated the search for alternative methods. In this respect, another possibility of dimension estimation is provided by the so-called *Local Intrinsic Dimensionality* (LID) [12, 25] (see also [5]). The idea is to extract some information from (local) matrices associated with a phase space trajectory, and to reduce the problem of dimension esti-

mation to that of a rank determination. More precisely, the algorithm proceeds as follows: (i) given an embedding dimension p, a number M of test points is selected at random on the attractor; (ii) for each of these points $\mathbf{X}_i$, the q-nearest neighbors $\mathbf{X}_{i(q)}$ are retained and organized in a ($p \times q$) matrix

$$\mathbf{Y}(i) = (\mathbf{X}_{i(1)} - \mathbf{X}_i, \mathbf{X}_{i(2)} - \mathbf{X}_i, ... \mathbf{X}_{i(q)} - \mathbf{X}_i)^T \quad (17)$$

whose rank is estimated by SVD, and (iii) the LID is then obtained as the average of these local ranks over the M chosen points:

$$\hat{D} = \frac{1}{M} \sum_{i=1}^{M} \text{rank} \left(\mathbf{Y}(i)\mathbf{Y}^T(i) \right) \qquad (18)$$

It is to be noted that, whereas GPA is well-founded from a theoretical point of view, LID is more difficult to justify (see [20]). The main justification of LID stems from [12] in terms of equivalence between the number of independent parameters which are necessary in either a first order Taylor expansion of the non-linear dynamics at a given point $\mathbf{X}_i$ or a local Karhunen-Loève expansion of the associated matrix $\mathbf{Y}(i)$. From a practical point of view, there are experimental evidences for a good agreement between LID and GPA estimates, with the advantage, for LID, of a very easy implementation at a low computational cost. Figure 5 provides results comparable to those of figure 4, but obtained with LID instead of GPA (the rank is estimated from a threshold fixed arbitrarily at 80 percent of the total energy). Apart from the gain in terms of computational load, it can be checked that LID suffers from deficiencies similar to GPA when one wants to use it as a discrimator "chaos vs noise".

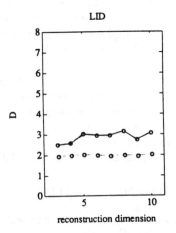

Figure 5. Results of the Local Intrinsic Dimensionality algorithm on a chaotic signal (Rössler, dotted line) and a stochastic process (Wiener-Lévy, full line).

4.3. FOURTH-ORDER EXTENSIONS

Both GPA and LID are very sensitive to additive noise and their performance is severely degraded, even at reasonable signal-to-noise ratios. Essentially motivated by a SNR improvement of the LID algorithm when the data points are corrupted by some additive Gaussian noise, a "higher-order version" of LID (referred to as HOLID) has recently been proposed [26]. It consists in introducing a fourth-order cumulant matrix $\mathbf{M}$:

$$\mathbf{M}_{ij} = \mathcal{E}\{x_i^3 x_j\} - 3\mathcal{E}\{x_i^2\}\mathcal{E}\{x_i x_j\} \qquad (19)$$

where $\mathcal{E}$ stands for the expectation operator and x_i for a component of the embedding vectors. Estimation is then performed on a neighborhood of q local data points, rank being deduced from the SVD of $\mathbf{M}$.

More generally, LID is by construction an eigen-method which only tracks *uncorrelated* (i.e. *linearly independent*) components in the embedding, while forcing the orthogonality of the decomposition vectors. This amounts to say that the coordinates of (17) are decomposed according to

$$\mathbf{X}_{i(k)} - \mathbf{X}_i = \sum_{j=1}^{p} \mu_{jk} \mathbf{Y}_j(\mathbf{X}_i), \qquad (20)$$

where the μ_{jk} are uncorrelated and the $\mathbf{Y}_j$ orthogonal (Karhunen-Loève expansion). Coming back to the interpretation of degrees of freedom in terms of unrelated coordinates of the state vector, it is more appealing to look for an *independent* component analysis, without imposing any orthogonality condition between the corresponding vectors, i.e. to only impose the requirement that the μ_{jk} be independent until fourth-order. A global fourth-order matrix, generalizing (19), can therefore be constructed, containing all possible cumulants from a set of p-dimensional vectors. Forcing all of these cumulants to be zero forms the basis of a new higher order version of LID (referred to as LID4), which has been proposed and applied in [19].

Another appealing approach consists in considering the problem in terms of source separation, in the spirit of [7]. This corresponds again to a decomposition of the form (20) under a constraint of statistical independence up to fourth-order, but this leads to a new distribution of the *energy* of the local observations. Figure 6 illustrates the effectiveness of this approach for improving the discrimation chaos vs noise, in cases where second-order methods fail.

As for the second-order LID algorithm, a weakness of this higher-order method is the need to estimate the effective rank of higher-order matrices, what is generally driven from empirical thresholding methods. Nevertheless, the first investigations conducted

up to now suggest that, in comparable circumstances, fourth-order algorithms tend to provide a result more related to the number of degrees of freedom involved in the dynamics than to the geometrical dimension of the attractor.

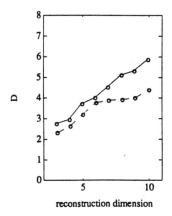

Figure 6. Results of 4th order extension of the Local Intrinsic Dimensionality algorithm on a chaotic signal (Rössler, dotted line) and a stochastic process (Wiener-Lévy, full line).

5. CONCLUSION

The purpose of this paper was to show that some more insight can be gained in the characterization of chaotic signals by using higher-order based techniques. For instance, although it has been stressed that the problem of *detecting* chaos (or, in other words, of testing for determinism) cannot be restricted to an *estimation* problem based on chaos-oriented algorithms, fourth-order estimation algorithms have been shown to outperform second-order (classical) ones within a detection context.

The few examples presented here are however far from exhausting the potential usefulness of higher-order techniques within the context of chaotic signal processing. We can mention, e.g., that only *low-dimensional* chaos has been considered, although increasing the number of degrees of freedom also leads to important problems. Fully developed turbulence is a physical example of such a situation, and its characterization relies heavily on higher-order concepts such as the so-called "structure functions" [21]. (By definition, a structure function measures the higher-order variance of the increment of a velocity field, and its scaling behavior (in the small scale limit) directly provides informations about deviations from Gaussianity in the fluctuations.)

From another point of view, it is clear that, beyond analysis, challenging problems of *prediction* are offered by chaotic signals and that non-linear models are to be explored further from such a perspective [6].

6. REFERENCES

[1] H.D.I. Abarbanel, "Chaotic Signals and Physical Systems", IEEE Int. Conf. on Acoust., Speech and Signal Proc. ICASSP-92, San Francisco, pp. IV.113–IV.116, 1992.

[2] A.M. Albano, J. Muench and C. Schwartz, "Singular Value Decomposition and the Grassberger-Procaccia Algorithm." Phys. Rev. A, Vol. 38, No. 6, pp.3017–3026, 1988.

[3] A.M. Albano, A. Passamante and M.E. Farrell, "Using Higher-Order Correlations to Define an Embedding Window", Physica D, Vol. 54, pp. 85–97, 1991.

[4] P. Bergé, Y. Pomeau and C. Vidal, Order within Chaos, Wiley-Interscience, New-York, 1984.

[5] D.S. Broomhead and G.P. King, "Extracting Qualitative Dynamics from Experimental Data.", Physica D, Vol. 20, pp.217–236, 1986.

[6] M. Casdagli, "Nonlinear Prediction of Chaotic Time Series", Physica D, Vol. 35, pp. 335–356, 1989.

[7] P. Comon, "Independent Component Analysis", Int. Signal Proc. Workshop on Higher Order Statistics, Chamrousse (France), pp. 111-120, 1991.

[8] J.F. Cardoso and P. Comon, "Tensor-Based Independent Component Analysis", EUSIPCO-90, Barcelona (Spain), pp. 673-676, Elsevier, 1990.

[9] P. Duvaut, "Non Linear Filtering in Signal Processing", Int. Signal Proc. Workshop on Higher Order Statistics, Chamrousse (France), pp. 41-50, 1991.

[10] J.P. Eckmann and D. Ruelle, "Ergodic Theory of Chaos and Strange Attractors", Rev. Mod. Phys., Vol. 57, No. 3, pp. 617–656, 1985.

[11] P. Flandrin and O. Michel, "Chaotic Signal Analysis and Higher Order Statistics", EUSIPCO-92, Brussels (Belgium), pp. 179–182, Elsevier, 1992.

[12] K. Fukunaga and D.R. Olsen, "An Algorithm for Finding Intrinsic Dimensionality of Data", IEEE Trans. on Comp., Vol. C-20, No. 2, pp. 176–183, 1971.

[13] A. Fraser and H.L. Swinney, "Independent Coordinates for Strange Attractors from Mutual Information.", Phys. Rev. A, Vol. 33, No. 2, pp. 1134–1140, 1986.

[14] A. Fraser, "Information and Entropy in Strange Attractors.", IEEE Trans. on Info. Theory, Vol. IT-35, No. 2, pp. 245-262, 1989.

[15] A. Fraser, "Reconstructing Attractors from Scalar Time Series: A Comparison of Singular system and Redundancy criteria.", Physica D, Vol. 34, pp. 391–404, 1989.

[16] P. Grassberger and I. Procaccia, "Measuring the Strangeness of Strange Attractors", Physica D, Vol. 9, pp. 189–208, 1983.

[17] M.G. Kendall, The Advanced Theory of Statistics - Vol. I, C. Griffin and Co. Ltd, London, 1952.

[18] W. Liebert and H.G. Schuster, "Proper Choice of the Time Delay for the Analysis of Chaotic Time Series", Phys. Lett. A, Vol. 142, No. 2–3, pp. 107–111, 1989.

[19] O. Michel and P. Flandrin, "An Investigation of Chaos-Oriented Dimensionality Algorithms Applied to AR(1) Processes", IEEE Int. Conf. on Acoust., Speech and Signal Proc. ICASSP-92, San Francisco, 1992.

[20] O. Michel and P. Flandrin, "Local Minimum Representation of a System for Estimating the Number of its Degrees of Freedom", IEEE-SP Workshop on Higher-Order Statistics, South Lake Tahoe (CA), 1993.

[21] J.F. Muzy, E. Bacry and A. Arnéodo, "Multifractal Formalism for Fractal Signals: the Structure-Function Approach versus the Wavelet-Transform Modulus-Maxima Method", Phys. Rev. E, Vol. 47, No. 2, 1993.

[22] A.V. Oppenheim, G.W. Wornell, S.H. Isabelle and K.M. Cuomo, "Signal Processing in the Context of Chaotic Signals", IEEE Int. Conf. on Acoust., Speech and Signal Proc. ICASSP-92, San Francisco, pp. IV.117–IV.120, 1992.

[23] A.R. Osborne and A. Provenzale, "Finite Correlation Dimension for Stochastic Systems with Power-law Spectra", Physica D, Vol. 35 pp. 357–381, 1989.

[24] T.S. Parker and L.O. Chua, Practical Numerical Algorithms for Chaotic Systems, Springer-Verlag, New York, 1989.

[25] A. Passamante, T. Hediger and M. Gollub, "Fractal Dimension and Local Intrinsic Dimension", *Phys. Rev. A*, Vol. **39**, No. 7, pp. 3640-3645, 1989.

[26] A. Passamante and M.E. Farrell, "Characterizing Attractors Using Local Intrinsic Dimension *via* Higher-Order Statistics", *Phys. Rev. A*, Vol. **43**, No. 10, pp. 5268-5274, 1991.

[27] D. Ruelle, *Chaotic Evolution and Strange Attractors*, Cambridge Univ. Press, 1989.

[28] D. Ruelle, "Deterministic Chaos: the Science and the Fiction", *Proc. Roy. Soc. London*, Vol. **A427**, pp. 241–248, 1990.

[29] R. Shaw, "Strange Attractors, Chaotic Behavior, and Information Flow". *Z. Naturforschung*, Vol. **36A**, No.1, pp. 80–112, 1981.

[30] F. Takens, "Detecting Strange Attractors in Turbulence", *Lecture Notes in Mathematics*, Vol. **898**, pp. 366–381, 1981.

[31] J. Theiler, "Some Comments on the Correlation Dimension of $1/f^\alpha$ noise", *Phys. Lett. A*, Vol. **155**, No. 8-9, pp. 480–493, 1991.

[32] R.C.L. Wolff, "A Note on the Behaviour of the Correlation Integral in the Presence of a Time Series", *Biometrika*, Vol. **77**, No. 4, pp. 689–697, 1990.

CHAOTIC SIGNALS AND SYSTEMS FOR COMMUNICATIONS

Kevin M. Cuomo   *Alan V. Oppenheim*

Research Laboratory of Electronics
Massachusetts Institute of Technology
77 Massachusetts Avenue, Cambridge, MA 02139

ABSTRACT

Chaotic systems provide a rich mechanism for signal design and generation for communications and a variety of signal processing applications. Because chaotic signals are typically broadband, noise-like, and difficult to predict they potentially can be utilized in various contexts for masking information-bearing waveforms and as modulating waveforms in spread spectrum systems. In this paper, we propose and demonstrate with a working circuit two approaches to communications based on synchronized chaotic signals and systems. In the first approach a chaotic masking signal is added at the transmitter and regenerated and subtracted at the receiver. The second approach utilizes modulation of the coefficients of the chaotic system in the transmitter and corresponding detection of synchronization error in the receiver to transmit binary-valued bit streams. We demonstrate both approaches using a transmitter circuit with dynamics that are governed by the chaotic Lorenz system. A synchronizing receiver circuit which exploits the ideas of synchronized chaotic systems is used for signal recover.

1. INTRODUCTION

Chaotic systems are nonlinear deterministic systems which can exhibit erratic and irregular behavior. The limiting trajectories of dissipative chaotic systems are attracted to a region in state space which forms a set having fractional dimension and zero volume. Trajectories on this limiting set are locally unstable, yet remain bounded within some region of the system's state space. These sets are termed "strange attractors" and exhibit a sensitive dependence on initial conditions in the sense that any two arbitrarily close initial conditions will lead to trajectories which rapidly diverge.

A particular class of chaotic systems possesses a self-synchronization property [1, 2]. A chaotic system is self-synchronizing if it can be decomposed into at least two subsystems: a drive system and a stable response subsystem(s) that synchronize when coupled with a common drive signal. For some synchronizing chaotic systems the ability to synchronize is robust. For example, the chaotic Lorenz system

This work was sponsored in part by the Air Force Office of Scientific Research under Grant Number AFOSR-91-0034-A, in part by a subcontract from Lockheed Sanders, Inc. under ONR Contract Number N00014-91-C-0125, and in part by the Defense Advanced Research Projects Agency monitored by the Office of Naval Research under Grant N00014-89-J-1489. K. M. Cuomo is supported in part through the MIT/Lincoln Laboratory Staff Associate Program.

is decomposable into two separate response subsystems that will each synchronize to the drive system when started from any initial condition. This property leads to some interesting applications, such as spread spectrum communication and signal masking as discussed in [3, 4].

In section 2 we describe the synchronizing characteristics of the Lorenz system of equations and their implementation as an analog circuit. In section 3 we discuss and demonstrate the implementation of the chaotic signal masking technique introduced in [3, 4] utilizing the Lorenz circuit. In section 4 we discuss and demonstrate an approach to binary communication utilizing coefficient modulation in the Lorenz circuit.

2. THE CIRCUIT EQUATIONS

The Lorenz system consists of a set of autonomous ordinary differential equations having a three-dimensional state space. These equations arise in the study of thermal convection [5] and are given by

$$
\begin{aligned}
\dot{x} &= \sigma(y - x) \\
\dot{y} &= rx - y - xz \\
\dot{z} &= xy - bz
\end{aligned}
\tag{1}
$$

where σ, r, and b are constant coefficients of the system. For our investigations, we use the values $\sigma = 16, r = 45.6$, and $b = 4$ which places the Lorenz system in a chaotic regime.

An interesting property of equation (1) is that it is decomposable into two stable subsystems [1, 2]. Specifically, a stable (x_1, z_1) response subsystem can be defined by

$$
\begin{aligned}
\dot{x}_1 &= \sigma(y - x_1) \\
\dot{z}_1 &= x_1 y - b z_1
\end{aligned}
\tag{2}
$$

and a stable (y_2, z_2) response subsystem by

$$
\begin{aligned}
\dot{y}_2 &= rx - y_2 - xz_2 \\
\dot{z}_2 &= xy_2 - bz_2 \ .
\end{aligned}
\tag{3}
$$

Equation (1) can be interpreted as the drive system since its dynamics are independent of the response subsystems. Equations (2) and (3) represent dynamical response systems which are driven by the drive signals $y(t)$ and $x(t)$ respectively. The eigenvalues of the Jacobian matrix for the (x_1, z_1) response subsystem are equal to $(-\sigma, -b)$. Since they are both negative, $|x_1 - x|$ and $|z_1 - z| \to 0$ as $t \to \infty$. Also, it can be shown numerically that the Lyapunov exponents of the (y_2, z_2) response subsystem are both negative and thus $|y_2 - y|$ and $|z_2 - z| \to 0$ as $t \to \infty$.

The two response subsystems can be used together to regenerate the full-dimensional dynamics which are evolving at the drive system. Specifically, if the input signal to the

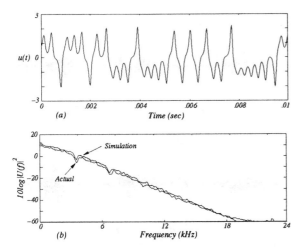

Figure 1: *Circuit Data: (a) A sample function of $u(t)$. (b) Averaged power spectrum of $u(t)$.*

(y_2, z_2) subsystem is $x(t)$, then the output $y_2(t)$ can be used to drive the (x_1, z_1) subsystem and subsequently generate a "new" $x(t)$ in addition to having obtained, through synchronization, $y(t)$ and $z(t)$. It is important to recognize that the two response subsystems given by equations (2) and (3) can be combined into a single system having a three-dimensional state space. This produces a full-dimensional response system which is structurally similar to the drive system (1). Further discussion of this result is given below where we describe the circuit implementation.

A direct implementation of equation (1) with an electronic circuit presents several difficulties. For example, the state variables in equation (1) occupy a wide dynamic range with values that exceed reasonable power supply limits. However, this difficulty can be eliminated by a simple transformation of variables. Specifically, we define new variables by $u = x/10, v = y/10$, and $w = z/20$. With this scaling, the Lorenz equations are transformed into

$$
\begin{aligned}
\dot{u} &= \sigma(v - u) \\
\dot{v} &= ru - v - 20uw \\
\dot{w} &= 5uv - bw \ .
\end{aligned} \tag{4}
$$

This system, which we refer to as the transmitter, can be more easily implemented with an electronic circuit because the state variables all have similar dynamic range and circuit voltages remain well within the range of typical power supply limits. We emphasize that our analog circuit implementation of (4) is exact, and not based on a piecewise linear approach as was used in [6].

To illustrate the chaotic behavior of the transmitter circuit, an analog-to-digital (A/D) data recording system was used to sample the appropriate circuit outputs at a 48 kHz rate and with 16-bit resolution. Figure 1(a) and (b) show a sample function and averaged power spectrum corresponding to the circuit waveform $u(t)$. The power spectrum is broad-band which is typical of a chaotic signal. Figure 1(b) also shows a power spectrum obtained from a numerical simulation of the Lorenz equations. As we see, the performance of the circuit and the simulation are consistent. Figure 2(a) and (b) show the circuit's chaotic attractor projected onto the uv-plane and uw-plane respectively. These data were obtained from the circuit using the stereo recording capability of the A/D system. Specifically, x-axis signals were applied to the left channel and y-axis signals were ap-

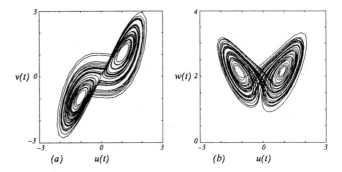

Figure 2: *Circuit Data: (a) Chaotic attractor projected onto uv-plane. (b) Chaotic attractor projected onto uw-plane.*

plied to the right channel, and then simultaneously sampled at a 48-kHz rate and with 16-bit resolution. The circuit's attractor is consistent with numerical simulation. A more detailed analysis of the transmitter circuit is given in [7].

A full-dimensional response system which will synchronize to the chaotic signals evolving at the transmitter (4) is given by

$$
\begin{aligned}
\dot{u}_r &= \sigma(v_r - u_r) \\
\dot{v}_r &= ru - v_r - 20uw_r \\
\dot{w}_r &= 5uv_r - bw_r \ .
\end{aligned} \tag{5}
$$

This system is referred to as the "u-drive" system or as the receiver in light of the various communications applications made possible using this system. For simplicity in notation we will refer to the transmitter state variables collectively by the vector $\mathbf{d} = [u, v, w]$ and to the receiver variables by the vector $\mathbf{r} = [u_r, v_r, z_r]$ when convenient.

It is straightforward to show analytically that synchronization in the Lorenz system is a global property of the nonlinear error dynamics between the transmitter and receiver. First, the dynamical errors, $\mathbf{e}$, are defined as

$$
\mathbf{e} = \mathbf{d} - \mathbf{r} \ .
$$

Under the condition of perfect coefficient matching between the transmitter and receiver a set of equations which govern the error dynamics are given by

$$
\begin{aligned}
\dot{e}_1 &= \sigma(e_2 - e_1) \\
\dot{e}_2 &= -e_2 - 20u(t)e_3 \\
\dot{e}_3 &= 5u(t)e_2 - be_3 \ .
\end{aligned} \tag{6}
$$

The origin of the error system is asymptotically stable provided that $\sigma, b > 0$. This result follows by considering the three-dimensional Lyapunov function defined by $E(\mathbf{e}, t) = \frac{1}{2}(\frac{1}{\sigma}e_1^2 + e_2^2 + 4e_3^2)$. The time rate of change of $E(\mathbf{e}, t)$ along trajectories is given by

$$
\begin{aligned}
\dot{E}(\mathbf{e}, t) &= \frac{1}{\sigma}e_1\dot{e}_1 + e_2\dot{e}_2 + 4e_3\dot{e}_3 \\
&= -(e_1 - \tfrac{1}{2}e_2)^2 - \tfrac{3}{4}e_2^2 - 4be_3^2 \\
&< 0
\end{aligned} \tag{7}
$$

which shows that $E(\mathbf{e}, t)$ decreases for all points in the systems state space. Furthermore, since the divergence of the vector field of (6) is a negative constant, equal to $-(\sigma + b + 1)$, it follows that any error volume will go to zero exponentially fast.

3. CHAOTIC SIGNAL MASKING

In this section, we discuss and demonstrate with a working circuit, chaotic signal masking. Our objective is to demonstrate the signal masking idea described in [3, 4] and to

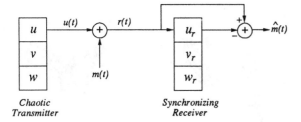

Figure 3: *Chaotic Signal Masking System.*

further illustrate that synchronizing chaotic systems offer potential opportunity for novel approaches to secure communications. In signal masking, a noise-like masking signal is added at the transmitter to the information-bearing signal $m(t)$ and at the receiver the masking is removed. The basic idea is to use the received signal to regenerate the masking signal at the receiver and subtract it from the received signal to recover $m(t)$. This can be done with the synchronizing receiver circuit since the ability to synchronize is robust, *i.e.* is not highly sensitive to perturbations in the drive signal and thus can be done with the masked signal. While there are many possible variations, consider, for example, a transmitted signal of the form $r(t) = u(t) + m(t)$. It is assumed that for masking, the power level of $m(t)$ is significantly lower than that of $u(t)$. The basic strategy then is to exploit the robustness of the synchronization using $r(t)$ as the synchronizing drive at the receiver. The dynamical system implemented at the receiver is

$$\begin{aligned}
\dot{u}_r &= 16(v_r - u_r) \\
\dot{v}_r &= 45.6r(t) - v_r - 20r(t)w_r \\
\dot{w}_r &= 5r(t)v_r - 4w_r \ .
\end{aligned} \qquad (8)$$

If the receiver has synchronized with $r(t)$ as the drive, then $u_r(t) \simeq u(t)$ and consequently $m(t)$ is recovered as $\hat{m}(t) = r(t) - u_r(t)$. Figure 3 illustrates the approach.

In [3] the feasibility of the approach was demonstrated through numerical simulation with almost perfect signal recovery. Using the working transmitter and receiver circuits, we demonstrate the performance of this system in figure 4 with a segment of speech from the sentence "He has the bluest eyes". The waveforms were obtained by sampling the appropriate circuit outputs at a 48 kHz rate and with 16-bit resolution. Figure 4(a) and (b) show the original speech, $m(t)$, and the recovered speech signal, $\hat{m}(t)$, respectively. Clearly, the speech signal has been recovered. Although more distortion is evident in the recovered waveform with the actual circuit implementation as compared with the numerical simulation in [3], the output is very intelligible in informal listening tests. Figure 5 illustrates that the power spectra of the chaotic masking signal, $u(t)$, and the speech are highly overlapping with an average signal-to-masking ratio of approximately -20dB.

4. CHAOTIC DIGITAL COMMUNICATION

In this section, we propose the use of synchronized chaotic systems to transmit and recover binary-valued bit streams. Synchronized chaotic systems are well suited to this application because the chaotic signals they produce have noise-like characteristics and the receiver is robust to uncertainties in the transmitter's initial condition.

The error dynamics of the Lorenz u-drive system are exponentially stable provided that the transmitter and receiver coefficients are identical. This suggests a way in

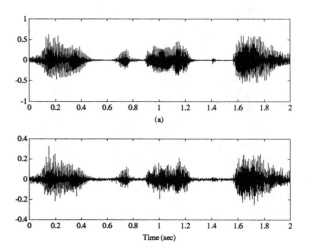

Figure 4: *Circuit Data: Speech Waveforms: (a) Original. (b) Recovered.*

which an information-bearing waveform could be embedded in a chaotic carrier and extracted at the receiver. The basic idea is to modulate a transmitter coefficient with the information-bearing waveform and to transmit the chaotic drive signal. Because of the modulation embedded in the carrier a time-varying coefficient mismatch exists between the transmitter and receiver. Upon reception, the coefficient mismatch will produce a synchronization error between the received drive signal and the receiver's regenerated drive signal with an error signal amplitude that depends on the modulation present. Using the synchronization error the coefficient mismatch can be detected and exploited in various ways for information transfer.

This modulation/detection process is illustrated in figure 6. In this figure, the coefficient "b" of the transmitter equations (4) is modulated by the information waveform, $m(t)$. The coefficients σ and r could also be used as the modulation coefficient, however, in [7] we show that there are some advantages to choosing b as the modulation coefficient. The information is carried over the channel by the chaotic signal $u_m(t)$ and the received signal, $r(t) = u_m(t) + n(t)$, serves as the driving input to the receiver. At the receiver the modulation is detected by forming the difference between $r(t)$ and the reconstructed drive signal, $u_r(t)$. If we assume that the signal-to-noise ratio of $r(t)$ is large, the error signal $e_1(t) = r(t) - u_r(t)$ will have a small average power if no modulation is present. However if, for example, the information waveform is a binary-valued bit stream, with a "1" representing a coefficient mismatch and a "0" representing no coefficient mismatch, then $e_1(t)$ will be relatively large in amplitude during the time period that a "1" is transmitted and small in amplitude during a "0" transmission. The synchronizing receiver can thus be viewed as a form of matched filter for the chaotic transmitter signal $u(t)$.

For purposes of demonstrating the technique, we use a square-wave for the information-bearing waveform as illustrated in figure 7(a). The square-wave produces a variation in the transmitter coefficient "b" with the zero-bit and one-bit coefficients corresponding to $b(0) = 4$ and $b(1) = 4.4$ respectively. The resulting modulated drive signal, $u_m(t)$, is used as the drive input to the synchronizing receiver system as depicted in figure 6. For transmission privacy it is important that the characteristics of the drive signal not be significantly altered by the presence of the modulation. A comparison of the averaged power spectrum of the drive signal with and without the embedded square-wave present shows that the power spectra are very similar and the presence of the embedded square-wave is not at all obvious [7]. Figure 7(b) shows the synchronization error power, $e_1^2(t)$, at the output of the receiver circuit. As expected, the coefficient mismatch between transmitter and receiver produces significant synchronization error power during a "1" transmission and very little error power during a "0" transmission. Also evident from this figure is the fast response time of the receiver at the transitions between the zero and one bits. Figure 7(c) illustrates that the square-wave modulation can be reliably recovered by lowpass filtering the synchronization error power waveform and applying a threshold test.

5. CONCLUSIONS

In this paper, we described and demonstrated with a working circuit, two approaches to private communications based on chaotic signals and systems. Using a signal masking approach we have shown that analog signals can be privately transmitted and recovered at the intended receivers. Also, signals represented as binary-valued bit streams can be privately communicated by modulating a transmitter coefficient with the information-bearing waveform and detecting the information with a synchronizing receiver circuit. These approaches were demonstrated using a transmitter circuit with dynamics that are governed by the chaotic Lorenz system. A synchronizing receiver circuit which exploits the ideas of synchronized chaotic systems was used for signal recover. We are actively investigating these methods as well as alternative approaches to secure communications based chaotic signals and systems.

REFERENCES

[1] L. M. Pecora and T. L. Carroll, "Synchronization in Chaotic Systems," *Phys. Rev. Lett.*, vol. 64, no. 8, pp. 821-824, Feb. 1990.

[2] T. L. Carroll and L. M. Pecora, "Synchronizing Chaotic Circuits," *IEEE Transactions on Circuits and Systems*, vol. 38, no. 4, pp. 453-456, April 1991.

[3] K. M. Cuomo, A. V. Oppenheim, and S. H. Isabelle, "Spread Spectrum Modulation and Signal Masking Using Synchronized Chaotic Systems," *MIT Research Laboratory of Electronics Technical Report 570*, Feb. 1992.

[4] A. V. Oppenheim, G. W. Wornell, S. H. Isabelle, and K. M. Cuomo, "Signal Processing in the Context of Chaotic Signals," *Proc. IEEE ICASSP-92*, Mar. 1992.

[5] E. N. Lorenz "Deterministic Nonperiodic Flow," *Journal of the Atmospheric Sciences*, vol. 20, pp. 130-141, 1963.

[6] R. Tokunaga, M. Komuro, T. Matsumoto, and L. O. Chua "Lorenz Attractor From an Electrical Circuit With Uncoupled Continuous Piecewise-Linear Resistor," *Intl. Jrnl. of Ckt. Theory and Applications*, vol. 17, pp. 71-85, 1989.

[7] K. M. Cuomo and A. V. Oppenheim "Synchronized Chaotic Circuits and Systems for Communications," *MIT Research Laboratory of Electronics Technical Report 575*, Nov. 1992.

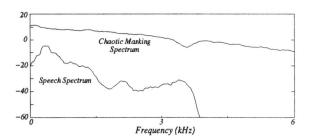

Figure 5: *Circuit Data: Power Spectra of Chaotic Masking and Speech Signals.*

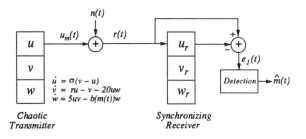

Figure 6: *Chaotic Communication System.*

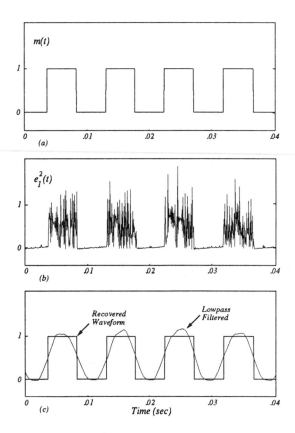

Figure 7: *Circuit Data: (a) Modulation Waveform. (b) Synchronization Error Power. (c) Recovered Modulation Waveform.*

Chapter 2: Adaptive and Learning Systems

Adaptive and Learning Systems

A model that consistently improves through use, through learning or adaptation, can provide better results than an accurate but fixed model of a signal. This becomes particularly important in situations such as language modeling or image recognition in which a tractable data set provides only sparse coverage of a very large state space. Artificial neural networks provide a net of interconnected simple nodes, each of which sums a number of weighted inputs and passes the results through a non-linearity, such as a threshold. The weights are adapted during use, and the network can provide robust classification of distributions generated by nonlinear processes. Neural networks are particularly useful if the data model is not known, and they can generalize to examples not yet seen that are drawn from the same probability distribution.

In a manner similar to neural networks, one may model a language learning task as a system of linear equations, where each equation establishes a sentence - action pair and each variable associates a word with an action. This approach is described in "Adaptive Language Acquisition Using Incremental Learning" (paper 2.1). Related work expands these systems from considering only the input message to also including the state of the surrounding world, as measured through an additional sensor channel. "Visual Focus of Attention in Adaptive Language Acquisition" (2.2) combines a typed keyboard message source with visual input to more accurately mimic the human learning process of combining information from multiple sources.

The concept of fuzziness, i.e., allowing a pattern to have partial membership in each of several classes, is combined with a neural network system in "Fuzzy-Decision Neural Networks" (2.3). This technique is shown to more accurately cope with marginal training patterns that lie within the overlap of two categories.

Hidden Markov models (HMM) have provided the basis for some of the best-performing systems for speech recognition. They place the bulk of the necessary computation in the up-front training processing, where real time is less critical than in the classification of unknown utterances during operation. Most training of HMMs assume a particular model structure (number of states, allowed transitions, etc.) and use the training data to set model parameters within that structure. "An Algorithm for the Dynamic Inference of Hidden Markov Models (DIHMM)" (2.4) proposes a means to incorporate the learning of the HMM structure to achieve reliable models on limited training data. It reports an improvement in recognition performance over fixed-model HMM training techniques.

Finally, "A Self-Learning Neural Tree Network for Recognition of Speech Features" (2.5) provides for the non-uniform correlation among speech sounds, exemplified by the fact that speech sounds are not all equally confusable to the human listener. It establishes an intra-class correlation among speech phones and reports results using a standard phonemic database.

Suggested Additional Readings:

[1]   B. K. Natarajan, *Machine Learning: A Theoretical Approach.* San Mateo, CA: Morgan Kaufman, 1991.

[2]   D. R. Hush and B. G. Horne, "Progress in Supervised Neural Networks," *IEEE Signal Processing Magazine,* vol. 10, no. 1, Jan. 1993.

[3]   R. P. Lippmann, "An Introduction to Computing with Neural Nets," *IEEE ASSP Magazine,* vol. 4, no. 2, pp. 4-22, April 1987.

[4]   P. Placeway, R. Schwartz, P. Fung, and L. Nguyen, "The Estimation of Powerful Language Models from Small and Large Corpora," *Proc. IEEE International Conference on Acoustics, Speech, and Signal Processing 93* (Minneapolis, MN, USA) April 1993, vol. 2 pp. 33-36.

[5]   L. Z. Wang and J. V. Hanson, "Competitive Learning and Winning-Weighted Competition for Optimal Vector Quantizer Design," *Neural Networks for Signal Processing III - Proc. of the 1993 IEEE-SP Workshop* 1993, pp. 50-59.

[6]   H. Ishibuchi, K. Nozaki, and N. Yamamoto, "Selecting Fuzzy Rules by Genetic Algorithm for Classification Problems," *Second IEEE International Conference on Fuzzy Systems,* (San Francisco, CA, USA) Mar. - Apr. 1993, pp. 1119-1124.

[7]   T. I. Boubez and R. L. Peskin, "Wavelet Neural Networks and Receptive Field Partitioning," *1993 IEEE International Conference on Neural Networks* (San Francisco, CA, USA) Mar. - Apr. 1993, pp. 1544-1549.

[8]   K. R. L. Godfrey and Y. Attikiouzel, "Self-Organized Color Image Quantization for Color Image Data Compression," *1993 IEEE International Conference on Neural Networks* (San Francisco, CA, USA) Mar. - Apr. 1993, pp. 1622-1626.

[9]   E. Wilson, S. Umesh, and D. W. Tufts, "Multistage Neural Network Structure for Transient Detection and Feature Extraction," *Proc. IEEE International Conference on Acoustics, Speech, and Signal Processing 93* (Minneapolis, MN, USA) April 1993, vol. 1 pp. 489-492.

ADAPTIVE LANGUAGE ACQUISITION
USING INCREMENTAL LEARNING

K. Farrell[1], R.J. Mammone[1], and A.L. Gorin[2]

[1]CAIP Center, Rutgers University, Piscataway, New Jersey 08855
[2]AT&T Bell Laboratories, Murray Hill, New Jersey 07974

ABSTRACT

An incremental approach to solving an algebraic formulation of the language acquisition problem is presented. This problem consists of solving a system of linear equations, where each equation represents a sentence/action pair and each variable denotes a word/action association [1]. The algebraic model for language acquisition has been shown [1] to provide advantages over the relative frequency estimate models when dealing with small-sample statistics. In this paper, two incremental methods are investigated to solve the system of linear equations. The incremental methods provide a *regularized* solution that is shown experimentally to be advantageous over the pseudo-inverse solution for classifying test data. In addition, the methods are more efficient with respect to computational and memory requirements.

1 INTRODUCTION

The methods for adaptive language acquisition presented in [2, 3, 4] rely on mutual information estimates that represent word/action associations. The word/action associations can be computed using smoothed relative frequency estimates, i.e., using the number of occurrences of a given word and the number of times an action occurred for sentences containing that word. That method for determining word/action associations is *context independent*, meaning that the update of a word/action association for a given word is independent of the other words that occur along with it in the sentence. This trait is undesirable and leads one to formulate the problem of *focused learning* [1], referring to the ability to concentrate on words that convey most of the meaning of a sentence.

As an example of focused learning, consider two sentences from the Inward Call Manager database [2], which consists of requests to an operator in a department store that may be routed to the *furniture*, *clothing*, or *hardware* departments. The two example sentences are: "*I'm looking for a mauve sweater*", and "*I need a new etarge*", where the underline denotes a new vocabulary word. The smoothed mutual information method will tend to create an equal level

of association of the words *mauve* and *etarge* with their corresponding classes, namely *clothing* and *furniture*. This is due to the algorithm only using the information that it has seen *mauve* occur once for the *clothing* category, and *etarge* occur once for *furniture* category. However, the desired response of an algorithm would be to create a relatively small association of *mauve* with clothing and a relatively large association of *etarge* with furniture. Intuitively, this motivates the use of an algorithm whose update for the word/action association is proportional to the error signal. In the above example the first sentence would probably be classified correctly, thus having a small error, whereas the second sentence would probably be misclassified, hence having a large error. Formulating the language acquisition problem as a system of linear equations [1], i.e., algebraic learning, is our proposed means of incorporating the error signal in the update for the word/action association.

Algebraic learning consists of modeling the language acquisition problem as a system of linear equations, where each equation represents a sentence/action pair and each variable denotes a word/action association. The pseudo-inverse solution for this system of linear equations has been found to provide connection weights that are less sensitive to small numbers of samples than are the smoothed relative frequency estimates. The pseudo-inverse solution in [1] was computed using a singular value decomposition (SVD). However, the direct computation of the pseudo-inverse has some limitations. It is costly to use for updating weights and is found to provide suboptimal performance for cross-validation.

In this paper, we present two methods of incremental learning to overcome the drawbacks of the pseudo-inverse solution. The following section reviews how the language acquisition problem can be formulated as a system of linear equations. The incremental methods for solving this system are then described in Section 3. Experimental results are provided for the text-based Inward Call Manager [2] system in Section 4 and in Section 5 the conclusions and summary of this paper are given.

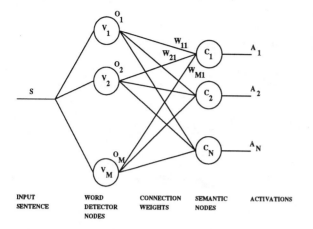

INPUT SENTENCE   WORD DETECTOR NODES   CONNECTION WEIGHTS   SEMANTIC NODES   ACTIVATIONS

Figure 1: Single layer neural network for sentence/action mapping

2 PROBLEM FORMULATION

Our most basic language acquisition model can be interpreted as a single layer neural network, as illustrated in Figure (1). The input nodes V_i are word detectors whose outputs O_i represent the probability that a word is in sentence s. In this text-based experiment, the output of the word detectors is either 0 or 1. Note that the number of input nodes M increases in time as the vocabulary grows. The connection weights w_{ij} represent the word/action association and was defined in [2] as the *mutual information* between words and actions. The output nodes C_j correspond to the set of N possible actions and are evaluated as the inner product of O and W_j. The action A is chosen according to which output node C_j has the largest value.

It was shown in [1] that the language acquisition model described above can be characterized as a set of linear equations. In particular, given a sentence/action pair for the jth sentence, define a vector λ_j as:

$$\lambda_j = <00...0100...10> \qquad (1)$$

where a one or zero in position i indicates the presence or absence of word i. A straight-forward generalization will use word probabilities in place of the $0s$ and $1s$ [5]. Note that the dimension of λ is equal to the number of words in the vocabulary, namely M, which grows over time. An equation derived for a sentence/action pair is thus:

$$<\lambda_j, W^k> = A_j^k = \left\{ \begin{array}{l} 1 \ if \ \lambda_j \ \in \ action \ k \\ 0 \ if \ \lambda_j \ \notin \ action \ k \end{array} \right\} . \quad (2)$$

This equation is independent of word order in a sentence, and can thus be characterized as a "bag of words" model. In [1], it was shown how to extend this model to depend on word order. By using the formulation in equation (2) for each sentence in the

system, a system of linear equations can now be expressed as:

$$\Lambda W^k = A^k, \quad 1 \le k \le N. \qquad (3)$$

In equation (3), Λ is an M by P matrix, whose row vectors are that as given in equation (1). The variable P represents the number of sentences, which also grows over time, W^k is an M dimensional vector containing the connection weights for the kth action, and A^k is a P dimensional vector representing the labels for the kth action. In this experiment, we consider the case where the labels $A^k \in \{1, 0\}$ denote whether or not the sentence corresponds to the kth action.

The pseudo-inverse solution to equation (3) is denoted as:

$$W^{k\dagger} = \Lambda^\dagger A^k, \qquad (4)$$

where the pseudo-inverse matrix $\Lambda^\dagger$ for equation (4) can be computed using the SVD of Λ. For example, the SVD decomposes the Λ matrix as a product of unitary matrices and and a diagonal matrix containing the singular values of Λ. The pseudo-inverse matrix $\Lambda^\dagger$ is computed by inverting the nonzero singular values and letting the zero singular values stay zero. In practice, singular values below some threshold are considered zero and are not inverted.

3 INCREMENTAL METHODS

Two incremental methods are evaluated for solving the system of equations in (3) for W^k. The first method minimizes the quantity:

$$E_k^2 = \sum_{j=1}^{P} [A_j^k - <\lambda_j, W^k>]^2, \qquad (5)$$

where j is the sentence index. This quantity is minimized for each of the k systems, hence the superscript k is omitted in the remaining discussion for simplicity of notation. The method used to minimize the error quantity in equation (5) is known as the row-action projection (RAP) [6] algorithm, and can be used to incrementally calculate the pseudo-inverse solution of equation (3). The RAP algorithm for the system in equation (3) is implemented by using the update equation:

$$W^{(i+1)} = W^{(i)} + \mu \frac{\epsilon_j}{\|\lambda_j\|} \frac{\lambda_j^T}{\|\lambda_j\|} \qquad (6)$$

where

$$\epsilon_j = A_j - <\lambda_j, W^{(i)}> . \qquad (7)$$

In expressions (6) and (7), the superscript i denotes the iteration, the vector λ_j refers to the jth row vector of the matrix Λ, ϵ_j is called the error, and μ is a gain parameter, which is usually chosen between zero and two. Intuitively, the weight vector W in equation (6) is updated by projecting in the direction $\lambda_j^T/\|\lambda_j\|$, by an amount given by $\epsilon_j/\|\lambda_j\|$. The choice of μ contributes to the trade-off between rate of the convergence and the accuracy of the solution. The RAP method minimizes the error quantity E in equation (5) by cycling over the P equations until the error is below some threshold. Each cycle over the P equations will henceforth be referred to as an epoch.

Asymptotically, the RAP algorithm will converge to the pseudo-inverse solution of equation (4). However, in the short term, the RAP algorithm de-emphasizes the inversion of small singular values and provides a *regularized* inverse solution as described in [6]. The resulting singular value taper is given by:

$$\sigma_i = \frac{1}{\sigma_\Lambda}\left[1 - (1 - \frac{\lambda}{N}\sigma_\Lambda^2)^{l+1}\right], \qquad (8)$$

where σ_Λ and σ_i are the singular values of the data matrix and its inverse, N is the dimension of the solution, and l is the iteration index for the block of P equations. The regularized solution represented by equations (6) and (7) is more robust to noise than the pseudo-inverse solution [6]. This regularization tends to improve the classification of test data since the inversion of small eigenvalues, as performed in the SVD computation of the pseudoinverse solution, tends to overfit the training data.

A second approach is a nonlinear incremental method, which minimizes the error quantity:

$$E_k^2 = \sum_{j=0}^{P}[A_j^k - f(<\lambda_j, W^k>)]^2. \qquad (9)$$

The function $f()$ used in equation (9) is the sigmoid activation function [7]:

$$f(<\lambda_j, W>) = \frac{1}{1 + e^{-<\lambda_j, W>}} = y_j. \qquad (10)$$

The effect of the nonlinear activation is to replace the error term ϵ_j in equation (7) with $\tilde{\epsilon}_j$, which is given as:

$$\tilde{\epsilon}_j = y_j(1 - y_j)(A_j - y_j). \qquad (11)$$

In equation (9), since $A_j^k \in 0, 1$ and $0 < f() < 1$, the error for the jth sentence is bounded by $0 < |E_j| < 1$. Thus, large errors of the same sign are deemphasized. The use of the sigmoid activation function introduces a different error norm that is more robust to this type of error [8].

4 EXPERIMENTAL RESULTS

The two incremental methods were applied to the text-based Inward Call Manager [2] database. The database consists of requests to a department store that may fall under the three categories of *furniture*, *clothing*, or *hardware*. This system consists of 1105 sentences comprised of the first sentence in each dialogue of a natural language experiment. The vocabulary size of the 1105 sentences is 1356. All experiments reported in this paper use the first 800 sentences for training, which contains 1122 vocabulary words.

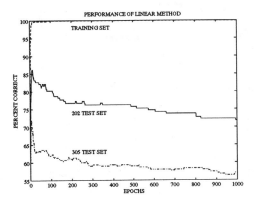

Figure 2: Linear Method

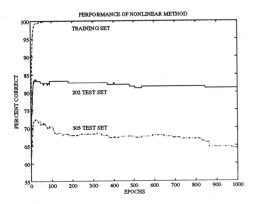

Figure 3: Nonlinear Method

Figures (2) and (3) show the learning curves for the two algorithms. These plots illustrate the performance when the system is trained on 800 sentences and tested with 1) the training set, 2) the test set (305 sentences), and 3) the subset of the test set (202 sentences) containing known *salient* words. For example, in the test sentence, *I'm looking for a table*, if *table* was encountered in the training set, then this sentence would be labeled as containing a known *salient* word.

The peak classification performance for both methods and test sets occurs at roughly 15 epochs. At the peak operating point, the linear method correctly classifies 99% of the training set, 85% of the 202 test set, and 72% of the 305 test set. The nonlinear method correctly classifies 98% of the training set, 84% of the 202 test set, and 72% of the 305 test set. The pseudoinverse as computed with an SVD (thresholding singular values less than 0.1) correctly classifies 99% of the training set, 69% of the 202 test set, and 57% of the 305 test set. These results are summarized in Table 1 along with the smoothed mutual information (SMI) estimates.

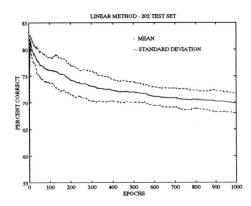

Figure 4: Linear Method - Random Training

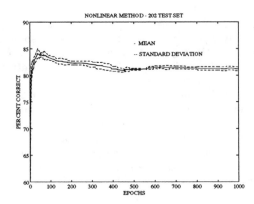

Figure 5: Nonlinear Method - Random Training

Table 1: Classification Performance

Method	Training	Test (305)	Test (202)
SMI	93%	73%	83%
SVD	99%	57%	69%
Linear-peak	99%	72%	85%
Nonlin-peak	98%	72%	84%
Linear-1000	99%	57%	72%
Nonlin-1000	99%	64%	81%

Though the peak performance of both incremental algorithms is roughly equal, the nonlinear method maintains its classification performance as opposed to the linear method, whose performance degrades after numerous iterations. An additional experiment was performed, where the 800 training sentences are randomly ordered prior to estimating the performance. Note that this does not change the final solution to the system of equations. However, the path that the weights take towards their optimum will be different and hence effect performance. Eight random orderings were tested, whose mean and one standard deviation for classification performance on the 202 test set are shown in Figures (4) and (5). The standard deviation at the 1000th epoch is 0.375 for the nonlinear method and 1.866 for the linear method. Hence, the nonlinear method is more robust than the linear method with respect to order sensitivity.

5 CONCLUSION

Two incremental methods have been evaluated on the adaptive language acquisition problem. For the 305 test set, the incremental methods correctly classify 72% of the test sentences at their peak operating point. When evaluated on a subset of the test set containing known salient words, the incremental methods correctly classify 84% of the test sentences. The nonlinear incremental method is found to maintain its level of generalization while the linear method is more vulnerable to overtraining. The non-linear method is also less sensitive to equation ordering. Both methods perform significantly better than the pseudo-inverse, which correctly classifies about 57% and 69% of the sentences in the 305 and 202 test sets, respectively. The performance of the incremental methods is similar to that of the smoothed mutual information method for the test set, but yields better performance for the training set.

REFERENCES

[1] N. Tishby and A. Gorin. Algebraic learning of statistical associations for language acquisition. In *Neural Networks for Speech and Image Processing*. Chapman Hall, 1993.

[2] A.L. Gorin, S.E. Levinson, A.N. Gertner, and E.R. Goldman. On adaptive acquisition of language. *Computer, Speech, and Language*, pages 101–132, Apr. 1991.

[3] A.L. Gorin, S.E. Levinson, and A.N. Gertner. Adaptive acquisition of spoken language. In *Proceedings IEEE ICASSP 1991*, Toronto, May 1991.

[4] L.G. Miller and A.L. Gorin. A structured network architecture for adaptive language acquisition. In *Proceedings IEEE ICASSP 1992*, San Francisco, CA, Mar. 1992.

[5] A.L. Gorin, L.G. Miller, and S.E. Levinson. Some experiments in spoken language acquisition. In *Proceedings IEEE ICASSP 1993*, Minneapolis, Mn, Apr. 1993.

[6] R.J. Mammone. *Computational Methods of Signal Recovery and Recognition*. Wiley, New York, NY, 1992.

[7] D.E. Rumelhart and J.L. McClelland. *Parallel Distributed Processing*. MIT Cambridge Press, Cambridge, Ma, 1986.

[8] P.J. Huber. *Robust Statistics*. Wiley, New York, NY, 1981.

VISUAL FOCUS OF ATTENTION IN ADAPTIVE LANGUAGE ACQUISITION

Ananth Sankar and Allen Gorin

AT&T Bell Laboratories
Murray Hill, NJ 07974

ABSTRACT

In our research on Adaptive Language Acquisition, we have been investigating connectionist systems that learn the mapping from a message to a meaningful machine action through interaction with a complex environment. Previously, the only input to these systems has been the message. However, in many devices of interest, the action also depends on the state of the world, thereby motivating the study of systems with multi-sensory input. In this work, we describe and evaluate a device which acquires language through interaction with an environment which provides both keyboard and visual input. In particular, the machine action is to focus its attention, by directing its eyeball toward one of many blocks of different colors and shapes, in response to a message such as \Look at the red square". The attention focus is controlled by minimizing a *time-varying potential function* that correlates the message and visual input. This correlation is factored through color and shape *sensory primitive subnetworks* in an *information-theoretic connectionist network*, allowing the machine to generalize between different objects having the same color or shape. The system runs in a *conversational mode* where the user can provide clarifying messages and error feedback until the system responds correctly. During the course of performing its task, a vocabulary of 431 words was acquired from 11 users in over 1000 unconstrained natural language conversations. The average number of inputs for the machine to respond correctly was only 1.4 sentences, and it retained 98% of what it was taught.

1. INTRODUCTION

In previous work [1, 2, 3], adaptive language acquisition systems have been explored which are based on connectionist models that learn the mapping from an input message to a meaningful machine response. In all those systems, the input to the machine was a typed or spoken message whose purpose was to induce the machine to perform some action. In many devices of interest, however, the machine response to the message also depends on the *state of the world*. This leads us to consider systems with *multi-sensory input*, i.e., input from both a user and other sources. In this paper, in particular, we investigate a *visual focus of attention task* in a blocks world, where the machine receives both a message and visual input. The machine action is to focus its attention on one of many blocks of different colors and shapes in response to an input message. Figure 1 shows a blocks-world example where the eyeball is shown responding correctly to the input message, \Look at the red square". The device acquires language by learning the associations between the input message and the visual input. We introduce the concept of *sensory primitive subnetworks* in a connectionist network through which the associations between the message and visual input is factored. This structured network enables rapid learning and generalization for such devices. A *time-varying potential function* is described which controls the attention focusing of the machine. We present the results of an experimental evaluation based on over 1000 typed conversations with 11 users over a three month period.

2. BACKGROUND

At present, automatic speech recognition technology is based upon constructing models of the various levels of linguistic structure assumed to compose spoken language. These models are either constructed manually

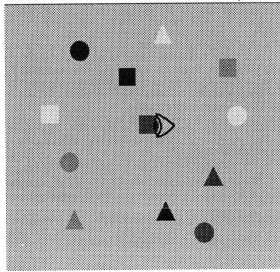

MESSAGE: LOOK AT THE RED SQUARE

Figure 1. A blocks-world example. The eyeball focuses its attention on a red square in response to the input message.

or automatically trained by example. A major impediment is the cost, or even the feasibility, of producing models of sufcient delity to enable the desired level of performance.

An alternative proposed in [1] is to build a device that acquires the necessary linguistic skills *during the course of performing its task*. A rst principle in [1] is that the primary function of language is to communicate meaning, so that language acquisition involves gaining the capability of decoding that meaning. The proposed mechanism is a connectionist network with information-theoretic weights, that learns the mapping from an input message to a meaningful machine response.

A *second principle* is that language is acquired by interaction with a complex environment. This led to the investigation of a connectionist network embedded in a feedback control loop, where the environment is a human user who provides both an input message and a semantic-level error signal as to the appropriateness of the machine's response. The machine learns to respond correctly by strengthening or weakening the connections between words and actions based on the error signal. The connection weights were dened to be the *mutual information* between words and actions, and were updated using smoothed relative frequency estimates.

The utility of these principles was rst studied for a text-based 3-action task that is described in detail in [1], and for spoken input in [2]. However, as a task grows in size and complexity, it is reasonable to question the capability of the machine to learn these complex mappings. While there does, indeed, exist a body of theory addressing such questions [4, 5], the results tend to be asymptotic in nature, requiring large numbers of examples for learning to occur. In contrast, a striking feature of human language

0-7803-0946-4/93 $3.00 © 1993 IEEE

acquisition is our ability to make sweeping generalizations from small numbers of observations. Thus, though a homogeneous network might, given sufcient data, be capable of learning the associations between messages and meaningful responses for complex tasks, we are not satised with such asymptotic results. An alternative is embodied in a *third principle*, that the device architecture should be well-matched with its environment and input/output periphery. Mechanisms for matching a network architecture to its output-periphery structure were explored for a data retrieval task in [3], and for matching environmental structure in the input language for an airline information system in [6]. In this paper, we present a method to match the network architecture to the device's multi-sensory input-periphery structure.

3. FOCUS OF ATTENTION MECHANISM

The vehicle for our investigations into multi-sensory input devices is a simulated blocks-world device, described below. We construct a device whose response to the visual and message inputs has the following innate properties. A *rst property* is that the device eyeball is attracted to bright and moving objects, with movement being more attractive. A *second property* is that the device becomes \bored" after looking at an object for a while, and then moves on to some other object. These two properties enable the device to explore the scene and learn through unsupervised methods, even in the absence of message inputs. A *third property* is that the device is attracted to objects whose visual input is strongly associated with the message input. While learning, the device builds these inter-sensory associations which result in a correct response. A *fourth property* is that the device is repelled from objects that the user deems to be inappropriate. The user conveys this information by typing \o", a prefabricated error signal, as the rst word of a sentence. We now describe these innate properties in greater detail.

1. *The eyeball is attracted to bright and moving objects*, where movement is more attractive than brightness. Thus, the user can catch the machine's attention by moving an object, at the same time providing a message, such as *This is a blue ball*", to teach the machine about that object. This property is implemented via a function $A_i(t)$, given by

$$A_i(t) = \frac{\left(m_i(t) + v_i^\alpha(t)\right)}{\sqrt{d_i(t)^2 + \epsilon}}, \qquad (1)$$

where, at time t, $m_i(t)$ and $v_i(t)$ are the intensity and speed, of object i, $d_i(t)$ is the distance between object i and the eyeball position, $\left(x(t), y(t)\right)$, and ϵ is a positive constant. $v_i(t)$ is raised to a constant power, α, which can be adjusted to make speed or movement more attractive than intensity. The distance term, $d_i(t)$, in the denominator, reects the fact that the device is more strongly attracted to objects that are close to the center of its eld of view.

2. *The device becomes bored after looking at an object for a while*, and then moves on to a different object. This causes the eyeball to explore the scene and gather information from many different objects as part of its innate behavior. Indeed, during such exploration, the network could learn what colors and shapes exist in its environment without human supervision via clustering different colors and shapes in its visual feature space. We postpone the investigation of such learning to future work. The boredom property is implemented via the function B_i, given by

$$B_i = \exp\left(-\lambda_2 \sum_{\tau=t-T_0}^{t} e^{-\lambda_1 d_i(\tau)}\right), \qquad (2)$$

where T_0 is a time window, and λ_1 and λ_2 are constants which affect the rate at which the device gets bored of an object. This function can be understood by imagining that the eyeball has been positioned at object i for T seconds. Since, during this time, T, the eyeball is close to object i, d_i would be small or close to 0. Thus, from Equation 2, we see that

$$B_i \approx e^{-\lambda_2 T}. \qquad (3)$$

This shows that B_i decays exponentially with T, the time for which the eyeball is focused on object i, causing it to lose interest in that object.

3. *The device is attracted to objects whose visual input is strongly associated with the message input.* The device acquires language by learning these associations. This property is implemented via a function U_i, which is a measure of the strength of the association between the message input and the visual input from object i. This function is discussed in greater detail in Section 4..

4. *Finally, the device is repelled from objects that are determined to be inappropriate by the user.* This is implemented via a function E_i, which is an encoding of the semantic-level error signal and is given as follows. Normally E_i is equal to zero. However, suppose the machine focuses its attention on object i in response to a message and the user decides that the response is inappropriate. The user then starts his next message input with the word \o", a prefabricated error signal indicating that object i is incorrect. E_i is then set to a large positive number. The user may also reward the device by typing *Good*", a prefabricated reward signal, causing the network weights to be updated as described in [1].

The properties described above are combined into a potential function, whose value at time t and position (x_i, y_i) of object i is given by

$$P_i(x_i, y_i, t) = -A_i B_i U_i + E_i, \qquad (4)$$

The eyeball is moved toward the object that has the smallest value of P_i. In other words, this is the object that exerts the maximum force on the eyeball. This object is given by

$$i_{min} = \arg\min_i P_i. \qquad (5)$$

The eyeball is moved toward the position $(x_{i_{min}}, y_{i_{min}})$ with a step size that decreases with the distance, $d_{i_{min}}$, between the eyeball and the object, i_{min}. This step size is given by

$$= 1 - e^{-\lambda_4 d_{i_{min}}}. \qquad (6)$$

Although the innate properties described above are psychologically plausible, we do not claim biological validity for the implementation.

4. NETWORK STRUCTURE

We now describe the network which implements the eyeball control. Following our third principle, we reect the device's input-periphery structure in the network architecture. The input periphery of our system has structure in that there is both message and visual input. Furthermore, the device can sense two visual features{color and shape. The visual input structure is reected in the network through the introduction of *sensory primitive nodes* corresponding to color and shape. These are organized into two independent subnetworks as shown in Figure 2. This is similar to the structure of the product network described in [3]. This method can be extended in a straightforward manner if the number of input features is greater than 2, by simply adding one new subnetwork for each new feature. For example, if the visual features corresponded to color, shape, and size, we would have three subnetworks.

In Figure 2, A_i, B_i, and E_i refer to the component functions in the potential function of Equation 4. The outputs, U_{C_i} and U_{S_i}, are measures of the strength of the associations between the message input and the visual input corresponding to the color and shape, respectively, of object i. The term U_i in Equation 4 is now given by adding U_{C_i} and U_{S_i}:

$$U_i = U_{C_i} + U_{S_i}. \qquad (7)$$

Figure 3 shows the details of the color subnetwork. The shape subnetwork is similar. The rst layer nodes are the vocabulary word detector nodes, v_m. The next layer of nodes are the color sensory primitive nodes, c_k. Initially, the subnetwork has no word or color sensory primitive nodes. As new words and colors are experienced, the device builds the color subnetwork by adding new word and color sensory primitive nodes.

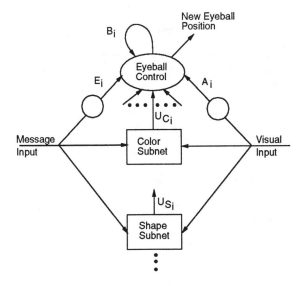

Figure 2. Network Architecture

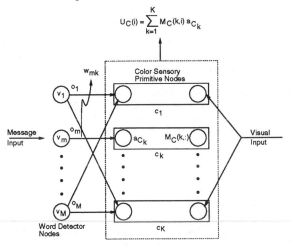

Figure 3. Color Subnetwork

If a word, v_m, in the lth sentence, $S(l)$, is not a vocabulary word, then a new word detector node is grown. The outputs of the word detector nodes, $o_m(l)$, are then given by

$$o_m(l) = \left\{ \begin{array}{ll} 1 & \text{if } v_m \in S(l) \\ 0 & \text{otherwise} \end{array} \right. . \tag{8}$$

Thus, $o_m(l)$ detects the presence of the word, v_m, in the sentence, $S(l)$.

The weight between the word, v_m, and the color sensory primitive, c_k, is the mutual information, $I(c_k, v_m)$, given by

$$w_{mk} = I(c_k, v_m) = \log \frac{P(c_k \mid v_m)}{P(c_k)}, \tag{9}$$

where $P(c_k \mid v_m)$ is the the probability that the appropriate color is c_k, given that the input word is v_m. Each sensory primitive node also has a bias term associated with it, given by

$$w_k = \log P(c_k). \tag{10}$$

where $P(c_k)$ is simply the probability of the appropriate color being c_k. As shown in [1], the weight and bias terms can be updated sequentially by smoothed relative frequency methods.

The output of the color sensory primitive node due to the lth sentence, $S(l)$ is given by

$$a_{C_k}(l) = f(\sum_{m=1}^{M} o_m w_{mk} + w_k), \tag{11}$$

where M is the number of vocabulary words w_{mk} and w_k are the weights and the bias of sensory primitive node k, which are dened in Equations 9 and 10, and $f(.)$ is the sigmoid function given by

$$f(x) = \frac{1}{1 + e^{-x}}. \tag{12}$$

A conversation segmentation algorithm [7] is used to segment the sentences, $S(l)$, into conversations where, within a conversation, the user's goal for the machine response is assumed to be constant. The outputs, $a_{C_k}(m)$, after the mth sentence in the conversation, are linearly combined with the outputs in response to the previous sentences, to produce a resultant output given by

$$A_{C_k}(m) = \gamma(m)a_{C_k}(m) + (1 - \gamma(m))A_{C_k}(m-1), \tag{13}$$

where $\gamma(m)$ is a smoothing factor. In our implementation we choose $\gamma(m) = 1/m$.

The output of the color sensory primitive node, c_k, due to the visual input from object i is given by

$$M_C(k, i) = \left\{ \begin{array}{ll} 1 & \text{if color of object } i = \text{color } k \\ 0 & \text{otherwise} \end{array} \right. . \tag{14}$$

This function simply detects the color k in object i. The outputs due to the message, given by Equation 13, and due to the visual input, given by Equation 14 are combined to give the output, U_{C_i} as follows:

$$U_{C_i} = \sum_{k=1}^{K} M_C(k, i)A_{C_k}. \tag{15}$$

It can be seen from Equation 15 and Equation 4 that the function, $U_C(i)$, decreases the value of $P_i(x_i, y_i, t)$ if the color of object i is that connoted by the message input. This is exactly what is needed, since the eyeball is moved toward the object with the minimum value of P_i.

5. EXPERIMENTAL RESULTS

We collected over 1000 typed conversations from 11 users who interacted with the system over a three month period. The sentences were *unconstrained in both grammar and vocabulary*. Each user could change the visual scene at any time by adding, deleting, or moving objects using a graphics software package which underlies the Blocks-world simulation [8, 9].

5.1. Vocabulary

The system started with *no* vocabulary for its task. However the words \Wo" and \Good" were preprogrammed to provide the negative and positive reinforcement as to the appropriateness of the device's response. The device acquired 431 words over the course of 1045 conversations. Figure 4 shows that the vocabulary continues to grow with new conversations and does not level off, even for this rudimentary task. This is because the users were constantly attempting to both teach the machine and test it through the use of novel and sometimes highly imaginative words. We also plot the frequency of a word against its rank in Figure 5 on a log-log graph. The linear behavior, known as Zipf's Law, shows that most words occur infrequently. This conrms the importance of continuing investigations into algorithms which learn from small numbers of examples. We also found that the semantically meaningful words for this task were spread out over all word frequencies, many of them occurring only a few times, and a few, like \Red", \Green", \Circle", and \Square" occurring many times. As shown in the results of the next section, the performance of our algorithm under small sample conditions is quite good.

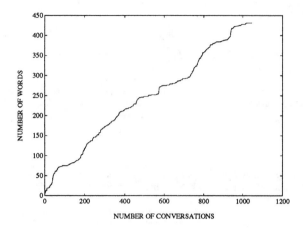

Figure 4. Vocabulary Growth

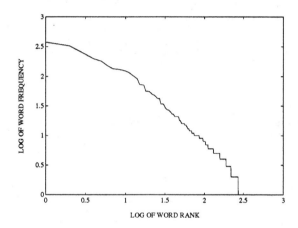

Figure 5. Rank-Frequency Plot

5.2. Conversation Length

Using the conversation segmentation algorithm [7], we found that the average conversation length was only 1.4 sentences with a standard deviation of 0.8. During the interactions, 95% of the conversations converged with a correct machine response, and in most cases a correct response (convergence) was achieved within 2 to 3 sentences even if an error was made in the response to the rst sentence. This demonstrates the error-recovery capability of the dialogue feedback control mechanism. Table 1 shows $P(m \leq k)$, the probability of converging within k sentences, for $k = 1$ to 5. In Figure 6, we plot $\log P(k) = \log \left(1 - P(m \leq k) \right)$ against k. The linear behavior of the plot shows that $P(m \leq k)$ exponentially approaches 1. This is as predicted by a convergence model described in [7].

5.3. Retention

We also measured the retention of the system, i.e., how well does it remember what it was taught. This is done by evaluating the correctness of the responses of the nal system to the rst sentences of each conversation where the system responded correctly in the original interactions. Thus, this is a measure of the nal system's performance on a subset of the training set. However the visual scenes presented during testing were new. We found that the system retained 98% of what it was taught.

6. CONCLUSION

We presented a study of an adaptive language acquisition system that has multi-sensory input as opposed to just a message input. We introduced the concept of sensory primitive subnetworks in a connectionist network, reecting the structure in the input periphery of the machine in order to

Conversation length, k	1	2	3	4	5
Probability of convergence, $P(k)$, within k sentences	0.69	0.88	0.95	0.98	0.99

Table 1. Convergence Probabilities

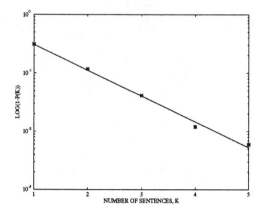

Figure 6. This plot shows that the probability of convergence approaches 1 exponentially

provide improved generalization. A time-varying potential function was described which controls the machine response to the environment. Our results show that the average conversation length was only 1.4 sentences, and the machine retains 98% of what it is taught. We view these encouraging experimental evaluations as early results in a long term investigation of such language acquisition systems.

REFERENCES

[1] A. Gorin, S. Levinson, A. Gertner, and E. Goldman, \Adaptive Acquisition of Language," *Computer Speech and Language*, vol. 5, pp. 101{132, 1991.

[2] A. Gorin, S. Levinson, and A. Gertner, \Adaptive Acquisition of Spoken Language," in *Proceedings IEEE International Conference on Acoustics, Speech, and Signal Processing*, vol. 2, pp. 805{808, 1991.

[3] L. Miller and A. Gorin, \A Structured Network Architecture for Adaptive Language Acquisition," in *Proceedings IEEE International Conference on Acoustics, Speech, and Signal Processing*, vol. 1, pp. 201{204, 1992.

[4] G. Cybenko, \Approximation by Superposition of a Sigmoidal Function," *Mathematics of Control, Signals and Systems*, vol. 2, pp. 303{314, 1989.

[5] E. Baum, \What Size Net Gives Valid Generalization?," *Neural Computation*, vol. 1, no. 1, pp. 151{160, 1989.

[6] A. Gertner and A. Gorin, \Adaptive Language Acquisition for an Airline Information Subsystem." to appear in *Neural Networks for Speech and Vision Applications*, ed. R. Mammone, Chapman&Hall, 1993.

[7] A. Sankar and A. Gorin, \Visual Focus of Attention in Adaptive Language Acquisition." to appear in *Neural Networks for Speech and Vision Applications*, ed. R. Mammone, Chapman&Hall, 1993.

[8] D. Blewett, S. Anderson, M. Kilduff, and M. Wish, \X widget based software tools for unix," in *Proceedings of USENIX*, January 1992.

[9] D. Blewett, S. Anderson, and M. Kilduff, \Virtual widgets and distributed processing using the x toolkit," in *Proceedings of Xhibition*, June 1992.

AN ALGORITHM FOR THE DYNAMIC INFERENCE OF HIDDEN MARKOV MODELS (DIHMM)

Philip LOCKWOOD, Marc BLANCHET

MATRA COMMUNICATION

Corporate Scientific and Technical Division
Rue J.P.Timbaud, 78392 BOIS D'ARCY cedex, BP 26, FRANCE

ABSTRACT

The DIHMM algorithm performs a robust estimation of the HMM topology and parameters. It allows a better control of the speech variability within each state of the HMM, yielding enhanced estimates. The DIHMM parameters (number of states, structure of the Gaussian mixture density functions (modes), transition matrix) are obtained from the training data via probabilistic grammatical inference techniques, welded in a Viterbi-like training framework. Experimental results on various databases indicate a global improvement of the recognition rates in adverse environments: the results averaged on three databases show an increase of 12.8% on raw data and 2.4% when using NSS [11].

I INTRODUCTION

HMMs are one of the most efficient modeling techniques used in automatic speech recognition, but their ability to derive reliable models depends on the accuracy and quality of the parametric estimations.

If, given a HMM topology (a set of states and transitions), standard algorithms exist to estimate these parameters from training data, most of the commonly adopted topologies are very crude and must be hypothesised by the system conceptor. The modeling capacities of HMMs are therefore not fully exploited.

It has been shown that models including accurate phonetic knowledge yield better performance [2], [3]. The way this knowledge is included in the system needs some type of expertise (phonetic transcription at least). Sometimes it is not easy to derive this knowledge automatically (eg. case of proper names), and it is not obvious that the derived models used to represent each phonetic component will be fully adapted to the particular modeling problem. Ideally, the algorithm should be able to learn this information from the observed data.

There are several ways to approach the problem. The first would be based on the LHMM framework presented in [12] by Pepper and Clements. In this case, the HMM model for each word would be chosen sufficiently large, using a duration model allowing high transition orders. The training scheme should naturaly converge to the optimal topology, provided that a significant amount of training material is available so as to reach a stabilised scheme. The method we propose follows the same philosophy but the training, performed incrementaly, should be less sensitive to lack of data.

Algorithms for the Dynamic inference of Markov networks have been proposed in the literature [4], [14]. These are adaptations or extensions of grammatical inference techniques [7]. Casacuberta and Vidal [4] have extended the approach to estimation of HMM parameters. A similar scheme was proposed by Falaschi and Pucci [6].

In [6], Falaschi et al. use a dynamic programming algorithm to align the (coded) utterances of a word, and build a lattice from the correspondances obtained. The lattice is pruned in a second stage. The scheme proposed by Casacuberta et al. [4] performs also in two stages. Estimation of a network using a grammatical inference specific scheme. The second stage prunes the network to come out with a more compact HMM-like model. The path probability criterion is used for pruning. There are two problems with this approach: the first is management of the network (complexity, restriction to the inference of finite state networks with no loops, need for a compiler to linearise the network), the second is the pruning strategy itself, as the difficulty will be to remove only the redundancy of the representation.

The "Hidden" structure of HMMs is attractive due to the "smoothing effect" (all states need not be accurately observed). Complexity is another aspect since the Markov networks generated are usually much more complex: large number of states, important overhead introduced by the data structure needed to manage the network.

The problem adressed here is the extension of standard HMM training schemes to the problem of learning the topology of the HMM (number of states, allowable transitions between states, number of modes in the density functions) from the observations. Such training techniques lead to models able to represent information more efficiently. It has a direct impact on performance in the context of the experiments reported here where our main problems are first to create models from a limited amount of training data, and second to have a better duration modeling as this is an important feature for speech recognition in mismatch conditions [11]. The DIHMM algorithm directly builds a compact HMM-model able to handle loop transitions or other durational models. The proposed scheme infers a linearised left-right HMM: in other words, no particular method to linearise the network is required.

II DYNAMIC INFERENCE SCHEME: THE DIHMM ALGORITHM

The dynamic inference algorithm presented hereafter (DIHMM) first finds (infers) the underlying topology of the HMM by performing substitutions, deletions or insertions, by solving successive string-to-network alignment problems: it then optimises the parameters of the DIHMM using standard Baum-Welch or Segmental K-Means (Viterbi) algorithms [9].

The DIHMM algorithm estimates all the parameters of a HMM: model topology (number of states, allowable transitions between states), the duration in each state, and the density functions (including the number of modes of the Gaussian mixture pdfs), using as only knowledge source, the available training material. To approach this, we propose an algorithm that performs in two stages. The first is the estimation of the structure or topology of the network, which includes an explicit network reestimation stage. The output of this stage is also a HMM with initialised parameters, both for durations and observations. The second stage is the optimisation of the estimations.

The flow diagram of the scheme is detailed in fig. 1. First the DIHMM network has to be initialised (step 1). Each utterance is then used successively to infer the network (step 2). The network is reestimated after each alignment (step 2). The last step is the refinement of the estimations using standard schemes (step 3). Figure 2 illustrates the basic operations generated by the the alignment stage: a substitution will add a mixture component, an elision will create an arc, and an insertion will add a state.

Figure 1
Flow diagram of the DIHMM algorithm

For each unit or word:
STEP 1: **Initialisation of the topology** from the first utterance
  - Creation of "start" and "stop" non emitting states
  - Coding of the utterance
  - Initialization of the model
STEP 2: **Dynamic inference of the topology** from all the other utterances and network reestimation
  - loop on utterances 2 to N:
    - Coding of the utterance
    - Alignment of the utterance to the model
    - Reestimation of the topology
STEP 3: **Optimisation of the HMM parameters** by a Viterbi (or Baum Welch) algorithm.

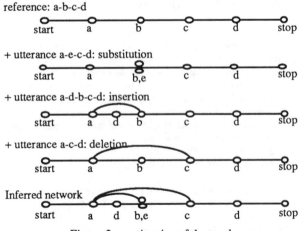

reference: a-b-c-d

+ utterance a-e-c-d: substitution

+ utterance a-d-b-c-d: insertion

+ utterance a-c-d: deletion

Inferred network

Figure 2: reestimation of the topology

Coding: This is the discretisation stage. Its aim is to derive a robust representation of the speech frames. It provides an inference guide by dividing the utterances into pseudo-phonetic sections. The output of this module is a set of labels (states) associated to functions (as in the tied-mixture model [8]) and a set of state durations computed from label repetitions.

Figure 3 -a
The DIHMM algorithm: string to network alignment

N: number of "states" of the input utterance;
M: number of states of the DIHMM
C[,]: array of cumulated costs
BACKI[,], PI, PSI: backpointers in the utterance string,
BACKJ[,], PJ, PSJ: backpointers in the DIHMM network

Initialization matrix of accumulated costs

BACKI[i,j] = 0, BACKJ[i,j] = 0 **for** all i,j, C[0,0] = 0

First line and column of C

FOR i = 1..N
  C[i,0] = C[i-1,0] + insertion-cost
FOR j = 1..M
  C[0,j] = Min{ C[0,k] } + elision-cost,
      k
      for every states k predecessors of state j

Alignment stage

FOR i = 1..N   {Input string}
  FOR j = 1..M   {Network states}
    IF the label of i is one of the component of state j,
      dist = 0, **ELSE** dist = 1
    C1 =   C[i-1,j] + dist * insertion-cost
    C2 = Min { C[i-1,k] } + dist * substitution-cost
         k
    C3 = Min { C[i,k] } + dist * deletion-cost
         k

    C[i,j] = Min{C1, C2, C3}

    IF C[i,j] = C1
      BACKI [i,j] = i-1   {Insertion}
      BACKJ [i,j] = j

    IF C[i,j] = C2
      BACKI [i,j] = i-1   {Substitution}
      BACKJ [i,j] = ArgMin{ C[i-1,k] }
                    k
    IF C[i,j] = C3
      BACKI [i,j] = i   {Elision}
      BACKJ [i,j] = ArgMin{ C[i,k] }
                    k
  ENDFOR
ENDFOR

Backtracking to find the optimal alignment and
determination of the network modifications

PI = N, PJ = M, PSI = BACKI [N,M], PSJ = BACKJ [N,M]
Lengthpath = 0
WHILE the path is not traced back completely
  Keep track of PI, PJ, and the type of modification:
    IF (PI>PSI, PJ>PSJ, dist(PI,PJ) = 0) matching PI/PJ
    IF (PI>PSI, PJ>PSJ, dist(PI,PJ) > 0) substitution PI/PJ
    IF (PI>PSI, PJ=PSJ) insertion PI/PJ
    IF (PI=PSI, PJ>PSJ) deletion PI/PJ
  incrementation of Lengthpath
  PI=PSI, PJ=PSJ, PSI=BACKI[PI,PJ], PSJ=BACKJ[PI,PJ]
(end of while)

Initialisation: The topology is initialised from the first utterance: each label of the utterance becomes a state of the HMM; if the same label occurs successively, only one state is created and the duration (label repetition) is recorded. The mixture component associated with the label and the duration information initialise the HMM state.

Alignment: Optimal string-to-network alignment is the basis of the inference, which modifies the DIHMM so as to include the new utterance. Costs are assigned to matches, substitutions, deletions and insertions, and a dynamic programming scheme computes the optimal alignment yielding network modifications.

Reestimation: The network is reestimated taking into account the basic operations issued from the alignment. These will create skips, or add states and/or transitions. The duration information is also reestimated.

The DIHMM inference and network reestimation stages are detailed in figure 3-a and -b respectively.

An important advantage of DIHMM is that <u>the inferred topology remains linear during all the inference stage</u>. The initialisation introduces a first order linear HMM, and afterwards each reestimation stage maintains the linearity: a substitution modifies only the structure of the state's mixture by adding or updating a Gaussian mixture component, an insertion adds a new state between state i-1 and state i, a deletion adds a new transition between state i-1 and state i. So none of these modifications breaks the linearity of the network. Figure 4 below illustrates for a particular word the inferred topology (state durations are not represented on the figure).

Figure 4: topology of the word "lipka" of the MATRA database. Each circle represent one Gaussian mixture component for a state

The topology inferred for a 43-word vocabulary and the 4 speakers of the speaker-dependent MATRA database (one of the databases used in these experiments) is illustrated in fig 5. On average, we obtain 18 states and 28 Gaussian mixture components per word.

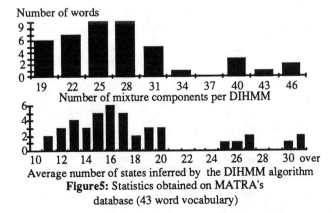

Figure5: Statistics obtained on MATRA's database (43 word vocabulary)

At the end of the inference stage, a HMM is obtained. An Estimate/Maximise scheme will fulfill the estimation process.

III HMM PARAMETERS ESTIMATION AND DURATION MODELING

The HMM parameters can be refined by using the traditional estimation algorithms (Baum-Welch algorithm, or segmental K-Means algorithm).

For duration modeling, two types of models can be obtained from the inference. It can be represented by self-transitions and the reestimation during inference if that of [9]. But as the DIHMM algorithm gives states with very short as well as very long durations, exponential duration laws are not adapted. We can seek a more efficient model, and for instance an expanded-state model is more appropriate [5]. These duration probabilities are computed during the inference stage, by counting the number of utterances giving a duration t to the state j (the same label on the test string is observed "t" times). For every t and j a duration histogram is built, which gives all the parameters of a Fergusson model. To overcome the poor estimation of such duration models, especially for speaker-dependent applications with a limited amount of training, a smoothing of these histograms is performed to which is appended an additional loop as shown in fig. 6. Even smoothed, duration parameters remain poorly estimated, but the duration is in fact modelled in context [13].

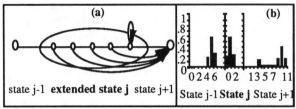

Figure 6: The Normalised histograms (b) map a Fergusson model (a). The duration is represented by a sub HMM.

IV EXPERIMENTAL CONTEXT

HMM recogniser: The speech parameters (Mel-cepstrum) are computed every 16 ms on a 32 ms window. In this implementation, the word-HMM is a linear sequence of 16 states, each state being represented by a Gaussian pdf (diagonal). A pooled fixed variance is used [11].

Databases: Three of the databases recorded within the ARS* ESPRIT project are used. The first two (MATRA, ENST) include 43 words, while the latter 34 (CSELT). The utterances recorded in silence conditions (car stopped) are used for training the system, the others for testing.

- The MATRA database contains 4 speakers, 2 female and 2 male; each speaker pronounced a set of 4 utterances in silence, 4 at 90 km/h and 4 at 130 km/h recorded in a medium range car. The sampling frequency is 8kHz.

- The ENST database contains 10 repetitions recorded in silence and 15 at 110 km/h in a low/medium-range car, for a male speaker. The sampling rate is 16 kHz. This database is also called ENST1 in some other work [1].

- The CSELT database contains 10 repetitions recorded with car stopped but in a noisy car environment, 5 at 70 km/h and 5 at 130 (recordings made in a low/medium-range car) and 4 speakers. The sampling rate is 16 kHz.

V RESULTS

Results obtained with standard HMMs and DIHMMs are compared on a speaker-dependent task where we have mainly two problems: limited amount of training material, and tests performed in noise and mismatch conditions. Results are given on raw data or after Non linear Spectral Subtraction (NSS:[11]).

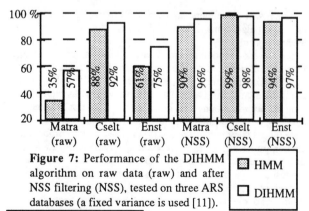

Figure 7: Performance of the DIHMM algorithm on raw data (raw) and after NSS filtering (NSS), tested on three ARS databases (a fixed variance is used [11]).

*ACKNOWLEDGEMENTS: Part of this work was performed with the support of the CEC under ESPRIT p2101 ARS project.

The DIHMM scheme performs as well as or better than the standard technique due mainly to a better representation of the duration and observation information. If the number of observations available to estimate the parameters within a HMM state is reduced, the influence of outliers will be minimised within the DIHMM framework. The drawback in this case can be an increase of the number of states of the model (especially with the expanded states duration modeling). In this case, the complexity can be reduced, for example by introduction of a beam in the recognition process. DIHMM increase the best recognition rates on the MATRA database from 90.4% to 96.1% with fixed variance, and from 94.6% to 97.9% with smoothed variance [10], [11]. Averaged on the three ARS* databases, we observe a rise of 12.8% on raw data and 2.4% after NSS. The root-based front-end has not been used in these tests [1].

VI CONCLUSION

This paper adressed the problem of automatic inference of the structure of Hidden Markov Models. A new scheme is proposed, with the following properties:

- The estimation stage (topology and other parameters) is data driven, through probabilistic grammatical inference schemes combined with Baum-Welch or Viterbi algorithms;
- The algorithm computes directly a linearised HMM;
- State duration is modelled by smoothed Fergusson models.

DIHMM has proved its superiority over standard HMMs in various conditions, especially in noise. It also clearly improves the best performances of the recognition system: averaged on the three databases, the improvement is of 12.8% on raw data and of 2.4% with NSS.

Some open issues still remain such as the necessity of embedding further network inference (step 2-fig. 1) with Viterbi training (steps 3 fig. 1). A clear advantage is that this would remove the influence of the order in which we input the utterances. An estimation loop on steps 2-3 would reduce the entropy of the DIHMM networks.

Finally, the main motivation for proposing such inference model is the developement of global training techniques where a single DIHMM would be created for the whole vocabulary. The DIHMM algorithm could also be a valuable tool on a number of other speech problems such as: subword modeling, discriminant training, or statistical language modeling.

[1] P. ALEXANDRE, J. BOUDY, P. LOCKWOOD, "Root homomorphic deconvolution schemes for speech recognition in car-noise environments", ICASSP, 1993.

[2] L.R. BAHL, J.R. BELLEGARDA, P.V. DE SOUZA, P.S. GOPALAKRISNAN, D; NAHAMOO, M; PICHENY, "A new class of fenonic Markov word models for large vocabulary continuous speech recognition", ICASSP, 1991.

[3] R. CARDIN, Y. NORMANDIN, R. DE MORI, "High performance connected digit recognition using codebook exponents", ICASSP, 1992.

[4] F.CASACUBERTA, E.VIDAL, B.MAS, H.RULOT, "Learning the structure of HMM's through grammatical inference techniques", ICASSP, 1990.

[5] A.COOK, M.J.RUSSEL, "Improved duration modelling in HMM using series-parallel configurations of states", Proceedings of the Institute of Acoustics Autumn Conference on Speech and Hearing 1986.

[6] A. FALASCHI, M. PUCCI, "Automatic derivation of HMM alternative pronounciation network topologies", Proc. EUROSPEECH 91.

[7] K.S. FU, "A Step Towards Unification of Syntactic and Statistical Pattern Recognition", IEEE-Transactions on PAMI, Vol. PAMI-8, NO. 3, May 1986.

[8] X.D. HUANG, M.A. JACK, "Semi-continuous hidden Markov models for speech signals", Computer, Speech and Language, Vol. 3, No. 3, 1989.

[9] B-H.JUANG, L.R.RABINER, "The segmental K-Means algorithm for Estimating parameters of Hidden Markov Models", IEEE Trans. on ASSP vol.38, no.9, Sept.1990.

[10] P.LOCKWOOD, J.BOUDY, M.BLANCHET, "Non-Linear spectral suntraction (NSS) and HMMs for robust speech recognition in car noise environments", ICASSP, 1992.

[11] P.LOCKWOOD, J.BOUDY, "Experiments with a Non-Linear Spectral Subtractor (NSS), Hidden Markov Models and the Peojection, for robust speech recognition in cars", Speech Comm., vol. 11, No. 2-3, June 1992.

[12] D.J. PEPPER, M.A. CLEMENTS, On the phonetic structure of a large hidden Markov model", Proc. IEEE-ICASSP Conf., 1991.

[13] J. PICONE, "On modeling duration in context", ICASSP, 1989.

[14] M.G. THOMASON, E. GRANUM, "Dynamic Programming Inference of Markov Networks from Finite Sets of Sample Strings", IEEE Trans. on PAMI, Vol. PAMI-8, NO. 4, July 1986.

FUZZY-DECISION NEURAL NETWORKS*

J.S. Taur and S.Y. Kung

Princeton University

ABSTRACT

In a decision-based neural network(DBNN)[5], the teacher only tells the correctness of the classification for each training pattern. In dealing with practical classification applications where significant overlap may exist between categories, a special care is needed to cope with the "marginal" training patterns. For these situations, a soft decision is more appropriate. This motivates a fuzzy-decision neural network(FDNN) which incorporates a penalty criterion into the DBNNs. Following [2], a penalty function is proposed which treats the errors with equal penalty once the the magnitude of error exceeds certain threshold. Theoretically, the FDNNs are less biased and they yield the minimum error rate when the number of the training patterns is very large. Simulation results confirm that the FDNN works more effectively than the DBNN when the the training patterns are not separable.

1. FUZZY DECISION NEURAL NETWORKS

1.1 Fuzzy Decision Neural Networks

In practice, some of the patterns on the borders between two neighboring categories may overlap in certain grey areas. In this case, a technique based on a somewhat fuzzy decision appears to be more suitable. In order to provide some tolerance, the defecting and undecided patterns must be suppressed or even ignored. To illustrate this point, let us study the following example.

To show how the noise can affect the decision boundary, two-dimensional artificial data are created in the following simulation. For each of the two classes, 200 patterns are drawn randomly from the following distributions.

$$
\begin{aligned}
p_1(x,y) &= 0.9 \times N(x,3,s^2)N(y,2,s^2) \\
&\quad + 0.1 \times N(x,3,5*s^2)N(y,2,5*s^2) \\
p_2(x,y) &= 0.9 \times N(x,1,s^2)N(y,2,s^2) \\
&\quad + 0.1 \times N(x,1,5*s^2)N(y,2,5*s^2)
\end{aligned}
$$

where $N(t,\mu,s^2)$ is a Gaussian with mean μ and variance s^2. (Here, $s = 0.15$.) The training patterns contain additive noises, represented by the second Gaussian terms in

---

*This research was supported in part by Air Force Office of Scientific Research under Grant AFOSR-89-0501A.

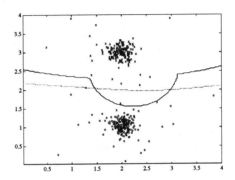

Figure 1: The solid decision boundary is obtained by a DBNN with two subclusters. The dotted decision boundary is obtained by using FDNN with $\xi = 1$.

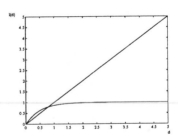

Figure 2: The difference between the loss functions of a hard-decision DBNN (linear line) and a fuzzy-decision neural network.

the foregoing distributions. Due to the noise distribution, the two classes overlap in the feature space(cf. Figure 1.) The decision boundary obtained by the DBNN is shown in Figure 1(solid line).

A more elegant approach is by imposing a proper penalty function to all the "bad" decisions and find the solution which minimizes the total penalty. This formulation leads to the *Fuzzy Decision Neural Network* (FDNN). It provides the flexibility of allowing a "soft" or "fuzzy" decision, as opposed to the hard (yes/no) decision in the decision based neural network (DBNN). In the DBNN, the teacher only tells the correctness of the classification for each training pattern. A decision-based learning rule is effective for clearly separable pattern categories, however, it is no longer so when encountering overlapping regions or fuzzy decision boundaries. Therefore, there must exist different degrees of error associated with each decision, for example, marginally erroneous, erroneous, and extremely erroneous. For these purposes, a proper penalty should be assessed according to the degree of error. Sometimes, even a marginally correct pattern will have to be slightly penalized.

Following the notations used in DBNN, suppose that $S = \{x^{(1)}, \ldots, x^{(M)}\}$ is a set of given training patterns; and the model function for the class Ω_i is denoted $\phi(x, w_i)$ for $i = 1, \ldots, L$. Following the original idea proposed in [3, 2], a cost function which provides a practical measure of the severeness of misclassification will be introduced. Suppose that $x^{(m)} \in \Omega_i$ and $j = arg\max_{j \neq i} \phi(x^{(m)}, w_j)$, then the measure d is defined by

$$d = d^{(m)}(x^{(m)}; w) = -\phi_i(x^{(m)}; w_i) + \phi_j(x^{(m)}; w_j) \quad (1)$$

where w denotes all the involving weight vectors. The larger the d is the greater the error is.

For minimum error classification, the following are two possible penalty functions:

$$\ell(d) = \frac{1}{1 + e^{-d/\xi}} \quad (2)$$

$$\ell(d) = \begin{cases} (d + \xi/2)/\xi & -\xi/2 < d < \xi/2 \\ 1 & d \geq \xi/2 \\ 0 & d \leq -\xi/2 \end{cases} \quad (3)$$

Finally, the overall cost-function can be defined as follows:

$$E(w) = \sum_{m=1}^{M} \{\ell(d^{(m)})\} \quad (4)$$

In particular, when the parameter ξ approaches zero, the three loss functions will approach the step function

$$\ell(d) = \begin{cases} 0 & d < 0 \\ 1 & d \geq 0. \end{cases} \quad (5)$$

The cost function as chosen will in turn lead to a minimum-error classifier, since the misclassification measure is exactly the recognition error count. Nevertheless, in its more general form, the cost function will still represent a good approximation of total number of errors in the training samples.

Now the decision-based learning rule can be reformulated as a gradient descent method to minimize the overall cost function. By incorporating the penalty criterion into the DBNNs[5], we obtain the following learning rule:

Algorithm 1 (Fuzzy Decision Learning Rule)
Suppose that the mth training pattern $x^{(m)}$ presented is known to belong to class Ω_i; and that the leading class for the pattern is denoted by an integer j. ($j = arg\max_{j \neq i} \phi(x^{(m)}, w_j)$)

Reinforced Learning:
$$w_i^{(m+1)} = w_i^{(m)} + \eta \ell'(d_i) \nabla \phi(x^{(m)}, w_i)$$
Antireinforced Learning: $\quad (6)$
$$w_j^{(m+1)} = w_j^{(m)} - \eta \ell'(d_i) \nabla \phi(x^{(m)}, w_j)$$

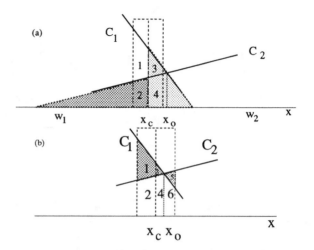

Figure 3: Decision boundaries for FDNNs and DBNNs.

where $\ell'(d_i)$ is the derivative of the penalty function evaluated at d_i. ∎

As a comparison, here we present the corresponding loss function adopted in the (hard) decision-based neural network:

$$\ell(d) = \zeta d$$

The two loss functions are shown in Figure 2. As shown in Figure 2 the two curves indicate the cost (penalty) of errors. Since a linear penalty function would impose too excessive penalty for the patterns with large error, an sigmoid function is relatively more appealing. It effectively treats the errors with equal penalty once the magnitude of error exceeds certain threshold. (This is in accordance with the so-called "minimum-error-rate" classifier defined by Duda[1].)

The FDNN can be applied to both the static and temporal recognition problems. The possible model functions for the static problems include LSE and likelihood function. Examples for the temporal problems are likelihood function, prediction error, or DTW distance as model functions. In other words, the FDNN formulation can be blended into traditional recognizer designs. For examples, a prediction-error based classifier, a hidden Markov model (HMM), or a dynamic time warping (DTW) based system. For more details, see [4].

1.2 Decision Boundaries of FDNNs

In the following, let us use a one-dimensional example to illustrate some theoretical observations. The block-adaptive updating is adopted in the discussion. We assume that the number of training samples is very large and the loss function is defined in Eq. 3. Also the gradient is normalized. We note that the DBNNs are biased classifiers while the FDNNs are unbiased with a proper window size.

As shown in Figure 3, two thick straight lines represent parts of the probability density functions of two different classes. According to [1], the decision boundary of the minimum error rate classifier is the intersection point of the two probability density functions. The optimal and current decision boundary are denoted as x_o and x_c respectively. The window of $\ell'(d)$ for the FDNN in Eq. 6 is shown in the dotted rectangular which is symmetric around x_c. The area under the probability density function within the window is the probability of the patterns to be in that interval and is divided into several regions marked by numbers. Assume that the model function is defined as $\phi_i = -(x - w_i)^2$, where w_i is the reference point used to represent that class. By taking the gradient of the model function and then normalizing it, the effect of each training sample in the window is moving the reference point one unit to the left or to the right.

As illustrated in Figure 3(a), when DBNNs converge, the two shaded regions must have the same area. Thus it is easy to see that DBNNs are biased.

As shown in Figure 3(a), when the window does not cover the correct decision boundary x_o, the regions marked 1,2,3, and 4 in class C_1 will move the reference points w_1 and w_2 to the right by the reinforced and anti-reinforced updating. Therefore the decision boundary will move to the right. On the other hand, the regions 2 and 4 in class C_2 will move the decision boundary to the left. The net effect of these two type of influence is that decision boundary moves to the right, which is the correct direction. The amount of movement is proportional to the sum of the areas of regions 1 and 3.

When x_o is within the window (Figure 3(b)), by the same argument as the above, we can show that when the network converges, the influence from these two types will reach an equilibrium. That is, the sum of the areas of region 1 and 3 should be the same as the area of the region 5. For simplicity, let us assume that the probability density functions around the intersection can be approximated by two straight lines. Then the two shaded triangles are similar to each other. The only chance for them to have the same area is when $x_c = x_o$. Thus the FDNNs are not biased and they yield the minimum error rate when the number of the training patterns is very large. If the window size is large, the FDNNs will be more prone to bias since the straight line approximation no longer holds. In the beginning phase , the window size should be large so that the decision boundary can move more rapidly to the neighborhood of the correct solution. The window size should gradually decrease so that the bias can be gradually reduced.

When the number of samples is not very large, the variance of the decision boundary for the FDNN with a small window may be large since there are not enough patterns included in the window. In this sense the selection of window size will depend on the tradeoff between bias and variance. In the previous discussion, the window does not have to be rectangular. As long as the the window (i.e. the derivative of the loss function) is symmetric around zero, e.g. Eq. 2, the above argument still holds.

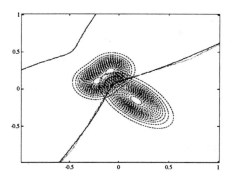

Figure 4: Two Gaussian-mixture distributions and their decision boundary.

ξ value	1	0.1	0.01	0.001
errors	5.5%	5.5%	5.23%	5.17%

Table 1: For the gaussian-mixture problem, the optimal error percentage is 5.3%.

2. SIMULATION RESULTS

In order to evaluate the performance of the FDNNs, three two-dimensional artificial data are studied. In all the examples, the FDNN with sigmoid loss function and elliptic basis function is adopted.

Noisy Data Again let us consider the previous example. In the simulation, $\xi = 1$ is chosen and the decision boundary obtained is shown in Figure 1(dotted line). Compared with the decision boundaries in Figure 1, the one obtained by FDNN is smoother and closer to the optimal boundary which is the straight line $y = 2$. are renormalized.)

Gaussian-Mixture Problem In this example, we consider the classification of two nonseparable training data sets with the Gaussian-Mixture distributions. The distributions and decision boundary created by Bayes decision rule is shown in Figure 4 (solid line) and a (training) classification error of 5.3% (for 3000 training samples) is obtained. In this simulation, if a 3-cluster DBNN with a positive vigilance[5] is used the best classification error rate for the training set is 5.53%. The simulation results for the FDNN with different ξ are shown in Table 1 and the estimated decision boundary for $\xi = 0.01$ is depicted in Figure 4 (dotted line). In the figure we can see that the estimated boundary is very close to true one and the error rate is smaller than the "optimal" value. This is because only the finite samples are used in the training and the network is biased toward the samples.

ξ value	10	1	0.1	0.01
errors	1%	0.1%	0.1%	1.1%

Table 2: For the two triangle problem, the optimal error percentage is 0% which can be obtained by DBNNs.

Two Triangle Problem To test how FDNNs perform when the data is separable, let us consider the problem of classifying two classes divided by the diagonal in a square box. For each class, 500 training patterns are generated randomly with a uniform distribution in its triangular region. The simulation results are listed in Table 2 for different ξ values. Although these two classes are separable, FDNNs can not get 100% classification rate. In this case, DBNNs obtain the perfect classification in just a few iterations.

3. DISCUSSION

From the above simulations, we found that FDNNs perform better than the DBNNs when there exists overlap between classes or when the data are noisy. If the data are noisy, the DBNNs will try to classify the unwanted noisy pattern and the decision boundary will become undesirably irregular (c.f. Figure 1). On the other hand, FDNNs take part of the major distribution into consideration and thus they are less sensitive to noise.

Our simulations indicate that the shape of loss function does not make significant difference. The loss function in Eq. 3 has also been studied, the classification rates are similar as those when Eq. 2 is adopted. However, if the loss function in Eq. 3(the dashed line in Figure 5(a)) is shifted $\xi/2$ to the right (the dashed line in Figure 5(b)), the classification error rate goes up. Although it still approximates the step function when ξ goes to zero, the performance degrades. The derivatives of these two loss functions are depicted in Figure 5. The loss function in Figure 5 (b) is similar to the DBNNs in that the correct patterns (those with negative d) are not included in the updating. The difference is that the degree of error is considered the same when the value d is larger than ξ. On the other hand, the loss function in Figure 5 (a) is very similar to the situation of the DBNNs with a positive vigilance[5], i.e. the pattern is considered correctly classified when the model function is larger than the rest by at least an amount of $\xi/2$. It is crucial to include the marginal correct patterns in the training. This attribute allows the decision boundary to be influenced by most of the data instead of by the misclassified noisy patterns only.

If the classes are separable and noise level is low, then the DBNNs are more effective. Although the classification rates are very similar, DBNNs converge much faster than the FDNNs. This is because the DBNNs concentrate on the misclassified patterns only and they can settle the dispute between classes more quickly.

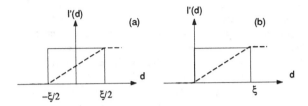

Figure 5: Derivative plots for two loss functions. The dashed lines are the corresponding loss functions.

From Table 1 and 2 we note that the performance of the FDNNs depends on the ξ value. Different ξ values are required to obtain the best results for different data sets. It controls the slope of the loss function and the area of the effective region. Note that the updating rule in Eq. 6 involves the derivative of the loss function. Since the value of the derivative vanishes when the difference of the model functions is large, there is equivalently an effective region in which the patterns have certain influence on the decision boundary. Larger ξ means more data are covered in both sides of the decision boundary and involved in the training procedure. One heuristic approach to choose a proper ξ is to compute the center of each class and then calculate the difference of the model function of centers. Then the difference is used to guide a selection of a suitable ξ. The value ξ is chosen to be a monotonic function of the difference. In this case the value ξ will not depend on the magnitude of the data. In our experiments, this approach always deliver the satisfactory performance. Based on the simulations, the FDNNs appear to be very effective in the noisy data environments.

References

[1] R. O. Duda and P. E. Hart. *Pattern Classification and Scene Analysis*. John Wiley, New York, 1973.

[2] B. H. Juang and S. Katagiri. Discriminative learning for minimum error classification. *IEEE Transactions on Signal Processing*, in press.

[3] S. Katagiri, C. H. Lee, and B. H. Juang. Discriminative multi-layer feed-forward networks. In B. H. Juang, S. Y. Kung, and C. A. Kamm, editors, *Neural Networks for Signal Processing, I*. Proceedings of the 1991 IEEE Workshop, Princeton, NJ, 1991.

[4] S. Y. Kung. *Digital Neural Networks*. Prentice Hall, Englewood Cliffs, NJ, 1993.

[5] S. Y. Kung and J. S. Taur. Hierarchical Perceptron (hiper) networks for classifications. In S. Y. Kung, F. Fallside, J. A. Sorensen, and C. A. Kamm, editors, *Neural Networks for Signal Processing, II*. Proceedings of the 1992 IEEE Workshop, Helsingoer, Denmark, 1992.

A SELF-LEARNING NEURAL TREE NETWORK FOR RECOGNITION OF SPEECH FEATURES

Mazin G. Rahim

CAIP Center, Rutgers University, Piscataway, N.J. 08855-1390

Abstract

This paper presents a Self-Learning Neural Tree Network (SL-NTN) for classification of speech features into phones. The SL-NTN employs a farthest-neighbor fuzzy-clustering algorithm to establish the intra-class correlation among speech phones, thus, splitting the phones in such a way to maximize the recognition performance while reducing the computational complexity. When evaluated on the 61 phones of the TIMIT database, the SL-NTN has shown to provide an 'optimal' trade-off between computational complexity and recognition performance. It also provides insight towards the interrelationship among the applied speech patterns.

1 Introduction

There are two issues to be considered when applying Neural Networks (NNs) for recognition of speech features. The first issue is that conventional NNs tend not to take advantage of the non-uniform correlation that exists among speech sounds. Instead, all sounds are considered equally confused. However, it is well known that the human ear can perceive certain sounds more distinctly than others (e.g., /aa/ & /b/ vs /p/ & /b/). The second issue is that conventional NNs require the network architecture to be specified prior to training. The architecture is typically found through trial and error. This results in an added computational complexity and a sub-optimal recognition performance.

To accommodate the varying complexity of different classification problems, Sankar et al [4] and Stromberg et al [5] proposed to combine the concept of decision trees with feed-forward neural networks to form a tree neural net. The authors evaluated their implementations on a vowel classification task and demonstrated an improvement in performance over classical decision tree methods (e.g., CART). Their recognition performance is comparable to that of conventional NNs, but with a significant reduction in computational complexity.

For performing more complex classification tasks, we recently proposed a binary tree architecture of NNs, referred to as the Neural Tree Network (NTN) [3]. The NTN was tested for recognition of 36 isolated phonemes. It demonstrated a recognition performance equivalent to that of conventional NNs, but with a reduction in computational complexity by a factor exceeding 8. The NTN was also found to capture important phonemic correlations which are known to exist in the human auditory system.

This paper presents a new and a more generalized self-learning approach towards designing a NTN. The Self-Learning NTN (SL-NTN) provides an 'optimal' trade-off between computational complexity and recognition performance. It also provides insight towards the interrelationship among the applied patterns. Experimental results are presented demonstrating the use of the SL-NTN for recognition of the 61 English phones of the TIMIT database [2].

The outline of this paper is as follows. Section 2 presents the spectrum analysis method. Section 3 describes the construction, training and recognition steps of the SL-NTN. Experimental results are presented in Section 4, and Section 5 gives the main conclusions of this study.

2 Spectrum Analysis

The task is to perform speaker-independent phone recognition on the 61 phones of the TIMIT database. Speakers are considered from two different accentual regions; New-England and Northern. This represents 114 training speakers and 37 different testing speakers. The feature extraction method is described as follows.

1. Each manually segmented phone is represented by 12 cepstral coefficients, $c_i(t)$; 12 cepstral

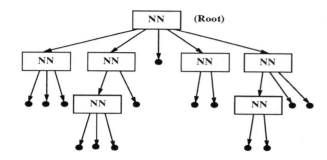

Figure 1: A schematic diagram of a NTN

derivatives, $\Delta c_i(t)$; 1 energy coefficient, $e(t)$; 1 energy derivative, $\Delta e(t)$. $c_i(t)$ are obtained through a 12th order Hamming-windowed LPC analysis performed on the middle 30 msec segment of the phone. $\Delta c_i(t)$ are approximated by a first-order polynomial applied over a 5-frame window, at a frame rate of a 100 frame/s. $e(t)$ is computed by taking the logarithm of the average frame energy, and $\Delta e(t)$ is computed in a similar way to $\Delta c_i(t)$.

2. In order to allow greater robustness, both $c_i(t)$ and $\Delta c_i(t)$ are multiplied by a weighting lifter:

$$w_i = 1 + 6\sin\frac{i\pi}{12} \qquad 1 \le i \le 12.$$

3. A scaling factor is applied to the components in order to equalize their variance. The combined cepstral and delta cepstral vectors, energy and delta energy form a 26-dimensional vector:

$$O(t) = (\{w_i c_i(t)\}, \alpha\{w_i \Delta c_i(t)\}, \beta e(t), \gamma \Delta e(t)).$$

4. Previous studies carried out by the author, as well as by other researchers [1], show that applying a Discrete Cosine Transform (DCT) to the feature vector, $O(t)$, and using a 16-dimensional DCT vector, as opposed to using the 26-dimensional vector $O(t)$, causes no degradation in the recognition performance. Therefore, each phone is represented by a vector $dct(t)$ $(=DCT[O(t)])$.

Applying the above strategy resulted in about 43,000 patterns (i.e., feature vectors) for training and about 14,000 patterns for testing (the number of patterns *per* phone ranged between 10 for /eng/ to 2,000 for /ix/).

3 SL-NTN: Self-Learning Neural Tree Network

A schematic diagram of a NTN is shown in Figure

1. At each node of the tree, a NN is trained to classify the input data into one of several sets of phones. These sets are determined by a clustering algorithm which splits the phones according to their relative confusion with each other. The NTN is self-learning since no external knowledge regarding the nature of the phones is required, thus, it is referred to as the Self-Learning NTN (SL-NTN). The SL-NTN establishes the intra-phone correlation independently and splits the phones in such a way so as to maximize the recognition performance while reducing the computational complexity.

The first step in designing the SL-NTN requires an estimate of the correlation among the applied phones (i.e., intra-phone correlation). This estimate is utilized in the clustering algorithm to allow a split into multiple branches at each node of the tree. The intra-phone correlation is obtained by training a conventional NN at the root level of the tree to classify the input patterns into their appropriate phones. The NN 'learns' to map a 16-dimensional DCT vector (i.e., $dct(t)$) into one of 61 phones. Following the training phase, the percentages of patterns confused among the different phone classes (i.e., confusion matrix) provide the intra-phone correlation.

The second step in designing the tree involves the application of the clustering algorithm. In [3], a farthest-neighbor clustering algorithm was utilized for splitting phonemes into two sets only. In this study, a more generalized clustering algorithm is introduced in which phones are being divided into two or more sets, imposing maximal inter-set correlation and minimal intra-set correlation. The number of sets formed by the clustering algorithm, at any node of the tree, is determined by an error threshold, T.

A flow diagram of the clustering algorithm is shown in Figure 2. Initially, two phones are selected (S_1^* and S_2^* - referred to as the 'nuclei') which are 'least' confused (i.e, farthest apart) according to the confusion matrix, D:

$$(S_1^*, S_2^*) = \arg \min_{i=1,N; j>i} \bar{d}(S_i, S_j),$$

where N is the number of available phones, and $\bar{d}(S_i, S_j)$ is the average percentage confusion error between the ith and j^{th} phones. Using D, two clusters are formed (i.e., M=2) by classifying all the phones into either S_1^* or S_2^*. Further clusters may be formed by defining new nuclei:

$$S_{M+1}^* = \arg \min_{j=1,M; i=1,N; S_i \ne S_j^*} \bar{d}(S_i, \{S_j^*\}),$$

where $\{S_j^*\}$ is the jth cluster including the class (nucleus), S_j^*. Every time a new cluster is created, two conditions are verified:

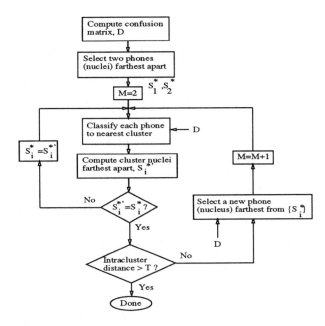

Figure 2: Flow diagram of the clustering algorithm

1. The intra-cluster distance

$$G = \sum_{i,j=1,M,i\neq j} \bar{d}(\{S_j^*\}, \{S_i^*\})$$

is minimum, given the required number of clusters and the intra-phone correlation. Otherwise, an iterative procedure is performed in which (M-1) of the nuclei are fixed and a search is carried out to find a new nucleus that satisfies the above condition.

2. The condition, G < T, is always satisfied. Otherwise, the clustering algorithm is terminated.

Because of the high intra-phone correlation among specific phones, it is not reasonable to expect new clusters to be created without increasing the value of G. To alleviate this problem, a type of 'fuzzy' clustering is introduced in which highly confused phones are shared among several clusters [3].

The building process of the SL-NTN continues by assigning a NN for each cluster and repeating the two design steps until all phones are classified appropriately. The designed SL-NTN is then trained to perform an M-way classification at each of its nodes. The number of sets, M, varies between 2 and N depending on the number of generated clusters at the individual tree nodes. Since the intra-set (intra-cluster) correlation is minimized by the clustering algorithm, NNs of the SL-NTN are now expected to be trained with minimal confusion error.

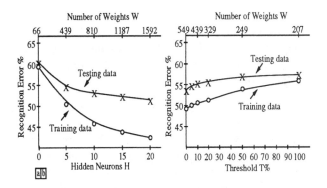

Figure 3: The percentage recognition error of the SL-NTN and the number of weights, W, when (a) varying H and setting T=5%, and (b) varying T and setting H=5

Two constraints are imposed when training the designed SL-NTN; (a) the number of hidden neurons of the tree NNs are set equal to that used while building the tree; (b) the 'optimal' weighting coefficients from the input layer to the hidden layer, obtained while building the tree, are used to initialize the SL-NTN prior to training. Initializing the variables of the tree in this manner helps to speed-up convergence.

When performing recognition of speech features using the SL-NTN, the input pattern is initially applied to the NN at the root level of the tree. The appropriate NN at the next higher level is then activated and the pattern is applied to its input layer. The pattern propagates through the tree until it is classified as one of the phones.

4 Experimental Results

Two criteria are used to evaluate the performance of the SL-NTN; percentage of recognition error, and computational complexity. The former is defined as the number of misclassified patterns, whereas the latter is measured in terms of the average number of weights, W, activated during the recognition phase. Two factors contribute to the value of W: (1) the error threshold, T, and (2) the number of hidden neurons, H, of each NN. Figure 3a shows the percentage recognition error of the SL-NTN when varying H from 0 to 20, and setting T to 5%. Figure 3b shows the recognition error when setting H, for each NN, to 5 and varying the threshold, T, from 0% to 100%.

The plots demonstrate that increasing the value of H for each network results in a reduction in the recognition error but at the expense of increasing W (increasing H from 5 to 20 reduces the error by 2.5% and increases W by a factor of about 4). Also, the recognition error seems to increase with increasing T, but with a reduction in W (increasing T from 5% to 100% increases the error by 2.5% but reduces W by a factor of over 2).

Further experiments conducted using a binary SL-NTN (i.e., T=0%) indicate that introducing 'fuzzy' clustering results in a 1% reduction in error at the expense of increasing W by a factor of 1.2. For comparison, a conventional NN with 100 hidden neurons (i.e., W=7861) is evaluated as having a 54% recognition error. This indicates that the SL-NTN reduces the computational complexity by a factor exceeding 10. Generally, our experiments suggest that if a conventional NN requires $V \times V$ weights to obtain a specific percentage accuracy, then the SL-NTN would need to activate approximately $V \cdot \log_2 V$ weights to achieve a similar accuracy. Also, the multi-level architecture of the tree is desired for hardware implementation (only one NN plus a look-up table to store the weights).

An important observation, which confirms previous findings in [3], suggests that the SL-NTN provides intra-set correlations which exist in the human auditory system. For example, when setting T=5% and H=5, the NN at the root node produces a separation into 4 sets of phones, namely, (a) silence and short bursts, (b) high voicing, (c) low voicing and nasals, and (d) affricates, fricatives and stop consonants. On the other hand, phones which are highly confused, even to the human ear (e.g., /s/ & /z/), are found to be grouped together at top levels of the tree. Generally, phones at low levels of the tree are grouped according to their manner of articulation whereas those at high levels of the tree are grouped according to their place of articulation. Recognition results demonstrating the performance of the SL-NTN on broad phonemic sets, as defined by the TIMIT database, and on dominant phonemic sets obtained at the root level are presented in Table 1.

5 Conclusions

A self-learning approach to designing neural tree networks is presented in this paper. The SL-NTN has been evaluated on a classification task involving 61 English phones. Results show that the SL-NTN provides (1) 'optimal' trade-off between computational complexity and recognition performance, and (2) insight towards the correlation among the phones. The ability of the SL-NTN to perform recognition on unlabeled speech data is currently being investigated.

Table 1: Recognition performance on broad phonemic sets, and on dominant phonemic sets at the root level

Phonemic Set	Training%	Testing%
silence & short bursts	91.3	89.9
fricatives	78.5	77.2
stop consonants	68.1	69.6
vowels and diphthongs	83.3	82.9
semi-vowels & liquids	68.9	67.5
affricates	52.0	38.6
nasals	56.0	55.4
Average	78.0	77.5

Dominant Set	Training%	Testing%
silence & short bursts	82.4	83.9
stop cons., affr. & fric.	91.8	92.1
high voicing	82.4	83.9
low voicing and nasals	87.1	85.7
Average	83.0	82.6

6 Acknowledgments

This research was made possible through the support of the New Jersey Commission on Science and Technology, and the member corporations of the CAIP Center at Rutgers University.

References

[1] D. J. Burr, "Comparison of Gaussian and Neural Network Classifiers on Vowel Recognition using the Discrete Cosine Transform," *IEEE ICASSP-92*, San Francisco, 1992.

[2] L. F. Lamel, R. H. Kessel and S. Seneff, "Speech Database Development: Design and Analysis of the Acoustic-Phonetic Corpus," *Proc. Speech Recognition Workshop (DARPA)*, 1986.

[3] M. Rahim, "A Neural Tree Architecture for Phoneme Classification with Experiments on the TIMIT Database," *IEEE ICASSP-92*, San Francisco, 1992.

[4] A. Sankar and R. Mammone, "Speaker Independent Vowel Recognition Using Neural Tree Networks," *IJCNN*, Seattle, July 1991.

[5] J-E Stromberg, J. Zrida and A. Isaksson, "Neural Trees: Using Neural Trees in a Tree Classifier Structure," *IEEE ICASSP-91*, pp. 137-140, Toronto, 1991.

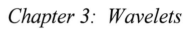

Chapter 3: Wavelets

Wavelet signal analysis creates a set of basis functions and transforms that describe signals as localized in both space and time. It introduces a multi-resolution approach that uses short analysis windows for high frequencies with long analysis windows for low frequencies. This analysis approach emphasizes the areas of change in a signal without expending undue computation to areas that are not changing. In this way, wavelet analysis is similar to the human perceptual capabilities that emphasize transients (such as edges and sharp frequency contrasts) over more slowly-varying frequency regions.

This chapter is only one of several places in which this book discusses wavelets. Architectures for wavelet processing are described in Chapter 7, and their application to image processing is described in Chapter 9. This chapter begins with a comprehensive tutorial that originally appeared in *Signal Processing Magazine*. "Wavelets and Signal Processing (paper 3.1) provides an excellent foundation for exploring the wavelet topics in the remainder of this chapter and elsewhere in this book.

"Speech and Image Signal Compression with Wavelets" (3.2) describes the application of wavelets to speech coding, comparing their performance and complexity to code-excited linear prediction (CELP). It describes wavelets applied to image coding, where their performance is compared to that of the discrete cosine transform (DCT). It is found that image and speech coding with wavelets achieves comparable performance at a given bit rate to DCT and CELP respectively, but with less computational complexity.

The concept of time-varying transforms, whose basis functions are tailored to the short-term properties of the signal, has routinely been applied to various block transforms. Paper 3.3, "A Class of Time-Varying Wavelet Transforms," extends this technique to wavelets, employing two different filter bank approaches. The tree structured filter bank is shown to provide good signal reconstruction but requires a brief transition period to switch from one wavelet transform to another. The parallel filter bank avoids this transition interval, but at the penalty of greater filter complexity.

Morphology-based algorithms apply non-linear operations to the analysis of signals and are suitable for realization in efficient parallel processing structures. "A Morphological Wavelet Transform" (3.4) extends the use of morphology, already successful in areas of image enhancement and pattern recognition, to duplicating the function of a wavelet filter bank, in both one and two dimensions.

Suggested Additional Readings:

[1]   G. W. Wornell, "Wavelet-Based Representation for the $1/f$ Family of Fractal Processes," *Proc. IEEE,* vol. 81, no. 10, pp. 1428-1450, Oct. 1993.

[2]   J. Serra, *Image Analysis and Mathematical Morphology.* London: Academic, 1982.

[3]   F. Hlawatsch and G. F. Boudreaux-Bartels, "Linear and Quadratic Time-Frequency Signal Representations," *IEEE Signal Processing Magazine,* vol. 9 no. 2, pp. 21-67, April 1992.

[4]   M. A. Colestock, "Wavelets - A New Tool for Signal Processing Analysts," *AIAA/IEEE Digital Avionics Systems Conference*, pp. 54-59.

[5]   T. I. Boubez and R. L. Peskin, "Wavelet Neural Networks and Receptive Field Partitioning," *1993 IEEE International Conference on Neural Networks* (San Francisco, CA, USA) Mar. - Apr. 1993, pp. 1544-1549.

[6]   R. Orr, C. Pike, M. Bates, M. Tzannes, and S. Sandberg, "Covert Communications Employing Wavelet Technology," *The Twenty-Seventh Asilomar Conference on Signals, Systems, and Computers* (Pacific Grove, CA, USA) Nov. 1994, pp. 523-527.

[7]   S. Mallat and Z. Zhang, "Adaptive Time-Frequency Transform," *Proc. IEEE International Conference on Acous - tics, Speech, and Signal Processing 93* (Minneapolis, MN, USA) April 1993, vol. 3 pp. 241-244.

[8]   J. Götze and M. Sauer, "Unitary Image Transforms and Their Implementation," *IEEE Pacific Rim Conference on Communications, Computers and Signal Processing* (Victoria, BC, Canada) May 1993, pp. 55-57.

Wavelets and Signal Processing

OLIVIER RIOUL and MARTIN VETTERLI

avelet theory provides a unified framework for a number of techniques which had been developed independently for various signal processing applications. For example, multiresolution signal processing, used in computer vision; subband coding, developed for speech and image compression; and wavelet series expansions, developed in applied mathematics, have been recently recognized as different views of a single theory.

In fact, wavelet theory covers quite a large area. It treats both the continuous and the discrete-time cases. It provides very general techniques that can be applied to many tasks in signal processing, and therefore has numerous potential applications.

In particular, the Wavelet Transform (WT) is of interest for the analysis of *non-stationary* signals, because it provides an alternative to the classical Short-Time Fourier Transform (STFT) or Gabor transform [GAB46, ALL77, POR80]. The basic difference is as follows. In contrast to the STFT, which uses a single analysis window, the WT uses short windows at high frequencies and long windows at low frequencies. This is in the spirit of so-called "constant-Q" or constant relative bandwidth frequency analysis. The WT is also related to time-frequency analysis based on the Wigner-Ville distribution [FLA89, FLA90, RIO90a].

For some applications it is desirable to see the WT as a signal decomposition onto a set of basis functions. In fact, basis functions called *wavelets* always underlie the wavelet analysis. They are obtained from a single prototype wavelet by dilations and contractions (scalings) as well as shifts. The prototype wavelet can be thought of as a bandpass filter, and the constant-Q property of the other bandpass filters (wavelets) follows because they are scaled versions of the prototype.

Therefore, in a WT, the notion of *scale* is introduced as an alternative to frequency, leading to a so-called *time-scale representation*. This means that a signal is mapped into a time-scale plane (the equivalent of the time-frequency plane used in the STFT).

There are several types of wavelet transforms, and, depending on the application, one may be preferred to the others. For a continuous input signal, the time and scale parameters can be continuous [GRO89], leading to the Continuous Wavelet Transform (CWT). They may as well be discrete [DAU88, MAL89b, MEY89, DAU90a], leading to a Wavelet Series expansion. Finally, the wavelet transform can be defined for discrete-time signals [DAU88, RIO90b, VET90b], leading to a Discrete Wavelet Transform (DWT). In the latter case it uses multirate signal processing techniques [CRO83] and is related to subband coding schemes used in speech and image compression. Notice the analogy with the (Continuous) Fourier Transform, Fourier Series, and the Discrete Fourier Transform.

Wavelet theory has been developed as a unifying framework only recently, although similar ideas and constructions took place as early as the beginning of the century [HAA10, FRA28, LIT37, CAL64]. The idea of looking at a signal at various scales and analyzing it with various resolutions has in fact emerged independently in many different fields of mathematics, physics and engineering. In the mid-eighties, researchers of the "French school," lead by a geophysicist, a theoretical physicist and a mathematician (namely, Morlet, Grossmann, and Meyer), built strong mathematical foundations around the subject and named their work "*Ondelettes*" (Wavelets). They also interacted considerably with other fields.

The attention of the signal processing community was soon caught when Daubechies and Mallat, in addition to their contribution to the theory of wavelets, established connections to discrete signal processing results [DAU88], [MAL89a]. Since then, a number of theoretical, as well as practical contributions have been made on various aspects of WT's, and the subject is growing rapidly [WAV89], [IT92].

The present paper is meant both as a review and as a tutorial. It covers the main definitions and properties of wavelet transforms, shows connections among the various fields where results have been developed, and focuses on signal processing applications. Its purpose is to present a simple, synthetic view of wavelet theory, with an easy-to-read, non-rigorous flavor. An extensive bibliography is provided for the reader who wants to go into more detail on a particular subject.

NON-STATIONARY SIGNAL ANALYSIS

The aim of signal analysis is to extract relevant information from a signal by transforming it. Some methods make *a priori* assumptions on the signal to be analyzed; this may yield sharp results if these assumptions are valid, but is obviously not of general applicability. In this paper we focus on methods that are applicable to any general signal. In addition, we consider invertible transformations. The analysis thus unambiguously represents the signal, and more involved operations such as parameter estimation, coding and pattern recognition can be performed on the "transform side," where relevant properties may be more evident.

Such transforms have been applied to *stationary* signals, that is, signals whose properties do not evolve in time (the notion of stationarity is formalized precisely in the statistical signal processing literature). For such signals $x(t)$, the natural "stationary transform" is the well-known Fourier transform [FOU88]:

$$X(f) = \int_{-\infty}^{+\infty} x(t)\, e^{-2j\pi f t} dt \qquad (1)$$

The analysis coefficients $X(f)$ define the notion of global frequency f in a signal. As shown in (1), they are computed as inner products of the signal with sinewave basis functions of infinite duration. As a result, Fourier analysis works well if $x(t)$ is composed of a few stationary components (e.g., sinewaves). However, any abrupt change in time in a non-stationary signal $x(t)$ is spread out over the whole frequency axis in $X(f)$. Therefore, an analysis adapted to *nonstationary* signals requires more than the Fourier Transform.

The usual approach is to introduce time dependency in the Fourier analysis while preserving linearity. The idea is to introduce a "local frequency" parameter (local in time) so that the "local" Fourier Transform looks at the signal through a window over which the signal is approximately stationary. Another, equivalent way is to modify the sinewave basis functions used in the Fourier Transform to basis functions which are more concentrated in time (but less concentrated in frequency).

SCALE VERSUS FREQUENCY

The Short-Time Fourier Transform: Analysis with Fixed Resolution.

The "instantaneous frequency" [FLA89] has often been considered as a way to introduce frequency de-

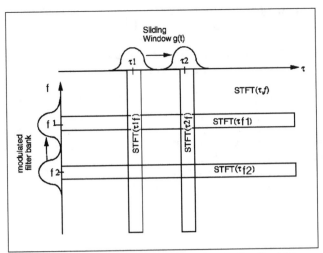

Fig. 1. Time-frequency plane corresponding to the Short-Time Fourier Transform. It can be seen either as a succession of Fourier Transforms of a windowed segment of the signal (vertical stripes) or as a modulated analysis filter bank (horizontal stripes).

and assume it is stationary when seen through a window $g(t)$ of limited extent, centered at time location τ. The Fourier Transform (1) of the windowed signals $x(t) g^*(t - \tau)$ yields the Short-Time Fourier Transform (STFT)

$$\text{STFT}(\tau, f) = \int x(t) \, g^*(t - \tau) \, e^{-2j\pi f t} \, dt \qquad (2)$$

which maps the signal into a two-dimensional function in a time-frequency plane (τ, f). Gabor originally only defined a synthesis formula, but the analysis given in (2) follows easily.

The parameter f in (2) is similar to the Fourier frequency and many properties of the Fourier transform carry over to the STFT. However, the analysis here depends critically on the choice of the window $g(t)$.

Figure 1 shows vertical stripes in the time-frequency plane, illustrating this "windowing of the signal" view of the STFT. Given a version of the signal windowed around time t, one computes all "frequencies" of the STFT.

An alternative view is based on a filter bank interpretation of the same process. At a given frequency f, (2) amounts to filtering the signal "at all times" with a bandpass filter having as impulse response the window function modulated to that frequency. This is shown as the horizontal stripes in Fig. 1. Thus, the STFT may be seen as a modulated filter bank [ALL77], [POR80].

From this dual interpretation, a possible drawback related to the time and frequency resolution can be shown. Consider the ability of the STFT to discriminate between two pure sinusoids. Given a window function $g(t)$ and its Fourier transform $G(f)$, define the "bandwidth" Δf of the filter as

pendence on time. If the signal is not narrow-band, however, the instantaneous frequency averages different spectral components in time. To become accurate in time, we therefore need a *two-dimensional* time-frequency representation $S(t,f)$ of the signal $x(t)$ composed of spectral characteristics depending on time, the local frequency f being defined through an appropriate definition of $S(t,f)$. Such a representation is similar to the notation used in a musical score, which also shows "frequencies" played in time.

The Fourier Transform (1) was first adapted by Gabor [GAB46] to define $S(t,f)$ as follows. Consider a signal $x(t)$,

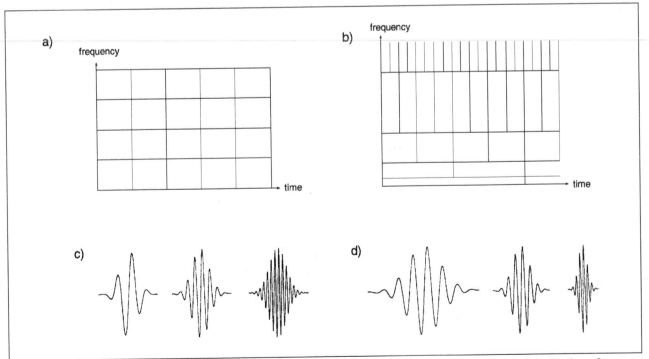

Fig. 2. Basis functions and time-frequency resolution of the Short-Time Fourier Transform (STFT) and the Wavelet Transform (WT). The tiles represent the essential concentration in the time-frequency plane of a given basis function. (a) Coverage of the time-frequency plane for the STFT, (b) for the WT. (c) Corresponding basis functions for the STFT, (d) for the WT ("wavelets").

$$\Delta f^2 = \frac{\int f^2 \, |G(f)|^2 \, df}{\int |G(f)|^2 \, df} \qquad (3)$$

where the denominator is the energy of $g(t)$. Two sinusoids will be discriminated only if they are more than Δf apart (This is an rms measure, and others are possible). Thus, the resolution in frequency of the STFT analysis is given by Δf. Similarly, the spread in time is given by Δt as

$$\Delta t^2 = \frac{\int t^2 |g(t)|^2 \, dt}{\int |g(t)|^2 \, dt} \qquad (4)$$

where the denominator is again the energy of $g(t)$. Two pulses in time can be discriminated only if they are more than Δt apart.

Now, resolution in time and frequency cannot be arbitrarily small, because their product is lower bounded.

$$\text{Time – Bandwidth product} = \Delta t \, \Delta f \geq \frac{1}{4\pi} \qquad (5)$$

This is referred to as the uncertainty principle, or Heisenberg inequality. It means that one can only trade time resolution for frequency resolution, or *vice versa*. Gaussian windows are therefore often used since they meet the bound with equality [GAB46].

More important is that once a window has been chosen for the STFT, then the time-frequency resolution given by (3), (4) is *fixed* over the entire time-frequency plane (since the same window is used at all frequencies). This is shown in Fig. 2a, while Fig. 2c shows the associated basis functions of the STFT. For example, if the signal is composed of small bursts associated with long quasi-stationary components, then each type of component can be analyzed with good time resolution or frequency resolution, but not both.

The Continuous Wavelet Transform: A Multiresolution Analysis.

To overcome the resolution limitation of the STFT, one can imagine letting the resolution Δt and Δf vary in the time-frequency plane in order to obtain a multi-resolution analysis. Intuitively, when the analysis is viewed as a filter bank, the time resolution must increase with the central frequency of the analysis filters. We therefore impose that Δf is proportional to f, or

$$\frac{\Delta f}{f} = c \qquad (6)$$

where c is a constant. The analysis filter bank is then composed of band-pass filters with constant relative bandwidth (so-called "constant-Q" analysis). Another way to say this is that instead of the frequency responses of the analysis filter being regularly spaced over the frequency axis (as for the STFT case), they are regularly spread in a logarithmic scale (see Fig. 3). This kind of filter bank is used, for example, for modeling the frequency response of the cochlea situated in the inner ear and is therefore adapted to auditory perception, e.g. of music: filters satisfying (6) are naturally distributed into octaves.

When (6) is satisfied, we see that Δf and therefore also Δt changes with the center frequency of the analysis filter. Of course, they still satisfy the Heisenberg inequality (5), but now, the time resolution becomes arbitrarily good at high frequencies, while the frequency resolution becomes arbitrarily good at low frequencies. For example, two very close short bursts can always be eventually separated in the analysis by going up to

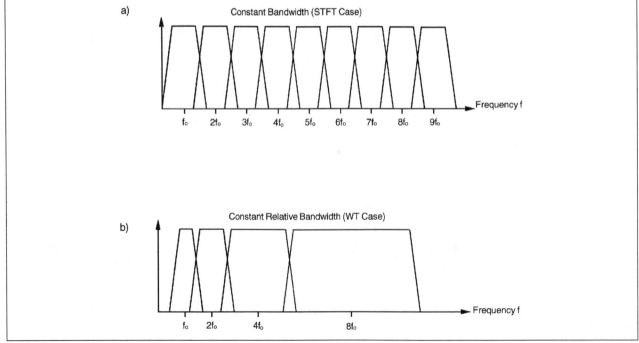

Fig. 3. Division of the frequency domain (a) for the STFT (uniform coverage) and (b) for the WT (logarithmic coverage).

higher analysis frequencies in order to increase time resolution (see Fig. 2b). This kind of analysis of course works best if the signal is composed of high frequency components of short duration plus low frequency components of long duration, which is often the case with signals encountered in practice.

A generalization of the concept of changing resolution at different frequencies is obtained with so-called "wavelet packets" [WIC89], where arbitrary time-frequency resolutions (within the uncertainty bound (5)) are chosen depending on the signal.

The *Continuous Wavelet Transform* (CWT) exactly follows the above ideas while adding a simplification: all impulse responses of the filter bank are defined as *scaled* (i.e. stretched or compressed) versions of the same prototype $h(t)$, i.e.,

$$h_a(t) = \frac{1}{\sqrt{|a|}} \ h(\frac{t}{a})$$

where a is a *scale factor* (the constant $1/\sqrt{|a|}$ is used for energy normalization). This results in the definition of the CWT:

$$\text{CWT}_x(\tau, a) = \frac{1}{\sqrt{|a|}} \int x(t) \ h^*\left(\frac{t-\tau}{a}\right) dt \qquad (7)$$

Since the same prototype $h(t)$, called the *basic wavelet*, is used for all of the filter impulse responses, no specific scale is privileged, i.e. the wavelet analysis is self-similar at all scales. Moreover, this simplification is useful when deriving mathematical properties of the CWT.

To make the connection with the modulated window used in the STFT clearer, the basic wavelet $h(t)$ in (7) could be chosen as a modulated window [GOU84, GRO84, GRO89]

$$h(t) = g(t) \ e^{-2j\pi f_0 t}$$

Then the frequency responses of the analysis filters indeed satisfy (6) with the identification

$$a = \frac{f_0}{f}$$

But more generally, $h(t)$ can be any band-pass function and the scheme still works. In particular one can dispense with complex-valued transforms and deal only with real-valued ones.

It is important to note that here, the local frequency $f = a f_0$ has little to do with that described for the STFT: indeed, it is associated with the scaling scheme (see Box 1). As a result, this local frequency, whose definition depends on the basic wavelet, is no longer linked to frequency modulation (as was the case for the STFT) but is now related to time-scalings. This is the reason why the terminology "scale" is often preferred to "frequency" for the CWT, the word "frequency" being reserved for the STFT. Note that we define *scale* in wavelet analysis like the scale in geographical maps: since the filter bank impulse responses in (7) are dilated as scale increases, large scale corresponds to contracted signals, while small scale corresponds to dilated signals.

WAVELET ANALYSIS AND SYNTHESIS

Another way to introduce the CWT is to define *wavelets* as basis functions. In fact, basis functions already appear in the preceding definition (7) when one sees it as an inner product of the form

$$\text{CWT}_x(\tau, a) = \int x(t) \ h_{a,\tau}^*(t) \ dt$$

which measures the "similarity" between the signal and the basis functions

The inner product is often used as a similarity measurement, and because both STFT's and CWT's are inner products, they appear in several detection/estimation problems. Consider, for example, the problem of estimating the location and velocity of some target in radar or sonar applications. The estimation procedure consists in first emitting a known signal $h(t)$. In the presence of a target, this signal will return to the source (received signal $x(t)$) with a certain delay τ, due to the target's location, and a certain distortion (Doppler effect), due to the target's velocity.

For narrow-band signals, the Doppler effect amounts to a single frequency shift f_0 and the characteristics of the target will be determined by maximizing the cross-correlation function (called "narrow-band cross-ambiguity function") [WOO53]

$$\int x(t)\, h(t-\tau)\, e^{-2j\pi f_0 t}\, dt = \text{STFT}(\tau, f)$$

For wide-band signals, however, the Doppler frequency shift varies in the signal's spectrum, causing a stretching or a compression in the signal. The estimator thus becomes the "wide-band cross-ambiguity function" [SPE67], [AUS90]

$$\frac{1}{\sqrt{|a|}} \int x(t)\, h\!\left(\frac{t-\tau}{a}\right) dt = \text{CWT}_x(\tau, a)$$

As a result, in both cases, the "maximum likelihood" estimator takes the form of a STFT or a CWT, i.e. of an inner product between the received signal and either STFT or wavelet basis functions. The basis function which best fits the signal is used to estimate the parameters.

Note that, although the wide-band cross-ambiguity function is a CWT, for physical reasons, the dilation parameter a stays on the order of magnitude of 1, whereas it may cover several octaves when used in signal analysis [FLA89].

$$h_{a,\tau} = \frac{1}{\sqrt{a}}\, h\!\left(\frac{t-\tau}{a}\right)$$

called *wavelets*. The wavelets are scaled and translated versions of the basic wavelet prototype $h(t)$ (see Fig. 2d).

Of course, basis functions can be considered for the STFT as well. For both the STFT and the CWT, the sinewaves basis functions of the Fourier Transform are replaced by more localized reference signals labelled by time and frequency (or scale) parameters. In fact both transforms may be interpreted as special cases of the cross-ambiguity function used in radar or sonar processing (see Box 2).

The wavelet analysis results in a set of wavelet coefficients which indicate how close the signal is to a particular basis function. Thus, we expect that any general signal can be represented as a decomposition into wavelets, i.e. that the original waveform is synthesized by adding elementary building blocks, of constant shape but different size and amplitude. Another way to say this is that we want the continuously labelled wavelets $h_{a,\tau}(t)$ to behave just like an *orthogonal basis* [MEY90]. The analysis is done by computing inner products, and the synthesis consists of summing up all the orthogonal projections of the signal onto the wavelets.

$$x(t) = c \iint_{a>0} \text{CWT}(\tau, a)\, h_{a,\tau}(t)\, \frac{da\, d\tau}{a^2} \qquad (8)$$

where c is a constant that depends only on $h(t)$. The measure in this integration is formally equivalent to $dt\, df$ [GOU84]. We have assumed here that both signal and wavelets are either real-valued or complex analytic so that only positive dilations $a > 0$ have to be taken into account. Otherwise (8) is more complicated [GRO84].

Of course, the $h_{a,\tau}(t)$ are certainly not orthogonal since they are very redundant (they are defined for continuously varying a and τ). But surprisingly, the reconstruction formula (8) is indeed satisfied whenever $h(t)$ is of finite energy and *band pass* (which implies that it oscillates in time like a short wave, hence the name "wavelet"). More precisely, if $h(t)$ is assumed sufficiently regular, then the reconstruction condition is $\int h(t)\, dt = 0$.

Note that the reconstruction takes place only in the sense of the signal's energy. For example, a signal may be reconstructed only with zero mean since $\int h(t)\, dt = 0$. In fact the type of convergence of (8) may be strengthened and is related to the numerical robustness of the reconstruction [DAU90a].

Similar reconstruction can be considered for the STFT, and the similarity is remarkable [DAU90a]. However, in the STFT case, the reconstruction condition is less restrictive: only finite energy of the window is required.

SCALOGRAMS

The *spectrogram*, defined as the square modulus of the STFT, is a very common tool in signal analysis because it provides a distribution of the energy of the signal in the time-frequency plane. A similar distribution can be defined in the wavelet case. Since the CWT behaves like an orthonormal basis decomposition, it can be shown that it is isometric [GRO84], i.e., it preserves energy. We have

$$\iint |\text{CWT}(\tau, a)|^2 \frac{d\tau,\, da}{a^2} = E_x$$

where $E_x = \int |x(t)|^2\, dt$ is the energy of the signal $x(t)$. This leads us to define the *wavelet spectrogram*, or *scalogram*, as the squared modulus of the CWT. It is a distribution of the energy of the signal in the time-scale

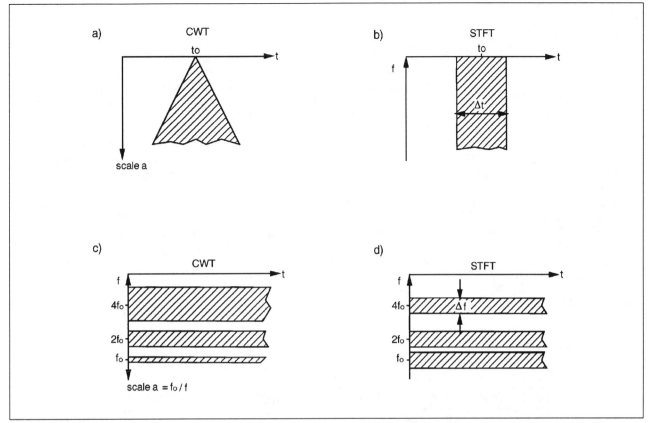

Fig. 4. Regions of influence of a Dirac pulse at $t=t_0$ (a) for the CWT and (b) for the STFT; as well as of three sinusoids (of frequencies f_0, $2f_0$, $4f_0$) for (c) the CWT and (d) the STFT.

plane, associated with measure $\dfrac{d\tau, da}{a^2}$, and thus expressed in power per frequency unit, like the spectrogram. However, in contrast to the spectrogram, the energy of the signal is here distributed with different resolutions according to Fig. 2b.

Figure 4 illustrates differences between a scalogram and a spectrogram. Figure 4a shows that the influence of the signal's behavior around $t = t_0$ in the analysis is limited to a cone in the time-scale plane; it is therefore very "localized" around t_0 for small scales. In the STFT case, the corresponding region of influence is as large as the extent of the analysis window over all frequencies, as shown in Fig. 4b. Moreover, since the time-scale analysis is logarithmic in frequency, the area of influence of some pure frequency f_0 in the signal increases with f_0 in a scalogram (Fig. 4c), whereas it remains constant in a spectrogram (Fig. 4d).

Both the spectrogram and the scalogram produce a more or less easily interpretable visual two-dimensional representation of signals [GRO89], where each pattern in the time-frequency or time-scale plane contributes to the global energy of the signal. However, such an energy representation has some disadvantages, too. For example, the spectrogram, as well as the scalogram, cannot be inverted in general. Phase information is necessary to reconstruct the signal. Also, since both the spectrogram and the scalogram are bilinear functions of the analyzed signal, cross-terms appear as interferences between patterns in the time-frequency or time-scale plane [KAD91] and this may be undesirable.

In the wavelet case, it has been also shown [GRO89]

that the *phase* representation more accurately reveals isolated, local bursts in a signal than the scalogram does (see Box 3).

To illustrate the above points, Fig. 5 shows some examples of spectrograms and scalograms for synthetic signals and a speech signal (see Box 3).

More involved energy representations can be developed for both time-frequency and time-scale [BER88, FLA90, RIO90a], and a link between the spectrogram, the scalogram and the Wigner-Ville distribution can be established (see Box 4).

WAVELET FRAMES AND ORTHONORMAL BASES

Discretization of Time-Scale Parameters

We have seen that the continuously labelled basis functions (wavelets) $h_{a,\tau}(t)$ behave in the wavelet analysis and synthesis just like an orthonormal basis. The following natural question arises: if we appropriately discretize the time-scale parameters a, τ, can we obtain a *true* orthonormal basis? The answer, as we shall see, is that it depends on the choice of the basic wavelet $h(t)$.

There is a natural way to discretize the time-scale parameters a, τ [DAU90a]: since two scales $a_0 < a_1$ roughly correspond to two frequencies $f_0 > f_1$, the wavelet coefficients at scale a_1 can be subsampled at $(f_0/f_1)^{\text{th}}$ the rate of the coefficients at scale a_0, according

BOX 3:
Spectrograms and Scalograms

We present in Fig. 5 spectrograms and scalograms for some synthetic signals and a real signal. The signals are of length 384 samples, and the STFT uses a Gaussian-like window of length $L = 128$ samples. The scalogram is obtained with a Morlet wavelet (a complex sinusoid windowed with a Gaussian envelope) of length from 23 to 363 samples.

The horizontal axis is time in both spectrograms and scalograms. The signal is shown on the top. The vertical axis is frequency in the spectrogram (high frequencies on top) and scale in the scalogram (small scale at the top). Compare these figures with Fig. 4, which indicates the axis system used, and gives the rough behavior for Diracs and sinewaves.

First, Fig. 5.1 shows the analysis of two Diracs and two sinusoids with the STFT and the CWT. Note how the Diracs are well time-localized at high frequencies

in the scalogram. Figure 5.2 shows the analysis of three starting sinusoids with different starting times (a low frequency starts first, followed by a medium and a high frequency sinewave). Figure 5.3 shows the transforms of a chirp signal. Again, the transitions are well resolved at high frequencies in the scalogram. Finally, Fig. 5.4 shows the analysis of a segment of speech signal, where the onset of voicing is seen in both representations.

Note that displaying scalograms is sometimes tricky, because parameters like display look-up tables (which map the scalogram value to a grey scale value on the screen) play an important but not always well understood role in the visual impression. Such problems are common in spectrogram displays as well.

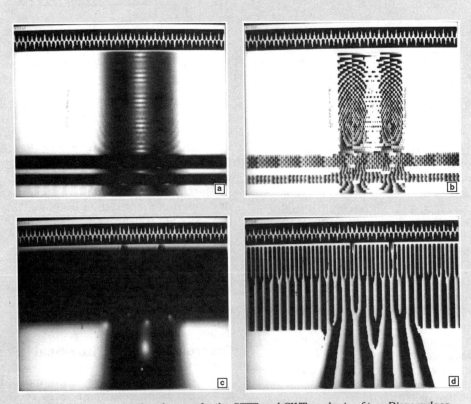

Fig. 5.1. Spectrogram and scalogram for the STFT and CWT analysis of two Dirac pulses and two sinusoids. (a) Magnitude of the STFT. (b) Phase of the STFT. (c) Amplitude of the WT. (d) Phase of the WT.

BOX 3: Spectrograms and Scalograms *(continued)*

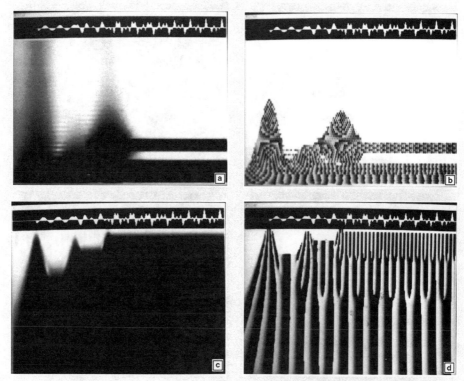

Fig. 5.2. Spectrogram and scalogram for the STFT and CWT analysis of three sinusoids with staggered starting times. The low frequency one comes first, followed by the medium and high frequency ones. (a) Magnitude of the STFT. (b) Phase of the STFT. (c) Amplitude of the WT. (d) Phase of the WT.

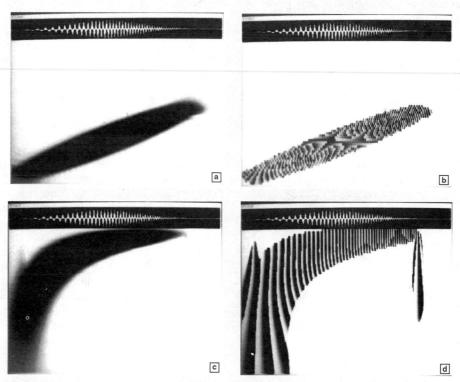

Fig. 5.3. Spectrogram and scalogram for the STFT and CWT analysis of a chirp signal. (a) Magnitude of the STFT. (b) Phase of the STFT. (c) Amplitude of the WT. (d) Phase of the WT.

BOX 3: Spectrograms and Scalograms (continued)

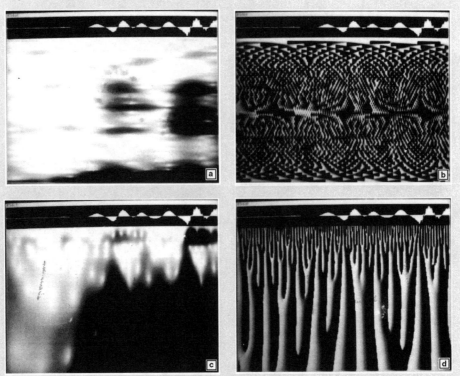

Fig. 5.4. Spectrogram and scalogram for the STFT and CWT analysis of a segment of speech, including onset of voicing. (a) Magnitude of the STFT. (b) Phase of the STFT. (c) Amplitude of the WT. (d) Phase of the WT.

to Nyquist's rule. We therefore choose to discretize the time-scale parameters on the sampling grid drawn in Fig. 7. That is, we have $a = a_0^j$ and $b = k\, a_0^j\, T$, where j and k are integers. The corresponding wavelets are

$$h_{j,k}(t) = a_0^{-j/2}\, h(a_0^{-j}\, t - kT) \tag{9}$$

resulting in wavelet coefficients

$$c_{j,k} = \int x(t)\, h_{j,k}^*(t)\, dt \tag{10}$$

An analogy is the following: assume that the wavelet analysis is like a microscope. First one chooses the magnification, that is, a_0^{-j}. Then one moves to the chosen location. Now, if one looks at very small details, then the chosen magnification is large and corresponds to j negative and large. Then, $a_0^j\, T$ corresponds to small steps, which are used to catch small details. This justifies the choice $b = k\, a_0^j\, T$ in (9).

The reconstruction problem is to find a_0, T, and $h(t)$ such that

$$x(t) \approx c \sum_j \sum_k c_{j,k}\, h_{j,k}(t) \tag{11}$$

where c is a constant that does not depend on the signal (compare with (8)). Evidently, if a0 is close enough to 1 (and if T is small enough), then the wavelet functions

are overcomplete. Equation (11) is then still very close to (8) and signal reconstruction takes place within non-restrictive conditions on h(t). On the other hand, if the sampling is sparse, e.g. the computation is done octave by octave ($a_0 = 2$), a true orthonormal basis will be obtained only for very special choices of h(t) [DAU90a, MEY90].

Wavelet Frames

The theory of wavelet frames [DUF52, DAU90a] provides a general framework which covers the two extreme situations just mentioned. It therefore permits one to balance (i) redundancy, i.e. sampling density in Fig. 7, and (ii) restrictions on h(t) for the reconstruction scheme (11) to work. The trade-off is the following: if the redundancy is large (high "oversampling"), then only mild restrictions are put on the basis functions (9). But if the redundancy is small (i.e., close to "critical" sampling), then the basis functions are very constrained.

The idea behind frames [DUF52] is based on the assumption that the linear operator $x(t) \to c_{j,k}$, where $c_{j,k}$ is defined by (10), is bounded, with bounded inverse. The family of wavelet functions is then called a frame and is such that the energy of the wavelet coefficients $c_{j,k}$ (sum of their square moduli) relative to that of the signal lies between two positive "frames bounds" A and B,

$$A\cdot E_x \leq \sum_{j,k} |\, c_{j,k}\, |^2 \leq B\cdot E_x$$

94

Box 4:
Merging Spectrogram, Scalogram, and Wigner Distribution into a Common Class of Energy Representations

There has been considerable work in extending the spectrogram into more general time-frequency energy distributions TF(τ, f). These all have the basic property of distributing the energy of the signal all over the time-frequency plane, i.e.,

$$\int \int TF\{\tau,f\}\, d\tau\, df = \int |x(t)|^2 dt$$

Among them, an alternative to the spectrogram for nonstationary signal analysis is the Wigner-Ville distribution [CLA80, BOU85, FLA89]

$$W_x(\tau,f) = \int x(\tau + \frac{t}{2})\, x^*(\tau - \frac{t}{2})\, e^{-2j\pi ft} dt$$

More generally, the whole class of time-frequency energy distributions has been fully described by Cohen [COH66], [COH89]: they can all be seen as smoothed (or, more precisely, correlated) versions of the Wigner-Ville distribution. The spectrogram is itself recovered when the "smoothing" function is the Wigner-Ville distribution of the analysis window!

A similar situation appears for time-scale energy distributions. For example, the scalogram can be written as [FLA90], [RIO90a]

$$|CWT\{\tau, a\}|^2 = \int \int W_x(t,v)\, W_h(\frac{t-\tau}{a},av)\, dt\, dv$$

i.e., as some 2D "affine" correlation between the signal and the "basic" wavelet's Wigner-Ville distribution. This remarkable formula tells us that there exist strong links between Wavelet Transforms and Wigner-Ville distributions. And, as a matter of fact, it can be generalized to define the most general class of time-scale energy distributions [BER88, FLA90, RIO90a], just as in the time-frequency case.

Figure 6 shows that it is even possible to go continuously from the spectrogram of a given signal to its scalogram [FLA90, RIO90a]. More precisely, starting from the Wigner-Ville distribution, by progressively controlling Gaussian smoothing functions, one goes through a set of energy representations which either tends to the spectrogram if regular two-dimensional smoothing is used, or to the scalogram if "affine" smoothing is used. This property may allow us to decide whether or not we should choose time-scale analysis tools, rather than time-frequency ones for a given problem.

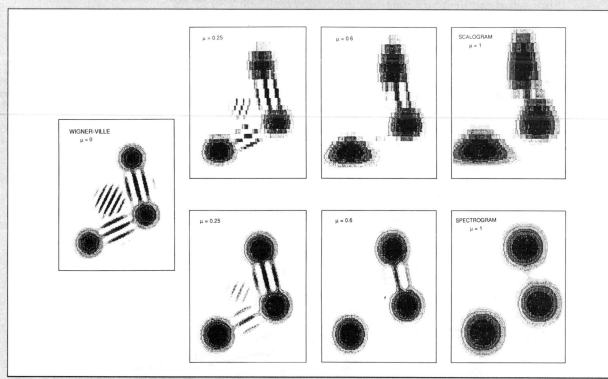

Fig. 6. From spectrograms to scalograms via Wigner-Ville. By controlling the parameter μ (which is a measure of the time-frequency extent of the smoothing function), it is possible to make a full transition between time-scale and time-frequency analyses. Here seven analyses of the same signal (composed of three Gaussian packets) are shown. Note that the best joint time-frequency resolution is attained for the Wigner-Ville distribution, while both spectrogram and scalogram (which can be thought of as smoothed versions of Wigner-Ville) provide reduced cross-term effects compared to Wigner-Ville (after [FLA90, RIO90a]).

where E_x is the energy of the signal $x(t)$.

These frame bounds can be computed from a_0, T and $h(t)$ using Daubechies' formulae [DAU90a]. Interestingly enough, they govern the accuracy of signal reconstruction by (11). More precisely, we have

$$x(t) \approx \frac{2}{A+B} \sum_j \sum_k c_{j,k} \, h_{j,k}(t)$$

with relative SNR greater than $(B/A+1)/(B/A-1)$ (see Fig. 8). The closer A and B, the more accurate the reconstruction. It may even happen that $A=B$ ("tight frame"), in which case the wavelets behave exactly like an orthonormal basis, although they may not even be linearly independent [DAU90a]! The reconstruction formula can also be made exact in the general case if one uses different synthesis functions $h'_{jk}(t)$ (which constitute the *dual frame* of the $h_{jk}(t)$s [DAU90a]).

Introduction to orthogonal wavelet bases

If a tight frame is such that all wavelets $h_{j,k}(t)$ (9) are necessary to reconstruct a general signal, then the wavelets form an *orthonormal* basis of the space of signals with finite energy [HEI90]. Recall that orthonormality means

$$\int h_{j,k}(t) \, h^*_{j',k'}(t) \, dt = \begin{cases} 1 & if \, j = j' \; and \; k = k' \\ 0 & otherwise \end{cases}$$

An arbitrary signal can then be represented exactly as a weighted sum of basis functions,

$$x(t) = \sum_{j,k} c_{j,k} \, h_{j,k}(t)$$

That is, not only the basis functions $h_{j,k}(t)$ are obtained from a single prototype function $h(t)$ by means of

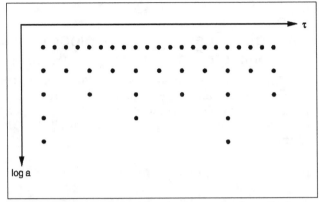

Fig. 7. *Dyadic sampling grid in the time-scale plane. Each node corresponds to a wavelet basis function $h_{j,k}(t)$ with scale 2^{-j} and shift $2^{-j}k$.*

scalings and shifts, but also they form an orthonormal basis. What is most interesting is that there do exist well-behaved functions $h(t)$ that can be used as prototype wavelets, as we shall see below. This is in sharp contrast with the STFT, where, according to the Balian-Low theorem [DAU90a], it is impossible to have orthonormal bases with functions well localized in time and frequency (that is, for which the time-bandwidth product $\Delta t \, \Delta f$ is a finite number).

Recently, the wavelet orthonormal scheme has been extended to synthesis functions $h'_{jk}(t) \neq h_{jk}(t)$, leading to so-called *biorthogonal* wavelet bases [COH90a], [VET90a], [VET90b].

THE DISCRETE TIME CASE

In this section, we first take a purely discrete-time point of view. Then, through the construction of iterated filter banks, we shall come back to the continuous-time

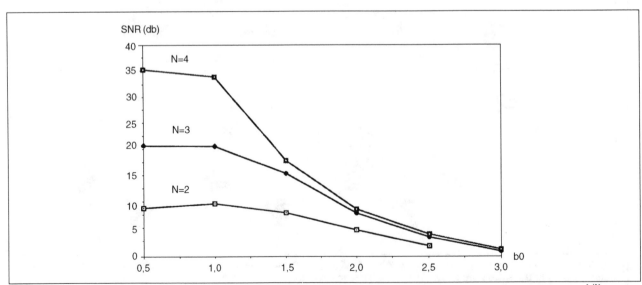

Fig. 8. *Reconstruction Signal/Noise Ratio (SNR) error after frame decomposition for different sampling densities $a_0 = 2^{1/N}$ (N = number of voices per octave), $b = a_0^j k \, b_0$ (after [DAU90a]). The basic wavelet is the Morlet wavelet (modulated Gaussian) used in [GRO89]. The reconstruction is done "as if" wavelets were orthogonal (see text), and its accuracy grows as N increases and b_0 decreases, i.e. as the density of the sampling grid of Fig. 7 increases. Therefore, redundancy refines the "orthogonal-like" reconstruction.*

Three Dimensional Displays of Complex Wavelet Transforms

As seen in Box 3, the wavelet transform using a complex wavelet like the Morlet wavelet (a complex sinusoid windowed by a Gaussian) leads to a complex valued function on the plane.

Phase information is also useful and thus, there is interest in a common display of magnitude and phase. This is possible by using height as magnitude and color as phase, leading to so-called "phasemagrams".

Two examples are shown here: a synthetic chirp in the upper figure (similar to the one in Fig. 5.3); and a triangle function below. In both cases, the discontinuous points are clearly identified at small scales (top of the figure). The chirp has two such points (beginning and end), while the triangle has three. At large scales, these signals look just like a single discontinuity, which is what an observer would indeed see from very far away. For the chirp, the phase cycles with increasing speed, as expected.

Signal analyses with a Morlet wavelet. The display shows magnitude as height and phase as color (phasemagram). The horizontal axis is time. Above) a synthetic chirp signal, with frequency increasing with time. Below) a triangle function.

case and show how to construct orthonormal bases of wavelets for continuous-time signals [DAU88].

In the discrete time case, two methods were developed independently in the late seventies and early eighties which lead naturally to discrete wavelet transforms, namely subband coding [CRI76], [CRO76], [EST77] and pyramidal coding or multiresolution signal analysis [BUR83]. The methods were proposed for coding, and thus, the notion of critical sampling (of requiring a minimum number of samples) was of importance. Pyramid coding actually uses some oversampling, but because it has an easier intuitive explanation, we describe it first.

While the discrete-time case has been thoroughly studied in the filter bank literature in terms of frequency bands (see e.g. [VAI87]), we insist here on notions which are closer to the wavelet point of view, namely those of *scale* and *resolution*. Scale is related to the size of the signal, while resolution is linked to the amount of detail present in the signal (see Box 1 and Fig. 9).

Note that the *scale* parameter in discrete wavelet analysis is to be understood as follows: For large scales, dilated wavelets take "global views" of a *subsampled* signal, while for small scales, contracted wavelets analyze small "details" in the signal.

The Multiresolution Pyramid

Given an original sequence $x(n)$, $n \in \mathbf{Z}$, we derive a lower resolution signal by lowpass filtering with a half-band low-pass filter having impulse response $g(n)$. Following Nyquist's rule, we can subsample by two (drop every other sample), thus doubling the scale in the analysis. This results in a signal $y(n)$ given by

$$y(n) = \sum_{k=-\infty}^{+\infty} g(k)\, x(2n - k)$$

The resolution change is obtained by the lowpass filter (loss of high frequency detail). The scale change is due to the subsampling by two, since a shift by two in the original signal $x(n)$ results in a shift by one in $y(n)$.

Now, based on this lowpass and subsampled version of $x(n)$, we try to find an approximation, $a(n)$, to the original. This is done by first upsampling $y(n)$ by two (that is, inserting a zero between every sample) since we need a signal at the original scale for comparison.

$$y'(2n) = y(n), \quad y'(2n+1) = 0$$

Then, $y'(n)$ is interpolated with a filter with impulse response $g'(n)$ to obtain the approximation $a(n)$,

$$a(n) = \sum_{k=-\infty}^{\infty} g'(k)\, y'(n - k)$$

Note that if $g(n)$ and $g'(n)$ were perfect halfband filters (having a frequency passband equal to 1 over the normalized frequency range $-\pi/2$, $\pi/2$ and equal to 0 elsewhere), then the Fourier transform of a(n) would be

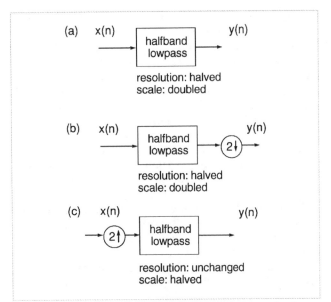

(a) x(n) → halfband lowpass → y(n)

resolution: halved
scale: doubled

(b) x(n) → halfband lowpass → (2↓) → y(n)

resolution: halved
scale: doubled

(c) x(n) → (2↑) → halfband lowpass → y(n)

resolution: unchanged
scale: halved

Fig. 9. Resolution and scale changes in discrete time (by factors of 2). Note that the scale of signals is defined as in geographical maps. (a) Halfband lowpass filtering reduces the resolution by 2 (scale is unchanged). (b) Halfband lowpass filtering followed by subsampling by 2 doubles the scale (and halves the resolution as in (a)). (c) Upsampling by 2 followed by halfband lowpass filtering halves the scale (resolution is unchanged).

equal to the Fourier transform of $x(n)$ over the frequency range $(-\pi/2, \pi/2)$ while being equal to zero elsewhere. That is, $a(n)$ would be a perfect halfband lowpass approximation to $x(n)$.

Of course, in general, $a(n)$ is not going to be equal to $x(n)$ (in the previous example, $x(n)$ would have to be a halfband signal). Therefore, we compute the difference between $a(n)$ (our approximation based on $y(n)$) and $x(n)$,

$$d(n) = x(n) - a(n)$$

It is obvious that $x(n)$ can be reconstructed by adding $d(n)$ and $a(n)$, and the whole process is shown in Fig. 10. However, there has to be some redundancy, since a signal with sampling rate f_s is mapped into two signals $d(n)$ and $y(n)$ with sampling rates f_s and $f_s/2$, respectively.

In the case of a perfect halfband lowpass filter, it is

clear that $d(n)$ contains exactly the frequencies above $\pi/2$ of $x(n)$, and thus, $d(n)$ can be subsampled by two as well without loss of information. This hints at the fact that critically sampled schemes must exist.

The separation of the original signal $x(n)$ into a coarse approximation $a(n)$ plus some additional detail contained in $d(n)$ is conceptually important. Because of the resolution change involved (lowpass filtering followed by subsampling by two produces a signal with half the resolution and at twice the scale of the original), the above method and related ones are part of what is called Multiresolution Signal Analysis [ROS84] in computer vision.

The scheme can be iterated on $y(n)$, creating a hierarchy of lower resolution signals at lower scales. Because of that hierarchy and the fact that signals become shorter and shorter (or images become smaller and smaller), such schemes are called signal or image pyramids [BUR83].

Subband Coding Schemes

We have seen that the above system creates a redundant set of samples. More precisely, one stage of a pyramid decomposition leads to both a half rate low resolution signal and a full rate difference signal, resulting in an increase in the number of samples by 50%. This oversampling can be avoided if the filters $g(n)$ and $g'(n)$ meet certain conditions [VET90b].

We now look at a different scheme instead, where no such redundancy appears. It is the so-called subband coding scheme first popularized in speech compression [CRI76, CRO76, EST77]. The lowpass, subsampled approximation is obtained exactly as explained above, but, instead of a difference signal, we compute the "added detail" as a highpass filtered version of $x(n)$ (using a filter with impulse response $h(n)$), followed by subsampling by two. Intuitively, it is clear that the "added detail" to the lowpass approximation has to be a highpass signal, and it is obvious that if $g(n)$ is an ideal halfband lowpass filter, then an ideal halfband highpass filter $h(n)$ will lead to a perfect representation of the original signal into two subsampled versions.

This is exactly one step of a wavelet decomposition using $\sin(x)/x$ filters, since the original signal is mapped into a lowpass approximation (at twice the scale) and

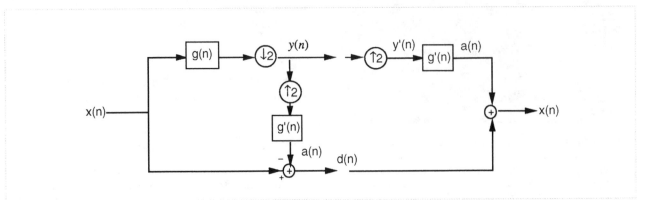

Fig. 10. Pyramid scheme. Derivation of a lowpass, subsampled approximation $y(n)$, from which an approximation $a(n)$ to $x(n)$ is derived by upsampling and interpolation. Then, the difference between the approximation $a(n)$ and the original $x(n)$ is computed as $d(n)$. Perfect reconstruction is simply obtained by adding $a(n)$ back.

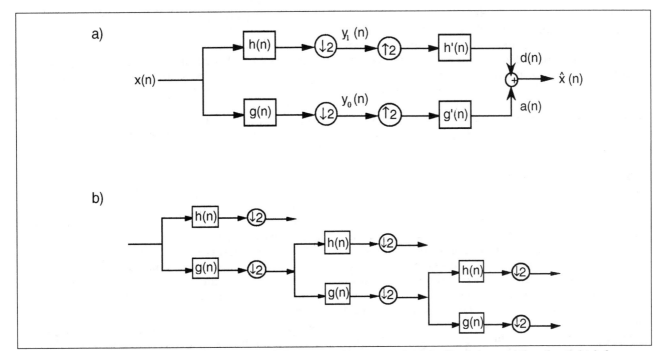

Fig. 11. Subband Coding scheme. (a) Two subsampled approximations, one corresponding to low and the other to high frequencies, are computed. The reconstructed signal is obtained by re-interpolating the approximations and summing them. The filters on the left form an analysis filter bank, while on the right is a synthesis filter bank. (b) Block diagram (Filter Bank tree) of the Discrete Wavelet Transform implemented with discrete-time filters and subsampling by two. The frequency resolution is given in Fig. 3b.

an added detail signal (also at twice the scale). In particular, using these ideal filters, the discrete version is identical to the continuous wavelet transform.

What is more interesting is that it is not necessary to use ideal (that is, impractical) filters, and yet $x(n)$ can be recovered from its two filtered and subsampled versions which we now call $y_0(n)$ and $y_1(n)$. To do so, both are upsampled and filtered by $g'(n)$ and $h'(n)$ respectively, and finally added together, as shown in Fig. 11a. Now, unlike the pyramid case, the reconstructed signal (which we now call $\hat{x}(n)$) is not identical to $x(n)$, unless the filters meet some specific constraints. Filters that meet these constraints are said to have *perfect reconstruction* property, and there are a number of papers investigating the design of perfect reconstruction filter banks [MIN85, SMI86, VAI88, VET86].

The easiest case to analyze appears when the analysis and synthesis filters in Fig. 11a are identical (within time-reversal) and when perfect reconstruction is achieved (that is, $\hat{x}(n) = x(n)$, within a possible shift). Then it can be shown that the subband analysis/synthesis corresponds to a decomposition onto an orthonormal basis, followed by a reconstruction which amounts to summing up the orthogonal projections. We will assume FIR filters in the following. Then, it turns out that the highpass and lowpass filters are related by

$$h(L - 1 - n) = (-1)^n g(n) \qquad (12)$$

where L is the filter length (which has to be even). Note that the modulation by $(-1)^n$ transforms indeed the lowpass filter into a highpass one.

Now, the filter bank in Fig. 11a, which computes convolutions followed by subsampling by two, evaluates inner products of the sequence $x(n)$ and the sequences $\{g(-n+2k), h(-n+2l)\}$ (the time reversal comes from the convolution, which reverses one of the sequences). Thus

$$y_0(k) = \sum_n x(n)\, g(-n + 2k)$$

$$y_1(k) = \sum_n x(n)\, h(-n + 2k)$$

Because the filter impulse responses form an orthonormal set, it is very simple to reconstruct $x(n)$ as

$$x(n) = \sum_{k=-\infty}^{\infty} \Big[\, y_0(k)\, g(-n + 2k) + y_1(k)\, h(-n + 2k)\,\Big] \qquad (13)$$

that is, as a weighted sum of the orthogonal impulse responses, where the weights are the inner products of the signal with the impulse responses. This is of course the standard expansion of a signal into an orthonormal basis, where the resynthesis is the sum of the orthogonal projections (see *Introduction to orthogonal wavelet bases* above).

From (12), (13) it is also clear that the synthesis filters are identical to the analysis filters within time reversal.

Such orthogonal perfect reconstruction filter banks have been studied in the digital signal processing literature, and the orthonormal decomposition we just indicated is usually referred to as a "paraunitary" or "lossless" filter bank [VAI89]. An interesting property of such filter banks is that they can be written in lattice form [VAI88], and that the structure and properties can be extended to more than two channels [VAI87, VAI89,

99

VET89]. More general perfect reconstruction (biorthogonal) filter banks have also been studied (see e.g. [VET86, VET90b, COH90a]). It has been also noticed [MAL89b, SHE90, RIO90b] that filter banks arise naturally when implementing the CWT.

Note that we have assumed linear processing throughout. If non-linear processing is involved (like quantization), the oversampled nature of the pyramid scheme described in the preceding section may actually lead to greater robustness.

The Discrete Wavelet Transform

We have shown how to decompose a sequence $x(n)$ into two subsequences at half rate, or half resolution, and this by means of "orthogonal" filters (orthogonal with respect to even shifts). Obviously, this process can be iterated on either or both subsequences. In particular, to achieve finer frequency resolution at lower frequencies (as obtained in the continuous wavelet transform), we iterate the scheme on the lower band only. If $g(n)$ is a good halfband lowpass filter, $h(n)$ is a good halfband highpass filter by (12). Then, one iteration of the scheme on the first lowband creates a new lowband that corresponds to the lower quarter of the frequency spectrum. Each further iteration halves the width of the lowband (increases its frequency resolution by two), but due to the subsampling by two, its time resolution is halved as well. At each iteration, the current high band portion corresponds to the difference between the previous lowband portion and the current one, that is, a passband. Schematically, this is equivalent to Fig. 11b, and the frequency resolution is as in Fig. 3b.

An important feature of this discrete algorithm is its relatively low complexity. Actually, the following somewhat surprising result holds: independent of the depth of the tree in Fig. 11b, the complexity is linear in the number of input samples, with a constant factor that depends on the length of the filter. The proof is straightforward. Assume the computation of the first filter bank requires C_0 operations per input sample (C_0 is typically of the order of L). Then, the second stage requires also C_0 operations per sample of its input, but, because of the subsampling by two, this amounts to $C_0/2$ operations per sample of the input signal. Therefore, the total complexity is bounded by

$$C_{total} = C_0 + \frac{C_0}{2} + \frac{C_0}{4} + \ldots < 2C_0$$

which demonstrates the efficiency of the discrete wavelet transform algorithm and shows that it is independent of the number of octaves that one computes. This bounded complexity had been noticed in the multirate filtering context [RAM88]. Further developments can be found in [RIO91a]. Note that a possible drawback is that the delay associated with such an iterated filter bank grows exponentially with the number of stages.

Iterated Filters and Regularity

There is a major difference between the discrete scheme we have just seen and the continuous time wavelet transform. In the discrete time case, the role of the wavelet is played by the highpass filter $h(n)$ and the cascade of subsampled lowpass filters followed by a highpass filter (which amounts to a bandpass filter). These filters, which correspond roughly to octave band filters, unlike in the continuous wavelet transform, are not exact scaled versions of each other. In particular, since we are in discrete time, scaling is not as easily defined, since it involves interpolation as well as time expansion.

Nonetheless, under certain conditions, the discrete system converges (after a certain number of iterations) to a system where subsequent filters are scaled versions of each other. Actually, this convergence is the basis for the construction of continuous time compactly supported wavelet bases [DAU88].

Now, we would like to find the equivalent filter that corresponds to the lower branch in Fig. 11b, that is, the iterated lowpass filter. It will be convenient to use z-transforms of filters, e.g. $G(z) = \sum_n g(n) z^{-n}$ in the following. It can be easily checked that subsampling by two followed by filtering with $G(z)$ is equivalent to filtering with $G(z^2)$ followed by the subsampling (z^2 inserts zeros between samples of the impulse response, which are removed by the subsequent subsampling). That is, the first two steps of lowpass filtering can be replaced by a filter with z-transform $G(z) \cdot G(z^2)$, followed by subsampling by 4. More generally, calling $G^i(z)$ the equivalent filter to i stages of lowpass filtering and subsampling by two (that is, a total subsampling by 2^i), we obtain

$$G^i(z) = \prod_{l=0}^{i-1} G(z^{2^l}) \tag{14}$$

Call its impulse response $g^i(n)$.

As i infinitely increases, this filter becomes infinitely long. Instead, consider a function $f^i(x)$ which is piecewise constant on intervals of length $1/2^i$ and has value $2^{i/2} g^i(n)$ in the interval $[n/2^i, (n+1)/2^i)$. That is, $f^i(x)$ is a staircase function with the value given by the samples of $g^i(n)$ and intervals which decrease as 2^{-i}. It can be verified that the function is supported on the interval $[0, L-1]$, where L is the length of the filter $g(n)$. Now, for i going to infinity, $f^i(x)$ can converge to a continuous function $g_c(x)$, or a function with finitely many discontinuities, even a fractal function, or not converge at all (see Box 5).

A necessary condition for the iterated functions to converge to a continuous limit is that the filter $G(z)$ should have a sufficient number of zeros at $z = -1$, or half sampling frequency, so as to attenuate repeat spectra [DAU88, DAU90b, RIO91b]. Using this condition, one can construct filters which are both orthogonal and converge to continuous functions with compact support. Such filters are called *regular*, and examples can be found in [DAU88, COH90a, DAU90b, RIO90b, VET90b]. Note that the above condition can be interpreted as a *flatness* condition on the spectrum of $G(z)$ at half sampling frequency. In fact, it can be shown

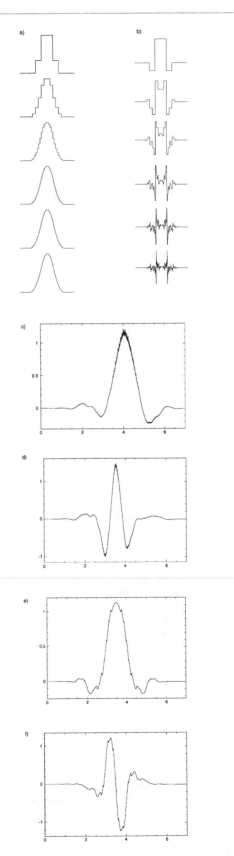

Box 5:
Regular Scaling Filters

It is well known that the structure of computations in a Discrete Wavelet Transform and in an octave-band filter bank are identical. Therefore, besides the different views and interpretations that have been given to them, the main difference lies in the filter design. Wavelet filters are chosen so as to be *regular*. Recall that this means (with the same notation as used in the main text sections on iterated filters), that the piecewise constant function associated with the discrete wavelet sequence $h_j(n)$ of z-transform $G^j(z)H(z^{2j})$ converges (e.g. pointwise), as j indefinitely increases, to a regular limit function $h_c(x)$. Equivalently, the piecewise constant function associated with the discrete "scaling" sequence $g_j(n)$ of z-transform $G^j(z)$ converges to a regular limit function $g_c(x)$. By "regular" we mean that the continuous-time wavelet $h_c(x)$ (or the scaling function $g_c(x)$) is at least continuous, or better, once or twice continuously differentiable. The regularity order is the number of times $h_c(x)$ (or $g_c(x)$) is continuously differentiable. Figures 12a and 12b show two examples, one where $g_c(x)$ is almost three times continuously differentiable and another where $g_j(n)$ diverges with fractal behavior.

Note that there are a number of classical filters, designed for two-band filter banks, which, unlike wavelet filters, are *not* regular. Figures 12c through 12f show two well-known examples: a Johnston filter [JOH80], and a Smith and Barnwell filter [SMI86]. The latter allows perfect reconstruction, while the former does not. Figure 12d shows that the Smith and Barnwell discrete sequences $h_j(n)$ do not tend to regular limit functions, but rather diverge. This is not surprising since the necessary condition that the low-pass filter has a zero at half the sampling frequency is violated (although this filter has 40 dB attenuation in the stop band [SMI86]). This eventually results, when j increases, in small, but rapid oscillations in $h_j(n)$. As for the Johnston filter (Figs. 12e and 12f), it can be shown that the wavelet limit function is continuous but not differentiable.

For wavelet filters, the more regular the limit function, the faster the convergence to this limit [RIO91b] —- and in practice the convergence is very fast. This justifies the study of the limit $h_c(x)$, which is almost attained after a few octaves of a logarithmic decomposition. Since an error in a wavelet coefficient (due e.g. to quantization) results, after reconstruction, in an overall error proportional to a discrete wavelet $h_j(n)$, regularity seems a nice property, e.g., to avoid visible distortion on a reconstructed image [ANI90].

From equations (12), (15) and (16), the knowledge of $g(n)$ suffices to determine the limit $h_c(x)$. Several methods have been developed to estimate the regularity order of $h_c(x)$ from the coefficients $g(n)$. Most are based on Fourier transform techniques [DAU88, COH90b]. Recently, time-domain techniques have been developed which provide optimal estimates [DAU90c, RIO91b].

Fig. 12. (a) Iterated low-pass filter $g_j(n)$ with $g(n) = (1,3,3,1)$, converges to a regular, smooth function. (b) Iterated low-pass filter with $g(n) = (-1,3,3,-1)$ converges to a fractal function (see text). (c) Smith and Barnwell 8-tap lowpass filter [SMI86], iterated 8 times. Divergence occurs, due to rapid oscillations in the temporal waveform. (d) Corresponding continuous-time wavelet (after [RIO90b]). (e) Johnston 8-tap lowpass filter [JOH80], iterated 8 times. The limit function is not differentiable. (f) Corresponding continuous-time wavelet.

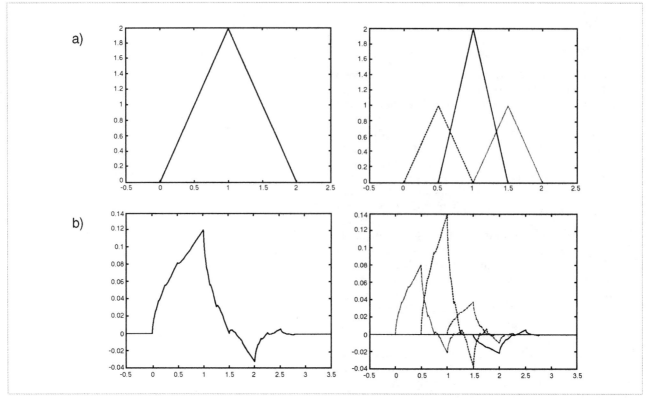

Fig. 13. Scaling functions satisfying two-scale difference equations. (a) the hat function. (b) the D_4 wavelet obtained from a 4-tap regular filter by Daubechies.

[AKA90], [SHE90] that the well-known Daubechies orthonormal filters [DAU88] are deduced from "maximally flat" low-pass filters [HER71]. Note that there are many other choices that behave very differently in terms of phase, selectivity in frequency, and other criteria (see e.g. [DAU90b]). An important issue related to regular filter design is the derivation of simple estimates for the regularity order (see Box 5).

It is still not clear whether regular filters are most adapted to coding schemes [ANI90]. The minimal regularity order necessary for good coding performance of discrete wavelet transform schemes, if needed at all, is also not known and remains a topic for future investigation.

Scaling Functions and Wavelets Obtained from Iterated Filters

Recall that $g_c(x)$ is the final function to which $f^i(x)$ converges. Because it is the product of lowpass filters, the final function is itself lowpass and is called a "*scaling function*" because it is used to go from a fine scale to a coarser scale. Because of the product (14) from which the scaling function is derived, $g_c(x)$ satisfies the following two scale difference equation [DAU90c]:

$$g_c(x) = \sum_{n=-\infty}^{\infty} g(n)\, g_c(2x-n) \qquad (15)$$

Figure 13 shows two such examples. The second one is based on the 4-tap Daubechies filter which is regular and orthogonal to its even translates [DAU88].

So far, we have only discussed the iterated lowpass and its associated scaling function. However, from Fig. 11b, it is clear that a bandpass filter is obtained in the same way, except for a final highpass filter. Therefore, in a fashion similar to (15), the wavelet $h_c(x)$ is obtained as

$$h_c(x) = \sum_{n=-\infty}^{\infty} h(n)\, g_c(2x-n) \qquad (16)$$

that is, it also satisfies a two scale equation.

Now, if the filters $h(n)$ and $g(n)$ form an orthonormal set with respect to even shifts, then the functions $g_c(x-l)$ and $h_c(x-k)$ form an orthonormal set (see Box 6). Because they also satisfy two scale difference equations, it can be shown [DAU88] that the set $h_c(2^{-i}x-k)$, $i,k \in \mathbf{Z}$, forms an orthonormal basis for the set of square integrable functions.

Figure 14 shows two scales and shifts of the 4-tap Daubechies wavelet [DAU88]. While it might not be obvious from the figure, these functions are orthogonal to each other, and together with all scaled and translated versions, they form an orthonormal basis.

Figure 15 shows an orthogonal wavelet based on a length-18 regular filter. It is obviously a much smoother function (actually, it possesses 3 continuous derivatives).

Finally, Fig. 16 shows a biorthogonal set of linear phase wavelets, where the analysis wavelets are orthogonal to the synthesis wavelets. These were obtained from a biorthogonal linear phase filter bank with length-18 regular filters [VET90a, VET90b].

We have shown how regular filters can be used to

The concept of multiresolution approximation of functions was introduced by Meyer and Mallat [MAL89a, MAL89c, MEY90] and provides a powerful framework to understand wavelet decompositions. The basic idea is that of successive approximation, together with that of "added detail" as one goes from one approximation to the next, finer one. We here give the intuition behind the construction.

Assume we have a ladder of spaces such that:

$$...\subset V_2 \subset V_1 \subset V_0 \subset V_{-1} \subset V_{-2} \subset ...$$

with the property that if $f(x) \in V_i$ then $f(x-2^{-i}k) \in V_i$, $k \in \mathbf{Z}$, and $f(2x) \in V_{i-1}$. Call W_i the orthogonal complement of V_i in V_{i-1}. This is written

$$V_{i-1} = V_i \overset{\perp}{\oplus} W_i \qquad \text{(B6.1)}$$

Thus, W_i contains the "detail" necessary to go from V_i to V_{i-1}. Iterating (B6.1), one has

$$V_{i-1} = W_i \overset{\perp}{\oplus} W_{i+1} \overset{\perp}{\oplus} W_{i+2} \overset{\perp}{\oplus} W_{i+3} \overset{\perp}{\oplus} ... \qquad \text{(B6.2)}$$

that is, a given resolution can be attained by a sum of added details.

Now, assume we have an orthonormal basis for V_0 made up of a function $g_c(x)$ and its integer translates. Because $V_0 \in V_{-1}$, $g_c(x)$ can be written in terms of the basis in V_{-1}, i.e., (15) is satisfied:

$$g_c(x) = \sum_n c_n g_c(2x-n)$$

Then it can be verified that the function $h_c(x)$ (16) (with the relation (12)) and its integer translates form an orthonormal basis for W_0. And, because of (B6.2), $h_c(x)$ and its scaled and translated versions form a wavelet basis [MAL89a, MAL89c, MEY90].

The multiresolution idea is now very intuitive. Assume we have an approximation of a signal at a resolution corresponding to V_0. Then a better approximation is obtained by adding the details corresponding to W_0, that is, the projection of the signal in W_0. This amounts to a weighted sum of wavelets at that scale. Thus, by iterating this idea, a square integrable signal can be seen as the successive approximation or weighted sum of wavelets at finer and finer scale.

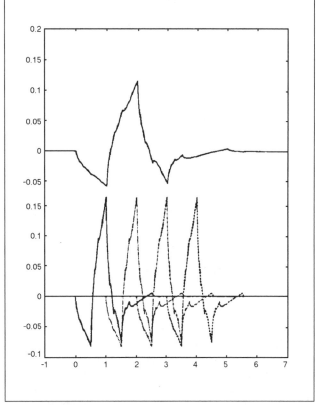

Fig. 14. Two scales of the D_4 wavelet and shifts. This set of functions is orthogonal.

APPLICATIONS OF WAVELETS IN SIGNAL PROCESSING

From the derivation of the wavelet transform as an alternative to the STFT, it is clear that one of the main applications will be in non-stationary signal analysis. While conceptually, the CWT is a classical constant-Q analysis, its simple definition (based on a single function rather than multiple filters) allows powerful analytical derivations and has already lead both to new insights and new theoretical results [WAV89].

Applications of wavelet decompositions in numerical analysis, e.g. for solving partial differential equations, seem very promising because of the "zooming" property which allows a very good representation of discontinuities, unlike the Fourier transform [BEY89].

Perhaps the biggest potential of wavelets has been claimed for signal compression. Since discrete wavelet transforms are essentially subband coding systems, and since subband coders have been successful in speech and image compression, it is clear that wavelets will find immediate application in compression problems. The only difference with traditional subband coders is the fact that filters are designed to be regular (that is, they have many zeroes at $z = 0$ or $z = \pi$). Note that although classical subband filters are not regular (see Box 5 and Fig. 12), they have been designed to have good stopbands and thus are close to being "regular", at least for the first few octaves of subband decomposition.

generate wavelet bases. The converse is also true. That is, orthonormal sets of scaling functions and wavelets can be used to generate perfect reconstruction filter banks [DAU88, MAL89a, MAL89c].

Extension of the wavelet concept to multiple dimensions, which is useful, e.g. for image coding, is shown in Box 7.

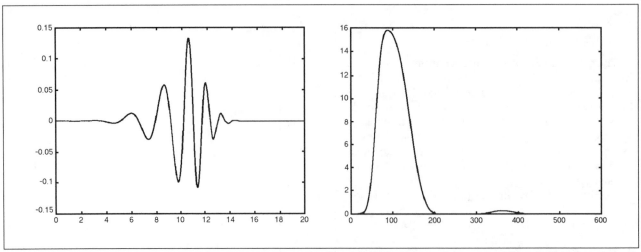

Fig. 15. Orthonormal wavelet generated from a length-18 regular filter [DAU88]. The time function is shown on the left and the spectrum is on the right.

It is therefore clear that drastic improvements of compression will not be achieved so easily simply because wavelets are used. However, wavelets bring new ideas and insights. In this respect, the use of wavelet decompositions in connection with *other* techniques (like vector quantization [ANI90] or multiscale edges [MAL89d]) are promising compression techniques which make use of the elegant theory of wavelets.

New developments, based on wavelet concepts, have already appeared. For example, statistical signal processing using wavelets is a promising field. Multiscale models of stochastic processes [BAS89], [CHO91], and analysis and synthesis of $1/f$ noise [GAC91], [WOR90] are examples where wavelet analysis has been successful. "Wavelet packets" [WIC89], which correspond to arbitrary adaptive tree-structured filter banks, are another promising example.

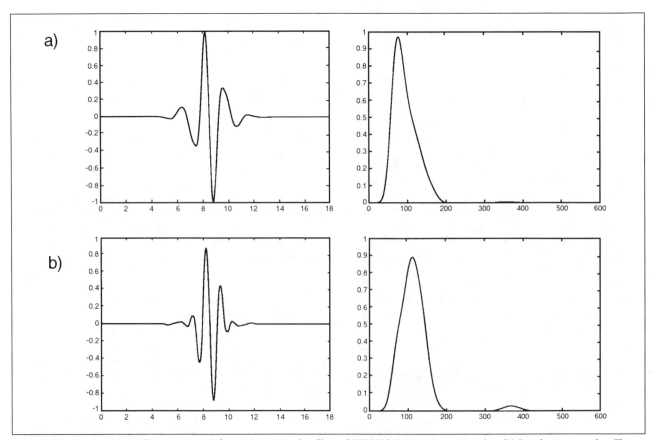

Fig. 16. Biorthogonal wavelets generated from 18-tap regular filters [VET90b]. (a) Analysis wavelet. (b) Synthesis wavelet. The time function is shown on the left and the spectrum is on the right.

Box 7: Multidimensional filter banks and wavelets

In order to apply wavelet decompositions to multi-dimensional signals (e.g., images), multidimensional extensions of wavelets are required. An obvious way to do this is to use "separable wavelets" obtained from products of one-dimensional wavelets and scaling functions [MAL89a, MAL89c, MEY90]. Let us consider the two-dimensional case for its simplicity. Take a scaling function $g_c(x)$ (15) and a wavelet $h_c(x)$ (16). One can construct for two-dimensional functions :

$$g_c(x,y) = g_c(x) \cdot g_c(y)$$

$$h_c^{(1)}(x,y) = g_c(x) \cdot h_c(y)$$

$$h_c^{(2)}(x,y) = h_c(x) \cdot g_c(y)$$

$$h_c^{(3)}(x,y) = h_c(x) \cdot h_c(y)$$

which are orthogonal to each other with respect to integer shifts (this follows from the orthogonality of the one dimensional component). The function $g_c(x,y)$ is a separable two-dimensional scaling function (that is, a lowpass filter) while the functions $h_c^{(i)}(x,y)$ are "wavelets". The set $\{h_c^{(i)}(2^i x\text{-}k, 2^j x\text{-}l), \ i = 1,2,3 \text{ and } j,k,l \in \mathbf{Z}\}$ forms an orthonormal basis for square integrable functions over $\mathbf{R}^2$. This solution corresponds to a separable two-dimensional filter bank with subsampling by 2 in each dimension, that is, overall subsampling by 4 (see Fig. 17).

More interesting (that is, non-trivial) multidimensional wavelet schemes are obtained when non-separable subsampling is used [KOV92]. For example, a non-separable subsampling by 2 of a double indexed signal $x(n_1, n_2)$ is obtained by retaining only samples satisfying:

$$\begin{pmatrix} n_1 \\ n_2 \end{pmatrix} = \begin{pmatrix} 1 & 1 \\ 1 & -1 \end{pmatrix} \begin{pmatrix} u_1 \\ u_2 \end{pmatrix}, \quad u_1, u_2 \in \mathbf{Z} \tag{B7.1}$$

The resulting points are located on a so-called quincunx sublattice of $\mathbf{Z}^2$. Now, one can construct a perfect reconstruction filter bank involving such subsampling because it resembles its one-dimensional counterpart [KOV92]. The subsampling rate is 2 (equal to the determinant of the matrix in (B7.1)), and the filter bank has 2 channels. Iteration of the filter bank on the lowpass branch (see Fig. 18) leads to a discrete wavelet transform, and if the filter is regular (which now depends on the matrix representing the lattice [KOV92]), one can construct non-separable wavelet bases for square integrable functions over $\mathbf{R}^2$ with a resolution change by 2 (and not 4 as in the separable case). An example scaling function is pictured in Fig. 19.

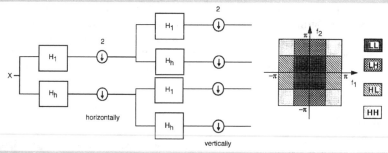

Fig. 17. Separable two-dimensional filter bank corresponding to a separable wavelet basis with resolution change by 4 (2 in each dimension). The partition of the frequency plane is indicated on the right. H_l and H_h stand for low-pass and high-pass filter, respectively.

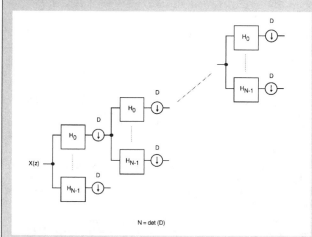

Fig. 18. Iteration of a non-separable filter bank based on non-separable subsampling. This construction leads to non-separable wavelets.

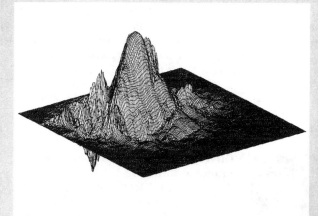

Fig. 19. Two-dimensional non-separable orthonormal scaling function [KOV92] (orthogonality is with respect to integer shifts). The resolution change is by 2 ($\sqrt{2}$ in each dimension). The matrix used for the subsampling is the one given in (B7.1).

CONCLUSION

We have seen that the Short-Time Fourier Transform and the Wavelet Transform represent alternative ways to divide the time-frequency (or time-scale) plane. Two major advantages of the Wavelet Transform are that it can zoom in to time discontinuities and that orthonormal bases, localized in time and frequency, can be constructed. In the discrete case, the Wavelet Transform is equivalent to a logarithmic filter bank, with the added constraint of regularity on the lowpass filter.

The theory of wavelets can be seen as a common framework for techniques that had been developed independently in various fields. This conceptual unification furthers the understanding of the mechanisms involved, quantifies trade-offs, and points to new potential applications. A number of questions remain open, however, and will require further investigations (e.g., what is the "optimal" wavelet for a particular application?).

While some see wavelets as a very promising brand new theory [CIP90], others express some doubt that it represents a major breakthrough. One reason for skepticism is that the concepts have been around for some time, under different names. For example, wavelet transforms can be seen as constant-Q analysis [YOU78], wide-band cross-ambiguity functions [SPE67, AUS90], Frazier-Jawerth transforms [FRA86], perfect reconstruction octave-band filter banks [MIN85, SMI86], or a variation of Laplacian pyramid decomposition [BUR83], [BUR89]!

We think that the interest and merit of wavelet theory is to unify all this into a common framework, thereby allowing new ideas and developments.

ACKNOWLEDGMENTS

The authors would like to thank C. Herley for many useful suggestions and for generating the continuous STFT and WT plots; and Profs. F. Boudreaux-Bartels, M.J.T. Smith and Dr. P. Duhamel for useful suggestions on the manuscript. We thank B. Shakib (IBM) for creating the three-dimensional color rendering of phase and magnitude of wavelet transforms (so-called "phasemagrams") used in the cover picture and elsewhere; C.A. Pickover (IBM) is thanked for his 3D display software and J.L. Mannion for his help on software tools. The second author would like to acknowledge support by NSF under grants ECD-88-11111 and MIP-90-14189.

Olivier Rioul was born in Strasbourg, France on July 4, 1964. He received diplomas in Electrical Engineering from the Ecole Polytechnique, Palaiseau, France, and from Telecom University, Paris, in 1987 and 1989, respectively.

Since 1989, he has been with the Centre National d'Etudes des Télécommunications (CNET), Issy-Les-Moulineaux, France, where he is completing work in the Ph.D. degree in Signal Processing at Télécom University, specializing in wavelet theory, image coding, and fast signal algorithms.

Martin Vetterli was born in Switzerland in 1957. He received the Dip. El.-Ing. degree from the Eidgenössische Technische Hochschule Zürich, Switzerland, in 1981; the Master of Science degree from Stanford University, Stanford CA, in 1982; and the Doctorat ès Science Degree from the Ecole Polytechnique Fédérale de Lausanne, Switzerland, in 1986.

In 1982, he was a Research Assistant at Stanford University, and from 1983 to 1986 he was a researcher at the Ecole Polytechnique. He has worked for Siemens and AT&T Bell Laboratories. In 1986, he joined Columbia University in New York where he is currently associate professor of Electrical Engineering, member of the Center for Telecommunications Research, and codirector of the Image and Advanced Television Laboratory.

He is a senior member of the IEEE, a member of SIAM and ACM, a member of the MDSP committee of the IEEE Signal Processing Society, and of the editorial boards of *Signal Processing* and *Image Communication*. He received the Best Paper Award of EURASIP in 1984 for his paper on multidimensional subband coding, and the Research Prize of the Brown Boveri Corporation (Switzerland) in 1986 for his thesis. His research interests include multirate signal processing, wavelets, computational complexity, signal processing for telecommunications and digital video processing.

REFERENCES

[AKA90] A.N. Akansu, R.A. Haddad, and H. Caglar, "The Binomial QMF-Wavelet Transform for Multiresolution Signal Decomposition," submitted *IEEE Trans. Signal Proc.*, 1990.

[ALL77] J.B. Allen and L.R. Rabiner, "A Unified Approach to Short-Time Fourier Analysis and Synthesis," *Proc. IEEE*, Vol. 65, No. 11, pp. 1558-1564, 1977.

[ANI90] M. Antonini, M. Barlaud, P. Mathieu and I. Daubechies, "Image Coding Using Vector Quantization in the Wavelet Transform Domain," in *Proc. 1990 IEEE Int. Conf. Acoust., Speech, Signal Proc.*, Albuquerque, NM, Apr.3-6, 1990, pp. 2297-2300.

[AUS90] L. Auslander and I. Gertner, "Wide-Band Ambiguity Function and a.x+b group," in Signal Processing, Part I: Signal Processing Theory, L. Auslander, T. Kailath, S. Mitter eds., Institute for Mathematics and its Applications, Vol. 22, Springer Verlag, New York, pp.1-12, 1990.

[BAS89] M. Basseville and A. Benveniste, "Multiscale Statistical Signal Processing," in Proc. 1989 IEEE Int. Conf. Acoust., Speech, Signal Proc., Glasgow, Scotland, pp. 2065-2068, May 23-26, 1989.

[BER88] J. Bertrand and P. Bertrand, "Time-Frequency Representations of Broad-Band Signals," in *Proc. 1988 IEEE Int. Conf. Acoust., Speech, Signal Proc.*, New York, NY, Apr.11-14, 1988, pp.2196-2199.

[BEY89] G. Beylkin, R. Coifman and V. Rokhlin, "Fast Wavelet Transforms and Numerical Algorithms. I," submitted, Dec. 1989.

[BOU85] G. F. Boudreaux-Bartels, "Time-Varying Signal Processing Using the Wigner Distribution Time-Frequency Signal Representation," in *Adv. in Geophysical Data Proc.*, Vol. 2, pp. 33-79, Jai Press Inc., 1985.

[BUR83] P.J. Burt and E.H. Adelson, "The Laplacian Pyramid as a Compact Image Code," *IEEE Trans. on Com.*, Vol. 31, No.4, pp. 532-540, April 1983.

[BUR89] P.J. Burt, "Multiresolution Techniques for Image Representation, Analysis, and 'Smart' Transmission," Proc. SPIE Conf. on Visual Communication and Image Processing., pp. 2-15, Philadelphia, PA, Nov. 1989.

[CAL64] A. Calderón, "Intermediate Spaces and Interpolation, the Complex Method," Studia Math., Vol. 24, pp. 113-190, 1964.

[CHO91] K.C. Chou, S. Golden, and A.S. Willsky, "Modeling and Estimation of Multiscale Stochastic Processes," in Proc. *1991 IEEE Int. Conf. Acoust., Speech, Signal Proc.*, Toronto, Ontario, Canada, pp. 1709-1712, May 14-17, 1991.

[CIP90] B.A. Cipra, "A New Wave in Applied Mathematics," *Science, Research News*, Vol. 249, August 24, 1990.

[CLA80] T.A.C.M. Classen and W.F.G. Mecklenbräuer, "The Wigner Distribution — A Tool for Time-Frequency Signal Analysis, Part I, II, III," *Philips J. Res.*, Vol.35, pp. 217-389, 1980.

[COH66] L. Cohen, "Generalized Phase-Space Distribution Functions," *J. Math. Phys.*, Vol. 7, No. 5, pp. 781-786, 1966.

[COH89] L. Cohen, "Time-Frequency Distribution - A Review," *Proc. IEEE*, Vol. 77, No. 7, pp. 941-981, 1989.

[COH90a] A. Cohen, I. Daubechies and J.C. Feauveau, "Biorthogonal Bases of Compactly Supported Wavelets," to appear in *Comm. Pure and Applied Math.*

[COH90b] A. Cohen, "Construction de Bases d'Ondelettes Hölderiennes," *Revista Matematica Iberoamericana*, Vol.6, No. 3 y 4, 1990.

[CRI76] A. Croisier, D. Esteban, and C. Galand, "Perfect Channel Splitting by Use of Interpolation, Decimation, Tree Decomposition Techniques," Int. Conf. on Information Sciences/Systems, Patras, pp. 443-446, Aug. 1976.

[CRO76] R.E. Crochiere, S.A. Weber, and J.L. Flanagan, "Digital Coding of Speech in Subbands," *Bell Syst. Tech. J.*, Vol.55, pp.1069-1085, Oct.1976.

[CRO83] R.E. Crochiere, and L.R. Rabiner, Multirate Digital Signal Processing, Prentice-Hall, Englewood Cliffs, NJ, 1983.

[DAU88] I. Daubechies, "Orthonormal Bases of Compactly Supported Wavelets," *Comm. in Pure and Applied Math.*, Vol.41, No.7, pp.909-996, 1988.

[DAU90a] I. Daubechies, "The Wavelet Transform, Time-Frequency Localization and Signal Analysis," *IEEE Trans. on Info. Theory*, Vol. 36, No.5, pp.961-1005, Sept. 1990.

[DAU90b] I. Daubechies, "Orthonormal Bases of Compactly Supported Wavelets II. Variations on a Theme," submitted to *SIAM J. Math. Anal.*

[DAU90c] I. Daubechies and J.C. Lagarias, "Two-Scale Difference Equations II. Local Regularity, Infinite Products of Matrices and Fractals," submitted to *SIAM J. Math. Anal*, 1990.

[DUF52] R.J. Duffin and A.C. Schaeffer, "A Class of Nonharmonic Fourier Series," *Trans. Am. Math. Soc.*, Vol. 72, pp. 341-366, 1952.

[EST77] D. Esteban and C. Galand, "Application of Quadrature Mirror Filters to Split-Band Voice Coding Schemes," Int. Conf. Acoust., Speech, Signal Proc., Hartford, Connecticut, pp. 191-195, May 1977.

[FLA89] P. Flandrin, "Some Aspects of Non-Stationary Signal Processing with Emphasis on Time-Frequency and Time-Scale Methods," in [WAV89], pp.68-98, 1989.

[FLA90] P. Flandrin and O. Rioul, "Wavelets and Affine Smoothing of the Wigner-Ville Distribution," in *Proc. 1990 IEEE Int. Conf. Acoust., Speech, Signal Proc.*, Albuquerque, NM, April 3-6, 1990, pp. 2455-2458.

[FOU88] J. B. J. Fourier, "Théorie Analytique de la Chaleur," in *Oeuvres de Fourier*, tome premier, G.Darboux, Ed., Paris: Gauthiers-Villars, 1888.

[FRA28] P. Franklin, "A Set of Continuous Orthogonal Functions," *Math. Annal.*, Vol.100, pp.522-529, 1928.

[FRA86] M. Frazier and B. Jawerth, "The φ-Transform and Decomposition of Distributions," *Proc. Conf. Function Spaces and Appl.*, Lund 1986, Lect. Notes Math., Springer.

[GAB46] D. Gabor, "Theory of Communication," *J. of the IEE*, Vol.93, pp.429-457, 1946.

[GAC91] N. Gache, P. Flandrin, and D. Garreau, "Fractal Dimension Estimators for Fractional Brownian Motions," in Proc. *1991 IEEE Int. Conf. Acoust., Speech, Signal Proc.*, Toronto, Ontario, Canada, pp. 3557-3560, May 14-17, 1991.

[GOU84] P. Goupillaud, A. Grossmann and J. Morlet, "Cycle-Octave and Related Transforms in Seismic Signal Analysis," *Geoexploration*, Vol.23, pp.85-102, Elsevier Science Pub., 1984/85.

[GRO84] A. Grossmann and J. Morlet, "Decomposition of Hardy Functions into Square Integrable Wavelets of Constant Shape," *SIAM J.Math.Anal.*, Vol.15, No.4, pp.723-736, July 1984.

[GRO89] A. Grossmann, R. Kronland-Martinet, and J. Morlet, "Reading and Understanding Continuous Wavelet Transforms," in [WAV89], pp.2-20, 1989.

[HAA10] A. Haar, "Zur Theorie der Orthogonalen Funktionensysteme," [in German] *Math. Annal.*, Vol. 69, pp. 331-371, 1910.

[HEI90] C.E. Heil, "Wavelets and Frames," in Signal Processing, Part I: Signal Processing Theory, L. Auslander, et al. eds., IMA, Vol. 22, Springer , New York, pp.147-160, 1990.

[HER71] O. Herrmann, "On the Approximation Probem in NonRecursive Digital Filter Design," *IEEE Trans. Circuit Theory*, Vol. CT-18, No. 3, pp. 411-413, May 1971.

[IT92] *IEEE Transactions on Information Theory*, Special issue on Wavelet Transforms and Multiresolution Signal Analysis, to appear January, 1992.

[JOH80] J.D. Johnston, "A Filter Family Designed for Use in Quadrature Mirror Filter Banks," in *Proc. 1980 IEEE Int. Conf. Acoust., Speech, Signal Proc.*, pp. 291-294, Apr. 1980.

[KAD91] S. Kadambe and G.F. Boudreaux-Bartels, "A Comparison of the Existence of 'Cross Terms' in the Wavelet Transform, Wigner Distribution and Short-Time Fourier Transform," Submitted *IEEE Trans. Signal Proc.*, Revised Jan. 1991.

[KOV92] J. Kovacevic and M. Vetterli, "Non-separable Multidimensional Perfect Reconstruction Filter Banks and Wavelet Bases for R^n," *IEEE Trans. on Info. Theory*, Special Issue on wavelet transforms and multiresolution signal analysis, to appear Jan. 1992.

[LIT37] J. Littlewood and R. Paley, "Theorems on Fourier Series and Power Series," *Proc. London Math. Soc.*, Vol.42, pp. 52-89, 1937.

[MAL89a] S. Mallat, "A Theory for Multiresolution Signal Decomposition: the Wavelet Representation," *IEEE Trans. on Pattern Analysis and Machine Intell*, Vol.11, No. 7, pp.674-693, July 1989.

[MAL89b] S. Mallat, "Multifrequency Channel Decompositions of Images and Wavelet Models," *IEEE Trans. Acoust., Speech, Signal Proc.*, Vol. 37, No.12, pp.2091-2110, December 1989.

[MAL89c] S. Mallat, "Multiresolution Approximations and Wavelet Orthonormal Bases of $L^2(R)$," *Trans. Amer. Math. Soc.*, Vol.315, No.1, pp.69-87, September 1989.

[MAL89d] S. Mallat and S. Zhong, "Complete Signal Representation with Multiscale Edges," submitted to *IEEE Trans. Pattern Analysis and Machine Intell*, 1989.

[MEY89] Y. Meyer, "Orthonormal Wavelets," in [WAV89], pp. 21-37, 1989.

[MEY90] Y. Meyer, *Ondelettes et Opérateurs*, Tome I. *Ondelettes*, Herrmann ed., Paris, 1990.

[MIN85] F. Mintzer, "Filters for Distortion-Free Two-Band Multirate Filter Banks," *IEEE Trans. on Acoust., Speech, Signal Proc.*, Vol.33, pp.626-630, June 1985.

[POR80] M.R. Portnoff, "Time-Frequency Representation of Digital Signals and Systems Based on Short-Time Fourier Analysis," *IEEE Trans. on Acoust., Speech, Signal Proc.*, Vol.28, pp.55-69, Feb. 1980.

[RAM88] T.A. Ramstad and T. Saramäki, "Efficient Multirate Realization for Narrow Transition-Band FIR Filters," in *Proc. 1988 IEEE Int. Symp. Circuits and Systems*, Helsinki, Finland, pp. 2019-2022, 1988.

[RIO90a] O. Rioul and P. Flandrin, "Time-Scale Energy Distributions: A General Class Extending Wavelet Transforms," to appear in *IEEE Trans. Signal Proc.*

[RIO90b] O. Rioul, "A Discrete-Time Multiresolution Theory Unifying Octave-Band Filter Banks, Pyramid and Wavelet Transforms," submitted to *IEEE Trans. Signal Proc.*

[RIO91a] O. Rioul and P. Duhamel, "Fast Algorithms for Discrete and Continuous Wavelet Transforms," submitted to *IEEE Trans. Information Theory*, Special Issue on Wavelet Transforms and Multiresolution Signal Analysis.

[RIO91b] O. Rioul, "Dyadic Up-Scaling Schemes: Simple Criteria for Regularity," submitted to *SIAM J. Math. Anal.*

[ROS84] A. Rosenfeld ed., Multiresolution Techniques in Computer Vision, Springer-Verlag, New York 1984.

[SMI86] M.J.T. Smith and T.P. Barnwell, "Exact Reconstruction for Tree-Structured Subband Coders," *IEEE Trans. on Acoust., Speech and Signal Proc.*, Vol. ASSP-34, pp. 434-441, June 1986.

[SHE90] M.J. Shensa, "The Discrete Wavelet Transform: Wedding the à Trous and Mallat Algorithms," submitted to *IEEE Trans. on Acoust., Speech, and Signal Proc.*, 1990.

[SPE67] J.M. Speiser, "Wide-Band Ambiguity Functions," *IEEE Trans. on Info. Theory*, pp. 122-123, 1967.

[VAI87] P.P. Vaidyanathan, "Quadrature Mirror Filter Banks, M-band Extensions and Perfect-Reconstruction Techniques," *IEEE ASSP Magazine*, Vol. 4, No. 3, pp.4-20, July 1987.

[VAI88] P.P. Vaidyanathan and P.-Q. Hoang, "Lattice Structures for Optimal Design and Robust Implementation of Two-Band Perfect Reconstruction QMF Banks," *IEEE Trans. on Acoust., Speech and Signal Proc.*, Vol. ASSP-36, No. 1, pp.81-94, Jan. 1988.

[VAI89] P.P. Vaidyanathan and Z. Doganata, "The Role of Lossless Systems in Modern Digital Signal Processing," *IEEE Trans. Education*, Special issue on Circuits and Systems, Vol. 32, No.3, Aug. 1989, pp.181-197.

[VET86] M. Vetterli, "Filter Banks Allowing Perfect Reconstruction," *Signal Processing*, Vol.10, No.3, April 1986, pp.219-244.

[VET89] M. Vetterli and D. Le Gall, "Perfect Reconstruction FIR Filter Banks: Some Properties and Factorizations, *IEEE Trans. on Acoust., Speech Signal Proc.*, Vol.37, No.7, pp.1057-1071, July 1989.

[VET90a] M. Vetterli and C. Herley "Wavelets and Filter Banks: Relationships and New Results," in *Proc. 1990 IEEE Int. Conf. Acoust., Speech, Signal Proc.*, Albuquerque, NM, pp. 1723-1726, Apr. 3-6, 1990.

[VET90b] M. Vetterli and C. Herley, "Wavelets and Filter Banks: Theory and Design," to appear in *IEEE Trans. on Signal Proc.*, 1992.

[WAV89] *Wavelets, Time-Frequency Methods and Phase Space*, Proc. Int. Conf. Marseille, France, Dec. 14-18, 1987, J.M. Combes et al. eds., Inverse Problems and Theoretical Imaging,

Springer, 315 pp., 1989.

[WIC89] M.V. Wickerhauser, "Acoustic Signal Compression with Wave Packets," preprint Yale University, 1989.

[WOO53] P.M. Woodward, *Probability and Information Theory with Application to Radar*, Pergamon Press, London, 1953.

[WOR90] G.W. Wornell, "A Karhunen-Loève-like Expansion for 1/f Processes via Wavelets," *IEEE Trans. Info. Theory*, Vol. 36, No. 4, pp.859-861, July 1990.

[YOU78] J.E. Younberg, S.F. Boll, "Constant-Q Signal Analysis and Synthesis," in Proc. 1978 IEEE Int. Conf. on Acoust., Speech, and Signal Proc., Tulsa, OK, pp. 375-378, 1978.

EXTENDED REFERENCES

History of Wavelets: see [HAA10, FRA28, LIT37, CAL64, YOU78, GOU84].

Books on Wavelets: (see also [WAV89, MEY90])

I. Daubechies, *Ten Lectures on Wavelets*, CBMS, SIAM publ., to appear.

Wavelets and their Applications, R.R. Coifman, I. Daubechies, S. Mallat, Y. Meyer scientific eds., L.A. Raphael, M.B. Ruskai managing eds., Jones and Bertlel pub., to appear, 1991.

Tutorials on Wavelets: (see also [FLA89, GRO89, MAL89b, MEY89])

R.R. Coifman, "Wavelet Analysis and Signal Processing," in Signal Processing, Part I: Signal Processing Theory, L. Auslander et al. eds., IMA, Vol. 22, Springer, New York, 1990.

C.E. Heil and D.F. Walnut, "Continuous and Discrete Wavelet Transforms," *SIAM Review*, Vol. 31, No. 4, pp 628-666, Dec. 1989.

Y. Meyer, S. Jaffard, O. Rioul, "L'Analyse par Ondelettes," [in French]*Pour La Science*, No.119, pp.28-37, Sept 1987.

G. Strang, "Wavelets and Dilation Equations: A Brief Introduction," *SIAM Review*, Vol. 31, No. 4, pp. 614-627, Dec. 1989.

Mathematics, Mathematical Physics and Quantum Mechanics: (see also [GRO84, DAU88, MEY90])

G. Battle, "A Block Spin Construction of Ondelettes, II. The Quantum Field Theory (QFT) Connection," Comm. Math. Phys., Vol. 114, pp. 93-102, 1988.

W.M. Lawton, "Necessary and Sufficient Conditions for Constructing Orthonormal Wavelet Bases," Aware Tech. Report # AD900402.

P.G. Lemarié and Y. Meyer, "Ondelettes et Bases Hilbertiennes," [in French] *Revista Matematica Iberoamericana*, Vol.2, No.1&2, pp.1-18, 1986.

T. Paul, "Affine Coherent States and the Radial Schrödinger Equation I. Radial Harmonic Oscillator and the Hydrogen Atom," to appear in *Ann. Inst. H.Poincaré* .

H.L. Resnikoff, "Foundations or Arithmeticum Analysis: Compactly Supported Wavelets and the Wavelet Group," Aware Tech. Report # AD890507.

Regular Wavelets: see [DAU88, DAU90b, DAU90c, COH90b, RIO91b]

Computer-Aided Geometric Design using Regular Interpolators:

S.Dubuc, "Interpolation Through an Iterative Scheme," *J. Math. Analysis Appl.*, Vol.114, pp.185-204, 1986.

N. Dyn and D. Levin, "Uniform Subdivision Schemes for the Generation of Curves and Surfaces," Constructive Approximation, to appear.

Numerical Analysis: (see also [BEY89])

R.R. Coifman, "Multiresolution Analysis in Nonhomogeneous Media," in [WAV89], pp. 259-262, 1989.

V. Perrier, "Toward a Method to Solve Partial Differential Equations Using Wavelet Bases," in [WAV89], pp. 269-283, 1989.

Multiscale Statistical Signal Processing: see [BAS89, CHO91].

Fractals, Turbulence: (see also [GAC91, WOR90])

A. Arnéodo, G. Grasseau, and M. Holschneider, "Wavelet Transform of Multifractals," *Phys. Review Letters*, Vol.61, No.20, pp.2281-2284, 1988.

F. Argoul, A. Arnéodo, G. Grasseau, Y. Gagne, E.J. Hopfinger, and U. Frisch, "Wavelet Analysis of Turbulence Reveals the Multifractal Nature of the Richardson Cascade," *Nature*, Vol.338, pp.51-53, March 1989.

One-Dimensional Signal Analysis: (see also [GRO89], [WIC-89])

C. D'Alessandro and J.S. Lienard, "Decomposition of the Speech Signal into Short-Time Waveforms Using Spectral Segmentation," in *Proc. 1988 IEEE Int. Conf. Acoust., Speech, Signal Proc.*, New York, Apr.11-14, 1988, pp.351-354.

S. Kadambe and G.F. Boudreaux-Bartels, "A Comparison of Wavelet Functions for Pitch Detection of Speech Signals," in *Proc. 1991 IEEE Int. Conf. Acoust., Speech, Signal Proc.*, Toronto, Ontario, Canada, pp. 449-452, May. 14-17, 1991.

R. Kronland-Martinet, J. Morlet, and A. Grossmann, "Analysis of Sound Patterns Through Wavelet Transforms," *Int. J. Pattern Recognition and Artificial Intelligence*, Vol.1, No.2, pp. 273-302, pp.97-126, 1987.

J.L. Larsonneur and J. Morlet, "Wavelets and Seismic Interpretation," in [WAV89], pp.126-131, 1989.

F. B. Tuteur, "Wavelet Transformations in Signal Detection," in *Proc. 1988 IEEE Int. Conf. Acoust., Speech, Signal Proc.*, New York, NY, Apr. 11-14, 1988, pp.1435-1438. Also in [WAV89], pp. 132-138, 1989.

Radar/Sonar, Ambiguity Functions: see e.g. [AUS90, SPE67, WOO53]

Time-Scale Representations: see [BER88, FLA89, FLA90, RIO90a].

Filter Bank Theory: (see also [CRI76, CRO76, EST77, CRO83, MIN85, SMI86, VAI87, VAI88, VAI89, VET86, VET89])

J.D. Johnston, "A Filter Family Designed for Use in Quadrature Mirror Filter Banks," *Proc. ICASSP-80*, pp.291-294, April 1980.

T.Q. Nguyen and P.P. Vaidyanathan, "Two-Channel Perfect-Reconstruction FIR QMF Structures Which Yield Linear-Phase Analysis and Synthesis Filters," *IEEE Trans. Acoust., Speech, Signal Processing*, Vol. ASSP-37, No. 5, pp.676-690, May 1989.

M.J.T. Smith and T.P. Barnwell, "A New Filter Bank Theory for Time-Frequency Representation," *IEEE Trans. on Acoust., Speech and Signal Proc.*, Vol. ASSP-35, No.3, March 1987, pp. 314-327.

P.P. Vaidyanathan, *Multirate Filter Banks*, Prentice Hall, to appear.

Pyramid Transforms: see [BUR83, BUR89, ROS84].

Multidimensional Filter Banks: (see also [KOV92])

G. Karlsson and M. Vetterli, "Theory of Two-Dimensional Multirate Filter Banks," *IEEE Trans. on Acoust., Speech, Signal Proc.*, Vol.38, No.6, pp.925-937, June 1990.

M. Vetterli, "Multi-Dimensional Subband Coding: Some Theory and Algorithms," *Signal Processing*, Vol. 6, No.2, pp. 97-112, Feb. 1984.

M. Vetterli, J. Kovacevic and D. Le Gall, "Perfect Reconstruction Filter Banks for HDTV Representation and Coding," *Image Communication*, Vol.2, No.3, Oct.1990, pp.349-364.

E. Viscito and J. Allebach, "The Analysis and Design of Multidimensional FIR Perfect Reconstruction Filter Banks for Arbitrary Sampling Lattices," *IEEE Trans. Circuits and Systems*, Vol.38, pp.29-42, Jan. 1991.

Multidimensional Wavelets: (see also [ANI90, KOV92, MAL89a, MAL89b, MAL89c, MAL89d]).

J.C. Feauveau, "Analyse Multirésolution pour les Images avec un Facteur de Résolution 2," [in French], Traitement du Signal, Vol. 7, No. 2, pp. 117-128, July 1990.

K. Gröchenig and W.R. Madych, "Multiresolution analysis, Haar bases and self-similar tilings of R^n," submitted to *IEEE Trans. on Info. Theory*, Special Issue on wavelet transforms and multiresolution signal analysis, Jan.1992.

SPEECH AND IMAGE SIGNAL COMPRESSION WITH WAVELETS

W. Kinsner and A. Langi

Department of Electrical & Computer Engineering
University of Manitoba
Winnipeg, Manitoba, Canada R3T 2N2
eMail: kinsner@ee.UManitoba.CA
Tel.: (204) 474-6490; FAX: (204) 275-0261

ABSTRACT

This paper presents a review and some experimental results of a time-frequency multiresolution analysis based on wavelets, as it applies to speech/audio and image/video signal compression. It compares the wavelet analysis to the traditional short-window techniques used in signal compression. The performance of the discrete wavelet transform in terms of the bit rates and signal quality is comparable to other good techniques such as the discrete cosine transform (DCT) for images and code-excited linear predictive coding (CELP) for speech, but with much less computational burden. Experiments with Lena's image and Daubechies 4-coefficient wavelet show that truncation of wavelet coefficients as high as 90% still produces 30dB peak signal-to-noise ratio (PSNR) quality. This is better than DCT. In an experiment on a male spoken sentence, the scheme reaches 12.82 dB segmental signal-to-noise ratio (SEGSNR) at less than 4.8 kbit/s rate. In comparison, the state-of-the-art CELP coding at 4.8 kbit/s can attain SEGSNR between 10-13 dB. Other experiments with images and Haar 2-coefficient wavelet are also highlighted.

1. INTRODUCTION

1.1 Why Is Coding of Speech and Image Special?

Speech and vision belong to the fundamental human activities involved in the perception of and interaction with the surrounding world. Both speech and images are highly redundant in time (speech, wideband speech, and music) and space (still images) or even space and time (moving images or video). Consequently, significant compression of the signals can be achieved by either scalar compression or vector quantization, with the latter giving improved performance for a given bit rate, or a lower bit rate for a given performance [GeGr92].

The analysis of speech and images is not simple because speech is nonstationary in time while images are nonperiodic in space. The problem is compounded by very rapid transitions that occur in stop consonants in speech and edges in images. The application of standard spectral techniques to such signals is not appropriate.

Speech and image coders and decoders can be treated jointly because the reconstructed signals are judged subjectively by the human aural and visual systems. Therefore, perceptual limitation of the ear and eye should be included in the encoding process to improve performance of the encoding schemes. For example, noise does not affect the perceived speech and image uniformly; instead the degradation is a function of noise frequency. This is in contrast to the older schemes based on the signal alone. Thus, digital coders must be evaluated based not only on their *bit rate*, the *delay* associated with the coding time, and the *complexity* in terms of processing and storage, but also on the *quality* of the reconstructed signal based on perceptual limitations.

1.2 Traditional Spectral Analysis

Wavelets are emerging as a competitor to Fourier, Walsh, Haar and other orthogonal bases whenever either nonstationary (short-window) or transient signal analysis must be performed. This applies to all signals that are nonstationary in time (e.g., narrow-band speech and wide-band audio) or in space (e.g., still images) or even space and time (e.g., video). Wavelets can also compete whenever some data must be either updated frequently or guessed (erroneous or missing data). This is possible because wavelet analysis allows *zooming* on segments of data or details in data (local analysis), thus concentrating on the window of interest, or avoiding errors and gaps in data. The zooming property stems from the variable-resolution property of wavelets, which is not available in Fourier analysis (fixed-resolution and infinite extent of complex exponentials).

Transform techniques such as the discrete Fourier transform (DFT) or discrete cosine transform (DCT), and subband techniques such as the conjugate quadrature-mirror filterbank (QMF) are suitable for stationary signal analysis. However, they are not suitable for analysis of nonstationary signals such as speech and audio (time-frequency) or images (space-frequency), or video (time-space-frequency), and for nonlinear perceptual distortion criteria. Consequently, these techniques have recently been extended to QMF trees with unequal-bandwidth branches and subband-DFT hybrids. A much more promising approach to time-frequency analysis is offered by wavelets [Kins91].

1.3 General Features of Wavelets

The basis vectors of a DFT are the sinusoids and cosinusoids of various frequencies and constant time support. This makes them dissimilar for different frequencies. On the other hand, wavelets are self similar in that they preserve their shape for low and high frequencies by *dilating*, i.e., either stretching the time base at low frequencies or contacting the time base for higher frequencies. This time-frequency characteristic of wavelets makes them naturally suitable not only for many signals whose short-time bursts involve high frequencies, and long-term events contain lower frequencies, but also for time varying events. Wavelets are also naturally suitable for images with either very uneven intensity histograms or largely uniform texture except for sharp transitions between them.

2. SHORT-WINDOW FOURIER ANALYSIS

An analysis of nonstationary signals requires a time (or space) window sliding over the signal in order to produce frequency analysis that is time (or space) dependent. This approach is called the windowed Fourier transform (WFT). Thus, instead of the usual Fourier transform $\hat{f}(\omega)$ of a signal $f(x)$

$$\hat{f}(\omega) = \int_{-\infty}^{+\infty} f(x)\, e^{-j\omega x}\, dx \tag{1}$$

a frequency transform at a given time (or space) is now appropriate

$$\hat{f}(\omega,u) = \int_{-\infty}^{+\infty} f(x)\, g_{\omega,u}(x)\, dx \tag{2}$$

where ω is the angular frequency, and u is the location of the window in time or space x. The simplest form of g is the product of the original complex exponential with a window

$$g_{\omega,u}(x) = g(x-u)\, e^{-j\omega x} \tag{3}$$

This expression generates a family of waveforms, as shown in Fig. 1. It is seen that although the waveforms have the same bandwidth, they are not self similar because the number of cycles in the *fixed* window increases with frequency. This is the *Gabor transform* (in recognition of Denis Gabor's early use of this form with a Gaussian window in his work on radar signals). This transform may lead to instabilities when reconstructing the original signal. A thorough study of completeness and stability of a discrete WFT is presented by Daubechies [Daub88] and Mallat [Mall89b]. If the Gabor transform function is modified to include dilation, then the function is called the *Gabor wavelet* (notice that the two functions are entirely different).

3. WAVELET ANALYSIS

3.1 Mallat Wavelet and Multiresolution Analysis

The problems with the WFT may be alleviated by the wavelet transform, as formulated by Grossman, Meyer, Coifman, Mallat and Daubechies. The self similarity can be achieved if the complex exponential is replaced by a dilation operator, s, acting on a mother (prototype or seed) wavelet $\psi(t)$ [Mall89b]

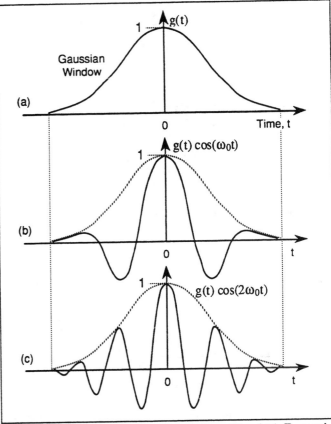

Fig. 1. Gabor wavelet. (a) Gaussian window. (b) and (c) Two real-part wavelets with single and double frequencies.

$$\psi_{s,u}(x) = \sqrt{s}\, \psi[\frac{(x-u)}{s}] \tag{4}$$

where s is the scaling factor, and u is the spatial displacement of the wavelet. Such wavelets must satisfy several conditions, including finite energy, continuity, integrability, no dc component, and minimum time-bandwidth product [Tute88], [Mall89b]. Figure 2a shows the mother wavelet (for s=1 and x−u=0). A dilated child wavelet is shown in Fig. 2b (s<1), and a contracted wavelet is shown in Fig. 2c (s>1).

Substituting Eq. 4 into 2 results in a wavelet transform (WT) denoted by Wf(s,u)

$$Wf(s,u) = \int_{-\infty}^{+\infty} f(x)\, \sqrt{s}\, \psi[\frac{(x-u)}{s}]\, dx \tag{5a}$$

or in the equivalent inner product form

$$Wf(s,u) = \langle f(x),\, \psi_s(x-u)\rangle \tag{5b}$$

The WT can be viewed as a filtering of the f(x) with a bandpass filter whose impulse response is related to $\psi_s(x-u)$.

It is important to understand how wavelets are scaled in the frequency domain. The Fourier transform of the wavelet given by Eq. 4 is

$$\hat{\psi}_s(\omega) = \frac{1}{\sqrt{s}}\hat{\psi}(s\omega) \tag{6}$$

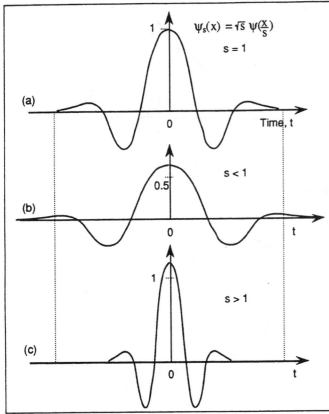

Fig. 2. Mallat wavelet in space (or time) domain. (a) Mother wavelet. (b) Dilated wavelet. (c) Contracted wavelet.

Figure 3 shows how the amplitude and bandwidth vary with the scale factor (though the bandwidth is the same for all wavelets on a logarithmic scale). It is seen that the resolution of the WT varies in the spatial and frequency domains. This is in contrast with the AFT which has a fixed resolution in both domains [Mall89b]. This variable-resolution property allows the WT to focus on irregularities in a signal and represent them locally.

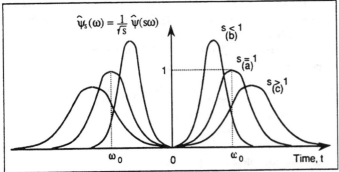

Fig. 3. Mallat wavelet in frequency domain. (a) Mother wavelet. (b) Dilated wavelet. (c) Contracted wavelet.

To illustrate this variable resolution, let us assume that σ_x be the standard deviation of the $|\psi(x)|^2$ around zero, and σ_ω be the rms bandwidth around ω_0. The spatial resolution is scaled by σ_x/s, while the frequency resolution is scaled as $s\sigma_\omega$, as shown in Fig. 4. For s<1 the spatial resolution is large, but the frequency resolution is narrow so that coarse features (context) are shown. For s>1, the spatial resolution decreases and the frequency resolution increases, so finer details can be seen in the signals (zooming). This

multiresolution representation provides a coarse-to-fine processing strategy and a broad framework for signal analysis and compression.

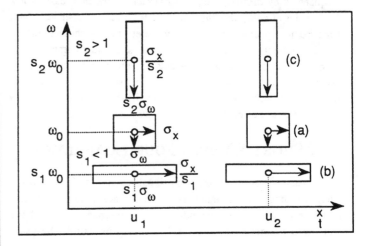

Fig. 4. Resolution cells in the phase-space for the Mallat wavelet. (a) Mother wavelet. (b) Dilated wavelet. (c) Contracted wavelet.

3.2 Daubechies Wavelet

Another wavelet suitable for locating abnormality in ECG signals was used by Tuteur [Tute88]. Many other wavelets have been proposed, including Haar, Stromberg, Meyer, Lemarie, Battle, and Daubechies. For example, a family of basis functions can be obtained from [Daub88]

$$h_{s,u}(x) = a_0^{-s/2}\, h(\frac{x}{a_0^s} - b_0 u) \quad \text{for } a_0>1,\ b_0\neq0 \tag{7}$$

where typically $a_0=2$ and $b_0=1$. The scale factor $a_0^{-s/2}$ is used to conserve the Euclidean L_2 norm of the wavelet. For large positive s, $h_{s,u}(t)$ is dilated by a_0^s and shifted by large steps $a_0^m b_0$, while for negative s, it is contracted and shifted in small steps. Thus, the wavelet translation is sharp in time at high frequencies, and is sharp in frequency at low frequencies (with corresponding loss of frequency and time resolution, respectively).

Wavelet packet analysis is a generalization of the Mallat multiresolution decomposition based on wavelet series. Any nonstationary signal can be viewed as a superposition of different wavelet packets occurring on different time scale and different times (or spacial scales at different locations). An analysis of such a signal separates it into constituent wavelet packets and arranges them according to a descending order. The matching process is based on the lowest entropy (highest efficiency of expansion) which also leads to an automatic segmentation of the signal. The signal can be reconstructed with a minimum number of wavelets. The "best basis" can be used to compress the temporal signal for transmission and storage.

3.3 Data Compression

The goal of signal compression is to represent a signal using the smallest number of data bits. In both *lossless* and *lossy* compression, it is important to obtain a highly-structured signal representation [Kins91]. In such a representation, we can decompose a signal into (i) a signal structure, and (ii) its parameters. The *structure* represents a class of signals, while the *parameters* represent *distinct features* of the signal to distinguish it from other signals within the same class. Assuming that the structure is known, we can represent the signal using the parameters which now require fewer number of bits that the original signal. During reconstruction, we incorporate the parameters into the structure to reconstruct the signal in its original form.

For lossy compression, a signal compression scheme relies on *vector* or *scalar quantization* of the parameters. Here, an efficient representation comes from a *coarse* quantization, that is the parameters are approximated by a few, finitely many values (or arrays of values) from a table. This approach usually results in a better compression efficiency than the lossless compression. However, there is a trade-off between a coarse quantization and the reconstruction quality (under a given criterion). A coarser quantization may result in poorer quality.

To deal with this trade-off, it is important to have parameters with a smaller dynamic range prior to their quantization. The values of such parameters span a narrow range, such that a coarse quantization still gives small quantization error, resulting in an acceptable reconstruction quality.

3.4 Application to Image and Sound Compression

The discrete wavelet transform (DWT) has been applied to many applications, particularly to image compression [e.g., ABMD90]. For example, Stephane Mallat at Courant Institute of Mathematical Sciences at New York University [Mall89a], [Mall89b] uses wavelets to compress 2D images by 40:1, with reconstructed images that are nearly indistinguishable from the original. He provides a spectrum of resolutions in a wavelet orthonormal basis, capable of representing an image very compactly. The representation is midway between the Fourier domain and space domain. He also points out how fractals could be analyzed with wavelets, using their self-similarity.

Lewis and Knowles [LeKn90] have refined the Mallat multiresolution technique for real-time moving images, with 78:1 compression ratio without image degradation. Their method has lower computational requirements than the discrete cosine transform (DCT), and is more suited to VLSI implementation.

Coifman at Yale University [Cipr90] uses packet wavelets for acoustic signal compression. In addition to data compression, wavelets can be applied to music and speech synthesis, improved medical imaging (CAT scan and nuclear-magnetic resonance), seismic exploration, and vibration analysis.

Daugman, Division of Applied Sciences, Harvard University, [Daug88] uses a neural network capable of transforming 2D discrete signals into generalized nonorthogonal 2D Gabor representations for image analysis, segmentation and compression with ratios up to 20:1. He also presents [Daug89] image compression to below 1 bpp (bit/pixel) using Gabor wavelets. Porat and Zeevi [PoZe89] discuss texture processing using a finite set of Gabor wavelets. Mallat [Mall89b] has pointed out several drawbacks of the Gabor transform when applied to image analysis.

In addition to the DWT, continuous wavelet transform (CWT) can be advantageous to some applications. An efficient algorithm for the computation of CWT for the analysis of bandpass signals that are scale-time perturbed is given in [GoBu90]. The algorithm uses the DWT to compute the CWT. A relationship of wavelets to filter banks as applied for nonstationary signals is discussed in [VeHe90].

3.5 New Developments

Wavelet Packet Analysis. The wavelet packet analysis with the best basis can also be used for parameter extraction for pattern recognition and diagnosis. For example, a standard procedure is to compare an ECG recording of 50 cycles with another recording in order to detect significant changes. Although this can be done optimally with factor analysis (Karhunen-Loeve K-L bases) which is the most efficient basis for capturing, on the average, most of the energy of an arbitrary sample, the computational effort is prohibitive. The reduction in complexity in terms of a single statistical best basis allows the factor analysis to be performed in real time [CoMW92]. This approach can also be used for pitch detection of speech [KaBB91], acoustic signal compression [Wick92] with 25 to 38% better compression than wavelet transform, and to second-generation (i.e., with structural and perceptual properties) image compression [FrMa92].

Wavelets and Fractals. Self similarity is the feature of fractals, both deterministic and stochastic. Deterministic fractals have the property that there exists a similarity (or even identity) between an object (the entire signal or its fragment) and a magnified version of itself. This also applies to stochastic fractals. Self similarity can then be regarded as a form of redundancy with respect to scale. Thus, finding appropriate affine transformations for the object may result in a very large compression ratios. Barnsley and co-workers have developed a theory of Iterated Function Systems (IFS) [Barn88] and Recurrent Iterated Function Systems [BaJa88], capable of representing complex shapes and textures with very few parameters (compressions from 100:1 to 10,000:1). The process of finding the IFS for a given signal is computationally very difficult. Since wavelets also possess the inherent self similarity, the wavelet transform, combined with the multiresolution analysis, may be suitable for a more systematic approach to the inverse problem of finding the IFS [FrDu90].

4. EXPERIMENTAL RESULTS

To demonstrate the potential of wavelets in signal compression, we now show a simple, yet useful and complete scheme of image and speech compression. The scheme is simple as it requires a few processes, and uses a simple Daubechies 4-coefficient analyzing wavelet, as suggested in [Pres91]. Yet, it significantly compresses images (less than 2 bits/pixel for 30 dB peak signal-to-noise ratio, PSNR), and is complete enough to show basic principles of using wavelets for signal compression.

4.1 An Example of Image Compression

In our experiment, we notice that the wavelet transform provides image representation in a form that is suitable for coarse quantization. As a sample image, we used a 256×256 pixel image of Lena (lena.img), with a total number of pixels of 65536, each 8 bit grey-level resolution. The wavelet transform of such a number of pixels produces a set of the same number (65536) of wavelet coefficients. The coefficients are exact representation, that is the inverse transform of them gives the exact image back.

In that form, the coefficients do not give efficient representation. However, we can now apply quantization on the coefficients. Our study of value distribution of the wavelet coefficients reveals that most of them are small values (close to zero); i.e., 80.92% of the coefficients have absolute values less or equal than 5% of the maximum absolute values of all the coefficients. Further observation shows that those small values can be considered insignificant with respect to the image reconstruction quality (in term of PSNR). To show this, we replace all small values with zero, and use the resulting set of coefficients to reconstruct the image. By varying the value of truncation, which is the percentage of coefficients that have been replaced by zero, we can obtain corresponding PSNR of the reconstructed image, which is shown in Fig. 5. Notice that truncation of the coefficients as high as 90% still produces 30dB PSNR quality. This means that only 10% of the coefficients are significant for that level of image quality.

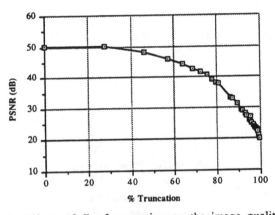

Fig. 5. Effects of % of truncation on the image quality for lena.img. This shows that even by replacing 90% of wavelet representation with 0, the quality is still good (≥ 30 dB).

What does this mean from a compression perspective? Knowing that the truncated coefficients have a structure with such an unbalanced distribution of values, we can immediately use *statistical approach* of data compression, such as *Shannon-Fano* technique [Kins91]. This technique assigns fewer number of bits for more frequently occurring values, resulting in a smaller total number of bits. Here, we simply apply the technique (using the popular PKZIP) to a file containing the truncated wavelet coefficients of the image. Figure 6 shows the relationship between the values of truncation percentage and the compression efficiency (in a term of bits/pixel). Here, 90% truncation translates to 2 bits/pixel compression. By combining Fig. 5. and Fig. 6, we obtain the relationship between the compression efficiency and the reconstruction quality, shown in Fig 7. As expected, we can see the trade-off between the compression efficiency and the image quality.

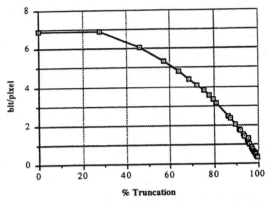

Fig. 6. Effect of the truncation to PKZIP compression for lena.img.

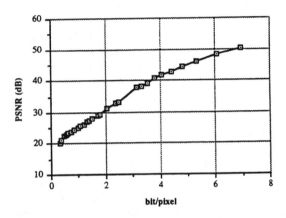

Fig. 7. Quality of the image for different compression rates for lena.img.

The above results show the compression capability of the wavelet representation of signal. Furthermore, we observe two attractive features of the scheme. First, although sets of the highly-truncated coefficients produce distorted images, the noise is in a form of white spots that are spatially localized. A low-pass filter similar to that for salt-and-pepper noise can reduce the effect of this noise. Second, a *pyramidal algorithm* used to implement the forward and inverse wavelet transform [Pres91] is simple and fast, such as a real-time VLSI or digital signal processor implementation is possible (see [Know90]).

How does the scheme fit into the signal compression framework reviewed before? The use of wavelet representation implies that our signal structure is a *linear series representation* [Fran69] which uses a set of wavelets derived from a known *mother wavelet* as an *orthonormal basis*. Then, the parameters are the *coefficients* of the series representation obtained through a usual *orthogonal projection* procedure [Krey78]. In our example, the mother wavelet is the 4-coefficient Daubechies wavelet, and the parameters are the wavelet coefficients.

Within this framework, we can see several possible improvements. Firstly, we notice that the selection of the Daubechies wavelet as a mother wavelet is rather arbitrary. Since it is well known that the efficiency of series representation depends heavily on the selection of a basis, a more proper mother wavelet may result in a better compression performance. Secondly, the truncation approach in quantizing the wavelet coefficients is not well developed yet and is open for improvement. The above truncation is based simply on the magnitude of each individual coefficient, while ignoring the position of the coefficient in the *time-scale plane* [RiVe91], or the relation of the coefficient with its neighbours. Those ignored aspects may result in a better quantization scheme, because they provide a more suitable threshold criterion for truncation, or a vector quantization scheme. Furthermore, we may obtain a scheme with a better reconstruction quality by incorporating *human perceptual* aspect in our threshold criterion or the vector quantization scheme. Thirdly, the compression process above was based only on the unbalanced number of zero-valued coefficients. Here, we leave the non-zero valued coefficients as they are. More study on the dynamic of the non-zero valued coefficients may reveal some structure of them, which may lead to a more efficient compression scheme.

A number of experiment have recently been performed on Lena and baboon images, using the standard discrete cosine transform (DCT), as well as the Daubechies 4-coefficient and Haar 2-coefficient wavelets [Lega93]. The implementation is based on the core presented by Press *et al.* [PTVF92]. The Daubechies wavelet was superior to DCT for full image compression, but inferior for small subimages. In fact, the superiority increases with the size of images, while the DCT compression remains the same regardless of the image size. The superior performance can be attributed to the more regularly spaced coefficients. Threshold tests have shown that Daubechies wavelet coefficients can be stored in 11 bits, similarly to the DCT [GoWo92]. The Haar wavelet transform was inferior to both the DCT and Daubechies. However, it can provide good results for images with regular patterns.

In summary, we have seen an example of using wavelet transform for image compression. This example shows the promising performance of wavelet-based compression, in term of optimizing the compression efficiency and the reconstruction quality. The scheme uses series representation with a wavelet basis as the signal structure, and wavelet coefficients as the parameters. In this scheme, the parameters are so sparse such that a statistical compression can significantly compress them.

4.2 An Example of Speech Compression Using Wavelets

We also can compress speech using a similar wavelet scheme. Although we use a simple approach as at the image compression above, the scheme performs amazingly well. In an experiment on a male spoken sentence, the scheme reaches 12.82 dB segmental signal-to-noise ratio (SEGSNR) at less than 4.8 kbit/s rates. In comparison, today's state of the art coding at 4.8 kbit/s (code excited linear predictive coding, CELP [Lang92]) has SEGSNR between 10-13 dB. Although our result may vary for different speech signals, it shows the promising capability of wavelet-based speech coding.

The scheme relies on the sparsity of the transform results. As shown in the Fig. 8., most of the coefficients have small magnitude. More than 90% of the samples have magnitudes less than 5% of the

maximum magnitude. Since we use an orthonormal wavelet transform, these numbers mean most of the speech energy is in the high-valued coefficients, which are a few. This understanding allows us to treat the small-valued coefficients as insignificant data.

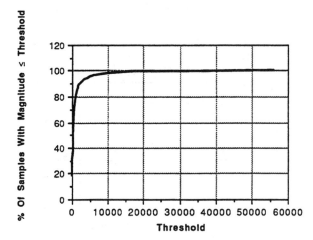

Fig. 8. Sparsity of the transform results. This figure shows that more than 90 % of the samples have magnitudes ≤ 5 % of the maximum magnitude.

To confirm that the small-valued coefficients are insignificant, we replace the small coefficients by zero (truncation), and use the results to reconstruct the speech. We then measure the closeness of the original and reconstructed speech by computing the segmental signal-to-noise ratio (SEGSNR), which is common in objective measurement of speech closeness. As shown in Fig. 9, only the high-valued coefficients are significant. With only 10 % of total coefficients, the scheme can provide 20 dB of SEGSNR.

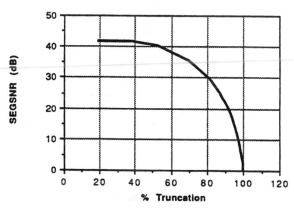

Fig. 9. Objective quality of truncated wavelet coefficients. This figure shows that small valued coefficients are redundant, since a SEGSNR of as high as 20 dB can be obtained from only 10% of the coefficients that have high magnitudes.

Consequently, compression can be achieved by first truncating the small-valued coefficients and then efficiently encoding them. As in the case of image compression above, the truncation causes unbalanced number of zero values in the set of coefficients. The next obvious step is to use statistical approach of compression to take advantages of such a value distribution. This scheme can be improved by many coding techniques. In our scheme, we encode consecutive zero-valued coefficients prior to applying a statistical compression. Here, a one-byte code and a one-byte number replace a consecutive zero-valued coefficients, where the number reflects the number of consecutive zero-valued coefficients. The scheme is able to reach a low bit rate of representation by increasing the percentage of truncated coefficients, as shown in Fig. 10.

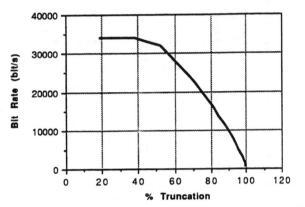

Fig. 10. Compression efficiency as a function of truncation percentage. Truncation by less than 95 % results in bit-rates less than 5000 bit/s.

We can now obtain the performance of the scheme by combining Fig. 9. and Fig. 10. As shown in Fig. 11, the scheme performs amazingly well, despite all the arbitrarily-selected techniques used in the scheme. The system achieves a SEGSNR of ≥ 12 dB at a rate less than 4.8 kbit/s. The performance is comparable to that of 4.8 kbit/s CELP, with only a fraction of CELP complexity. The performance may be better with careful selection of the techniques.

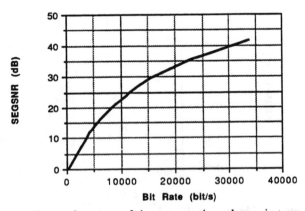

Fig. 11. The performance of the compression scheme, in term of objective quality (SEGSNR) and compression efficiency (bit rate). The scheme is comparable to CELP at a low bit rate. At a rate of 4.8 kbit/s, the scheme achieves > 12 dB SEGSNR.

In summary, this example shows the promising use of wavelet transforms for speech compression. The wavelets seem to be well suited for representing speech signals, in a sense that most of speech energy are in a few wavelet coefficients which have high values. The compression can then focus on obtaining and efficiently representing those coefficients. Our example shows one simple way of doing it, that is with truncating the small-valued coefficients, and using statistical approach to compress the results. This approach results in a high performance speech compression system.

5. CONCLUSIONS

The discrete wavelet transform (DWT) performs very well in the compression of images and speech. Its performance in terms of the bit rates and signal quality is comparable with other good techniques such as the discrete cosine transform (DCT) for images and code-excited linear predictive coding (CELP) for speech, but with much less computational burden. Much work must be done to improve the wavelet compression. More specifically, the scheme: could improved by: (i) finding the best mother wavelet for a certain type of signals, (ii) finding new truncation criteria or vector quantization based on time-scale, neighbourhood, and perceptual analysis, for better reconstruction quality, and (iii) finding the dynamic of the high-valued coefficients to have better compression efficiency.

Acknowledgements

This work was supported in part by the Natural Sciences and Engineering Research Council of Canada.

References

[ABMD90] M. Antonini, M. Barlaud, P. Mathieu, and I. Daubechies, "Image coding using vector quantization in the wavelet transform domain," *Proc. IEEE Intern. Conf. Acoustics, Speech & Sign. Processing*; ICASSP90 (Albuquerque, NM; Apr. 3-6, 1990), IEEE Cat. no. 90CH2847-2, vol. 4, pp. 2297-2300, 1990. (A new image vector quantization technique for wavelets.)

[Barn88] M.F. Barnsley. *Fractals everywhere*. San Diego, CA: Academic Press, 394 pp., 1988.

[BaJa88] M.F. Barnsley and A.E. Jacquin, "Application of recurrent iterated function systems to images," *SPIE*, vol. 1001 Visual Communications and Image Processing, 1988.

[Cipr90] B.A. Cipra, "A new wave in applied mathematics," *Science*, vol. 249, pp. 858-859, 24 Aug. 1990. (Describes recent developments in wavelets at Bell Laboratories (Ingrid Daubechies), Courant Institute, and New York University (Stephane Mallat), Yale University (Ronald Coifman).)

[CoMW92] R.R. Coifman, Y. Meyer, and V. Wickerhauser, "Wavelet analysis and signal processing," in *Wavelet and Their Applications*, M.B. Ruskai, G. Beylkin, R. Coifman, I. Daubechies, S. Mallat, Y. Meyer, and L. Raphael (eds.), pp. 153-178. Boston, MA: Jones and Bartlett, 474 pp., 1992. (A computer software package. Also the idea of the best wavelet basis for numerical computations.)

[Daub88] I. Daubechies, "Orthonormal bases of compactly supported wavelets," *Communic. in Pure and Applied Math.*, vol. 41, pp. 909-996, Nov. 1990. (A comprehensive exposure to wavelets.)

[Doug88] J.G. Daugman, "Complete discrete 2D Gabor transforms by neural networks for image analysis and compression," *IEEE Trans. Acoustics, Speech, Signal Processing*, vol. 36, pp. 1169-1179, Jul. 1988. (Similar to [Mall89a].)

[Daug89] J.G. Daugman, "Entropy reduction and decorrelation in visual coding," *IEEE Trans. Biomedical Eng.*, vol. 36, pp. 107-114, Jan. 1989. (Gabor's transform.)

[Fran69] L. E. Franks, *Signal Theory*. New Jersey: Prentice-Hall Inc., 317 pp., 1969.

[FrDu90] G.C. Freeland and T.S. Durrani, "IFS fractals and the wavelet transform," *ICASSP'90*, CH2847-2, 1990, pp. 2345-2348.

[FrMa92] J. Froment and S. Mallat, "Second generation compact image coding with wavelets," in *Wavelets – A Tutorial in Theory and Applications*, C.K. Chui (ed.), pp. 655-678. New York, NY: Academic Press, 723 pp., 1992. (Edge detection in images.)

[GeGr92] A. Gersho and R.M. Gray. *Vector Quantization and Signal Compression*. Boston, MA: Kluwer, 1992. (Both authors have contributed to the field significantly.)

[GoBu90] R.A. Gopinath and C.S. Burrns, "Efficient computation of the wavelet transforms," *Proc. IEEE Intern. Conf. Acoustics, Speech & Sign. Processing*; ICASSP90 (Albuquerque, NM; Apr. 3-6, 1990), IEEE Cat. no. 90CH2847-2, vol. 3, pp. 1599-1601, 1990.

[GoWo92] R.C. Gonzalez and R.E. Woods. *Digital Image Processing*. Reading, MA: Addison-Wesley, 1992.

[KaBB91] S. Kadambe and G.F. Boudreax-Bartels, "A comparison of a wavelet functions for pitch detection of speech signals," *Proc. IEEE Intern. Conf. Acoustics, Speech & Sign. Processing*; ICASSP91 (Toronto, ON; May. 14-17, 1991), IEEE Cat. no. 91CH2977-7, vol. 1, pp. 449-452, 1990.

[Kins91] W. Kinsner, "Review of data compression methods, including Shannon-Fano, Huffman, arithmetic, Storer, Lempel-Ziv-Welch, fractal, neural network, and wavelet algorithms," *Technical Report*, DEL91-1, Jan. 1991, 157 pp.

[Know90] G. Knowles, "VLSI architecture for the discrete wavelet transform," *Electronics Letters*, vol. 26, pp. 1184-1185, 19 Jul. 1990.

[Krey78] E. Kreyzig, *Introductory Functional Analysis with Application*. New York: John Wiley & Sons, 688 pp., 1978.

[Lang92] A. Langi, "Code-excited linear predictive coding for high-quality and low-bit rate speech," *M.Sc. Thesis*, Winnipeg, MB: Department of Electrical and Computer Engineering, University of Manitoba, Mar. 1992. (An implementation of near-real-time CELP at 4.8 kbps.)

[Lega92] T.L. Legal, "Comparison of transform coding techniques: Discrete Wavelet Transform versus Discrete Cosine Transform," *B.Sc. Thesis*, Winnipeg, MB: Department of Electrical and Computer Engineering, University of Manitoba, Mar. 1993. (Still image processing with Daubechies and Haar wavelets.)

[LeKn90] A.S. Lewis and G. Knowles, "Video compression using 3D wavelet transform," *Electronics Letters*, vol. 26, pp. 396-398, 15 Mar. 1990. (Their method is better than the discrete cosine transform.)

[Mall89a] S.G. Mallat, "A theory for multiresolution signal decomposition: The wavelet representation," *IEEE Trans. Pattern Analysis Machine Intelligence*, vol. 11, pp. 764-693, Jul. 1989. (Review of multifrequency channel decomposition of images and wavelet models.)

[Mall89b] S.G. Mallat, "Multifrequency channel decompositions of images and wavelet models," *IEEE Trans. Acoustics, Speech, Signal Processing*, vol. 37, pp. 2091-2110, Dec. 1989. (Similar to [Mall89a].)

[PoZe89] M. Porat and Y.Y. Zeevi, "Localized texture processing in vision: Analysis and synthesis in the Gaborian space," *IEEE Trans. Biomedical Eng.*, vol. 36, pp. 115-129, Jan. 1989. (Gabor's transform, not wavelets used.)

[Pres91] W. H. Press, "Wavelet Transforms," *Harvard-Smithsonian Centre for Astrophysics Preprint No. 3184*, 1991. (Available through anonymous ftp at 128.103.40.79, directory /pub).

[PTVF92] W.H. Press, S.A. Teukolsky, W.T. Vetterling, and B.P. Flannery. *Numerical Recipes in C*. (2nd ed) New York, NY: Cambridge University Press, 1992. (This second edition contains the wavelet transform.)

[RiVe91] O. Rioul and M. Vetterly, "Wavelets and signal processing," *IEEE Signal Processing Magaz.*, IEEE CT 1053-5888/91 pp. 14-38, Oct. 1991.

[Tute88] F.B. Tuteur, "Wavelet transformations in signal detection," *Proc. IEEE Intern. Conf. Acoustics, Speech & Sign. Processing*; ICASSP88 (New York, NY; Apr. 11-14, 1988), IEEE Cat. no. 88CH2561-9, vol. 3, pp. 1435-1438, 1988. (Wavelets applied to the detection of ventricular delayed potentials, VLP, in ECG.)

[VeHe90] M. Vetterli and C. Herley, "Wavelets and filter banks: Relationship and new results," *Proc. IEEE Intern. Conf. Acoustics, Speech & Sign. Processing*; ICASSP90 (Albuquerque, NM; Apr. 3-6, 1990), IEEE Cat. no. 90CH2847-2, vol. 3, pp. 1723-1726, 1990.

[Wick92] M.V. Wickerhauser, "Acoustic signal compression with wavelet packets," in *Wavelets – A Tutorial in Theory and Applications*, C.K. Chui (ed.), pp. 679-700. New York, NY: Academic Press, 723 pp., 1992. (Wavelet packets applied to speech.)

A CLASS OF TIME-VARYING WAVELET TRANSFORMS *

Iraj Sodagar *Kambiz Nayebi* *Thomas P. Barnwell, III*

Digital Signal Processing Laboratory
Georgia Institute of Technology
School of Electrical Engineering
Atlanta, Georgia 30332

ABSTRACT

In this paper, the concept of the time-varying discrete-time wavelet transform is introduced. In the time-varying wavelet transform, the basic tiling of the time-frequency plane stays unchanged and only the properties of the analysis wavelets are changed in time. Our primary objective in designing such analysis-synthesis structures is to insure that the transform is invertible at all times. Two design procedures are developed based on the time-varying filter bank design. The first approach uses a tree structure of time-varying 2-band filter bank and the second approach is based on parallel nonuniform filter banks. The concept can easily be extended to general tree-structures and wavelet packets.

1. INTRODUCTION

In the past, time-frequency signal processing has often been used to decompose time-varying signals. The short-time Fourier transform (STFT) and the Gabor transform are the classical linear time-frequency transforms with time and frequency shifted basis functions. The wavelet transform, which has been extensively studied over the past few years, is a recent member of the linear time-frequency transform family.

A more flexible way of exploiting the time-varying nature of many types of signals is by using *time-varying* transforms in which the properties of the transformation are changed to match the short-term properties of the signal. The ultimate goal of such systems is to use adaptive transforms in which the time-frequency tiling and basis functions are tailored to the properties of the signal. This is similar to the time-varying block transforms, such as the Karhunen-Loeve transforms, in which the transformation is data dependent.

This paper introduces the concept of the time-varying wavelet transform as a special case of a more general class of time-varying transforms. This class of transforms is capable of realizing time-varying tilings of the time-frequency plane. In [1], authors have shown an example of such a time-varying tiling using tree structures. Our primary objective in designing such maximally decimated transformations is to keep the time-varying transform invertible at all times.

*THIS WORK IS SUPPORTED IN PART UNDER THE JOINT SERVICES ELECTRONICS PROGRAM CONTRACT #DAAL-03-90-C-0004.

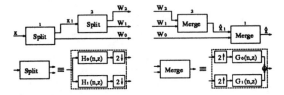

Figure 1: **Block diagram of the 2-stage tree structure used for DTWT and its inverse.**

In this paper, the main focus is on the time-varying discrete wavelet transforms (DTWT). In such transforms, the basic tiling of the time-frequency plane stays unchanged and only the properties of the analysis wavelets such as its regularity or its frequency characteristics are changed in time.

Knowing the relationship between two-band filter banks and the DTWT [2–4], the time-varying DTWT can be studied in the context of time-varying filter banks. In time-varying filter banks [5, 6], system filters are changed in time and the exact reconstruction property of the system is always preserved. In this paper, we apply time-domain formulation of filter banks for the design of time-varying DTWT. The proposed time-varying DTWT is achieved by switching among a finite number of time-invariant DTWT.

The paper is structured as follows. First, we study two different analysis-synthesis structures for implementing time-invariant DTWT and their properties. Next, we present the time domain formulation for time-varying filter banks. Finally, the time-varying DTWT is studied in the context of both a tree structure and a parallel implementation of DTWT.

2. THE DISCRETE WAVELET TRANSFORM

The discrete-time wavelet transform (DTWT) is usually implemented using an octave-band tree structure. Each stage of the tree structure is an identical 2-band decomposition and only the low frequency band of this 2-band system is further divided. The inverse transform can be implemented in a similar structure using the corresponding 2-band synthesis section. Figure (1) shows a 2-stage tree structure.

If the basic 2-band system provides perfect reconstruction (PR), the DTWT is invertible. A desired DTWT can be obtained by designing the 2-band filter bank with the required properties. As an example, orthogonal DTWTs can be obtained by using paraunitary 2-band systems. In the same way, a time-varying DTWT may be realized by using a time-varying 2-band system.

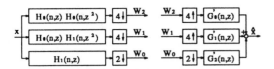

Figure 2: **Block diagram of the parallel synthesis structure used for 2-level DTWT and its inverse.**

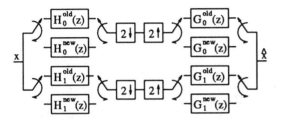

Figure 3: **Block diagram of direct switching of filter banks between two 2-band systems.**

For a finite number of splitting stages, such a tree structure can also be presented in the form of a parallel nonuniform filter bank. The analysis filters of this structure are obtained by convolving the 2-band analysis filters and their upsampled versions. Figure (2) shows the equivalent parallel form of Figure (1). It is worth noting that the parallel and the tree structure implementations of the time-varying DTWT have quite different properties. This will be discussed later in this paper.

3. TIME VARYING FILTER BANKS

The principles of time-varying filter bank structures are presented in [5,6]. In such structures, the time and frequency resolutions of the system are changed in time and the PR property is always preserved. The basic idea is to determine the set of necessary and sufficient conditions to provide PR at all times. These conditions can be expressed in matrix form as:

$$\mathbf{A}_n \mathbf{s}_n = \mathbf{b}_n \qquad (1)$$

where n is a time index. $\mathbf{A}_n$ contains the analysis filter coefficents which produce the existing samples in the synthesis section at time n. $\mathbf{s}_n$ contains the synthesis filter coefficents and the vector $\mathbf{b}_n$ is the ideal impulse response [6]. We only consider the case in which the analysis section is switched from one filter bank to another. For this case, a set of time-varying synthesis filters must be found such that they satisfy the above equations. Since such synthesis systems will not always exist for any arbitrary set of analysis systems, a redesign of the analysis systems may be required.

Direct switching between two different 2-band systems is shown in Figure (3). The two sets of PR filter banks are designed with 8-tap analysis and synthesis filters. The frequency characteristics of the analysis filters are shown in Figure (4). As is well known, the input/output relation of a time-varying system can be expressed by its time-varying impulse response. Figure (5a) shows the impulse response of the above system around the transition period. In this figure, $h(n,m)$ is the response at time n to a unit input at time m. For a PR system, $h(n,m)$ has a height of one

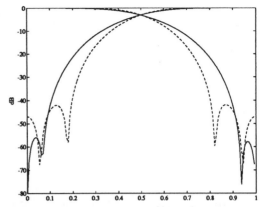

Figure 4: **Frequency characteristics of two independent 2-band perfect reconstructing filter banks. Full lines show the first set and dotted lines show the second set.**

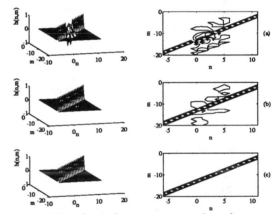

Figure 5: **The impulse response of a time-varying 2-band filter bank around transition. The time-varying impulse responses are shown in surface plots at left and in contour plots at right. (a)Direct switching. (b) LS design. (c) PR design.**

along a diagonal line and zero everywhere else in the (m,n)-plane. As is shown in Figure (5a), this system provides PR at samples before and after reconstruction transition period, but not during this period. This is because during the transition, some samples used by the new synthesis filters are generated by the old analysis filters. The transition period in this example has a length of 6 samples. One trivial solution for avoiding switching distortion is to run both filter banks during transition period and transmit the extra samples. To provide PR without transmitting extra samples, the analysis and synthesis sections should be designed for the time-varying case. Using the independently designed analysis sections of Figure 4, we can approximately solve the equations of each state using the least square (LS) method. Figures (5b) shows the impulse response of the LS solution. The impulse response of the system is significantly improved compared to direct switching, but still PR is not provided. For PR design, the independent filter banks are used as the starting point in a design procedure including

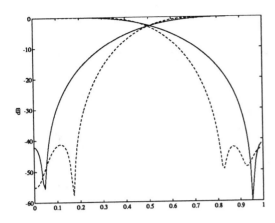

Figure 6: **The analysis filter responses of two 2-band systems with PR transition period. Full lines show the first set and dotted lines show the second set.**

optimization. Figures (5c) shows the impulse response of the PR time-varying system designed for above example. Figure (6) show frequency responses of the corresponding analysis filters. As it can be seen, the resulting analysis filters are different than the initial filters.

4. THE TIME VARYING DTWT

A simple form of the time-varying DTWT is achieved by changing the basis functions among a finite number of choices. Each of these time-invariant transforms can be implemented using a separate filter bank. Hence, switching between these filter banks results in the time-varying DTWT. The time-invariant DTWT was implemented in the form of a tree-structure and a parallel nonuniform filter bank in the first section. Here, we extend each of these structures to time-varying case to obtain the time-varying DTWT.

Time-Varying Tree Structures The first approach uses the octave-band tree structure. Changing the basis functions of DTWT is equivalent to changing the splitting units in the tree structure. So, to design a time-varying DTWT using the tree structure, one can design the basic time-varying 2-band system with the necessary constraints and use it in the tree structure. Here, we study the time-varying 2-level tree structure of Figure 1 as a simple example. Figure (7) shows a schematic diagram of the signals in the analysis section of this tree structure. All analysis filters are assumed to be 8-tap filters. In this figure, x is the input signal. x_1 is the lowpass output of the first stage and W_0, W_1, W_2 are the DTWT outputs. At time $n = 0$, the *old* analysis filters are switched to the *new* ones. Each of the old and new analysis sections produces a distinct DTWT. The empty and solid dots in this figure show the samples which are generated by the old and the new analysis sections respectively. Before time 0, all outputs are generated by the old analysis section and belong to the old DTWT. At time 0, all filter coefficients are changed, so at this time and after, x_1 and W_0 are generated by the new analysis section. But at the input of the second stage, 1 sample is generated by the new filter and 7 samples are generated by the old filter. The old input samples of the second stage are gradually replaced by the outputs generated by the new filter. At time 16, all input samples of the second stage are generated by the new analysis filters of the first stage. Therefore, the

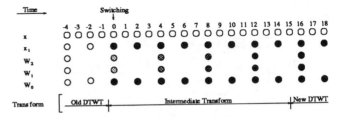

Figure 7: **Diagram of a 2-level DTWT outputs during the transform transient period.**

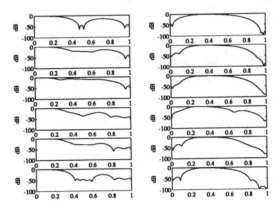

Figure 8: **The frequency response of W_2 (left column) and W_1 (right column) of the 2-level time-varying DTWT around transition. Top: the old DTWT. Middles: Intermediate transform. Bottom: the new DTWT.**

tree structure outputs belong to the new DTWT at and after this time. At the period $n = 0, \cdots, 14$, W_1 and W_2 are produced form a mixture of the old and the new x_1's, and they do not belong to either the old or the new DTWTs. These samples are shown in gray dots in Figure (7). In this period, which is called *transform transition period*, the transform is not a 'true' DTWT. As it is shown in Figure (8), the intermediate transform has frequency characteristics close to the old and the new DTWTs. For this example, we have used the time-varying 2-band system of Figure (6). In general, for a tree structure with M stages and filters of length N, the maximum number of intermediate samples at each output of the analysis section is $N - 1$. In practice, with the finite number of splitting stages, not having a true DTWT in the transition periods does not seem to be a major disadvantage.

From reconstruction point of view, a PR time-varying tree structure can be obtained by using PR time-varying split-merge pairs. This is because each split-merge pair in a tree structure is equivalent to a pure delay and if it is replaced by a another PR split-merge pair with equal delay, the system remains PR at all times. Therefore, the time-varying DTWT is invertible at all times when it is implemented in the form of tree-structure using a PR time-varying 2-band filter bank.

Time-Varying Parallel Structures Another approach for designing the time-varying DTWT is to implement it as a time-varying corresponding parallel structure. Chang-

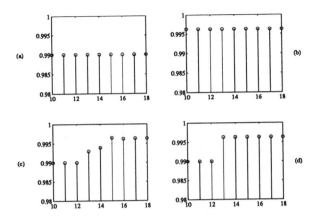

Figure 9: The W_2 output for a pulse in the tree structure and the parallel filter bank around transition: (a)The old DTWT. (b) The new DTWT. (c)Changing transform using tree structure. (d) Changing transform using parallel structure.

ing the basis functions of DTWT is equivalent to changing the analysis section in the parallel nonuniform filter bank. Switching the analysis sections results in changing the DTWT immediately in the sense that there is no intermediate transform during switching. This is because signal decomposition is performed in a single stage. As it is shown in Section 3, direct switching usually results in a substantial amount of reconstruction distortion during the transition period. To reduce the amount of distortion in the transition periods, the required synthesis sections can be obtained by solving equations (1) using the LS method. To provide PR at all times, the analysis filters should be redesigned using equations (1) as discussed in Section 3. Therefore, time-varying DTWT design can be considered as a time-varying nonuniform filter bank problem with some extra constraints. These constraints provide the relations between the analysis filters and the basis functions of DTWT.

In the time-varying system, the tree structure and the parallel form are not equivalent, because convolution is a time-invariant operator. So the synthesis filters obtained from the time-varying 2-band systems can not provide perfect reconstruction in the time-varying parallel filter bank. Figure (9) demonstrates the difference between outputs of transform around transition periods in the tree structure and the parallel implementation. A time-varying DTWT is implemented in the forms of Figures (1) and (2) using the analysis filters shown in Figure (6). Figure 9 shows $W_2(m)$ output in both implementations. The input signal is chosen to be an step function with an step size of 1. The transform is switched at $m = 13$. As it is shown, the output of the parallel implementation is immediately changed form the old transform to the new one, but the output of the tree structures includes 4 intermediate samples between the old and the new transform. After $m = 17$, the outputs of both structure are same.

For a M-level DTWT (equivalently a M-stage tree structure), the parallel structure has $M+1$ bands. The maximum length of filters in this structure is $(2^M-1)(N-1)+1$, where N is the length of filters in the basic 2-band system. The decimation ratios varies from 2^M in the lowest band to 2 in

the highest. Therefore, switching a wavelet transform with N_1-tap analysis filters to another transform with N_2-tap analysis filters, needs a reconstructing transient period of length $(2^M - 1)(N_2 - 1) - 1$. Note that the basic time-varying 2-band system has a transition period of length $N_2 - 2$. Since the number of necessary and sufficient constraints for PR is a function of length of transition period [6], the number of equations increases exponentially with M. Considering the number of free variables of analysis section in both tree structure and parallel implementation is same ($2(N_1 + N_2)$), designing a time-varying DTWT using parallel structure is much more difficult than using the tree structure.

5. CONCLUSION

In this paper, the time-varying DTWT is introduced. It was shown that the time-varying DTWT can be implemented using tree and parallel structures. Although the tree structure can provide good reconstruction, changing from one wavelet transform to another requires a transition period. On the other hand, time-varying parallel structures provide immediate switching from one wavelet transform to another but designing transforms with exact reconstructing transition period is more difficult in this implementation. Both methods can also be used to design time-varying wavelet packets in a similar way.

REFERENCES

[1] C. Herley, J. Kovacevic, K. Ramchandran, and M. Vetterli, "Arbitrary orthogonal tilings of the time-frequency plane," *IEEE-SP International Symposium on Time-Frequency and Time-Scale Analysis*, October 1992.

[2] I. Duabechies, "Orthonormal bases of compactly supported wavelets," *Comm. in Pure and Applied Math.*, vol. 41, pp. 909–996, 1988.

[3] M. Vetterli and C. Herley, "Wavelets and Filter Banks: Relationships and New Results," *Proceedings of the International Conference on Acoustics, Speech, and Signal Processing*, pp. 1723 – 1726, April 1990.

[4] S. Mallat, "A theory of multiresolution signal decomposition: The wavelet representation," *IEEE Trans. on Pattern Analysis and Machine Intell.*, vol. 11, pp. 674–693, 1989.

[5] K. Nayebi, T. P. Barnwell, and M. J. T. Smith, "Analysis-synthesis systems with time-varying filter bank structures," *Proceedings of the International Conference on Acoustics, Speech, and Signal Processing*, March 1991.

[6] K. Nayebi, I. Sodagar, and T. P. Barnwell, "The wavelet transform and time-varying tiling of the time-frequency plane," *IEEE-SP International Symposium on Time-Frequency and Time-Scale Analysis*, October 1992.

A MORPHOLOGICAL WAVELET TRANSFORM

H. Cha

Advanced TV and Display Lab.
A/V R&D Center
Samsung Electronics Co.
Suwon, Kyungki-Do, Korea 441-742

L. F. Chaparro

Real Time Signal Processing Laboratory
Department of Electrical Engineering
University of Pittsburgh
Pittsburgh, PA 15261

ABSTRACT

Using adaptive structuring functions, we develop a morphological interpolation that allows a signal representation similar to the given by the wavelet bank of filters. With morphological operators the interpolation problem is reduced to solving non–linear equations iteratively to get an approximate expansion of a sampled signal in terms of the structuring functions. We obtain a pyramid-like structure to decompose the signal into smoothed and detail components at different scales, just as in the wavelet representation. The use of non–linear filters in our algorithm reduces the computational complexity associated with the decomposition and synthesis. Our representation is valid for one- and two-dimensional signals. In the two-dimensional case we consider the non–unique ordering of the structuring functions, and the variety of possible sampling, decimation and interpolation procedures. We illustrate our one- and two-dimensional representations by means of examples.

1. INTRODUCTION

The bank of filters implementation of the multiresolution signal decomposition based on the discrete wavelet transform [5] has great significance in the representation of signals. However, it requires appropriate choice of wavelets and compensation of the linear phase introduced by the filtering. Moreover its extension to two-dimensions is complex.

Recently, there has been great interest in applying morphology in the processing and representation of signals. Morphology-based algorithms are generally efficient since morphological operators are non–linear and well suited for parallel implementation. We propose a simple morphology–based technique analogous to the wavelet bank of filters representation and valid for one- and two-dimensional signals. This work extends the morphological sampling of Haralick and his co-workers [4], the signal decomposition using simple functions proposed by Pitas and Venetsanopoulos [7, 8], and the use of multiple structuring functions in filtering by Song and Delp [9]. Our algorithm applies an interpolation procedure based on morphological opening with structuring functions obtained from orthogonal polynomials defined on a window. The one-dimensional algorithm

is easily extendable to two-dimensions once the generation and ordering of two-dimensional structuring functions as well as two-dimensional multirate considerations are taken care of [1, 10].

It will be shown by means of examples that our procedure provides results similar to those obtained using the wavelet representation. However, our algorithm reduces the computational complexity by using non-linear filters, permits a perfect reconstruction with no need for linear phase compensation, provides flexibility in the choice of structuring functions and can be easily extended to two-dimensions.

2. ONE-DIMENSIONAL MORPHOLOGICAL WAVELET TRANSFORM

2.1. Morphological Interpolation

In [4] it is shown that under special conditions a morphological sampling theorem permits the reconstruction of sampled signals. We show in this section, that under general conditions one can develop a morphological interpolation method that permits a good reconstruction of sampled signals by adapting the structuring functions.

Consider a signal $f(n) \geq 0$, $n \in K$, and let $f(m)|_s$ be the sampled signal defined for $m \in K \cap S$ where S is the sampling domain. The positivity condition is not restrictive in image processing applications and in general can be satisfied by appropriate shifting, but it is required since the umbra concept used in gray–level morphology does not permit us to differentiate strictly positive signals from the rest. Thus, if $K = \{n | 0 \leq n \leq N-1\}$ and $S = \{m | m = nL, 0 \leq n \leq (N-1)/L\}$, the sampled signal $f(m)|_s$ will be equal to $f(n)$ every L samples. Defining the integer $\theta = (N-1)/L$ as the order of the interpolation, then given the sampling rate L we determine the number of points needed for the interpolation. Thus if $L = 2$, a second order ($\theta = 2$) interpolation requires five points.

Given a set of real-valued polynomials defined on K, we generate positive and normalized structuring functions $\{k_i(n) | 0 \leq i \leq N-1; n \in K; 0 \leq k_i(n) \leq 1\}$, and define

$$f^c(n) = \max_{n \in K}[f(n)] - f(n).$$

The following interpolation algorithm gives us a minimal reconstruction of $f(n)$:

$$z_0(n) = \begin{cases} f(n)|_s & n \in K \cap S \\ \text{undefined} & \text{otherwise} \end{cases}$$

$$\begin{aligned} f_i(n) &= z_i(n) \circ a_i k_i(n) \\ z_{i+1}(n) &= z_i(n) - f_i(n)|_s, \ n \in K \cap S \end{aligned}$$

where $i = 0, 1, \cdots, \theta$ and $\circ$ is morphological opening. To get a maximal reconstruction, we apply the above procedure to $f^c(n)$. This is equivalent to performing opening on the umbra of $f(n)$ and then on the umbra of $f^c(n)$ which is the complement of the umbra of $f(n)$. The interpolated signal is found to equal

$$\begin{aligned} \tilde{f}(n) &= \sum_{i=0}^{\theta} f_i(n) \\ &= \sum_{i=0}^{\theta} z_i(n) \circ a_i k_i(n) \end{aligned} \qquad (1)$$

whenever the corresponding interpolation error

$$e(n) = z_0(n) - \tilde{f}(n)$$

is smaller than the interpolation error, defined in a similar way, for $f^c(n)$.

In the examples, we use a second-order interpolation and the structuring functions are derived from the discrete Legendre orthogonal polynomials [6]. The first three of the five structuring functions are used in the interpolation. It should be mentioned that in the second order interpolation, it is not necessary to process both $f(n)$ and $f^c(n)$ if one uses the convexity of the signal. If the signal is convex $f(n)$ is processed, otherwise $f^c(n)$ is processed.

2.2. Properties and Calculation of $\{a_i\}$

If in (1) we choose the $\{a_i\}$ so that

$$z_i(n) \circ a_i k_i(n) = a_i k_i(n)$$

then that equation resembles a polynomial interpolation. As in [2] the following are properties of the procedure that permit us to find the $\{a_i\}$.

Proposition 1. Given $z(n), n \in K \cap S$, and the structuring function $k(n), n \in K$, then

$$z(n) \circ k(n) = \min_{\ell \in K \cap S}[z(\ell) - k(\ell)] + k(n)$$

Proposition 2. If $z_i(n) \circ a_i k_i(n) = a_i k_i(n)$, $a_i \geq 0$, then:

$$\begin{aligned} (i) & \quad 0 \leq z_{i+1}(n) \leq z_i(n) \\ (ii) & \quad \{a_i\} \text{ are unique} \\ (iii) & \quad a_0 = \min_{n \in K \cap S}[z_0(n)] \\ (iv) & \quad a_i \leq \max_{n \in K \cap S}[z_i(n)] \end{aligned}$$

According to the above propositions, to find the a_i that satisfies the non–linear equation $z_i(n) \circ a_i k_i(n) = a_i k_i(n)$ we minimize the following error:

$$\begin{aligned} \varepsilon_i(n) &= a_i^q k_i(n) - z_i(n) \circ a_i^q k_i(n) \\ &= - \min_{\ell \in K \cap S}[z_i(\ell) - a_i^q k_i(\ell)] \end{aligned}$$

where a_i^q is the estimate of a_i at the q-th iteration. Since $\varepsilon(n)$ is constant, we propose the following iterative procedure to find a_i:

$$\begin{aligned} a_i^0 &= \max_{n \in K \cap S} z_i(n) \\ \varepsilon_i^q &= a_i^q k_i(n) - z_i(n) \circ a_i^q k_i(n) \\ a_i^{q+1} &= a_i^q - \varepsilon_i^q \end{aligned}$$

It can be easily shown the error is monotonically decreasing and it tends to zero as $a_i^q \to a_i$.

When using orthogonal polynomials to generate the structuring functions $\{k_i(n)\}$ we need to consider the shifted and scaled orthogonal polynomials $\{g_i(n)\}$ as well as the complimentary set $\{1 - g_i(n)\}$. This is due to the positivity of the $\{a_i\}$. To determine whether $g_i(n)$ or $1 - g_i(n)$ is to be chosen as $k_i(n)$ in the representation, we calculate the corresponding reconstruction errors and choose the one that gives the smaller one [2].

2.3. Morphological Wavelet Representation

The wavelet bank of filters given by [5] separates the signal into low and high frequency components at different bands. The analysis and synthesis is accomplished using linear filtering and decimation/interpolation. An analogous morphological implementation is obtained by using the above interpolation algorithm along with decimation and interpolation.

The complexity of our algorithm increases proportionally to the order θ. For a given order, increasing L degrades the quality of the interpolation but improves the data compression. The analysis and synthesis procedures are illustrated in Fig. 1 and 2. The signal is sampled with a rate L and processed by the morphological interpolator Ψ which behaves as a non–linear smoother. Both smoothed and detail signals are then decimated by a factor of L.

Thus, using a second-order interpolator and $L = 2$ the interpolated signal is given by

$$\Psi(f_i(n)|_s) = \{f_i(0), \tilde{f}_i(1), f_i(2), \tilde{f}_i(3), \cdots\}$$

where $\tilde{f}$ is the interpolated value, and the detail signal is

$$\begin{aligned} d_i(n) &= f_i(n) - \Psi(f_i(n)|_s) \\ &= \{0, f_i(1) - \tilde{f}_i(1), 0, f_i(3) - \tilde{f}_i(3), \cdots, \}. \end{aligned}$$

The downsampled signal is denoted as $f(n) \downarrow$ and $\Psi(f_i(n)|_s) \downarrow$ is the decimated smoothed signal. The decimated detail signal is then defined as

$$d_i(n) \downarrow = \{f_i(1) - \tilde{f}_i(1), f(3) - \tilde{f}_i(3), \cdots\}$$

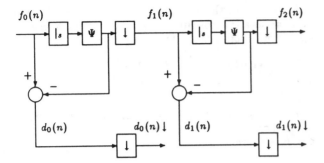

Figure 1. Morphological signal decomposition

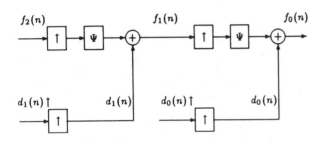

Figure 2. Morphological signal synthesis

where we take odd terms instead of even terms. Basically, $d_i(n)$ is the interpolation error and is the equivalent of the high-frequency component in the wavelet implementation. Using a dual procedure we can synthesize the signal exactly.

3. TWO-DIMENSIONAL MORPHOLOGICAL WAVELET TRANSFORM

The real advantage of our procedure is seen clearly in the two-dimensional case. The extension of the wavelet representation to two–dimensions is difficult due to the need to generate bivariate wavelets and the greater variety of sampling, decimation/interpolation procedures. In [5] separable two-dimensional filters are used and thus the filtering, decimation and interpolation procedures are one-dimensional applied column-wise and row-wise. The case of non-separable filters makes the procedure much more complex.

3.1. Two-dimensional Morphological Interpolation

Sampling in two-dimensions can be done in many ways: row by row (or column by column) – as in the one-dimensional case – which we call the row/column sampling or by the quincunx sampling method [10, 1].

For an $(M + 1) \times (N + 1)$ image $\{f(m, n)\}$, the row

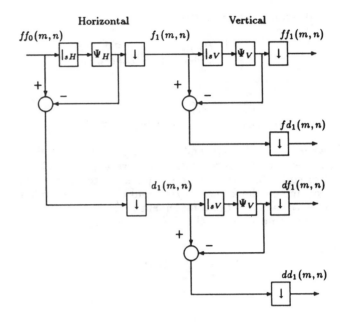

Figure 3. Two-dimensional morphological signal decomposition with row sampling

sampling method yields as $f(m, n)|_s$

$$\begin{pmatrix} f(0,0) & * & f(0,2) & * & \cdots & f(0,N) \\ f(1,0) & * & f(1,2) & * & \cdots & f(1,N) \\ \vdots & \vdots & \vdots & \vdots & \ddots & \vdots \\ f(M,0) & * & f(M,2) & * & \cdots & f(M,N) \end{pmatrix}$$

where $*$ correspond to the disregarded samples. Mathematically, the row sampling is equivalent to a lattice transformation [10] between the integer lattice $\{(n_1, n_2)\}$ to the sampling lattice $\{(m_1, m_2)\}$:

$$\begin{bmatrix} m_1 \\ m_2 \end{bmatrix} = \begin{bmatrix} 1 & 0 \\ 0 & 2 \end{bmatrix} \begin{bmatrix} n_1 \\ n_2 \end{bmatrix}$$

The quincunx sampling [10] corresponds to a transformation from the integer lattice to a sampling lattice given by

$$\begin{bmatrix} m_1 \\ m_2 \end{bmatrix} = \begin{bmatrix} 1 & -1 \\ 1 & 1 \end{bmatrix} \begin{bmatrix} n_1 \\ n_2 \end{bmatrix}$$

The quincunx sampling of an $(M + 1) \times (N + 1)$ image gives as $f(m, n)|_s$

$$\begin{pmatrix} f(0,0) & * & f(0,2) & \cdots & f(0,N) \\ * & f(1,1) & * & \cdots & * \\ f(2,0) & * & f(2,2) & \cdots & f(2,N) \\ \vdots & \vdots & \vdots & \ddots & \vdots \\ f(M,0) & * & f(M,2) & \cdots & f(M,N) \end{pmatrix}$$

where as before $*$ stands for the omitted sample.

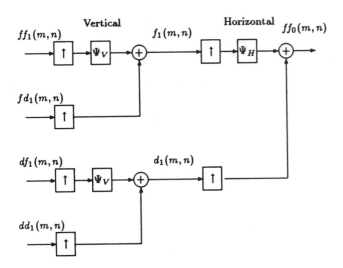

Figure 4. Two-dimensional morphological signal synthesis with row sampling

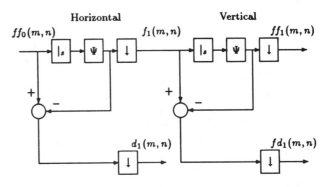

Figure 5. Two-dimensional morphological signal decomposition with quincunx sampling

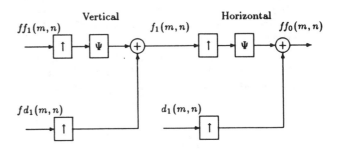

Figure 6. Two-dimensional morphological signal synthesis with quincunx sampling

The interpolation for the row/column sampling can be done using the one-dimensional interpolators discussed before. Although in the quincunx sampling one-dimensional interpolators can also be used, it is possible to develop a two–dimensional interpolation procedure using bivariate structuring functions, similar to the one-dimensional. The structuring functions are generated as the product of one-dimensional ones. Ordering of these functions is arbitrary since there is no unique ordering in the two–dimensional space.

3.2. Two-dimensional Morphological Wavelet Representation

The analysis and synthesis procedures for row/column sampling are illustrated in Fig. 3 and 4. The interpolator Ψ is one-dimensional and processes the image horizontally (vertically) corresponding to the row (column) sampling.

If we choose the quincunx sampling Ψ is a two-dimensional interpolator and processes the image block by block. The structures for the analysis and the synthesis are shown in Fig. 5 and 6.

The decimation in a 3×3 image block to obtain $f_i(m,n)$ and $d_i(m,n)$ is done by taking the even samples $m_0 + n$ of the signals for $m_0 = 0, 1, 2, \cdots$. Fixing m and letting n vary we are processing the data horizontally. The number of rows of the processed image remains the same, but the number of columns are reduced in half. We proceed similarly to obtain the vertically decimated signals $ff_{i+1}(m,n)$ and $fd_{i+1}(m,n)$ which would reduce the number of rows by half, and keeps the same number of columns.

4. EXAMPLES

In Figures 7 we show the representation of a one-dimensional signal using our algorithm and second order interpolator with $L = 2$. We compare it with the

wavelet representation obtained using Daubechies [3] wavelet of length 8. The two decompositions are quite similar, however with our algorithm the signal components are not delayed and the detail signals show more clearly the local maxima (which can be used in edge detection or transient identification).

To illustrate the application of our two-dimensional method for data compression, we consider Lenna's image shown in Figure 8. The image processed with the row sampling is shown in Figure 9. The image processed with the quincunx sampling is shown in Figure 10. To provide us a means of comparison the decomposed image using Daubechies' 8 tap wavelet and row sampling is shown in Figure 11.

5. CONCLUSIONS

In this paper, we have a new representation of one- and two-dimensional signals using gray-level morphology that performs similarly as the wavelet bank of filters. However, our representation uses non–linear filters and the implementation relies on maximum and minimum functions instead of inner products or convolutions. Synthesis with our algorithm is guaranteed to be exact because the procedure compensates for the interpolation error.

REFERENCES

[1] Adelson, E. H. and Simoncelli, E. "Orthogonal Pyramid Transforms for Image Coding," *Proc. SPIE, Visual Comm. and Image Processing II*, pp. 50-58, 1987.

[2] Cha, H. and Chaparro, L.F. "A Morphological Polynomial Transform," *Proc. IEEE ICASSP-93*, pp. v-173-76, Minneapolis, Apr. 1993.

[3] Daubechies, I. "Orthonormal Bases of Compactly Supported Wavelets," *Comm. Pure and Applied Math.*, vol. 41, pp. 909-996, 1988

[4] Haralick, R. M., *et. al* "The Digital Morphological Sampling Theorem," *Trans. Acoustics, Speech, and Signal Processing*, vol. ASSP-37, no. 12, pp. 2067-2090, Dec. 1989.

[5] Mallat, S. G. "A Theory for Multiresolution Signal Decomposition: The Wavelet Representation," *IEEE Trans. Pattern Analysis and Machine Intelligence*, vol. PAMI-11, No. 7, pp. 674-692, Jul. 1989.

[6] Neuman, C. P. and Schonbach, D. I. "Discrete (Legendre) Orthogonal Polynomials," *International Journal for Numerical Methods in Engineering*, vol. 8, pp. 743-770, 1974.

[7] Pitas, I. and Venetsanopoulos, A. N. "Morphological Shape Decomposition," *IEEE Trans. Pattern Analysis Machine Intell.*, vol. PAMI-12, pp. 38-45, Jan. 1990.

[8] Pitas, I. and Venetsanopoulos, A. N. *Nonlinear Digital Filters: Principles and Applications*, Kluwer Academic Publishers, 1990.

[9] Song, J. and Delp, E. J. "The Analysis of Morphological Filters with Multiple Structuring Elements," *Computer Vision, Graphics, and Image Processing*, vol. 50, pp. 308-328, 1990.

[10] Vaidyanathan, P.P. *Multirate Systems and Filter Banks*, Prentice-Hall, 1993.

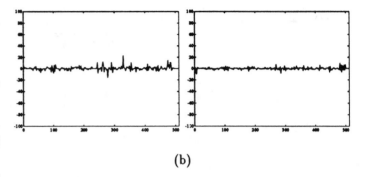

(b)

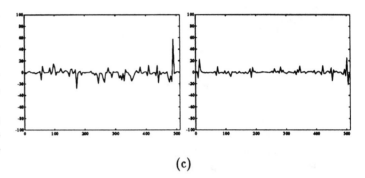

(c)

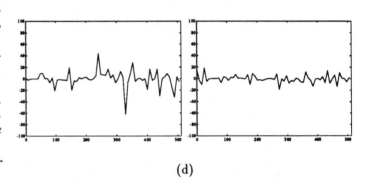

(d)

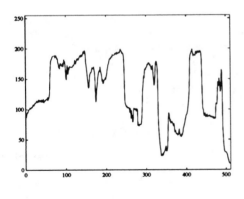

(a)

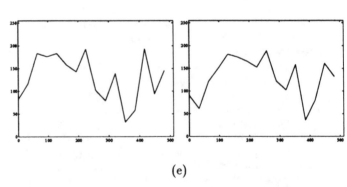

(e)

Figure 7. (a) Original signal. Detail signals of the morphological and wavelet decompositions, respectively: (b) first stage; (c) second stage; (d) third stage; and (e) smooth signal components of fifth stage. All components are displayed with same resolution

Figure 8.

Figure 10.

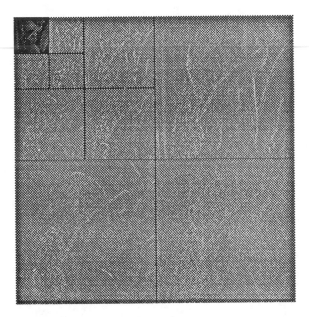

Figure 9.

Figure 11.

Chapter 4: Vector Quantization

Vector Quantization

Vector quantization maps a sequence of feature vectors to a digital symbol that indicates the identity of the most similar vector to the input vector from among those in a precomputed dictionary. This symbol is transmitted as a lower-bit-rate representation of the input vector and is then decoded by using the symbol to index into another copy of the dictionary, or codebook, of reproduction feature vectors. In an alternative view, the symbol sequence that arises from a vector quantized waveform can itself be used as a representation of the signal for recognition processing. The generation of the codebook, the ordering of its elements to allow efficient search, and the restricting of the search space to reduce the number of searches and accelerate the search process are described in this chapter.

Neural computing techniques are used in "Code-Excited Neural Vector Quantization" (paper 4.1), in which the combinatorial explosion of complexity that occurs for vector quantizers with large numbers of vector dimensions or high coding rate (large codebooks) is alleviated by the use of parallel computation of neural networks.

A computational bottleneck that limits real-time throughput for vector quantization is finding the nearest neighbor to the point in feature space that corresponds to an incoming vector from among large number of codebook points in a multidimensional space. "Algorithms for Fast Vector Quantization" (4.2) discusses and compares three nearest-neighbor search algorithms and applies them to representative data for a variety of vector dimensions and codebook sizes. "A Non-Metrical Space Search Algorithm for Fast Gaussian Vector Quantization" (4.3) presents a way to speed vector quantization by avoiding the comparison of the incoming vector to a substantial portion of the codebook. While past work has described methods to restrict the search space that result in occasionally missing the best-fitting code vector, this paper includes techniques that provide a result identical to full-search result.

Generating the codebook for a vector quantizer requires finding the collection of vectors as centroids to represent the signal source that minimizes total quantization distortion across the set of training signals. Iterative techniques such as the Generalized Lloyd Algorithm monotonically reduce the distortion for the training set upon each iteration, but they may converge on a suboptimal result by becoming trapped in a local minimum. Simulated annealing has been applied to such optimization problems to perturb the converging temporary result out of a local minimum, as exemplified by "Codebook Optimization by Cauchy Annealing" (4.4).

Suggested Additional Readings:

[1]   C. Constantinescu and J. A. Storer, "Improved Techniques for Single-Pass Adaptive Vector Quantization," *Proc. IEEE,* vol. 82, no. 6, pp. 933-939, June 1994.

[2]   R. M. Gray, "Vector Quantization," *IEEE ASSP Magazine,* vol. 1 no. 2, pp. 4-29, April 1984.

[3]   D. M. Monro and B. G. Sherlock, "Optimum DCT Quantization," *Proc. Data Compression Conference '93* 1993, pp. 188-194.

[4]   L. Owsley and L. Atlas, "Ordered Vector Quantization for Neural Network Pattern Classification," *Neural Networks for Signal Processing III - Proc. of the 1993 IEEE-SP Workshop* 1993, pp. 141-150.

[5]   R. M. Gray, K. L. Oehler, K. O. Perlmutter, R. A. Olshen, "Combining Tree-Structured Vector Quantization ith Classification and Regression Trees," *The Twenty-Seventh Asilomar Conference on Signals, Systems, and Computers* (Pacific Grove, CA, USA) Nov. 1994, pp. 1494-1498.

[6]   G. Poggi and E. Sasso, "Codebook Ordering Techniques for Address-Predictive VQ," *Proc. IEEE International Conference on Acoustics, Speech, and Signal Processing 93* (Minneapolis, MN, USA) April 1993, vol. 5 pp. 586-589.

[7]   K. C. Adkins, B. B. Bibyk, and R. T. Kaul, "Associative Computation Circuits for Real-Time Processing of Satellite Communications and Image Pattern Classification," *1993 International Symposium on Circuits and Systems* (Chicago, IL, USA) May 1993, pp. 227-230.

[8]   E. S. H. Cheung and A. G. Constantinides, "Fast Nearest Neighbor Search Algorithms for Self-Organising Map and Vector Quantization," *The Twenty-Seventh Asilomar Conference on Signals, Systems, and Computers* (Pacific Grove, CA, USA) Nov. 1994, pp. 946-950.

[9]   P. Knagenhjelm, "How Good Is Your Index Assignment?" *Proc. IEEE International Conference on Acoustics, Speech, and Signal Processing 93* (Minneapolis, MN, USA) April 1993, vol. 2 pp. 423-426.

CODE-EXCITED NEURAL VECTOR QUANTIZATION

Zhicheng Wang, John V. Hanson
Department of Electrical and Computer Engineering
University of Waterloo
Waterloo, Ontario, Canada N2L 3G1

Abstract

The Generalized Lloyd Algorithm (GLA), better known as the Linde-Buzo- Gray (LBG) algorithm is the most widely used technique in classical vector quantization (VQ) for speech or image signal compression. However, the encoding complexity of the algorithm grows exponentially with the product of coding rate and vector dimension, which prohibits applying the technique to tasks with moderate to large encoding rates or vector dimensions. This paper presents a new VQ scheme which overcomes the successive search coding computation of traditional techniques by using a quasi-parallel mapping technique and makes VQ practical for higher encoding rates and/or vector dimensions. Neural computing techniques are used to implement parallel encoding and decoding mappings in VQ and the developed algorithm is applied to quantizing Gauss-Markov processes and artificial data sources. Comparisons of performance with the LBG algorithm are given.

1 Introduction

In traditional Vector Quantization (VQ), the information is processed successively along a single line which leads to a processing bottleneck in the quantization search phase. The basic VQ design technique, the LBG algorithm– a clustering approach, is inherently limited for practical applications by its design and encoding complexity. More specifically, both the design and the encoding complexity grow exponentially with the product of coding rate and vector dimension and the storage requirement for the codebook increases linearly with the product. Although some vector quantization techniques such as tree-structured VQ, multistage VQ, gain/shape VQ, etc.[1], have been proposed to mitigate the complexity barrier at the expense of codebook storage, they

inevitably compromise the optimality of direct "full-search" VQ.

It is of interest to develop VQ techniques that do not suffer an exponential growth in complexity as a function of the rate-dimension product and hence are instrumentable for moderate to large encoding rates or vector dimensions.

Some research has been done on vector quantization with neural networks such as learning vector quantization (LVQ) [2] which is actually a generalized LBG algorithm, and the multi-layer net as front-end [3] which maps a high dimensional signal vector into a reduced dimensionality feature vector. This may then be vector quantized with a lower complexity than the direct quantization of the original vector. However, both still suffer the nearest-neighbor successive search.

Neural networks offer high overall data throughput rates due to parallel data processing when embedded in a hardware implementation. Based upon neural computing techniques, we have developed a code-excited neural vector quantization (CENVQ) algorithm in which the quantization consists of two mappings – encoding mapping and decoding mapping. Each is performed in parallel in the neural network. These two mappings are stored in the form of weights in the network which can be much more efficient than the storage of the conventional codebook.

In our vector quantization learning algorithm, the individual distortion $d(X_{nk}, Y_n)$ is minimized by taking into account the whole distortion E, improving the quantizer performance. Moreover, the minimization is inherently related to the partition. The centroid Y_n is a function of the inputs passing a partition, not just an arithmetic mean as in the LBG algorithm, and can be arbitrarily approximated based upon the minimization of the distortion measure.

In our neural vector quantizer, an input vector is directly mapped to a corresponding codeword with parallel computation. Hence the encoding in the

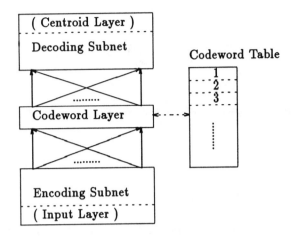

Figure 1: Block Diagram of CENVQ Model

CEVNQ system is much faster than in conventional VQs. In particular it is a breakthrough in VQ practical applications for moderate to large coding rates or vector dimensions.

2 The CENVQ System

The architecture of the neural quantizer is shown in Figure 1. It is a multilayer neural system which consists of an encoding subnet with weight set W_e, a codeword layer with r units, and a decoding subnet with weight set W_d, and a codeword table. The centroid (output) layer is the same as the input layer. The codeword layer is between the encoding and the decoding subnet. The interconnection schemes can be simply feedforward interfield, or include lateral connections within a layer.

The architecture is determined by the number $\{N_l, 1 \leq l \leq L\}$ of processing elements per layer. Here, $N_1 = N_L = N$. The elements of the lth layer are connected to the previous layer and their outputs $o_i^{(l)}, 1 \leq i \leq N_l$ is determined by

$$o_i^{(l)} = f^{(l)}\left(\sum_{j=1}^{N_{l-1}} w_{ij}^{(l)} o_j^{(l-1)} + \sum_{k=1, k \neq i}^{N_l} s_{ik} o_k^{(l)} + b_i^{(l)}\right) \quad (1)$$

where $\{b_i^{(l)}\}$ are the biases, $\{s_{ik}\}$ the strengths of the lateral connections within a layer, and $1 \leq i, k \leq N_l$, $2 \leq l \leq L$. However, $N_{l_c} = r$ elements at the codeword layer $l = l_c$. If the $\{s_{ik}\}$ are all zero, the network becomes feedforward. For hidden layers ($2 \leq l \leq L - 1$), the activation functions $\{f^{(l)}\}$ are

nonlinear and could be the same for all hidden layers. The hyperbolic tangent function is chosen as the activation function for the hidden layers. The output layer contains the same number of elements N as the input space dimension, and we use a linear activation function for the output (centroid) layer.

Assume that the input vector $X = (x_1, \ldots, x_k) \in \Re^k$, the output vector (centroid) $Y = (y_1, \ldots, y_k) \in \Re^k$, and the vector of the codeword layer is $C \in R^r$, that is, $X = O^{(1)}, Y = O^{(L)}$ and $C = O^{(l_c)}$. The system is a realization of two mappings, one being an encoding (Θ) mapping from a continuous input space to a discrete codeword space,

$$C = \Theta(X, W_e), \quad (2)$$

The other is a decoding (Φ) mapping from a codeword space to an output (reproduction) space,

$$Y = \Phi(C, W_d). \quad (3)$$

The decoding subnet is driven by a codeword vector selected from the codeword table. The input vector is compared with the output vector Y. For each input X, a vector C with minimum of d(X,Y) is selected by means of searching the codeword table.

$$C^* = min_C^{-1} d(X, \Phi(C, W_d)). \quad (4)$$

The total distortion over a data set is:

$$E = \sum_X d(X, Y) = \sum_X d(X, \Phi(C, W_d)). \quad (5)$$

The goal of the decoding subnet is to find the minimum E via W_d

$$W_d^* = min_{W_d}^{-1} E(X, C, W_d). \quad (6)$$

The encoding mapping is realized based on the decoding mapping result which is the assignment of a codeword to each input vector. During the training of the encoding mapping, the total distortion at the codeword layer is:

$$E_c = \sum_X d(C, O^{l_c}) = \sum_X d(C, \Theta(X, W_e)). \quad (7)$$

The encoding subnet learns the coding mapping by the minimization of E_c via W_e

$$W_e^* = min_{W_e}^{-1} E_c(C, X, W_e). \quad (8)$$

Both mappings are learned by taking into account the unknown probability density function of the data set through the weight updating. Therefore, the optimal polytopes of a partition can be any form approximated by the mappings.

3 The CENVQ Learning Algorithm

Multilayer neural networks have been applied to data compression, especially image data compression [6]. But in previous data compression using multilayer network architectures, internal representations were used as reduced dimensionality feature vectors. Hence, the exponential growth in complexity remained unsolved, although the vector dimension was reduced.

In our CENVQ learning algorithm, we use the codeword vectors at the codeword layer as fundamental entities to be determined by searching in the codeword space. But training the weights in the encoding and decoding subnets uses a descent gradient method similar to the error back-propagation algorithm or its generalization to recurrent neural networks [7] if the $\{s_{ik}\}$ are not zero. For the Φ mapping, we use codewords from a codeword space to excite the decoding subnet from the codeword layer as inputs. There is an iteration procedure for searching the best codewords corresponding to input vectors and for updating the weights of the decoding subnet to improve the Φ mapping. However the weights update with a whole span through the data set to catch its statistical character. After this procedure, the relationship between the codewords and training data vectors has been established, and this can be used to train the encoding subnet for the Θ mapping.

The learning process has four stages:

1. Initialize the weights in both encoding and decoding subnets with random values within a small value range.

2. For each input vector of the complete training data set, choose the best codeword vector by searching in the codeword space, based on the objective function of distortion $d(X, \Phi(C, W_d))$,

3. Train the decoding subnetwork to update the Φ mapping with a descent gradient algorithm similar to the generalized error back-propagation algorithm. If the average distortion for the whole training set is below a threshold, go to the stage 4. Otherwise, go to stage 2.

4. Using the relationship established in the previous stage, train the encoding subnet for the Θ mapping with the training data vectors as inputs and the corresponding codewords as outputs.

The iterations of stage 2 and 3, the codeword vector selection and weight updating influence each other. The best codeword for a input vector at this interation could be different from the previous codeword.

| Partition | Average Distortion | |
	F-SVQ	CENVQ
1	0.099738	Nil
2	0.084598	0.084595
3	0.083374	0.083374

Table 1: CENVQ and LBG with different partitions for a artificial data source

| Bits/vector | SNR (db) | |
	F-SVQ	CENVQ
1	3.277	3.277
2	6.015	5.925
3	8.278	8.203
4	10.384	10.245

Table 2: Performance comparisions between CENVQ and LBG algorithm

With the nth iteration, equations (4) and (6) become

$$C^*(n) = min_{C(n)}^{-1} d(X, \Phi(C^*(n), W_d(n))) \quad (9)$$

$$W_d(n+1) = min_{W_d}^{-1} E(X, C^*(n), W_d(n)) \quad (10)$$

The storage requirement is very high when the training data set is large: codewords of all training vectors are stored. However we can have a much smaller table used for learning and substitute its entries with unused training vectors iteratively.

4 Simulation Results

The simple but nontrivial example demonstrated in [4] which used an artificial data source is presented with our CENVQ model. The coding rate is 2 bits/vector. The results with different partitions are shown in Table 1, which indicate that the CENVQ gives better partitions. In the LBG algorithm, a partition is determined by an initial codebook. However, in CENVQ, a partition is influenced by the learning rate and the system always tends to converge to a partition with the least total distortion.

We present experimental results for a Gauss-Markov source, which $\{x_n\}$ is defined by the difference equation

$$x_{n+1} = ax_n + G_n, \quad (11)$$

where $\{G_n\}$ is a zero mean unit variance i.i.d. Gaussian time series. This source is of interest since it is a commonly used model of real data, used for comparing data compression systems. We here consider only the case a=0.9 of a highly correlated source, the correlation coefficient being typical of speech and scanned image data.

The code-excited neural vector quantizers were made for the vector dimension N=5 with rate of R=1, 2, 3, 4 bits per vector. The neural quantizers are symmetric about a codeword layer. The resulting SNRs are listed in Table 2. The performance of both algorithms is almost the same. Hence, the CENVQ can at least be the same optimality as the direct "full-search" VQ technique and better than the other vector quantization techniques indicated in [1]. The CENVQ can achieve better results if the algorithm is modified to escape from poor local minima as presented in [9]. The CENVQ can overcome the empty cell problem in traditional VQ algorithms and the neural vector quantizer presented in [8].

Usually we choose one hidden layer for encoding and decoding subnets, respectively, which can realize the approximate nonlinear Θ and Φ mappings [5]. However the size of hidden layers was chosen experimentally in our simulations and it is not always true that the larger size gives better performance. A pruning technique could be used to design optimal size networks for best performance and least computational expense.

5 Conclusions

A new quantization technique has been developed based upon the parallel computation of neural networks, in which the computation and storage requirements are not influenced by the quantization task directly. However they are determined by the size of the quantizer network, whose performance is influenced by the size of hidden layers. We conjecture that the CENVQ has better generalization capability than LBG algorithm. Although the CENVQ is quite computationally intensive in the design phase, this is a secondary concern since the complexities of the two parallel mappings in the actual encoding and decoding operations are not affected. Actually, the encoding operation is carried out in parallel with hardware implementation, breaking through the computational barrier of traditional vector quantization techniques. Furthermore, the decoding mapping can be substituted with table-look-up procedures in the decoding operation. In summary, this paper demonstrates a VQ algorithm which uses parallel mapping to overcome the successive search coding computational barrier and makes VQ practical for applications involving high coding rates and/or vector dimensions.

References

[1] R.M.Gray, "Vector Quantization", *IEEE ASSP Magazine*, pp.4-29, April 1984.

[2] T.Kohonen, "Improved Versions of Learning Vector Quantization", *Proceedings of 1990 IEEE International Joint Conference on Neural Networks*, vol.I, pp.545-550, 1990.

[3] N.Sonehara, M.Kawato, S.Miyake, and K.Nakane, "Image Data Compression Using a Neural Network Model", *Proceedings of International Joint Conference on Neural Networks*, vol.II, pp.35-40, 1989.

[4] Y.Linde, A.Buzo, and R.M.Gray, "An Algorithm for Vector Quantizer Design", *IEEE Trans. Commun.*, vol.COM-28, pp.84-95, Jan.1980.

[5] K.Funahashi, "On the Approximate Realization of Continuous Mappings by Neural Networks", *Neural Networks*, Vol.2, pp.183-192,1989.

[6] A.Namphol, M.Arozullah, and S.Chin, "Higher Order Data Compression with Neural Networks", *Proceedings of 1991 IEEE International Joint Conference on Neural Networks*, vol.1, pp.55-59, Seattle, WA, July 1991.

[7] F.J.Pineda, "Generalization of Back-Propagation to Recurrent Neural Networks", *Physical Review Letters*, vol.59, no.19, pp.2229-2232, November 1987.

[8] Z.Wang and J.V.Hanson, "A Neural Vector Quantizer", *Proceedings of 1992 IEEE International Symposium on Circuits and Systems*, vol.1, pp.351-354, San Diego, CA, May 1992.

[9] Z. Wang and J.V.Hanson, "Design Optimization of Code-Excited Neural Vector Quantizers", submitted to *The 1993 IEEE International Conference on Neural Networks*, San Francisco, CA, March 28-April 1, 1993.

[0]ITRC and NSERC support for this work is gratefully acknowledged.

Algorithms for Fast Vector Quantization

Sunil Arya

Department of Computer Science
University of Maryland, College Park, Maryland, 20742

David M. Mount

Institute for Advanced Computer Studies and
Department of Computer Science
University of Maryland, College Park, Maryland, 20742

1   Introduction

The nearest neighbor problem is to find the point closest to a query point among a set of n points in k-dimensional space. Finding the nearest neighbor is a problem of significant importance in many applications. One important application is vector quantization, a technique used in the compression of speech and images [9]. Samples taken from a signal are blocked into vectors of length k. Based on a training set of vectors, a set of codevectors is first precomputed. The technique then encodes each new vector by the index of its nearest neighbor among the codevectors.

The rate r of a vector quantizer is the number of bits used to encode a sample and it is related to n, the number of codevectors, by:

$$n = 2^{rk}$$

For fixed rate, the performance of vector quantization improves as dimension increases but, unfortunately, the number of codevectors grows exponentially with dimension. There have been two major approaches to deal with this increase in complexity. The first approach is to impose structure on the codebook, so that the nearest neighbor or an approximation to it can be found rapidly [9]. Some deterioration in performance occurs because the imposition of structure results in a non-optimal codebook. The second approach is to preprocess the unstructured codebook so that the complexity of nearest neighbor searching is reduced [8], [7], [5]. However, the complexity of practical algorithms that fall under this approach increases very rapidly with dimension, and their running time is little better than brute-force linear search for rate about 1 bit per sample or less (i.e., for $n \leq 2^k$).

In this paper we show that if one is willing to relax the requirement of finding the true nearest neighbor, it is possible to achieve significant improvements in running time and at only a very small loss in the performance of the vector quantizer. We present three algorithms for nearest neighbor searching:

(1) the standard k-d tree search algorithm [4] similar to the version proposed by Sproull [11], but with the improvement of replacing distance estimates with exact distance values,

0-8186-3392-1/93 $3.00 © 1993 IEEE

(2) a new search algorithm, which we call priority k-d tree search, in which cells of the search tree are visited in priority order of distance from the query point, and

(3) a neighborhood graph search algorithm [2] in which a directed graph is constructed for the point set and edges join neighboring points.

We performed numerous experiments on these algorithms on point sets from various distributions, and in dimensions ranging from 8 to 16. We employed a rate of 1 bit per sample for these experiments. We studied the running times of these algorithms measured in various ways (number of points visited, number of floating point operations). We also measured the performance of these algorithms in various ways (relative error, signal to noise ratio, and probability of failing to find the true nearest neighbor). Our studies show that, for many distributions in high dimensions, the latter two algorithms provide a drastic reduction in running time over standard approaches with very little loss in performance.

2 Nearest neighbor searching using k-d trees: standard approach

Bentley introduced the k-d tree as a generalization of the binary search tree in higher dimensions [3]. Each internal node of the k-d tree is associated with a hyperrectangle and a hyperplane orthogonal to one of the coordinate axis, which splits the hyperrectangle into two parts. These two parts are then associated with the two child nodes. The process of partitioning space continues until the number of data points in the hyperrectangle falls below some given threshold. These hyperrectangles are called buckets and the corresponding nodes are the leaf nodes of the tree. Data points are only stored in the leaf nodes, not in the internal nodes.

Friedman, Bentley and Finkel [7] gave an algorithm to find the nearest neighbor using optimized k-d trees that takes $O(\log n)$ time in the expected case, under certain assumptions on the distribution of data and query points. The internal nodes of the *optimized k-d* tree split the set of data points lying in the corresponding hyperrectangle into two equal parts, along the dimension in which the data points have maximum spread. The algorithm works by first descending the tree to find the data points lying in the bucket that contains the query point. Then it examines surrounding buckets if they intersect the sphere S centered at the query point and having radius equal to the distance between the query point and the closest data point visited so far. The lower and upper limits of the corresponding hyperrectangle (*bounds array*) can be explicitly stored in each of the internal nodes of the k-d tree, to help prune the search so that buckets that do not intersect the sphere S are not examined. Sproull [11] showed that there was no need to explicitly store the bounds array, and gave an alternative version to save space and speed up the search. Sproull's version examines buckets if they intersect the smallest hypercube enclosing sphere S. We refine the implementation further so that buckets are examined only if they actually intersect sphere S (we call this the *distance* refinement, and refer to the algorithm incorporating this refinement as the *standard* approach). This implies a very significant saving in high dimensions because of the large difference in the volume of a sphere and the volume of the enclosing hypercube in these dimensions. For example, given 65,536 points in 16 dimensions from an uncorrelated Gaussian source, the numbers of points visited with and without this refinement are 14,500 and 50,000, respectively. These averages were computed over 25,000 query points, also from the same source. For these and all other experiments described in this paper, we used optimized k-d trees with one data point per bucket and measured distances in the Euclidean norm.

Briefly, the algorithm incorporating the distance refinement works as follows. At each leaf node visited we compute the squared distance between the query point and the data point in the bucket and update the nearest neighbor if this is the closest point seen so far. At each internal node visited we first search the subtree whose corresponding hyperrectangle is closer to the query point. Later, we search the farther subtree if the squared distance between the query point and the closest point visited so far exceeds the squared distance between the query point and the corresponding hyperrectangle. To facilitate the computation of the squared distance between the query point and the hyperrectangle, we maintain a variable *cpdist2* which keeps track of this quantity. We also maintain an array *cp* to keep track of the distance between the query point and the hyperrectangle along each dimension. As the k-d tree is traversed, it is possible to update *cpdist2* and the appropriate element of *cp* in constant time for each node visited. For more details, the interested reader is referred to [1].

3   Nearest neighbor searching using k-d trees: priority approach

The standard k-d tree algorithm usually comes across the nearest neighbor much before the search terminates. One may view the extra search as the price to pay to guarantee that the nearest neighbor has been found. If we are willing to sacrifice this guarantee, then the complexity can be reduced by interrupting the search before it terminates (say, after a fixed number of points have been visited). In this case, it is desirable to order the search so that buckets more likely to contain the nearest neighbor are visited early on. This suggests a variant of the standard k-d tree algorithm that visits the buckets of the k-d tree in increasing order of distance from the query point.

The algorithm maintains a priority queue of subtrees, where the priority of a subtree is inversely related to the distance between the query point and the hyperrectangle corresponding to the subtree. Initially, we insert the root of the k-d tree into the priority queue. Then we repeatedly carry out the following procedure. First, we extract the subtree with highest priority from the queue. Then we descend this subtree to find the bucket closest to the query point. We update the nearest neighbor if the data point in the bucket is the closest point visited so far. As we descend the subtree, for each node u that we visit we insert u's sibling into the priority queue. The algorithm terminates when the priority queue is empty, or sooner, if the distance from the query point to the hyperrectangle corresponding to the highest priority subtree is greater than the distance to the closest data point. The code for an implementation of this algorithm is given in [1].

To justify the correctness of this algorithm, observe first that we may regard a subtree as a set of buckets that are contained in the hyperrectangle associated with the subtree. Then the claim is that at the beginning of each iteration of the above procedure, the following invariant holds: the subtrees in the priority queue are mutually disjoint (have no bucket in common) and their union is the set of buckets not yet visited. This claim rests on the observation that at each iteration a subtree T is removed from the priority queue and replaced by (0 or more) subtrees whose union is the set of all the buckets contained in subtree T minus the bucket in subtree T that is visited next. The correctness of the algorithm follows easily from the above claim and the fact that the subtree with the highest priority contains the bucket closest to the query point.

4 Neighborhood graphs

We give here a brief overview of an approach to nearest neighbor searching based on the notion of neighborhood graphs, which was introduced in [2]. A neighborhood graph is a connected graph (directed or undirected) whose vertices are the set of data points, such that two points are adjacent to one another if they satisfy some local criterion. For example, the Delaunay triangulation is an undirected neighborhood graph in which two points are adjacent if there is a sphere passing through the two points that contains no other point in its interior.

Given a neighborhood graph we can search for the nearest neighbor of a query point using a greedy strategy. We start the search with the data point p from the bucket of the k-d tree containing the query point. We repeatedly carry out the following steps. We *expand* the point p, by which we mean that we compute the distance to the query point for all those neighbors of point p that have not yet been expanded. Among such neighbors, we expand the point that is closest to the query point. We continue to expand points in this manner until we arrive at a point all of whose neighbors have already been expanded (the search is said to have reached an *impasse*), or the number of points visited by the algorithm exceeds some prespecified cut-off value. Then we end the search and output the closest data point visited.

The neighborhood graph we use for nearest neighbor searching is quite similar to the relative neighborhood graph (RNG) [12]. In the RNG, two points p and r are adjacent if there is no point that is simultaneously closer to both points than they are to one another. The modified graph we build is equivalent to a graph presented by Jaromczyk and Kowaluk [10] which was used as an intermediate result in their construction of the RNG. It is based on the following pruning rule. For each point p in the data set, we consider the remaining points in increasing order of distance from p. We remove the closest point x from this sequence, create a directed edge from p to

x, and remove from further consideration all points s such that $dist(p, s) > dist(x, s)$. This process is repeated until all points are pruned. Figure 1 shows an example of this process applied to a point p in three stages; at each stage, a new edge is directed from p to one of its neighbors. The three neighbors of point p in the neighborhood graph are x, y and z. This variant, called the RNG*, can be computed in $O(n^2)$ time, where n is the number of points. Details on the degree of the RNG* and intuitive reasons for its appeal in nearest neighbor searching can be found in [2].

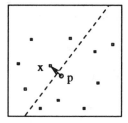 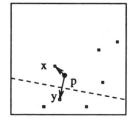 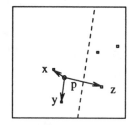

Figure 1: Sparse neighborhood graph (RNG^*).

Although the worse case behavior of the greedy algorithm can be quite bad, our experimental studies indicate that in high dimensions the search quickly zeroes in to find the nearest neighbor and only rarely reaches an *impasse* before finding the nearest neighbor.

5 Empirical Analysis

Before running the experiments, we optimized many aspects of the code for all three algorithms. We mention two of these optimizations. First, the well-known *partial distance* optimization was implemented for all three algorithms: as we compute the squared distance between the query point and the data point by summing the contribution from each dimension, we exit the loop when the accumulated sum of the squares becomes too large [11]. This optimization diminished the total number of floating point operations at only a small increase in the number of comparisons. Second, for the RNG*-search, we saved the results of the partial distance computations so that they could be used again if the same point was encountered on expanding several different points.

We studied how the performance of the vector quantizer changes as a function of the complexity of the algorithms. We focused on rate 1 bit per sample in dimensions ranging from 8 to 16. For each of the three algorithms the search is interrupted if the number of points visited by the algorithm reaches a certain threshold (*cut-off value*), and the closest point visited until then is taken as the output of the algorithm. By varying this cut-off value the complexity of each algorithm can be changed. We used 25,000 query points for each experiment and recorded the following for each query point at each cut-off (here d_e is the distance between the query point and the data point output by the algorithm and d_n is the distance between the query point and its nearest neighbor):

141

- The number of data points visited by the algorithm. A data point is said to be *visited* if the algorithm accesses its coordinates. Each data point is counted at most once in this total. If the algorithm terminates before the cut-off is reached, then this quantity is the same as the number of points visited until termination, otherwise it is the cut-off value.

- The number of floating point operations per sample. In this we included all floating point additions, subtractions, multiplications and comparisons (except comparisons with zero) performed by the algorithm, not just those involved in computing distances between points. The total number of floating point operations is divided by the dimension to get the per sample average.

- Whether the true nearest neighbor has already been found.

- The error-factor which is defined as $(d_e - d_n)/d_n$.

- The distortion per sample which is defined as $(d_e)^2/k$, where k is the dimension.

Three measures of performance were computed at each cut-off — the signal-to-noise ratio (SNR), the average error-factor and the miss probability (probability of failing to find the nearest neighbor). The SNR is defined as $10\log_{10}(V/D)$, where V is the variance of the samples and D is the average distortion per sample. All these averages are taken over the entire set of query points. Of these measures of performance, SNR is the most significant one for vector quantization. The other two measures are here principally to aid in a better empirical understanding of the algorithms.

Two measures of complexity were computed at each cut-off — the average number of points visited and the average number of floating point operations per sample. The number of points visited is a useful quantity to study, but since the algorithms have different overheads, the number of floating point operations is more directly related to the complexity of the algorithms. Our studies indicate that the number of floating point operations is a reasonable measure of the search time, and can be used to compare the algorithms.

We conducted experiments using the Gaussian and the Laplacian sources. Both uncorrelated and correlated sources were used. For the correlated sources, we used 0.9 as the correlation coefficient. All the sources had zero mean and unit variance.

In dimension 16 we used codebooks consisting of 65,536 codevectors generated by the k-d tree based Equitz algorithm [6]. We sped up Equitz algorithm in several ways and, for uncorrelated sources, instead of building balanced k-d trees as is customary, we partitioned the hyperrectangles corresponding to the internal nodes such that a random number of points were contained in each part. This led to codebooks of better quality. The size of the training set used was 32 times the size of the codebook. For our experiments the training set and test set were different.

Figure 2 shows the variation of the SNR with the average number of floating point operations per sample, for the uncorrelated Gaussian, uncorrelated Laplacian, correlated Gaussian, and correlated Laplacian sources, respectively. For a more careful

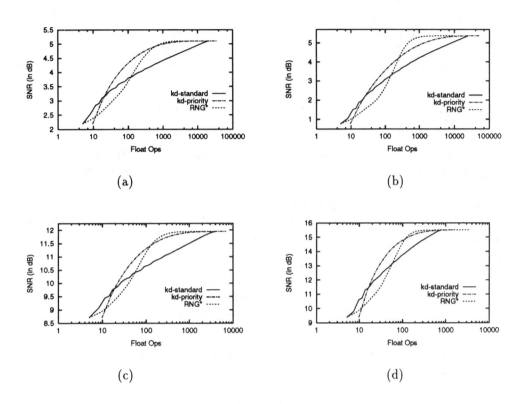

Figure 2: SNR vs. Average floating point operations per sample. (a) Uncorrelated Gaussian. (b) Uncorrelated Laplacian. (c) Correlated Gaussian. (d) Correlated Laplacian.

DISTRIBUTION	SNR-MAX	KD-STANDARD	KD-PRIORITY	RNG*
Uncorrelated Gaussian	5.11	12000	1100	850
Uncorrelated Laplacian	5.36	18500	4500	850
Correlated Gaussian	11.94	2500	550	300
Correlated Laplacian	15.51	650	400	200

Table 1: Floating point operations per sample to achieve SNR within 0.1 dB of SNR-MAX.

DISTRIBUTION	SNR-MAX	KD-STANDARD	KD-PRIORITY	RNG*
Uncorrelated Gaussian	5.11	19000	5000	2000
Uncorrelated Laplacian	5.36	24000	15000	2000
Correlated Gaussian	11.94	3700	1700	600
Correlated Laplacian	15.51	800	950	450

Table 2: Floating point operations per sample to achieve SNR within 0.01 dB of SNR-MAX.

study of the high performance region (say, less than 0.1 dB deterioration from the performance obtained by full exhaustive search, SNR-MAX), we make two tables. Table 1 compares the three algorithms in terms of the average number of floating point operations per sample needed to achieve SNR within 0.1 dB of SNR-MAX. Table 2 shows the same needed to achieve SNR within 0.01 dB of SNR-MAX. Due to lack of space, we summarize the key observations for dimension 16 in the high performance region:

- RNG*-search is the fastest of the three algorithms followed by the priority k-d tree algorithm. Both these algorithms often achieve very significant speed-ups, sometimes by a factor of over 10, compared to the standard k-d tree algorithm.

- The complexity of RNG*-search ranges from being just a little better than the priority k-d tree algorithm to being much better, sometimes achieving speed-ups by a factor of over 5.

- All three algorithms achieve significant speed-ups over full exhaustive search, with negligible loss in performance (less than 0.01 dB). RNG*-search achieves massive speed-ups by a factor of over 100 compared to full exhaustive search. Even the standard k-d tree algorithm achieves speed-ups by a factor of over 8 compared to full exhaustive search.

- The two k-d tree algorithms require similar storage, while the RNG*-search may require about twice as much (depending on the source and the implementation). Regarding dimension as fixed and ignoring the space needed for the data points, the storage requirements of all three algorithms is $O(n)$. The constant of proportionality for the k-d tree algorithm is largely independent of dimension, while for the RNG*-search it shows a moderately exponential growth. For the uniform distribution, empirical studies show that the degree of RNG* grows roughly as $2.90(1.24^k)$ in the asymptotic case, and as $1.46(1.20^k)$ when the number of points grow as 2^k [2]. The cost of building the RNG* is $O(n^2)$ while that of building the k-d tree is $O(n \log n)$. This is an enormous difference because the number of points is so large.

For the uncorrelated Gaussian source, Figure 3 shows how the average error-factor and the probability of failing to find the nearest neighbor vary with the average

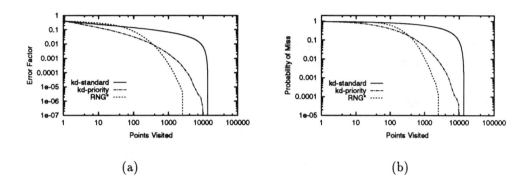

(a)   (b)

Figure 3: Uncorrelated Gaussian. (a) Average error-factor vs. Average points visited.
(b) Probability of missing the nearest neighbor vs. Average points visited.

number of points visited. These graphs show that both these quantities fall much more
rapidly for the RNG*-search than for the priority k-d tree algorithm. This suggests
that the RNG*-search would enjoy very significant advantage over the priority k-d
tree algorithm in any application where these quantities are critical.

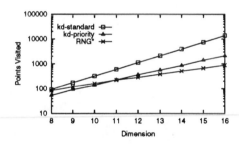

Figure 4: Uncorrelated Gaussian: Average points visited to achieve error-factor 0.001.

We also conducted experiments for the uncorrelated Gaussian source in several
other dimensions ranging from 8 to 16, using random codebooks. In each case we
used a rate of 1 bit per sample. In all these dimensions we found that the priority
k-d tree algorithm and the RNG*-search are much faster than the standard k-d tree
algorithm. In dimensions below 11 we found the priority k-d tree algorithm to be
faster than the RNG*-search, while in dimensions above 11 the RNG*-search is faster
in the high performance region. This can be seen in Figure 4 which shows, in various
dimensions, the average number of points visited by the three algorithms to achieve
an average error-factor of 0.001.

6 Acknowledgements

We would like to thank Nam Phamdo for many helpful discussions on the subject of generating codebooks, and Nariman Farvardin for useful suggestions.

References

[1] S. Arya and D. M. Mount. Algorithms for fast vector quantization. Technical Report CS-TR-3017, University of Maryland Institute for Advanced Computer Studies (UMIACS), January 1993.

[2] S. Arya and D. M. Mount. Approximate nearest neighbor queries in fixed dimensions. In *Proc. 4th ACM-SIAM Sympos. Discrete Algorithms*, 1993.

[3] J. L. Bentley. Multidimensional binary search trees used for associative searching. *Communications of the ACM*, 18(9):509–517, September 1975.

[4] J. L. Bentley. K-d trees for semidynamic point sets. In *Proc. 6th Ann. ACM Sympos. Comput. Geom.*, pages 187–197, 1990.

[5] D. Y. Cheng and A. Gersho. A fast codebook search algorithm for nearest-neighbor pattern matching. In *Proc. IEEE Int. Conf. Acoust., Speech, Signal Processing*, volume 1, pages 265–268, April 1986.

[6] W. H. Equitz. A new vector quantization clustering algorithm. *IEEE Transactions on Acoust., Speech and Signal Process.*, 37(10):1568–1575, October 1989.

[7] J. H. Friedman, J. L. Bentley, and R.A. Finkel. An algorithm for finding best matches in logarithmic expected time. *ACM Transactions on Mathematical Software*, 3(3):209–226, September 1977.

[8] K. Fukunaga and P. M. Narendra. A branch and bound algorithm for computing k-nearest neighbors. *IEEE Transactions on Computers*, 24:750–753, July 1975.

[9] A. Gersho and R. M. Gray. *Vector Quantization and Signal Compression*. Kluwer Academic, 1991.

[10] J. W. Jaromczyk and M. Kowaluk. A note on relative neighborhood graphs. In *Proc. 3rd Ann. ACM Sympos. Comput. Geom.*, pages 233–241, 1987.

[11] R. L. Sproull. Refinements to nearest-neighbor searching in k-dimensional trees. *Algorithmica*, 6, 1991.

[12] G. T. Toussaint. The relative neighborhood graph of a finite planar set. *Pattern Recognition*, 12(4):261–268, 1980.

A NON-METRICAL SPACE SEARCH ALGORITHM FOR FAST GAUSSIAN VECTOR QUANTIZATION†

E.G. Schukat-Talamazzini, M. Bielecki, H. Niemann, T. Kuhn, S. Rieck

Universität Erlangen-Nürnberg, Lehrstuhl für Informatik 5 (Mustererkennung)
Martensstraße 3, 8520 Erlangen, Germany
E-mail: schukat@informatik.uni-erlangen.de

Abstract

We present three algorithms to speed up full covariance multivariate Gaussian vector quantizers. The speed-up is achieved by avoiding the distance calculation for a considerable number of codebook classes at each input frame. In two cases, this pruning is guided by thresholds $U_{\kappa\lambda}$ which are computed for each class pair in a preprocessing stage. The computation of the $U_{\kappa\lambda}$'s from the codebook parameters by the gradient projection method leads to an admissible search strategy. To our knowledge, there is no formerly published algorithm that can be proved admissible for Gaussian codebooks. Two of the proposed search procedures trade accuracy for speed. Both of them allow more than fivefold speed-up vector quantization at very low frame error rates, and without any degradation of word accuracy.

1 Introduction

During the last decade, it has been a standard approach in automatic speech recognition to represent phonetic speech units using hidden Markov models (HMM's). In the case of HMM's with discrete or semi-continuous [9] probability density functions (PDF's) the mapping of the parametric input to discrete categories is performed by a vector quantizer (VQ) [7]. The distance function $u_\kappa(x)$ which measures the dissimilarity between a real-valued input vector x and the VQ codebook class Ω_κ is usually chosen as the (squared) Euclidean distance

$$u_\kappa(x) = (x - \mu_\kappa)^{\mathrm{T}}(x - \mu_\kappa) \qquad (1)$$

or the negative logarithm of a class-dependent multivariate Gaussian PDF

$$u_\kappa(x) = \log(|2\pi K_\kappa|) + (x - \mu_\kappa)^{\mathrm{T}} K_\kappa^{-1}(x - \mu_\kappa) \quad (2)$$

with mean vectors μ_κ and covariance matrices K_κ. The VQ stage of a recognizer forms a serious computional bottleneck because its cost (in terms of the number of multiplications per input frame) is roughly

proportional to the number of codebook classes and to the square of the input vector dimension, if Eq. (2) is considered. For example, if $K = 256$ codebook classes and feature vectors of dimension $d = 24$ are required, which are typical values for a speech recognition application, the computation amounts to

$$K \cdot \left(1 + d + (d+1) \cdot \frac{d}{2}\right) = 83\,200 \qquad (3)$$

multiplications every 10 milliseconds which is prohibitive for a real-time system without special hardware.

Algorithms to speed up VQ's: In the literature a lot of approaches to speed up VQ's are described. The basic technique is to avoid a full search during the classification of an input frame, i.e., to make sure that distance calculations have to be performed only for a possibly small fraction of the codebook classes. To achieve that, (binary) decision trees can be built up during or after the codebook design phase [3, 19]. Without relying on specifically structured codebooks a fast preselection of class candidates may be obtained using projection or hypercube methods [4] or constructing sequential classifiers [12]. A dynamic pruning of codebook classes from further consideration can take place during distance computation (partial distances [1], hypercubes [18, 17], minimax method [4]) or after distance computation (branch and bound [13], triangle inequality pruning [14]).

Our purpose here was to speed up the VQ component of a semi-continuous HMM recognizer. The VQ codebook is of Gaussian type and the codebook parameters are subject to Baum-Welch reestimation [9]. As a consequence, algorithms which require a specialized codebook design are not applicable because the codebook parameters are given in advance. Furthermore, many of the approaches cited above explicitly choose the dissimilarity function $u_\kappa(x)$ to be the Euclidean distance, or at least they assume the form $u_\kappa(x) = d(\mu_\kappa, x)$ where $d(\cdot, \cdot)$ is a metric. This, however, obviously does not apply to Gaussian quantizers (Eq. (2)).

Several of the above search strategies are approximate: they make quantization faster but, in general, *admissibility* is lost, i.e., the accelerated version cannot

† This work was partly funded by the Commission of the EC under ESPRIT contract P 2218 (*SUNDIAL*).

be guaranteed to produce output identical to the full search case. To our knowledge, there is no published algorithm for Gaussian codebooks that has been proved admissible. Of course, a slight distortion of the quantization process can be tolerated, namely if the impact on the recognition accuracy is neglegible.

The remainder of this article is organized as follows. In Sec. (2) we will generalize the TIP (triangle inequality pruning) algorithm to the case of Gaussian PDF codebooks. Two non-metrical space search (NMSS) algorithms will be developed: the first one (*"geometrical"* pruning VQ) inherits the admissibility from the TIP algorithm, whilst the second one (*"empirical"* pruning VQ) trades accuracy for speed. In Sec. (3) we describe the (non-admissible) *"sequential"* pruning VQ, which is in essence a multi-stage Gaussian classifier. Finally, in Sec. (4) the three competing approaches are quantitatively evaluated on a large set of speech data with respect to computation time and quantization accuracy.

2 Non-metrical space search

The TIP algorithm [14] relies on the assumption that the distance function can be expressed as $u_\kappa(x) = d(x, \mu_\kappa)$ where $d(\cdot, \cdot)$ is a metric and μ_κ is the centroid of codebook class Ω_κ, $\kappa = 1, \ldots, K$. From the triangle inequality

$$d(x, \mu_\kappa) + d(x, \mu_\lambda) \geq d(\mu_\kappa, \mu_\lambda) =: 2 \cdot U_{\kappa\lambda} \quad (4)$$

we obtain the codebook class pruning rule

$$u_\kappa(x) < U_{\kappa\lambda} \quad \longrightarrow \quad \boxed{\text{prune class } \Omega_\lambda} \quad (5)$$

This rule states that if the current input vector x is sufficiently close to class Ω_κ, the competing class Ω_λ can already be excluded without any distance computation (cf. Fig. (1)). This can be seen as follows: assume $x \in \Omega_\lambda$. Then we we have in particular $u_\lambda(x) < u_\kappa(x)$ and hence

$$u_\kappa(x) + u_\lambda(x) < 2 \cdot u_\kappa(x) < 2 \cdot U_{\kappa\lambda} \quad (6)$$

which is in contradiction with Eq. (4).

The threshold matrix $U = (U_{\kappa\lambda})_{\kappa,\lambda=1,\ldots,K}$ is calculated in a preprocessing step from the codebook centroids.

For an arbitrary VQ characterized by the distance functions $u_\kappa(x)$, $\kappa = 1, \ldots, K$ we would also like to precompute a threshold matrix U with the pruning rule (Eq. (5)) still valid. For the reason of optimal acceleration, the thresholds $U_{\kappa\lambda}$ should be maximal with that property. The best pruning matrix conceivable is certainly that particular one defined by Eq. (7):

$$U_{\kappa\lambda}^{(1)} = \min_{x \in \Omega_\lambda} u_\kappa(x) \quad (7)$$

$$U_{\kappa\lambda}^{(2)} = \min_{x \in \omega_\lambda} u_\kappa(x) \quad (8)$$

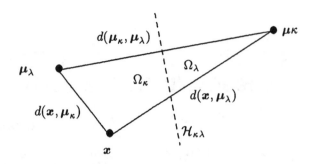

Figure 1: The principle of triangle inequality pruning

$$U_{\kappa\lambda}^{(3)} = \min_x \{ u_\kappa(x) \mid u_\kappa(x) \geq u_\lambda(x) \} \quad (9)$$

$$U_{\kappa\lambda}^{(4)} = \min_{x \in \mathcal{H}_{\kappa\lambda}} u_\kappa(x) \quad (10)$$

Unfortunately, for the interesting case of $u_\kappa(x)$ being Gaussian log likelihoods, to our knowledge there is no closed nor iterative solution for Eq. (7). A first way out [10] is to choose *empirical* pruning thresholds $U_{\kappa\lambda}^{(2)}$ by substituting the pattern class Ω_κ in Eq. (7) by a hopefully representative sample ω_κ. However, since $U_{\kappa\lambda}^{(1)} \leq U_{\kappa\lambda}^{(2)}$ this solution offers additional pruning power at the cost of strict admissibility.

Geometrical pruning: Thus we decided to compute the *geometrical* pruning thresholds $U_{\kappa\lambda}^{(3)}$, too. As we have $U_{\kappa\lambda}^{(3)} \leq U_{\kappa\lambda}^{(1)}$, geometrical pruning is admissible, but probably somewhat less efficient as compared with empirical pruning. Eq. (9) is an instance of a non-linear optimization problem with non-linear restrictions [8] and it can be solved by the method of iterative gradient projection [15]. Details of this procedure for the special case of Gaussian likelihoods (Eq. (2)) can be found in [12]. In practice, we computed the coefficients $U_{\kappa\lambda}^{(4)}$ of Eq. (10), where $\mathcal{H}_{\kappa\lambda} = \{ x \mid u_\kappa(x) = u_\lambda(x) \}$ indicates the set of boundary vectors between classes Ω_κ and Ω_λ. It was shown in [2] that $U_{\kappa\lambda}^{(4)} = U_{\kappa\lambda}^{(3)}$ is valid if only the means μ_κ, μ_λ are separated by the class boundary $\mathcal{H}_{\kappa\lambda}$. Thus, for the non-pathological case we obtain $U_{\kappa\lambda}^{(3)}$ simply through minimizing $u_\kappa(x)$ within the class boundary.

Search algorithm: To summarize, the entire NMSS algorithm proceeds as follows: In the preprocessing step, the pruning matrix $U^{(3)}$ is computed from the codebook PDF parameters, yielding typically 20 iterations of gradient projection per threshold. Alternatively, $U^{(2)}$ requires a set of design data to be analyzed, using the standard vector quantizer without pruning.

During analysis at each time frame a candidate list is initialized to include all codebook classes. Then repeatedly the next candidate is popped from the list, its likelihood score is computed, and the remainder of the list is checked for classes which can be precluded from distance calculation using the pruning rule (Eq. (5)). Finally, when the list is empty, we select the best-scoring class for output.

An important issue is the choice of the *candidate list order*. We want to assure that in most cases the list starts with the correct, or at least a well-scoring class to gain satisfactory savings. Assuming speech data as input, acoustic properties (and so our feature vectors) tend to move relatively slowly in time which is due to the bounded velocity of human articulatory movements. With this in mind we chose a candidate list order varying dynamically in time, putting the recent winner class on the top whenever a frame classification has been completed. Preliminary experiments [2] have shown the superiority of this procedure to static or random ordering policies.

3 Sequential pruning

The process of vector quantization may also be decomposed into a sequence of *"filters"*

$$\xrightarrow{\mathcal{C}_0} \boxed{\mathbf{VQ}_1} \xrightarrow{\mathcal{C}_1} \ldots \xrightarrow{\mathcal{C}_{i-1}} \boxed{\mathbf{VQ}_i} \xrightarrow{\mathcal{C}_i} \ldots \xrightarrow{\mathcal{C}_{I-1}} \boxed{\mathbf{VQ}_I} \xrightarrow{\mathcal{C}_I}$$

where the quantizer $\mathbf{VQ}_i$ computes probability scores for each class in its input candidate list $\mathcal{C}_{i-1}$ and produces an output candidate list $\mathcal{C}_i$ by selecting the most promising classes by application of a beam pruning criterion. The entry $\mathcal{C}_0$ is the entire set of class labels, and $\mathcal{C}_I$ consists of the single best-scoring class of the final VQ stage. The $\mathbf{VQ}_i$'s are designed as Gaussian quantizers using the initial d_i coefficients of the input vectors for probabilistic scoring, where d_i increases with i, and $d_I = d$. Consequently, the accuracy as well as the computational demand grows from the entry to the exit of the quantizer chain. Thus, a sequential VQ is completely specified by fixing the number of stages I, and, for each $1 \leq i \leq I$, the stage-dependent subspace dimensions d_i beam widths b_i (the latter of which is a rather delicate task).

4 Quantitative evaluation

We evaluated the VQ speed-up algorithms using multi-speaker 16 kHz continuous speech data, parametrically represented by 24 features (12 *mel*-cepstrum coefficients [5] plus 12 temporal derivatives [6]) per 10 msec frame, and three different Gaussian PDF codebooks. The latter were designed using a training set of 11 hours of speech data (100 utterances by each of 31 female and 48 male speakers). $\mathbf{CB}_{220}$ ($\mathbf{CB}_{64}$) was created by means of a 220-class (64-class) LBG design [11]. The "real-world" vector quantizer $\mathbf{CB}_{\text{SCHMM}}$ was initialized by merging 44 phone-specific Gaussian 5-mixtures, giving 220 classes. Afterwards, $\mathbf{CB}_{\text{SCHMM}}$ was integrated into the semi-continuous HMM recognizer ISADORA [16], and three codebook reestimations were run.

We compared three VQ speed-up approaches, that is, geometrical (GP), empirical (EP), and sequential (SP) pruning. Moreover, for each of the pruning strategies different settings of its relevant control parameters (GP: no. of gradient projection iterations; EP: amount of data to adapt $\mathbf{U}^{(2)}$; SP: see Sec. (3)) were tested. The competing algorithms were analyzed on an independent test set (100 utterances by 1 female and 3 male speakers; 0.5 hours of speech) regarding two groups of performance criteria.

	PDFs calc's	CPU time s/up	RTF	frame error	word acc.
	100.0	1.0	6.03	0.00	94.88
iter's	GEOMETRICAL				
1	62.3	1.04	5.75	4.39	94.63
3	65.7	0.98	6.10	2.16	—
21	68.4	0.94	6.38	1.20	94.88
utter's	EMPIRICAL				
79	11.3	7.06	0.85	3.28	94.50
237	14.4	5.16	1.16	1.08	94.88
790	17.7	4.15	1.45	0.38	94.88
$b_1 = b_2 = ?$	SEQUENTIAL				
0.1	6.6	8.52	0.70	13.01	93.63
0.01	15.4	4.86	1.23	1.41	95.13
0.001	24.0	3.41	1.76	0.17	94.88
0.0001	31.7	2.67	2.25	0.04	94.88
0.00001	38.3	2.25	2.68	0.02	94.88

Table 1: Performance of the $\mathbf{CB}_{\text{SCHMM}}$ quantizer

Their computational complexity was measured in terms of the CPU time elapsed on a 25 MIPS DECstation 5200, the relative speed-up (s/up) w.r.t. the full search VQ, and the real-time factor (RTF; CPU seconds per second of speech). Additionally, we recorded the percentage of full PDF calculations necessary, as compared to the exhausive search case. This latter figure is completely independent of any hardware/software implementation issues.

The degree of admissibility was assessed on the level of input frames by computing the percentage of input vectors with VQ class labels coinciding with the "true" (full search) labels. On the word level, we examined the influence of lacking admissibility on the overall speech recognition performance, given by the word accuracy. The recognition performance measured by the word accuracy is computed on a vocabulary of 162 words. In preliminary experiments with Gaussian quantizers for single-speaker data we observed speed-ups of about 3 (GP), 10 (EP, 6.9% frame error), and 6 (SP, 2.3% frame error). Even larger speed-up could be gained with Euclidean distance classifiers. When moving to the multi-speaker scenario, the pattern changes as is shown in Tab. (1) for the 220-class SCHMM codebook. Geometrical pruning saves one third of the PDF calculations, but due to the overhead introduced by the search strategy virtually no practical advantage in terms of speed remains.

Much more, however, can be gained if speed is traded for admissibility. With empirical and sequential pruning this trade-off can be controlled by adjusting the amount of training data or the width of preselection. For example, if designed using 3×79 training utterances, EP has to run only 14.4% of all PDF calculations, yielding a fivefold acceleration at 1% frame error, without any degradation of word recognition accuracy.

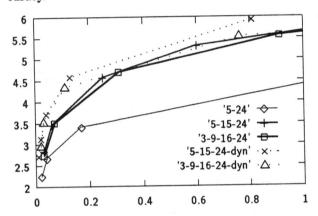

Figure 2: Frame error (x-axis) versus speed-up (y-axis) for different SP quantizer designs.

The sequential VQ was designed two-stage ($d = (5, 24)$, $b_1 = b_2$), and the results listed in Tab. (1) for varying beam width indicate a performance very similar to that of EP. However, we felt that the option to adjust the d_i's and b_i's provided some additional room for improvement in sequential pruning. The plotted curves in Fig. (2) show the behaviour of different sequential quantizers in the *accuracy/speed*-plane. The d_i's are given in the panel. 'dyn' refers to designs with exponentially increasing b_i's, as opposed to constant beam widths. The 3-stage dynamical beam version speeded up the quantization process by a factor of four at a frame error of less than 0.1%.

Note the surprisingly high frame error figures of the GP runs in Tab. (1). The admissibility guaranteed by the geometrical approach appears purely academic, given that even 21 threshold iterations did not lead us to a satisfactory frame accuracy. A closer look at the LBG codebooks $\mathbf{CB}_{64}$ and $\mathbf{CB}_{220}$, however, which achieved zero error after the 18th iteration, gives a possible explanation: unlike the LBG design, SCHMM codebook reestimation do not necessarily result in well-separated pattern space partitions. Rather, we very frequently observed instances of the pathological situations mentioned already in Sec. (2), where codebook classes Ω_κ do not even include their mean vectors μ_κ. Apart from the GP accuracy, we didn't observe any significant deviations between the performance of $\mathbf{CB}_{\text{SCHMM}}$, $\mathbf{CB}_{64}$, and $\mathbf{CB}_{220}$ codebooks, and thus omit the latter's results.

5 Conclusion

We presented three algorithms to speed up full covariance multivariate Gaussian vector quantizers. The speed-up is achieved by avoiding the distance calculation for a considerable number of codebook classes at each input frame. GP is a generalization of the triangle inequality pruning algorithm and guarantees admissibility for Gaussian VQ's, too. It provides moderate acceleration for LBG codebook, but fails in more general domains. EP as well as SP trade accuracy for speed. Both of them allow VQ speed-ups of five and more at very low frame error rates, and without any degradation of word accuracy.

References

[1] C.-D. Bei and R.M. Gray. An Improvement of the Minimum Distortion Encoding Algorithm for Vector Quantization. *IEEE Trans. on Communications*, COM-33:1132–1133, 1985.

[2] M. Bielecki. *Schnelle Suchverfahren zur Vektorquantisierung*. Technical Report, Diplomarbeit IMMD5 (Mustererkennung), Universität Erlangen, 1992.

[3] D.-Y. Cheng and A. Gersho. A Fast Codebook Search Algorithm for Nearest-Neighbor Pattern Matching. In *Proc. Int. Conf. on Acoustics, Speech, and Signal Processing*, pages 6.14.1–6.14.4, 1986.

[4] D.-Y. Cheng, A. Gersho, B. Ramamurthi, and Y. Shoham. Fast Search Algorithms for Vector Quantization and Pattern Matching. In *Proc. Int. Conf. on Acoustics, Speech, and Signal Processing*, pages 9.11.1–9.11.4, 1984.

[5] S.B. Davis and P. Mermelstein. Comparison of Parametric Representation for Monosyllabic Word Recognition in Continuously Spoken Sentences. *IEEE Trans. on Acoustics, Speech and Signal Processing*, ASSP-28(4):357–366, 1980.

[6] S. Furui. Speaker-Independent Isolated Word Recognition Using Dynamic Features of Speech Spectrum. *IEEE Trans. on Acoustics, Speech and Signal Processing*, ASSP-34(1):52–59, 1986.

[7] R.M. Gray. Vector Quantzation. *IEEE ASSP Magazine*, 1(2):4–29, 1984.

[8] R. Horst. *Nichtlineare Optimierung*. Hanser Verlag, München, 1979.

[9] X.D. Huang and M.A. Jack. Semi-Continuous Hidden Markov Models for Speech Signals. *Computer Speech & Language*, 3(3):239–251, 1989.

[10] A. Kaltenmeier. (personal communication). 1991.

[11] Y. Linde, A. Buzo, and R.M. Gray. An Algorithm for Vector Quantizer Design. *IEEE Trans. on Communications*, 28(1):84–95, 1980.

[12] H. Niemann. *Klassifikation von Mustern*. Springer Verlag, Berlin, Heidelberg, New York, 1983.

[13] H. Niemann and R. Goppert. An Efficient Branch-and-Bound Nearest Neighbour Classifier. *Pattern Recognition Letters*, 7:67–72, 1988.

[14] M. Orchard. A Fast Nearest-Neighbor Search Algorithm. In *Proc. Int. Conf. on Acoustics, Speech, and Signal Processing*, pages 2297–2300, 1991.

[15] J. B. Rosen. The Gradient Projection Method for Nonlinear Programming. Part II. Nonlinear Constraints. *SIAM J. Appl. Math.*, 9:514–532, 1961.

[16] E.G. Schukat-Talamazzini and H. Niemann. ISADORA — A Speech Modelling Network Based on Hidden Markov Models. *Computer Speech & Language*, (submitted), 1993.

[17] M. Soleymani and S. Morgera. A Fast MMSE Encoding Technique for Vector Quantization. In *Proc. Int. Conf. on Acoustics, Speech, and Signal Processing*, pages 656–659, 1989.

[18] M. Soleymani and S. Morgera. An Efficient Nearest Neighbor Search Method. In *Proc. Int. Conf. on Acoustics, Speech, and Signal Processing*, pages 677–679, 1987.

[19] X. Wu. A Tree-Structured Locally Optimal Vector Quantizer. In *Proc. Int. Conf. on Pattern Recognition*, pages 176–181, 1990.

CODEBOOK OPTIMIZATION BY CAUCHY ANNEALING

Zhicheng Wang and John V. Hanson
Department of Electrical and Computer Engineering
University of Waterloo
Waterloo, Ontario, Canada N2L 3G1
Telephone: 519-885-1211 ext.3896. Fax:519-746-3077

Abstract

The codebook design is a central issue of vector quantization (VQ) for speech or image signal compression. Simulated annealing (SA) has been applied to codebook design with encoder perturbations[3]. This paper presents a vector quantizer design method which associates the LBG algorithm with Cauchy annealing (CA) by decoder perturbations. This stochastic algorithm is able to find a globally optimal solution much faster than classical simulated annealing methods and remove the initial condition as the codebook design parameter. Moreover, deterministic Cauchy annealing algorithms have been proposed for computing complexity reduction and they are much more efficient than stochastic Cauchy annealing.

1 Introduction

The crux of a vector quantizer is the codebook. Given an adequate statistical specification of the source and a prescribed codebook size, the most challenging task is to design a codebook which contains the nearly globally-minimum-distortion collection of centroids that effectively represent the source to be encoded. Codebook design is a one-time off-line operation, so its complexity does not affect the performance of the real-time source coder (vector quantizer). However, the design complexity is of concern for moderate to large coding rates or vector dimensions with appropriate sizes of training sets, when the computation power is limited. Most design procedures today lead to a locally optimal or even suboptimal codebook that may be substantially inferior to the globally optimal solution. The traditional and most widely used VQ design technique is the LBG algorithm, otherwise known as the Generalized Lloyd Algorithm (GLA); an iterative heuristic optimization procedure in which an initial codebook is continually refined so that each iteration reduces the distortion involved in coding a given training set. Convergence is assured in the sense that the distortion of the sequence of trial codebooks is guaranteed to reduce monotonically with each iteration. However, since the multidimensional distortion function is nonconvex and the initial codebook is ad hoc, this algorithm easily gets trapped in local minima of the distortion surface, resulting in a suboptimal codebook. Hence, the initial codebook becomes crucial to the codebook design.

Simulated annealing has been successfully applied to a number of combinatorial optimization problems of high complexity. J.Vaisey and A.Gersho [3] combined the SA and LBG algorithms to design the codebook for a vector quantizer. In their algorithm, named the SA-GLA algorithm, they took the codebook design as a combinatorial optimization problem by perturbing the encoder of a vector quantizer. The motivation of our research is to improve the codebook design with the aim of finding a globally optimal codebook with less computational complexity and faster convergence. In this paper a fast simulated annealing, Cauchy Annealing, is applied to codebook design with decoder perturbations, and is treated as an optimization problem in a continuous space.

2 Cauchy Annealing

When the classical cost function has a single minimum, the conventional method can provide the unique ground state, and any method of gradient descent can approach the minimum. However, when the cost function has multiple extrema, a nonconvex optimization technique that allows tunneling and variable sampling and which accepts hill-climbing for escaping from local minima is required.

Simulated annealing is the artificial modeling of an annealing process which heats and then gradually cools a system to the freezing point to let the system achieve states with globally minimal energy. By appropriately defining an effective temperature for the multivariable system, simulated annealing can solve a wide collection of optimization problems. Kirkpatrick et al.[4] were the first to use simulated annealing to solve optimization problems. B.Hajek [5] surveyed the basic theory of simulated annealing in discrete space and in continuous space and their applications. A necessary and sufficient condition for the convergence to the global minimum has been proven in 1984 by Geman and Geman [6] for the classical simulated annealing based on a strictly local sampling. It is required that the time schedule of changing the fluctuation variance, described in terms of the artificial cooling temperature $T_g(t)$, be inversely proportional to a logarithmic function of time, given a sufficiently high initial temperature T_o:

$$T_g(t) = \frac{T_o}{\log(1+t)} \qquad (1)$$

A fast simulated annealing (FSA) technique is described by Szu [7], which uses both random walks and long jumps to generate new states and hence offers a faster convergence procedure than classical simulated annealing. Because states are generated with a Cauchy probability density in FSA, we name it as Cauchy Annealing (CA). The cooling schedule for CA is:

$$T_g(t) = \frac{T_o}{(1+t)} \qquad (2)$$

Proof of the cooling schedule for CA to insure a globally optimal solution was presented in [7]. The advantage of using the Cauchy distribution is that the ability to take advantage of locality is preserved, but the presence of a small number of very long jumps allows faster escape from local minima. CA turns out to be better than any algorithm based on any bounded variance distribution which is equivalent to the gaussian diffusion process from the Central Limiting Theorem (CLT).

The classical simulated annealing may characterized by: (i) bounded generating probability density (thermal diffusion), e.g., Gaussian probability density

$$G(X) = \frac{1}{2\pi\sqrt{T}} \exp(-\frac{X^2}{T}), \qquad (3)$$

(ii) an inversely logarithmic update cooling schedule, and (iii) the canonical hill-climbing acceptance probability

$$p(\Delta E) = \exp(-\Delta E/T), \qquad (4)$$

where $\Delta E = E_{t+1} - E_t$ is the increase of cost incurred by a transition.

The Cauchy annealing is defined by (i) the unbounded variance generating probability, say the Cauchy distribution:

$$G_c(X) = \frac{T(t)}{T^2(t) + X^2}, \qquad (5)$$

(ii) the update cooling schedule which may be inversely linear in time (equation(2)), (iii) the hill-climbing acceptance probability (equation(4)). Alternatively the following probability could be used,

$$p(\Delta E) = \frac{1}{1 + \exp(\frac{\Delta E}{T})}. \qquad (6)$$

In a continuous space, assume a function V on R^n to be minimized. Simulated annealing is implemented as a diffusion with the following equation

$$X(t+1) = X(t) - \nabla V(X(t))\Delta t + \sqrt{2T}W(t) \qquad (7)$$

where W(t) is white noise. Cauchy annealing generates the solution of the stochastic differential equation as follows:

$$dX = -\nabla V(X(t))dt + T\tan(u(t)) \qquad (8)$$

where u(t) is a random variable uniformly distributed on $(-\frac{\pi}{2}, \frac{\pi}{2})$. The above two equations are gradient descent algorithms combined with the normal distribution and the Cauchy/Lorentzian distribution processes respectively.

3 Codebook Optimization by CA (COCA)

The codebook design can be viewed as an optimization problem where the objective is to find the codevectors of a codebook that best represent the training set in terms of the MSE. Since it is generally very difficult and impractical to find an analytically tractable model of a vector source and equally difficult to utilize an analytical model for codebook design, a training set of empirical data that statistically represents the source is an essential starting point for any codebook design algorithm.

The main features of the LBG algorithm are: (a) it uses a training set of typical data to define an implicit source probability density function, and (b) it performs an iterative operation where each iteration maps a codebook into a new and improved codebook. The LBG method is appealing because the iteration is based on the well known necessary conditions for optimality and results in a monotonically decreasing distortion, which, however, actually makes the method very easy to get stuck in poor local minima.

In the LBG method the "state" of the optimization can be viewed as a set of labels assigning each training vector to one of N clusters. The centroids of these clusters define the current codebook. Vaisey and Gersho applied the simulated annealing in finite state space to the LBG algorithm, in their term GLA, for codebook optimization. In their SA-GLA method the state perturbation was accomplished in the encoder by altering the labels associated with the training vectors, changing the clusters and hence the current codebook. Obviously, they considered the codebook updating as a transition in a discrete space by changing a training vector label (i.e. the clusters) and then calculating the corresponding centroids. Therefore, the computation requirement is very high because the combination in terms of the N clusters (i.e. centroids) of a codebook, the number of vectors in a training set, and training vectors in different clusters is large.

However, we directly perturb the decoder, the centroids in the continuous (state) space for the codebook updating. In this method, the clusters can be obtained by a nearest neighbor search based upon the current centroids. The total distortion of a codebook on a training set is a function of the centroids in the codebook:

$$E(t) = f(C_1(t), C_2(t), \ldots, C_N(t)) \qquad (9)$$

Although it is impractical to find this analytical function, the distortion with respect to the current centroids can be calculated by nearest neighbor clustering. Hence, instead of equation (8), we use

$$C_n(t+1) = C_n(t) + T\tan(u(t)). \qquad (10)$$

The codebook design procedure is described as follows. Begin by setting the effective value of T to an initial temperature T_0, and randomly set an initial codebook. Alter the centroids based on equation(10), and compute the resulting change in the distortion function, $\Delta E = E(t+1) - E(t)$. If $\Delta E \leq 0$, the perturbed codebook (centroids) is accepted as the new codebook. If $\Delta E > 0$, the perturbation is accepted with the probability $p(\Delta E) = \frac{1}{1+\exp(\frac{\Delta E}{T})}$. The perturbation continues until an equilibrium condition has been reached. The temperature T is then reduced to the next lower temperature in a predetermined sequence of temperatures, and perturbations are again carried out on the codebook. Cauchy annealing terminates when the convergence criterion is met and then the optimization algorithm becomes a greedy optimization algorithm.

In the procedure mentioned above, the conditions determining an annealing schedule are: (1)the initial temperature, (2)the temperature decrement, (3)the equilibrium conditions, and (4)the stopping, or convergence criterion. For an annealing schedule to be problem independent, the parameters in the four conditions should be determined by the algorithm itself and should not have any predefined values. (1) In simulated annealing, the initial temperature can be determined based on the condition proposed by White [8], $T_0 \gg \sigma$, where σ is the standard deviation of the distortion distribution. For Cauchy annealing, we choose $T_0 \simeq \sigma$ because of the Lorentzian wings of the Cauchy probability density that imply occasional long jumps among local sampling. Therefore, from the starting temperature we can predict that the optimization procedure by Cauchy annealing is shorter in the temperature range (implying faster in time) than by simulated annealing. Actually, we can also take the variance of the coded training set, V_s, as a reference as σ to choose T_0. We propose $T_0 \simeq V_s$ for simulated annealing and $T_0 \ll V_s$ for Cauchy annealing. (2) Huang,et.al. [9], proposed that the temperature should decrease by:

$$T_{t+1} = T_t \exp(-\frac{\lambda T_t}{\sigma}). \qquad (11)$$

where $\lambda \leq 1$. Kirkpatrick [4] suggested an intuitive temperature reducing method: $T_n = (0.9)^n T_0$, which many have used without observing any difficulty in convergence of the simulated annealing for most of normal problems. For our algorithm, we can use a factor less than 0.9 in the intuitive temperature decreasing sequence. (3) The equilibrium condition is met intuitively when a fixed number of attempts are made or a minimum number of attempts are accepted. (4) The stopping criterion is that all the accessible codebooks at that temperature have comparable distortions. The temperature is then set to zero. At last, we can consider that we have obtained a starting codebook near a globally minimal distortion after the annealing procedure and then use a greedy algorithm like the LBG algorithm to obtain the globally optimal codebook. We call this

codebook design algorithm the Codebook Optimization by Cauchy Annealing, or COCA.

4 Codebook Optimization by Deterministic Annealing

As we can see in the previous section, the Cauchy annealing is computationally expensive for adaptive vector quantizer design and the major complexity of the Cauchy annealing approach is the randomness of walk, the computation of ΔE and the slowness of reaching thermal equilibrium. Some quasi-deterministic annealing approaches were proposed with setting the acceptance probability

$$p(\Delta E) = \begin{cases} 1 & \text{if } \Delta E < 0 \\ 0 & \text{otherwise} \end{cases} \qquad (12)$$

or/and adding the gradient into the annealing [10]

$$C_n(t+1) = C_n(t) + \alpha(t)[X - C_n(t)] + T\tan(u(t)). \qquad (13)$$

The gradient conscience avoids the blind random walk in annealing, resulting in approaching a globally optimal solution fast. Under the condition eq.(12) and the noise is Gaussian, the optimization method ensures convergence to a global minimum with probability 1 on the compact set [11]. Here we used Cauchy noise instead of Gaussian noise.

We also suggest a deterministic Cauchy annealing approach that eliminates the need to calculate ΔE and to reach thermal equilibrium for major complexity reduction. We propose the following changes in the Cauchy annealing to establish the deterministic Cauchy annealing procedure:
1) Accept every proposed random walk for a winning code vector to avoid computing ΔE by setting $p_a(\Delta E) = 1$;
2) Perform temperature update once for each sweep according to the temperature cooling schedule.

Although the stochastic principles are still incorporated in the proposed approach with eq.(10) for winning code vector update, the annealing schedule becomes deterministic after the above changes have been made, which is the reason we call it Deterministic Cauchy Annealing. No global asymptotic convergence theorem has been found for this approach. However, the justification is empirical success and the principles of the stochastic scheme.

5 Simulation Results

The simple but nontrivial example demonstrated in [2] is presented using the Codebook Optimization by Cauchy annealing. The results with the different initial codebooks (Table 1) have shown that with the LBG algorithm, the minima being reached are determined by their initial codebooks. However, using the algorithm presented in this paper, the global minimum is always reached no matter what the initial codebook.

Initial Condition	Average Distortion	
	LBG	COCA
1	0.099738	0.082710
2	0.084598	0.082710
3	0.083374	0.082710
4	0.082710	0.082710

Table 1: Comparison of performance between LBG and COCA algorithms

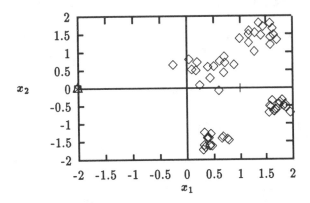

Figure 1: Codevectors of the LBG algorithm

The following example illustrates the initial condition problem for different design algorithms of vector quantizer. The data is generated from four Gaussian functions with different means and variances. In this example, all the initial codevectors were set to the same value of $(-2, 0)$.

As shown in Figure 1 for the LBG algorithm, the codevectors but one have not been updated at all, resulting in a variance-value high distortion. Even with Kohonen learning vector quantization in Figure 2, only two codevectors (units) won and were modified although, unlike the LBG algorithm, the Kohonen learning is not a "greedy" algorithm. The kohonen learning modifies the codebook on-line using

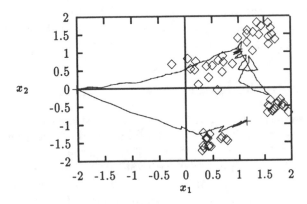

Figure 2: Codevectors update in Kohonen learning

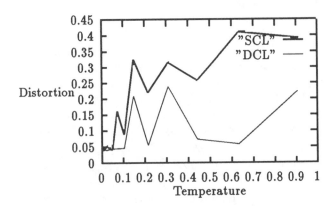

Figure 3: Annealing Curves. SCL stands for stochastic CA and DCL for quasi-deterministic CA

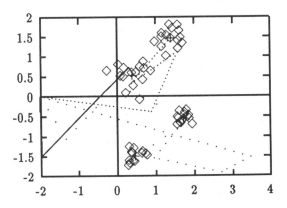

Figure 4: Codevectors update in deterministic Cauchy annealing

an instantaneous estimation of the gradient, known as the stochastic gradient. However, the Cauchy annealing algorithms can reach the globally optimal solution in Figure 4 and 5. From Figure 3, the quasi-deterministic Cauchy annealing converges faster than stochastic Cauchy annealing.

We present experimental results for Gauss-Markov sources. A Gauss-Markov source or first order autoregressive source $\{x_n\}$ is defined by the difference equation

$$x_{n+1} = ax_n + G_n, \qquad (14)$$

where $\{G_n\}$ is a zero mean unit variance i.i.d. Gaussian time series. This source is of interest since it is a commonly used model of real data, used for comparing data compression systems, and since its optimal performance (the rate-distortion function) can be expressed parametrically and either evaluated, approximated, or bounded [12]. We here consider only the case a=0.9 of a highly correlated source, the correlation coefficient being typical of speech and scanned image data.

The SNR's obtained from coding the training set using codebooks designed with the LBG algorithm and the

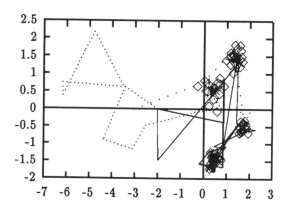

Figure 5: Codevectors update in stochastic Cauchy annealing

	SNR (db)	
Bits/vector	LBG	COCA
1	3.277	3.415
2	6.015	6.325
3	8.278	8.603
4	10.384	11.047

Table 2: Comparison of the performance between LBG and COCA algorithms

COCA algorithm are depicted in Table 2 for the vector dimension K=5 with coding rate of R=1, 2, 3, 4 bits per vector. The splitting algorithm [2] was applied on the training sequence to obtain the initial codebook for both algorithms. Obviously, the results of the COCA algorithm are better than those of the LBG algorithm. In our experiments, the times used in codebook design by Cauchy annealing are approximately a tenth less than those by simulated annealing. With regard to larger training sets or real image data, and codebooks with larger coding rates or/and longer vector dimensions, the COCA algorithms will show much advantage over the LBG algorithm in finding optimal codebook and work faster than the SA-GLA algorithm in codebook design.

6   Conclusions

New codebook design algorithms have been developed based upon a Cauchy annealing technique. The COCA algorithms are reliable methods to improve the quality of codebooks for vector quantization. The two important features of this technique are: (1) a global optimal codebook on a training set is consistently achieved regardless of the initial condition of the codebook, which is an improvement over the LBG algorithm which inevitably varies with the initial codebook; (2) faster cooling schedules and convergent rates and much lower computation complexities than codebook optimization algorithms using simulated anneal-

ing with encoder or decoder perturbations are achieved. Although the codebook design complexity of the COCA algorithms is higher than that of the LBG algorithm, the complexity of the actual coding operation is unaffected since the codebook is designed only once.

References

[1] R.M.Gray, "Vector Quantization", *IEEE ASSP Magazine*, pp.4-29, April 1984.

[2] Y.Linde, A.Buzo, and R.M.Gray, "An Algorithm for Vector Quantizer Design", *IEEE Trans. Commun.*, vol.COM-28, pp.84-95, Jan.1980.

[3] J. Vaisey and A.Gersho, "Simulated Annealing and Codebook Design", *Proceedings of 1988 International Conference on Acoustics, Speech and Signal Processing*, pp.1176-1179, April 1988.

[4] S.Kirkpatrick, C.D.Galatt, and M.P.Vecchi, "Optimization by Simulated Annealing", *Science*, vol.220, pp.671-680, May 13, 1983.

[5] B.Hajek, "A Tutorial Survey of Theory and Application of Simulated Annealing", *Proceedings of 24th Conference on Decision and Control*, pp.775-760, December 1985.

[6] S.Geman and D.Geman, "Stochastic Relaxation, Gibbs Distributions, and the Bayesian Restoration of Images", *IEEE Transactions on Pattern Analysis and Machine Intelligence*, PAMI-6, pp.721-741, 1984.

[7] H.Szu and R.Hartley, "Fast Simulated Annealing", *Physics Letters*, vol.122, no.3,4, pp.157-162, June 1987.

[8] S.White, "Concepts of Scale in Simulated Annealing", *Proceedings of the International Conference on Computer Design*, pp.646, 1984.

[9] M.F.Huang, F.Romeo, and A.Sangiovanni-Vincentelli, "An Efficient General Cooling Schedule for Simulated Annealing", pp.381, 1986.

[10] Z.Wang and J.V.Hanson, "Cauchy Learning Vector Quantization", *Proceedings of World Congress on Neural Networks*, vol.IV, pp.605-608, Portland, Oregon, July 1993.

[11] F.J.Solis and J.B.Wets, "Minimization by Random Search Techniques", *Mathematics of Operations Research*, vol.6, no.1, pp.19-30, February 1981.

[12] R.M.Gray and Y.Linde, "Vector quantizers and predictive quantizers for Gauss-Markov sources", *IEEE Trans. Commun.*, vol.COM-30, pp.381-289, Feb.1982.

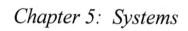

Chapter 5: Systems

Chapter 5 marks the transition from algorithms to architectures in this consideration of signal processing technology. It discusses pervasive system-wide issues that arise in signal processing implementations, while Chapter 6 examines reusable components that apply across many signal processing implementations, and Chapter 7 presents specific solutions to particular problems that can be incorporated as elements in an end-to-end solution.

This chapter begins with the consideration of the system-wide issue of power. Often postponed until late in the design when a physical designer begins developing a cooling method for the system, the advent of portable low power signal processing systems requires that power be included as a design criterion at the very early stages of design, even influencing the choice of algorithm. "Cost, Power and Parallelism in Speech Signal Processing" (paper 5.1) argues that traditional signal processing components emphasize speed over power, while portable applications instead require minimum power and just enough speed to achieve real-time operation. The paper reviews the techniques that can reduce system power at the earliest design stages and shows how parallel slower smaller processors can achieve real time operation at a significant power savings over traditional one-processor approaches. An independent view of design for minimal power consumption is provided in paper 5.2, "System Design of a Multimedia I/O Terminal," which demonstrates the application of many of the power-reduction techniques described in the preceding paper.

With the earlier introduction of total system considerations, the growth of concurrent engineering, and the increasing importance of trying an early version of a signal processing system in the context for which it is designed, there is an emphasis on rapid prototyping. "Fast Prototyping of Real Time Systems: A New Challenge?" (5.3) interconnects domain-specific simulation engines into a heterogeneous architecture to provide rapidly-realized real time simulation of the multimedia I/O terminal described in paper 5.2. Rapid attainment of a system prototype is also addressed in "Hardware - Software Trade-Offs for Emulation" (5.4), which explores tradeoffs between multiple digital signal processing chips and multiple field-programmable gate arrays (FPGAs) to achieve real-time emulation of a neural network exclusive-or function.

The general problem of matching algorithms to architectures is treated in "Optimal Mapping of DSP Applications to Architecture" (5.5). This paper addresses both the problem of synthesizing a digital signal processing architecture according to some optimality condition, and optimally mapping a DSP application to multiple programmable DSP chips. A method that includes chaining, loop winding, and the automatic selection of functional units promises to formalize the intuitive actions of a signal processing architecture designer.

The unfolding of loops in computational flow provides an in-line program. Depending upon how the degree of the unfolded version of a computation grows upon unfolding, the computation can be improved through the application of unlimited parallelism. "On Unlimited Parallelism of DSP Arithmetic Computation" (5.6) identifies the necessary and sufficient conditions for achieving unbounded improvement through increasing parallelism.

Suggested Additional Readings:

[1]   W. H. Wolf, "Hardware-Software Co-Design of Embedded Systems," *Proc. IEEE,* vol. 82, no. 7, pp. 965-989, July 1994.

[2]   S. Kleinfeldt, M. Guiney, J. K. Miller, and M. Barnes, "Design Methodology Management," *Proc. IEEE,* vol. 82, no. 2, pp. 231-249, Feb. 1994.

[3]   S. J. Lipoff, "Personal Communications Networks Bridging the Gap Between Cellular and Cordless Phones," *Proc. IEEE,* vol. 82, no. 4, pp. 564-571, April 1994.

[4]   K. G. Shin and P. Ramanathan, "Real-Time Computing: A New Discipline of Computer Science and Engineering," *Proc. IEEE,* vol. 82, no. 1, pp. 6-24, Jan. 1994.

[5]   J. W. S. Liu, W. K. Shih, K. J. Lin, R. Bettati, and J. Y. Chang, "Imprecise Computations," *Proc. IEEE,* vol. 82, no. 1, pp. 83-94, Jan. 1994.

[6]   United States Advanced Research Projects Agency, *Proceedings: 1st Annual RASSP Conference* (Arlington, VA, USA) August 1994.

[7]   F. B. Maciel, Y. Miyanaga, and K. Tochinai, "A Performance-Driven Approach to the High-Level Synthesis of DSP Algorithms," *1993 IEEE International Symposium on Circuits and Systems* (Chicago, IL, USA) May 1993, pp. 1658-1661.

[8]   H. Wang, N. Dutt, A. Nicolau, and I.-Y. S. Siu, "High-Level Synthesis of Scalable Architectures for IIR Filters Using Multichip Modules," *30th Design Automation Conference* (Dallas, TX, USA) June 1993, pp. 336-342.

[9]   D. Desmet and D. Genin, "ASSYNT: Efficient Assembly Code Generation for Digital Signal Processors Starting from a Data Flowgraph," *Proc. IEEE International Conference on Acoustics, Speech, and Signal Processing 93* (Minneapolis, MN, USA) April 1993, vol. 3 pp. 45-48.

COST, POWER, AND PARALLELISM
IN SPEECH SIGNAL PROCESSING

Richard F. Lyon

Apple Computer
One Infinite Loop
Cupertino, CA 95014

ABSTRACT

The cost of speech signal processing and other computationally intensive functions is increasingly influenced by power consumption as products are made smaller and more portable. That is, in portable products, weight and battery life are bigger issues than silicon area and total computational capability. A recent emphasis on the power problem within the VLSI signal processing community has led to an understanding of how parallelism can significantly reduce the cost of a system by greatly reducing clock speed, supply voltage, and power consumption, even though at the expense of silicon area and other measures of efficiency. Several different kinds and degrees of parallelism, including massive analog parallelism, should be considered in planning to reduce the total cost of speech signal processors. In this tutorial paper, recent and older ideas are reviewed with respect to their potential applicability to modern products, as well as with respect to their difficulties.

1. INTRODUCTION AND MOTIVATION

Speech processing and other signal processing systems have enough regularity and inherent parallelism that they are amenable to a wide range of optimization techniques. The VLSI academic and industrial communities have traditionally focussed much more emphasis on optimization of silicon area and clock speed than of power consumption, though with portable products becoming popular this trend is changing quickly. Analytical optimization techniques sometimes lead to general rules of thumb; for example, designers know to break up amplifiers and drivers into stages with gains not much above e to optimize delay [1]. But when the rules of thumb come from the optimization of a single variable, they will be at odds with the optimization of a more relevant joint cost function. Since system-level cost functions have been difficult to define at the chip design stage, and since good joint optimization tools have been difficult to get and to use, chip designers may still be applying rules of thumb that lead to very power-inefficient, and hence costly, product designs.

Historically, we have seen how applying huge numbers of transistors to a problem can reduce the cost of a system (as in the use of a DSP chip instead of a few inductors and capacitors to implement a filter). In the age of portable products, we should think in terms of applying huge numbers of transistors to reduce the size and weight of the battery. The examples surveyed in this tutorial support the idea of using parallelism of slow processing units as a power-saving alternative to the single fast central processor that characterizes many of today's desktop and portable products.

1.1. Power as a Primary Cost Factor

In today's desktop computers, the silicon chips and associated components tend to cost more than the power supply and cooling systems. While this may also be true in portable computing products, the weight and limited operating time of the portable power source implies another kind of cost to the user. The option of increasing the energy density by using more exotic batteries is available, but at a very high cost. In new generation RISC-based desktop systems, getting rid of the heat is an increasingly costly problem. Overall, it seems that electrical energy is very cheap, but storing and transporting it, and getting rid of its waste heat, are increasingly expensive in more compact and higher-performance products. Mead has argued that computer system cost has stayed proportional to power consumption over many generations of machines, and that we should expect this trend to continue [2].

In spite of this situation, we do not have good institutionalized ways of estimating and trading off the true costs associated with the power consumption of our computational subsystems. In this tutorial, we emphasize power as the primary cost factor that needs more explicit consideration at all levels of design, in hopes of raising the level of attention paid to power by chip and product designers.

To make things worse, other kinds of hard-to-evaluate costs are often involved in system optimization. For example, Narkiewicz and Burleson [3] present techniques that "allow tradeoffs between VLSI costs, performance and precision," while Orailoglu and Karri [4] "systematically explore the three-dimensional design space spanned by cost, performance, and fault-tolerance constraints." Apparently performance, precision, and fault-tolerance are important dimensions that, like power, haven't yet been integrated into "system cost functions."

1.2. Portable Speech Processors

Portable sound and speech products tend to have severe real-time computational requirements relative to other portable products, and hence a bigger problem with battery weight and running time. We need to develop a bag of tricks to attack this general problem at all levels.

An important property of speech processors is that they have a bursty load. Voice coders, recognizers, synthesizers, recorders, etc. only need to process speech when speech is present. An important source of power efficiency in such systems is the ability to shut down or slow down parts of the system when their capabilities are not needed. A less obvious power saving technique is to move some of the work from the fast real-time mode into a slower and more efficient background mode, for example in a voice-mail compressor. In section 2 we explain how a slow mode not only reduces the power level, but can also reduce the total energy consumption.

1.3. "Are DSP Chips Obsolete?"

In a recent paper, Stewart, Payne, and Levergood [5] posed the question of what a specialized DSP chip is good for in the age of fast RISC chips. Using the DEC Alpha architecture [6] as an example, they argue that it is generally preferred to let a fast RISC chip do the kinds of computational jobs, such as speech processing, that have been largely relegated to DSP chips during the last decade. Unfortunately, they totally ignore power as a cost factor, and propose a solution that is about an order of magnitude more energy hungry than the DSP alternative. We discuss how optimizing a chip for speed leads to its inefficient use of energy, again using the DEC Alpha RISC chip as an example.

1.4. Previous Surveys

There are several previous excellent surveys of power-saving techniques and the size-cost-speed-voltage-power trends and limits of CMOS technology (mostly from an academic perspective). An early study by Swanson and Meindl (Stanford) [7] explores CMOS logic structures at very low supply voltages. Mead and Conway (Caltech and Xerox) [8] provide in-depth chapters on the "Physics of Computational Systems", including technology scaling and energetics of very low voltage CMOS, as well as an introduction to the kinds of "Highly Concurrent Systems" architectures that can take advantage of low-speed low-power operation (with contributions from Kung and Leiserson of CMU and Browning and Rem of Caltech). Vittoz (Swiss CSEM and EPFL) describes "Micropower" techniques for portable electronic devices [9]. Mead (Caltech) [2] emphasizes power costs and the power

efficiency of special-purpose and analog signal processing implementations in "Neuromorphic" systems. Burr, Williamson, and Peterson (Stanford) [10] survey low-power issues in modern high-performance signal processing systems, emphasizing current projects at Stanford. Brodersen, Chandrakasan, and Sheng (Berkeley) [11] provide an excellent tutorial with specific examples of choices at many levels that can improve power efficiency in signal processors.

In the present tutorial, rather than introducing new ideas, we attempt to assemble ideas to support an attitude adjustment of design engineers, emphasizing the conclusions of Mead and Conway [8] over a decade ago, that "In real systems, the cost of power, cooling, and electrical bypassing often exceeds the cost of the chips themselves. Hence any discussion of the cost of computation must include the energy cost of individual steps of the computation process." and "Perhaps the greatest challenge that VLSI presents to computer science is that of developing a theory of computation that accommodates a more general model of the costs involved in computing."

2. POWER CONSUMPTION BASICS

Power consumption in modern CMOS circuits is usually dominated by dynamic power related to the charging and discharging of circuit nodes between the two power supply voltage levels that represent logic 1 and logic 0. Dynamic power can be expressed as:

$$P = fCV^2$$

where f is the clock frequency, C is the effective total capacitance being switched at the rate of one transition per clock cycle, and V is the supply voltage.

The effective total capacitance is computed as a weighted sum of all the node capacitances in the circuit, with each node's weight equal to its number of logic transitions per clock cycle, averaged over the conditions of interest. Clock nodes have two transitions per cycle, so they count double, while static logic nodes may average one-half transition per cycle or less; precharged logic nodes are typically somewhere in between.

2.1. Reducing Power

It might almost go without saying that power can be reduced in three ways:
- Reduce Frequency
- Reduce Capacitance
- Reduce Voltage

Importantly, the voltage factor applies twice, so any reduction in voltage is worth twice as much power savings as a proportionate reduction in frequency or capacitance. But the supply voltage may also be the most difficult parameter for the designer to change freely, due to system and compatibility constraints. Typically reducing the supply voltage will also require reducing the clock frequency, so other changes will be needed to maintain performance.

Reductions in effective switched capacitance can come from two main sources: physical-level and circuit-level optimizations that reduce node capacitance per logic gate, and logical/architectural reorganizations that reduce the number of nodes or gates being switched. Avoiding clock and data transitions in portions of the processor that are not needed for a particular part of the computation is one technique that pays off well. Making low-level "sizing" optimization criteria favor power rather than speed will result in smaller transistors, smaller cell layouts, and shorter wires, all contributing to lower capacitance.

Reductions in clock frequency are generally acceptable only if there is some other change that allows the system computational requirements to be met. For example, the average clock frequency can be reduced if the clock frequency is made variable and the computational requirements are variable. Or the clock frequency can be cut by a factor of N if N copies of the processor are used and can efficiently share the computational load. In this case, the silicon area and capacitance increase by a factor of N, but the lower clock rate will allow a lower supply voltage and a great power savings. This kind of trading of frequency and voltage against area and parallelism is a large under-exploited source of power savings.

CMOS circuits also dissipate power during switching due to the fact that the transistor charging a node doesn't quite turn off before the other transistor tries to discharge the node. This "crossover" or "short-circuit" power is again proportional to the effective switching frequency and to a "strength times slowness" factor that scales pretty much like the capacitance; but this power is also a very expansive function of supply voltage (between quadratic and exponential). If the supply voltage is reduced to around twice the typical transistor threshold, the crossover power involves only subthreshold conduction, and is largely negligible (as long as threshold magnitudes are large compared to kT/q); but for 3-V and 5-V chips it may be a significant power as well as a significant contributor to power supply noise spikes. DC power is also significant if threshold voltages are reduced to the point where the turned-off transistor has a substantial leakage. Burr and Peterson [12] analyze the contribution of crossover power and DC power at very low threshold and supply voltages, where they are important considerations in establishing limits to ultra low power operation.

2.2. Circuit Speed versus Voltage

Operating a system at a supply voltage higher than the clock frequency requires is a big waste of power. The system designer should think about scaling supply voltage and system clock frequency together for best efficiency. Using self-timed circuits or voltage-dependent clock oscillators, it is possible to make the system speed adapt to conditions such as dropping battery voltage, changing temperature, etc. For example, the technique of Von

Kaenel *et al.* [13] can be used to drop the supply voltage to match the logic speed, or can be turned around to adjust the logic speed to match the supply voltage.

The designer needs to know what to expect in the speed versus voltage tradeoff even when the system handles it automatically. The expected scaling between voltage and speed depends on what voltage range in considered; figure 1 characterizes some possibilities.

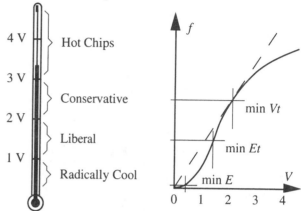

Figure 1. A crude characterization of power supply voltage ranges available to the chip and system designer, and an approximate speed-voltage curve (solid) shown with a linear approximation (dashed).

The speed of logic circuits has traditionally been approximated as linear with supply voltage, as in the dashed line in figure 1. This "rule of thumb" was reasonable for supply voltages large compared to a threshold voltage, in older technologies that did not show short-channel effects and velocity saturation within the typical range of supply voltages. According to this approximation, each 1% increase in speed costs a 1% increase in voltage, a 2% increase in switching energy, and a 3% increase in power.

Many designers have internalized a second-order correction derived from the standard quadratic MOS transistor model—speed varies as supply voltage minus threshold, and hence is more than proportional to supply voltage:

$$\frac{\Delta f}{f} > \frac{\Delta V}{V} \quad \text{"old scaling"}$$

This relation tells designers that reducing the supply voltage has a relatively large speed penalty—a rule that is no longer true at typical standard voltages.

A relevant cost function that can be optimized is the voltage-time product Vt, or equivalently Et^2, where switching energy E is proportional to V^2, and t is the logic delay, inversely proportion to f. According to "old scaling", this cost function is optimized by letting V increase toward infinity. In modern fine-line processes, velocity saturation is a dominant effect in the 3-V to 5-V region, so circuit speed actually varies less than linearly with supply voltage:

$$\frac{\Delta f}{f} < \frac{\Delta V}{V} \quad \text{"new scaling"}$$

The actual optimum of the Et^2 criterion will occur at the voltage that defines the boundary between the old scaling, where velocity saturation was negligible, and the new scaling, where it dominates the threshold effect. This critical voltage is decreasing as the technology scales, and is already below 3 Volts, according to the speed derating curve of the Philips HLL (High-speed Low-power Low-voltage) CMOS logic family [14]. Many designers have not yet internalized this new reality, so they continue to pay *more than* a 3% power increase for each 1% speed increase by keeping the supply voltage around 3 to 5 Volts.

Other sensible criteria that may be analytically optimized, as reviewed by Burr and Peterson [12], include the energy-time product Et and the switching energy E. Optimizing E puts zero weight on speed, which is probably not useful in current products, but is useful in deriving lower bounds to sensible supply voltages. At the optimum, where supply voltage is approximately equal to threshold voltage, power will be proportional to speed, but only very low speeds will be achievable. As the cost of transistors continues to fall, making up for the speed loss with increased parallelism will push efficient designs toward this limit.

Optimizing Et means that each 1% increase in speed is worth a 1% increase in switching energy and a 2% increase in power. Burr and Peterson [12] show that this criterion is optimized when the supply voltage is equal to three times the threshold voltage, ignoring velocity saturation and subthreshold effects. If velocity saturation is already significant at this voltage, the actual optimum voltage will be even lower.

If clock frequency varies linearly with voltage, power varies as the cube. In modern technologies, the power will be less than cubic at traditional high voltages, and more than cubic at sufficiently low voltages. But the large power effect at low voltages comes with a corresponding clock speed penalty. The switching energy or energy per clock cycle is still proportional to V^2. Therefore, if a system can be built with a slow low-voltage "background" mode and a faster "foreground" or "real-time" mode, computations that can be deferred to the slower mode can be done much more energy efficiently.

3. DIGITAL PARALLELISM OPTIONS

At the recent IEEE Workshop on VLSI Signal Processing, there was considerable excitement around the topic of low-power techniques. In addition to the tutorial of Brodersen *et al.* [11] mentioned above, there were several contributed papers on low-power techniques [15, 16], as well as a number of papers on architecture alternatives that provide a range of examples that the techniques could be applied to or evaluated on. For this tutorial, I draw digital parallelism examples mainly from that workshop and its predecessors, and from my own work. Varying styles and degrees of parallelism distinguish the architectural options.

One useful tool for coarse architecture-level evaluation of alternatives is the "power factor approximation" (PFA) method of Chau and Powell [17, 18], which provides a method of estimating the energy cost of a system based on the number and size of multiplication, storage, and communication operations, etc. Berkelaar and Theeuwen [19] provide a tool that directly manipulates designs to explore the area-power-delay space. In both of these approaches, however, since they do not consider supply voltage among the parameters under the control of the designer, some of their conclusions do not apply when supply voltage is jointly optimized with architecture. Alternatively, Chandrakasan *et al.* [15] specifically consider architecture and algorithm transformations jointly with voltage optimization to minimize power at a specified throughput, and find that the optimum supply voltage on real-world examples is often about 1.5 Volts, somewhat below the Et optimum.

3.1. Bit-Serial vs. Bit-Parallel

An obvious level of parallelism familiar in most digital computing machines is bit-level parallelism. Microprocessor chips have been advancing exponentially in this dimension, though at a relatively slow rate (4-bit to 64-bit machines in 20 years is a doubling every 5 years, which is quite slow compared to the rate of improvement of other metrics in this field). There is little reason to expect or want an increase in the size of arithmetic operands beyond 32, 64, or 80 bits, depending on the application, and therefore little opportunity for increased parallelism of this sort.

At the other end of the spectrum, bit-serial methods [Denyer&Renshaw 83?] are attractive because they reduce interconnect complexity and make it easier to employ other levels of parallelism. Lyon [20, 21, 22] has shown both dedicated (functionally parallel) and programmable (SIMD) bit-serial architectures for signal processing, and Wawrzynek and Mead have shown a configurable architecture [23, 24, 25]. The shallow logic depths and simple interconnect topologies of these designs were exploited for speed and area efficiency, but can also contribute to energy efficiency in configurations that meet the throughput needs at a lower clock rate and supply voltage. Summerfield and Lyon [26] showed that bit-serial dedicated signal processors built with standard cell automatic place and route tools can be much more locally connected, and hence more area and energy efficient, than typical bit-parallel designs.

When higher levels of parallelism are difficult to use, as in general-purpose von Neuman machines, parallelism at the bit level is essential to achieving high performance. Because of this fact, bit-parallel techniques have come to dominate the computing industry, including computer engineering education. Most architecture examples in the literature start from an implicit assumption of bit-level parallelism, but could in most cases be adapted to bit-serial approaches.

3.2. Programmable, Configurable, etc.

Programmability is the key to the cost effectiveness and proliferation of microprocessors, but comes at a high energy cost relative to more specialized or dedicated architectures. In portable products that process speech and other signals, some flexibility and re-programmability may still be valuable, but power should take on a much larger weight in the cost function. An attractive intermediate between dedicated fixed-function architectures and programmable fetch-execute architectures is the family of "configurable" architectures.

The configurable bit-serial architecture of Wawrzynek and Mead [23, 24, 25], mentioned above, implemented a variety of music synthesis algorithms about three orders of magnitude faster than a programmable microprocessor of similar technology, with much better energy efficiency.

Yeung and Babaey's configurable "nanoprocessor" based architecture [27] has been benchmarked at about a billion operations per second per chip, with 64 nanoprocessors per chip, in a variety of signal processing domains including speech recognition. Like Wawrzynek and Mead's, this architecture uses static scheduling of a signal flow graph to attain an effective compromise between generality and efficiency. But because each nanoprocessor includes a small control store, it is even more flexible.

Dedicated hardware for predetermined regular algorithms is the most efficient approach when it is applicable. Systolic arrays represent a maximally parallel approach to dedicated hardware for large signal processing problems, and are a natural candidate for low-voltage operation. Good reviews of systolic and array techniques are given by Rao and Kailath [28] and by Kung [29].

An example of the design of a VLSI digital audio equalizer is given by Slump et al. [30]. Two VLSI designs were completed and compared. Both are essentially dedicated chips, but with internal control programmability. The first try, based on a single multiplier, turned out to be inefficient due to the difficulty of controlling it in a way that would take advantage of all the regularities and efficiencies known at the algorithm level. The second version, using four smaller multipliers on one chip, did the job with less than half the silicon and at a lower clock rate, due largely to more flexible programming. The details are complex enough that we do not try to draw a simple conclusion, except to note that more multipliers does not necessarily mean higher cost. Architectural exploration is still needed to arrive at good designs.

3.3. RISC, VLIW, SIMD, MIMD, etc.

Within the realm of programmable computing machines, parallelism options abound. Multi-processor machines with single or multiple instruction streams (SIMD or MIMD) have been widely explored. SIMD machines have reduced control overhead, but MIMD machines are more general. Within single-processor machines, multiple function units may be organized using very-long-instruction-word (VLIW) or superscalar techniques. VLIW is more efficient, given a good compiler, but superscalar is compatible with pre-existing instruction-set architectures, so is a more popular technique—for example, the Alpha is a dual-issue machine, or superscalar up to two instructions per cycle. Explicit vector instructions are another possibility, either controlling a vector pipeline or a SIMD vector datapath.

All of these techniques have traditionally been explored for their speed potential. But because efficient use of parallelism allows a given throughput at a lower clock rate and therefore a lower supply voltage, the same techniques are promising for power reduction. The most effective alternative for sufficiently regular problem classes, as found in speech signal processing, will be those that reduce control complexity and overhead, such as explicitly vector oriented alternatives.

4. PHYSICAL LEVEL OPTIMIZATION

Many choices are available in converting a logic design to circuits and layouts. We review a few possibilities here in terms of their impact on speed, power, and area costs.

4.1. The Speed/Power Penalty

Circuits optimized for speed use much more power than more moderate designs, when measured at a fixed clock rate and supply voltage where both will operate correctly. Under such conditions, the more moderate design is preferred. But the dominant "Hot Chips" approach optimizes for speed, ignoring other system cost factors. In this subsection, we discuss several physical-level choices, and how optimizing them for speed increases power consumption and system cost.

Transistor sizing is a key optimization tool. For very high speed designs, it is generally necessary to use large transistors, so that stray and wire capacitance is reduced relative to active gate capacitance. But at the high-speed end of the range, a large increase in size, and thus total capacitance, is needed to get a modest speed improvement. That is, the energy of a computation increases much more rapidly than the speed as transistors are made larger, with diminishing returns at higher speed. Larger transistors also lead to larger cell layouts and longer interconnect wires, to compound the problem. Tools are available for optimizing transistor sizes for speed, but usually not for other criteria.

Semi-custom ASIC designs based on gate arrays or standard cells are particularly susceptible to the problem of excessive capacitance, both because transistors are generously oversized for speed and because wire routing capacitance is often far from optimal. Good gate array and standard cell families optimized for energy efficiency would be quite useful. Dedicating extra wiring channel space and feedthrough options could help reduce long wires, and thereby trade increased area for improved efficiency and speed. Using smaller transistors and better optimized macrocells could recoup the area cost while further improving the energy efficiency at the cost of speed.

In optimizing for speed, it is important to minimize clock skew. Clock gating logic that would be useful for reducing activity in unused portions of a circuit could potentially save lots of power, but would increase clock skew and therefore adversely impact speed. In the Alpha chip, since speed was paramount, it was decided to drive the clock signal ungated to almost everywhere that it was needed, irrespective of the energy wasted.

Some circuit forms that are designed for high speed, such as "true single phase clock" (TSPC) dynamic logic [31, 32], require fast clock transitions to operate correctly. Therefore, large clock driver transistors are needed, and a large capacitive load is presented to the clock pre-driver stage. Almost half the power consumed by the Alpha is in clock distribution, and since it uses TSPC and is optimized for speed, almost half of that in the pre-drivers.

The combination of non-gated clocks and the TSPC logic form has allowed the Alpha to run at 200 MHz and claim the "Hot Chip" prize of its day, but at a tremendous cost in energy efficiency. The packaging, cooling, and high-speed interconnect issues in any product that uses the Alpha near its rated speed will dominate the system cost. Products that use the Alpha well below its rated speed, to save cost, will still suffer a substantial penalty from the fact that the chip was optimized for speed, even if they run at a reduced voltage.

In general, an improvement in circuit speed can be converted to an improvement in power by reducing the supply voltage to bring the speed back down. However, as argued above without proof, this speed-optimization approach still results in a design way off the optimum for other more realistic cost functions.

4.2. The Area/Power Symbiosis

In many cases, optimizing a circuit for area, rather than for speed, will lead to reduced computation energy with only a modest speed penalty. Smaller transistors lead to smaller cells and shorter wires, for less total capacitance, as mentioned above. Circuit choices that minimize the number of transistors may also reduce the effective capacitance being switched.

A popular technique in nMOS for saving transistors and area is the use of pass-transistor logic [8]. Simple single-ended n-type pass-transistor logic has not been used much in CMOS because the typical CMOS inverter characteristic is not compatible with the reduced logic-high level at the output of a pass transistor: it is difficult to achieve a low logic threshold and a low DC current. An "Ultra Low Power" technique proposed by Lowy and Tiemann [16] solves this problem by a simple increase in the magnitude of p-type thresholds (and optionally also a reduction in n-type thresholds), thereby achieving an excellent combination of the advantages of nMOS and

CMOS. Their technique is particularly useful when the circuit must operate at conventional supply voltages and be compatible with conventional CMOS circuits. It is unclear whether it works well or has compelling advantages at very low supply voltages.

Another pass-transistor logic form termed "complementary pass-transistor logic" (CPL) has been described by Yano et al. [33]. They used cross-coupled pullups to restore logic levels after complementary pass-transistor paths, rather than modifying inverter thresholds, so more transistors are needed than in simple pass-transistor logic. According to Brodersen *et al.* [11], this circuit form, if used with reduced n-type transistor thresholds, is a full order of magnitude more energy efficient, in an adder application, than differential cascode voltage switch logic (DCVSL), with other logic forms studied falling in between (Lowy and Tiemann's new logic form was not included, but may be even better).

Another dual-rail version of pass-transistor logic, termed complementary set-reset logic (CSRL), has been proposed by Wawrzynek and Mead [23]; the pass-transistor paths need not be complementary, but instead represent set and reset conditions for registers. In many applications CSRL circuits are almost as compact as nMOS-like circuits, but simpler and more robust, requiring no process modifications. CSRL is not known for its high speed, but Mead's group and others use this versatile form extensively in conjunction with micropower analog designs [34].

In some pass-transistor circuit forms, pass transistors may be used to implement dynamic storage. In that case, as with any dynamic logic form, leakage will determine a lower bound on clock speed and hence on power, unless other methods are employed to "staticize" the design. In CSRL, shift registers are inherently fully static, but dynamic storage is used in registers that neither set nor reset.

5. "MICROPOWER" TECHNIQUES

Reducing power consumption by more than a few orders of magnitude will require more extensive changes in our thinking. We briefly comment on both digital and analog "micropower" approaches.

5.1. Low Switching Energy

The analyses of Burr and Peterson [35] show that 3-dimensional MCM packaging of high-performance signal processing systems will require operation near the switching energy optimum, with transistor threshold voltages agressively reduced into the 100–200 mV range, due to power and cooling costs. This approach currently appears radical, but it is well motivated by technology trends and projected needs in real applications. More moderate threshold reductions will work well with most of the other power-saving methods discussed here.

In a new paper, Younis and Knight [36] show how to implement "Charge Recovery Logic" in CMOS to recover an increasing fraction of switching energy at decreasing clock rates. This approach appears even more radical, but looks like a good way to achieve a flexible speed-power tradeoff, with power porportional to speed squared, without reducing the power supply voltage.

5.2. Massive Analog Parallelism

The massively parallel "neural" micropower analog signal processing approach of Mead [37] has been applied to speech signal processing by Lyon and Mead and their colleagues [38, 39, 40, 41, 42]. Compared to the custom digital bit-serial approach of Summerfield and Lyon [26], the analog approach uses about an order of magnitude less silicon area and several orders of magnitude less power, but delivers much less precision.

Mead and Faggin [2, 43] have argued that for low-precision sensory front ends, where data variability and noise are inherently large compared to circuit component variability and noise, analog approaches show a significant advantage.

In terms of silicon area for state storage, the crossover between analog and digital approaches can be estimated based on the number of bits of precision needed. For some number of bits, N, the area of N digital memory cells will roughly match the area of an analog state storage capacitor and related circuits big enough for N bits of precision. We estimate that N is around 8 bits today. The additional local circuitry and power needed to process the stored state will almost always favor analog, since digital representations require lots of switching of lots of bits, while micropower analog approaches only incrementally change stored state voltages. Analog approaches may be preferred in terms of power even up to 12 or more bits of precision.

Analog techniques that use high-performance operational amplifiers are not as energy efficient as the micropower techniques, but may sometimes still be competitive with digital approaches.

6. LOW-POWER CHALLENGES

Implementing the ideas presented in this tutorial presents the chip designer and system designer with a number of challenges, some of which will best be met by the consensus of the industry through new standards and conventions.

Power supply voltage is a tool of great leverage for power reduction, but is not something that is easy to change continuously, or separately on different subsystems. Particularly in battery operated products, the power supply voltage must be planned to meet voltages attainable from one or several electrochemical cells, of a type selected for a variety of complex reasons. If the product can be operated without a voltage regulator, there will be greater savings, but then a more variable supply voltage must be tolerated, possibly along with a variable clock speed. If multiple chips are needed, the supply voltage must be selected to be within the intersection of their operating voltage regions—nearly impossible with today's chips if you want to operate at an efficient low voltage—or parts of the system need to operate at different voltages. If part of the system is operated at a lower voltage by dropping the supply through

a dissipative series regulator, then the factor of V^2 savings that was expected at the chip level will not be realized at the system level—only one factor of V will apply.

Specialization is another powerful tool for power and silicon area savings. A general-purpose von Neuman machine has a computational overhead of several orders of magnitude over a more specialized or dedicated signal processing architecture, while a programmable DSP chip falls somewhere between. But specialization can be costly by limiting the capabilities of a system, or by requiring additional design time. Efficient configurable or programmable architectures with a reasonable degree of specialization for signal processing applications are needed, but there is not yet a consensus as to which architectures to pursue beyond the single-chip DSP. Specializations that rely on parallelism with low control overhead, as in SIMD signal processors, offer a combination of advantages.

Analog micro-power techniques are tremendously power efficient relative to computing with digital representations of signals, but the kinds of accuracy, repeatability, and testability that we have come to expect from digital approaches are difficult or impossible to attain. This does not mean we are pessimistic on the outlook for massively parallel analog subsystems. Rather, we need to solve some problems and identify appropriate applications, in order to find good commercial success possibilities such as the one Synaptics found for optical character reading. Faggin and Mead [43] discuss this challenge.

Dynamically controlling the power-speed tradeoff is another important challenge. Making an existing portable computer architecture smart enough to judge the user's priorities may be quite difficult. Giving the user control over the tradeoff may be easier. Just as users of videotape have accepted the job of choosing 2-hour, 4-hour, or 6-hour recording mode, the user of a portable product could choose 2-hour, 4-hour, or 6-hour mode. A user stuck on an airplane would appreciate the option. If the product is a voice recorder, the tradeoff may influence sound quality; for other applications, it may influence responsiveness or other factors. The user will learn to make the tradeoff.

Can an existing chip spec'd at 200 MHz at 3.3 Volts and 30 Watts be operated an order of magnitude slower with two orders of magnitude power reduction, as expected around the Et optimum? Such scaling appears to be realized in Philips HLL CMOS [14]. Will the Alpha run at 20 MHz with a 1.1 Volt power supply? We suspect not. Typically somewhere in a chip will be unusual circuits with an extra threshold drop, which will kill performance rapidly as supply voltage is reduced. Chip designers need to learn that such circuits, which are motivated by optimizing performance at a particular supply voltage, take away a powerful power saving tool from the system designer.

7. CONCLUSIONS

There are a number of techniques available for reducing power consumption and system cost in speech signal processing products, and in other computing systems, as demanded especially by high-performance portable applications. Reducing the power supply voltage is the most important technique available for reducing the switching energy of logic functions. Using smaller transistors also reduces switching energy. Both of these techniques result in slower operation, so performance may need to be made up by parallelism. For parallelism to be effective, it needs to computationally efficient and power efficient, but also flexible enough for the range of tasks that it may be employed to implement. The extra silicon usage of this "low and slow" approach [44] will be a small price to pay for system-level savings.

The key impediment to taking advantage of cost-saving and power-saving techniques is the difficulty in changing to a new way of thinking about performance tradeoffs as the economics and technology of microelectronics continue to evolve. System designers and chip designers need to work together to accelerate progress.

In summary, we offer the following list of **top ten** [45] **standard ways to waste power and increase system cost:**

10. Do everything with a single central programmable processor, running fast enough for the worst-case computational demand.

9. When the processor is not needed to do real work, keep it running fast idle cycles.

8. Run the clock as fast as the chips allow, even when the application doesn't demand it.

7. Make all memory references across a central high-speed heavily-loaded system bus.

6. Optimize circuits for speed, using low fanouts, large transistors, and lots of pipelining.

5. Don't tolerate the additional clock skew of gated clocks to control which processing units are in use.

4. Use register circuits that require fast clock edges to operate correctly.

3. Use standard power supply voltages, regulated from a higher battery voltage.

2. Use digital signal representations wherever possible.

And finally, the number one way that you as a custom chip designer can waste power and increase cost in a portable speech processor:

1. To avoid being outsmarted by a clever system designer, make sure your chip won't operate correctly at a low voltage or a low clock rate.

REFERENCES

[1] A. M. Mohsen and C. A. Mead, "Delay Time Optimization for Driving and Sensing of Signals on High-Capacitance Paths of VLSI Systems," *IEEE J. Solid-State Circ.* **14**, pp. 462-470, 1979.

[2] C. Mead, "Neuromorphic Electronic Systems," *Proc. IEEE* **78**, pp. 1629–1636, 1990.

[3] J. Narkiewicz and W. P. Burleson, "VLSI Performance/Precision Tradeoffs of Approximate Rank-Order Filters," *VLSI Signal Processing, V* (K. Yao, R. Jain, and W. Przytula, eds.) pp. 185–194, IEEE 1992.

[4] A. Orailoglu and R. Karri "A Design Methodology for the High-Level Synthesis of Fault-Tolerant ASICs," *VLSI Signal Processing, V* (K. Yao, R. Jain, and W. Przytula, eds.) pp. 417–426, IEEE 1992.

[5] L. C. Stewart, A. C. Payne, and T. M. Levergood, "Are DSP Chips Obsolete?," *Intl. Conf. on Signal Processing Applications and Technology,* pp. 178–187, DSP Associates, Boston, 1992.

[6] D. W. Dobberpuhl *et xxii al.* "A 200-MHz 64-b Dual-Issue CMOS Microprocessor," *IEEE J. Solid-State Circ.* **27** pp. 1555–1567, 1992.

[7] R. M. Swanson and J. D. Meindl, "Ion-Implanted Complementary MOS Transistors in Low Voltage Circuits," *IEEE J. Solid-State Circ.* **7**, pp. 146–153, 1972.

[8] C. Mead and L. A. Conway, *Introduction to VLSI Systems,* Addison-Wesley, 1989.

[9] E. A. Vittoz, "Micropower Techniques," in *Design of MOS VLSI Circuits for Telecommunications* (Y. Tsividis and P. Antognetti, eds.), Prentice-Hall, 1985.

[10] J. Burr, P. R. Williamson, and A. Peterson, "Low Power Signal Processing Research at Stanford," *3rd NASA Symposium on VLSI Design,* pp. 11.1.1–11.1.12, Moscow, Idaho, Oct., 1991.

[11] R. Brodersen, A. Chandrakasan, and S. Sheng, "Low-Power Signal Processing Systems," *VLSI Signal Processing, V* (K. Yao, R. Jain, and W. Przytula, eds.) pp. 3–13, IEEE 1992.

[12] J. Burr and A. Peterson, "Ultra Low Power CMOS Technology," *3rd NASA Symposium on VLSI Design,* pp. 4.2.1–4.2.13, Moscow, Idaho, Oct., 1991.

[13] V. Von Kaenel, P. Macken, and M. G. R. Degrauwe, "A Voltage Reduction Technique for Battery-Operated Systems," *IEEE J. Solid-State Circ.* **25** pp. 1136–1140, 1990.

[14] Philips Semiconductors, "Fast, low-power HLL & LV-HCMOS logic families… …for systems with 1.2 V to 3.6 V supplies," undated product literature.

[15] A. Chandrakasan, M. Potkonjak, J. Rabaey, and R. Brodersen, "An Approach for Power Minimization Using Transformations," *VLSI Signal Processing, V* (K. Yao, R. Jain, and W. Przytula, eds.) pp. 41–50, IEEE 1992.

[16] M. Lowy and J. J. Tiemann, "Ultra-Low Power Digital CMOS Circuits," *VLSI Signal Processing, V* (K. Yao, R. Jain, and W. Przytula, eds.) pp. 31–40, IEEE 1992.

[17] S. R. Powell and P. M. Chau, "Estimating Power Dissipation of VLSI Signal Processing Chips: The PFA Technique," *VLSI Signal Processing, IV* (H. S.

Moscovitz, K. Yao, and R. Jain, eds.), pp. 250–259, IEEE 1990.

[18] S. R. Powell and P. M. Chau, "A Model for Estimating Power Dissipation in a Class of DSP VLSI Chips," *IEEE Trans. Circ. and Sys.* **38** pp. 646–650, 1991.

[19] M. R. C. M. Berkelaar and J. F. M. Theeuwen, "Real Area-Power-Delay Trade-Off in the *Euclid* Logic Synthesis System," *Custom Integrated Circuits Conference,* pp. 14.3.1–14.3.4, IEEE, 1990.

[20] R. F. Lyon, "A Bit-Serial VLSI Architectural Methodology for Signal Processing," in *VLSI 81 Very Large Scale Integration* (J. P. Gray, ed.), Academic Press, 1981.

[21] R. F. Lyon, "FILTERS: An Integrated Digital Filter Subsystem" and "MSSP: A Bit-Serial Multiprocessor for Signal Processing," in *VLSI Signal Processing: A Bit-Serial Approach,* P. B. Denyer and D. Renshaw, Addison-Wesley, 1985.

[22] R. F. Lyon, "MSSP: A Bit-Serial Multiprocessor for Signal Processing," in *VLSI Signal Processing* (P. Cappello et al., eds.), IEEE Press, 1984.

[23] J. Wawrzynek and C. Mead, "A New Discipline for CMOS Design: An Architecture for Sound Synthesis," in *Chapel Hill Conference on Very Large Scale Integration,* H. Fuchs, ed., Computer Science Press, 1985.

[24] J. Wawrzynek and C. Mead, "A VLSI Architecture for Sound Synthesis," in *VLSI Signal Processing: A Bit-Serial Approach,* P. B. Denyer and D. Renshaw, Addison-Wesley, 1985.

[25] J. Wawrzynek and C. Mead, "A Reconfigurable Concurrent VLSI Architecture for Sound Synthesis," *VLSI Signal Processing, II* (S. Y Kung, R. E. Owen, and J. G. Nash, eds.), pp. 385–396, IEEE Press, 1986.

[26] C. D. Summerfield and R. F. Lyon, "ASIC Implementation of the Lyon Cochlea Model," *Proc. IEEE International Conference on Acoustics, Speech, and Signal Processing,* pp. V-673–676 1992.

[27] A. K. W Yeung and J. M. Rabaey, "A Data-Driven Architecture for Rapid Prototyping of High Throughput DSP Algorithms," *VLSI Signal Processing, V* (K. Yao, R. Jain, and W. Przytula, eds.) pp. 225–234, IEEE 1992.

[28] S. K. Rao and T. Kailath, "Regular Iterative Algorithms and Their Implementation on Processor Arrays," *Proc. IEEE* **76,** pp. 259–269, 1988.

[29] S. Y. Kung, *VLSI Array Processors,* Prentice-Hall, 1988.

[30] C. H. Slump, C. G. M. van Asma, J. K. P. Barels, and W. J. A. Brunink, "Design and Implementation of a Linear-Phase Equalizer in Digital Audio Signal Processing," *VLSI Signal Processing, V* (K. Yao, R. Jain, and W. Przytula, eds.) pp. 297–306, IEEE 1992.

[31] Y. Ji-ren, I. Karlsson, and C. Svensson, "A True Single Phase Clock Dynamic CMOS Circuit Technique," *IEEE J. Solid-State Circ.* **22,** pp. 899–901, 1987.

[32] J. Yuan and C. Svensson, "High-speed CMOS Circuit Technique," *IEEE J. Solid-State Circ.* **24,** pp. 62–70, 1989.

[33] K. Yano, T. Yamanaka, T. Nishida, M. Saito, K. Shimohigashi, and A. Shimizu, "A 3.8-ns CMOS 16x16-b Multiplier Using Complementary Pass-Transistor Logic," *IEEE J. Solid-State Circ.* **25,** pp. 388–395, 1990.

[34] C. A. Mead and T. Delbrück, "Scanners for Visualizing Activity of Analog VLSI Circuitry," *Analog Integrated Circuits and Signal Processing* **1,** pp. 93–106, 1991.

[35] J. Burr and A. Peterson, "Energy Considerations in Multichip Module-based Multiprocessors," *IEEE Intl. Conf. on Computer Design,* pp. 593–600, Oct., 1991.

[36] S. Younis and T. Knight, "Practical Implementation of Charge Recovering Asymptotically Zero Power CMOS," *1993 Symposium on Integrated Systems* (C. Ebeling and G. Borriello, eds.), Univ. of Washington, in press (and personal communication).

[37] C. Mead, *Analog VLSI and Neural Systems,* Addison-Wesley, 1989.

[38] R. F. Lyon and C. Mead, "An Analog Electronic Cochlea," *IEEE Trans. ASSP* **36,** pp. 1119–1134, 1988.

[39] R. F. Lyon, "CCD Correlators for Auditory Models," *25th Asilomar Conference on Signals, Systems and Computers,* IEEE Computer Society Press, 1991.

[40] C. A. Mead, X. Arreguit, and J. Lazzaro, "Analog VLSI Model of Binaural Hearing," *IEEE Trans. Neural Networks* **2,** pp. 230–236, 1991.

[41] Lazzaro, J. and Mead, C., "Silicon models of auditory localization," in *An Introduction to Neural and Electronic Networks* (S. F. Zornetzer, J. L. Davis, and C. Lau, eds.), Academic Press, 1990.

[42] L. Watts, D. Kerns, R. Lyon, and C. Mead, "Improved Implementation of the Silicon Cochlea," *IEEE J. Solid State Circ.* **27,** pp. 692–700, May 1992.

[43] F. Faggin, "VLSI Implementation of Neural Networks," in *An Introduction to Neural and Electronic Networks* (S. F. Zornetzer, J. L. Davis, and C. Lau, eds.), Academic Press, 1990.

[44] G. Gilder, *MICROCOSM,* pp. 145–148, Simon and Schuster, NY, 1989.

[45] D. Letterman *et al., The "Late Night with David Letterman" Book of Top Ten Lists,* Simon & Schuster, 1990.

System Design of a Multimedia I/O Terminal

Anantha Chandrakasan, Tom Burd, Andy Burstein,
Shankar Narayanaswamy, Samuel Sheng, Robert Brodersen
University of California at Berkeley

A key to the success of future Personal Communications Systems is going to be the end-user devices or terminals that will allow users to have untethered access to fixed multimedia information servers. This paper addresses the key issues behind the design of a multimedia I/O terminal currently being implemented. The portability requirement places severe constraints on the power consumption requiring a system level approach to power reduction. This includes partitioning of computation and optimizing algorithms, architectures, circuit styles and technology. In addition to reducing the power consumption and meeting bandwidth requirements, portable operation imposes additional challenges on the system design in order to deal with high bit error rates of wireless communications.

1. INTRODUCTION

Over the past several years, the area of personal communications has seen explosive growth in the number of services and types of technologies that have become readily available. In voiceband communications, systems such as mobile analog cellular telephony, radio pagers, and cordless telephones have become commonplace, despite their limited nature and sometimes poor quality of transmission. In portable computation, "notebook" computers are boasting capabilities far in excess of the desktop machines of five years ago, with multi-MIPS RISC-based portable workstations quickly emerging. However, in spite of the myriad of technologies that are already available, there has been little by way of integration of these diverse services, merging computation and communications in a portable unit. Thus our vision of a future personal communications system currently centers on such integration of services, providing ubiquitous access to data and communications accessed through a specialized, wireless multimedia terminal.

As discussed in [1], our view of the wireless terminal is that it will be in full duplex communication with a networked base station, which serves as a gateway between the wired and wireless mediums. Thus, through the base station, a user accesses desired services over the high-speed communications backbone, including communicating with another person also linked into the network. However, this idea can be extended to a user communicating not only with another person, but with network "servers." Since the data bandwidth of future fiber-optic networks is easily in excess of 10 gigabits/sec, these centralized servers can provide an extensive variety of information services to users. Some key features of a personal communications system will likely include:

■ Access to large commercial databases that will contain various types of information, such as news, financial information, traffic data, etc.

■ Access to digital video databases containing both entertainment and educational media.

■ Simplified entry mechanisms such as voice-recognition and handwriting-recognition interfacing to access the above functions [2].

This paper will focus on the critical design issues behind the multimedia terminal. This includes:

■ Portability which places constraints on the size, weight and power consumption.

■ Wireless communications which imposes a constraint on the available communication bandwidth and the need to design for high BER (Bit Error Rates).

Optimizing performance in each of these areas is crucial in meeting the performance requirements of the overall system, in providing a small, lightweight terminal for personal communications.

2. TERMINAL PORTABILITY

The portability requirement of the terminal will place severe constraints on the size, weight, and power consumption. The various strategies used to reduce the power consumption at the system level will be discussed.

Power availability

One of the primary objectives in the design of portable systems is power reduction, which is required to maximize run time with a minimum of size and weight allocated to batteries. Unfortunately, the improvement in battery technology over the last 30 years, has been modest, with the capacity (Watt-hours/Pound) for Nickel-Cadmium cells increasing only by a factor of two, to the present value of 20 Wh/lb. The only likely new technology, which will take over from NiCd over the next few years is Nickel-Metal Hydride, since it alone provides a combination of enhanced performance (30-35 Wh/lb) while being environmentally safe [3]. However, projections for the next 5 years, for even this new technology, predict an improvement of only an additional 30%.

We can thus make a reasonable estimate of the power available for future portable systems. For example, assuming that 4 pounds is an upper limit on the weight of the portable device, 1/2 of which is the battery; with 8 hours of run time, the average power used for all functions must be less than 8 watts. The amount which can be used for computation is even less than this limited amount in those applications which require a high quality video display. Though present displays (or more accurately their backlight) would consume this entire power budget, it seems reasonable to expect that in the next 5 years that display power levels of several watts will be achieved, leaving only 2-3 watts for implementing all the computation and I/O support in our relatively heavy portable unit.

It might be hoped that the improvements in fabrication technology might be the answer to providing low power computation. Unfortunately, the recent "advances" in microprocessor design, which have reported single chip power dissipation levels upwards of 30 watts, demonstrate that full exploitation of higher clock rates results in dramatic *increases* in power dissipation, not the needed reductions. However, ultra low voltage operation with an architectural design strategy can reduce the power requirements by an *order of magnitude* [4]. Yet, even this will not be enough to make possible the most exciting new future applications of portable electronics; they will additionally require the use of the future, ubiquitous wireless communication networks.

Partitioning of Computation

The problem of implementing low power, general purpose processing (X-server, word processing, etc.) is primarily being addressed by power down strategies of subsystems, with a limited amount of voltage scaling [5]. This is effective to some extent, but because of the difficulty in determining when an application is really idle, there also needs to be modifications to the software and operating system to fully utilize processor inactivity. Unfortunately, a primary figure-of-merit for computers, that use one of the industry standard microprocessors, has become the clock rate, which results in architectures which are antagonistic to low power operation. On the other hand, parallel architectures, operating at reduced clock rates, can have the *same* effective performance in terms of instructions/second at substantially reduced power levels.

Another strategy to reducing the power required for general purpose computation is to effectively **remove** it from the portable. This is accomplished by providing communications capability to computational resources, either at a fixed base station at the other end of a wireless link or back on the network at available compute servers. This is simply the use of the well known client-server architecture, in which the portable unit degenerates to an I/O server (e.g. a wireless X-terminal). The basic strategy we use for text/graphics support is to transmit pen data over the uplink, perform the processing (X-server as well as client applications) on the base station or on a remote compute server sitting on the wired backbone and transmit incremental bit-maps from the X-server over the down-link. Hence most of the processing for supporting text/graphics is removed from the terminal, leaving behind only the text/graphics frame-buffer. Similarly, certain application specific tasks such as speech and handwriting recognition are moved to the backbone. PCM sampled speech (64Kbits/sec) is transmitted back and forth over the wireless link. Since video is accessed over a wireless network with limited available spectrum (< 2Mbits/sec allocated per user), video is transmitted in a compressed format and therefore decompression has to be performed on the terminal. Figure 1a shows the layout of a master controller chip designed to perform the interface functions to various I/O devices including a serial port (wireless side), a pen input, a speech codec and a text/graphics display. The controller runs at a supply voltage of 1.5V (1.2µm CMOS technology, 9.1mmx9.1mm) and uses level conversion pads to communicate with I/O devices running at higher voltages (5V).

Optimizing Algorithms

Minimizing the operation count is important in minimizing the switching events and hence the power consumption. In order to illustrate how algorithms can be optimized for low-power, lets look at the video decompression module. Various algorithms including transform coding and vector quantization have been widely explored and used for image compression. Transform based compression using the Discrete Cosine Transform (DCT) has been especially popular and forms the basis for most image compression standards (JPEG and MPEG). One feature of the DCT is that it is symmetric, i.e the coder and decoder have equal complexity, as shown in Table 1. Another approach is to use Vector Quantization, which has a complex coder but a very simple memory-lookup decoder, making it well suited for single-encoder, multiple-decoder systems. For this application, the VQ solution provides a means of implementing real-time decompression using very little computation and power.

| Method | Computation Requirements | |
	coder	decoder
DCT fast algorithm	4 multiply 8 additions 6 mem. access	4 multiply 8 additions 6 mem. access
Vector Quantization (Tree search - differential CB)	8 multiply 8 additions 8 mem. access	1 mem. access

Table 1: Complexity of compression algorithms.

Optimizing Architectures

The main contribution to power consumption in CMOS circuits is attributed to the charging and discharging of parasitic capacitors that occurs during logical transitions. The average switching energy of a CMOS gate (or the power-delay product) is equal to $C_{avg} \cdot V_{dd}^2$, where C_{avg} is the average capacitance being switched per clock, and V_{dd} is the supply voltage. The quadratic dependence of energy on voltage makes it clear that operating at the lowest possible voltage is most desirable for optimizing the energy per computation; unfortunately, reducing the supply voltage comes at a cost of a reduction in the computational throughput [4].

One way to compensate for these increased delays is to use architectures that reduce the speed requirements of operations while keeping throughput constant. To illustrate this point consider the text/graphics frame-buffer. The frame-buffer is written into by the X-server over the RF link and is read out to the display. While the write from the X-server is both asynchronous and slow, the read is synchronized to the refresh rate of the display (60Hz). Since the read operations occur much more frequently than the write operations, the system is optimized for the read operation. The conventional approach is to read data in a serial fashion; however, by exploiting the fact that read is sequential in nature, several pixels can be read in parallel (32 pixels in our system), hence reducing the speed requirements on the memory. This way, we can run the frame-buffer at supply voltages in the range of 1.2V-1.5V (and hence very low power levels) while meeting the high refresh rate requirements. Note that the X-server write is random, and hence the server has been modified to make the write sequential (hence enabling a low-power implementation).

Circuit Style and Technology

There are a number of options available in choosing and optimizing the basic circuit approach and topology for implementing various logic and arithmetic functions. We have developed a cell-library optimized for low-power operation. The main features of this library are: low-voltage operation (1.2-1.5V), reduced number of transistors to implement various functions, using minimum sized transistors as much as possible, reducing the average switching activity (e.g. reducing glitching), and optimizing layout styles to minimize the interconnect routing lengths and their associated parasitic capacitances. For example, the SRAM uses self-timed circuitry to reduce glitching activity, activates only one block at a time to reduce the capacitance switched, uses reduced voltage swings on the bit-lines (0.5V), operates at a supply voltage of 1.5V and consumes about 1mW at 10Mhz operation. This is *orders* of magnitude lower than existing current day solutions. Figure 1b shows a chip photo of the SRAM. For packaging, we are using Multi-chip Modules and Chip-on-Board which drastically reduces the interchip interconnect capacitance.

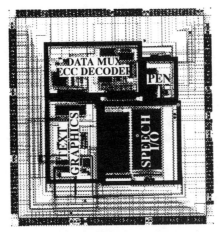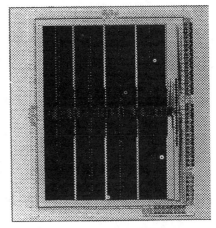

Figure 1a: Master controller chip to interface to Pen, Speech and Display.
Figure 1b: 64Kbit SRAM for the text/graphics frame-buffer.

3. WIRELESS COMMUNICATIONS

Wireless communication implies a limitation on the available communication bandwidth and the need to design for high bit error rates. These two constraints will be discussed in this section.

Spectral Efficiency

In order to support the various types of multimedia data such as video (transmitted in a compressed format), text / graphics (transmitted in incremental bit-maps) and speech, we need about 2Mbit/sec per user on average, and therefore spectrum usage is of great concern. Our approach to solve this problem is to utilize cellular networking techniques to achieve spatial reuse. Unlike conventional sized cells, which have a radius on the order of kilometers, we are using very small picocells (with a radius on the order of 5m). Using picocells implies that we will need about 100Mhz of RF spectrum to support a high density of users (~8-10 users in a 5m radius) while using conventional sized cells would require more than a Ghz of RF spectrum.

There is another significant advantage to picocellular wireless systems: since the transmit power is scaled down as the cells move closer together to reduce interference, power consumed in the portable's transmitter to drive the antenna is correspondingly reduced. Whereas existing cellular systems utilize 1 watt transmit power for voiceband RF links in 5 mile cells, a picocellular system with 5 meter cells will require less than one milliwatt to maintain the link.

Design for High Bit Error Rates (BER)

Wireless communications naturally implies a noisy transmission medium. To compare, wired ethernet achieves BER far lower than 10^{-12}, due to the extremely clean nature of the channel. For our system, with all users simultaneously communicating with full power, we have estimated a worst case BER of 10^{-3}.

Given this BER, consider the robustness of the text/graphics scheme which transmits incremental bit-maps over the RF link. The protocol used is to transmit a base address, length, and bit-mapped data. So errors show up in two forms; errors in the bit-mapped data which appear as dots on the screen, as seen from Figure 2b, and errors in the control information (address and length) which appear as lines on the screen. Our approach is to protect the control information, and leave the bit-mapped data unprotected. We see that even for a very high BER of 10^{-3}, we still get "acceptable" performance. Bandwidth can be traded for better quality by providing extra error protection for the control data.

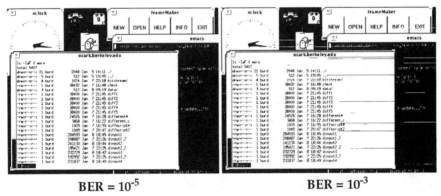

BER = 10^{-5} BER = 10^{-3}

Figure 2: Approach to high BER; transmit more "data" and minimal "control"

For video, there are three key features that are used to make it robust at high bit error rates:

- A ping-pong frame-buffer scheme that provides immunity against bursty type errors. i.e the previous frame is repeated in case of packet loss.

- Vector quantization decoding which localizes errors in space. i.e an error in the codeword that is transmitted appears as a corrupted 4x4 block on the screen and does not corrupt the entire screen.

- Intra-frame compression which localizes errors in time. i.e errors do not accumulate over time.

4. SUMMARY

A key to the success of future Personal Communications Systems is going to be the end-user devices or terminals that will allow users to have untethered access to fixed multimedia information servers. The portability requirement imposed by these terminals will place severe constraints on the total power consumption. To minimize the power consumption, it is necessary to exploit the availability of communications between the terminal and computational resources in the base-station and high speed backbone network, making the terminal an I/O device. Other techniques for achieving low-power include optimizing algorithms for minimal switching activity, using parallel architectures that enables slow and low-voltage operation, optimizing memory structures for reduced capacitance, and utilizing low-power cell libraries. The digital hardware for the terminal runs at a supply voltage of 1.5V and uses level-conversion circuits to communicate to commercial I/O devices running at higher supply voltages.

In addition to reducing the power consumption and meeting bandwidth requirements, wireless operation imposes an additional constraint, namely noise. Our approach to provide robustness against high BERs is to transmit I/O information and minimal "control" data through system level optimizations.

References

[1] R. W. Brodersen, A. Chandrakasan, and S. Sheng, "Low-Power Signal Processing Systems", 1992 VLSI Signal Processing Workshop, pp. 3-13.
[2] A. Burstein, A. Stoelze, R.W. Brodersen, "Using Speech Recognition in a Personal Communications System," in Proc. 1992 International Communications Conf., Chicago, IL, June 1992.
[3] J. S. Eager, "Advances in Rechargeable Batteries Spark Product Innovation", Proc. 1992 Silicon Valley Computer Conference, Santa Clara, CA, August, 1992, pp. 243-253.
[4] A. Chandrakasan, S. Sheng, R.W. Brodersen, "Low-Power CMOS Digital Design", IEEE Journal of Solid-State Circuits, Vol. 27, No. 4, pp. 473-484, Apr. 1992.
[5] D. Dahle, "Designing High Performance Systems to Run from 3.3V or Lower Sources", Silicon Valley Personal Computer Conference, pp. 685-691, 1991.

Acknowledgments

This project is funded by ARPA. The authors would like to thank Randy Allmon, Prof. Jan Rabaey, and Brian Richards for their invaluable contributions.

EECS Department
Cory Hall,
Graduate Student Box #65,
University of California
Berkeley, CA 94720
USA
e-mail: anantha@eecs.berkeley.edu

Fast Prototyping of Real Time Systems: A New Challenge?

Jan M. Rabaey

As electronic systems become more complex and gather more and more functionality, the attention of the design automation community is gradually drifting away from the chip prototyping problem and starts to focus on the development of system design technology. One might wonder how system prototyping differs from the traditional chip design problem. In this paper, we will demonstrate that there are indeed some fundamental differences, which will effect the design methodology at both the specification, simulation and implementation levels. On the other hand, we will also argue that a large part of the chip design methodology is valuable at the system level as well.

1. INTRODUCTION

When arguing about system design, a single question always comes to mind: What is really understood under the term *system* and where does it differ from, for instance, an integrated circuit. Answering this question brings us a long way towards understanding the difference between system and traditional integrated circuit prototyping. We have adopted the following definition of a system, which has, as will become obvious in the rest of the paper, a profound impact on the design methodology and tools: *a system is a self-contained entity, which is composed of a variety of sub-components with heterogeneous properties, communicating with each other using a variety of protocols.*

This definition has some important repercussions. First of all, no mention is made of a hardware platform. A system can span just a single integrated circuit, a printed circuit board, a card cage or a number of boxes. Secondly, the emphasis in system design is on heterogeneity. It brings together a number of widely varying components and architectures, some of them implemented in hardware, others realized in software, which use different means of communication and synchronization (for instance, synchronous or asynchronous, message passing based versus master-slave, etc.). This is in contrast to the traditional chip prototyping, which is based on a single architectural paradigm and uses a single communication mechanism, often being a synchronous, statically scheduled protocol.

179

Figure 1 shows an example of a complex system [1], which will be used as the driver example in the rest of the paper. It demonstrates the basic properties of a system in an ample way. The overall compute system consists of a number of components with varying architectures (compute servers, base stations or terminals), communicating with each other using a variety of protocols (wireless or wired). Each component on itself is once again a system. For instance, the infopad terminal consists of a number of elements, such as a radio modem or a video decompression unit. In this particular case, the units converse with each other using a buffered non-blocking message passing protocol.

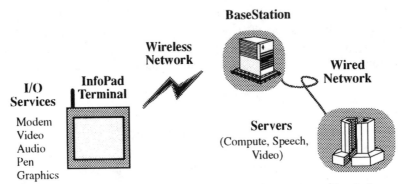

Figure 1 Infopad: a communication and computing system with wireless access.

This heterogeneity will be reflected throughout the design flow, presented in the following sections. We will consequently discuss issues regarding specification, simulation/emulation, design partitioning and implementation.

2. DESIGN SPECIFICATION/DESCRIPTION

As mentioned, a typical system merges a number of entities, whose underlying semantics are very diverging. Furthermore, these modules tend to communicate with each other using a variety of mechanisms. Believing that all these domains with their particular properties can be adequately modeled in a single environment is wishful thinking. From the VHDL experience, we have learned that meaningful design analysis and synthesis is only possible when the semantics of a domain and its interface with the surrounding environment are well understood.

Therefore, it is our belief that a heterogeneous specification environment is essential for the description of complex systems. In such an environment, each entity is modeled using an appropriate *domain*, which matches its underlying concepts. For instance, most signal processing algorithms are adequately modeled using (static or dynamic) data flow. A finite state machine model is a good match for a global controller and an event driven model fits a communication channel with its random nature.

The hierarchical nature of a system, furthermore, requires that these domains can be nested. For instance, our experience has shown that a *communicating processes* domain [2, 3] is often the best suited for the description of a system at the uppermost level. The individual components of the system are modeled as a set of concurrent processes, communicating with each other by message passing. Each of the components is be described in the appropriate semantic model (data-flow, sequential code, etc.) or, on its own turn, can be a complete system.

The idea of hierarchical, heterogeneous system modeling forms the core concept of the Ptolemy system [4], developed at U.C. Berkeley. It is implemented as an object-oriented framework, within which diverse models of computations can co-exist and interact.

As an example, Figure 2 shows a high level description of the Infopad system in the Ptolemy environment. It models the base station, the wireless channel and the wireless terminal as a set of processes. Depending upon the abstraction level, each pro-

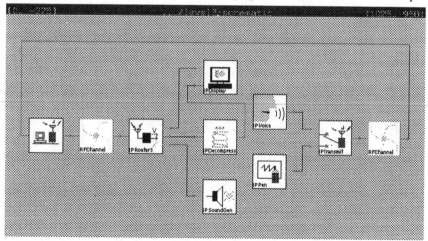

Figure 2 Communicating processes model of Infopad.

cess on its own is modeled in many different ways, either as an abstract model representing the I/O behavior only or, in more detail, by a complete sub-domain.

3. SIMULATION/EMULATION

Heterogeneous modeling results in leaner and more efficient simulation: as each sub-system is defined in its own domain with well understood semantics, a dedicated simulation engine can be used, which exploits the properties of that particular domain and avoids the overhead of the generic serve-all simulation approach. As an example, efficient architecture simulation of a micro-processor is possible when adopting a single evaluation per clock cycle regime. A simulator based on this concept is easily about a 100 times faster than a VHDL simulator, which can do exactly the same, but has to carry the extra overhead of being a full fledged event driven simulator.

The clear definition of the communication semantics, both within and between domains has another important advantage. The simulation and analysis of complex system, such as the multi-media computation and communication system of Figure 1, requires a tremendous amount of computation cycles, for instance to determine network bottlenecks or to analyze the effects of channel errors. Often, we would like to accelerate the simulation by distributing it over multiple workstations or by swapping in the actual hardware, when available. As an example, as the video server is implemented on an existing RAID server, video streams can be delivered directly from that server over the network instead of using a file based approach.

Such a hardware-software co-simulation is more easily achieved when the communication mechanism is well understood. In the communicating processes domain, for instance, a system-node can be migrated from the simulator to actual hardware, by replacing the simulation model of the node by a wrapper process, which behaves exactly like the original node from an input-output perspective, but has no other internal function but to transmit and receive the messages to/from the accelerator hardware [3].

This concept is demonstrated in the following example, which analyzes the basestation-terminal link of the infopad system. The basestation, router and audio sub-systems are implemented in software, while the video part of the terminal is implemented on an accelerator board, consisting mainly of a TMS320C30 (Figure 4). The link between the host workstation and the accelerator board is rather complex and proceeds over the ethernet to the VME-bus via a bridge, implemented on a single board micro-processor (Figure 3). It is worth mentioning that the migration of the code from the software domain to the hardware accelerator is totally automatic and all necessary software processes (at the host workstation, bridge and accelerator site) are synthesized from the defined communication protocol and its properties. Figure 4 shows a model of the simulation. The messages gen-

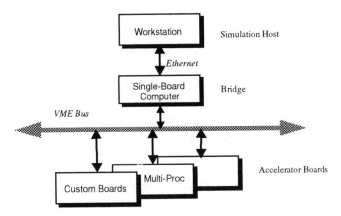

Figure 3 Hardware-Software Co-Simulation Environment.

erated by the remote process to monitor the simulation are shown in a separate window.

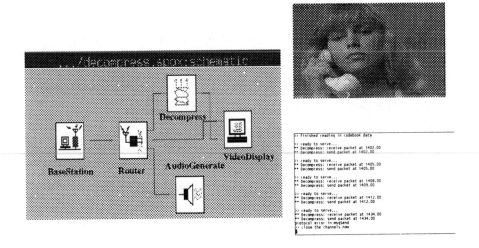

Figure 4 Hardware-Software Co-Simulation of Infopad.

The hardware-software co-simulation approach can result in *an incremental proto-typing* design methodology. The major hurdle to the realization of this concept is the long latency of contemporary computer networks. Efforts towards the realization of low-latency networks are under way, however [5].

4. DESIGN PARTITIONING

One of the key challenges in the system design process is the partitioning of the design into sub-components and the choice of the appropriate architectural and hardware platform for each of them. This task is often quoted to be *intuitive*, but actual design experiences have shown the opposite. For instance, in the infopad design, moving functionality from the terminal to the base-station or the servers can reduce the power consumption of the hand-held terminal, but might result in bandwidth bottlenecks or reduced quality.

The major problem with system level partitioning is a lack of information on how a high level decision impacts the design cost and system performance. To decide if a certain module should be implemented in software or hardware, one needs an approximative idea with regards to the performance, cost or power consumption for both alternatives. As the actual architecture might not been known at that time or as a complete design synthesis of every alternative might take too much time, there exists a dire need for effective techniques to generate an reliable performance model of a module (area, time, power), given an architectural template.

One would hope that a single estimation or performance analysis tool might be sufficient to cover the complete architectural and algorithmic space. Unfortunately, that is not the case. As the estimator is a simplified version of the actual synthesis/ compilation process, its mechanisms will be influenced by the nature of the architecture, the properties of the application and the cost factor to be optimized. In all, four classes of performance analysis tools can be identified [6]:

- *Deterministic* estimators, based on a simplified synthesis process.
- *Stochastic* Estimators, which obtain a statistical mode from the analysis of a large set of design examples.
- *Benchmarking*. This approach is especially effective for black box synthesis/ compilation tools.
- *Profiling* (simulation based) to anticipate the impact of data dependencies.

What estimation approach to use depends upon the specification paradigm, the target architecture and nature of the mapping process. Published results for each of the estimator classes indicate that average prediction accuracies within 5-10% are definitely within reach. The availability of efficiently generated performance models is the first step towards either an interactive or automatic system partitioning and design style selection environment.

5. IMPLEMENTATION

From an implementation perspective, we can divide the components into two classes: *Data Processing Modules*, which implement the various application specific algorithms and the *Interconnect Modules,* which implement the communication protocols linking the Data Processing Modules [2].

As obvious from the preceding discussions, no single architectural template will cover all the potential implementations of the Data Processing Modules. A mix of structural and behavioral module generators, hence, have to be provided. At this level, we can borrow extensively from the design approaches and methodologies used in custom design: for instance, database and design management tools, used for ASIC's are readily adapted to the system level. The concepts of design re-use and modular design, which made complex ASIC design possible in the first place, are extendable as well.

For example, a library of parameterized hardware modules can be provided for commonly used entities, such as microprocessor or signal processor sub-systems, memory modules, data acquisition modules and bus interfaces. Parameters can be the number of processors in a multi-processor module, the size of the memory or the number of ports in an interface macro-element. Besides generating the netlist, the module generator might also define the placement, invoke board or MCM routers and even might synthesize some of the glue logic. In the SIERA system [7], a structure description language (called *sdl*) with powerful parametrization facilities is used to describe the module topology. An example of such a module is shown in Figure 5a, which displays an instance of a uni-processor module (build around the TMS320C30). The SIERA module library contains elements such as a VME slave module, a static RAM block and a multi-processor sub-system.

Other module generators start from a behavioral description. Existing ASIC synthesis tools, such as C2Silicon [8] and HYPER [9] are readily available as system level building block generators. An example of a module, implementing a robot controller and generated using the C2Silicon and LAGER [10] systems, is shown in Figure 5b. These blocks are easily incorporated into the global system, once a black box model, specifying the input/output interface, is available. One can even go one step further: combining a software compiler with a pre-existing processor board or a parameterized processor module, results in a *"software"* module generator. This demonstrates that there is no conceptual difference between hardware and software from a system perspective. Concepts such as modularity and reusability are valid in both cases. The key to the integration between hardware and software is a clear understanding of the interface semantics.

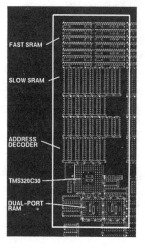
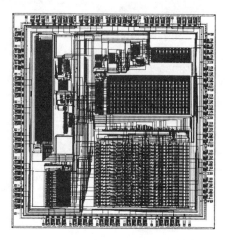

(a) A uni-processor structural module.   (b) C2Silicon Generated Processor

Figure 5 System level module generators.

The latter will have a distinct impact on the nature of the Interconnect Modules. Depending upon the protocol chosen and the hardware platform, an interconnect module can vary from pure wires, a composition of glue logic, a complex hardware module containing buffer elements and complex control or a simple piece of software, implementing a semaphore and running on a processor module. A interconnect generator can, hence, vary from the very simple to the complex. For instance, on a typical integrated circuit, the communication protocol is implicit and based on a clock-driven, statically scheduled event ordering. A message passing protocol can easily be overlaid on top of the VME slave module, mentioned earlier as well as implemented as an RPC call, as was discussed earlier. Event driven protocols, such as often occur in board level design, require the generation of an asynchronous control module [11]. These protocols, and the associated generation, can be pretty complex, as is illustrated in Figure 6, which shows the protocol governing the write from a processor to a memory module.

6. SUMMARY

The central concept in system prototyping is heterogeneity, which is reflected at all levels of the design process. A clear understanding of the semantics of each subsystem and its relations with the environment results in modular, hierarchical design process, not that distinct from the well-established chip prototyping methodology.

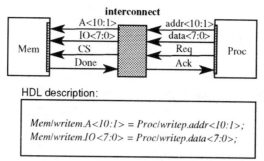

HDL description:

Mem/writem.A<10:1> = Proc/writep.addr<10:1>;
Mem/writem.IO<7:0> = Proc/writep.data<7:0>;

Figure 6 Processor to Memory Interface (Write Operation) -
Protocol Specification.

References

[1] R. W. Brodersen, A. Chandrakasan, and S. Sheng. *Technologies for Personal Communications*. 1991 Symposium on VLSI Circuits. Digest of Technical Papers, May 1991, pp. 5-9.

[2] M.B. Srivastava, R.W. Brodersen, "Rapid-Prototyping of Hardware and Software in a Unified Framework," *Proc. of ICCAD-91*, November 1991.

[3] S. Lee and J. Rabaey, "A Hardware-Software Co-Simulation Environment", *Proc. Hardware-Software Co-design Workshop*, Cambridge, October 1993.

[4] J. Buck, S. Ha, E.A. Lee, D. Messerschmitt, "Ptolemy: A Framework for Simulating and Prototyping Heterogeneous Systems," *International Journal of Computer Simulation*, special issue on "Simulation and Software Development."

[5] D. Patterson and D. Culler, "A Distributed Super-Computer Server", Internal Report Univ. Of California, Berkeley.

[6] I, Verbauwhede and J. Rabaey, ".Synthesis for Real Time Systems: Solutions and Challenges", Journal on VLSI Signal Processing, Nov. 1993.

[7] J. Sun, et al., "SIERA: A CAD Environment for Real Time Systems", Proc. Physical Design Workshop, May 1991

[8] L. Thon, K. Rimey, L. Svensson, "From C to Silicon," Chapter 17 in R. W. Brodersen (ed.). *Anatomy of Silicon Compiler*. Kluwer Academic Publishers, 1992.

[9] C. Chu et al., "HYPER: An Interactive Synthesis Environment for Real Time Applications", *Proc. ICCD 1989*, Cambridge, 1989.

[10] R. W. Brodersen (ed.). *Anatomy of Silicon Compiler*. Kluwer Academic Publishers, 1992

[11] J. Sun and R. Brodersen, "Design of System Interface Modules", Proc. ICCAD 1992, pp. 478-481, Nov. 1991.

Acknowledgments

Major funding for this project is provided by ARPA. The author appreciates the contributions of Mani Srivastava, Seungjun Lee, Alfred Yeung, Jane Sun, Miodrag Potkonjak, Lisa Guerra, Edward Lee and Robert Brodersen.

EECS Department
University of California
Berkeley, CA 94720
USA
e-mail: jan@zabriskie.berkeley.edu

Hardware-Software Trade-Offs for Emulation

M. Adé, P. Wauters, R. Lauwereins, M. Engels, J.A. Peperstraete

This article discusses the hardware versus software trade-offs for the functional and bit-compliant emulation of digital systems. It considers two classes of components: processors (software emulation) and field programmable gate arrays (hardware emulation). In a case study, a neural exclusive-or function is emulated on both platforms. This leeds to the conclusion that many applications will need a heterogeneous emulation environment.

1. INTRODUCTION

We define *emulation* as the execution of an application algorithm in real-time — i.e. at the speed determined by the application itself or by the environment into which the application will be placed — before the final implementation platform is constructed. The goal of emulation is to verify whether the envisioned implementation of the application meets the initial requirements before actually doing the implementation. This lets us spot difficulties as early in the design phase as possible, and hence reduces development costs and time-to-market. Depending on the level in the design process at which the emulation is done, we discriminate functional, bit-compliant, architecture-compliant and timing-compliant emulation.

- *Functional* emulation allows to verify the correctness of the chosen algorithm, to compare alternative algorithms, and to tune algorithmic parameters[1]. At this level, computational accuracy nor architectural or timing aspects are taken into account.

- In *bit-compliant* emulation, all computations are performed using the same word-length as in the final implementation. Also the rounding, truncation and overflow/underflow handling mechanisms mimic those of the final implementation. Its goal is to analyse finite word-length effects and possibly to reduce the word-length as much as allowed by the initial requirements.

- *Architecture-compliant* emulation uses the same architectural entities as the final implementation. As a consequence, we can verify whether the architecture chosen to implement the algorithm still meets the requirements.

- In *timing-compliant* emulation, all signals have the same timings as in the final implementation. Its goal is to check and correct timing problems before the actual product is built.

Further in this paper, we will not consider architecture-compliant neither timing-compliant emulation, but will restrict ourselves to the analysis of the properties of functional and bit-compliant emulation engines.

2. COMPONENTS FOR EMULATION

This section first discusses the requirements that functional and bit-compliant emulation pose to the emulation engine. Based on these requirements, two possible emulation platforms emerge: multiprocessors (software emulation) and field programmable gate arrays (hardware emulation). Next, we review the properties of both platforms.

Properties of emulation

For functional and bit-compliant emulation, architectural neither timing aspects are taken into account. The *architecture* used is hence *irrelevant*.

The most important property of both emulation set-ups is their *re-programmability*, for the components used as well as for the interconnection between the components. Building a dedicated bread-boarded prototype for each application is too time consuming and not flexible enough to perform the typical what-if analysis that is due in this early stage of the design.

The requirement that the emulation engine executes the application algorithm in real-time leads to set-ups which consist of *multiple components* interconnected by a network to exchange intermediate results. Indeed, the fact that emulation components necessarily are more general purpose than the final implementation chips, induces that less functionality of the application will fit on a single emulation component than on a single implementation chip.

Input/output (I/O) re-configurability is a highly desirable but difficult to achieve feature. Depending on the environment into which the final implementation will be placed, analogue-to-digital converters, phase-locked-loops or voltage converters must be present at the inputs or outputs. In practice, vendors of emulation engines shield themselves from this problem by letting the user build a dedicated I/O bread-board.

For bit-compliant emulation, an additional property of the emulation machine is that all computations have to be done with exactly the *same word-length* and rounding effects as in the final implementation.

Based on these requirements, two possible emulation components emerge: microprocessors (μPs) and field programmable gate arrays (FPGAs). Microprocessors have a fixed hardware architecture; the application algorithm that is executed on this architecture is determined by a program stored in PROM or RAM memory (software emulation). The architecture of FPGAs on the other hand is not fixed but determined by the configuration, stored in PROM or RAM memory (hardware emulation); the application algorithm is embedded into this architecture. Next sections will review the properties of both platforms.

Properties of a multiprocessor emulation set-up

• Each μP possesses an arithmetic logic unit (ALU) with a *fixed word-length*. Bit-compliant computations that involve words with a smaller word-length hence need additional operations to modify the result. This is especially true for bit-serial computations.

• Since the execution of an algorithm requires multiple computations and the μP typically has only one ALU, these computations are executed *sequentially*. Each μP possesses a sequential controller to impose the order indicated in the program.

• Typically, each μP has one or more external data and address buses to which additional data or program memory can be attached. Modern μPs allow between 16K and 4G words of *external memory*.

• When interconnecting multiple μPs together, they generally operate *asynchronously* with respect to each other, since they have a different clock oscillator. Even when using a single clock, it is difficult to let them work in lock step, since it is hard to let all μPs come out of reset state on the same clock phase.

• Most μPs have poor provisions for interconnecting them via a static *interconnection network* (apart from the Inmos Transputer series and the Texas Instruments TMS320C40). Dedicated hardware will be needed to give each μP a direct connection to a number of neighbouring nodes.

Properties of a field programmable gate array set-up.

• Hardware schemes are built up by tying together bit-oriented logic blocks. FPGAs hence have *no fixed word-length* and bit-compliant emulation is easily guaranteed. The larger the word-length to be implemented, the more chip area is consumed.

• Each separate computation of an algorithm is mapped to dedicated hardware, and may be executed *concurrently*. A sequential controller is hence not necessary.

• Typically, FPGAs only have a very *small amount of on-chip RAM* available (< 1 Kbit). If additional memory is necessary, an external memory interface could be implemented at the cost of consuming a rather large chip area and many I/O pins.

• Multiple FPGAs may work *asynchronously as well as synchronously* with respect to each other. When no additional pipeline stages may be introduced into the design, the maximum reachable clock frequency of a synchronous design will be lower for a multiple-FPGA solution than for a single-FPGA due to the slower off-chip communication.

- The communication between two tasks that are mapped onto two different FPGAs may be implemented by just interconnecting the corresponding external I/O pins. Dedicated communication hardware is hence not required. However, when not enough I/O pins are available, time-multiplexing has to be introduced at the cost of a substantial use of chip area.

3. CASE STUDY OF A HARDWARE-SOFTWARE TRADE-OFF

In this section, we compare the implementation of a neural exclusive-or function on DSPs and FPGAs. We first give a short overview of the application. In the next subsections we present two implementations of this neural exclusive-or, respectively on a multi-DSP and multi-FPGA system. The last section compares these implementations and makes some general conclusions about the hardware-software trade-off.

An N-input exclusive-or is a generalised symmetric functions, i.e. a function for which the output only depends on a norm of its inputs [4]. As shown in [5], generalised symmetric functions can be realised by means of a feed-forward network of linear-threshold elements, for which the weights and thresholds are fixed at the construction. A neuron with n binary inputs, digital weights w_{ij} and thresholds t_i, is described by equation (1). The output of the neuron equals y_i. The number of neurons in the network is thus $1 + \lfloor (N+1)/2 \rfloor \cdot 2$. Figure 1 shows an example where N equals 3.

$$y_i = \begin{cases} 1 & \text{when } \sum_j w_{ij} x_j \geq t_i \\ 0 & \text{otherwise} \end{cases} \tag{1}$$

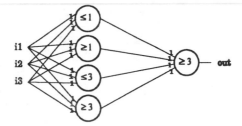

Figure 1. Neural exclusive-or with N=3

Implementation on a multi-DSP

A first implementation platform is a pipeline of up to 4 Motorola 56001 DSPs. This processor is well suited for our application because its ALU contains a multiply-accumulate unit (MAC), which exactly reflects the computations of the neural activation function.

For the implementation of the neural network we followed the approach of [6, 2] : the application is statically (at compile-time) assigned, routed, and scheduled. Due to this, the run-time overhead is minimised and the main program for every processor reduces to a simple sequencer. This sequencer calls a macro for every neuron and also performs the synchronisation on external inputs and outputs.

In general, intra-processor communication between the macros is organised through software buffers. However, for the neural network all buffers can be realised as a single memory location.

For the inter-processor communication we use the serial line of the processor. Since only one line is available between two processing elements, all communication is multiplexed. The communication is handled via send, receive, and transfer tasks, scheduled along with the other tasks, and using the software polling mechanism. When no direct hardware link is available, communication is routed through intermediate processors, by means of transfer tasks, which combine a send and a receive function. Neglecting the small overhead for the interrupt routine that starts and ends the application, the number of synapses in function of N for a DSP 56001 running at 27 MHz is shown in figure 2.

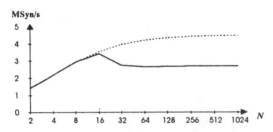

Figure 2. : Performance of a DSP in function of N

As can be seen from this plot, the upper limit for the performance is 4.5 MSyn/s. For N=12, the performance is 3.31 MSyn/s. For N=15, internal memory of the DSPs will be insufficient to hold all data since the number of synapses increases with N. Each external memory reference requires two extra cycles and parallel moves are impossible. Instead of 6 cycles for processing a synapse with internal data, it would take 10 cycles with external data. Therefore, performance will degrade starting at $N = 16$.

In our particular application, using more DSPs of the serial pipeline causes a deterioration of performance. This is due to the fact that all neurons of the first layer need all inputs. With the serial lines that take 96 cycles for transferring a 24-bit word, this communication constitutes the bottleneck of the application.

Implementation on a multi-FPGA

The second implementation is performed on a Virtual ASIC machine from InCA Ltd. with 8 Xilinx 3090 FPGAs that can be configured by the user. The FPGA has 320 bit-oriented Configurable Logic Blocks (CLBs) and 144 bi-directional and tri-state I/O pins. Each CLB may implement any combinatorial function of at most 5 inputs and contains 2 flip-flops. The interconnect between the CLBs is programmable. All configuration data are stored in on-chip RAM.

We assume that a direct mapping of the nodes of the network fits in the machine. Hence, no time multiplexing of neurons must be considered and all available parallelism of the application is exploited. Inter-chip and intra-chip communication is synchronous and performed by means of a global clock.

The area requirements increase quadratically with N, since not only the number of neurons increases linearly with N, but also the area occupied by each neuron.

Confronting these results with the 320 CLBs of one XC3090 FPGA learns that theoretically a network with at most $N = 16$ can be implemented on a single FPGA. However, practical implementations show that routing congestion limits N to 12, while only 43% of the available CLBs are in use. For larger N, the design has to be partitioned over several FPGAs. Due to limitations on area usage, pin count, delays, etc., the optimal number of 43% area usage will be reduced to 30%. For the VA with 8 FPGAs this results in a maximum N of 26. With several VAs, we can go further. Because of pin limitations, the largest neuron that fits in one FPGA has approximately 96 inputs. Since we do not allow to split one neuron over several FPGAs, $N=96$ is considered to be the absolute maximum. Figure 3 shows the overall performance of the implementation of the neural exclusive-or on a multi-FPGA in function of N.

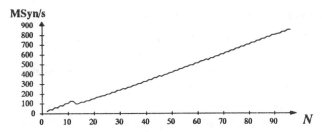

Figure 3. Performance of a (multi-)FPGA in function of N

A timing analysis of the single FPGA implementations showed that a 15 MHz clock can be reached for a XC3090PG175-100 device. When more FPGAs are used, off-chip delays are introduced, which are substantially larger than the on-chip delays. With direct connections available, a maximum delay of 100ns seems to be feasible. A 10 MHz clock is thus assumed for multi-FPGA implementations, which is for N larger than 12.

Trade-off analysis

As can be seen from the previous sections, FPGAs are far better suited for emulating a feedforward network of linear threshold elements than DSPs. Emulating an exclusive-or function with $N=12$ inputs, a single DSP obtains 3.31 MSyn/s, and a single FPGA 117 MSyn/s. This difference becomes even larger when N increases, since multiple FPGAs can execute the algorithm in parallel. Due to the high communication overhead when interconnecting multiple DSPs via their built-in serial lines, a single DSP still outperforms a parallel computer solution. Four main reasons for these differences may be seen:

- The binary logic of the application requires only a small amount of chip area on an FPGA, while the DSP cannot take advantage of its 24 bit architecture.

- The application itself contains a high degree of parallelism. This can readily be exploited even on a single FPGA, while a DSP has to stick to sequential execution due to its single ALU.

- The application only needs a very limited amount of memory storage. Since the weights are known at implementation time, they may be hard coded in the combinatorial logic on an FPGA without consuming scarce flipflops.

- Due to the small granularity of the application, communication becomes relatively more important. This prohibits us from taking advantage of a multiprocessor implementation. The same property on the other hand lets us scatter the small-grained computation parts over multiple FPGAs, while the small word-length avoids us from running out of I/O pins.

Next aspects also influence the trade-off, but were not present in the case study :

- Multi-rate applications as well as asynchronous activation of sender with respect to receiver necessitate the use of a buffer between sender and receiver. Implementing buffers on FPGAs consumes a large amount of the scarcely available memory bits. DSPs have ample memory to implement buffers.

- In general, the execution order of the various tasks assigned to the same device, may be determined at compile-time (static scheduling) or at run-time (dynamic scheduling). For hardware emulation, a dynamic schedule only requires a local test per task. For software emulation, a dynamic schedule requires the execution of multiple test and jump instructions at run-time. Asynchronous applications, which evidently lead to dynamic schedules, hence suffer from a rather large run-time scheduling overhead with software emulation.

• When the application is too large to fit on the available multi-FPGA emulation engine, multiplexing should be considered. To obtain a substantial gain in area, there should be many similar tasks. Indeed, besides an extra controller, the task itself will also take more area. Due to the sequential execution of multiple tasks on the same hardware, only a small performance benefit will be left compared to the much simpler DSP implementation.

• When one task of the application is too large to fit on a single FPGA, it has to be split over multiple FPGAs. The interconnectivity within a task is mostly much denser than between tasks. Splitting a single task over multiple FPGAs therefore often leads towards a shortage in I/O pins.

• For applications with several parallel inputs of many bits, pin limitations might prohibit implementation on a multi-FPGA. Here, multiplexing of some channels could be considered. If feasible, then at least extra control and multiplexing-demultiplexing hardware is required.

• The single built-in serial communication line of the DSP processors is clearly too slow for use as communication medium in a parallel computer. By building dedicated hardware, connected to the memory interface bus of the processor, fast interprocessor communication could be achieved. This will probably pose an intolerable overhead for small messages.

• Bit-compliant emulation fits better on FPGAs where the word-length can be chosen the same as in the final implementation, than on DSPs where additional instructions are necessary to modify computation results in a bit-compliant way.

4. CONCLUSION

Emulation of digital systems on the functional or bit-compliant level requires an emulation engine consisting of multiple processors and/or multiple field programmable gate arrays. Parts of the application that have a large word-length, need a large memory storage or buffer size, are favourably emulated in software on processors. Parts that have a small computation as well as communication granularity, have a high degree of parallelism or require dynamic run-time scheduling are better emulated in hardware on programmable gate arrays. Many applications possess parts that are well suited for software emulation as well as parts suited for hardware emulation. This leads to heterogeneous emulation engines. In [7] we showed, for instance, that a mixture of FPGAs and DSPs is essential for realizing a complete prototype of a digital radio receiver, because of the combination of a high frequency bit-serial synchronization part with computation intensive, word-wide FFT-calculations. The software environment used for prototyping hence needs to support this heterogeneity. Currently, no such tools are commercially available, but research tools are in development [8].

References

[1]   Lauwereins R., Engels M., Peperstraete J.A., Steegmans E., Van Ginderdeuren J., "GRAPE: a CASE Tool for Digital Signal Parallel Processing", IEEE ASSP magazine, special issue on Computer Aided Signal Processing Tools for Research, Vol. 7, No. 2, April 1990, pp. 32-43.

[2]   Engels M., "Design and Programming of Multiprocessors for Real-time Digital Signal Processing", Ph.D. dissertation, K.U.Leuven-ESAT, March 1993.

[3]   Anonymous, "The Programmable Gate Array Data Book", Xilinx Inc., 1992, pp 2/1-2/60.

[4]   Siu K.-Y., Bruck J., "Neural Computation of Arithmetic Functions", Proceedings of the IEEE, Vol. 78, No. 10, October 1990, pp. 1669-1675

[5]   R. Lauwereins, J. Bruck, "Efficient Implementation of a Neural Multiplier", Proceedings of the Int. Conf. on Microelectronics for Neural Networks, Kyrill & Method Verlag, Ed. U. Ramacher, U. Ruckert, J. Nossek, Oct. 16-18, 1991, Munchen, Germany, pp.217-230.

[6]   Engels M., Lauwereins R., Peperstraete J.A., "A Static Routing and Synthesis Tool based on a Heuristic Search Algorithm", Proc. of the Workshop on Parallel and Distributed Processing WP&DP'91, to be published in Elsevier Science Publishers series, Ed. Boyanov, Sofia, Bulgaria, April 13-19, 1991.

[7]   Engels M., Lauwereins R., Peperstraete J.A., "Rapid Prototyping for DSP Systems with Multiprocessors", IEEE Design & Test of Computers, June 1991, pp. 52-62.

[8]   Lauwereins R., Engels M., Peperstraete J.A., "GRAPE-II: A Tool for the Rapid Prototyping of Multi-Rate Asynchronous DSP Applications on Heterogeneous Multiprocessors", Proceedings of the Third International Workshop on Rapid System Prototyping, IEEE Computer Society Press, Ed. Nick Kanopoulos, Research Triangle Park, Nord Carolina, USA, June 23-25, 1992, pp. 24-37.

Acknowledgements

This work has partly been sponsored by the Belgian Government under the Concerted Action on Applicable Neural Networks and under IUAP-50, and by the EC under Esprit project 6800 (Retides).

Marleen Ade, Piet Wauters, Rudy Lauwereins[1], Marc Engels, J.A. Peperstraete
Katholieke Universiteit Leuven
Departement Elektrotechniek
K. Mercierlaan 94
3001 Heverlee
Belgium
e-mail : Marleen.Ade@esat.kuleuven.ac.be

---

[1]Senior Research Assistant of the Belgian National Fund for Scientific Research

OPTIMAL MAPPING OF DSP APPLICATIONS TO ARCHITECTURES

Catherine H. Gebotys and Robert J. Gebotys

Department of Electrical and Computer Engineering,
University of Waterloo, Waterloo, Ontario. N2L 3G1 Canada

Emerging programmable VLSI technologies have increased both the interest in and the industrial impact of CAD tools which map digital signal processing (DSP) applications to custom architectures. This paper presents for the first time an optimization approach to synthesizing DSP-specific architectures utilizing programmable VLSI technologies. A new integer programming (IP) model is presented that supports simultaneous scheduling, allocation, and retiming or loop winding. The IP model is used to map a DSP application to a high speed application-specific architecture and to map multiple DSP applications to a programmable architecture. The same model can also be used to map an application to multiple chips. Results show that the optimization approach synthesizes architectures that are 10% to 24% faster with up to 12% higher throughputs than previously published architectures. This research provides Industry with 1) a DA tool for optimal mapping of DSP applications to high performance architectures that can readily take advantage of emerging programmable VLSI technologies, and 2) a methodology to support optimal mapping of large DSP applications to multiple VLSI programmable chips.

1. Introduction

Since field programmable VLSI technologies are a factor of three slower than mask programmable technologies, architectural synthesis tools must look for increases in speed by extracting parallelism from the application, selecting optimal functional units, minimizing latency or execution time, and maximizing throughput. DSP applications can be represented by large grain data flow graphs (LGDFG) where each node represents a DSP block, such as a filter. The DSP block can in turn be represented by a fine grain DFG (FGDFG), where nodes are code operations such as a multiplication.

Although research in architectural synthesis has been active and developing over the last 10 years, it has had very little impact in industry. One of the reasons contributing to this problem is that few synthesizers have been able to produce high quality architectures. In order to synthesize high quality architectures, retiming, loop winding, functional pipelining and exploration of different types of functional units must be supported during application-specific architectural synthesis. Retiming or loop winding is necessary to maximize the throughput of recursive or non-recursive filters respectively. However for improvements beyond the above the synthesizer should be able to schedule and allocate architectures for unfolded loops. It has been proven that unfolding of any FGDFG by an optimum unfolding factor will transform it into a perfect rate DFG, or one that can be scheduled

rate-optimally[1] . The selection of different types of functional units from the library (including multicycled, pipelined, or chained functional units) must also be supported to synthesize high speed architectures. An architectural synthesis tool that supports functional pipelining, retiming, loop winding, and selection of different types of functional units is a necessity for high speed DSP applications.

The interest in high level synthesis tools has recently increased due to the emergence of low cost VLSI specifically field programmable gate arrays technologies. However since the FPGAs are slower than mask programmable gate arrays the high level synthesis must be optimal and wider design exploration (including clock period) must be performed in order to find good architectures which can utilize this technology. The LGDFGs[2] , representing large DSP systems, must also be mapped onto architectures of multiple (heterogeneous) DSP chips. Thus high level synthesis tools should be robust enough to support scheduling and allocation of the LGDFGs as well as the FGDFG. One of the earliest tasks to be performed in high level synthesis is that of scheduling and allocation of functional units. In fact it is well known that these early decisions have the largest impact on the final VLSI implementation. These early tasks are highly interdependent, thus in order to obtain high quality architectures one must solve them simultaneously. Unfortunately the architectural synthesis problem is believed to be NP-hard, since many of its subtasks have been defined as NP-complete. NP-hard means that there will exist some problems which require exponential time to solve. However the research presented in this paper shows that many problems can be solved very quickly to global optimums.

The objective of architectural synthesizers is to transform a FGDFG (or input algorithm or behavior) into a hardware architecture that satisfies performance constraints (throughput, latency) and minimizes cost (area). Synthesizers must minimize the latency (or execution time) given area constraints or minimize the chip area given constraints on the latency and throughput.

2. Problem Description and Previous Research

The following two problems, describe an important part of the architectural synthesis problem that will be examined in this paper. We will represent the DSP application by a FGDFG where nodes are the code operations (such as addition, multiplication) and the arcs are data transfer between operations.

Problem 1. *By mapping each operation (task) of the FGDFG (LGDFG) to a time or cstep, produce a schedule and synthesize an (multichip) architecture that for a given throughput, satisfies latency and area constraints, and supports selection of different* *types of functional units including chained functional units.*

An extension of this problem is to schedule multiple FGDFGs simultaneously to allocate architectures which can execute each FGDFG separately (programmable architecture) or multiple FGDFGs simultaneously (which we define as an interleaved architecture).

Problem 2. *By mapping each operation (task) of the FGDFG (LGDFG) to a cstep, produce a schedule and synthesize an (multichip) architecture that maximizes throughput while satisfying latency, and area constraints. Throughput is maximized by simultaneously performing loop winding or retiming.*

Problem 1 is called simultaneous scheduling and allocation, however its extension allows programmable and interleaved architectures to be synthesized. Problem 2 is called simultaneous scheduling, allocation, and loop winding or retiming. Alternatively in problem 1, the problem of minimizing area, given constraints on execution time will also be addressed. We define chained functional units as having no registers to hold the variable output from the first functional unit and input to the second operation.

Previous IP approaches[3-5] to simultaneous scheduling and allocation do not perform retiming or loop winding. Also previous research[6] has not investigated scheduling and allocation of LGDFGs or synthesis of programmable or interleaved architectures. The MILP model in [5] supported selection of chained types of functional units, however only very small examples could be solve in long cpu times due to the complexity of the model. A percolation based scheduler, PBS, in [7] and a pipeline scheduler, PLS,[8] both provide a heuristic approach to retiming during scheduling for synthesis of architectures. A range-chart-guided iterative algorithm for scheduling data flow graphs on multiprocessors has been developed[9] however it has not been applied to the synthesis of architectures.

A new model for synthesizing DSP-specific VLSI architectures, is introduced. The basis of this model is an extension of the previously researched IP approach[10] to simultaneous scheduling and resource allocation. Using polyhedral theory[11] , constraints were formulated to reduce the size of the search space and the problem was solved using the branch and bound algorithm. In this paper a new IP model is introduced to specifically support for the first time (1) retiming or loop winding simultaneously with scheduling and allocation, (2) synthesis of programmable and interleaved architectures, and (3) synthesis of multichip architectures for LGDFGs. This research breaks new ground by (1) providing industry with synthesized high performance architectures for DSP applications (through optimal retiming and loop winding), and (2) synthesizing programmable and interleaved architectures which are capable of executing a number of filters simultaneously or one at a time.

The following terminology will be used in this paper: The set of nodes in the data flow graph (DFG) is called K. k = a code operation of the input algorithm, $k \varepsilon K$. A partial order between k_1 and k_2 is represented by $k_1 <\bullet k_2$, or in other words code operation k_1 produces (or outputs) data that is used by code operation k_2. The variables of the model are $x_{j,k,t}$. When $x_{j,k,t} = 1$, code operation k is assigned to time (cstep) j ($j\varepsilon Z$, set of integers) and operation k is assigned to functional unit type t

($t\varepsilon Z$). $t_z\varepsilon T(k_z)$ means that code operation k_z can be implemented by functional unit type t_z. For example a multiplication operation, k, can be implemented by $t=1$ a two-cycle multiplier and $t=2$ a pipelined multiplier, $T(k)=\{1,2\}$ (further details will be provided later in the paper concerning the use of t for chained operations). The notation $j_z \varepsilon R(k_z)$ means that $asap(k_z) \leq j_z \leq alap(k_z)$ (where $asap/alap(k_z)$ is the as soon/late as possible csteps for operation k_z). The notation for retiming and loop winding will be presented in section 4. The variable I_t represents the number of functional units of type t. For simplicity we will assume that $t_z \varepsilon op(C_z, L_z)$ which means that the type of functional unit t_z requires C_z csteps to produce output data and can accept new input data at a minimum of L_z csteps. For example a single cycle type of functional unit is $t\varepsilon op(1,1)$ and a two-cycle pipelined type of functional unit is $t\varepsilon op(2,1)$. We use the notation $time(k_1,k_2) \leq or \geq or =T$ to represent ($\sum\limits_{j} x_{j,k_2} - \sum\limits_{j} x_{j,k_1} \leq or \geq or =T$) the maximum, minimum or fixed time constraint between the two operations. Note that $time(k_1,k_2) \leq T$ is equivalent to $time(k_2,k_1) \geq -T$. For functional pipelining or pipelining of the algorithm, we define Te as the latency or number of csteps between the start of the algorithm to the completion of the algorithm. We define f as the number of csteps between successive initiations of the algorithm (often called pipeline latency or iteration period bound). Any variable with $^{LB/UB}$ will denote the value of the lower bound (LB) or upper bound (UB) of the variable.

3. The IP Model for Problem 1

The integer programming model for simultaneous scheduling and allocation of functional units of the DSP architecture is given below with extension to programmable and interleaved architectural synthesis. The basic model which performs scheduling and allocation is first presented followed by the constraints for chaining of functional units, functional pipelining, and allocation of programmable and interleaved architectures.

The code operation assignment constraint, (1.1), ensures that each code operation of the input algorithm will be assigned to one cstep and one type of functional unit. The type of functional unit constraint, (2.1), calculates the number of functional units of a particular type to be allocated. The data precedence constraint, (3.1), prevents an operation k_1 from being scheduled after operation k_2 whenever $k_1 <\bullet k_2$ and t_1 is not a chained type of functional unit. The precedence constraint is equivalent to the minimum timing constraint $time(k_1,k_2) \geq C_1$, $t_1\varepsilon T(k_1)$, and more details can be found in[3] . This constraints supports multicycle operations as well as pipelined operations. In order to support chaining of operations the constraints described in the next section are required.

$$\sum_{t\varepsilon T(k)} \sum_{j\varepsilon R(k)} x_{j,k,t} = 1, \forall k \qquad (1.1)$$

$$\sum_{j1=j}^{j1=j+(L_1-1)} \sum_{k_1,t_1 \varepsilon T(k_1)} x_{j1,k_1,t_1} \le I_{t_1}, \forall j,t_1 \qquad (2.1)$$

$$\sum_{t_1 \varepsilon T(k_1)} \sum_{\substack{j_1 \varepsilon R(k_1), j-(C_1-1) \le j_1}} x_{j_1,k_1,t_1} + \qquad (3.1)$$

$$\sum_{t_2 \varepsilon T(k_2)} \sum_{\substack{j_2 \le j \\ j_2 \varepsilon R(k_2)}} x_{j_2,k_2,t_2} \le 1, \forall k_1 <\bullet k_2, j \varepsilon R(k_2) \bigcap (R(k_1)+C_1-1)$$

Chaining

Consider a single cycle chained type of functional unit, called an adder-adder (aa), composed of two additions $+_1<\bullet+_2$. The variables which represent selection of the adder-adder and operation scheduling, are $x_{j,+_1,t_1}$ and $x_{j,+_2,t_2}$, where $t_1=t_{aa1}$ and $t_2=t_{aa2}$ represent the first and the second addition operations in the adder-adder respectively. Different values of t are used to ensure that no more than two successive adders are chained together (or scheduled in the same cstep). The following constraint ensures that the adder-adder will execute in one cstep, time$(+_1,+_2)=0$ $\forall +_1<\bullet+_2, t_1=t_{aa1}, t_2=t_{aa2}$. This (fixed) timing constraint is represented in the IP model as time$(+_1,+_2)\ge0$ and time$(+_1,+_2)\le0$ $\forall t_1=t_{aa1}, t_2=t_{aa2}$ along with the following inequalities
$x_{j_1,+_1,t_{aa1}}+ \sum_{j\neq j_1, j \varepsilon R(+_2)} x_{j,+_2,t_{aa2}} \le 1, \quad x_{j_1,+_2,t_{aa2}}+ \sum_{j\neq j_1, j \varepsilon R(+_1)} x_{j,+_1,t_{aa1}} \le 1$
$\forall j_1, +_1<\bullet+_2$. However each addition operation may also be implemented with a one-cycle adder type of functional unit. Therefore the remaining constraints are time$(k_1,+_2)\ge C_1$, $\forall k_1<\bullet+_2, t_1\neq t_{aa1}$, and time$(+_1,k_2)\ge C_1, \forall +_1<\bullet k_2, t_2\neq t_{aa2}$.

Another example of a chained functional unit, called a multiplier-adder (ma), is a multiplication chained with an addition, $*_1<\bullet+_2, t_1=t_2=t_{ma}$ (for multiplier-adder type), where $C_{ma}=2$. Thus the following timing constraints are used time$(*_1,+_2)\ge1 \forall *_1<\bullet+_2, t_1=t_2=t_{ma}$, time$(*_1,k_2)\ge C_1 \forall *_1<\bullet k_2$, $t_2\neq t_{ma}$ (this includes $t_1=t_{ma}$), and time$(k_1,+_2)\ge C_1 \forall k_1<\bullet+_2$, $t_1\neq t_{ma}$ or $t_2\neq t_{ma}$. The IP approach supports chaining of two or more operations to define any type of functional unit. Unlike previous research, no preprocessing of operations is required in order to simultaneously select chained types of functional units.

Functional Pipelining

Functional pipelining for a fixed initiation rate (every f csteps) can be supported through modification of constraint (2.1). It is replaced by $\sum_{n=f}^{n=p} \sum_{j1=j}^{j1=j+(L_1-1)} \sum_{\substack{k_1,j1+nf \varepsilon R(k_1) \\ t_1 \varepsilon T(k_1)}} x_{j1+nf,k_1,t_1} \le I_{t_1}, \forall t_1$. In this

inequality, the filter is initiated every f csteps, where Te/f = p and Te is the latency of the FGDFG.

Cost Function

Both an area cost function and latency cost function can be formulated and minimized. For the latency formulation we use an extra operation called end, which must succeed all operations in the FGDFG. This operation, end, marks the end of the loop or FGDFG. The latency objective function to be minimized is then formulated as $Minimize \sum_{j \varepsilon R(end)} (j-1)*x_{j,end,0}$ which equals Te (where subscript t is assigned to zero since it is not necessary).

An area cost function can be formulated as $Minimize \sum_t area(t)*I_t$, where area(t) is the area of the functional unit of type t.

Programmable Architectures

Now consider the case where we wish to synthesize an architecture that is capable of executing a number of different types of filters. The problem is to find out what types of functional units and how many of each to use in the architecture along with a schedule for each filter on this architecture. The IP model simultaneously schedules and allocates a programmable architecture capable of executing all of the filters. For example let us assume that there are f FGDFGs, $K=\{K_1,...,K_f\}$, each representing a different filter. Constraints (1.1) and (3.1) are formed for all $k \varepsilon K$ however a modified version of constraint (2.1) specifically constraint (2.1*) given below, is used. Inequality (2.1*) ensures that the number of functional units of a type is the maximum number required over the complete set of filters.

$$\sum_{j1=j}^{j1=j+(L_1-1)} \sum_{\substack{k_1 \varepsilon K_f \\ t_1 \varepsilon T(k_1)}} x_{j1,k_1,t_1} \le I_{t_1}, \forall K_f,j,t_1 \qquad (2.1*)$$

Interleaved Architectures

Assume that time interleaved data samples of several input signals are available (similar to word-serial time division multiplexing). Furthermore assume that each input signal, s_i, must be processed by a filter, f_i. The problem is to generate an architecture that executes a number of filters, FGDFGs, simultaneously. Each filter is functionally pipelined with initiation rate equivalent to the rate at which its input data is arriving. For example consider two input signals, s_0 and s_1, which arrive alternatively at successive csteps. In the IP model the complete set of operations from both filters are used, $k \varepsilon K = \{K_0, K_1\}$ where each filter is scheduled starting at cstep one. Therefore at the even numbered csteps samples of the input signal s_0 arrive and at the odd numbered csteps samples of the input signal s_1 arrives. Furthermore if the filters, f_0 and f_1, are functionally pipelined, then data for each filter arrives every two csteps (f=2). To model this the following constraints (2.1.1*,2.2.1*) are used in place of constraint (2.1):

$$\sum_{n=1}^{n=p} \sum_{j1=j}^{j1=j+(L_1-1)} \sum_{\substack{k_1 \varepsilon K_1 \\ t_1 \varepsilon T(k_1)}} x_{(j1+2n),k_1,t_1} + \qquad (2.1.1*)$$

$$\sum_{n=1}^{n=p} \sum_{j1=j}^{j1=j+(L_1-1)} \sum_{\substack{k_1 \varepsilon K_0 \\ t_1 \varepsilon T(k_1)}} x_{(j1+2n+1),k_1,t_1} \le I_{t_1}, \forall j,t_1$$

$$\sum_{n=1}^{n=p} \sum_{j1=j}^{j1=j+(L_1-1)} \sum_{\substack{k_1 \varepsilon K_1 \\ t_1 \varepsilon T(k_1)}} x_{(j1+2n+1),k_1,t_1} + \qquad (2.2.1*)$$

$$\sum_{n=1}^{n=p} \sum_{j1=j}^{j1=j+(L_1-1)} \sum_{\substack{k_1 \varepsilon K_0 \\ t_1 \varepsilon T(k_1)}} x_{(j1+2n),k_1,t_1} \le I_{t_1}, \forall j,t_1$$

In inequality (2.1.1*) all the odd csteps of f_0 are scheduled with the even csteps of f_1 and in (2.2.1*) all the odd csteps of f_1 are scheduled with the even csteps of f_0. The resulting architecture simultaneously processes the two incoming signals. In this way the hardware can be shared in a more optimal manner than implementing a separate architecture for each filter. For example, during one cstep f_0 may use one multiplier and f_1 may use two multipliers and one adder. During another cstep f_0 may use two multipliers and f_1 may use one multiplier and one adder. The synthesized interleaved architecture would only need three multipliers and one adder as opposed to two separate architectures, one with two multipliers and one adder for f_1, and the other architecture with two multipliers for f_0. The IP model can synthesize an optimal architecture for simultaneous execution of both filters while meeting throughput and latency requirements.

4. The IP Model for Problem 2

The following notation in addition to the notation presented in section 2 will be used to present the loop winding/retiming IP model. Let l represent the loop iteration number, $l \varepsilon Z$, $l \varepsilon 1,2,...,LI$, where LI is the maximum number of loop iterations. For example assume a loop was unrolled three times, at one cstep operations from loop iteration 3 would be assigned l=1, from loop iteration 2 would be assigned l=2 and from loop iteration 1 would be assigned l=3. To model loop winding or retiming we assume that there are again K code operations and each can be assigned to only one loop iteration. Therefore the variables for this model are $x_{j,k,l}$, where $x_{j,k,l}=1$ means that code operation k is assigned to cstep j and assigned to loop iteration l. Precedence constraints are now dependent upon which loop iteration is assigned to a code operation. So $a^{l1} <\bullet b^{l2}$ means that data produced by code operation a from loop iteration $l1$ is used as input to code operation b of loop iteration $l2$. If $l1 > l2$ then the DSP application is a recursive filter and retiming is being performed by the IP model. If $l1 \leq l2$ for all precedence constraints then the DSP application is non-recursive and loop winding is being performed.

Consider the following example where $a^{l0} <\bullet b^{l1}$, $l0 \leq l1$, and $b^{l3} <\bullet c^{l2}$, $l2 < l3$. Assume that LI=2 J=10 and the set of valid csteps for c to be scheduled in is R(1,c)={1,2,3,4,5} and operation a and b can be scheduled in R(1,a)={7,8,9,10} and R(1,b)={8,9,10}. Normally the as late as possible time for operation a would be 9 since it must preced operation b, however a and b can be retimed so the latest cstep in l=1 becomes 10. Therefore if operation a and/or b are retimed they can be scheduled in csteps R(2,b) = {1,2,3,4} and R(2,a)={1,2,3}. To compute the asap and alap values for each l, we first establish J and LI. We set the maximum number of csteps to (LI)(J) and compute the asap and alap values for each operation. Then we divide the region into LI sets of csteps, each comprising of J csteps and use these sets to obtain the asap and alap values for each code operation in each loop iteration, or R(l,k).

The following equations define the IP model which simultaneously performs scheduling, hardware allocation and retiming or loop winding. Equation (1.2) ensures each code operation is assigned to only one cstep and one loop iteration. Equation (2.2) defines the precedence constraint for operation k1 of loop iteration l1 which outputs data to operation k2 of loop iteration l2. For the case where $l1 \leq l2$ each of the four terms from left to right

are also illustrated from left to right in figure 1b) as four rectangular boxes. The constraint ensures that at most only one of k1 or k2 can be scheduled at one of these csteps located in the box. A feasible schedule is shown as a circle k1 (representing scheduled cstep location for k1) joined to the circle k2 with an arc (to represent data transfer). This is generated for all j. For the case of l1>l2 this constraint generates a different set of boxes shown in figure 1c), again with corresponding terms and boxes. Equation (3.2) ensures that all code operations are scheduled at least one cstep before the end operation which is used to determine the value of f (or pipeline latency).

Equation (4.2) ensures that data precedence constraints are maintained across loop iterations. The cstep jend is defined as jend $=f+1= \sum_j x_{j,end,0}$. This also describes how we can formulate our cost function which maximizes throughput ie. f by $minimizing \sum_j (j-1)x_{j,end,0}$. An example is illustrated in figure 1a) where $* <\bullet +$ and the multiplier has a $C_*=2$, and the multiplier is scheduled at f-1 and the adder at j=1. For example if the multiplier is scheduled at the last cstep in l=1 (contributing one to left hand side of constraint (4.2)) then the adder can at the earliest be scheduled at cstep j=2 (outside of the boxed region and therefore constraint (4.2) is feasible). This is necessary since the two cycle multiplier is scheduled across the loop iteration, where one cycle is j=f in l=1 and the next cycle is at j=1 in l=2.

Equation (5.2) determines the number of functional units allocated in the architecture. Equation (6.2) allocates the correct number of functional units as well except i checks for multicycle operations which may be scheduled across loop iterations.

$$\sum_l \sum_{j \varepsilon R(l,k)} x_{j,k,l}=1, \forall k \qquad (1.2)$$

$$\sum_{l4<l1j4\varepsilon R(l4,k2)} x_{j4,k2,l4} + \sum_{\substack{j2 \varepsilon R(l1,k1) \\ j2 \geq j-(C_{k1}-1)}} x_{j2,k1,l1} + \qquad (2.2)$$
$$\sum_{\substack{j1 \leq j \\ j1 \varepsilon R(l2,k2)}} x_{j1,k2,l2} + \sum_{l3>l2}\sum_{j3 \varepsilon R(l3,k1)} x_{j3,k1,l3} \leq 1. \ \forall j,k1^{l1} <\bullet k2^{l2}$$

$$\sum_{\substack{j1 \leq j \\ j1 \varepsilon R(0,end)}} x_{j1,end,0} + \sum_{\substack{j2 \geq j \\ j2 \varepsilon R(l,k)}} x_{j2,k,l} \leq 1. \ \forall k,l,j \qquad (3.2)$$

$$x_{(jend-c),k1,l} + \sum_{l2<lj3\varepsilon R(l2,k2)} x_{j3,k2,l2} + \qquad (4.2)$$
$$\sum_{l1>lj1\varepsilon R(l1,k1)} x_{j1,k1,l1} + \sum_{(C_{k1}-c)\geq j2 \geq 1} x_{j2,k2,l} \leq 1.$$
$$\forall k1^{l1} <\bullet k2^{l2}, C_{k1} \geq c \geq 1$$

$$\sum_{k \varepsilon T(t)} \sum_l \sum_{\substack{(j-Lk+1)\leq j1 \leq j \\ j1 \varepsilon R(l,k)}} x_{j1,k,l} + I_t^{LB} \sum_j x_{j,end,0} \leq I_t. \forall j,t \qquad (5.2)$$

$$\sum_{k \varepsilon T(t)} \sum_l (\sum_{Lk-1-j \geq c \geq 0} x_{(jend-1-c),k,l} + \sum_{(Lk-1)\geq c \geq 0,j-c \geq 1} x_{j-c,k,l}) \qquad (6.2)$$
$$\leq I_t. \forall j,t$$

These data precedence constraints (3.1),(2.2),(3.2), and (4.2) are facet-generating (for a subset of the problem[10]). In other

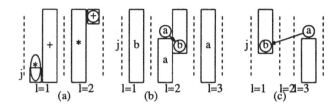

Figure 1. *Scheduling operations simultaneously with loop winding. Code operations in (a) are * < •+, and in (b),(c) are a < •b. Constraint (2.2) is illustrated in 1b) and 1c) and (4.2) is illustrated in 1a), where c=1. Only one operation can be scheduled within the boxed regions (csteps increase from top to bottom of figure).*

words they contribute significantly to decreasing the size of the search space and explains why many IP synthesis problems can be solved in very good execution times. Register allocation constraints and conditional code support and timing constraints for the new model are obtained by substituting $\sum_l x_{j,k,l}$ for $x_{j,k}$ in the previously researched constraints[10].

Methodology

The formulation for area is the same as that presented in section 3. This can be used to place a constraint on area while one is maximizing throughput using this IP model (presented in section 4). One can also put a constraint on Te while minimizing f (or maximizing throughput ie. $minimize \sum_j (j-1)x_{j,end,0}$). This is performed by calculating the asap and alap csteps for each code operation so that the Te constraint is met. For example if the constraint is $Te \leq Tmax$, then we choose LI and J so that LI(J-1) $\leq$ Tmax $\leq$ LI(J). The asap and alap are calculated for a total of Tmax csteps and R(l,k) are derived as described earlier. During the design exploration our methodology starts with LI=2 and minimizes f. Then we increase LI by one until there is no further change in f (and Tmax constraints are still met, LI(J-1) $\leq$ Tmax $\leq$ LI(J)). These series of IP solutions obtain the maximum throughput given constraints on area and Te. In the next phase of the methodology, one can then fix the maximum throughput and minimize the area and alternatively the Te using the IP model described in section 3. This methodology provided very good results as we shall see in the next section for a discrete cosine transform DSP application.

5. Synthesizer Results and Discussion

To illustrate the IP model based architectural synthesizer for architectural exploration, we have chosen the fifth order elliptic wave filter (EWF), 16 point FIR filter (FIR), and AR lattice filter (AR) from the HLSW benchmark suite[12], a discrete cosine transform (DCT, realistic industrial example[13]), and an unfolded large grain data flow graph (LG) taken from [1]. The clock period exploration which spans multicycled to chained operations, and retiming support with the IP model is illustrated with the recursive EWF example. The functional pipelining and synthesis of both programmable and interleaved architectures is demonstrated with functionally pipelined AR and FIR filters. The LGDFG support provided by the model is demonstrated by scheduling and allocating multichip architectures with the LG

example. All filters were solved also with the loop winding and retiming IP model for solving problem 2. In all cases the lower bounds on variables, such as Te, f, and I_t, were obtained by solving the relaxed linear program, requiring normally less than one cpu second. The IP solver was GAMS/ZOOM which uses a branch and bound algorithm running on a MIPS RC6280 workstation. The IP model size is given by the number of variables (Var), and number of constraints (Eqn).

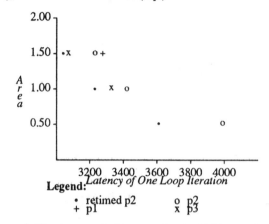

Figure 2. *The Area vs. Delay curve for the EWF example. Different clock period illustrated in figure 3 are used and retiming is explored.*

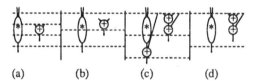

Figure 3. *Clock periods of p1=156ns (a), p2=190ns (b), p3=256ns (c), and 380ns (d) are indicated with vertical space between dashed horizontal lines. The different types of functional units, such as the three cycle multiplier in (a) or the two chained adders in (c), are also illustrated.*

Figure 3 illustrates the four clock periods. These were selected based on the multiplier and adder obtained from [14]. The multiplier has an area of 49K mil^2 and a delay of 375 ns. The adder has an area of 1200 mil^2 and a delay of 151 ns and registers have a delay of 5 ns. The vertical length of the operations in figure 3 is proportional to its propagation delay. The distance between the dashed horizontal lines is proportional to the clock period. For example in figure 3c), in one clock period two successive addition operations can be performed. If we assume the propagation delay to be less than the sum of the propagation delays through two adders[15] we obtain a clock period of 256ns (151+151/3+5). Retiming improved the latency of the EWF (for the same area) in one case from 3990ns to 3610ns. In another case exploration of different clock periods improved the latency of the EWF (for the same area) from 3230ns to 3072ns where the improvement is due to chaining operations and thus increasing the clock period. The architectures were synthesized in reasonable cpu times ranging from 1 to 9 cpu seconds to solve the IP problems with 150 to 384 binary variables. Retimed schedules are

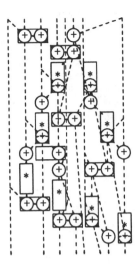

Figure 4. *Example of an EWF schedule with one adder-adder and two multiplier-adder functional units allocated.*

Table 1. IP Synthesis Results for Programmable and Interleaved Architectures

IP Model			f	*	+	Te	
Var	Eqn	CPU (sec)				AR	FIR
Programmable Architectures							
419	426	0.49	2	8	8	11	10
523	538	3.23	3	6	5	12	10
Interleaved Architectures							
419	426	4.78	4	6	7	16	16

also illustrated in the area delay curve in figure 2.

To give an example of the speed advantage the IP approach provides one can map the solutions to the library described in [16] where areas of *, + are 17000, 1500 units and delays of *, + are 80ns, 55 ns (and registers have delay of 2ns). The minimum delay for the EWF in[16] (using a fixed clock period of 100ns) was 1800ns with an area of 21500 units. The IP approach's delay, which uses period 4 in figure 3 which is 112ns for this library, was 1456ns (13 csteps) with an area of 21500 units.

The EWF was also used to demonstrate how the IP model can support automatic selection of chained operations. The following types of functional units were allowed: two-cycle multipliers, two-cycle multiplier-adders, one-cycle adders, and a one-cycle adder-adders. The IP solver minimized the area of the architecture by selecting and allocating functional units simultaneously with scheduling operations, given an upper bound of 14 csteps on the total execution time. Figure 4 shows the IP solution requiring one adder-adder and two multiplier-adders. This architecture was globally optimized in 13.6 cpu seconds, requiring 202 variables and 447 inequalities. Note that the single addition operations shown in figure 4 can be implemented with the second addition in the adder-adder or with the addition in one of the multiplier-adders.

To illustrate the robustness of the IP model the LG example was simultaneously scheduled and allocation of multichips was performed. This example had tasks which ranged from 2 csteps

in duration to 20 csteps in duration (C=2,..20 in $k\varepsilon op(C,L)$). The IP problem had 214 variables, 396 inequalities however only (108 iterations) 0.28 cpu seconds were required to solve it to a global optimum. Only three chips were allocated with a total latency of 104 csteps.

Table 1 illustrates the synthesis of programmable and interleaved architectures. Both the functionally pipelined FIR and AR filter examples were used. The l column refers to the number of csteps between successive initiations of the filter. In the last row the l=4 utilized incoming data samples every 4 csteps for each filter: ie. at cstep 1 data sample # 1 arrives for the AR filter, at cstep 3 data sample # 1 arrives for the FIR filter, at cstep 5 data sample # 2 arrives for the AR filter, etc. Both programmable and interleaved architectures for these two filters could be allocated simultaneously with scheduling in very fast cpu times.

Tables 2, 3 and 4 illustrate the loop winding IP model (presented in section 4) results for the filters which solve problem 2. The retimed EWF example is shown in table 2 and compared with the percolation based scheduler, PBS, results[7]. Each IP solution required less than 7 cpu seconds to solve and had up to 250 binary variables and up to 450 constraints. Allocation of two cycle pipelined multipliers (*p, L=1,C=2) as well as two cycle multipliers (*, L=2,C=2) were solved. In table 3 the loop winding of the FIR, AR, and DCT examples are shown, with constraints on the number of adders (+), subtractors (-) and two cycle multipliers (*). The first two DCT results were solved using an advanced optimization package called LAMPS on a IBM RISC 6000. This was necessary due to the memory requirements for the IP model which had over 1000 binary variables and over 2000 constraints. In table 4 the results of the DCT from table 3 were used to fix f, for solving problem 1 using the IP model presented in section 3, which was used to minimize area and then minimize Te. The final resulting architectures are listed and compared with PLS results[8].

Table 2. Comparison of Retiming Results for the EWF Example.

			f (Pipeline latency)	
*p	*	+	IP	PBS[7]
	1	2	19	21
	2	2	17	18
	2	3	16	18
	3	3	16	17
1		2	18	19
1		3	16	18
2		2	17	18
2		3	16	17
3		3	16	17

6. Conclusions

In this paper we have defined a new IP model, that supports simultaneous loop winding, retiming, and synthesis of programmable or interleaved architectures. This is in contrast to [10] that used a different IP model which could not retime, loop wind, select types of functional units or chain operations. For the first time we can optimally retime or loop wind simultaneously with scheduling and allocation. By supporting chained, multicycled, or pipelined functional units, the IP approach has synthesized up to 24% faster architectures than previous research[16]. The IP

Table 3. Loop Winding with IP Model

Example	+	-	*	f	LI	Var,Eqn,CPUsec
FIR	2	2	8	2	2	198v,280e,0.3s
	3	3	5	2	2	110v,145e,0.3s
	4	4	4	4	4	255v,318e,0.6s
	8	8	3	5	5	254v,336e,0.3s
AR	4	4	4	4	4	296v,418e,3s
DCT	1	1	1	16	2	1010v,1954e,32s†
	1	1	2	13	2	785v,1167e,78s†
	2	2	2	8	2	410v,570e,74s
	3	3	3	6	2	410v,570e,1s
	4	4	4	4	3	262v,383e,2s
	6	6	6	3	3	179v,296e,0.3s

† *solved with LAMPS IP solver on IBM RISC6000 model 350*

Table 4. DCT Comparison of IP with PLS[8]

+	-	*	IP		PLS[8]	
			Te	f	Te	f
7	7	8	9	2	-	-
5	5	6	9	3	10	3
4	4	4	10	4	11	4
3	3	3	11	6	-	-
2	2	2	11	8	12	8

approach synthesized architectures with up to 12% higher throughputs than previous research[7] (see table 2) and up to 11% improvement in latency (see table 4). Globally optimal schedules and allocations which minimize an area cost function or latency have been synthesized in practical execution times. The IP model is robust enough to handle multiple filters to synthesize programmable architectures. The solution to IP problems with over 500 binary variables, to solve simultaneous scheduling and allocation for two filters required under 5 cpu seconds. Finally IP problems whose operations have large latencies were solvable in practical cpu times using the same general IP model, thus supporting scheduling and allocation of LGDFGs. Although the worst case complexity is exponential (since we are solving an IP problem), we have found that many problems can be solved extremely quickly to global optimums. IP models with over 1000 variables and over 2000 constraints have been solved in under one cpu minute. As input algorithms become larger we can partition the behavior or take advantage of the mathematical flexibility of the IP approach to model systems with hierarchy and regularity.

In summary this research for the first time synthesizes programmable architectures in practical cpu times, supports simultaneous loop winding or retiming, and automatically selects different types of functional units. This is important for industry since these early decisions made during high level synthesis have the greatest impact on the final design. In contrast to other synthesizers we can obtain exact globally optimal schedules, allocations and selections of types of functional units, that support retiming. Previous synthesizers could at best guarantee a locally optimal solution, which may not satisfy stringent latency or area constraints of new field programmable gate array technologies. The mathematical basis of the approach provides the flexibility of modeling complex constraints (such as retiming, loop winding) which are critical to achieving high performance. The polyhedral theory behind this mathematical model explains why we can solve many problems in practical cpu execution times. Finally the methodology forms a DA tool that can be brought to market quickly since it based on highly reliable mathematical software. For the first time this research 1) provides industry with a DA tool for synthesizing high performance DSP VLSI architectures targeted for emerging programmable technologies, and 2) a new methodology that can support DSP architectural design exploration to minimize area-delay costs. A new IP model has been developed for synthesis of multichip systems[17] (which accounts for a communication delay for data transfers between different chips), and simultaneous selection of clock periods[18].

C.H.Gebotys would like to thank Dr.D.J.Burns, Dean of Engineering, for his generous financial support. R.J.Gebotys would like to thank the University of Waterloo (where he is currently on sabbatical, on leave from Wilfrid Laurier Univ.) for its hospitality. This research is supported in part by grants from NSERC, ITRC, and MICRONET.

References

1. K.K.Parhi and D.G.Messerschmitt, "Static Rate-Optimal Scheduling of Iterative Data-Flow Programs via Optimum Unfolding," *IEEE Trans. on Computers*, **40**(2) pp. 178-194 (Feb 1991).

2. E.A.Lee and D.G.Messerschmitt, "Static Scheduling of Synchronous Data Flow Programs For Digital Signal Processing," *IEEE Trans. on Computers*, **C-36**(1) pp. 24-35 (Jan 1987).

3. C.H.Gebotys and M.I.Elmasry, *Optimal VLSI Architectural Synthesis: Area, Performance, Testability*, Kluwer Academic Publishers (1992).

4. C-T Hwang, J-H. Lee, and Y-C. Hsu, "A Formal Approach to the Scheduling Problem in High-Level Synthesis," *IEEE Transactions on CAD*, **10**(4) pp. 464-475 (1991).

5. L. Hafer and A. Parker, "A formal Method for the Specification, Analysis and Design of Register-Transfer-Level Digital Logic," *IEEE Transactions on Computer Aided Design of Circuits and Systems*, **CAD-2**(1) pp. 4-17 (Jan 1983).

6. M. McFarland, A. Parker, and R. Camposano, "Tutorial on High-Level Synthesis," *Design Automation Conference*, pp. 330-336 (1988).

7. R.Potasman, J.Lis, A.Nicolau, and D.Gajski, "Percolation Based Synthesis," *IEEE/ACM Design Automation Conference*, pp. 444-449 (1990).

8. C.-T.Hwang, Y.-C.Hsu, and Y.-L.Lin, "Scheduling for Functional Pipelining and Loop Winding," *IEEE/ACM Design Automation Conference*, pp. 764-769 (1991).

9. S.M.Heemstra de Groot, S.H. Gerez, and O.E. Herrmann, "Range-Chart-Guided Iterative Data-Flow Graph Scheduling," *IEEE Trans. on Circuits and Systems-I*, **39**(5) pp. 351-364 (May 1992).

10. C.H. Gebotys and M.I. Elmasry, "Optimal Synthesis of High-Performance Architectures," *IEEE Journal of Solid-State Circuits*, **27**(3)(March 1992).

11. G.L. Nemhauser and L.A. Wolsey, *Integer and Combinatorial Optimization*, Wiley Interscience (1988).

12. hlsw and B. Mayo (Coordinator), *High-Level Synthesis Workshop Clearinghouse, email: hlsw-request@decwrl.dec.com*. 1988.

13. D.J.Mallon and P.B.Denyer, "A New Approach to Pipeline Optimization," *European Design Automation Conference*, pp. 83-88 (March 1990).

14. A.Parker, "Tutorial on High Level Synthesis," *Can. Conf. on VLSI*, (1990).

15. S.Note, F.Catthoor, G.Goossens, and H.DeMan, "Combined Hardware Selection and Pipelining in High-Performance Data-Path Design," *IEEE Trans. on CAD*, **11**(4) pp. 413-423 (April 1992).

16. L.Ramachandran and D.D.Gajski, "An Algorithm for Component Selection in Performance Optimized Scheduling," *IEEE International Conf. on CAD*, pp. 92-95 (1991).

17. C.H.Gebotys, "Optimal Synthesis of MultiChip Architectures," *IEEE/ACM International Conference on CAD*, (November 1992).

18. C.H.Gebotys, "Synthesizing Embedded Speed-Optimized Architectures," *IEEE Journal of Solid-State Circuits*, (March 1993).

On Unlimited Parallelism of DSP Arithmetic Computations

Miodrag Potkonjak
C&C Research Laboratories, NEC USA, Princeton

Jan Rabaey
Dept. of EECS, University of California, Berkeley

ABSTRACT

We present a conceptually simple and practical technique which identifies necessary and sufficient conditions for achieving unlimited improvement in the throughput of a general DSP arithmetic computation (DAC). For the cases when necessary conditions are fulfilled, this technique is a basis for an approach which transforms an arbitrary DAC so that unlimited parallelism is realized. An important special case, linear computations, is addressed using a specially tuned version of the general algorithm so that the introduced latency and hardware overhead are minimized. The effectiveness of the new approach is demonstrated on several real life examples.

1. MOTIVATION AND PRIOR ART

We address the computational concurrency of DSP arithmetic computations (DACs), computations which are performed on a semi-infinite stream of input data using additions, subtractions, multiplications, and divisions. Because of the iteration bound present in recursive DACs, the throughput for this class of computations is often limited by technology and processing constraints. Although several special classes of recursive DACs have been successfully transformed so that arbitrarily high sampling rates can be achieved, a consistent framework for determining when the application of transformations leads to unlimited parallelism has not been addressed.

Rapid progress in semiconductor technology, IC design methodology and CAD tools, removed many of the initial constraints which imposed area as the primary constraint, and elevated throughput as the main criterion for design comparison. Since the mid sixties [Hoc65, Sto73, Kog73], there have been numerous successful attempts at optimizing several classes of recursive computations for asymptotically unlimited throughput. While the first researchers concentrated their effort in the field of numerical solutions of differential equations, and later on for several important classes of DSP computations (e.g. some classes of adaptive linear filters and dynamic programming like computation such as Viterbi decoders) it has been shown that arbitrarily high sampling rates can be achieved by restructuring the computations using special sets of transformations [Gol68, Cha69, Vol70, Bur71, Men87, Mes88, Par89, Fet90, Lin91]. Recently, Lin and Messerschmitt [Lin91] achieved important theoretical results, showing

that an arbitrary finite state machine has unlimited concurrency. However, their method is only applicable to systems with a moderate number of binary states, which practically rules out its application on even very small numerical computations. This leaves open the important question: "which DSP arithmetic computations have structures suitable for the exploration of unlimited parallelism?"

The major result of this paper is a technique which identifies when a DAC can be transformed so that arbitrarily high speed-up is possible. For instances of a DAC where necessary conditions are satisfied, a procedure which speeds it up is presented. The procedure is built by combining the concept of temporal loop unrolling (known in the DSP and design literature under a number of different formulations such as look-ahead, K-slow, recursive doubling and block filtering techniques) with common subexpression replication and a well organized set of algebraic transformations. A more detailed description of the work presented here can be found in [Pot93].

2. MAIN IDEA

The main idea behind this new procedure for the unlimited improvement of throughput is conceptually simple, yet effective. The approach is to build on the top of a combination of work done in theoretical computer science in the area of parallel arithmetic computations [Bre74, Mul76, Val83, Mil87, Mil88, May90] and work which addresses the use of computational transformations in the compiler and high level synthesis domain [Pot92]. These results are combined with loop unfolding techniques, techniques widely used in the numerical analysis and VLSI DSP communities. For the general case, we use the results of Valiant et. al [Val83] and Miller, Ramachandran and Kaltofen [Mil88, Kal86]. In some cases, the algorithms presented by Brent [Bre74] and Muller and Preparata [Mul76] are sufficient and introduce lower hardware overhead. Finally, for the important case of linear computations we rely on the recently introduced maximally fast implementation algorithm [Pot92]. We will concentrate in the rest of the paper on the restricted case when the DAC is division free. However, by using either [Kal86] or [Str73], DACs with divisions can be handled in the same manner. The key result on which we are basing the new approach is that the computation of an arbitrarily organized computation of a polynomial of formal degree d (see next section) which uses n operators, can be computed in time O ($\log n$ ($\log n$ + $\log d$)).

From now on, we will refer to those procedures as algebraic speed-up procedures. There are two major obstacles which limit the performance of the algebraic speed-up procedures when applied to DSP arithmetic computations. First, DSP computations are normally represented as directed acyclic graphs (DAGs) with multiple outputs (often including the states). Second, the algebraic speed-up procedures will only achieve major improvements for large values of n, or, in other words, it is important that the DACs are as large as possible. Both considerations can be elegantly and efficiently addressed using computational transformations.

Common subexpression replication makes it possible to transform an arbitrary DAG into a tree, where each output depends only on a partial subset of the inputs and where fanout to multiple outputs has been eliminated. Common subexpression replication acts as an enabling transformation for the application of the speed-up procedures. To create large graphs (as is beneficial for the algebraic speed-up procedures), the time loop is unrolled a number of times. In this way, arbitrary large graphs and hence speed-ups can be obtained. This is at the expense of extra hardware. It will be proven however that the cost increase is controlled by well defined bounds. The interesting feature of this procedure is that it can be applied on general DACs, and there it is a generalization of a previously published technique [Pot92], which only handles linear computations. For a special case of linear computation, an additional use of grouping and pipelining of all primary inputs and computing them using the Horner scheme results in smaller hardware overhead and better trade-off between latency and throughput [Pot93].

The presented optimization procedure is illustrated for the 2nd order IIR direct form II filter (Figure 1), as shown in Figures 2-4. Figure 2 shows the initial structure of the filter, Figure 3 the 2 times unfolded filter and Figure 4 the structure after the application of common subexpression replication, followed by the algebraic speed-up procedure. The critical path of the final implementation is only 3, although two iterations are performed. Since the initial critical path equaled 4, the throughput is improved by a factor of 2.67.

3. MAJOR RESULT

3.1. When does a DAC have an unlimited parallelism implementation?

An algebraic straight-line program consists of a sequence of assignment statements executed in a specified order. The expressions on the right-hand side are expressions which involve arithmetic operators (+ , - ,* , /) that are applied either on an input variable or variable to which value is assigned in an earlier assignment. Straight-line programs are often represented as DAGs with a single output node. Since both representations, when only (+, - , *) are used (when division is also used see [Str73]), compute some polynomial, the formal degree of a computation represented as a straight-line program or a DAG is defined as the degree of the computed polynomial [May90]. Similarly, for each node of the corresponding DAG, the formal degree can be defined. Note that when a node is + or -, its degree is equal to the maximum degree of its inputs, when a node is *, the degree is the product of the degrees of its input degrees. Theorem 1 states a simple and computationally efficient test to detect whether a given DAC can be implemented at arbitrarily high speed [Pot93].

Theorem 1: The throughput of a recursive DAC can be unlimitedly improved if and only if the formal degree of the N-times unrolled initial iteration is growing at a sublinear rate compared to N.

The required testing of the rate of growth of the formal degree can be easily achieved by analyzing feedback dependencies of all multiplications [Pot93].

3.2 Algorithms for achieving unlimited parallelism

The Algorithm for the arbitrary speed-up of an arithmetic DSP computation can be described using the following pseudo-code:

1. Unfold the basic iteration N times;

2. Organize the arithmetic expression for each output as a tree using common subexpression replication;

3. Using the algebraic-speed-up procedure, reduce the critical path of the unfolded expressions for all outputs;

4. Pipeline all computation outside recursive loops;

5. Minimize the resulting graph using common subexpression elimination.

The power of the algorithm is expressed by the following theorems. The detailed proofs of all theorems can be found in a technical report [Pot93].

Theorem 2: The throughput of a recursive DAC which satisfies the conditions from Theorem 1 can be increased arbitrarily.

Theorem 3: The ratio of the transformed to the initial AT^2 product grows only as the square of logarithm.

Theorem 4: The proposed algorithm produces the design which has asymptotically the minimum critical path among all designs with the same latency.

3.3 Linear Computation

For the important and widely used special case of linear

206

Before Unfolding:

$$x_1 = In_1 + a*x_0 + b*y_0$$

$$y_1 = x_0 \qquad\qquad\qquad\qquad (a)$$

$$Out_1 = x_1 + c*x_0 + d*y_0$$

After Unfolding:

$$x_1 = In_1 + a*x_0 + b*y_0 \qquad x_2 = In_2 + a*x_1 + b*y_1$$

$$y_1 = x_0 \qquad\qquad\qquad\qquad y_2 = x_1 \qquad\qquad\qquad (b)$$

$$Out_1 = x_1 + c*x_0 + d*y_0 \qquad Out_2 = x_2 + c*x_1 + d*y_1$$

After Transformations:

$$x_2 = In_2 + a*In_1 + (a*a + b)\,x_0 + a*b*y_0$$

$$y_2 = In_1 + a*x_0 + b*y_0 \qquad\qquad\qquad\qquad (c)$$

$$Out_2 = In_2 + (a+c)\,In_1 + (a*a + b + c*a + d)\,x_0 + (a*b + c*b)\,y_0$$

FIGURE 1. Functional dependences for 2nd order IIR filter before and after transformations

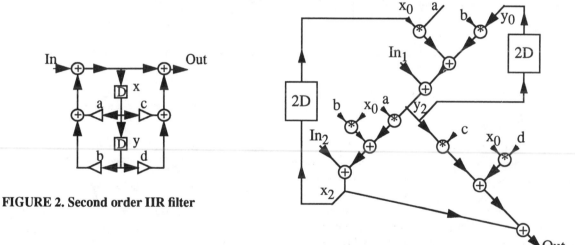

FIGURE 2. Second order IIR filter

FIGURE 3. Flow graph of the unfolded IIR filter

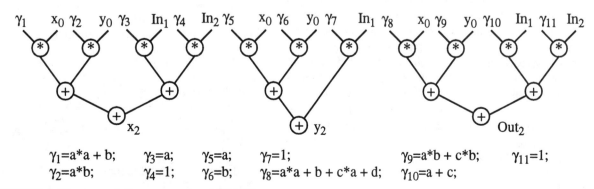

$\gamma_1 = a*a + b; \qquad \gamma_3 = a; \qquad \gamma_5 = a; \qquad \gamma_7 = 1; \qquad\qquad \gamma_9 = a*b + c*b; \qquad \gamma_{11} = 1;$

$\gamma_2 = a*b; \qquad\quad \gamma_4 = 1; \qquad \gamma_6 = b; \qquad \gamma_8 = a*a + b + c*a + d; \qquad \gamma_{10} = a + c;$

FIGURE 4. Flow graph for restructured IIR filter.

computation, by using the maximally fast implementation of linear programs in the third step of the new algorithm, a significant improvement over the previously published result can be achieved, as illustrated with the following two theorems [Pot93]:

Theorem 5: The throughput of a recursive linear DAC can be increased arbitrarily so that the latency increases at the same rate as the throughput.

Theorem 6: The ratio of the initial and the transformed *AT* for a linear DAC is constant.

3.4. Experimental Results

The effectiveness of the procedure is illustrated in Table 1 for two linear examples (51st order FIR filter and 8th order Avenhaus IIR filter) and two non-linear examples (a Volterra filter and an Adaptive FIR filter).

Unfolding Factor Example	Initial	2	5	10	50	100
51 FIR	1	14.6	36.5	73	365	730
8 Avenhaus	1	3.6	9.0	18	90	180
Volterra	1	1.25	2.9	4.4	18	33
Adaptive FIR	1	3.9	8.1	15	61	110

Table 1: The throughput improvement for four examples

4. CONCLUSION

Necessary and sufficient conditions are derived for identifying when an arbitrary DAC can be transformed using unfolding so that the effective throughput is arbitrarily high. An efficient algorithm has been presented, which transforms an arbitrary DSP arithmetic computation which satisfies the mentioned criteria, so that the throughput of the transformed implementation is arbitrarily high. From a theoretical point of view, the importance of the new approach is in the re-establishment of loop unfolding as a powerful and general optimization technique. It also opens new directions for the exploration of loop unfolding by combining it with other transformations. From the practical side, since the hardware overhead is bounded and of an acceptable nature, the approach enables the speed-up of many real-life DSP numerical computations. For the important special case of linear computation, the new procedure results in solutions where the AT product is constant, regardless of speed-up and latency is traded at a linear rate for throughput.

5. REFERENCES

[Bre74] R.P. Brent: "The parallel evaluation of general arithmetic expression", Journal of ACM, Vol. 21, No. 2, pp. 201-206, 1974.

[Bur71] C.S. Burrus: "Block Implementation of Digital Filters", IEEE Trans. on Circuits Theory, Vol. 18., No. 6, pp. 697-701.

[Cha69] D. Chanoux: "A method of Digital Filter Synthesis", M.S. thesis, MIT, Cambridge, MA, May 1969.

[Fet90] G. Fetwies, H. Meyr: "A 100 Mbit Viterbi Decoder Chip: Novel Architecture and its Realization", IEEE Internation Conference on Communications, Vol. 2, pp. 463-467, 1990.

[Gol68] B. Gold, K.L. Jordan: "A Note on Digital Filter Synthesis", Proc. of IEEE, pp. 1717-1718, 1968.

[Hoc65] R. Hockney: "A fast direct solution of Poisson's equation using Fourier analysis", Journal of ACM, Vol. 12, No. 1, pp. 95-113, 1965.

[Kog73] P.Koge, H. Stone: "A parallel algorithm for the efficient solution of a general class of recurrence equations", IEEE Trans. of Comp., Aug 1973.

[Lin91] H. Lin, D.Messerschmitt: "Finite State Machine has Unlimited Concurrency", IEEE Trans. on CAS, Vol. 38, No. 5, pp. 465-475, 1991.

[May90] E.W. Mayr: "Theoretical Aspects of Parallel Computation", in "VLSI and Parallel Computation", ed. by R. Suaya, G. Birtwistle, Morgan Caufman, Palo Alto, CA 1990.

[Men87] T. Meng, D.G. Messerschmitt: "Arbitrarily high sampling rate adaptive filters", IEEE Trans. ASSP, apr. 1987.

[Mes88] D.G. Messerschmitt: "Breaking the recursive bottleneck", in "Performance Limits in Communication Theory and Practice", J.K. Skwirzinsky, Ed., Kluwer, Amsterdam, 1988.

[Mil87] G.Miller, S.-H. Teng: "Dynamic parallel complexity of computational circuit", 19th ACM Symposium on Theory of Computing, pp. 254-263, 1987.

[Mil88] G. Miller, V. Ramachandran, E. Kaltofen: "Efficient parallel evaluation of straight-line code and arithmetic circuits", SIAM Journal of Computing, Vol. 17, No. 4, pp. 687-695, 1988.

[Mul76] D.E. Muller, F.P. Preparata: "Restructuring of Arithmetic Expressions For Parallel Evaluation", Journal of ACM, pp. 534-543, 1976.

[Par89] K. K. Parhi, D. Messerschmitt: "Pipeline interleaving and parallelism in recursive Filters, Parts 1 and 2", IEEE Trans. on ASSP, Vol. 37, pp. 1099-1117, 1117-1134, 1989.

[Pot92] M. Potkonjak, J. Rabaey: "Maximally Fast and Arbitrarily Fast Implementation of Linear Computations", ICCAD-92, pp. 304-308, 1992.

[Pot93] M. Potkonjak, J. Rabaey: "On Unlimited Parallelism in DSP Arithmetic Computations", Technical Report 5510-093-001, NEC USA, Princeton, 1993.

[Sto73] H. Stone: "An efficient algorithm for the solution of a tridiagonal linear system of equations", Journal of ACM, Vol. 20, No. 1, pp. 27-38, 1973.

[Str73] V. Strassen: "Vermeidung von divisionen", Journal of Reine U. Angew. Math., Vol. 264, pp. 182-202, 1973.

[Val83] L. Valiant, S. Skyum, S. Berkowitz, C. Rackof: "Fast parallel computation of polynomials using few processors", SIAM Journal of Computing, Vol. 12, No. 4, pp. 641-644, 1983.

[Vol70] H.B. Voelcker, E.E. Hartquist: "Digital Filtering via Block Recursion", IEEE Trans. on Audio Electroacous., Vol. 18. No. 2, pp. 169-176.

Chapter 6: Components

Components represent the underlying building blocks from which processors are built. The ability to perform high-speed multiplications has been a long-standing challenge to real-time signal processing, as evidenced by the body of algorithms that has arisen to provide good performance without multiplications. Multiplier circuits are subject to tradeoffs of space against time -- fully parallel multipliers provide a result in a single clock cycle but require high power and area, while serial multipliers are small but require many clock cycles to compute a product. "Serial - Parallel Multipliers" (paper 6.1) reviews a variety of serial, parallel, and mixed multiplier circuits and compares their delay and projected size within an integrated circuit layout.

A comparison between serial and parallel processing techniques is offered by "Variable Word Length DSP Using Serial-by-Modulus Residue Arithmetic" (6.2). For situations in which the instantaneous dynamic range within each computation can be established for all expected signals (a demanding condition most easily met in more regular signal processing functions like filters), a variable word length and residue arithmetic can reduce processor complexity.

As real time signal processing requirements exceed the capacity of a uniprocessor, the use of multiple signal processors becomes necessary. The pervasive problem of multiprocessor communication is addressed in "Design and Implementation of an Ordered Memory Access Architecture" (6.3), which imposes the regularity of synchronous data flow upon interprocessor communication to achieve efficiency, simplicity, and scalability. Interprocessor communication is also addressed in "Sizing of Communication Buffers for Communicating Signal Processors" (6.4). As an advance from established methods that use lifetime analysis techniques to size buffers for non-conditional interactions, this paper uses dynamic data flow to generate a network of finite state machines to allow the accommodation of conditional behavior in intercommunicating processes. In another version of the space-time tradeoff, memory access speed may be balanced against the size of random access memory. "A Two-Level On-Chip Memory for Video Signal Processor" (6.5) uses an upper level memory to buffer the processor input/output from a high-bandwidth data bus, followed by a low-level memory that supports the demands of parallelism imposed by a pipeline video signal processor chip.

The skew of clock signals (e.g., the difference in arrival time of the same clock pulse at different parts of a synchronous circuit) becomes a limiter of performance as digital systems increase in size and complexity. The use of a clock input circuit that blocks further clock signals for the remainder of a clock period has been applied to reduce clock skew in "Synchronous Clocking Schemes for Large VLSI Systems" (6.6).

Suggested Additional Readings:

[1]   K. C. Adkins, B. B. Bibyk, and R. T. Kaul, "Associative Computation Circuits for Real-Time Processing of Satellite Communications and Image Pattern Classification" *1993 International Symposium on Circuits and Systems* (Chicago, IL, USA) May 1993, pp. 227-230.

[2]   N. T. Thao and M. Vetterli, "A Deterministic Analysis of Oversampled A/D Conversion and Sigma-Delta Modulation," *Proc. IEEE International Conference on Acoustics, Speech, and Signal Processing 93* (Minneapolis, MN, USA) April 1993, vol. 3 pp. 468-471.

[3]   H. Choi, W. P. Purleson, and D. S. Phatak, "Optimal Wordlength Assignment for the Discrete Wavelet Transform in VLSI," *VLSI Signal Processing VI* (The Netherlands) November 1993, pp. 325-333.

[4]   D. J. Ostrowski, P. Y. K. Cheung, and K. Roubaud, "An Outline of the Intuitive Design of Fuzzy Logic and its Efficient Implementation," *Second IEEE International Conference on Fuzzy Systems* (San Francisco, CA) Mar. - Apr. 1993, pp. 184-189.

Serial-Parallel Multipliers

Hawkins H. Yao and Earl E. Swartzlander, Jr.

Department of Electrical and Computer Engineering
University of Texas at Austin
Austin, Texas 78712

Abstract

Digital filters and other signal processing applications often involve a large number of multiplications. For many of these applications, the algorithm is sufficiently parallel that an implementation with many serial-parallel multipliers may be more efficient than implementation that uses a few parallel multipliers. This paper examines the design of several serial-parallel multipliers, compares and contrasts the different characteristics and improvements made on these multipliers. Comparison is also made between serial and parallel multipliers to give a rough estimate of the size versus speed trade off for using serial-parallel multipliers.

1.0 Introduction

Multipliers are often the central concern in implementing a digital signal processing system. The design of multipliers has, therefore, been an area of intense study. Over the years, many different design variations have been proposed and improved. Frequently, two main approaches are taken in attempts to speed up the multiplication operations of the arithmetic unit in these systems. One approach is to minimize the delay of the multiplier by parallelizing the additions of the partial products which improves the computational turn-around time and thereby allows more operands to be acted upon in a unit of time. The other scheme seeks to parallelize the algorithm in an attempt to increase throughput. Serial-parallel multipliers are particularly attractive for the latter purpose due to their relatively small size which allows many more of these multipliers to be integrated in a given area than parallel multipliers.

In applications involving a large number of multiplications, concurrent executions of the multiplication operations on separate serial multiplying units can significantly improve the applications' throughput. A large number of parallelizable multiplication operations can be found in many applications. Examples include signal and image processing, pattern recognition, and applications requiring the computation of inner products, as suggested by Swartzlander, Gilbert, and Reed [1].

This paper examines several serial-parallel multipliers, compares the cost and speed of these serial-parallel multipliers and contrast them against parallel multipliers.

2.0 Evolution of serial-parallel multipliers

The early serial-parallel multipliers can be traced to those in digital data processing systems and digital filters where data is sampled and fed into the system serially. Serial arithmetic units can be conveniently exploited to transform the data as it is being input into the system. Oftentimes, a fixed operation is to be performed on a huge stream of data. This prompted the development of the Serial-Parallel Multiplier, as shown in Figure 1, where a fixed coefficient Y is applied in parallel and the multiplication is performed as the bit-serial stream of data X is fed across the multiplier. As X_0 is fed into the multiplier, all the partial product terms associated with X_0 are formed, including P_0, which immediately obtained at the output. The interstage latches save the other terms to be added on the next cycle. After all N bits of X are input, 0 is input for K more cycles to drain the latches. A Serial-Parallel multiplier takes N + K clock cycles to compute the N + K bit product of a N bit multiplicand and a K bit multiplier or coefficient, where each cycle is the delay time for a full adder and a latch.

The Serial-Parallel multiplier evolved into the Serial Pipeline multiplier (Figure 2) with added circuitry to perform truncation. The coefficient bits and the control signals for performing truncation are both applied in parallel. The truncation property often does not present any serious limitation as the precision of the product in most systems generally needs to be kept at the same precision as that of the multiplicand or multiplier. The Pipeline Multiplier computes an N bit product in N + 1 clock cycles, a higher throughput than that of the Serial-Parallel Multiplier achieved at the expense of lower precision. Although the

Pipeline Multiplier requires fewer clock cycles, a long combinatorial path through all the adders to the output presents a long propagation delay that significantly limits the allowable clock cycle rate.

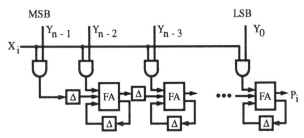

FIGURE 1. Serial-Parallel multiplier.

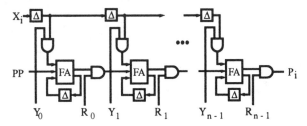

FIGURE 2. Serial Pipeline multiplier.

This is improved upon by Richard Lyon [2] with the addition of interstage latches. To further generalize the function of the multiplier to allow for a variable multiplier/coefficient, Lyon also added latches for the coefficient Y as well as those for the control signal so that all signals are input serially. Although the long combinatorial delay in eliminated, the multiplier requires twice as many cycles to complete a multiplication. A single module of the multiplier in shown in Figure 3.

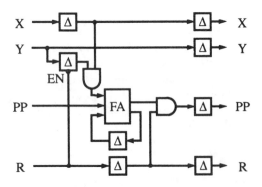

FIGURE 3. Enhanced Pipeline Multiplier.

3.0 Two's complement serial-parallel multipliers

All the multipliers described thus far process unsigned (i.e., positive) numbers. Multiplication of signed numbers can only be performed with additional logic to convert

two's complement form to sign-magnitude form. After the product is computed, additional post-processing logic is

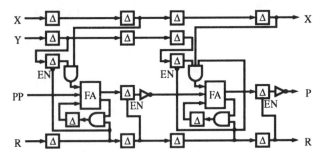

FIGURE 4. Post complementing stages for the enhanced serial pipeline multiplier.

also needed to convert the product back to its two's complement form. These additional logic blocks are, by no means, trivial extension to the existing logic. Lyon simplifies the design by modifying the last stage and introducing an additional post-complementing stage. The resulting last two stages of the multiplier, as shown in Figure 4, performs fully two's complement multiplication. Lyon proceeded to perform yet an additional improvement by extending the design to use Booth's recoding algorithm. This involves expanding on the complementing stage and

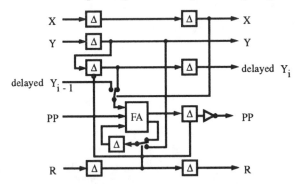

FIGURE 5. A single module of the bit serial multiplier using Booth recoding.

using the resultant module as the basic module for all the coefficient bits. The complementing module is modified to use multiplexors for selecting a preset carry and delayed y_i. This modification converts the carry save adder to a borrow-save subtracter that implements the subtraction of partial products. The resulting multiplier consists of identical repeating modules for which a single module is shown in Figure 5.

4.0 Two's Complement Quasi-Serial Multiplier

The Quasi-Serial Multiplier [5] was developed as an alternative to the add-and-shift scheme commonly implemented at its time. The quasi-serial scheme uses an N-bit parallel counter for reducing the partial product terms. As each bit is shifted and fed into the multiplier, a column of partial product terms are formed by the AND gates. The parallel counter performs an N to $\log_2 N$ bit reduction, with the carries saved in the carry register for the next cycle. The delay of each cycle is determined by the combinatorial parallel counter which is $2 \log_2 (N+1) - 3$ times the delay of a full adder..

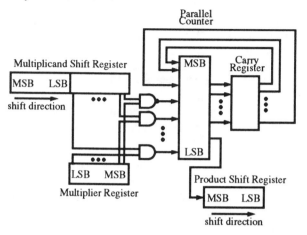

FIGURE 6. The Two's Complement Quasi-Serial Multiplier.

The Quasi-Serial Multiplier was later enhanced to handle two's complement numbers [6]. The Two's Complement Quasi-Serial Multiplier uses an expanded multiplicand register to hold an N-bit multiplicand that is sign-extended to N + K bits in the multiplication process, where K is the size of the multiplier or coefficient. The scheme also uses a NAND gate for the first product bit and requires presetting the carry register to 1 in the multiplication process. The Two's Complement Quasi-Serial Multiplier takes N + K clock cycles to form an N + K two's complement product.

5.0 A more modular and symmetrical bit serial multiplier

A different approach for a more symmetrical and modular design is later proposed for bit-serial multiplier and is due to the works of Buric and Mead [3]. To accommodate for two's complement operands, the original design consists of four input adders in which one input is a special "subtract" signal to be applied simultaneously with the most significant bits of the multiplicand and multiplier. The four to two adder for which the overflow is discarded implements a borrow save subtracter when the "subtract" signal is asserted. This design is later modified slightly to elimi-

nate the unusual adder design and simplify the operation of the multiplier [4].

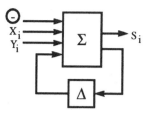

FIGURE 7. Four-to-two bit serial adder/subtracter.

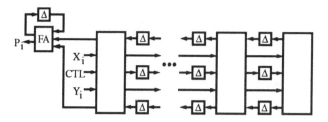

FIGURE 8. Two's complement Bit Serial Multiplier developed by Buric and Mead.

The multiplier consists of N identical bit multiplying modules, inter-stage flip flops, and an additional bit Serial adder with a carry-save flip flop, as shown in Figure 8, a magnified view of a single stage of the multiplier is shown in Figure 9. Conceptually, the operation of the multiplier can best be illustrated by its mechanism of generating the partial product bit matrix (see Figure 10).

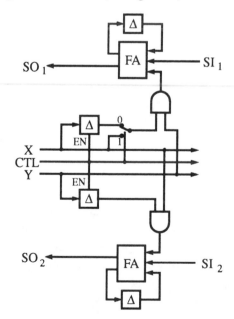

FIGURE 9. Magnified view of a module of the bit serial multiplier.

The multiplier forms the product by partitioning the generation and summation of the partial products along the diagonal of the partial product bit matrix. The bits on the diagonal proper are apportioned to the x partition and, thus, explain the necessary asymmetry between the upper and lower sections of a multiplying module. On each clock cycle, partial product bits in each partition are formed in the pattern shown in Figure 10.

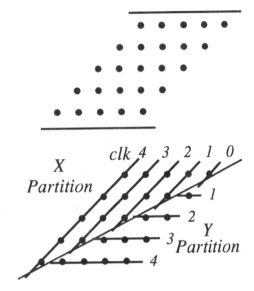

FIGURE 10. Partial product forming and partitioning in the bit serial multiplying modules.

Shown on the sides are the clock cycles when the oblique rows of partial product bits are formed. In a multiplying module, the multiplexor selecting between the value from the x latch and the value from the x bus provides a short path bypassing the latch before a bit of x is latched into the flip flop. This has the effect of aligning the extra partial bit products in the x partition one clock cycle ahead of the corresponding y partition. The extra serial adder adds the results from the two partitions to obtain the final product.

The operation of the multiplier is as follows: The least significant bits of the operands are applied simultaneously with the initial "start" signal of the control bus. A bit of the product is formed after a delay for the signals to propagate through the multiplexors and adders of one stage of the multiplier. A clock is then applied to latch the carries and advance the control signal to the next stage. The control signal is reset to zero for the next and subsequent cycles. At the start of the subsequent clock cycles, the next pair of bits from each operand is applied to the x and y buses. The process repeats for 2N-1 clock cycles for the 2N bits of the sign extended operands, producing a two's complement product of 2N-1 bits.

6.0 Comparison of the multipliers

As a comparison, the characteristics of the serial-parallel multipliers analyzed in this paper are summarized in Table 1. These characteristics are also to be compared with those of the parallel multipliers.

Table 1 shows the multipliers as constructed to perform multiplication of two N bit operands. Although the serial-parallel multipliers may have drastically different designs, they share similar delay and complexity characteristics. The only exception is the serial pipeline multiplier. Because of its long combinatorial path, its delay is $O(N^2)$ to the word size. Delay and size characteristics are tabulated for word sizes from 16 to 64 on Table 1. From Table 1, it is evident that even though Lyon's enhanced two's complement multiplier requires less delay between

TABLE 1. Comparison of the various multipliers.

Multiplier	Allowable Format	Product Precision	Clock Cycle† (t_{cyc})	Total Delay	Complexity
Sequential Serial-parallel Multipliers :					
Serial Parallel	positive	2N	$\Delta_G+\Delta_{FA}+\Delta_{setup}+\Delta_{hold}$	$2N\,t_{cyc}$	O(N)
Serial Pipeline	positive	N - 1	$N(2\Delta_G+\Delta_{FA})+\Delta_{setup}+\Delta_{hold}$	$N\,t_{cyc}$	O(N)
Lyon's two's complement	two's complement	N - 1	$\Delta_G+\Delta_{FA}+\Delta_{setup}+\Delta_{hold}$	$2(N+2)$ t_{cyc}	O(N)
Lyon's Booth Recoding	two's complement	N - 1	$\Delta_{FA}+2\Delta_{mux}+\Delta_{setup}+\Delta_{hold}$	$2N\,t_{cyc}$	O(N)
Two's complement Quasi-serial	two's complement	2N	$\Delta_G+k\Delta_{FA}+\Delta_{setup}+\Delta_{hold}$‡	$2N\,t_{cyc}$	O(N)
Buric & Mead's Bit Serial	two's complement	2N - 1	$\Delta_G+2\Delta_{FA}+\Delta_{mux}+\Delta_{setup}+\Delta_{hold}$	$2N\,t_{cyc}$	O(N)
Combinatorial Multipliers :					
Array	positive	2N	N/A	O(N)	$O(N^2)$
Wallace/ Dadda	positive	2N	N/A	O(logN)	$O(N^2)$

† Δ_G denotes the delay through a two input AND gate. Δ_{FA} denotes the delay through a full-adder. $\Delta_{setup}+\Delta_{hold}$ denotes the setup and hold time for the latches.

‡ $k = \log_2(N+1) - 3$, the delay through the parallel counter.

clocks, its requirement of twice as many clock cycles to complete a multiplication makes its total delay approximately equal to the Buric and Mead's bit serial multiplier. However, The higher precision of Buric and Mead's multiplier makes the latter more attractive for applications where a large number of individual products are accumulated. As a figure of merit, a delay-size product is computed for each multiplier and tabulated in Table 1. This figure provides a simple mechanism for comparing the multipliers taking the approximate delay and size together.

7.0 Summary and conclusions

Of all the serial-parallel multipliers, Buric and Mead's bit serial multiplier offers some of the most desirable features. It is the most modular and symmetrical of all the two's complement serial-parallel multipliers. This significantly simplifies the layout of the multiplier. Additionally, its ability to handle two's complement numbers without any additional logic makes it easily expandable. The multiplier can be expanded to any bit size by simply attaching more modules to the end. However, it is not as favorable if the delay-size product alone is used for evaluation. For what the Mead and Buric Multiplier offers in symmetry and layout simplicities it lacks in the overall desirable feature of low delay and size.

Comparing all multipliers in general, it is apparent that all serial-parallel multipliers have delay that is a linear function of word size. The serial-parallel multipliers also have order N complexity in terms of the occupied area. Inherent to the serial operation is the required setup and hold time of the latches that is absent from purely combinatorial multipliers. This makes the computational delay of the serial-parallel multipliers longer than combinatorial multipliers even for combinatorial multiplier with order N delay. However, the serial-parallel multipliers compare much more favorably with parallel multipliers in size. The

TABLE 2. Delay and size characteristics.

Multiplier	Approximate Delay†			Approximate Size‡		
	16	32	64	16	32	64
Sequential Serial-parallel Multipliers :						
Serial Parallel	160	320	640	50	100	200
Serial Pipeline	1280	5120	20480	50	100	200
Lyon's two's complement	210	400	780	170	310	600
Lyon's Booth Recoding	320	640	1280	200	400	770
Two's Complement Quasi-serial	550	1480	3720	30	50	80
Buric & Mead's Bit Serial	360	710	1410	170	330	650
Combinatorial Multipliers :						
Array	100	190	380	260	1030	4100
Wallace/Dadda	30	40	50	250	1000	4000

†. Delay is computed assuming a unit delay for gates, 3 gate delay for adders and multiplexors, 1 gate delay for setup and hold times.

‡. Size is computed assuming a unit size for adders and latches.

parallel multipliers typically have N^2 complexity whereas the serial ones occupy an area proportional to the word size. This is especially significant when a large number of independent multiplication operations is to be performed. The same area occupied by one parallel multiplier can hold several serial-parallel multipliers, all can be utilized in parallel.

TABLE 3. Delay-size product.

Multiplier	Delay-Size Product		
	16	32	64
Sequential Serial-parallel Multipliers :			
Serial Parallel	8000	32000	128000
Serial Pipeline	64000	512000	4096000
Lyon's two's complement	35700	124000	468000
Lyon's Booth Recoding	64000	256000	985600
Two's Complement Quasi-serial	16500	74000	297600
Buric & Mead's Bit Serial	61200	234300	916500
Combinatorial Multipliers :			
Array	26000	195700	1558000
Wallace/Dadda	7500	40000	200000

References

[1] Earl E. Swartzlander Jr., Barry K. Gilbert, Irving S. Reed, "Inner Product Computers," *IEEE Transactions on Computers*, Vol C-27, pp. 21-31, 1978.

[2] Richard F. Lyon, "Two's Complement Pipeline Multipliers," *IEEE Transactions on Communications*, pp. 418-425, 1976.

[3] M. R. Buric and C. A. Mead, "Bit-Serial Inner Product Processors in VLSI," *Proceedings of Caltech Conference on VLSI*, pp. 155-164, 1981.

[4] William A. Rogers, CHIEFS user manual, p. 33, August 1986.

[5] Earl E. Swartzlander Jr., "The Quasi-Serial Multiplier," *IEEE Transactions on Computers*, Vol C-22, pp. 317-321, 1973.

[6] T. G. McDaneld and R. K. Guha, "The Two's Complement Quasi-Serial Multiplier," *IEEE Transactions on Computers*, Vol C-24, pp. 1233-1235, 1975.

[7] Earl. E. Swartzlander Jr., "Parallel Counters," *IEEE Transactions on Computers*, Vol C-22, pp. 1021-1024, 1973.

Variable Word Length DSP Using Serial-by-Modulus Residue Arithmetic

W. Kenneth Jenkins and Andrew J. Mansen
Department of Electrical and Computer Engineering and
The Coordinated Science Laboratory
1308 West Main St.
Urbana, Illinois 61801 U.S.A.

ABSTRACT
The concept of a variable word length sum-of-products signal processing kernel is developed based on a serial-by-modulus residue number system (RNS) architecture. Due to the fact that the RNS is not a weighted number representation, if the instantaneous dynamic range requirement can be estimated it is possible to perform the computation with only enough residue digits to provide the necessary dynamic range. Although it is difficult to estimate instantaneous dynamic range requirements for arbitrary computational processes encountered in general purpose computers, in well defined signal processing tasks, such as the sum-of-products kernel found in digital filters and vector processors, it may be possible to take advantage of a variable word length design to improve computational efficiency for special-purpose processors.

1. INTRODUCTION
It has been shown in the past that residue number system (RNS) designs can be used effectively for realizing important sum-of-products DSP kernels, such as a finite convolution characteristic of FIR digital filters [1]. The modular structure of RNS arithmetic induces a natural modularity in the hardware, so a system can be partitioned to fit conveniently onto integrated circuits to achieve good parallelism and testability in a VLSI design. Redundancy can be incorporated into the RNS digits to allow error detection, location, and correction [2]. In recent years RNS arithmetic has been used more for DSP applications that are dominated by repetitive multiply-accumulate cycles than for general-purpose computing where many types of diverse operations tend to be equally important.

Because of its carry-free structure, RNS arithmetic has attracted considerable attention for high speed computation, especially when high precision (large word length) is required. More recently, researchers have studied processor designs based on serial-by-modulus (SBM) RNS concepts [3,4]. In a SBM-RNS architecture, the residue digits are processed sequentially in circuits that handle only one modular operation at a given time, thereby sacrificing speed for circuit simplicity. This paper extends the SBM-RNS concept to achieve adaptive word length processing. In a sum-of-products DSP processor, it is proposed to monitor the dynamic range requirements as an on-line process, and to use only enough dynamic range to accommodate the instantaneous need. It is believed that further development of this concept will lead to reduced computation and increased processing speeds in certain types of batch processing applications where excess dynamic range can be removed adaptively.

2. SBM-RNS ARCHITECTURES
In a SBM-RNS architecture the residue digits $x_1, \ldots, x_L$ are transmitted through the processor serially, i.e., x_1 arrives first, then x_2, etc. Assume that the moduli are b-bit integers, so that there are "b" parallel bits throughout the entire processor. In rough terms, the SBM RNS architecture operates L times slower than a fully parallel design, although the hardware complexity is substantially reduced. Serial-by-modulus arithmetic, similar in principle to bit-serial binary arithmetic that was popular in the early days of digital filters, represents a compromise between the slower speed of bit-serial binary arithmetic and the high cost of bit parallel binary arithmetic.

An SBM-RNS architecture using prime moduli requires an efficient realization for both mod (p_i-1) logarithmic addition (multiplication) and mod p_i addition (accumulation). A typical design uses a standard 2's-complement adder whose output addresses a ROM correction table, which produces the correct modular sum from the binary sum. When the adder implements an index addition the inverse logarithmic mapping is encoded into the same correction table so translation from index form to residue form does not require any additional hardware. Consider an RNS defined by the four 5-bit moduli 31, 29, 23, and 19. When two residues are added in a 2's-complement adder, the result may require as many as six bits, i.e., one additional "buffer" bit is needed to represent all possible results. Furthermore, since there are four moduli, it will take two "label" bits to uniquely identify each residue so that it can be unambiguously identified with the correct timing interval. In total each residue, together with its timing label, can be encoded with 8-bits, as illustrated in the block diagram of Fig. 1. The ROM in Fig. 1 consists of four independent correction tables corresponding to the four different m_i, i=1, 2, 3, and 4. The 2-bit timing label, appended to the corresponding residue by two extra bits, serves as the two highest order bits of the address, so the correct look-up table is automatically addressed.

Since the label bits are attached to each residue digit (as opposed to being generated locally by a synchronized

0-7803-0946-4/93 $3.00 © 1993 IEEE

clock), simple comparitors can be used at strategic locations to check that no timing skew has occurred and to guarantee that digit synchronization is properly maintained throughout the processor. A design based on the four moduli listed above realizes more than 18 bits of effective word length in a module that has only an 8-bit word length. Not only has the silicon area (and corresponding power requirement) been significantly reduced as compared to the digit parallel form, but the module can be more easily tested due to its short word length. Of course, the effective speed of the processor is reduced by an approximate factor of four. Note that two modules similar to the one shown in Fig. 1 (one for multiplication and one for addition) form the basis of a sum-of-products DSP kernel. The sum-of-products concept is illustrated in Fig. 2 where the standard IA cell is programmed for $\text{mod}(p_i-1)$ addition and the standard A cell is programmed for $\text{mod } p_i$ addition.

3. VWL FILTER DESIGNS

The SBM-RNS architecture can be used as the basis for the design of a variable word length (VWL) processor. Consider the situation in which a large amount of stored dated must be processed at high speed by a sum-or-products operator, such as a digital correlator or an FIR digital filter. This scenario often occurs in video processing, where frames are buffered and stored in order to give the processor working time to complete the processing on a frame. Assume for this discussion that the processor is designed to have a total of L moduli, where for convenience L is taken to be an odd number. According to RNS theory, some of the residue terms representing a number X may be redundant and, therefore, discarded. For simple RNS operations such as addition, subtraction, and multiplication, none of the residue digits corresponding to different moduli interact with each other. Therefore, redundant digits do not need to be computed since their results are not required. Which terms are in excess depend solely on the final output of the computation and not on intermediate results. According to ring theory, for computations involving addition and multiplication, the range of an intermediate result is not used to calculate necessary word-length. So, for a product of sums or a sum-of-products, only the size of the final result determines how many residues are necessary for that computation.

To implement a VWL filter, certain assumptions must be made in order to adaptively change the number of residue terms. Since the output of the current calculation is unknown, it is not certain how many digits are necessary; thus, the previous output(s) must be used. It is assumed that the output signal is slowly changing such that the current computation would use approximately the same number of terms that were needed in the previous calculation. A set of buffer residues must be added at times to make sure the output always remains valid. Since this can take away some of the efficiency of the algorithm, as small a buffer as possible should be used.

Several different methods of determining the buffer are discussed in the following section.

3.1. Buffering Methods

In case the current output exceeds the range of the minimum set of moduli needed for the previous computation, one or more extra terms may be added to prevent overflow. The added buffer reduces efficiency but is necessary in some situations. This section outlines several buffering methods.

3.1.1. Constant Buffer

With a constant buffer, once the minimum number of terms is determined by the result of the previous calculation, one more residue term is added to prevent overflow in the next calculation. This creates a buffer range that is added to the current range so that the new result can exceed the current range and still remain valid. Even with a buffer, overflow can occur if the output increases too abruptly. If the signal is not changing slowly enough, then the current output could possibly jump over the buffer range and create an overflow condition. If two redundant terms are added, instead of only one, the probability of overflow is further reduced.

3.1.2. Adaptive Buffer

One drawback to the constant buffer method is that extra terms may not always be necessary and will only slow down the system. For instance, the output could be far from the end of a range and not changing fast enough to go into the buffer range. This means that no buffer term would be necessary until the output gets closer to the upper edge of the current range. If the constant buffer method were used, the extra term(s) would unnecessarily be computed. However, there are instances with any signal when a buffer is needed because the previous output is near the edge of the current range and, no matter how slowly the signal is fluctuating, the next may spill into a larger range. A method that could adapt to these situations should improve the efficiency of the filter by eliminating unwanted calculations.

An adaptive algorithm proposed in this work determines when to use a buffer by calculating how close the previous output is to the edge of the next larger range. If it far from the boundary then no buffer is added. If it is close, then two buffer terms are added. If it is between these extremes then only one buffer term is added. The current range is taken to be the minimum range necessary to represent the previous output. The closeness ranges are fixed in size and do not fluctuate when the current range changes.

3.1.3. Linear Prediction

The buffering methods described above depend on the assumption that the output is slowly changing so that the current output will not be far from the previous one. This is not necessarily a good assumption and may cause the output to jump over the buffer. Also, they do not take into account if the output is decreasing towards a smaller range. They only assume that the output will be larger than the previous and, therefore, will need a larger range. If a prediction algorithm is added to the above methods, then better tracking should occur. Some of the experiments documented in [4], as well as the example presented in the next section, use a simple two-term linear prediction method prior to computing the buffer. In this case, a linear prediction of the current output from previous outputs is used to estimate of the range requirement of current output.

3.2 Computer Experiments

Computer experiments were performed in software written in C that was designed to emulate an SBM FIR filter. The program takes a specified set of strictly prime moduli, computes the necessary correction tables, and stores them in arrays which act as the hardware ROM. The filter coefficients and the input data are read from files in binary form; binary-to-residue and residue-to-binary conversions are performed inside the program. By passing different parameters from the command line, the simulator is designed to perform any of the buffering algorithms previously discussed.

Data analysis was also performed by the simulator. For each output computed, a tabulation of the actual number of residues used was saved in a file. From this, the average number of residues for the entire batch of data was computed, along with a histogram showing the statistical distribution of the number residues were used. Furthermore, the locations of and the total number of overflow errors were tracked throughout the process. To show how an optimum VWL algorithm could perform, the minimum number of necessary residue terms was stored for each output computed. An average is taken to determine the optimum figure of merit.

Due to space limitations it is possible to present here results from only one experiment, although many experiments were performed and are fully documented in [4]. The results presented here are for the seven moduli {31, 29, 23, 19, 17, 13, 11}, which has a dynamic range of approximately 30 bits. When the number of moduli required is less than maximum, the subset chosen is taken from the largest modulus to the smallest so that the broadest sub-range is retained. For example, if only three terms are needed then {31, 29, 23} are used.

All the different buffering methods were simulated by first estimating the current output by simply using the previous output, or by linear prediction using the previous two output samples. For each estimated output, the minimum number of residues necessary to represent that number is computed, and the buffer terms are then added to that number. For the constant buffer case, the buffer size is defined to be either one or two, depending on the method chosen. With the adaptive buffer method, the two closeness ranges are passed as parameters to the simulator for buffer selection.

The test data used as an input signal for processing by the VWL filer in this example was a neural pattern recorded from a single brain cell of a rat, with a record of 1000 samples. This particular test signal was chosen because the signal contains neural spikes that require a large dynamic range, and also because there is considerable "idle" time between the spikes during which the VWL algorithm can reduce the computational range. The filter characteristic used in the VWL filter experiment was a low-pass FIR filter designed using the Hamming window.

One particular experiment is summarized in Table 1, where results are listed for a constant buffer algorithm with one term, a constant buffer algorithm with two terms, and an adaptive buffer algorithm with different choices for two closeness ranges, as listed. In this experiment the filter coefficients and the input were scaled so that the range of the output did not exceed the range of five moduli. However, the number of residue digits that are actually used can be as high as seven. The low-pass filter used in these comparisons had a normalized cutoff of 0.1 rads.

It can be seen in Table 1 that as the closeness ranges are reduced in size, the average number of moduli used by the adaptive algorithm decreases. However, as the number decreases, so does the average dynamic range, until the number of overflow errors begins to increase rapidly. Note that the linear estimation scheme is more conservative in the sense that for the same closeness ranges it maintains a lower error rate. But it also does not reduce the average number of residues as much. These results are not intended to be conclusive, but rather to illustrate how an adaptive word length design might function. More research will be needed to determine the effectiveness of VWL designs in practical problems.

ACKNOWLEDGEMENTS

The research is supported by the National Science Foundation under grant number NSF MIP 91-00212.

REFERENCES

[1] M. A. Soderstrand, W. Kenneth Jenkins, Graham A. Jullien, and Fred J. Taylor, eds., *Residue Number System Arithmetic: Modern Applications in Digital Signal Processing*, IEEE Press, New York, NY, 1986.

[2] H. Krishna, K.-Y. Lin, and J.-D. Sun, "A coding theory approach to error control in redundant residue number systems - Part I: Theory and single error correction," *Transactions on Circuits and Systems*, vol. 39, no. 1, pp 8-17, January 1992.

[3] W. K. Jenkins and S. F. Lao, "The design of an integrated RNS digital filter module based on serial-by-modulus residue arithmetic," *Proceedings of the 1987 International Conference on Signals, Systems, and Computers*, Port Chester, New York, pp 634-638, October 1987.

[4] A. J. Mansen, "Variable Word-length DSP Using Serial-by-Modulus Residue Arithmetic, M.S. Thesis, Department of Electrical and Computer Engineering, University of Illinois, Urbana-Champaign, May 1993.

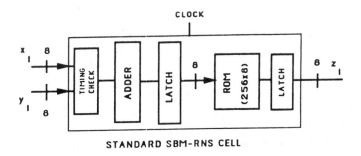

Fig. 1 Block diagram of a serial-by-modulus computational element.

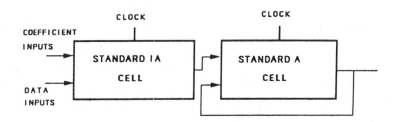

Figure 2. Two SBM computational elements forming a sum-of-products DSP module.

Table 1. Results of Using the 7-term VWL RNS Filter on Neural Data

Buffer Method	w/out Linear Estimation		w/ Linear Estimation	
	Errors	Ave. # of Res.	Errors	Ave. # of Res.
Constant Buffer				
1 Term	0%	5.204	0%	5.220
2 Term	0%	6.202	0%	6.215
Adaptive Buffer				
$10^6, 10^7$	0%	5.940	0%	5.951
$10^5, 10^7$	0%	5.479	0%	5.526
$10^4, 10^7$	0%	5.273	0%	5.298
$10^3, 10^7$	0%	5.214	0%	5.228
$10^2, 10^7$	0%	5.205	0%	5.219
$10, 10^7$	0%	5.204	0%	5.218
$1, 10^7$	0%	5.204	0%	5.218
$10^5, 10^6$	0%	5.219	0%	5.265
$10^4, 10^6$	0%	5.013	0%	5.036
$10^3, 10^6$	0%	4.953	0%	4.966
$10^2, 10^6$	0%	4.944	0%	4.957
$10, 10^6$	0%	4.943	0%	4.956
$1, 10^6$	0%	4.943	0%	4.956
$10^4, 10^5$	1%	4.558	0%	4.614
$10^3, 10^5$	2%	4.495	0%	4.542
$10^2, 10^5$	2%	4.481	0%	4.532
$10, 10^5$	2%	4.479	0%	4.531
$1, 10^5$	2%	4.479	0%	4.531
$10^3, 10^4$	17%	4.137	2%	4.322
$10^2, 10^4$	18%	4.121	2%	4.311
$10, 10^4$	18%	4.119	2%	4.310
$1, 10^4$	18%	4.119	2%	4.310
$10^2, 10^3$	48%	3.696	5%	4.245
$10, 10^3$	51%	3.649	6%	4.244
$1, 10^3$	51%	3.649	6%	4.244

DESIGN AND IMPLEMENTATION OF AN ORDERED MEMORY ACCESS ARCHITECTURE[1]

S. Sriram, and E. A. Lee

EECS Department, UC Berkeley, Berkeley CA94720.
(sriram@ohm.berkeley.edu, eal@ohm.berkeley.edu)
December 15, 1992

ABSTRACT

This paper describes a multiprocessor machine for real-time Digital Signal Processing that uses commercial programmable DSP chips. The architecture is a shared memory, single shared bus parallel processor designed to run signal processing tasks that can be statically scheduled. The design is based on the architecture proposed in [1]. A prototype has since been built. The implementation details and performance results are discussed here.

1 REAL-TIME SIGNAL PROCESSING USING MULTIPLE DSP CHIPS

DSPs are ideal for medium throughput applications, such as speech and digital audio. However, when attempting to meet real time constraints, one is often faced with the task of squeezing the application to fit onto relatively few instructions available per sample on a processor.

For example, consider processing audio at a 44 KHz sampling rate on a processor with a 60 ns cycle time. Audio samples arrive once every 227 ms. Hence there are only about 380 instructions available to process each sample while maintaining real time constraints.

Use of multiple DSPs is thus an attractive option for increasing the amount of processing that can be done per sample. The utility of a multi-DSP machine is determined largely by two issues. First is the overhead associated with communication and synchronization between processors and second is the available software environment for programming the multi-processor machine. Several multi-DSP architectures have been demonstrated by the research community. Special hardware is usually employed for reducing communication overhead. The SMART array [2] for example is a distributed shared memory machine with custom VLSI parts for maintaining coherence. The iWARP architecture [3] employs separate communication and computation agent, implemented using custom VLSI. In [4], the authors describe a DSP96002 based multi-DSP system with an "intelligent communication controller" implemented on a gate array for communicating through a high speed bus.

Our approach is to reduce overhead by restricting ourselves to a certain subclass of DSP applications. By sacrificing generality, we acheive efficient communication between processors with relatively simple hardware. Such an approach is ideal for low cost implementation of embedded applications that fall into this subclass, on multiple DSPs. Also, software environment used for programming the multi-processor machine is closely tied to the hardware methodology employed.

2 SYNCHRONOUS DATA FLOW

We address DSP applications that can be specified as a data-flow graph, with nodes (actors) being individual tasks and directed arcs between them representing flow of data (tokens). Synchronous Data Flow (SDF) refers to a subclass of data flow graphs where the actors lack data dependency in their firing patterns [5]. Thus in SDF graphs the number of tokens produced and consumed in each of the output and input arcs of each actor is constant and fixed at compile time. In exchange for this restriction several nice properties are obtained for these graphs. In particular, self-timed scheduling, i.e. assigning actors to processors and specifying their firing order, can be done efficiently at compile time. Nearly optimal static periodic multi-processor schedules can be obtained for these graphs [6]. Also, given approximate execution times for the actors, one can constrain the order in which processors would need to access shared resources at run time without unduly sacrificing performance.

It is seen that a large set of DSP algorithms fall into the SDF paradigm. In our group at Berkeley, we have implemented Gabriel, a block diagrammatic software environment for DSP based on the SDF model [7]. Ptolemy, the next generation system developed by our group, handles several different models, SDF being one of them [8]. Both systems can generate executable code for processors from block diagram specifications. Parallel scheduling algorithms have been implemented to schedule a flow graph onto multiple processors. Each processor gets a subset of all the nodes present in the flow graph, and it executes them in the prescribed order. When data tokens need to flow between processors, interprocessor communication

---

1. This work was supported by grants from SRC (grant number 92-DC-008), Motorola Inc., and ArielCorp.

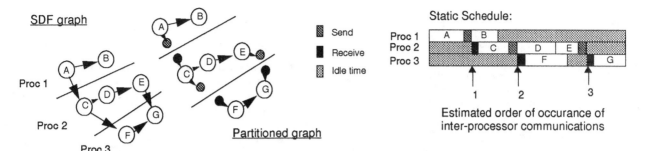

FIGURE 1. Illustration of the scheduling mechanism

primitives are inserted in the code of the sender and the receiver. The dataflow graphs we deal with are typically fine to medium grained, with each node representing no more than few tens of instruction cycles. Data going from one processor to another is usually one word at a time, in other words the data is not 'packetized' into blocks. Figure 1 illustrates the multiprocessor scheduling mechanism.

Under the software environment above we describe an architecture for synchronous data flow.

3 ARCHITECTURE

3.1 Motivation

Our goal was to design parallel hardware that would mesh well with the software methodology to be used on top of it. The result, hopefully, would be an architecture optimized, both in terms of performance as well as cost, for the restricted set of applications (in this case SDF) that we were concentrating on. We look at a single bus shared memory architecture, mainly because it results in simpler hardware and an easier partitioning problem. For such an architecture, communication of data through shared memory involves bus and memory contention. Usually there is a bus arbiter to resolve simultaneous bus access requests. Memory accesses, then, must be synchronized using semaphore mechanisms. All this adds to the inter-processor communication (IPC) overhead, not to mention the bus bandwidth wasted on unsuccessful semaphore checks.

In our lab we had a four processor Motorola DSP56000 architecture based on the "bus arbiter and semaphore" mechanism. Because of the associated overhead, 30 instruction cycles were required to transfer a word between processors. For digital audio, 30 cycles represents about 8% of the processing time available on a processor, in the scenario of Section 1. This implies that the scheduler must use as few IPCs as possible, thus severely restricting the parallelism that could potentially be exploited.

In principle we could reduce the IPC overhead by managing semaphores in hardware, using multi-ported memory and so on, under the penalty of complex and expensive hardware. Instead, noting that the above approach is too

general for our purposes, we look at an alternative solution.

3.2 Ordered Transaction Principle

Suppose we knew the exact run time behavior of each of the task nodes, down to the clock cycle. Also suppose that all processors run synchronously. We could then determine at compile time the instruction each processor executes on each clock tick (the *fully-static* approach in [1]). This approach would obviate bus-memory contention and result in near zero overhead IPC. However it would be impractical, since we usually do not know execution times of nodes to the degree required here.

Now, suppose we retain only the *order* of execution of nodes and the *order* in which shared memory is accessed, based on the static schedule. Since the application fits the SDF model, imposing this constraint does not sacrifice efficiency, provided that approximate execution times of the nodes are known. The approach is robust even for inaccurate execution time information. Performance efficiency however, may suffer in this case. Run time imposition of an order pre-determined at compile time obviates bus and memory contention, as in the fully-static case. No bus arbitration is required, and semaphores are not needed. A central controller simply maintains the pre-determined order by granting the bus to each processor on its turn. In the best case, if the scheduler has done a good job, the processor already has the bus when it needs it and can go ahead with its shared memory transaction. On the flip side, if the scheduler does not do a good job, a processor may have to wait until the bus is granted to it, or even worse, a processor may not yet be ready when it is granted the bus, thus blocking other processors that need the bus.

This approach is hardware lean: bus arbitration logic already present on the processors can be used. The access controller is simply a counter addressing memory that is downloaded with the schedule at compile time. No other hardware is required to implement the scheme. Moreover, in the best case it takes only 3 instruction cycles per transaction, which is about as well as we can hope to achieve.

3.3 Hardware Implementation

As a proof-of-concept prototype, we have built a 4 processor single bus shared memory architecture, called the ordered memory access architecture (OMA) board, with hardware support for ordering shared memory transactions (Figures 2 & 3). In our implementation, we use 4 Motorola DSP96002 processors on a single printed circuit board, operating with a 33MHz common clock. Even though our techniques can be applied to most off-the-shelf DSP chips, we chose the DSP96002, because its dual bus 32-bit architecture lends itself naturally to our specific application. A processing element consists of a DSP96002 with up to 256Kbytes of onboard zero wait state static RAM on one port (local memory); the other port of all processors are connected together to form the shared bus. There is provision for up to 512Kbytes of onboard static shared RAM.

The transaction controller is implemented as a presettable counter on a Xilinx FPGA. This was done with a view towards future experimentation with the transaction ordering mechanism. The access order list is extracted from the schedule and downloaded into the transaction controller memory. The schedule counter steps through this list at runtime. Stored entries in the schedule RAM correspond to processor IDs. This output is latched and decoded to obtain bus grant signals for the processors. The schedule counter is incremented each time the bus is released by a processor. Host interface and a simple I/O mechanism is also implemented on the Xilinx array. Re-use of the gate array has saved us a fair amount of glue logic, thus making the board less complex. Buffers for connecting multiple boards are included. We can configure multiple boards such that they form a single bus, or we could have cleaved busses with communication between busses implemented on the Xilinx chip.

The host to this board is a DSP56000 Sbus based card from Ariel Corp. This card is used to control the OMA board as well as download code from a SPARCstation running UNIX. The 56000 also acts as a interrupt processor,

providing the OMA board with real time I/O. Currently, we feed the OMA board with data from a compact disc (CD) player and obtain data from it at CD rates.

The OMA board is a 10 layer through-hole PCB, with dimensions of 11" by 7". It was designed and laid out using the tools developed here at UC Berkeley [9] and was fabricated by MOSIS.

4 RESULTS

The 4 processor prototype has been tested and functional correctness has been verified. We have been able to achieve processor to processor to communication over the shared bus with a cost of 3 instruction cycles. The multiprocessor system has also been integrated into Ptolemy and Gabriel, for automatic scheduling, code generation and downloading from a Sun Sparc workstation. Some of the applications that have been tested on the OMA board are as follows.

4.1 1024 point complex FFT

Data is assumed to be present in shared memory. The transform coefficients are written back to shared memory. A single 96002 processor on the board performs the transform in 3.0 milliseconds (ms). With all four processors, it takes 1.0 ms. Each processor performs a 256 point FFT from the first stage of a radix 4 computation. This example is communication intensive: the throughput is limited by the available bus bandwidth.

4.2 Music synthesis

A synthesis algorithm for plucked strings was implemented using the Karplus-Strong algorithm. Synthesis of each voice involves a noise source, a pulse generator, a delay, a filter operation and scaling operations. A single processor could fit 7 voices in real time (44 KHz sampling

FIGURE 2.   Block diagram of four processor OMA board

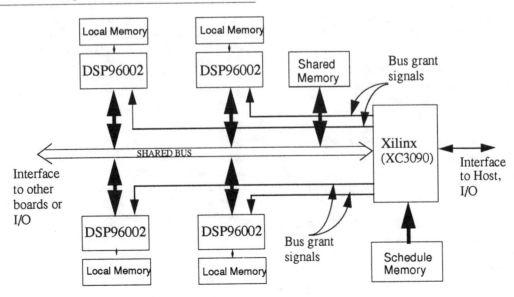

rate). Partitioned accross 4 processors, with 15 IPCs, we could fit in 28 voices. This example is not communication intensive; the low overhead IPC mechanism easily absorbs the extra cycles associated with IPC.

4.3 QMF filter bank

A filter bank was implemented to decompose audio from a CD player into 5 bands. The resynthesis bank was also implemented together with the decompostion part. This involved 16 multirate filters (18 taps each). There were 85 IPCs in the final schedule .

5 CONCLUSION

We have presented the design and implementation of a multi-DSP architecture that acheives low overhead inter-processor communication with low hardware complexity. Program entry, scheduling and code generation for this can be done under the Ptolemy environment. Some applications have been run on the board. We plan to expand the existing I/O capability of the board by adding peripheral modules. This will enable us to evaluate the architecture under several other applications.

Imposition of a single fixed order at run time means that no data dependency can be tolerated. To run non-SDF graphs on the OMA architecture, we use a presettable counter as the transaction controller. Any processor that has posession of the shared bus can make the transaction controller jump to another bus access schedule by presetting the schedule counter. Conditional branches can thus be handled by computing access schedules for each branch outcome and switching between them at run time. Compilation process for this scheme under the Ptolemy environment will be the subject of future research. We can then

evaluate how well this approach to handling limited data dependency works in practice.

6 REFERENCES

[1] J. C. Bier, S. Sriram, and E. A. Lee, "A Class of Multiprocessor Architectures for Real-Time DSP", *VLSI Signal Processing IV*, 1990.

[2] W. Koh, "A Reconfigurable Multiprocessor System for DSP B ehavioural Simulation", Ph.D. Thesis, ERL, UC Berkeley, June 1990.

[3] S. Borkar *et. al.*, "iWarp: An Integrated Solution to High-Speed Parallel Computing", *Proceedings of Supercomputing 1988 Conference*, Orlando, Florida.

[4] A. Gunzinger *et. al.*, "Architecture and Realization of a Multi Signalprocessor System", *Proceedings of the International Conference on Application Specific Array Processors*, Berkeley, California, August, 1992.

[5] E. A. Lee, and D. G. Messerschmitt, "Synchronous Data Flow", *IEEE Proceedings*, Sept. 1987.

[6] G. C. Sih, and E. A. Lee, "Multiprocessor Scheduling to Account for Inter-processor Communication", Ph.D. Thesis, ERL, UC Berkeley, April 1991.

[7] J. C. Bier *et. al.*, "Gabriel: A Design Environment for DSP", *IEEE Micro Magazine*, Oct. 1990, Vol. 10.

[8] J. Buck, S. Ha, E. A. Lee, and D. G. Messerschmitt, "Ptolemy: A Framework for Simulating and Prototyping Heterogenous Systems", Invited paper in the *International Journal of Computer Simulation*, to appear.

[9] M. B. Srivastava, "Rapid Prototyping of Hardware and Software in a Unified Framework", Ph.D. Thesis, ERL, UC Berkeley, June 1992.

[10] J. Buck, and E. A. Lee, "The Token Flow Model", *Data Flow Workshop*, Hamilton Island, Australia, May 1992.

FIGURE 3. Photograph of the board

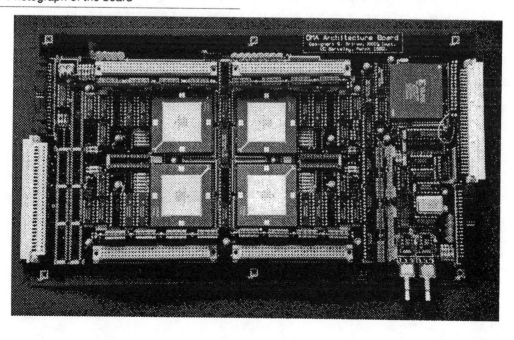

Sizing of Communication Buffers for Communicating Signal Processes

Tilman Kolks, Bill Lin and Hugo de Man

In this paper, a method is presented for the minimal sizing of communication buffers in an environment of communicating signal processes. Already existing methods based upon lifetime analysis techniques are mainly suited for sizing buffers in a network of processes with *non*-conditional interactions. Here we consider a more general problem where the execution of each process depends inherently on *complex interactions* with other processes as well as on its *internal conditional behavior*. To address the buffer sizing for this general case, we propose the use of *implicit state enumeration* techniques. A dynamic data flow graph (DDF) is used as the model to describe the entire system of communicating processes. This DDF is translated into a network of finite state machines that forms a state space on which methods based on implicit state space exploration are applied. To reduce the size of the state space, abstraction techniques are used to produce a simplified version in which irrelevant information is removed. The feasibility of the proposed approach is demonstrated by a practical signal processing example.

1. INTRODUCTION

Large systems are normally described as a collection of smaller, interacting subsystems, also called *processes*. The usual computational model in signal processing applications to describe such a system is a *dynamic data flow graph*, e.g. in [11]. The processes in such a system are subgraphs of the data flow graph.

In the actual implementation of the system, the individual processes are implemented independently from each other using their own appropriate architectural style. *Interfaces* are required between the subsystems that exchange data to ensure that their I/O behavior is attuned to each other. This interfacing becomes a part of the specification of the final system. An example of such a set of communicating processes on a single chip is a mobile terminal transmitter [1]. This example is used as a case study in this paper.

Much effort has been devoted to automated synthesis of the processes, e.g. in

[2]. However, less attention has been paid to automatic synthesis of interfaces through which processes communicate. One of the important issues in the design of interfaces is to determine the **size of intermediate buffers** in the interface required to temporarily store data to be transferred. This problem is addressed in this paper, where we consider a buffer to be a storage unit with **once in once out** characteristics.

In literature, only the sizing of buffers for the restricted types of non-conditional processes have been treated [3, 4, 5]. The sizing problem of communication buffers becomes more difficult in situations where processes contain conditionals that influence the communication behavior. In the telecommunication domain, networks of control dominated concurrent processes do occur as shown in figure 1. Each process

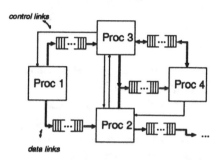

Figure 1. Processes interacting through communication buffers.

in such a network can has complex conditional behavior. In addition, the processes interact with each other. As a result, execution depends on these interactions because of control and data dependencies and techniques to size buffers based upon lifetime analysis (e.g. [3]) are no longer applicable.

To solve the buffer sizing problem for this case, we propose the use of methods based upon implicit state enumeration techniques [6, 7]. We start from a dynamic data flow graph to describe the system and determine the buffering capacity of the edges of the graph by transforming the graph into a network of communicating finite state machines by viewing each edge as a *counter*-like state machine. On this set of state machines, state space exploration algorithms can be applied to determine the required minimum sizes of the buffers. The most important aspect of this approach is its ability to handle conditional paths in communicating processes.

In the following sections we discuss our approach in detail, starting with an introduction of the techniques that are used for implicit state space exploration. It is followed by an outline of the dynamic data flow model (DDF) which we use to describe the overall system. Then the transformation step of the DDF into a single sequential machine is described after which we propose our method to solve the synthesis problem. Finally, we discuss a case study.

2. PRELIMINARIES

Binary Decision Diagrams

A binary decision diagram (BDD) [8] is a canonical graph representation of Boolean functions. It is a directed acyclic graph in which each internal node in the graph is associated with a Boolean decision variable and has two outgoing arcs. The reader is referred to [8] for details. BDD's hold key properties that make them a better representation for reasoning about Boolean functions than previously used representations, like the sum-of-products representation. In addition to the canonical property, BDD's are in most cases much more compact than any sum-of-products denoting the same function.

Characteristic Functions

BDD's can also be used to manipulate finite sets efficiently. Any subset A of B^n can be denoted by a unique Boolean function $\chi_A : B^n \to B$. This function is called the **characteristic function** of A and is defined in the following way: $\chi_A(\mathbf{x}) = 1$ if and only if $\mathbf{x} \in A$. A characteristic function is a very interesting implicit representation of Boolean sets because there is a direct correspondence between set operators and the logical operators. Since a characteristic function is a Boolean function, it can be represented with a binary decision diagram and the set operations can be performed directly on the BDD representation.

In addition to the basic propositional logic operations, the existential operator ($\exists$) is required for the implicit manipulation of sets. This Boolean quantifier can also be computed directly on a BDD representation. The existential quantification of a set of Boolean variables $X = \{x_1, \ldots, x_n\}$ with respect to the Boolean formula f can be evaluated as

$$\exists X(f) = \exists x_1(\exists x_2(\ldots \exists x_{n-1}(\exists x_n(f))\ldots)),$$
$$\exists x_i(f) = f_{x_i} \vee f_{\overline{x_i}},$$

where $f_{x_i}(f_{\overline{x_i}})$ denotes the **cofactor** of formula f with respect to a literal $x_i(\overline{x_i})$. $f_{x_i}(f_{\overline{x_i}})$ is a new formula obtained by substituting $1(0)$ in place of the variable x_i in the formula f.

Transition Relation

Consider a finite state machine represented by a 6-tuple $(I, O, S, \delta, \lambda, \sigma)$ with input and output alphabet I resp. O. $\delta : I \times S \to S$ is the next-state function, $\lambda : I \times S \to O$ is the output function and σ is the initial state(set). We can use a characteristic function to describe the next-state function of a state machine [6, 7] where $I \subseteq B^n, S \subseteq B^t$ are binary encoded . By viewing the next-state function as a ternary relation $T \subseteq I \times S \times S$ we can build a characteristic function for it. The characteristic function that encodes the relation T is a boolean function $T(x, i, y) : \{0, 1\}^{n+2t} \rightarrow \{0, 1\}$ called the **transition relation**. This transition relation indicates that next state y can be reached in exactly 1 transition from state x under input i.

The transition relation of a state machine can be used to compute the set of states R^* reachable by some transitions from an initial state(set) σ. This set is called the **reachable state set**. A procedure to compute the reachable state set is called **implicit state space enumeration** [6, 7]

A transition relation, together with the initial state(set) completely models an FSM and can be used to traverse the state space of the machine efficiently.

3. DYNAMIC DATA FLOW MODEL

Signal processing systems are frequently described using a dynamic data flow graph (DDF) [10, 11] or using the more restricted type of synchronous data flow graph (SDF). Here we use the DDF as a system model which allows for expressing conditionals and multi rate systems.

Formally we describe a DDF G by a 3-tuple $< V, E, R >$. V is the set of nodes (operations) of G, $E \subseteq V \times V$ is the set of directed edges of G. A function R associates with each node $v \in V$ a set of 3 rules: a firing, consumption, and production rule. The possible conditional behaviour of an operation node is expressed with its rules. Edges carry data in a *first in, first out* fashion. The data on an edge e is represented by a *stream* of tokens, $Str(e)$. The tokens in the stream may be labelled by values that can influence the execution of rules of an operation. To distinguish between these types, we speak of an *interpreted* and an *non-interpreted* stream.

Operation nodes that represent a constant value can be assumed to regenerate a token whenever it is needed by a successor node. Operation nodes can be used for the refinement or abstraction of other data flow graphs. The firing of a node in a DDF takes unit time.

In figure 2, an example of a DDF with 4 operations is shown as an illustration.

The problem of sizing buffers between processes in a system relates to determining the minimum sizes of the streams associated to edges. In the sequel, we will therefore not make a distinction between the terms buffer and stream.

4. SYNTHESIS PROCEDURE

The overall synthesis procedure to determine these sizes can be summarized as follows:

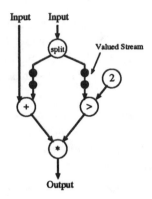

Figure 2. An example of a DDF.

1. model each operation node and stream as a finite state machine, which can be implicitly represented as a transition relation using BDD's

2. compute the set of reachable states (represented as a BDD) from the set of transition relations using implicit enumeration techniques

3. for each stream, derive the set of reachable states (represented as a BDD) for the corresponding state machine

4. from the BDD representation of the reachable states of the buffer, determine the minimum buffer size required to implement the corresponding communication buffer

Each of these steps will now be described in detail.

5. DYNAMIC DATA FLOW GRAPH TRANSFORMATION

In transforming a DDF into a set of interacting finite state machines, a key observation is that streams can be modelled by *counter-like* structures. We will informally describe those structures.

A non-interpreted stream $Str(e)$ is modelled by an FSM $B_e^0 \equiv \langle \mathcal{I}_e, \mathcal{O}_e, \mathcal{Q}_e, \delta_e, \lambda_e, q_e \rangle$ represents a counter. $\mathcal{I}_e \equiv \{r_e^1 \cdots r_e^n, w_e^1 \cdots w_e^m\}$ representing the read and write operations on the buffer imposed *solely* by the production and consumption rules of nodes in the DDF that are adjacent to e. $\mathcal{Q}_e \equiv \{0, 1, \ldots, maxsize_e\}$ and $\mathcal{O}_e \equiv \{over_e\}$ indicates that the buffer can "overflow" in a next write to the buffer. q_e presents the initial buffer offset which normally equals zero. The next-state function implements the counter increment and decrement operation triggered by the inputs.

To model the n-bit wide *interpreted* stream $Str(e)$ of an edge e, we first create a machine that models a one-bit FIFO: $B_e^1 \equiv \langle \mathcal{I}_e^1, \mathcal{O}_e^1, Q_e^1, \delta_e^1, \lambda_e^1, q_e^1 \rangle$. Again, $\mathcal{I}_e^1$ is solely determined by the adjacent nodes in the DDF. The machine stores 1-bit values written to it, and removes entries that are read. Also here an output function is provided that indicates that the machine can overflow in a next write action. From this machine an n-bit wide interpreted stream is formed by taking the synchronous product B_e^n over n state machines B_e^1.

Finally, a node v in a DDF is converted to a state machine $P_v \equiv \langle I_v, O_v, Q_v, \delta_v, \lambda_v, q_v \rangle$. This machine has one state, $Q_v = \{q_v\}$, and contains two self loops. One transition has as input condition the firing rule of $R(v)$ and as output the composition of the production and consumption rule. The other transition fires when the firing rule is not satisfied and does not affect any stream contents.

Then, a dynamic data flow G is transposed into a network of state machines by replacing every node $v \in V$ by a state machine P_v, and replacing every edge $e \in E$ by a state machine $B_e^{n(e)}$ of appropriate bit-width $n(e)$ ($n \geq 0$).

5.1. Reachable Buffer States

In the previous subsection, a network of state machines is formed. A transition relation $T_j(x_j, i_j, y_j)$ is build for every operation node (process) P_j as well as a transition relation $\hat{T}_e(\hat{x}_e, \hat{i}_e, \hat{y}_e)$ for every buffer $B_e^{n(e)}$. The synchronous transition relation $T_C(x, i, y)$ of the synchronous product machine can be formed from these transition relations [7]

$$T_C(x, i, y) \equiv \prod_{j \in V} T_j(x_j, i_j, y_j) \times \prod_{e \in E} \hat{T}_e(\hat{x}_e, \hat{i}_e, \hat{y}_e) \tag{1}$$

As mentioned in section 2, this transition relation $T_C(x, i, y)$ can be used to obtain the reachable state set $R_*(x_1, \ldots, x_{|V|}, \hat{x}_1, \ldots, \hat{x}_{|E|})$, where x_j indicates a state of a process P_j and $\hat{x}_k$ indicates a state of a buffer $B_k^{n(k)}$. In this case x_j is always the same state since a process P_j only contains one state.

By the FSM description of a buffer, a fixed number of states is implied for it. Since we do not know in advance how large the buffers may become, we *expand* the buffers dynamically during the state space traversal. For this purpose, we make use of the output function $over_e$ of a buffer $B_e^{n(e)}$. If such an output indicates that the preimposed limit of the buffer size can be reached, we expand the buffer by adding an additional state variables in the transition relation. The additional variables are also added to reachable state set computed so far. In this way we are sure that a buffer is always large enough.

5.2. Minimum Buffer Size

From the reachable state set $R_*(x_1, \ldots, x_N, \hat{x}_1, \ldots, \hat{x}_K)$ of the product machine, the set $Q_*(\hat{x}_1, \ldots, \hat{x}_K)$ of all *reachable buffers states* can be derived by existentially quantifying over all non-buffer state variables:

$$Q_*(\hat{x}_1, \ldots, \hat{x}_K) = \exists S \, R_*(x_1, \ldots, x_N, \hat{x}_1, \ldots, \hat{x}_K), \quad S = \{x_1, \ldots, x_N\} \quad (2)$$

To derive the set of reachable states for a particular buffer B_i, we can existentially quantify the other buffer state variables as follows:

$$Q_*(\hat{x}_i) = \exists S \, Q_*(\hat{x}_1, \ldots, \hat{x}_K), \quad S = \{\hat{x}_j \mid \hat{x}_j \neq \hat{x}_i\} \quad (3)$$

$Q_*(\hat{x}_i)$ represents the set of reachable states for buffer $B_i^{n(i)}$. From this set, the largest reached buffer size can be determined which represents the minimum buffer size. We refer to [12] for further details.

6. OTHER PROGRAMMING MODELS

Other programming models than a dynamic data flow graph may be used to describe systems. Two such models are the synchronous data flow graph (SDF) and the communicating finite state machines model.

The SDF model is more restricted than the DDF model in that it does not allow for conditionals to be expressed. The translation from a DDF to a set of interacting finite state machines as described is therefore also applicable for a SDF model. Due to this restriction, only non-interpreted streams have to be dealt with and only counters are required to replace the streams.

The other model of communicating finite state machines is the a more general one. In this model, processes are described by arbitrary finite state machines [12]. Also here our translation step is easily adapted by replacing the finite state description of an operation in a DDF by a finite state machine of the communicating finite state machines model. We use this more general model in our application study.

7. A CASE STUDY

To demonstrate our method, we considered a part of the mobile terminal transmitter [1]. The part consists out of three parts as is shown in figure 3. The model of communicating FSM's was used to describe this example.

The first part of the example is an A/D converter operating at 8 kHz. It produces data consumed by the Voice Coder (Vocoder) [9] that operates at 5.4 MHz. In between, two buffers are needed: one for the data that is read by the vocoder, the other in case the first buffer is read by the vocoder and the ADC produces a sample. It is then

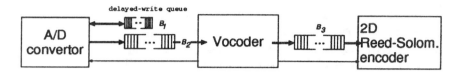

Figure 3. Partial Block diagram of Mobile Terminal Transmitter.

temporarily stored in the delayed-write buffer and later rerouted to the buffer. The data produced by the vocoder is encoded by a bidimensional Reed-Solomon encoder operating at 15 kHz. Again a buffer is needed for the data exchange. This buffer is a blocking queue, resulting in the Reed-Solomon encoder to enter an idle state if it tries to read data from the buffer when it is empty. Operation is continued if data is again present in the buffer.

The Vocoder is a complex signal processing algorithm, requiring over 100,000 clock cycles per execution frame. This implies that for a complete pass over the state space of the vocoder itself at least as many iterations are required, making this method prohibitively expensive. The use of abstraction is of vital importance in these applications. Since the data-path of the design does not influence the behavior on the buffers it can be abstracted. In addition we were able to perform abstraction on the state space to a significant extent while maintaining the Control Flow behavior of the design.

The abstraction enabled us to perform the sizing of the different buffers as well as verifying access properties and underflows for this example in short run times [12], using the methods described in this paper.

8. CONCLUSIONS

We have presented a method that is able to synthesize communication buffers in an environment of communicating signal processes. This method, unlike previous proposed methods, enables us to take into account conditional paths in the processes. A dynamic data flow model is used to describe the system, but other models are applicable as well. It was argued and shown that the use of abstraction in describing the processes as finite state machines is mandatory and possible so that the method can be successfully applied in a signal processing environment.

References

[1] L. Philips, I. Bolsens, B. Vanhoof, J. Vanhoof and H. De Man Silicon Integration of Digital User-end Mobile Communication Systems In *Proc. IEEE Int. Conf. on Communications*, 1993.

[2] H. de Man, F. Catthoor, G. Goossens, J. Vanhoof, J. L. Van Meerbergen, S. Note, J. A. Huisken. Architecture-Driven Synthesis Techniques for VLSI Implementation o f DSP Algorithms. *Proc. of the IEEE*, 78(2):319–335, February 1990.

[3] J. Decaluwe, J. M. Rabaey, J. L. Van Meerbergen, and H. J. De Man. Interprocessor Communication in Synchronous Multiprocessor Digital Signal Processing Chips. *IEEE Transactions on Acoustics, Speech and Signal Processing*, 37(12):1816–1828, December 1989.

[4] T. Amon and G. Borriello. Sizing Synchronization Queues: A Case Study in Higher Level Synthesis. In *Proc. of the 28th Design Automation Conference*, pages 690–693, 1991.

[5] J Huisken, A. Delaruelle, B. Egberts, P. Eckhout, and J. Van Meerbergen. Synthesis of Synchronous Communication Hardware in a Multiprocessor Architecture. In *Proceedings of the 5th High Level Synthesis Workshop*, March 1991.

[6] O. Coudert, C. Berthet, and J. C. Madre. Verification of Sequential Machines Using Boolean Functional Vectors. In *IMEC-IFIP Int. Workshop on Applied Formal Methods for Correct VLSI Design*, pages 111–128, November 1989.

[7] H. Touati, H. Savoj, B. Lin, R. K. Brayton, and A. Sangiovanni-Vincentelli. Implicit State Enumeration of Finite State Machines Using BDD's. In *Proc. of Inter. Conf. on Computer-Aided Design*, November 1990.

[8] R. Bryant. Graph-Based Algorithms for Boolean Function Manipulation. In *IEEE Trans. on Computers*, volume C-35, pages 677–691, August 1986.

[9] B. Vanhoof, I. Bolsens, S. De Troch, L. Philips, J. Vanhoof, and H. De Man. Design of a Voice Coder with the Cathedral-II Silicon compiler. In *Proc. of the EuroAsic Conference*, pages 23–25, Feb 1993.

[10] R. Lauwereins, M. Engels and J. Peperstraete. GRAPE-II: A Tool for the Rapid Prototyping of Multi-Rate Asynchronous DSP Applications on Heterogeneous Multiprocessors. In *Proc. of the 3d Workshop on Rapid System Prototyping*, June 1992.

[11] J. Buck, S. Ha, E. Lee and D. Messerschmitt. Ptolemy: A Framework for Simulating and Prototyping Heterogeneous Systems. In *Inter. Journal of Computer Simulation (to be published)*.

[12] T. Kolks, B. Lin and H. De Man. Sizing and Verification of Communication Buffers for Communicating Processes. In *Proc. of the Inter. Conference on Computer-Aided Design*, November 1993.

Acknowledgements

The authors would like to thank Gjalt de Jong and Eric Verlind for in depth discussions on the subject.

Tilman Kolks
IMEC vzw.
B-3001 Leuven
Belgium
e-mail: kolks@imec.be

A Two-level On-chip Memory for Video Signal Processor

Chia-Hsing Lin, Jen-Sheng Hung and Chein-Wei Jen

Institute of Electronics, National Chiao Tung University
Hsinchu 30050, Taiwan, ROC

ABSTRACT

In this paper, we propose an on-chip memory architecture for video signal processor(VSP). According to the nature of different data locality in video source coding applications, the memory adopts a novel two-level scheme for making trade-off between capacity and flexibility. The upper level -- Memory A provides enough storage capacity to reduce the impact on the limitation of chip I/O bandwidth, and the lower level -- Memory B provides enough data parallelism and flexibility to meet the requirements of multiple re-configurable pipeline function units in a single VSP chip.

We have finished the design of a prototype memory design using 1.2-μm SPDM SRAM technology and will fabricate it through TSMC, in Taiwan.

1. INTRODUCTION

In video signal processing applications, the diverse complex algorithms and huge data requirement make those applications computation-intensive. To fit the real time demand, parallel processing scheme, such as a video signal processor (VSP) with multiple processing elements working concurrently, is necessary. However, the chip I/O bandwidth will limit the utilization of processing units if it cannot meet with the computation rate. Therefore, an on-chip memory with enough capacity and an adequate structure is necessary to maximize the utilization of processing elements and reduce the demand of high chip I/O bandwidth.

Capacity and flexibility are the major considerations for the on-chip memory structure in order to reduce the impact on the limitation of chip I/O bandwidth and to meet the access requirements of multiple re-configurable processing elements in a single chip VSP. The on-chip memory design with dual-port cells' scheme may satisfy the capacity requirement; however, it is difficult for multiple processing elements to access the data in the same location without introducing processor stalls if VSP doesn't provide complex routing and addressing scheme. On the other hand, a multi-port memory structure may be a candidate to support enough parallelism. However, as the number of ports increases, the area growths make it is hard to provide enough storage capacity. Therefore, an on-chip memory structure should be able to make the trade-off between size and parallelism for on-chip memory.

In this paper, a two-level memory architecture that can fit the requirements of on-chip memory is proposed. The upper level -- Memory A provides enough storage capacity to release VSP chip from high I/O bandwidth demand, and the lower level -- Memory B supports enough data parallelism to meet the demand of multiple pipeline processing elements in a single VSP chip. In Section 2, We will discuss the memory model that is suitable for VSP under two critical considerations: capacity and parallelism. According this model, we will establish a two-level approach to on-chip memory and describe it in section 3. A prototype design of 3.5-kbyte that applied the two-level concept will be presented in Section 4. The conclusions are summarized in Section 5.

2. MEMORY MODEL FOR VIDEO APPLICATIONS

Both providing enough storage capacity and supporting data parallelism are the goals of on-chip memory design in order to reduce the impact on the limitation of chip I/O bandwidth and to meet the access requirements of multiple re-configurable pipeline processing elements in a single chip VSP. The memory, such as those in VSPs designed by Yamauchi[4], Murakami[5], and Tamitani[6], consists of relatively smaller dual-port cells, may meet the size requirement; however, multiple processing units can not access the data in the same location concurrently if VSP doesn't provide complex routing and addressing scheme. On the other hand, a multi-port memory structure may be a way to support enough parallelism. However, as the number of ports increases, the area growths make it hard to embed storage with enough capacity in a single chip.

It is obviously that we should make a trade-off between size and flexibility. After observing the frequently used block operations in video source coding algorithms such as motion vector calculation, 2-D convolution and VQ, etc., it is shown that there exists different data locality in those video applications. This seems to mean hierarchical memory design is useful. Therefore, we proposed a two-level model of on-chip memory(Fig. 1): The upper level -- Memory A, stores most of the would-be repeatedly used data as economic as possible, and the lower level -- Memory B, represents a moving window capable of offering large data parallelism. This memory structure is suitable for block-matching algorithms used for ME or VQ because we can easily map search area and template onto two memory levels. Furthermore, there is also no need of extra routing mechanism to perform bit-reverse, butterfly, or transpose if memory B has sufficient read ports and write ports. Thus, algorithms like DCT, FFT, and filtering are also well performed.

Obviously, multi-port RAM cells are needed to satisfy the desire of multiple processing elements for large amount of data provided Memory B. However, problems involved in the realization of multi-port memory are: (i) To concurrently enable different cells of a memory bank, it must duplicate the decoder, and therefore the word lines as well. (ii) It must duplicate both bit lines and sense amplifiers while the number of read ports increases. (iii) Each cell should have adequate driving capability to cope with the worst case situation, which occurs at a time that one cell read by all ports simultaneously[1]. (iv) There needs at least one more transistor as control for each additional port.

Since both word lines and bit lines are duplicated, the increase of chip area owed to these two parts is expected to be $O(N^2)$, where N is the number of ports. That is, the area cost will become high if the required number of ports is large, which in turn worsen the cycle time of memory. This makes this scheme be not suitable for a VSP with large number of embedded processing elements. It is also less expansible if the progress of VLSI technology makes it possible to embed more PEs in single chip.

3. THE PROPOSED TWO-LEVEL MEMORY STRUCTURE

On the basis of the above discussion of the difficulty in the realization of real multi-port memory, we use one-read-one-write memory banks to emulate such memory. The basic methodology is to duplicate all possibly used items into the corresponding memory banks. Fig.2 and Fig.3 show the applications of this architecture in butterfly-style memory access and blocking-matching, respectively.

In Fig.2, data x0 through x7 are copied into all eight bank-B's, and the butterfly style operation can be finished without the need of special routing.

In Fig.3, the block matching operation is demonstrated. At clock 0, the frame pixels are loaded into bank-A's, which is constituted of four one-read-one-write memory banks (a1, a5, a9, a13, ... in one bank, a2, a6, a10, ... in another bank, and so on) and the template pixels from t1 to t16 are scattered into four of the eight bank-B's. Assume we are now calculating the absolute-difference values of the second row of both blocks: at clock 1 pixel a19 to a22 are duplicated into the other four of bank-B's, and four absolute-difference operations can be executed simultaneously at the following clock (clock 2).

The proposed memory for four processing elements is shown in Fig.4. Two storage levels -- Memory A and Memory B constitute this memory. Each level consists of multiple one-read-one-write memory banks that are organized to be a quit flexible multi-port structure. The Memory A, which should provide enough storage capacity, behaves as the input buffer memory. Because of the random access characteristic in diverse motion vector detection algorithms, we think that it is appropriate for this level to have the size equal than the search area (used for ME or VQ) plus required capacity for pre-loading data of next search block in order to prevent unnecessary processor stall. On the other hand, the memory B, which should provide enough data parallelism, can offer up to sixteen 8-bit parallel read ports alone with four 16-bit parallel write ports for four function units. The size of this level should depend on read/write ratio between Memory B and PEs. Four 32b buses are used to support the computation rate between those two levels. We also use several bus switches to control the connection-disconnection of different buses. Therefore, through different setting of bus switches, the Memory B can be further divided into two independent 8-read-2-write or four independent 4-read-1-write memories with equal size, and each has the same size as the original 16-read-4-write structure. This memory can be organized as a 16-one-read-4-one-write structure, too, and the total size is sixteen times the 16-read-port-4-write-port case.

Fig.5 shows the details of one memory bank B. SEL2's decide the data writing pattern of bank-B's. Table.1 lists the enable line(s) chosen by SEL2. By way of proper writing sequence as well as the setting of SEL1's and SEL2's, this memory can be organized with fair elasticity. Table.2 lists some of the examples.

Switches in Fig.4 decide the break or concatenation of buses, which can divide bank B's up to four groups. Data from bank A's, direct input or processing elements (PE's) are written to bank B's according to the setting of SEL1's and SEL2's. If the configuration is chosen that each input data will appear in all selected N read columns, those N columns compose an N-read-M-write memory, where M may vary from one to four depending on how those columns are arranged.

The memory we proposed can be used for video signal processor with embedded multiple re-configurable pipeline function units to fit data flow patterns in algorithms. We may choose the (read ports)/(write ports) ratio and data precision according to the average input/output ratio and accuracy requirement. Compared with the memory structure that uses shared bus's structure and explicitly routing mechanism[4], the memory we proposed is more changeable under different considerations. There is also no need of explicit output buffer memory[4,6] and of extra routing mechanism because of the flexibility of Memory-B. The area growth ratio of this memory structure is $O(N)$ instead of $O(N^2)$ like real multi-port memory.

4. IMPLEMENTATION

We have finished and fully simulated a prototype design of proposed memory structure with $1.2~\mu$ SPDM technology and will fabricated it through TSMC, in Taiwan. With the limitation in die size, we only implemented the half the required memory size for D = 16 and B = 16, that is, 1.5K (bytes) for Memory-A and 2K (bytes) for Memory-B. Each Memory-B bank consists of sixteen 16bytes' sub-banks like Fig.5. However, we place those sub-banks to be a sixteen-by-one linear array instead of a four-by-four mesh in actual layout so that we could eliminate unnecessary global routing and to share the write word line. The penalty of that is the need of additional column decoders and more heavy word line loading. By separating the read-write circuitry, the required one-read-one-write memory banks can be implemented. The prototype consists of eight-transistor, two-port, static-RAM cells and occupies 7.2 x 9.3 mm^2 chip. The simulation results using SPICE is shown in Fig.6, where four cycles: Write '1', Read '1', Write '0', Read '0' are performed. The layout of the core circuit is shown in Fig.7.

5. CONCLUSION

Observing the features of different data locality in video source coding algorithms, we proposed a two-level on-chip memory to moderate the conflict between capacity and flexibility existed in Video Signal Processor with multiple embedded processing elements. The memory adopted a scheme that duplicates all possibly used items into the corresponding memory banks to emulate the real hard-to-implement multi-port storage. It consists of one-read-one-write dual-port modules that can be set up different configurations according the ratio of Read/Writ desired by PE's. Therefore, it is very flexible and can fully support VSP applications. The memory organization is independent of the physical implementation of the one-read-one-write memory blocks.

A prototype 3.5-kbyte memory with two-level structure was designed and will be fabricated with 1.2-μm CMOS technology.

6. REFERENCES

[1] R. D. Jolly, "A 9-ns, 1.4-Gigabyte/s, 17-Ported CMOS Register File." *IEEE JSSC*, Vol. 26, NO. 10, pp.1407-1412. Oct. 1991.

[2] K.-I. Endo T. Matsumura, J. Yamada, "A Flexible Multiport RAM Compiler for Data Path." *IEEE JSSC*, Vol. 26, NO. 3, pp. 343-348, Mar. 1991.

[3] K.-I. Endo, T. Mastumura, J. Yamada, "Pipelined, Time-Sharing Access Technique for an Integrated Multiport Memory." *IEEE JSSC*, Vol. 26, NO 4, pp. 549-554, Apr. 1991.

[4] H. Yamauchi, Y. Tashiro, T. Minami, Y. Suzuki, "Architecture and Implementation of a Highly Parallel Single-Chip Video DSP." *IEEE Trans. Circuit and System for Video Technology*, Vol. 2, NO. 2, pp. 207-220, Mar. 1992.

[5] T. Murakami, K. Kamizawa, M. Kameyama, S.I. Nakagawa, "A DSP Architectural Design for Low Bit-Rate Motion Video Codec." *IEEE Trans. Circuit Syst.*, vol. 36, NO. 10, pp. 1267-1274, Oct. 1989.

[6] I. Tamitani, H. Harasaki, T. Nishitani, Y. Endo, "A Real-Time HDTV Signal Processor: HD-VSP." *IEEE Trans. Circuit and System for Video Technology*, vol. 1, NO. 1, pp. 35-41, Mar. 1991.

[7] H. B. Bakoglu, "Circuit, Interconnections, and Packaging for VLSI", Chap. 4, Addison-Wesley, 1990.

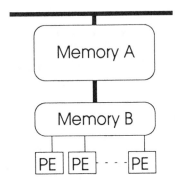

Figure 1. Two-level Memory Model for VSP with multiple processing units

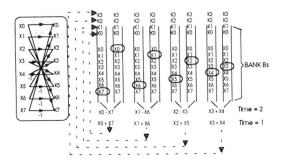

Figure 2. Butterfly-style memory access

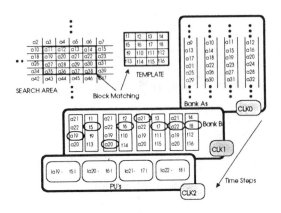

Figure 3. Blocking Matching for ME or VQ

All memory cells are dual-ported

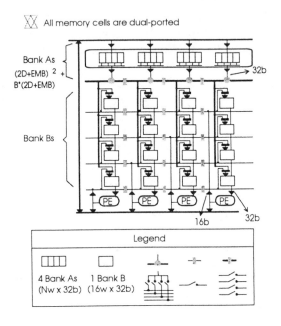

Figure 4. The proposed Memorry Organization

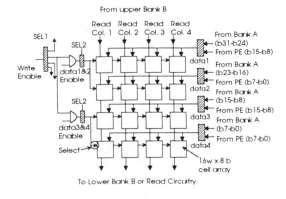

Figure 5. Single Memory Bank B

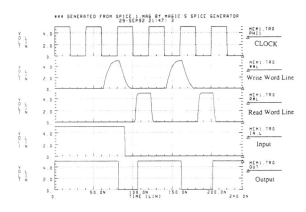

Figure 6. SPICE simulation results -- Four cycles: Write 1, Read 1, Write 0, Read 0 are shown respectively

setting sequence of SEL2	Result data in respevtive columns			
	col. 1	col. 2	col. 3	col. 4
00→01 →10→ 11	data4	data3	data2	data1
00→01	data2	data2	data1	data1
00→10	data2	data1	data2	data1
00→11	data2	data1	data1	data1

Table.2 Final results in the four columns of a B memory bank while control signal of SEL2 is change sequentially, and has a duration 1 clock.

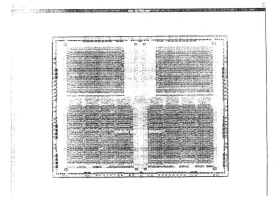

Figure 7. The layout of the prototype memory chip

Column (s) selected	1	2	3	4
SEL2=00	O	O	O	O
SEL2=01	O	O		
SEL2=10	O		O	
SEL2=11	O			

Table.1 The enabled column(s) in a B memory bank corresponding to different control signals of SEL2

Synchronous Clocking Schemes For Large VLSI Systems

Ahmed El-Amawy and Umasankar Maheshwar

Department of Electrical and Comptuer Engineering
Louisiana State University, Baton Rouge, LA 70803-5901

Abstract

Recently a novel clock distribution scheme called Branch-and-Combine(BaC) has been proposed. The scheme guarantees constant skew bound irrespective of the size of the clocked network. It utilizes simple nodes to process clock signals. The paper uses a VLSI model to compare the properties of the new scheme to those of the well established H-Tree approach. Our study considers clocking 2-D processor meshes of arbitrary sizes. We evaluate and compare the relevant parameters of both schemes in a VLSI layout context. We show that for each BaC network, there is a certain threshold size after which it outperforms the corresponding tree network. We utilize parameters such as *clock skew, link costs, node costs* and *area efficiency* as the basis for comparison.

1. Introduction

In synchronous control schemes, a single clock source is typically employed to provide a continuous stream of clocking events spaced equally in time(voltage transitions, pulses, .. etc.). Ideally, a clocking event should reach and affect all data processing units (processors or elements) at the same time. However, due to several factors such as different threshold voltages, different path lengths and buffer delays, the events usually reach the elements at different times. The difference in arrival times is called *clock skew*. The effects of clock skew on system performance have been studied in [1]-[3]. The problem clock skew poses is that it directly relates to the clocking rate. Larger skew requires slower clock rate for the system to operate correctly [1], [2]. If system A which is skew-free can safely operate with a minimum clock period of T, then a similar system B which suffers from skew $\leq \tau$ must use a clock period $T + \tau$. Clearly the problem is more significant in larger systems.

In this paper we utilize a VLSI model to evaluate and compare H-Tree and BaC clocking networks. For simplicity we only consider clocking a 2-D mesh of data processors. We evaluate the target networks based on a set of parameters such as link costs, clock skew, node costs and VLSI area. In Section 2 we review and point out certain basic features of Section 3 contains a comparative analysis of both clocking schemes. Section 4 contains our conclusion.

2. Basics of the Two Schemes

2.1 Tree clocking networks

In [2], Fischer and Kung introduced a model (Called Summation Model) in which delay variations from link to link (of equal length) are not ignored. They used the mdoel to analyze a tree clocking network and proved that clock skew is $\Omega(N)$, where N is the size (number of nodes) of the tree. In [4] Kugelmass and Steiglitz developed two probabilistic models in which the propagation delay on every source to processor path is the sum of independent contributions which are identically distributed. Their metric free model predicts skew upper bound to grow as $\Theta(N^{1/4} \log N^{1/4})$ where N is the number of clocked processors. In a VLSI layout context, a metric-free model would not be applicable, however.

Consider clocking a 2-D mesh of processors in a VLSI context using a binary tree. The leaves are assumed to clock the processors in one-to-one fashion. Hence, the leaves must be embedded in a 2-D mesh and their distances to the root should be equal as much as possible to minimize skew. These requirements are satisfied by an H-tree layout, [2].

Now we consider a *metric* model. The layout shown in Figure 1 is that of an H-Tree with 64 leaves clocking a (square) 2-D mesh with 64 processors. All nodes in the H-tree layout are identical containing a single buffer each. The leaves of the tree are marked with circles to indicate that they directly clock the pro-

† Supported in part by NSF grant no. MIP9117206 and in part by NSF/LEQSF under contract 1992-3-ADP-04.

cessors in the data network. The index within each leaf node indicates the processor clocked by the node. The intermediate nodes are marked by black spots whereas grid points which have not been used in the layout are marked with x's. It can be seen that the link length between a parent and its child is not constant. We now model the network as follows: 1) Delay in a link between two adjacent grid points = λ (grid points may be adjacent vertically or horizontally). 2) Delay in a node = λ (we consider the link delay to be equal to the delay in a single buffer). 3) $\lambda = x \pm a$ where a is the factor due to process parameter variations; and 4) Delay in a link is directly proportional to the length of the link. We assume the delay over a unit length is λ

Lemma 1: Under the above metric model, an Tree clocking network using H-Tree layout, clock skew is $O(\sqrt{N})$

Proof: It can be seen from the layout in Figure 1 that the link lengths between a parent and its children doubles every two levels as we traverse up the tree from the leaves. For the sake of convenience and without loss of generality, we assume that the size of the 2-D mesh = $mn = N$ is a power of 2. That is $N = 2^k$, where k is an integer. Therefore the size of the tree is $2N-1$ and the number of leaves is N. A tree with N leaves has $\log N + 1$ levels and the length of the path from the root to any leaf is $\log N$. Hence if $\log N$ is even, then the delay through this path will be given by the expression

$$(1+1+2+2+........+2^{\frac{\log N}{2}}+2^{\frac{\log N}{2}})(\lambda)+(\log N)(\lambda)$$

This expression reduces to [6]

$$(4\sqrt{N}-2+\log N)(\lambda)$$

Hence the worst case clock skew is given by

$$(4\sqrt{N}-2+2\log N)2a = (8\sqrt{N}-4+2\log N)a$$

For the case when $\log N$ is odd, it can be easily shown that skew is given by

$$(12\sqrt{\frac{N}{2}}-4+2\log N)a$$

Hence using H-Tree layout, Tree clocking network has a clock skew of $O(\sqrt{N})$. $\square$

2.2 BaC Networks

Branch and Combine (BaC) clocking has recently been introduced by El-Amawy [1], [7]. The most attractive property of BaC clocking is its ability to guarantee constant skew upper bound between any pair of directly communicating processors, regardless of network size. This is achieved by employing simple

clock nodes in the clock distribution network whose main function is to process clock signals such that skew is controlled within known bounds. The nodes also are responsible for clocking the data processors. Although a clock node could clock more than processor, we assume here that each node is assigned to a unique processor and each processor is assigned a unique node.

The principle on which BaC clocking is based is that the graph underlying the clock network is cyclic in nature. Each pair of adjacent nodes must be included in a cycle of finite length a node typically will have more than one input. Only the first arriving pulse (on any input) during any clocking event will trigger the node. Once triggered the node remains unresponsive (inert) to further inputs for a period $T_h > \Delta$, where Δ is the delay through a node and one of its output links. This guarantees that the node will be triggered once (by the first arriving pulse) per clocking event and that all subsequent input pulses belonging to the same event will be absorbed (since they are known to arrive within T_h units of time from the first [1]). The above also guarantees that the skew between the outputs of any two nodes in the same cycle is $\leq (L-1)\Delta$.

Figure 2 shows two examples of BaC networks for clocking a 2-D mesh of processors. These networks have been called the F_2 and F_4 networks, respectively [1], [7]. Note that F_x refers to a network where node Fan-in Fan-out=x. It has been shown in [1], [7] that for the F_2 and F_4 networks the maximum skew between the outputs of two adjacent nodes is 3Δ and Δ, respectively where $\Delta = $ (node + link) delay. In BaC networks node delay typically amounts to two gate delays [7], [1]. A third network (F_3) for clocking a 2-D mesh was reported in [1]. However since that network lies midway between the F_2 and F_4 networks, we will not consider it here. Hence for BaC networks $\Delta = 4\lambda$ since each link is assumed to span a distance of two grid points and node delay is 2λ [1]. Thus for the F_2 and F_4 networks maximum skew is 12λ and 4λ, respectively.

3. Comparative Analysis

To judge the merits/demerits of different clocking schemes in a VLSI context, it is necessary to evaluate their characteristic parameters and to work out cost-performance analysis under exactly the same set of conditions. In our study we assume that both clocking networks are designed to clock the same data network. Therefore the locations of the nodes which directly clock the processors are fixed. That is, if we superimpose a BaC network layout on the H-Tree layout, then the nodes of the BaC network will exactly lie on the nodes which directly clock the processors in the H-Tree layout (these are the circled nodes in Figure 1). We will perform simple comparative analysis based on *Clock skew, Link costs, Node costs, Area efficiency, and Maximum edge length*. Before we get into the analysis, we establish a relationship between the values of x and a where $\lambda = x \pm a$. We introduce a new parameter k called the *variation ratio*, where $k = x/a$. We will see shortly in the analysis that this factor plays an important role in network evaluations.

3.1 Clock Skew

We call into focus the expressions for clock skew given earlier. Skew comparisons between the two BaC networks and the H-Tree are listed in Tables 1 and 2. For each table the value of clock skew is calculated for different values of k and N. $k = 1$ implies that delay variation is 100% and $k = 10$ implies that variance is 10%. In an actual VLSI implementation we consider the range $1 \leq k \leq 10$ to represent a valid and complete interval for all practical purposes. Our aim now is to identify the threshold level at which clock skew of the H-Tree clocking network becomes greater than that of a BaC network. We call this size the *threshold size* indicated by $N_{cs}(k,f)$ which depends on both k, the variation ratio, and f, the fan-out.

From Tables 1 and 2 we can see that $N_{cs}(10,2) = 256$, $N_{cs}(10,4) = 32$ and $N_{cs}(10,2) = 256$. Note that $N=256$ corresponds to a 2-D mesh of size 16×16 which is a reasonably small size. When we look at the threshold sizes, we notice that they are not large and therefore, it is reasonable to assume that to clock medium to large data networks, BaC networks always outperform the H-Tree in terms of clock skew.

3.2 Link Costs

If the underlying technology used in the VLSI implementation is the same for both networks, we can assume that the cost of a link is proportional to its length. Let the cost of one link(one unit of the grid) be l_c. Let $\log N$ be even. Then $l_c \times$. Total-link-length will give the cost of the network. In the F4 network [1], [7] it can be seen that there are 4 nodes with node degree 4, $4(\sqrt{N} - 2)$ nodes with node degree 6 and the rest have node degree 8. Therefore total number of links
$= 4 \times 4 + 4 \times (\sqrt{N} - 2) \times 6 + (N - 4 - (4 \times (\sqrt{N} - 2))) \times 4/2$.
This works out to $4 \times (N - \sqrt{N})$. Therefore, for the

F4 network Total cost $= 4 \times (N - \sqrt{N}) \times l_c$. For the F2 network Total cost $= 2 \times (N - \sqrt{N}) \times l_c$.

In the case of H-Tree layout we have already observed that link lengths double every two levels. But it is also true that the number of links reduces by a factor of 2 every level. Therefore if we write the values in the form of a series, starting from the leaves and going up level by level the picture looks like
Link length $\rightarrow$ 1 1 2 2 4 4 8 8.....
No. of links $\rightarrow N\dfrac{N}{2} \dfrac{N}{4} \dfrac{N}{8} \dfrac{N}{16} \dfrac{N}{32} \dfrac{N}{64} \dfrac{N}{128}$.....
As the cost of the links is the summation of the cost of links at each level, we get
Link costs $= (N +$
$\dfrac{N}{2} + \dfrac{N}{2} + \dfrac{N}{4} + \dfrac{N}{4} + \dfrac{N}{8} + \dfrac{N}{8} + \dfrac{N}{16} ..... + \sqrt{\dfrac{N}{2}} + \sqrt{N})$.
The expression reduces to $3(N - \sqrt{N})$. It can be seen from the plots shown in Figure 3 that, the function for the Link cost for the Tree clocking network grows faster than that for the F2 network but slower than that for the F4 network. It can also be observed that the Link cost for any of the networks is $\Theta(N)$.

3.3 Node Costs

In the Tree clocking network, each node consists of a single buffer. In BaC networks, each node is likely to consist of 10-20 gates [1], depending on the specific design. From the structure of the two networks it can be seen that the total number of nodes in a tree clocking network $= 2N - 1$ and the total number of nodes in a BaC network $= N$. Though the number of nodes in the tree clocking network is double that of BaC networks, cost of a node in BaC network offsets this factor. Hence we can say that the total node cost of the tree clocking network will be less than that of BaC networks, perhaps by a factor of 10.

3.4 Area Efficiency

Conventionally, area efficiency of a layout is termed as the ratio of the number of grid points utilized by the topology of the network being embedded into the VLSI grid [5]. Here, the total number of grid points $= (2\sqrt{N} - 1)^2 = (4N - 4\sqrt{N} + 1)$. Therefore for the Tree clocking network, Area efficiency $= \dfrac{2N - 1}{4N - 4\sqrt{N} + 1} = 50\%$ when $N >>$. For the F2 and F4 networks, all the grid points are utilized as either nodes or as connection points. Therefore, for these BaC networks Area efficiency $= 1$ which means 100% utilization. Clearly BaC networks are much more area efficient compared to the Tree clocking network.

3.5 Maximum edge length

In a VLSI layout, it is very important to analyze the maximum edge length in the structure. The reason is that, VLSI design is limited by the fact that two pulses cannot physically exist on the same wire at the same time [2]. Due to this limitation, the clock rate of the network becomes dependent on the maximum edge length. A plot is shown in Figure 4 which illustrates this and compares the maximum clock rate for both the schemes. For the H-Tree, when network size becomes very large, the maximum clocking rate has to be lowered substantially to ensure correct operation of the network. This makes BaC networks more efficient in that regard since for BaC networks, the maximum clocking rate is independent of network size.

4. Conclusion

We have comapred BaC clocking to H-Tree clocking in a VLSI context. It has been shown that for clocking a 2-D mesh of processors there is a threshold size beyond which a BaC network will outperform the H-Tree in terms of skew. For small to medium sized networks (64-256 nodes), H-Tree may be a better choice. For larger sized networks BaC clocking outperforms the H-Three in terms of skew, maximum edge length and area efficiency. However, the H-Tree is less costly in terms of node costs for any size network. Insofar as link costs are concerned, both types of networks have comparable costs.

References

1. A. El-Amawy, "Clocking arbitrarily large computing structures under constant skew bound," *IEEE Transactions on Parallel and Distributed Systems*, vol. 4, no.3, pp 241-255, March 1993.

2. A.L. Fisher and H.T. Kung, "Synchronising Large VLSI Processor arrays,"' *IEEE Transactions on Computers,* vol.c-34: pp.734-740, August 1985.

3. M. Afghani and C. Svenson, "Performance of Synchronous and Asynchronous Schemes for VLSI Systems," *IEEE Transactions on Computers,* vol.41, no.7, July 1992.

4. S. Kugelmass and K. Steiglitz, "An Upper Bound on Expected Clock Skew in Synchronous Systems," *IEEE Transactions on Computers,* vol.39, no.12: pp. 1475-1477, Dec 1990.

5. H. Y. Youn and A. D Singh, "On Implementing Large Binary Tree Architecture in VLSI and WSI," *IEEE Transactions on Computers,* vol.38, no.4, April 1989.

6. M. R. Spiegel, "Mathematical handbook of formulas and tables," *McGraw Hill,* N.Y, 1986.

7. A. El-Amawy, "Branch-and-combine clocking of arbitrarily large computing networks," *Proc.Intl.Conf. on Parallel Procesing,* St.Charles, IL, pp.I-409-417, Aug 1991.

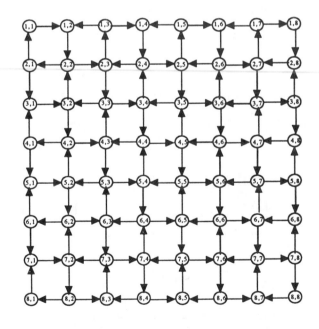

Figure 2(a): An F2 BaC clocking network for a 2D mesh of size 64.

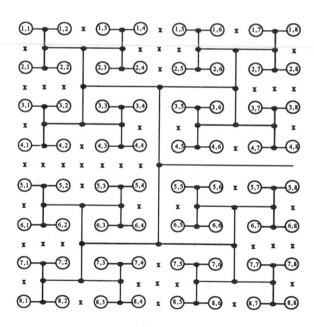

Figure 1: Tree clocking network for a 2D mesh of size using H-Tree layout.

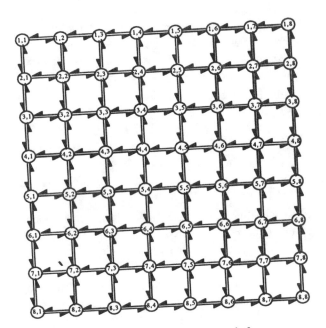

Figure 2(b): An F4 BaC clocking network for a

N→ k↓	1	2	4	8	16	32	64	128	256
1	4 / 8	10 / 8	14 / 8	24 / 8	36 / 8	54 / 8	72 / 8	106 / 8	140 / 8
2	4 / 12	10 / 12	14 / 12	24 / 12	36 / 12	54 / 12	72 / 12	106 / 12	140 / 12
3	4 / 16	10 / 16	14 / 16	24 / 16	36 / 16	54 / 16	72 / 16	106 / 16	140 / 16
4	4 / 20	10 / 20	14 / 20	24 / 20	36 / 20	54 / 20	72 / 20	106 / 20	140 / 20
5	4 / 24	10 / 24	14 / 24	24 / 24	36 / 24	54 / 24	72 / 24	106 / 24	140 / 24
6	4 / 28	10 / 28	14 / 28	24 / 28	36 / 28	54 / 28	72 / 28	106 / 28	140 / 28
7	4 / 32	10 / 32	14 / 32	24 / 32	36 / 32	54 / 32	72 / 32	106 / 32	140 / 32
8	4 / 36	10 / 36	14 / 36	24 / 36	36 / 36	54 / 36	72 / 36	106 / 36	140 / 36
9	4 / 40	10 / 40	14 / 40	24 / 40	36 / 40	54 / 40	72 / 40	106 / 40	140 / 40
10	4 / 44	10 / 44	14 / 44	24 / 44	36 / 44	54 / 44	72 / 44	106 / 44	140 / 44

Clock skew values are in units of "a" $\dfrac{\text{H-tree}}{\text{F4}}$

Table 1: Max skew for H-Tree Vs. F4.

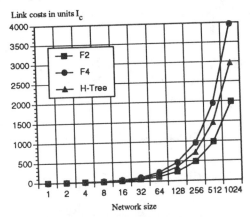

Figure 3: H-Tree Vs. BaC networks based on Link cost*

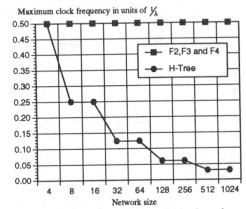

Figure 4: H-Tree Vs. BaC networks based on maximum clock frequency.

N→ K↓	1	2	4	8	16	32	64	128	256
1	4 / 24	10 / 24	14 / 24	24 / 24	36 / 24	54 / 24	72 / 24	106 / 24	140 / 24
2	4 / 36	10 / 36	14 / 36	24 / 36	36 / 36	54 / 36	72 / 36	106 / 36	140 / 36
3	4 / 48	10 / 48	14 / 48	24 / 48	36 / 48	54 / 48	72 / 48	106 / 48	140 / 48
4	4 / 60	10 / 60	14 / 60	24 / 60	36 / 60	54 / 60	72 / 60	106 / 60	140 / 60
5	4 / 72	10 / 72	14 / 72	24 / 72	36 / 72	54 / 72	72 / 72	106 / 72	140 / 72
6	4 / 84	10 / 84	14 / 84	24 / 84	36 / 84	54 / 84	72 / 84	106 / 84	140 / 84
7	4 / 96	10 / 96	14 / 96	24 / 96	36 / 96	54 / 96	72 / 96	106 / 96	140 / 96
8	4 / 108	10 / 108	14 / 108	24 / 108	36 / 108	54 / 108	72 / 108	106 / 108	140 / 108
9	4 / 120	10 / 120	14 / 120	24 / 120	36 / 120	54 / 120	72 / 120	106 / 120	140 / 120
10	4 / 132	10 / 132	14 / 132	24 / 132	36 / 132	54 / 132	72 / 132	106 / 132	140 / 132

Clock skew values are in units of "a" $\dfrac{\text{H-tree}}{\text{F2}}$

Table 2: Max skew for H-Tree Vs. F2.

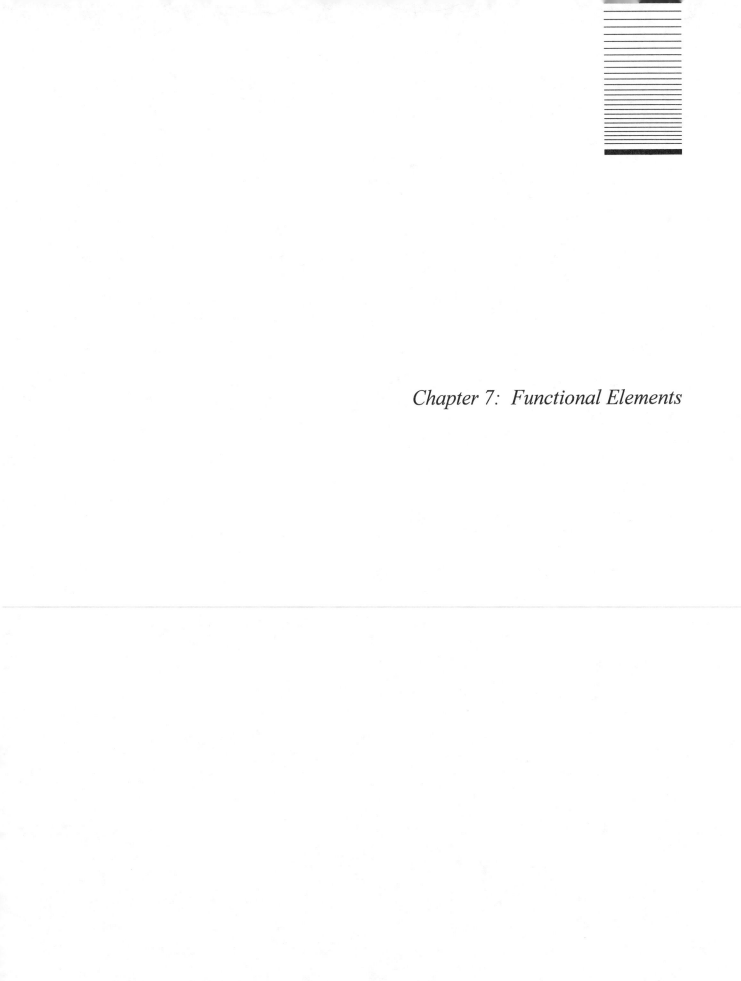

Chapter 7: Functional Elements

Functional Elements

"Functional elements" is the name applied to those implementations that execute an entire class of algorithms, an end-to-end algorithm, or a significant algorithm component. They are the most application-specific of the technology elements discussed so far, and they may be used in isolation or in combination with other such higher-level building blocks.

First among these is "High Speed Arithmetic Coder/Decoder Architectures" (paper 7.1), which introduces the interval tree search method as a means to realize arithmetic coders and examines the hardware complexity of several variations in architecture. The more common Huffman coding, used as part of the MPEG and JPEG standard image encoding schemes, is addressed in "Area Efficient VLSI Architectures for Huffman Coding" (7.2). The paper proposes a uniprocessor single chip approach and compares it to an existing implementation that uses massively parallel processing elements. The discussion of coding concludes with "An Efficient Prime-Factor Algorithm for the Discrete Cosine Transform and Its Hardware Implementations" (7.3). This paper describes a method to transform a large one-dimensional problem into a simpler multidimensional version by using a remapping of indices, and it illustrates the method by converting a 15-point discrete cosine transform to a 3x5 point two-dimensional transform.

Image analysis applications often rely upon accurate edge detection early in their processing. "Edge Detection Using Fine-Grained Parallelism in VLSI" (7.4) applies a fine-grained processor mesh to both the Laplacian and Sobel edge detection operators, following with a Hough transform to detect lines within an image. The Hough transform is also calculated in real time to detect road lane boundaries in "Compact Hardware Realization for Hough Based Extraction of Line Segments in Image Sequences for Vehicle Guidance (7.5). The single board system, which operates at up to 25 frames per second, has been used to guide a vehicle in real traffic conditions.

Comparing the similarity of two patterns can require the real-time stretching and compressing of the test pattern to best align its features to those of the reference pattern. "Implementing Dynamic Programming Algorithms for Signal and Image Processing on Array Processors" (7.6) examines dynamic time warping, boundary following, and the Viterbi algorithm under the assumption that the appropriate cost (distance) function is computed elsewhere and downloaded. These algorithms derive their implementation complexity from the fact that they are data dependent and require the tracking of several possible mapping paths until the comparison is complete and the winning path is selected. "Hybrid Survivor Path Architectures for Viterbi Decoders" (7.7) combines the register-exchange and trace-back methods of updating and decoding the survivor paths and estimates the implementational complexity.

The vector quantization method of signal coding requires the real-time conversion of an incoming sequence to the format of a codebook of precomputed vectors, the comparison of that input vector to each (or for limited search methods, many) of the codebook entries, and the output of the top candidate index, all within the period of an analysis frame that is short compared to the timescale of signal variations. A vector quantizer integrated circuit has been built that can accommodate a codebook of over 1024 vectors for real-time telecommunications-bandwidth speech processing, as described in "A High Performance Vector Quantization Chip" (7.8). Analog signal processing is used to implement an image vector quantizer chip in "An 8b CMOS Vector A/D Converter" (7.9).

The interest in wavelet transforms has led to application-specific integrated circuits for their real-time execution. A comprehensive review of systolic arrays, parallel filter banks, and single-instruction multiple data path arrays for performing both the discrete and continuous wavelet transform is provided in "Architectures for Wavelet Transforms (7.10). Efficient implementation of parallel filter banks has been addressed as part of subband coding of speech and images prior to the recent emphasis on wavelets. "Memory-I/O Tradeoff and VLSI Implementation of Lapped Transforms for Image Processing" (7.11) extends the work of subband coding architectures, exploring several architectural alternatives and proposing a single chip implementation capable of decoding for high-definition television (HDTV).

Like wavelets, fuzzy logic has begun to receive the attention of special-purpose architectural efforts. In "7.5 MFLIPS Fuzzy Microprocessor Using SIMD and Logic-in-Memory Structure" (7.12), a pair of integrated circuits have been designed and fabricated that partition the "if" portion and the "then" portion of fuzzy inference into separate chips. A capability of performing real time processing for up to 16 input variables, 960 rules, and 16 output variables is achieved.

The efficient programming of signal processing algorithms upon established architectures is not often addressed in a systematic manner. "Implementation of Several RLS Nonlinear Adaptive Algorithms Using a Commercial Floating Point Processor" (7.13) addresses this area for the Motorola DSP 56K floating point signal processor, describing how to perform real-time division and square roots, exploring various precision of floating point operations, and comparing several software architectures for the RLS adaptive filtering algorithm.

Suggested Additional Readings:

[1]   P. G. Howard and J. S. Vitter, "Arithmetic Coding for Data Compression," *Proc. IEEE,* vol. 82, no. 6, pp. 857-865, June 1994.

[2]   Y. H. Hu, "Cordic-Based VLSI Architectures for Digital Signal Processing," *IEEE Signal Processing Magazine,* vol. 9 no. 3, pp. 16-35.

[3]   E. A. Lee, "Programmable DSP Architectures: Part I," *IEEE ASSP Magazine,* vol 5, no. 4, pp. 4-19, Oct. 1988.

[4]   H. Choi, W. P. Purleson, and D. S. Phatak, "Optimal Wordlength Assignment for the Discrete Wavelet Transform in VLSI," *VLSI Signal Processing VI* (The Netherlands) November 1993, pp. 325-333.

[5]   D. J. Ostrowski, P. Y. K. Cheung, and K. Roubaud, "An Outline of the Intuitive Design of Fuzzy Logic and its Efficient Implementation," *Second IEEE International Conference on Fuzzy Systems* (San Francisco, CA) Mar. - Apr. 1993, pp. 184-189.

HIGH-SPEED ARITHMETIC CODER/DECODER ARCHITECTURES

Gireesh Shrimali and Keshab K. Parhi

Department of Electrical Engineering
University of Minnesota, Minneapolis, MN 55455

ABSTRACT

The state of art in data compression is arithmetic coding, not the better known Huffman method. To a unique data string, arithmetic coding technique assigns a unique fractional value on the number line between 0 and 1. The speed of this algorithm is limited because of its inherent recursive nature. In this paper we present the design of fast decoders using a novel *interval tree search* method. The decoder can be modeled as a FSM (finite state machine), enabling the application of *look-ahead* technique to achieve higher speeds. Look-ahead approach leads to slight degradation in performance (in terms of the adder/subtractor delay in the coder/decoder due to increased word lengths). We improve the performance of the decoder by using *redundant arithmetic*. The tree search method combined with redundant arithmetic and look-ahead leads to desired speed-ups without any degradation in performance.

1. INTRODUCTION

Arithmetic coding is the best available technique for reducing the encoded data rate [Elias, in an unpublished paper]. It dispenses with the restriction that symbols must translate into integral number of bits. By using real codeword lengths it achieves the theoretical entropy bound to compression rate.

The feedback associated with the arithmetic coder/decoder limits the speed of the algorithm. The computation associated with feedback loops imposes an inherent upper bound on the available iteration rate, referred to as the iteration bound [2]. The speed of the algorithm can be improved by modeling the coder/decoder as a finite state machine (FSM) and applying *look-ahead* techniques [3]-[6].

In this paper, we combine the techniques of *interval tree search*, *look-ahead*, and *redundant arithmetic* to design high speed architectures for arithmetic decoders. The tree search can be modeled as a FSM and can be implemented using combinational logic or PLAs [5]. The speed of the decoder can be improved by using look-ahead technique alongwith parallel processing. This leads to degradation in performance as the wordlength for the arithmetic operations increases. The decoder performance can be improved by using redundant arithmetic [7] in place of two's complement arithmetic. Since addition/subtraction in redundant arithmetic is independent of the wordlength [8]-[9], we achieve the required speed-up.

This research was supported by the Office of Naval Research under contract number N00014-91-J-1008.

The organization of this paper is as follows. Section 2 presents the basic arithmetic coding/decoding algorithms. Section 3 presents high speed encoder design, using look-ahead. Section 4 presents high speed decoder architectures, using interval tree search, look-ahead, and redundant arithmetic.

2. ARITHMETIC CODING

In arithmetic coding, a message is represented by an interval of real numbers between 0 and 1. As the message becomes longer, the interval needed to represent it becomes smaller, and the number of bits to specify that interval grows. Successive symbols of the message reduce the size of the interval in accordance with the symbol probabilities. The more likely symbols reduce the range less than the less likely symbols and hence add fewer bits to the message [10].

2.1. ENCODER DESIGN

By restricting the fractional parts of the code word lengths $(l(k)'s)$ to come from a limited set [11], we can achieve a finite state machine (FSM) formulation of the coding process [1]. Letting $s=s_1, s_2, \ldots, s_n, s_i \in \{0, 1, \ldots, N-1\}$ to be a string and k to be the next symbol to be coded, the coding process can be defined by

$$
\begin{aligned}
C[a, sk] &= C[a, s] + A[x(s), k]2^{-Y(s)} \\
x(sk) &= z[x(s), k] = F(1 + x - F(l(k))) \\
Y(sk) &= Y(s) + y[x(s), k] = I(l(k) + z[x, k]) \quad (1)
\end{aligned}
$$

where $x(\emptyset) = a$, $Y(\emptyset) = C(a, \emptyset) = 0$, and $\emptyset$ is the empty string. In addition, $z[x, k]$ and $y[x, k]$ represent the state transition and output functions, respectively, of the FSM. Here $F(.)$ and $I(.)$ denote the fractional and integer parts, respectively, of a real argument. The $A[x, k]'s$, also referred to as *augends*, depend on the present state x and the length of the present symbol. These real numbers (< 2), having r fractional bits, are calculated a-priori and stored in a look-up table [1].

2.2. DECODER DESIGN

The decoder tries to imitate what the encoder has gone through. The decoding process can be described by using a similar FSM to the one described by (1). For string s (section 2.1) consider the following state trajectory

$$
a = a_0 \xrightarrow{s_1/y_1} a_1 \xrightarrow{s_2/y_2} a_2 \ldots \xrightarrow{s_n/y_n} a_n = x(s)
$$

Also, from (1), by iteration we get

$$C[a,s] = A[a,s_1] + A[a_1,s_2]2^{-y[a,s_1]} + \cdots$$
$$+ A[a_{n-1},s_n]2^{-y[a,s_1]-\cdots-y[a_{n-1},s_n]}. \quad (2)$$

Under suitable selection of y's and A's, the first term in (2) will be greater than the sum of others. Then the left most symbol s_1 can be decoded for the largest index k for which

$$A[a,k+1] > C[a,s] \geq A[a,k]. \quad (3)$$

The decoding process can be continued by the equation

$$C[z[a,s_1],s_2,\ldots,s_n] = (C[a,s] - A[a,s_1])2^{y[a,s_1]} \quad (4)$$

where y and z have the same transitions as described by (1).

3. HIGH SPEED ENCODER USING LOOK-AHEAD

Since the basic decoding equations allow us to represent the encoder as a FSM, application of look-ahead to achieve higher speeds becomes easy [3]-[4]. Applying two steps of look-ahead to the coding equation (1), for symbols k and k', we get

$$C[a,skk'] = C[a,s] + 2^{-Y(s)}A[x(s),kk']$$
$$x(skk') = F(2 + x(s) - F(l(kk')))$$
$$Y(skk') = Y(s) + I(l(kk') + x(skk')). \quad (5)$$

where $A[x(s),kk'] \stackrel{def}{=} [A[x(s),k] + A[z[x(s),k],k']2^{-y[x(s),k]}]$ and $l(kk') \stackrel{def}{=} l(k) + l(k')$.

Now, by storing $A[x,kk']$ and $l(kk')$ for all values of k and k', we achieve an encoder representation which processes two symbols at a time with the same time complexity it used to encode one symbol with (1). Thus, we achieve a speed-up of two. By applying another stage of look-ahead to (5), i.e., by including symbol k'' in addition to k and k', we get the following representation.

$$C[a,skk'k''] = C[a,s] + 2^{-Y(s)}A[x(s),kk'k'']$$
$$x(skk'k'') = F(3 + x(s) - F(l(kk'k'')))$$
$$Y(skk'k'') = Y(s) + I(l(kk'k'') + x(skk'k'')) \quad (6)$$

where $l(kk'k'') \stackrel{def}{=} l(k) + l(k') + l(k'')$ and $A[x(s),kk'k''] \stackrel{def}{=} A[x(s),kk'] + A[z[x(s),k],k'],k'']2^{-(y[z[x(s),k],k']+y[x(s),k])}$. By storing $A[x(s),kk'k'']$ and $l(kk'k'')$ for all values of k, k' and k'', we achieve an encoder representation which processes three symbols at a time with the same time complexity it used to encode one symbol with (1). Thus, we achieve a speed-up of three. Higher speed-ups can be similarly achieved though with the increasing hardware penalty in proportion to N^M where N is the number of symbols in an alphabet and M is the steps of look-ahead.

4. HIGH SPEED DECODER DESIGN

The number of left most bits, of the code string, required for (4) is at most $r+1$. This enables the decoder to work on parts of code string instead of waiting for the complete code string to arrive, making real time processing possible. Decoding of a symbol consists of two parts. First, the decoder searches for symbol k by (3) and then modifies the code string using (4). A general decoder architecture is shown in Fig. 1. The buffer holds the bits of the code-string modified by the bit-stuffing removal circuit. The tree search operates on the first few bits of the $r+1$ bit code string part and on average tracks down the desired $A[x,k]$ in time much less than r. Once a symbol is decoded, the modification circuit processes the $r+1$ bits and modifies C, y, and x according to (4) and (1). The A's are obtained from the same table used by the encoder. Looking at (4), we see that $r+1-y$ bits remain after the subtraction operation. These are fed back to the buffer which appends y more bits to the right of these $r+1-y$ bits to obtain a code string of $r+1$ bits. The whole process is repeated for decoding of the next symbol.

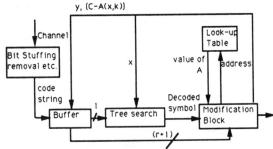

Figure 1. The Decoder Block Diagram

4.1. INTERVAL TREE SEARCH METHOD

The tree search uses different *interval trees* for different x's. Every node in the tree has an interval (range) associated with it. The range of a parent node is equally shared by its children. The transition between a parent and a child can be represented by a bit. Depending upon the present node, the next bit changes the range associated with the input bit stream (the more the depth of a node the less is the range associated with it). This concept can be used to identify the $R[x,k]$'s ($R[x,k] \stackrel{def}{=} [A[x,k+1],A[x,k])$) for decoding symbol k. The tree search stops once a node's range falls inside one of the R's. Such nodes are referred to as terminal nodes (or outputs) whereas the remaining nodes are referred to as intermediate nodes (or states).

In practice, we track down to a node whose range corresponds to parts of two adjoining R's. This combined interval, i.e., $R[x,k] \bigcup R[x,k+1]$, is much bigger than any of the component R's . Therefore, the depth of the interval tree becomes less and decoding is faster. The ambiguity arising from this approach is resolved by the modification circuit. Now it uses two similar circuits for k and $k+1$. By tree search we have ensured that $A[x,k] \leq C < A[x,k+2]$. If $C - A[x,k+1]$ is positive then symbol $k+1$ is decoded, otherwise symbol k is decoded. Fig. 2 shows the modified interval search tree for a typical four symbol alphabet. The terminal nodes can represent either one symbol or a pair of symbols. A symbol can be represented by many terminal nodes and the decoding length for the symbol (measured in bits) is taken as the average of the depths of all such nodes. The average decoding length per symbol for the tree is given by $\sum p(i)l_d(i)$ where $l_d(i)$ is the decoding length for symbol i. Note that the average decoding length of 3.22 bits/symbol is much less than the wordlength of 8.

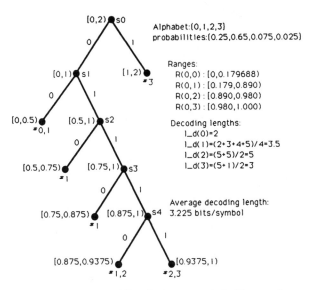

Alphabet:(0,1,2,3)
probabilities:(0.25,0.65,0.075,0.025)

Ranges:
R(0,0) : [0,0.179688)
R(0,1) : [0.179,0.890)
R(0,2) : [0.890,0.980)
R(0,3) : [0.980,1.000)

Decoding lengths:
l_d(0)=2
l_d(1)=(2+3+4+5)/4=3.5
l_d(2)=(5+5)/2=5
l_d(3)=(5+1)/2=3

Average decoding length:
3.225 bits/symbol

Figure 2. Binary Interval Tree (nodes are marked with ranges)

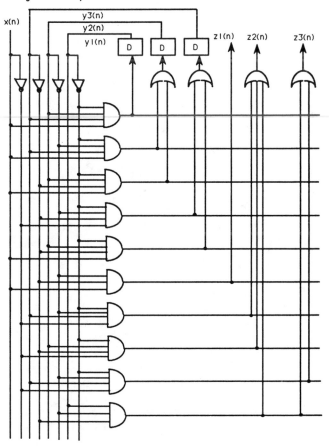

Figure 3. Minimally coded hardware architecture (y's and z's are coded bits representing states and outputs respectively)

The tree search can be implemented using a finite state machine where the states of the machine represent the intermediate nodes and outputs represent the terminal nodes (or symbols decoded) [5],[12]. Fig. 3 shows a minimal hardware implementation of the sequential decoder corresponding to the tree of Fig. 2 [13]. Note that the FSM has five states and four outputs and it processes one bit per cycle.

A simple mapping of the tree onto hardware would involve assigning AND gates to branch transitions and OR gates for the state s_0 and outputs. This approach does not minimize the hardware but gives a rough estimate just by looking at the tree under consideration. We will use this approach in later sections.

The time complexity of the search tree method is proportional to the average decoding length. The time complexity of the modification circuit is proportional to r, the codeword length. Since both circuits use combinatorial logic, the overall complexity can be taken as proportional to $l_{avg}(decoding) + r$, assuming that both use the same clock (i.e., these maintain the same bit processing time).

4.2. HIGH SPEED DECODER ARCHITECTURES USING LOOK-AHEAD

We observe that (4) and (1) describe the decoder as a finite state machine. This enables us to apply look-ahead to design high speed modification block in a way similar to section 3 [13].

4.2.1. HIGH SPEED TREE SEARCH

Since we use $A[x, kk'k'']$ and $A[x, kk']$ in place of $A[x, k]$, the search trees have to be constructed for composite A's. This increases the average decoding length (and time) for the binary tree. This problem can be alleviated using higher order trees [12]. A higher order tree processes multiple bits at a time and thus is faster than the binary tree. For e.g., a quaternary tree processes two bits at a time and is twice faster than the binary tree. An octal tree, processing three bits at a time, is three times faster. In general, the speed-up relates the average decoding times of the higher order tree used for the extended codes (using look-ahead) and the binary tree used for the original code. The speed-up for the search process can be defined as

$$Speed - Up = \frac{t_{avg}(2)}{t_{avg}(h)}$$
$$= \frac{l_{avg}(2)}{l_{avg}(h)/(\log_2(h) * n)} \quad (7)$$

where h and n denote the order of the higher order tree and number of symbols decoded simultaneously, respectively. The $t_{avg}(.)$'s and $l_{avg}(.)$'s denote the average (per symbol) decoding times and lengths for respective trees.

For example, for the alphabet used in Fig. 2 the binary tree has an average decoding length of 3.22 bits/symbol and it processes one bit at a time to decode a symbol. The quaternary tree (h=4) for the same alphabet for the extended code (one step of look-ahead) processes two bits at a time and has an average decoding length of 5.64 bits/(two symbols). The speed-up, therefore, is given by $\frac{t_{avg}(2)}{t_{avg}(4)} =$

$\frac{l_{avg}(2)}{l_{avg}(4)/(\log_2(4)*2)} = \frac{3.22}{5.64/(2*2)} = 2.28$. We observe that the speed of the search circuit increases by a factor greater than two and the speed of modification circuit increases by two; thus the overall speed-up for the decoder is at least two, as desired.

Similarly, the octal code ($h = 8$) for the same alphabet for the extended code, using two steps of look-ahead, processes three bits at a time to decode three symbols. The average decoding length is 8.55 bits/(three symbols), and the speed up is given by $\frac{l_{avg}(2)}{l_{avg}(8)} = \frac{l_{avg}(2)}{l_{avg}(8)/\log_2(8)*3} = 3.3947$. In this way, the speed of the search circuit increases by a factor greater than three and the speed of the modification block increases by three. Therefore, the overall speed up is at least three, as desired.

Since we use composite A's, there is a degradation in the speed of the modification circuit because of increased word lengths. The word lengths for the extended codes involving one and two steps of look-ahead now become $r + max(y)$ and $r + 2 * max(y)$ respectively, in place of r. As the time complexity of the modification circuit depends on the word length, increased word lengths result in increased delay in the arithmetic unit.

A solution to this problem is to use *redundant arithmetic* in place of ordinary arithmetic. This approach is described next.

4.2.2. REDUNDANT ARITHMETIC

In redundant arithmetic, each digit can take values from the set $\{-1,0,1\}$. Using MSD first approach [8]-[9], it can be seen that an adder/subtractor has a latency independent of the wordlength. The modification unit takes a constant time to operate on any number of bits and we achieve the desired speed-up every time look-ahead is applied. An efficient tree search method can be developed in a way similar to the one used in section 3. Since the digits come from a different set than $\{0,1\}$, every parent has three children and the ranges of the nodes are modified accordingly. The redundant number search tree corresponding to Fig. 2 is shown in Fig. 4.

By extending the idea of higher order trees to redundant arithmetic, we can develop 9-ary and 27-ary trees in analogy to the quaternary and octal trees used for 2's complement arithmetic. These trees would process two and three digits respectively, at a time and can be used to decode the extended codes. The 9-ary tree corresponding to the ternary tree in Fig. 4 is shown in Fig. 5. Notice that for every node, the three left and three right children are shared by its left and right neighbors respectively. In addition, every node has only seven children nodes since two pairs of branches go to the same nodes. Thus, every node has only one unique child.

Overall decoder time complexity using redundant arithmetic is given in table 1 alongwith speed-ups and comparisons to decoder time complexities for ordinary arithmetic. It is assumed that the original code uses a ternary tree for decoding and the extended codes use 9-ary trees for decoding. Ordinary arithmetic refers to two's complement arithmetic used in previous sections. The decoding times are presented in terms of number of internal cycles and we see that use of redundant arithmetic not only gives better performance than ordinary arithmetic for original code, but

also a speed-up of over two and over three using look-ahead. Note that computations of both sequential and extended decoders make use of redundant arithmetic.

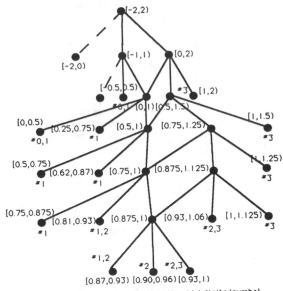

Average decoding length= 4.14 digits/symbol

Figure 4. Redundant arithmetic search tree

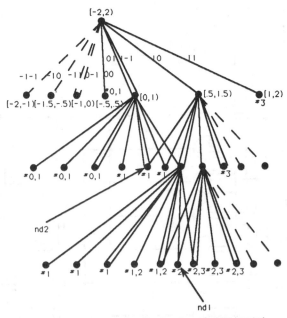

Average decoding length = 4.84 bits/symbol

Figure 5. 9-ary Interval Tree

A simple hardware mapping of these trees is possible using the approach described in section 4.1 and the hardware cost can be estimated. The number of AND and OR gates can be determined directly from the trees. Different trees use different types of multiple input gates. The number of inputs for the AND gates depends on the type of tree being used. In a ternary tree, a branch transition is indicated by a digit, and therefore, the AND gates employed for decoding a ternary tree would have two inputs each :one corresponding to the node and the other corresponding to the branch digit. Similarly, a 9-ary tree would require 3-input AND gates. The OR gates will be used for the s_0 state, outputs and the internal nodes having multiple inputs. Table 2 gives the hardware complexity for various codes. So as to get a rough comparison of the hardware costs, we try to approximate a p input gate by $(p-1)$ 2-input gates with a latency of $\lceil log_2 p \rceil$ 2-input gates. Here we assume a power-of-two decomposition of a multiple input gate. Table 3 shows the hardware complexity for the trees used for decoding various codes, in terms of 2-input gates. The reader should keep in mind that these figures give a very gross comparison of the hardware costs.

Table 1. Decoder time complexity for various codes using redundant arithmetic

expt. no.	code type	state x	Time complexity(per symbol)		
			search	modify	total
#1	original	0	4.14	3	7.14(11.22)
		1	3.21	3	6.21(10.38)
	extended	0	1.76	1.5	3.26
		1	1.62	1.5	3.12
	extended-2	0	1.42	1	2.42
		1	1.32	1	2.32
#2	original	0	6.33	3	9.33(21.74)
		1	6.03	3	9.03(21.85)
	extended	0	2.62	1.5	4.12
		1	2.49	1.5	3.99
	extended-2	0	2.32	1	3.32
		1	2.22	1	3.22

Table 2. Hardware complexity for various codes (x=0)

Experiment number	code type	hardware complexity(number of gates)	
		AND	OR
#1 (4 symb)	original	28(2)	1(15),5(2),1(5),1(6),1(8)
	extended	197(3)	1(82),28(2),27(3),1(4),3(6),6(7),3(10) 1(12),1(14)
	extended-2	912(4)	1(389),133(2),134(3),4(5),9(6),7(7) 6(8),11(9),7(10),2(11),7(12),8(13) 1(15)
#2 (16 symb)	original	115(2)	1(52),24(2),5(4),3(5),2(6),4(7),2(8)
	extended	2759(3)	1(1121),378(2),474(3),57(4),49(5), 36(6),31(7),27(8),15(9),18(10),8(11) 4(12),4(13),3(14),2(16)
	extended-2	44795(3)	1(18097),6050(2),7850(3),772(4), 755(5),638(6),593(7),401(8),368(9), 227(10),111(11),78(12),54(13), 31(14),14(15),5(16),3(17),3(18), 3(19),1(20),1(23)

Table 3. Hardware complexity for various codes for (x=0) (in terms of 2-input gates)

Experiment number	code type	hardware complexity(number of gates)	
		AND	OR
#1 (4 symb)	original	28	35
	extended	394	238
	extended-2	2796	1292
#2 (16 symb)	original	115	150
	extended	5518	3891
	extended-2	89590	63208

5. CONCLUSION

In this paper we presented systematic and fast architectures for arithmetic codes. These methods use interval tree search, look-ahead and redundant arithmetic. This approach is well suited for alphabets where number of symbols is low and tree search can be implemented using simple combinatorial logic. Design of more efficient architectures for arithmetic codes with large number of states is being currently investigated.

6. REFERENCES

[1] J. Rissanen, and G.G. Langdon, Jr., "Arithmetic Coding", *IBM J. Res. Develop.*, **23(2)**, March 1979, pp. 149-162.

[2] M. Renfors, and Y. Neuvo, "The Maximum Sampling Rate of Digital Filters under Hardware Speed Constraints", *IEEE Trans. on Circuits and Systems*, **28(3)**, March 1981, pp. 196-202.

[3] K.K. Parhi, "Algorithm Transformation Techniques for Concurrent Processors", *Proceedings of the IEEE*, (Special issue on supercomputer technology), **77(12)**, Dec. 1989, pp. 1879-1895.

[4] H.D. Lin, and D.G. Messerschmitt, "Improving the Iteration Bound of Finite State Machines", *in Proc. of IEEE Conf. on Computer Design*, Oct. 1989.

[5] K.K. Parhi, "High Speed VLSI Architectures for Huffman and Viterbi Decoders", *IEEE Trans. on Circuits and Systems*, **39(6)**, June 1992, pp. 385-391.

[6] P.M. Kogge, and H.S. Stone, "A Parallel Algorithm for the Efficient Solution of a General Class of Recurrence Equations", *IEEE Trans. on Computers*, **C-22**, Aug. 1973, pp. 786-792.

[7] A. Avizienis, "Signed Digit Number Representations for Fast Parallel Arithmetic", *IRE Trans. on Eletronic Computers*, **EC-10**, Sept. 1961, pp. 1389-400.

[8] S.C. Knowels et. al., "Bit Level Systolic Architectures for High Performance IIR Filters", *Journal of VLSI Sig. Proc.*, 1, 1989, pp. 9-24.

[9] H.R. Srinivas, and K.K. Parhi, "High Speed VLSI Arithmetic Processor Architectures Using Hybrid Number Representation", *Journal of VLSI Signal Processing*, 4, 1992, pp. 177-198.

[10] G.G. Langdon, Jr., "An Introduction to Arithmetic Coding", *IBM J. Res. Develop.*, **28(2)**, March 1984, pp. 135-149.

[11] R.M. Karp, "Minimum Redundancy Coding for the Discrete Noiseless Channel", *IRE Trans. on Info. Theory*, **IT-7**, Jan. 1961, pp. 27-38.

[12] K.K. Parhi and G. Shrimali, "A Concurrent Lossless Coder for Video compression", *in Proceedings of the IEEE VLSI Signal Processing Workshop*, Napa Valley, Oct. 1992.

[13] G. Shrimali and K.K. Parhi, "High-Speed Arithmetic Decoder Architectures", *in Proceedings of the Sixth SIAM Conference on Parallel Processing for Scientific Computing*, March 1993.

AREA EFFICIENT VLSI ARCHITECTURES FOR HUFFMAN CODING

Heonchul Park and Viktor K. Prasanna

Department of Electrical Engineering-Systems, EEB-243
University of Southern California
Los Angeles, CA 90089-2562

ABSTRACT

In this paper, we present simple and area efficient VLSI architectures for Huffman coding, an industrial standard proposed by MPEG, JPEG, and others. We use a memory of size $O(n \log n)$ to store a Huffman code tree, where n is the number of symbols. It requires few simple arithmetic operations on the chip for real-time encoding and decoding. Based on our scheme, we show a design for 8-bit symbols. The proposed design requires 256×9 and 64×18-bit memory modules to process 8-bit symbols. The chip occupies a silicon area of $3.5 \times 3.5 mm^2$ using 1.2 micron CMOSN standard library cells. Compared with, known parallel implementation [7] which requires upto 65536 PEs, the proposed architecture leads to a single PE design.

1. INTRODUCTION

Low bit rate coding is essential for image applications such as TV transmission, video conferencing, remote sensing via satellite, computer communication, facsimile transmission, etc.. Data compression techniques reduce the storage and communication channel bandwidth needed to process such signals. Most conventional data compression techniques require lossless coding method at the end of the encoder to obtain additional compression in image applications. In addition, some applications require lossless coding schemes. Several lossless coding schemes have been proposed for these purposes. One of these lossless coding schemes is Huffman coding (also known as optimal variable length coding) which provides optimal coding for a fixed length input. It is recommended as a standard coding method by MPEG, JPEG, and others for image applications [1, 6, 11].

Huffman coding assigns a 0/1 sequence for each input symbol such that the total average length of the

THIS RESEARCH WAS SUPPORTED IN PART BY NSF UNDER GRANT IRI-9145810 AND IN PART BY DEFENSE ADVANCED RESEARCH PROJECTS AGENCY UNDER CONTRACT F-49620-90-C-0087.

code is minimal based on a predetermined weight of each input symbol. The Huffman code is represented as a binary tree where each leaf of the tree represents a symbol. A symbol is encoded by the path from the root to the corresponding leaf. Thus, a symbol with larger weight has shorter path length.

Most known designs [2, 5] do not satisfy the real-time requirements, since these operate under 1.5 Mbps. Recently, several Huffman codec designs have been reported [7, 8] to support current and future terminals and disk controllers operating in real-time. In [7], several designs for Huffman codecs have been shown based on reversed Huffman tree. These attain upto 40 million symbols/sec of throughput. However, these require redesign of the architecture whenever a different tree is used, since the architecture is a direct mapping of the tree. It requires upto $O(n^2)$ PEs and most of the PEs remain idle during the computation, where n is the number of symbols. In [8], a single chip approach has been introduced. For encoding 7-bit ASCII symbols, most area of the chip is consumed by a memory of size 512×12 bits. The use of reversed tree in the design makes the decoding easier. However, the reversed tree results in more number of nodes than the original tree. In addition, there is no known worst case bound on the memory needed for storing the reversed tree.

We propose an area efficient VLSI architecture for Huffman coding which requires $2n \times (\log{}^1 n + 1)$ memory for $\log n$-bit symbol encoding and decoding using a simple mapping of the Huffman tree onto memory. Based on this, we obtain Huffman codec architectures for 8-bit symbols, which require memory of size 256×9 and 64×18 bits. The binary tree mapping employed in this design leads to fast encoding and decoding. It can support real-time operations required in MPEG rates. Our design requires a small amount of area for ALU and control unit, due to the simplicity of the algorithm. The needed processing power is competitive with the parallel implementation shown in [7].

The organization of this paper is as follows. In the

---

[1]Throughout this paper, logarithms are to base 2.

next section, our mapping of the tree onto memory, and encoding and decoding algorithms based on this mapping are discussed. A new VLSI architecture for Huffman coding is shown in Section 3. Concluding remarks are made in Section 4.

2. MAPPING HUFFMAN TREE ONTO STORAGE

In this section, we show a compact scheme for storing a Huffman tree, which also leads to simple and faster coding and decoding algorithms. Throughout this paper, we assume that a memory access, an arithmetic or logic operation on k-bit data, or a register transfer can be performed in unit time.

A binary tree with n leaves in a Huffman tree has $2n-1$ nodes. If a direct storage of the Huffman code for each symbol is used, then the storage for the complete tree would require $O(n^2/k)$ words of memory (in the worst case), assuming each word is k-bit wide. Since paths from the root to leaves in a Huffman tree may be of different length, this storage scheme can lead to wastage of memory. Using table look-up on the above storage scheme, encoding of a symbol takes $\lceil l/k \rceil$ time units, assuming l is the length of the encoded symbol. Alternately, the technique in [3] can be used. It uses memory of size $O(n \log n + 4n)$ bits ($n \log n$ bits are used to store symbols and $4n$ bits are used to store the nodes in the tree), then encoding of a symbol takes $O(n)$ time units [3], since this storage scheme does not have any extra information to help fast tree traversal.

In the rest of this paper, the *canonical* tree representation of Huffman tree is used. In this representation, all the leaves are to the left of all the internal nodes in each level of the tree. It is known that any Huffman tree can be transformed into a canonical tree without increasing the average code length [10].

2.1. MEMORY MAP FOR ENCODING

For encoding, the internal nodes are labeled top-down from 0 to $n-1$ and within each level the nodes are labeled left to right. The pointer to the parent of each symbol (which corresponds to a leaf node) is stored in locations 0 to $n-1$. The pointer of the i-th symbol is stored in the i-th location using $\log n$ bits. For each node, one additional bit is added to store the value of the edge between the node and its parent (which denotes if the symbol node is the left or the right child of its parent). The pointer of the i-th internal node is stored in location $(n+i)$. The stored pointer value is [label of the parent node - n]. The root of the tree has a *nil* pointer represented by "0". Thus, the total number of words of storage is $2n-1$ to represent a Huffman

tree with $2n-1$ nodes and each word consists of two fields: a $\log n$ bits pointer field and a 1-bit edge field. Total storage needed is $O(n \log n)$ bits. Notice that the values stored in the pointer field in memory locations n to $2n-2$ are in non-decreasing order (this fact is used to further compact the storage in Section 3).

Encoding of an input symbol is performed by a simple tree traversal starting from the leaf (corresponding to the symbol) to the root. The encoder obtains a symbol from the input buffer. Using the input symbol value as memory address, fetch a pointer j and 1-bit output. Note that the pointer of an internal node j is stored in the $(n+j)$-th location. Fetch the data in location $(n+j)$ and output the 1-bit data in the edge field. The output bit is pushed onto a stack. The pointer is updated using the data in the pointer field. Continue the above traversal, until $j = 0$, which denotes the root of the tree. Then, the output stored in the stack are popped out to an output buffer. This is necessary since the tree traversal is performed from a leaf to the root of the tree. The time for encoding is $O(l)$, where l is the length of the encoded output.

2.2. MEMORY MAP FOR DECODING

Throughout this paper, we assume the (decoded) symbols are represented by $\log n$ bits. For decoding, we employ a different memory map of the tree. All the nodes of the tree are labeled from 1 to $2n-1$, top-down and left to right within each level. Symbols are stored in an array A of size $n \times \log n$, such that the symbol in the i-th location has a lower node label than that of the symbol in the $(i+1)$-th location, $0 \leq i \leq n-2$. The number of leaves in the j-th level of the tree is stored in the j-th location of an array B. Thus, a memory of $2n \log n$ bits is needed for decoding.

Retrieval of symbols is as follows: Let P_T denote a pointer to a node in a level of the tree during the tree traversal. P_T has the position of the node from the leftmost node in the current level, where the leftmost node is designated to be at position "1". Also, assume that a variable C_s is used to accumulate the number of symbols in the tree up to the previous level of the traversal. Let C_L denote the current level of the tree traversal. The root is assumed to be at level 0. The tree traversal begins at the root. At this time, C_s and C_L are initialized to "0". Also, P_T is initialized to "1". The search continues until a leaf node is reached. If $P_T \leq B[C_L]$, then it implies that P_T points to a leaf and output $A[C_s + P_T]$. Otherwise, the tree traversal has to continue to the next level of the tree. C_s, C_L, and P_T are updated as follows:

- $C_s \leftarrow C_s + B[C_L]$.

- $P_T \leftarrow 2(P_T - \mathrm{B}[C_L]) + r - 1$. The rationale for this update is as follows. Every internal node has exactly two children. The children of the i-th node from the leftmost internal node at the current level are located at positions $(2i - 1)$ and $(2i)$ from the leftmost node in the next level. r is the input bit received from the communication channel during the current clock cycle.

- $C_L \leftarrow C_L + 1$.

Notice that all the above three update operations can be performed simultaneously during a clock cycle.

Compared with the known design in [8] which uses 512×12 bits of memory for 7-bit symbol coding, our design requires 256×8 bits of memory.

3. ARCHITECTURE AND IMPLEMENTATION

Based on the encoding and decoding algorithms discussed in Section 2, we show a compact design for Huffman codec. For sake of illustration and evaluation of the resulting design, we consider 8 bit symbols. 8-bit symbol encoding and decoding arises in many applications.

Consider a Huffman tree for 8-bit symbols. The tree has 256 leaves. We can store the Huffman tree in a memory of 511×9 bits as shown in Section 2. We can further reduce the memory requirement for storing the internal nodes by employing a simple coding scheme based on the observation that the values stored in the pointer field of the internal nodes are in non-decreasing order. The tree is divided into 64 groups of 4 nodes each, such that the node labeled i is in the $\lfloor i/4 \rfloor$-th group, $0 \le i \le 254$. In each group, 9 bits are assigned to the first node to store its 8-bit pointer value and 1-bit edge value. The other three nodes are coded into 3 bits each: two bits are used to store the difference between the pointer value of the first node in the group and its own and one bit is used to store the 1-bit edge value. In the labeling discussed in Section 2, the difference is always less than 4. These internal nodes are stored using 64×18 bits of memory. Note that the memory requirement to store the internal nodes is reduced by half, since four 9-bit words are now stored in two 9-bit words. First 8 bits are used to represent (un-coded) pointer value. The next 6 bits are used to store the three differences in pointer values. The last 4 bits are used to store the 1-bit output corresponding to the edges in the group.

Figure 1 shows an overall block diagram of our design. The main component in the architecture is two memory modules M_1 and M_2 of size 256×9 and 64×18

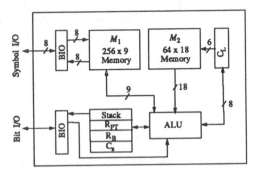

Figure 1: A block diagram of the codec

bits. For encoding, M_1 stores the pointer of the symbols and their edge values. Each location in M_2 stores the pointers and edges corresponding to a group (of 4 nodes) discussed above. For decoding, M_1 stores array A. Array B used in decoding is stored in M_2. Notice that, in practice, the length of an encoded symbol is at most 20. The less likely events are coded by an escape symbol followed by a fixed length code to avoid extremely long code words and to reduce the cost of implementation[1, 6]. Since in practical designs the length of a coded symbol is restricted to 20 which implies that the total number of levels in the tree is at most 20; the corresponding size of array B is 20×9 bits. Array B is stored in memory locations 0 to 19 in M_2. The contents of M_1 and M_2 are initialized before codec operation begins.

The ALU consists of two 8-bit adders and a comparator. The stack stores the (intermediate) encoded output and is used to reverse the data at the end of each encoding. A 20-bit long stack is sufficient for data reversal during encoding. 8-bit registers R_{PT} and R_B are used for tree traversal in decoding and are implemented using master-slave flip-flops. In addition, there are two counters. Counter C_s is used to store the length of the coded symbol in encoding. It is also used to store the number of symbols upto the previous level of the traversal and to output the symbol in decoding. Counter C_L is used to access M_2. Using bidirectional I/O ports (BIO), 8-bit input to be encoded and 8-bit output of the decoder share the same I/O ports. Also, a 1-bit bidirectional I/O port is used for bit serial output of the encoder and bit-serial output of the decoder.

If the length of the encoding is l, the algorithm takes at most $2l$ steps, assuming one clock cycle per level of the tree and l clock cycles to reverse the contents of the stack. Thus, the throughput of the codec is [clock speed/$(2 \times l/8)$]. For example, the throughput is over 40Mbps assuming 40 MHz two-phase clock and 50%

compression ratio.

In decoding, an 1-bit data is input to the chip during each clock cycle. An 8-bit symbol is output in bit-parallel mode once the search arrives at a leaf. Since it takes one clock cycle for each input bit, the decoder has 40Mbps throughput, assuming 40MHz two-phase clock.

The chip was implemented in 1.2 micron CMOSN process using CMOSN standard cells for registers, multiplexers, and logic gates. The memory modules were generated using CMOSN RAM generators. The design uses a 2-phase nonoverlapping clock. Cadence design tools running on SUN workstations were used for analyzing the entire design. The design approach was to draw schematic diagram using "EDGE" (Cadence CAD tool) and perform automatic placement and routing. The die size was found to be $3.5 \times 3.5mm^2$. Most silicon area is consumed by the memory. The longest path in the design was from R_d to the comparator. The delay time in the longest path was found to be less than $25ns$ using SPICE simulation. The number of I/O pins in the proposed design is 28 including signal and power ports. The number of I/O pins can be reduced into 20 by using bit serial input for memory update. The proposed design is more compact, compared with the known design [7] which requires $6.8 \times 6.9mm^2$ area using 2 micron SCMOS cell library. The layout shows that memory modules consume most of the silicon area. Notice that the design in [7] has been proposed for 7-bit symbols and it requires twice the memory required by our design (for 8-bit symbols). In addition, SCMOS standard cell requires much smaller area than CMOSN cell. Also, the design in [8] employs customized RAM cell.

4. CONCLUSION

We have shown a VLSI architecture for Huffman coding which requires a simple control logic and arithmetic unit. The most area consuming part is the memory. The average throughput for encoding/decoding based on our design for 8-bit symbols is 40Mbps, assuming 40 MHz two-phase clock and 50% compression ratio, the design can be implemented using less than 28 I/O pins including pins for data, control, and power. Notice that the design in [8] requires 1024×13 bits of memory for 8-bit symbol Huffman codec and also needs more than 55 I/O pins. Our design is competitive with the parallel implementation in [7] with respect to the throughput requirements. In addition, the proposed architecture can support MPEG data rates and data rates arising in many real-time applications.

Additional details of this paper can be found in [9].

Acknowledgment: We thank Mr. Dong-Hyun Heo at USC-ISI for his assistance in obtaining the layout and area estimate of our design.

5. REFERENCES

[1] E. A. Fox, "Advances in interactive digital multimedia systems," *IEEE Computer,* pp. 9-21, Oct. 1991.

[2] K. Hazboun and M. Bassiouni, "A multi-group technique for data compression," *Int. Conf. on Management of Data,* pp. 284-292, 1982.

[3] G. Jacobson, "Space-efficient static trees and graphs," *IEEE Symp. on Foundations of Computer Science,* pp. 549-554, 1989.

[4] G. Langdon, "An introduction to arithmetic coding," *IBM J. Res. Develop.,* pp. 135-149, 1984.

[5] R. Lea, "Text compression with an associative parallel processsors," *Comp. J.,* pp. 45-76, 1978.

[6] D. Le Gall, "MPEG: A video compression standard for multimedia applications," *Comm. of the ACM,* pp. 46-58, April, 1991.

[7] A. Mukherjee, N. Ranganathan, and M. Bassiouni, "Efficient VLSI designs for data transformation of tree-based codes," *IEEE Trans. on Circuits and Systems,* pp. 306-314, 1991.

[8] A. Mukherjee, J. W. Flieder, N. Ranganathan, "MARVLE: A VLSI chip for variable length encoding and decoding," *Manuscript,* Dept. of Computer Science and Engineering, Univ. of South Florida, 1992.

[9] H. Park and V. K. Prasanna, "Area efficient VLSI architectures for Huffman coding," *Manuscript,* Dept. of EE-Systems, University of Southern California, Oct. 1992.

[10] J. S. Vitter, "Design and analysis of dynamic Huffman codes," *Journal of the Association for Computing Machinery,* pp. 825-845, 1987.

[11] G. K. Wallace, "The JPEG still picture compression standard," *Comm. of the ACM,* pp. 30-44, April, 1991.

AN EFFICIENT PRIME-FACTOR ALGORITHM FOR THE DISCRETE COSINE TRANSFORM AND ITS HARDWARE IMPLEMENTATIONS

PeiZong Lee and Fang-Yu Huang

Institute of Information Science, Academia Sinica
Nankang, Taipei, Taiwan, R. O. C.

ABSTRACT

In this paper, we present a new prime-factor algorithm for the DCT. The input index mapping we adopt is the Ruritanian mapping, the output index mapping we employ is the same one as Lee's [Lee 89]. We also study hardware implementations for the prime-factor DCT. The methodology, which deals with general $(N_1 \cdot N_2)$-point DCTs, where N_1 and N_2 are mutually prime, is illustrated by converting a 15-point DCT problem into a (3×5)-point 2-D DCT problem.

1. INTRODUCTION

The prime-factor decomposition is a fast computational technique for many important digital signal processing operations, such as the convolution [Nussbaumer 81], the discrete Fourier transform (DFT) [Burrus 77, 87, 88, Burrus and Eschenbacher 81, Kolba and Parks 77, Rothweiler 82, Sorensen, Jones, Heideman, and Burrus 87, and Temperton 83, 85, 88], the discrete Hartley transform (DHT) [Chakrabarti and JáJá 90, Lun and Siu 92, and Sorensen, Jones, Burrus, and Heideman 85], and the DCT [Chakrabarti and JáJá 90, Lee 89, and Yang and Narasimha 85]. It has both theoretical and practical significance. Its main theoretical rationale is to convert a large-size one-dimension problem, by employing certain appropriate index mappings, into a multidimensional one. Then, we can deal with the resulting groups of small-size problems in each dimension.

For practical considerations, since in a typical DSP processor the memory for data storage is expensive and usually not large, it is more feasible to process a small-size problem one at a time. In addition, when this approach is combined with efficient short-length algorithms, such as Rader's algorithm [Nussbaumer 81], or Winograd type minimum multiplication algorithms in

This work was partially supported by the NSC under Grants NSC 81-0408-E-001-505.

DFT [Winograd 78], or Heideman's small odd-length DCT modules [Heideman 92], etc., it would be of practical interest in reducing the scalar multiplication complexity.

Although, previously Yang and Narasimha [Yang and Narasimha 85] have proposed a prime-factor DCT algorithm, its index mapping is very complicated. From our point of view, an index mapping not only should be easy to understand, but also should be efficient in running. Lee [Lee 89] has achieved this goal. However, his input index mapping is realized by constructing and combining two index tables, which could occupy additional memory space and would be infeasible in variable-size applications.

Chakrabarti and JáJá [Chakrabarti and JáJá 90] develop a systolic architecture for implementing Lee's algorithm. Because they want to compute the DCT from the DHT, they modify the index mappings, which are essentially the same as Lee's. However, they did not discuss the actual implementation for these index mappings.

In this paper, the input index mapping we adopt is the Ruritanian mapping, since its efficient realization can be based on the previous research efforts. In addition, the resulting algorithm complexity is by no means increased. As for the output index mapping, we employ the same one as Lee's, for which might be the most natural one in view of the DCT transform kernel's structure.

2. DERIVATION OF THE PRIME-FACTOR DECOMPOSITION

We briefly sketch our method. Consider the following simplified version of the IDCT (inverse discrete cosine transform):

$$x_n = \sum_{k=0}^{N-1} X_k \cos(\frac{\pi(2n+1)k}{2N}) , \qquad (1)$$

where X_k, $0 \leq k \leq N - 1$, is the DCT output data sequence, and x_n, $0 \leq n \leq N - 1$, is the IDCT output data sequence.

We want to convert Equation (1) into the form in Theorem 2 by taking appropriate input and output index mappings. The input index mapping connects the input index k, $0 \leq k < N$, to (k_1, k_2), $0 \leq k_1 < N_1$ and $0 \leq k_2 < N_2$. The output index mapping connects the output index n, $0 \leq n < N$, to (n_1, n_2), $0 \leq n_1 < N_1$ and $0 \leq n_2 < N_2$. Since the DCT is orthonormal, the forward transform also can be realized by taking the transpose of the inverse transform.

The input index mapping we adopt is the Ruritanian mapping, which was also used in the prime-factor DFT algorithms [Burrus 87, 88, Burrus and Eschenbacher 81, Kolba and Parks 77, and Rothweiler 82], and several researchers have studied its efficient implementation methods [Lun, Chan, and Siu 90, Wong, Chan, Lun, and Siu 90, Wong and Siu 89, 91]. That is,

$$\forall\, k \in [0, N - 1],\ \exists!\, (k_1, k_2) \in [0, N_1 - 1] \times [0, N_2 - 1],$$

such that

$$k = \psi(k_1, k_2) = (N_2 k_1 + N_1 k_2) \bmod N\,,$$

where $N = N_1 N_2$, N_1 and N_2 are relatively prime.

In other words, based on the Ruritanian mapping, the integers k on the interval $[0, N-1]$ correspond one-to-one to the lattice points in the region $[0, N_1 - 1] \times [0, N_2 - 1]$. For the sake of our derivation, these lattice points are divided into five disjoint groups.

Let

$$
\begin{aligned}
f(k_1, k_2) &= N_2 k_1 + N_1 k_2\,, \\
g(k_1, k_2) &= N_2 k_1 - N_1 k_2\,.
\end{aligned}
$$

Define

$E = \{(k_1, k_2) \,|\, k_1 = 0 \text{ or } k_2 = 0\}\,,$
$A = \{(k_1, k_2) \,|\, f(k_1, k_2) < N \text{ and } g(k_1, k_2) > 0;\ k_1 k_2 \neq 0\}\,,$
$B = \{(k_1, k_2) \,|\, f(k_1, k_2) < N \text{ and } g(k_1, k_2) < 0;\ k_1 k_2 \neq 0\}\,,$
$C = \{(k_1, k_2) \,|\, f(k_1, k_2) > N \text{ and } g(k_1, k_2) < 0;\ k_1 k_2 \neq 0\}\,,$
$D = \{(k_1, k_2) \,|\, f(k_1, k_2) > N \text{ and } g(k_1, k_2) > 0;\ k_1 k_2 \neq 0\}\,.$

Let

$$X_{(k_1, k_2)} = X_k, \text{ where } k = \psi(k_1, k_2)\,.$$

Now, we define $Y_{(k_1, k_2)} =$

$$
\begin{cases}
X_{(k_1, k_2)}, & \text{if } (k_1, k_2) \in E \\
X_{(k_1, k_2)} + X_{(k_1, N_2 - k_2)}, & \text{if } (k_1, k_2) \in A \\
X_{(k_1, k_2)} + X_{(N_1 - k_1, k_2)}, & \text{if } (k_1, k_2) \in B \\
-X_{(N_1 - k_1, N_2 - k_2)} + X_{(N_1 - k_1, k_2)}, & \text{if } (k_1, k_2) \in C \\
-X_{(N_1 - k_1, N_2 - k_2)} + X_{(k_1, N_2 - k_2)}, & \text{if } (k_1, k_2) \in D.
\end{cases}
\tag{2}
$$

Theorem 1 :

$$x_n = \sum_{k_1=0}^{N_1-1} \sum_{k_2=0}^{N_2-1}$$

$$Y_{(k_1, k_2)} \cos\left(\frac{\pi(2n+1)k_1}{2N_1}\right) \cos\left(\frac{\pi(2n+1)k_2}{2N_2}\right).$$

Next, we introduce the output index mapping, which was proposed by Lee [Lee 89]. For the completeness of this presentation, we list it here for easy reference.

The output index mapping: $\varphi(n) = (n_1, n_2)$, where

$$
\begin{aligned}
\bar{n}_1 &= n \bmod 2N_1 \\
\bar{n}_2 &= n \bmod 2N_2 \\
n_1 &= \begin{cases} \bar{n}_1, & \text{if } \bar{n}_1 < N_1 \\ 2N_1 - 1 - \bar{n}_1, & \text{otherwise} \end{cases} \\
n_2 &= \begin{cases} \bar{n}_2, & \text{if } \bar{n}_2 < N_2 \\ 2N_2 - 1 - \bar{n}_2, & \text{otherwise.} \end{cases}
\end{aligned}
\tag{3}
$$

Theorem 2 :

$$x_{(n_1, n_2)} = \sum_{k_1=0}^{N_1-1} \sum_{k_2=0}^{N_2-1}$$

$$Y_{(k_1, k_2)} \cos\left(\frac{\pi(2n_1+1)k_1}{2N_1}\right) \cos\left(\frac{\pi(2n_2+1)k_2}{2N_2}\right).$$

Prime-factor algorithm for the IDCT

Input: An N-point data sequence X_k, $0 \leq k < N$, where $N = N_1 N_2$, N_1 and N_2 are mutually prime.

Output: An N-point data sequence x_n, $0 \leq n < N$.

Step 1: Apply the Ruritanian mapping on X_k to construct an $(N_1 \times N_2)$-point data matrix $X_{(k_1, k_2)}$, where $0 \leq k_1 < N_1$ and $0 \leq k_2 < N_2$.

Step 2: Modify the data matrix $X_{(k_1, k_2)}$ to get the data matrix $Y_{(k_1, k_2)}$ according to Equation (2).

/* This is proved to be correct in Theorem 1. */

Step 3: /* Row-column evaluation */

1. Execute the N_1-point IDCT for each of the N_2 columns of $Y_{(k_1, k_2)}$, the result is $T_{(k_1, k_2)}$.

2. Execute the N_2-point IDCT for each of the N_1 rows of $T_{(k_1, k_2)}$, and the result is $x_{(n_1, n_2)}$, where $0 \leq n_1 < N_1$ and $0 \leq n_2 < N_2$.

/* This is proved to be correct in Theorem 2. */

Step 4: Apply the output index mapping in Equation (3) to get the final result x_n. □

k_2 k k_1	0	1	2	3	4
0	0	3	6	9	12
1	5	8	11	14	2
2	10	13	1	4	7

Table 1: The Ruritanian mapping: $k = (5k_1 + 3k_2)$ mod 15, when $N = 15$, and $N_1 = 3$, $N_2 = 5$.

n_2 n n_1	0	1	2	3	4
0	0	11	12	6	5
1	10	1	7	13	4
2	9	8	2	3	14

Table 2: The output index mapping when $N = 15$, and $N_1 = 3$, $N_2 = 5$.

Example : In this example, $N = 15$, and $N_1 = 3$, $N_2 = 5$. Figure 1 shows the signal flow graph for implementing the 15-point 1-D IDCT, which can be converted into the (3×5)-point 2-D IDCT. First, when given a 15-point 1-D input data sequence X_k, $0 \leq k < 15$, we transform this 1-D data sequence by the Ruritanian mapping to a (3×5)-point 2-D data matrix $X_{(k_1,k_2)}$, $0 \leq k_1 < 3$ and $0 \leq k_2 < 5$. Table 1 shows the Ruritanian mapping for this case. Next, we modify the 2-D data matrix $X_{(k_1,k_2)}$ to get the 2-D data matrix $Y_{(k_1,k_2)}$ according to Equation (2). Then, we apply the row-column evaluation for the computation: first, we deal with five 3-point IDCTs; then, take the transpose of the result; and then, deal with three 5-point IDCTs. Finally, the output data sequence $x(n)$, $0 \leq n < 15$, can be obtained by using the output index mapping in Equation (3). Table 2 shows the output index mapping for this case.

Note that, the signal flow graph in Figure 1 performs the IDCT if the signals flow from left to right, and performs the DCT if the signals flow from right to left.

3. HARDWARE IMPLEMENTATIONS

We consider hardware implementations for the prime factor DCT, too. We are especially interested in the hardware designs which are suitable for the VLSI (Very

Large Scale Integration) implementations. We have shown three hardware designs for the prime-factor DCT. The first one, which is a VLSI circuit fabricated directly according to the signal-flow graph as shown in Fig. 1, might be not easy to implement when the problem size becomes large. The second one is a linear systolic array implementation, which contains N processing elements and can solve an N-point DCT problem in $O(N)$ systolic steps. The third one is a mesh-connected systolic array implementation, which contains $O(N_1^2 + N_1 N_2 + N_2^2)$ processing elements and can solve the row-column evaluation in $O(N_1 + N_2)$ systolic steps, where $N = N_1 N_2$, and N_1 and N_2 are mutually prime. If we consider the VLSI performance measure AT^2 on the second and the third designs, then the latter one, which is $O((\max\{N_1, N_2\})^4)$, is better than the former one, which is $O(N^4)$, where A means the circuit area and T means the execution time. Figure 2 shows the complete data flow diagram of the mesh-connected systolic array implementation for computing the 15-point prime-factor IDCT.

Finally, we generalize Wang's method [Wang 82] to design an algorithm for computing the discrete sine transform (DST) from the DCT. We also design a prime factor DST algorithm based on the prime-factor DCT algorithm.

4. REFERENCES

[1] C. Chakrabarti and J. JáJá. Systolic architectures for the computation of the discrete Hartley and the discrete Cosine transforms based on prime factor decomposition. *IEEE Transactions on Computers*, C-39(11):1359–1368, November 1990.

[2] B. G. Lee. Input and ouput index mappings for a prime-factor decomposed computation of discrete Cosine transform. *IEEE Transactions on Acoustics, Speech, and Signal Processing*, 37(2):237–244, February 1989.

[3] P.-Z. Lee and F. Y. Huang. An efficient prime-factor algorithm for the discrete cosine transform and its hardware implementations. *Submitted for publication*, July 1992.

[4] H. V. Sorensen, D. L. Jones, M. T. Heideman, and C. S. Burrus. Real-valued fast Fourier transform algorithms. *IEEE Transactions on Acoustics, Speech, and Signal Processing*, ASSP-35(6):849–863, June 1987.

[5] P. P. N. Yang and M. J. Narasimha. Prime factor decomposition of the discrete Cosine transform. In *IEEE International Conference on Acoustics, Speech, and Signal Processing*, pages 772–775. IEEE, 1985.

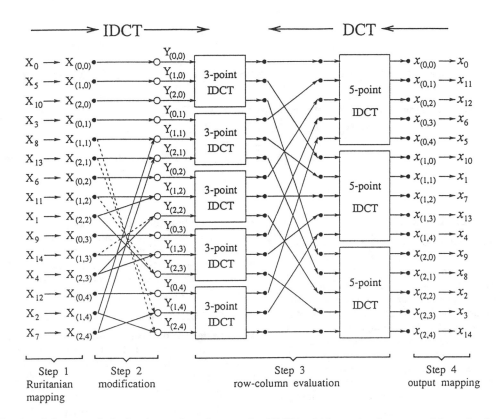

Fig. 1. The signal-flow graph for implementing the 15-point IDCT, which can be decomposed into the (3 × 5)-point 2-D IDCT. If signals flow from left to right, it performs the IDCT; if signals flow from right to left, it performs the DCT. Solid lines represent transfer factor 1, while dashed lines represent transfer factor −1. Circles o represent adders.

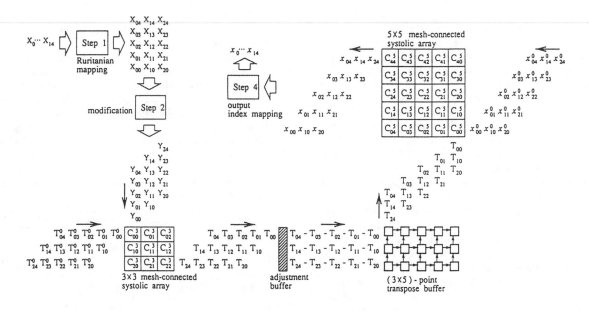

Fig. 2. The complete data-flow diagram of the mesh-connected systolic array implementation for computing the 15-point prime-factor IDCT.

EDGE DETECTION USING FINE-GRAINED PARALLELISM IN VLSI

Chetana Nagendra   *Manjit Borah*   *Mohan Vishwanath*   *Robert M. Owens*   *Mary Jane Irwin*

Department of Computer Science
Pennsylvania State University
University Park, PA 16802

ABSTRACT

This paper demonstrates an optimal time algorithm and architecture for edge detection in real time using fine grained parallelism. Given an image in the form of a two-dimensional array of pixels, this algorithm computes the Sobel and Laplacian operators for skimming lines in the image and then generates the Hough array using thresholding. Hough transforms for M different angles of projection are obtained in a fully systolic manner without using any multiplication or division. An implementation of the algorithm on the MGAP - a fine-grained processor array architecture being developed at the Pennsylvania State University, is shown which computes at the rate of approximately 75,000 Hough transforms per second on a 256 × 256 image using a 25 MHz clock. It is also shown that the algorithm can be easily extended to general case of Radon transforms.

1. INTRODUCTION

Image processing and computer vision problems are computation intensive by nature. During the last few years the trend has been towards making use of parallel computing to solve these problems resulting in significant speedups. Edge detection on a digital image is used in many applications like Computer aided Tomography, Robotics Vision etc. The idea underlying most edge detection techniques is the computation of a local derivative operator and then taking projection along different contours. Hough transform is a useful tool for obtaining projection data for binary images.

A large number of algorithms for edge detection have been developed for mesh architectures [4] [7] [9]. However, all these architectures use parallelism at a 'macro' level. They assume that each processor has a significant amount of computing power at the same time requiring availability of a large number of processors. Hence they are not ideally suited for VLSI. Therefore these algorithms need to be modified to make them suitable for fine-grained processors that cannot have a lot of computing power, but their availability in large numbers can be obtained in VLSI. Given an $N \times N$ image, we show a $\mathbf{T} = \mathbf{O}((N+M)\sqrt{p})$ [1] algorithm for computing Hough transforms at M different angles and $\mathbf{A} = \mathbf{O}(N^2 log N)$ implementation of the algorithm on the MGAP - a fine-grained processor array architecture being developed at Penn State University. The algorithm uses

---

[1] p is the operand precision

simple operations like add/subtract, compare etc. avoiding operations like multiplication, square-root that are expensive on fine grained processors. We show that it is possible to compute at the rate of approximately 75,000 Hough transforms per second on a 256 × 256 image using a clock rate of 25 MHz.

2. EDGE DETECTION ALGORITHM

We consider the problem of detecting lines in an image with $N \times N$ pixels on an $N \times N$ array of processors connected as a mesh. At the highest level, our algorithm consists of first skimming the image for lines and then applying the Hough transform on the binary array thus obtained.

2.1. Line-skimming Algorithm

In [2], Davies gives a fast and accurate scheme for skimming lines from a gray scale image which consists of the following steps.

1) A simple Differential Gradient is computed on the pixels using the Sobel masks, which provide sufficiently accurate estimates of the x and y components of the local intensity gradient $g = \sqrt{g_x{}^2 + g_y{}^2}$. A reasonable estimate of the gradient is obtained by $g = |g_x| + |g_y|$ which avoids the expensive *square root* operation [3].

2) This is followed by the Laplacian operator L which produces the second derivative of the gray level image and thus produces information about the dark and bright sides of an edge at the same time thinning down the line to one pixel.

3) Finally some thresholding function is applied on the resulting array to get a binary image which is called the Hough array. Hough transforms can now be applied on it to recognize the lines.

Algorithms for the Sobel and Laplacian operators on a mesh are given below. Note that they can be computed for all the pixels simultaneously.

Sobel (S_x)	Laplacian
Shift $p_{i,j}$ East	Shift $p_{i,j}$ East
Shift $p_{i,j}$ West	Shift $p_{i,j}$ West
$R_{i,j} = p_{i+1,j} - p_{i-1,j}$	$R_{i,j} = p_{i-1,j} + p_{i+1,j}$
Shift $R_{i,j}$ North	Shift $R_{i,j}$ North
Shift $R_{i,j}$ South	Shift $R_{i,j}$ South
$S_{i,j} = R_{i,j} * 2 + R_{i,j-1}$ $+ R_{i,j+1}$	$S_{i,j} = R_{i,j-1} + R_{i,j+1}$ $- p_{i,j} * 4$
$S_{x\ i,j} = S_{x\ i,j}/8$	$S_{x\ i,j} = S_{x\ i,j}/2$

In the above algorithms $p_{i,j}$ represents the pixel value in the processor $P_{i,j}$ and $p_{i+1,j}, p_{i-1,j}$ denote pixel values re-

$$S_x = \frac{1}{8} \begin{vmatrix} -1 & 0 & 1 \\ -2 & 0 & 2 \\ -1 & 0 & 1 \end{vmatrix} \quad S_y = \frac{1}{8} \begin{vmatrix} 1 & 2 & 1 \\ 0 & 0 & 0 \\ -1 & -2 & -1 \end{vmatrix} \quad L = \frac{1}{2} \begin{vmatrix} 1 & 0 & 1 \\ 0 & -4 & 0 \\ 1 & 0 & 1 \end{vmatrix}$$

Figure 1. Sobel & Laplacian Operators

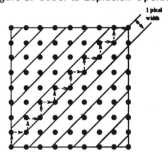

Figure 2. Hough Transform

ceived from the West and East neighbors respectively while $p_{i,j+1}$ and $p_{i,j-1}$ are the values received from the South and North neighbors respectively. Similar notations are used for $R_{i,j}, S_{i,j}$ and the variables in the other algorithms.

In the above we explained the computation of the S_x component of the Sobel operator. The S_y component can be computed in a similar manner by replacing the East and West shifts by North and South shifts. Then we compute $S_{i,j} = |S_x|_{i,j} + |S_y|_{i,j}$ at each processor.

Next, the thresholding can be achieved by broadcasting the threshold value to the processors and then each processor comparing the threshold value with its computed value and storing 1 or 0 accordingly.

2.2. Hough Transforms

The Hough transform of a binary image is a set of projections of the image taken from various angles. For a certain angle θ, the image can be viewed as being partitioned into parallel bands one-pixel wide running at the given angle of projection θ (Fig 2). The band with the highest accumulated sum is the best line segment in the image at that angle. For a given pixel we can determine the band it belongs to using the lower limit l and the upper limit u of the band (Fig 3). The difference in the values of l and u on successive pixel columns is $tan\theta$ (Fig 3).

We assume that θ lies between 0 and $\pi/4$. Lines with larger slopes are computationally more difficult and are avoided since they can be computed by transposing and/or rotating the image about the origin, the $X - axis$ or the $Y - axis$. These operations can be done in $O(N)$ time on the MGAP. The projections are taken in steps of δ starting from 0. The step size δ is usually $\pi k/4N$, where $k \in [1, N]$.

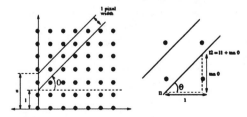

Figure 3. Lower and upper bounds of a band

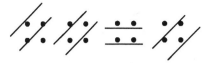

Figure 4. Possible groupings of pixels assuming $0 \leq \theta \leq \pi/4$

But in practice the step size is limited by the precision required for $tan\delta$ which is the unit of increment of the limit values in our algorithm.

Since the bands are one pixel wide and θ lies within the specified range, there will be at most two pixels of the same column in the same band (Fig 4). Thus the pattern of any band will be a series of horizontal pixels followed by a pixel in either the diagonal or the vertical direction. Our algorithm for edge detection is given below :

Algorithm : Hough Transform

For $\theta = 0$ to $\pi/4$ in steps of δ do
$P_{i,j}$:
1. Receive $sum, l_{i-1,j}, u_{i-1,j}, \& tan\theta$ from West neighbor
2. Compute sum and update l and u
   2.1. $sum_{i,j} = sum + p_{i,j}$
   2.2. $l_{i,j} = l_{i-1,j} + \tan\theta; u_{i,j} = u_{i-1,j} + \tan\theta$
3. Check if East neighbor is in the same band
   3.1. $l'_{i,j} = l_{i,j} + tan\theta$
   3.2. If $(l' > j)$
         Shift $l_{i,j}, u_{i,j} \& sum_{i,j}$ North
4. Only processors that received data from South in the prev step do the next step
   If $(j < u_{i,j-1})$
         $sum_{i,j} = sum + p_{i,j}$
5. All processors shift
   Shift $sum_{i,j}, l_{i-1,j}, u_{i-1,j}, \& tan\theta$ East

Note that the algorithm makes use of only simple operations like additions and comparisons. For step 4 we assume that each processor has a certain bit that is usually zero. This bit is set when data is received from the South. After step 4, it is reset once again.

3. ARCHITECTURE OF THE MICROGRAIN ARRAY PROCESSOR

Recently, fine grained architectures like MPP [1] and the Connection Machine [5] have gained popularity since they are specialized for problems that exhibit a high degree of parallelism. The MGAP [6] is a fine grain VLSI processor array suitable for solving image and signal processing problems. It consists of an $n \times n$ array of processing cells referred to as **digit processors** (Fig 4). A digit processor is just a small RAM and some multiplexers. Each processor is connected to its four nearest neighbors. An external controller broadcasts global control information to the processors and sends addresses to the memory unit so that every processor accesses the same memory location. However, it is possible for each processor to receive inputs from different neighbors by configuring them differently. It is possible to perform operations on a subset of the processors by conditionally enabling or disabling some of them. Algorithms for SIMD

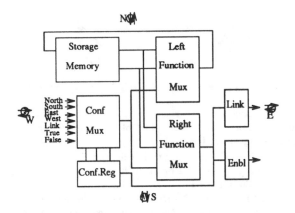

Figure 5. A single digit cell in the Micro Grain Processor

mesh architectures can be easily adapted to run efficiently on it. The strength of the MGAP lies in its flexibility and enormous amount of raw processing power.

Radix 4 fully redundant two's complement number representation is employed by the Micro Grain Processors. Hence each digit requires 3 bits of storage. p digit processors can be organized into a $\sqrt{p} \times \sqrt{p}$ subarray called **word processor** to perform arithmetic on p digit operands. It has been shown that the MicroGrain is capable of performing all the basic arithmetic operations on fixed and floating point numbers. The following table gives the number of cycles taken for some relevant operations.

Operation	Cycles
Word Shift	$2 + 3\sqrt{p}$
Digit Shift	$5 + 3k$
Addition	25
Subtraction	25
Comparison	$14\sqrt{p} + 33$
Multiplication	$O(p)$
Division	$O(p)$
Square-root	$O(p)$
2-D Sobel	$12\sqrt{p} + 140$
2-D Laplacian	$12\sqrt{p} + 120$

4. MAPPING OUR ALGORITHM ONTO THE MGAP

The above algorithm maps directly onto the MGAP. Initially the entire image is loaded into the processor array, one pixel element per word processor. The two Sobel operators are computed followed by the Laplacian operator and a thresholding function. They can be applied to every element in parallel, Sobel computation involves 8 word shifts and 7 addition/subtractions and Laplacian operator is computed with 4 shifts and 3 add/subtracts. The thresholding operation involves one comparison and broadcast. The divisions and multiplications in the Sobel and Laplacian operators are achieved by digit shifts.

Next, the external controller quantizes the image array into m bands depending upon θ. For each row, the controller introduces $sum_{i,j} = 0$, $l'_{i,j}$, $u'_{i,j}$ and $tan\theta$ into the array from the leftmost column. $l'_{0,j} = l_{0,j} - tan\theta$ and $u'_{0,j} = u_{0,j} - tan\theta$ is input so that the computation of l and u can be uniformly done by all the processors by adding $\tan\theta$.

The Hough transform algorithm cosists of 4 main phases :

1. Compute 1 [Step 2] : involves 3 additions = O(1).
2. Conditionally shift North [Step 3] : involves an addition, a comparison and 3 word shifts = $O(\sqrt{p})$.
3. Compute 2 [Step 4] : involves a comparison and an Addition = $O(\sqrt{p})$.
4. Unconditionally shift East [Step 7] : involves 4 word shifts = $O(\sqrt{p})$.

Since the computation in one step takes $O(\sqrt{p})$ cycles, and the maximum number of steps a sum needs to propagate is N, the total complexity of the algorithm for a given angle is $O(N\sqrt{p})$. Notice that data advances through the array one column at each step, from left to right. Therefore we can feed in the values for the next angle with one step delay between them. Thus M different projections on an image can be computed in $O((N + M)\sqrt{p})$ cycles, including the initial latency.

If sufficient precision is not used, propagation of error could take place while computing:
$$l_{i,j} = l_{i-1,j} + \tan\theta; \ u_{i,j} = u_{i-1,j} + \tan\theta \ \ldots (1)$$
Since there are exactly N rows and columns and $l, u \leq N$, there are only N^2 distinct values of l and u. Note that the precision in the algorithm is represented in terms of digits. Hence if d is the number of bits representing one digit of MGAP, then the computation in (1) can be represented with $p \leq \frac{log(N^3)}{d}$ digits. Thus p is $O(logN)$ and the running time of the algorithm is $O((N+M)\sqrt{p}) = O((N+M)logN)$.

The initial latency involves the time taken to move in the first image $(1 + N(1 + 3\sqrt{p}))$ which is 2561 cycles for a 256×256 image with $p = 9$. At a clock rate of 25MHz this is about 103 μsec. Pipelining is not only possible between various Hough transforms but also between successive images. As the last transform is being computed for some image the next image is moved in, column by column so that by the time we are done with one image, the next image is already loaded into the array. Thus the delay between successive images is only one word shift which is about 440 nsec. Computing the Hough array is estimated to take about 260 μs, and the delay between successive Hough transforms is estimated to be approximately 13 μs.

5. ALGORITHM FOR RADON TRANSFORMS

The general case of *Radon transform* is used for representation and manipulation of image data [8] and is helpful for efficient detection of shapes and many other applications. Computation of Radon transforms involve taking projections of a gray level image along different linear contours. Hough transform can be seen as a special case of Radon Transforms. Considering that each pixel value is represented by g digits, we will need $p' = p + g$ digits to represent the accumulated value along each line of projection. In reality, images of size more than 512×512 are not very common and 256 gray levels are used for most of the applications, therefore we need to double the precision in order to compute Radon transforms. Extending the algorithm to compute the Radon transform would require us to increase the precision by a factor of two which would in turn increase the time by a factor $\sqrt{2}$.

Noting that the gray values are represented by only p digits we can compute the Redon transform without doubling the precision, and instead accumulating the high order sum value in another variable at the expense of an increase in the time by a small amount. The $sum_{i,j}$ value is split into $suml_{i,j}$ and $sumh_{i,j}$ representing the low and high order sum digits respectively. The carry out of the addition of $p_{i,j}$ to $suml_{i,j}$ is added to $sumh_{i,j}$. The modified algorithm is given below.

Algorithm : Radon Transform

For $\theta = 0$ to $\pi/4$ in steps of δ do

$P_{i,j}$:

1. Receive $suml, sumh, l_{i-1,j}, u_{i-1,j}, \&\ tan\theta$ from West neighbor
2. Compute sum and update l and u
   2.1. $suml_{i,j} = suml + p_{i,j}$
   2.2. $sumh_{i,j} = sumh + Carry$
   2.2. $l_{i,j} = l_{i-1,j} + \tan\theta$; $u_{i,j} = u_{i-1,j} + \tan\theta$
3. Check if East neighbor is in the same band
   3.1. $l'_{i,j} = l_{i,j} + tan\theta$
   3.2. If $(l' > j)$
        Move $l_{i,j}, u_{i,j} \& suml, sumh$ North
4. Only processors that received data from South in the prev step do the next step
   If $(j < u_{i,j-1})$
        $suml_{i,j} = suml + p_{i,j}$
        $sumh_{i,j} = sumh + Carry$
5. All processors shift
   Move $suml, sumh, l_{i-1,j}, u_{i-1,j}, \&\ tan\theta$ East

Note that here we start with the gray level image itself and pipelining is between two Radon projections on the same image and also between loading two different images. The time between two Radon transforms on the same image is approximately $16\mu sec$.

6. CONCLUSION

We have presented an edge detection algorithm with real-time capability, that is suitable for a fine grain mesh architecture. It has the advantage of being fully systolic and uses short and simple data movements in the mesh. Many algorithms for computing the Hough transform exist. An algorithm has been implemented on the GAPP which takes 7 sec to run for a 512×512 image and about 0.3 sec to run for a 100×100 image [9]. A comparison of the time complexities of some other algorithms is given below.

GAPP [9]	$O(NM\xi)$†
Cypher et. al. [7]	$O((N + M)\xi)$†
Guerra et. al. [4]	$O((N + M)\xi)$†
MGAP	$O((N + M)\sqrt{p})$

† Since the first three algorithms do not take into consideration the complexity of the operations at each step, for comparison purposes we assume at least a ξ complexity per step because some of them might possibly require multiplication and division. It can be seen that our algorithm exhibits **optimal asymptotic complexity** [7]. Note that since multiplication is a relatively difficult operation in fine-grained processors, we have avoided it in the computation of the Hough transforms.

REFERENCES

[1] K.E. Batcher. Design of a massively parallel processor. *IEEE Trans. on COmputers*, 1980.

[2] E.R. Davies. Skimming technique for fast accurate edge detection. *Signal Processing*, 1992.

[3] R.C. Gonsalez and R.E. Woods. *Digital Image Processing*, chapter 7. Addison-Wesley, 1992.

[4] C. Guerra and S. Hambrusch. Parallel algorithms for line detection on a mesh. *Journal of Parallel and Distributed Computing*, 1989.

[5] W.D. Hillis. *The Connection Machine*. MIT Press, 1986.

[6] R.M. Owens and M.J. Irwin. A micro-grained vlsi signal processor. *Submitted to IEEE Trans. on Computers*, .

[7] J.L.C. Sanz R.E. Cypher and L. Snyder. The hough transform has o(n) complexity on n × n mesh connected computers. *SIAM Journal of Computing*, 1990.

[8] E.B. Sanz, J.L.C. Hinkle and A.K. Jain. *Radon and Projection Transform-Based Computer Vision*. Springer-Verlag, 1988.

[9] T.M. Silberberg. The hough transform on the geometric arithmetic parallel processor. *Proc. IEEE Workshop on Computer Architecture for Pattern Analysis and Image Data Base Management*, 1985.

COMPACT HARDWARE REALIZATION FOR HOUGH BASED EXTRACTION OF LINE SEGMENTS IN IMAGE SEQUENCES FOR VEHICLE GUIDANCE

Jörg Schönfeld, Peter Pirsch

University of Hannover, Laboratorium für Informationstechnologie
Schneiderberg 32, D–3000 Hannover 1, Germany

ABSTRACT

Important key symbols used in many image processing applications are straight line segments of edges and contours. The real time determination of straight line symbols in video sequences requires high computational rates which call for application specific circuits (ASICs). For a vision based vehicle guidance system a compact image processing unit has been developed which extracts straight lines as symbols for the street boundaries. This unit consists of two specially designed ASICs for the segmentation of symbols and a general programmable micro controller for high level processing. The implemented pipeline of algorithm comprises gradient operation, edge thinning, line verification and Hough transformation. Although the image processing unit is originally developed for vehicle guidance the hardware can be used for other industrial and robotic applications as well.

1. INTRODUCTION

Real time applications in the field of digital image processing require a large amount of computational power. The huge number of operations are in the order of mega operations per second or even giga operations per second. For this reason the hardware used for real time image processing tasks is in most cases of large size and power consuming. In order to provide the high computational rates often multiprocessor systems based on digital signal processors, array processors or transputers are used. Nevertheless there are many applications where a complete image processing unit in more compact hardware is requested. For this applications dedicated VLSI realizations are demanded. Due to the strong need of miniaturization and the high computational rates on the other hand the influence of algorithms for image processing onto the hardware have to be taken into account.

One of these applications is image processing for automatic vehicle guidance. Several activities in this special area are reported in literature [1][2][3]. But these investigations are in the field of algorithm research where programmable hardware like workstations and commercial multiprocessor systems are used. In opposite to this our approach was to realize a compact image processing hardware based on special VLSI circuits.

An automatic vehicle guidance system has to extract information for steering from a greyscale image provided by a camera. As displayed in Fig.1 the vision based system can be divided into the image processing unit which extracts relevant information of the greyscale image and a vehicle control unit which uses the image information to control the vehicle steering. This paper is addressed to the hardware realization of the image processing unit used for street boundary recognition. The main task of the image processing unit is data reduction. This is achieved by transformation of the pixel oriented image data to a list of line symbols for each image frame.

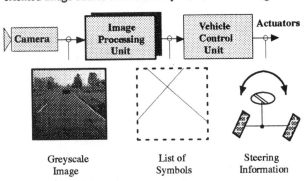

| Greyscale Image | List of Symbols | Steering Information |

Figure 1. Structure of an Automatic Vehicle Guidance System

Section 2 describes the pipeline of necessary image processing tasks. The algorithms are explained in section 3. Aspects of the hardware realization are given in section 4. The distribution of the processing tasks to ASICs and a micro controller are described in the sections 5 and 6. The developed image processing unit is presented in section 7. Section 8 provides an outlook to additional applications of the developed ASICs.

2. THE TASK OF THE IMAGE PROCESSING UNIT

Very important key symbols which are also used in many other applications are straight line segments of edges and contours. For this task a contour based processing pipeline of different algorithms has been developed consisting of the image processing steps gradient operation, edge thinning, line verification and Hough transformation. As a last step a rule based post processing is applied to the result of the Hough Transformation to reduce the line candidates from bundles of lines to just two lines representing the left and right street boundary. This gives also the basis for dynamic adaptation of the threshold in the Hough space.

In opposite to other works on special image processing hardware [8]–[13] not only a single algorithm but a pipeline of different algorithms matched to each other and tuned for real-time processing have been realized. The effects of the specific algorithms are displayed in Fig.2.

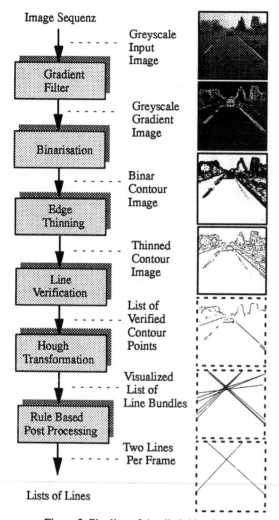

Image Sequenz

Gradient Filter — Greyscale Input Image

Binarisation — Greyscale Gradient Image

Edge Thinning — Binar Contour Image

Line Verification — Thinned Contour Image

Hough Transformation — List of Verified Contour Points

Rule Based Post Processing — Visualized List of Line Bundles

— Two Lines Per Frame

Lists of Lines

Figure 2. Pipeline of Applied Algorithms

3. THE ALGORITHMS

The first part of the pipeline is a **gradient operation** to determine the magnitude and the direction of the gradients.

G_y Mask
```
 1  1  1  1  1
 1  1  1  1  1
 0  0  0  0  0
-1 -1 -1 -1 -1
-1 -1 -1 -1 -1
```
G_x Mask
```
 1  1  0 -1 -1
 1  1  0 -1 -1
 1  1  0 -1 -1
 1  1  0 -1 -1
 1  1  0 -1 -1
```

Gradient Magnitude:
$$|G| = \frac{|G_x| + |G_y|}{2} \quad (1)$$

Gradient Direction:
$$G_{dir} = G_x / G_y \quad (2)$$

Figure 3. Gradient Filtering

The gradient operation is followed by the **edge thinning**. The edge thinning is performed by a maximum search inside a window of 7 x 1 pixel. The search window is oriented according to the gradient direction.

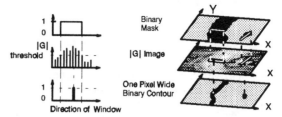

Figure 4. Binarisation and Edge Thinning

The result of the edge thinning is a one pixel wide binary contour image. In the next step the **line verification** removes small unconnected areas like texture and noise. The line verification is performed by a length count in a window perpendicular to the gradient direction. The size of the window is programmable from 7 x 1 pixels to 1 x 1 pixels. By the later option the size of the removed areas can be chosen.

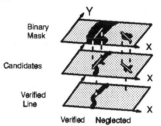

Figure 5. Line Verification

The fourth step in the pipeline is a modified **Hough transformation** converting the input data stream of unconnected contour points into a list of straight lines as output [4][5][6].

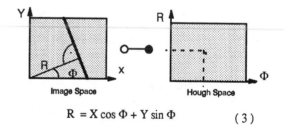

$$R = X \cos \Phi + Y \sin \Phi \quad (3)$$
$$\Phi = \arctan (G_x / G_y) \quad (4)$$

Figure 6. Hough Transformation

The evaluation of the Hough space is performed by the application of a threshold [$Thresh_{Hough}$] to the contents of the accumulation cells referred to as "weight" of the lines. This weight represents the number of contour points voting for a line. The threshold can be fixed or dynamically programmed to adapt the system to changes in the scene material. The result of the evaluation of the Hough space are bundles of lines. These line bundles represent the candidates for possible street boundaries (Fig. 2).

The last operation in the pipeline is the **rule based post processing** which is performed by a standard micro controller. The task is the reduction of the line bundles to the two most distinct lines representing the street boundaries.

In a first step a very simple street model is applied. The model says that there must be exactly one left and one right street boundary. The criterion for this is the range of the angle Φ.

Rule 1 : left boundary $\qquad$ $0° < \Phi < 90°$
Rule 2 : right boundary $\qquad$ $90° < \Phi < 180°$

Applying these rules the result of the hough transformation is separated into a left and a right line bundle. Inside these two bundles the lines with the highest weight are determined. By this second processing step two lines per frame are selected as the final result.

This model based approach is also used for the dynamic adaptation of the threshold for the Hough space evaluation. Therefore the bundle with the lower number of lines [n_{min}] is determined.

The applied calculation for adapting the threshold $Thresh_{Hough}$ is :

$3 > n_{min}$ $\qquad$ $Thresh_{Hough\ t+1} = Thresh_{Hough\ t} - 1$
$10 > n_{min}$ $\qquad$ $Thresh_{Hough\ t+1} = Thresh_{Hough\ t} + 1$
else $\qquad$ $Thresh_{Hough\ t+1} = Thresh_{Hough\ t}$

The rules of the post processing are kept as simple as possible to achieve a real time performance with reasonable expense.

4. FROM ALGORITHMS TO HARDWARE

The operations of the adapted image processing algorithms have been verified by simulations on workstations with real scene material. In the beginning of the hardware design phase the algorithms were simulated considering the envisaged hardware constrains in terms of accessible memory space, wordwidth and control signals in order to verify the later hardware behavior.

Due to the strong need of miniaturization the algorithms have been adapted to the requirements of the hardware realization. As recommended by Prewitt [14] simplified filter kernels with kernels coefficients 1,0,–1 were chosen for example. Furthermore the size of the kernels in our hardware is fixed to 5 x 5 coefficients as trade off between larger kernel sizes resulting in more line delays and smaller kernels which are more effected by noise and texture like structures.

As a crucial point the calculation of the gradient direction in the filter part gives the basis for the reduction of required computations in the following three processing steps of the pipeline. Therefore a divider with 50 nsec cycle time and three pipeline stages has been realized in hardware. The realized divider scheme is the Non-Restoring-Division [15] based on a Subtract-And-Shift operation.

5. TWO ASICS FOR LOW AND MEDIUM LEVEL PROCESSING

The main part of the developed image processing unit are two application specific VLSI circuits. To keep the hardware modular one ASIC performs the first three processing steps mentioned above and converts a greyscale image into a list of relevant contour points. This list is the input for the second ASIC which performs the Hough transformation to concatenate single contour points to parametrized lines. The two ASICs are manufactured in a 1.2 μm CMOS gate array technology. The complexity of the ASICs is 16k and 11k gates respectively.

6. THE TASKS OF THE MICRO CONTROLLER

Beside the realization of the rule based post processing described in section 3 the micro controller is used for additional tasks as well. One is to transmit start parameters to the ASICs after a reset or a power up. Those parameters are:

– the active window size, programmable from 16 x 16 pixels up to 512 x 512 pixels,
– the location of the active window in the video frame given by the start column and start row, and
– the search window size for line verification process.

All parameters can be fixed during processing or updated dynamically from frame to frame.

The most important task of the micro controller is to deliver the results of the image processing unit to the vehicle control unit. This communication is carried out in a hand shake mode .

7. THE IMAGE PROCESSING UNIT

Beside the two ASICs developed at the University of Hannover and the micro controller the image processing unit consists of an A/D–converter and memory devices necessary for data buffering and the realization of the Hough space. The introduced hardware performs the pixel oriented low level tasks, the medium level segmentation of the line symbols and the rule based post processing in real time. The entire image processing unit is realized on one printed circuit board . It has the dimension of 100×160 mm^2, a weight of 380 gr. and a power dissipation of less than 5 Watts. For an image format of 512 by 512 pixels and a frame rate of 25 Hz the board provides a computational power of 520 mega operations per second.

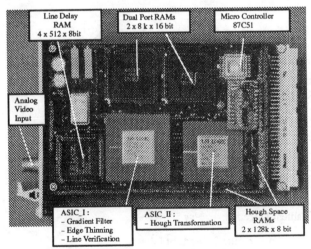

Figure 3. Single Board Image Processing Unit

The image processing unit can operate with fixed or self-adapting parameters or can be controlled by the high level processing part to achieve a dynamic adaptation during operation. By the dynamic controlling it becomes possible to modify threshold signals due to changing image material caused by different lighting conditions in real world application.

The latency of the processing pipeline is 1.1 frame times. Most of this time is required for the Hough transformation. It is unavoidable because the Hough space can only be analyzed after an entire image has been transformed.

The image processing unit (Fig.3) was developed and successfully integrated into an experimental host system of the Volkswagen AG to demonstrate the functionality under real traffic conditions.

8. FIELD OF APPLICATIONS

The two ASICs developed in this project can be used as a set or separately and have a restricted programmability in terms of image sizes, thresholds and parameters. By this it becomes possible to adapt them to different applications. For instance the results of the first ASIC can be used in combination with an object tracing algorithm or the Hough transformation ASIC as coprocessor in an image processing system.

Because straight line segments of edges and contours are important key symbols used in many applications this image processing hardware is not only of interest for vehicle guidance but for other industrial and robotic applications as well.

9. CONCLUSION

At the University of Hannover an image processing unit for vehicle guidance on motorways has been developed and tested under real traffic conditions. Due to the special restrictions to size and power consumption in vehicles applications specific CMOS circuits have been designed for the computation intensive image processing tasks. The ASICs have been combined with standard components as an A/D-converter, RAMs and a micro controller to build a single board image processing unit which can process images of 512 by 512 pixels and a frame rate of 25 Hz in real time.

ACKNOWLEDGEMENT

This work is part of PROMETHEUS project supported by the European automotive industry and the Bundesministerium für Forschung und Technik. Special thanks to the Volkswagen AG for providing the experimental host system.

REFERENCES

[1] P. Kahn, et.al. "A Fast Line Finder for Vision-Guided Robot Navigation" IEEE Trans. On Pattern Analysis and Machine Intelligence Vol 12 , No.11, November 1990, pp. 1098-1102.

[2] C. Thorpe, et.al. "Vision and Navigation for the Carnegie-Mellon Navlab" IEEE Trans. on Pattern Analysis and Machine Intelligence Vol 10 , No.3, May 1988, pp. 362-373.

[3] D. Kuan, et. al. "Autonomous Robotic Vehicle Road Following". IEEE Trans. On Pattern Analysis and Machine Intelligence, Vol. 10, No. 5, Sept. 1988, pp. 648-658.

[4] E.R.Davies " Occlusion Analysis for Object Detection Using The Generalized Hough Transformation" Signal Processing 1989/16, pp. 267-277, North-Holland.

[5] T.Risse, "Hough Transform for Line Recognition: Complexity of Evidence Accumulation and Cluster Detection" Computer Vision, Graphics and Image Processing, 1989/46, pp. 327-345.

[6] H. Koshimizu, et.al. "FIHT2 Algorithm: A Fast Incremental Hough Transform", IEICE Trans. on Communications Electronics Information and Systems, Vol.E74, No.10, Oct. 1991, pp. 3389 - 3392.

[7] H.M.Alnuweiri, et.al "Optimal Image Computations on reduced VLSI Architectures", IEEE Trans. on Circuits and Systems, Vol.36, No.10, Oct 1989.

[8] C.Y.Lee, et.al. " An Efficient Architecture For Real Time Edge Detection" , IEEE Trans. On Circuits and Systems, Vol. 36, No.20, Oct. 89. pp. 1350-1359.

[9] H.Ringshauser, et.al "Real-Time Computation of Image Structures for Different Resolutions" Frauenhofergesellschaft Reports 4 / 90 , pp. 45-50.

[10] D. Ben-Tzvi, et.al "Analog Implementation of the Hough Transform Suitable for Single Chip Implementation", IEEE Int. Sym. on Circuits and Systems May 1990, pp. 53-56.

[11] A.L. Fischer, et.al. "Computing the Hough Transform on a Scan Line Array Processor" IEEE Trans. on Pattern Analysis and Machine Intelligence Vol 11 , No.3, March 1989, pp. 262-265.

[12] LSI LOGIC "L64 250 Histogram / Houghtransform Processor" Data sheet, August 1989.

[13] PLESSEY "Digital Signal Processing IC Handbook" 1991.

[14] E.R. Dougherty, et.al "Matrix Structured Image Processing" Prentice-Hall, 1987.

[15] K. Wang "Computer Arithmetic Principles, Architecture and Design" Wiley&Son, 1979.

Implementing Dynamic Programming Algorithms for Signal and Image Processing on Array Processors

W.H. Chou, K.I. Diamantaras, and S.Y. Kung[†]

Abstract: This paper presents the implementation of a generic dynamic programming algorithm on array processors. A dynamic programming (DP) chip is proposed to speed up the processing of the dynamic programming tasks in many applications, including the Viterbi algorithm, the boundary following algorithm, the dynamic time warping algorithm, etc. By adopting a torus interconnection network, an internal/external dual buffer structure, and a multilevel pipelining design, a performance of several GOPS per DP chip is expected. Both the dedicated hardware design and the data flow control of the DP chip are discussed in this paper.

1. INTRODUCTION

Dynamic Programming [1] is a classical optimization technique. Although its complexity is much lower than the exhaustive search, it still demands a large computation power. The dedicated hardware design is usually needed to speed up the processing of the dynamic programming algorithm. For example, several chips [2, 3] designed for dynamic time warping can be considered as a special case of DP. However, the complicated design of their instruction sets to support the general cost function calculation hampers the efficient utilization of the chip area and limits the clock rate. Based on the observation that the complexity of DP is usually at least two orders higher than the cost functions in most applications, we assume that the host computer could calculate the cost functions and download them to the DP chips. A DP processor is fully customized for the dynamic programming task. Many DP processors can be put in one DP chip because of their simplicity and regularity. The torus interconnection structure makes it applicable to the 1D signal processing as well as the 2D image processing. The internal/external buffer design allows the data downloading, DP processing, and results uploading executed simultaneously. Pipelining at different levels is adopted, from the processor to bit level. The computation power of a single chip with tens of DP processors is expected to reach the order of GOPS.

A brief overview of the dynamic programming algorithm is given in this section. Sections 2 and 3 show the dynamic programming formulation of the Viterbi and the boundary following algorithms respectively. Section 4 presents the design of the DP processor and the architecture of the DP chip. The data flow control is discussed in Section 5.

Dynamic Programming Algorithm   DP is applicable to functions of the form

$$f(x_1, x_2, \cdots, x_T) = \sum_{t=1}^{T} s(x_t) + \sum_{t=1}^{T-1} h(x_t, x_{t+1}) \qquad (1)$$

where the variables $x_1, \cdots, x_T$ take on discrete values, $x_t \in \{a_1, a_2, \ldots, a_N\}$. The exhaustive search approach to optimizing f has very large computational complexity $O(N^T)$. DP takes advantage of the special structure of f to dramatically reduce this complexity. Let $f_t(x_t)$ be the partial cost for having traveled just t stages:

$$f_t(x_t) \equiv \min_{x_1, x_2, \ldots, x_{t-1}} [\sum_{i=1}^{t-1} h(x_i, x_{i+1}) + \sum_{i=1}^{t-1} s(x_i)] + s(x_t)$$

for all $t = 1, 2, \ldots, T$. Then the following recursive formulas can be used to compute the minimum:

$$f_1(x_1) = s(x_1) \qquad (2)$$

$$f_t(x_t) = \min_{x_{t-1}} [f_{t-1}(x_{t-1}) + h(x_{t-1}, x_t)] + s(x_t) \qquad (3)$$

$$\min_{x_1, x_2, \ldots, x_T} [f(x_1, x_2, \cdots, x_T)] = \min_{x_T} [f_T(x_T)] \qquad (4)$$

Furthermore, the global optimizing solution $\{x_1^*, x_2^*, \cdots, x_T^*\}$ can be calculated by maintaining – as the algorithm proceeds – a predecessor for every value of each variable. Precisely, for every value of x_t we define

$$\text{predecessor}(x_t) = x_{t-1}^*(x_t) = \arg \min_{x_{t-1}} [f_{t-1}(x_{t-1}) + h(x_{t-1}, x_t)] \qquad (5)$$

Equations (2)–(5) define the dynamic programming algorithm. The complexity of the algorithm is $O(N^2 T)$, as opposed to the $O(N^T)$ complexity of the exhaustive search. This is many orders of magnitude improvement, especially for the cases where N and/or T are large.

2. THE VITERBI ALGORITHM

The Viterbi algorithm [4] has been applied to many problems with the trellis structure. For example, the retrieving phase of the hidden Markov model [5] is a model-scoring problem which finds the most likely model with the highest probability $Pr(O|\Lambda)$, where $O = \{O(1), \ldots, O(T)\}$ is the observation data, $\Lambda = \langle A, B, \pi \rangle$, A is the state-transition matrix $\{a_{ij}\}$, B is the emission-distribution matrix $\{b_{ik} \equiv b_i(v_k)\}$, and π is the initial state distribution. The Viterbi algorithm is used to determine the most probable state sequence Q^* generating the observation sequence O. It consists of the following two procedures:

The forward evaluation procedure:

$$u_i(t+1) = \max_{1 \leq j \leq N} a_{ij}\delta_j(t) \tag{6}$$

$$\delta_i(t+1) = u_i(t+1)b_i(O(t+1)) \tag{7}$$

for $1 \leq t \leq T - 1$ and $1 \leq i \leq N$. Here $\delta_j(t)$ denotes the probability of the most likely subsequence ending at the candidate state j at the stage t. The initial conditions are set to $\delta_i(1) = \pi_i b_i(O(1))$. To facilitate the path backtracking, the indices $\{j_i^*(t)\}$ are introduced to keep track of the most probable incoming paths:

$$j_i^*(t) = \arg \max_{1 \leq j \leq N} a_{ij}\delta_j(t)$$

for $t = 1, \ldots, T - 1$.

The path backtracking procedure: Once the $\{j_i^*(t)| \ 1 \leq i \leq N, 1 \leq t \leq T - 1\}$ are obtained, the most probable path can be determined by path-backtracking.

Dynamic Programming Formulation The Viterbi algorithm can be rewritten as a dynamic programming algorithm by defining:

$$
\begin{aligned}
f(q_1, q_2, \ldots, q_T) &= \log Pr(O, Q|\Lambda) \\
&= \log[\pi_{q_1} b_{q_1}(O(1)) \times \prod_{t=1}^{T-1} a_{q_{t+1}q_t} b_{q_{t+1}}(O(t+1)] \\
&= \log \pi_{q_1} + \sum_{t=1}^{T} \log b_{q_t}(O(t)) + \sum_{t=1}^{T-1} \log a_{q_{t+1}q_t} \quad (8)
\end{aligned}
$$

Eq. (8) has the same format as Eq. (1) if $x_t = q_t$ for $1 \leq t \leq T$, and

$$s(x_t) = \begin{cases} \log \pi_{q_1} + \log b_{q_1}(O(1)) & \text{if } t = 1 \\ \log b_{q_t}(O(t)) & \text{otherwise} \end{cases} \tag{9}$$

$$h(x_t, x_{t+1}) = \log a_{q_{t+1}q_t} \tag{10}$$

The computational requirements for the cost functions of $s(x_t)$ and $h(x_t, x_{t+1})$ can be derived from Eq. (9) and Eq. (10) as

$$N(T + 1) \text{ log-operations } + N \text{ additions for } s(x_t) \tag{11}$$

and

$$N^2 \text{ log-operations for } h(x_t, x_{t+1}) \tag{12}$$

The computational requirement for the dynamic programming procedure is

$$N(N + 1)(T - 1) \text{ additions } + N^2(T - 1) \text{ max/min} \tag{13}$$

Assume that a HMM model has 30 states and the length of the observation sequence is 150. Equations (11)–(13) show that more than 98% of the computational load is due to the dynamic programming procedure.

3. THE BOUNDARY-FOLLOWING PROCEDURE

Dynamic programming can also be applied to many image processing problems. For example, Ballard and Brown [6] showed how to formulate the boundary-following procedure by dynamic programming. A local edge detection operator is applied to a gray-level picture to produce edge magnitude and direction information. One possible criterion for a "good boundary" is a weighted sum of high cumulative edge strength and low cumulative curvature, that is, for an n−pixel curve,

$$h(x_1, \ldots, x_n) = \sum_{k=1}^{n} s(x_k) + \alpha \sum_{k=1}^{n-1} q(x_k, x_{k+1}) \tag{14}$$

where $s(x_k)$ is the edge strength at the pixel x_k, α is a negative weight, and

$$q(x_k, x_{k+1}) = \text{ diff } [\phi(x_k), \phi(x_{k+1})]$$

is the difference of edge directions between two pixels, x_k and x_{k+1}. There is an implicit constraint that consecutive x_k's must be grid neighbors. Note that Eq. (14) has the same format as Eq. (1) and can be optimized by

$$\begin{aligned}
f_1(x_1) &= s(x_1) \\
f_k(x_k) &= \max_{x_{k-1}}[f_{k-1}(x_{k-1}) + \alpha q(x_{k-1}, x_k)] + s(x_k)
\end{aligned}$$

4. DP ARRAY PROCESSOR DESIGN

The array processor approach to the DP implementation is discussed in this section.

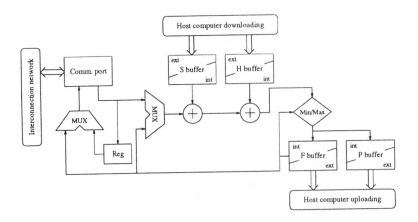

Figure 1. The block diagram of a DP processor.

DP Processor design   Figure 1 illustrates the block diagram of a DP processor. The $f_{t-1}(x_{t-1})$ value in Eq. (3) is collected either from the local memory or from the other processors. It is then added to $s(x_t)$ and $h(x_{t-1}, x_t)$. (Note that, $s(x_t)$ is a constant w.r.t. all possible states x_{t-1}, so it does not matter that whether it is added before or after the min/max operator.) The sum is compared with the stored max/min value. The max/min value and the predecessor information will be updated if necessary. In order to match the dynamic range of the data size in various applications, 32-bit fixed-point arithmetic units (two adders and one comparator) are used. The S and H buffers store the data of $s(x_t)$ and $h(x_{t-1}, x_t)$ respectively. Since the h function has higher dimensions than the s function, the H buffer is much larger than the S buffer. However, 2 Kbytes for the H buffer should be enough in most cases. The F and P buffers store the data of $f_t(x_t)$ and the predecessor information respectively. They have the same size as the H buffer. Each of the S, H, F, P buffers is comprised of two memory banks: the external buffer and the internal buffer. The external buffer serves as a medium between the DP processor and the host computer. The internal buffer supports the computation in each DP processor. These two sub-buffers can be switched in just one instruction cycle. Random accessibility to those buffers is preferred for the sake of flexibility. Nevertheless, by carefully arranging the data, a simpler sequentially accessible memory can be adopted. The *Reg* in Figure 1 is used to buffer some data during the communication.

DP Chip design   The DP processor is very simple and we can put many of them in a single chip. Figure 2 shows the block diagram of a DP chip

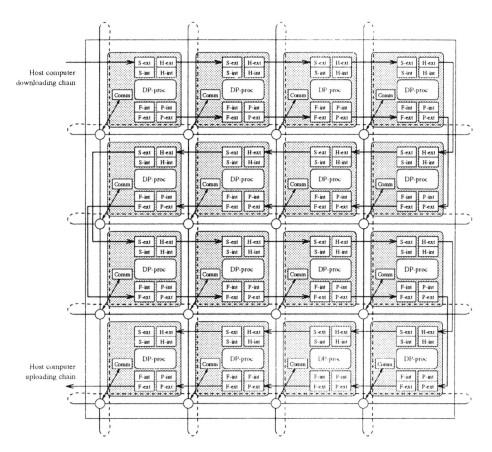

Figure 2. The block diagram of a DP chip with 4 × 4 DP processors. The dashed lines are the wrap-around links supported by the board.

with 4 × 4 DP processors. A torus topology is selected as the interconnection network because image processing applications usually involve 2D communications and it is easy to embed 1D communication structure in the torus network. Note that the interconnection topology inside the DP chip is a mesh with enhanced extension ability to support size scalability. The wrap-around links of the torus are supported by the board. The width of the communication channel is limited by the chip area as well as the allowable pin count. Byte-serial channel design for a 4 × 4 DP chip requires 128 data pins. If more DP processors are packed in one chip, the width of the communication channel may be further dropped to a nibble, or even a single bit. Some serial/parallel converters are needed when the full width cannot be supported. By analyzing the communication requirements in most cases, we use only one pair of input/output communication channels in each DP processor to reduce the complexity of the processor. The network switches are programmable for data routing. A complete dynamic programming task would be composed of the following three stages:

1. *Downloading:* The host computer calculates the cost functions and downloads the information into the DP chip.

2. *Dynamic Programming:* The DP chip performs the dynamic programming algorithm based on the provided cost functions.

3. *Uploading:* The results of dynamic programming are uploaded to the host computer when the DP is finished.

The dual buffer concept allows the DP processor to execute the dynamic programming algorithm with the internal buffers, and the host computer to download the new data and upload the old results from the external buffers simultaneously. The downloading chain and the uploading chain also increase the chip testability. The SIMD control is adopted in our design. The control words have a VLIW format to simplify the decoding circuit. The format contains a field for the arithmetic operation, a field for the switch configuration, a field for the downloading/uploading chain control, and, perhaps, a field for the RAM addressing and the processor masking mechanism.

5. DATA FLOW CONTROL

In this section, the data flow control for the Viterbi algorithm and the boundary following algorithm are presented.

SIMD scheduling for the Viterbi algorithm   The Viterbi algorithm requires every node to talk to each other. The ring structure is a straightforward result according to this concern. It is easy to embed the ring structure into a torus. However, the configuration of the switching nodes is not homogeneous, which may hamper the SIMD control. Fortunately, the torus structure can support a homogeneous broadcasting scheme. Figure 3(a) shows the communication timing for the upper-leftmost DP processor to broadcast its value to the other processors. The same communication timing can be applied to every node. It thus solves the inhomogeneous problem and gets rid of the ring design. Figure 3(b) illustrates the communication configuration of a DP processor at each time step. Note that, the data broadcasting of the 4×4 processors only takes 16 cycles which are the same as what the ring structure needs.

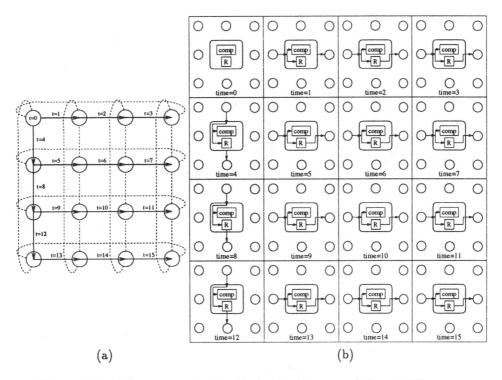

(a)   (b)

Figure 3. (a) The communication timing for the upper-leftmost DP processor to broadcast its value to the other processors. (b) The communication configuration of a DP processor at each time step. R stands for a register delay, while *Comp* is the computation circuitry.

Spiral scheduling for the boundary following algorithm   In the boundary-following algorithm, only the values of 8 neighbors need to be concerned. Figure 4(a) depicts the communications needed for the 8 neighbors to send their values to the center. There are 9 cycles required: 1 cycle for the center pixel itself and 8 cycles for the neighbors. Based on this communication pattern, Figure 4(b) shows the time-evolution of the communication configurations of one processor.

6. CONCLUSION

In this paper, a dedicated design for a generic dynamic programming algorithm is reported. By using a massively parallel architecture, we expect the performance to be at least 100 times faster than the conventional approach. Two examples in the signal and image processing areas demonstrate the po-

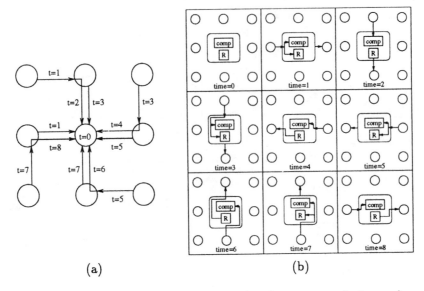

(a)   (b)

Figure 4. (a) The communication timing for the center node to receive all
the information from its 8 neighbors. This timing and communications pattern avoids diagonal connections. (b) The communication
pattern of a node with its 8 neighbors.

tential applications of the DP chip.

References

[1] R. Bellman, *Applied Dynamic Programming*, Princeton University Press, 1962.

[2] David J. Burr, Bryan D. Ackland, and Neil Weste, *Array Configurations for Dynamic Time Warping*, IEEE Transaction on ASSP, Vol. ASSP-32, No. 1:119-128, February 1984.

[3] Robert Kavaler, R. W. Brodersen, Tobias G. Noll, Menachem Lowy, and Hy Murveit, *A Dynamic Time Warp IC for a One Thousand Word Recognition System*, Proc. of IEEE ICASSP'84:25B.6.1-25B.6.4, 1984.

[4] A. J. Viterbi, *Error bounds for convolution codes and an asymptotically optimal decoding algorithm*, IEEE Transactions on Information Theory, IT-13:260-269, April 1967.

[5] L. R. Rabiner and B. H. Juang, *An introduction to hidden Markov models*, IEEE ASSP Magazine, 3(1):4-16, January 1986.

[6] D. H. Ballard and C. M. Brown, *Computer Vision*, Prentice Hall, 1982.

† Department of Electrical Engineering, Princeton University, Princeton, NJ 08544, USA.

Hybrid Survivor Path Architectures for Viterbi Decoders

Peter J. Black and Teresa H.-Y. Meng

Information Systems Laboratory, Stanford University, CA 94305

Abstract

A new approach to survivor path architectures for Viterbi decoders is proposed based on hybrid architectures that combine the classical register-exchange and trace-back methods. Two classes of hybrid architecture are proposed: the hybrid pretrace-back architecture and the hybrid trace-forward architecture. Pretrace-back is a preprocessing of the add-compare-select (ACS) decisions to increase the effective trace-back recursion rate, while trace-forward is a concurrent processing of the decisions to initialize the trace-back recursion. Both of these architectures can be implemented using a single compact memory, typically twice the survivor path length in size, without any loss in throughput compared to conventional trace-back architectures. Based on area estimates for common decoding problems the hybrid architectures reduce the required chip area by up to 40% compared with the popular k-pointer trace-back architecture.

1. Introduction

Survivor path decode is a commonly overlooked aspect of VLSI Viterbi decoder design. The chip area of state-of-the-art decoders is often dominated by survivor path memory and the throughput is often limited by the rate at which the survivor path sequence can be decoded. Hence innovations in either aspect would significantly impact the overall area and/or throughput of the design.

The two classical algorithms for survivor path storage and decode are the *register-exchange* method and the *trace-back* method [1]. Both algorithms require a recursive update which fundamentally limits the throughput. The focus of this paper is on the most difficult survivor path update problem; that is, architectures that can support a complete trellis update per clock cycle (i.e. fully parallel decoders). To date, trace-back is the preferred method for trellis complexities above 8 states due to reduced area and power dissipation.

In this paper a new approach to survivor path decode is proposed based on *hybrid* architectures. These hybrid architectures combine the high throughput, forward processing characteristics of the register-exchange method with the area-efficient, backward processing characteristics of the trace-back method, to yield overall area reductions while maintaining high throughput.

The organization of the paper is as follows. In Sections 2 and 3 the register-exchange and trace-back methods are reviewed. In Sections 4 and 5 the hybrid pretrace-back and hybrid trace-forward architectures, respectively, are described. In Section 6, the areas of the classical and hybrid architectures are compared and finally the paper is concluded with a summary.

2. Register Exchange Method

The register-exchange method is the simplest and most direct method of survivor path update and decode. Associated with every trellis state is a register which contains the survivor path

leading to that state. Each survivor path is uniquely specified by and stored as the sequence decisions along the survivor path. The decision sequence $d^S_{n-\Delta, n}$ of the survivor path to state S from time $n - \Delta$ to n is given by the recursive update

$$d^S_{n-\Delta, n} = (d^{S'}_{n-1-\Delta, n-1} \ll 1)\, d^S_n \qquad (1)$$

where S' is the predecessor state of S as determined by its decision d^S_n from the ACS update. The current state decision is used to select the predecessor state decision sequence which is left shifted to allow the current decision to be appended to the sequence. This update occurs concurrently for all states, hence the name register-exchange, since each update corresponds to an exchange of the register contents modulo the shift and append.

An example register-exchange network for a 4-state radix-2 (binary shift register) trellis is shown in Fig. 1. The decision sequence length Δ is limited to the survivor path length L. Under this condition all the survivor paths should merge with high probability and hence the oldest decision of any state can be chosen as the basis for the decoded output.

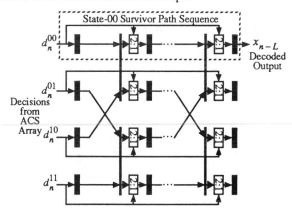

Fig. 1. Register-exchange network for a 4-state, radix-2 trellis (2:1 represents a 2 to 1 multiplexer).

3. Trace-Back Method

The trace-back method is a backward processing algorithm for survivor path update and decode. Such an algorithm is non-causal and requires the decisions to be stored in a decision memory prior to tracing back the survivor path. The trace-back recursion estimates the previous state S_{n-1} given the current state S_n according to the following update

$$S_{n-1} = f(S_n, d^S_n). \qquad (2)$$

The current state decision d^S_n is read from the decision memory using the current state S_n and time index n as an address. The function $f(\cdot)$ is dependent on the structure of the trellis. For the common radix-2 trellis the recursion simplifies to

$$S_{n-1} = d^S_n (S_n \gg 1), \qquad (3)$$

This work was supported in part by DARPA.

which corresponds to a 1-bit right shift of the current shift register state with input equal to the current state decision.

Practical trace-back architectures must achieve a throughput which is matched to the ACS update rate while minimizing the required decision memory. The one-pointer trace-back architecture described in the following subsection is a practical implementation of the architecture given in [2].

3.1. One-Pointer Trace-back Architecture

The one-pointer trace-back architecture decodes a block of decisions for each survivor path traced [1]. The decision memory is organized as a cyclic buffer and is partitioned into a write region and a read region as shown in Fig. 2. During each block decode phase, new decision vectors (one decision per state) are written to the write region while the survivor path is traced and decoded from the read region. The read region is divided into a merge block of length L and a decode block of length D. Trace-back of the merge block is used to generate an estimate of the initial state at the beginning of the decode block. Once a block is decoded it becomes the write block for the next phase and the other logical partitions appropriately shift.

To ensure continuous operation the trace-back operations must complete in the time it takes to fill the write region. The required decision memory length M as a function of the trace-back recursion rate TRR (i.e. the number of trace-back recursions per clock cycle) is given by

$$M = L(1 + \frac{2}{TRR - 1}) \quad \text{for } TRR > 1. \qquad (4)$$

The one-pointer trace-back architecture is simple and compact; however, its practicality is limited by the need to support recursion rates greater than one (typically 2 to 3). The classical solution to this problem is to run multiple trace-back processes concurrently as described in the following subsection.

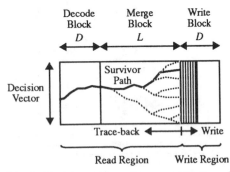

Fig. 2 Decision memory for one-pointer trace-back.

3.2. k-Pointer Trace-back Architecture

The k-pointer trace-back architecture is so named because it supports trace-back of k survivor paths concurrently [1,3]. Assuming each trace-back stream is iterated at the minimum trace-back recursion rate of one, a k-pointer trace-back architecture requires $2k$ independent memories each of length $L/(k-1)$, equivalent to a total memory length M of

$$M = L(2 + \frac{2}{k - 1}) \quad \text{for } k > 1. \qquad (5)$$

Increasing the number of trace-back pointers reduces the required decision memory; however, minimal memory does not equate to minimal area. As the memory length is reduced the

number of memory partitions is increased and hence the peripheral logic overhead is increased.

4. Hybrid Pretrace-back Architectures

The trace-back recursion rate of the one-pointer trace-back architecture can be reduced to one by applying lookahead to the trace-back recursion. In general the trace-back recursion from n to $n - \Delta$ can be collapsed in to a single trace-back recursion of the form

$$S_{n-\Delta} = F(S_n, d^S_{n-\Delta, n}), \qquad (6)$$

where $d^S_{n-\Delta, n}$, the survivor path sequence from state S_n, is thought of as a composite decision and $F(\cdot)$ is the collapsed trellis mapping function. Since the state sequence is unknown prior to trace-back, the composite decisions must be calculated for all states and stored in the decision memory. This lookahead computation is referred to as Δ-*levels of pretrace-back*[1].

For the common radix-2 trellis and 2-levels of pretrace-back the trace-back recursion simplifies to

$$S_{n-2} = d^S_{n-2, n}(S_n \gg 2), \qquad (7)$$

equivalent to trace-back of a radix-4 trellis using 2-bit decisions.

The hybrid Δ-level pretrace-back architecture is shown in Fig. 3. The pretrace-back computation can be implemented using a register-exchange network of length Δ. Every Δ iterations the survivor path of each state is output as the composite pretraced decision for that state. Using Δ-levels of pretrace-back, the trace-back recursion iterates from n to $n - \Delta$ in a single trace-back operation.

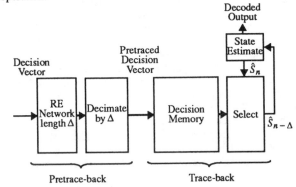

Fig. 3 Hybrid pretrace-back architecture.

4.1 Decision Memory Organization

The decision memory organization for the hybrid pretrace-back architecture is the same as the one-pointer architecture described in Section 3.1. The memory can be implemented using either single-ported or dual-ported memory.

Single-ported architectures exploit the fact that the pretraced decision vector stream is decimated by a factor of Δ. Decision vector write operations occur once every Δ cycles, leaving the other $\Delta - 1$ cycles for trace-back read operations. When using pretraced decisions, the trace-back recursion rate refers to the number of composite decision recursions per cycle as given by (6). The effective trace-back recursion rate $ETRR$ is the aver-

---

1. The concept of pretrace-back was published by the authors in the context of the radix-4 design presented in [4]. Independently, the equivalent of 8-levels of pretrace-back was published in [5].

age recursion rate for the original trellis and is given by

$$ETRR = (\Delta - 1) \times TRR. \qquad (8)$$

Although trace-back recursion rates greater than one can be supported, the reason for using pretrace-back is to reduce the required recursion rate to the minimum of one, thus maximizing throughput. In the following discussion TRR is assumed to be one. The total decision memory length M is given by

$$M = L\left(1 + \frac{2}{ETRR - 1}\right) \quad \text{for} \ ETRR > 1. \qquad (9)$$

In the context of pretraced decisions, a decision memory length of M implies a memory capable of storing all decisions over a time interval M, so the number of stored vectors is M/Δ.

Using dual-ported memory, read operations can occur on every cycle, hence $ETRR = \Delta$. Based on area estimates dual-ported architectures are approximately 20% larger than single-ported architectures due to the lower density memory.

4.2. Pretrace-back Unit

Implementation of pretrace-back using a register-exchange network relies on organizing the trellis into a constant geometry schedule. For shift register trellises, such a reorganization is not only possible for the full N-state trellis, but is also possible for the subtrellises over the pretrace-back interval of length Δ. By exploiting the subtrellis structure, the area complexity of the register-exchange interconnect is reduced from $\Omega(N^2\Delta)$ to $\Omega(N\Delta^2)$ and hence the approach scales well for arbitrary N.

An example of this reorganization for 3-levels of pretrace-back is shown in Fig. 4. In this case the 4-state subtrellis has the structure of a constant geometry schedule; however, each row of state nodes does not correspond to the same state. The pretrace-back (PTB) unit is implemented using a conventional register-exchange network, except that the decision controlling the multiplexers for each register row must follow the state sequence of the corresponding row in the subtrellis. For the 3-level pretrace-back example this is implemented by adding a 3-way switch to each decision input, and at the end of each 3 cycle period the registers contain the pretraced decisions for the states at time n as shown in Fig. 5. The complete N-state, 3-level PTB unit consists of $N/4$ 4-state, 3-level PTB units.

5. Hybrid Trace-forward Architectures

The one-pointer trace-back architecture requires trace-back recursion rates greater than one, because the survivor path is traced through the merge block to estimate the starting state for the decode block. Estimating this state a priori using a procedure referred to as *trace-forward*, trace-back of the merge block is avoided and the trace-back recursion rate is reduced to one.

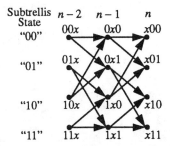

Fig. 4. 4-state subtrellis for a radix-2 trellis over the pretrace-back interval $n-3$ to n. State labels are binary and x represents the common state bits within the subtrellis.

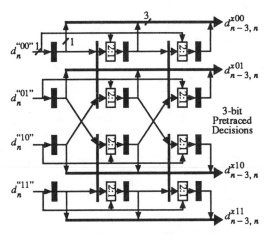

Fig. 5. 4-state, 3-level subtrellis pretrace-back unit.

In describing the trace-forward method it is useful to define the tail of a survivor path. Given a time reference m, every trellis state S at time $m + \Delta$ has an associated survivor path which traces back to some state at time m, referred to as the tail state $T^S_{m,m+\Delta}$. For $\Delta > L$ all the survivor paths should trace-back to the same state and hence any tail state can be used as a state estimate $\hat{S}_m$ for the trellis at time m.

The trace-forward method is a forward recursive procedure for calculating the tail states over the interval m to $m + \Delta$. Associated with every state is a register which contains the tail state for that state. The tail states are initialized at time m, at which point the survivor paths are of zero length, and hence the tail state and the current state are the same. This leads to the following initial condition for the trace-forward recursion

$$T^S_{m,m} = S. \qquad (10)$$

The trace-forward update is merely a selection operation, where the decision of the current state S is used to select the tail state of the predecessor state S' to form the updated tail state as follows

$$T^S_{m,m+\Delta} = T^{S'}_{m,m+\Delta-1}. \qquad (11)$$

This update occurs concurrently for all states and is similar in form to the register-exchange method without the shift.

The hybrid trace-forward architecture is shown in Fig. 6. The trace-forward unit is used to initialize the trace-back recursion by estimating the starting state of the decode block.

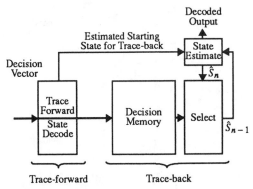

Fig. 6. Hybrid trace-forward architecture.

5.1. Decision Memory Organization

Organization of the decision memory is similar to the one-pointer trace-back architecture except that all blocks are of equal length L as shown Fig. 7. During each block decode phase, new decision vectors are written to the write region while the survivor path is traced and decoded from the read region. Concurrently, the new decisions are processed by the trace-forward unit, which estimates the state at the beginning of the each write block. Since the write block becomes the merge block of the next phase, this state is the required starting state for the next decode block.

At first glance the required decision memory length is $3L$; however, because the read and write rates are matched, the write block can be folded onto the decode block, thus reducing the memory length to $2L$. After a decision vector is read and decoded, its location is available for the next written vector.

5.2. Trace-forward Unit

The architecture of a trace-forward state decoder for a 4-state radix-2 trellis is shown in Fig. 8. At time m each tail state registers is initialized with the state index. On subsequent iterations the interconnect and multiplexers implement the recursive update under the control of the current decisions. To estimate the state at time m, the trace-forward unit is busy from m to $m+L$, at which point the state $\hat{S}_m$ can be decoded; hence state estimates are generated at intervals of L.

The basic building block is a trace-forward select (TFS) unit, which consists of a multiplexer and register as shown in Fig. 8. Since the topology of the trace-forward decoder is identical to the ACS array for the same decoder, area-efficient ACS topologies can be transformed into to area-efficient TFS topologies by simply substituting TFS units for ACS units.

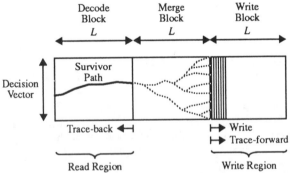

Fig. 7. Decision memory for hybrid trace-forward.

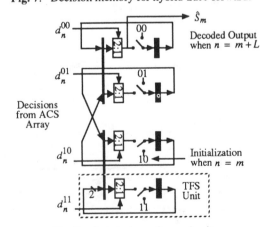

Fig. 8. 4-state, trace-forward unit.

6. Area Comparisons

The practical significance of the new hybrid architectures is evident when compared against the classical k-pointer architectures for two widely used trellises. The 64-state, radix-2 trellis is the basis of R=1/2 punctured convolutional codes and the 16-state, radix-4 is the basis of the 4D Wei trellis code. Based on area estimates, the minimal area architectures for these trellises are summarized in Table 1. In both cases, use of the hybrid architectures reduces the estimated area by approximately 40%.

Pretrace-back, which is decision based, is more sensitive to the number of bits per decision (radix). Trace-forward, which is state based, is more sensitive to the number of bits per state label (states). Based of area estimates the hybrid pretrace-back architecture is smaller for radix-2 trellises of more than 16-states, while the hybrid trace-forward architecture is smaller for fewer states and higher radix trellises.

Trellis	Architecture	Decision Memory M	Relative Area
64-state radix-2 (L=90)	hybrid pretrace-back	$2L$	0.66
	3-pointer trace-back	$3L$	1.0
16-state radix-4 (L=20)	hybrid trace-forward	$2L$	0.59
	2-pointer trace-back	$4L$	1.0

Table 1. Area estimates for trace-back architectures based on layout designed in the MOSIS scalable CMOS process.

7. Summary

A new approach to survivor path architectures is proposed based on hybrid architectures that combine the classical register-exchange and trace-back methods. Two classes of hybrid architecture are proposed: the *hybrid pretrace-back* architecture and the *hybrid trace-forward* architecture. Both architectures are based on the one-pointer trace-back architecture. Pretrace-back is a lookahead computation applied to decisions prior to trace-back, while trace-forward is a state estimate procedure used to initialize trace-back. Both of these architectures can be implemented using a single compact memory, typically $2L$ in length, without any loss in throughput compared to conventional trace-back architectures, and offer area reductions of up to 40%.

References

[1] G. Feygin and P. G. Gulak, "Survivor Sequence Memory Management in Viterbi Decoders," Technical Report CSRI-252, University of Toronto, Jan 1991.

[2] C. M. Radar, "Memory Management in a Viterbi Decoder," *IEEE Trans. Commun.*, Vol. COM-29, No. 9, Sep 1981.

[3] R. Cypher and C. B. Shung, "Generalized Trace-back Techniques for Survivor Memory Management in the Viterbi Algorithm," *Proc. GLOBECOM 90*, Vol. 2, pp. 1318-1322, Dec 1990.

[4] P. J. Black and T. H.-Y. Meng, "A 140Mb/s, 32-State, Radix-4 Viterbi Decoder," *ISSCC Dig. Tech. Papers*, pp. 70-71, Feb 1992.

[5] E. Paaske, *et al.*, "An Area-Efficient Path Memory Structure for VLSI Implementation of High Speed Viterbi Decoders," *INTEGRATION*, Vol. 12, No. 1, p. 79-91, Nov 1991.

A HIGH PERFORMANCE VECTOR QUANTISATION CHIP

M. Yan, Y. Hu and J.V. McCanny

A demonstrator chip is described for real time speech vector quantisation applications. The chip has been fabricated in a 1.2 micron CMOS technology and contains 38,000 transistors. It occupies a die area of only (4.8×4.6) mm^2 and can perform over 80 million multiply/ accumulate operations per second. The chip fully demonstrates the benefits which result from using a regular dedicated VLSI architecture.

1. INTRODUCTION

Vector Quantization (VQ) is an efficient coding technique where data compression is achieved by performing pattern matching processes [1]. However, the computational complexity involved is highly demanding, particularly when a full VQ codebook search is required to achieve optimal performance. Over the past number of years this has prompted research into special purpose VLSI architectures for implementing real-time VQ systems. Important contributions have been made by authors such as Kolagotla et al [2,3], Davidson and Gersho [4], Nelson and Read [5], Dianysian and Baker [6] and Ramamoorthy and Brahmaji [7]. The authors have also undertaken research in this field and in recent publications [8,9] have described the benefits of implementing a range of VQ architectures which are based on bit serial inner product array (IPA) processors.

The purpose of this paper is to describe a chip design based on such an architecture. The resulting device demonstrates the modular nature of such an architecture and its suitability in the design of high performance real-time systems.

2. ARCHITECTURE

The overall architecture of the VQ chip is illustrated schematically in Fig. 1. It consists of two word parallel, bit serial IPA circuits and associated comparator circuits. It also contains a data formatter which allows normal word serial, bit parallel data to be input to the chip. These are used to perform distortion measure computation and pattern matching, as described in detail elsewhere [8,9]. The associated support and control circuitry is minimal due to the highly regular nature of the underlying architecture. Much of this comprises clock buffers and clock distribution, data synchronisation and loading signals and an output enable signal. The resulting chip floorplan is shown in Fig. 2. All circuit

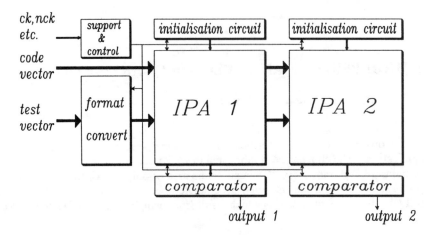

Figure 1 Chip architecture

blocks correspond to those in Fig. 1. In the diagram, *c,b* stands for clock buffer and *FCMP* refers to a comparator circuit. Upper and lower latch arrays (*ULA* & *LLA*) are used for synchronisation purposes.

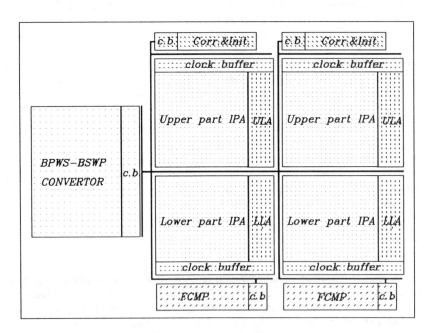

Figure 2 Chip floorplan

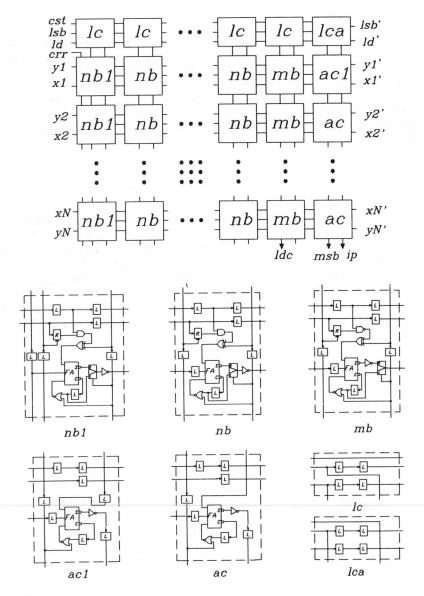

Figure 3 Inner product array and cell circuits

3. INNER PRODUCT ARRAY (IPA) CIRCUIT

The two inner product array blocks (Fig. 3) form the central part of the chip. These are cascaded so that the outputs data from the first IPA can be used as input to the second.

The IPA consists of N serial/parallel multipliers [10] where N is the vector dimension. It operates on continuous bit serial, word parallel input vectors and computes their inner products. On the fly word truncation (rounding) circuitry is included in the two IPA's. The IPA circuit computes the vector inner product (IP) values of the form $(\Sigma x \bullet y + z)$, where x, y are two vectors of dimension N and z is a constant. This is based on the observation that in VQ the popular squared error distortion can be replaced by an inner product computation [8,9,11].

4. COMPARATOR

The comparator circuit is used to compare all the inner product (distortion) values in order to determine the best match. Each comparator contains four n_s-bit delay latches, one n_s-bit shift register (n_s is the system wordlength) and a circuit for comparing values, labelled the *CU* in Fig. 4 . Two sets of delay latches are included, one set for storing the updated maximum IP value and the current input IP value from the IPA above (via *ip*), and the other for storing their associated addresses. The shift register (SR) is used to store the address of the final maximum IP value and its output (via *io*) is enabled by an active HIGH *shc* signal, as shown.

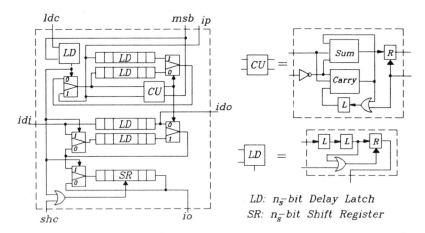

Figure 4 Comparator circuit

5. INITIALISATION CIRCUIT

The purpose of initialisation circuit is two fold. Firstly, as the multipliers in the IPA are based on the Baugh-Wooley two's complement multiplication algorithm [12], a fixed initial partial product (pp) term $(2^{ns-1} + 2^{ny-1})$ has to be added to

286

produce the final product, where n_y is the initial data wordlength for both x and y. Secondly, a constant, z, has also to be added to the partial IP values to generate the final modified IP values of the form $(\Sigma x \bullet y + z)$. The initialisation circuit is shown in Fig. 5.

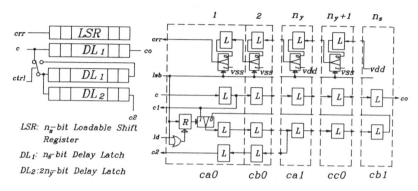

Figure 5 Initialisation circuit

As illustrated in Fig.5 ,the initial pp term, labled *crr* (correction term), is generated using an n_s-bit loadable shift register. The constant value (*c*) which normally corresponds to y is loaded into an n_s-bit delayed latch under the control of *lsb* and *ld* (load) signals. It is correctly timed by using extra latches and made available, via *c2,* to the accumulator of the IPA below. A delayed, unchanged output, *co*, is also provided to allow the loading of constants into successive IPA's.

6. DATA FORMAT CONVERTER

In many DSP systems, data is organised in a bit parallel, word serial (BPWS) manner. A data format converter has therefore been included to convert this into the bit serial, word parallel (BSWP), time skewed format required by the IPA's. This is illustrated in Fig. 6 along with appropriate cell logic circuits.

The circuit actually consists of two identical converters, as is apparent from the logic descriptions of the *MC* and *NC* cells These allow bit parallel data to be input at the same time as bit serial data is output. The bit parallel data are input to the conversion circuit via the data lines denoted by the row of *p's* at the bottom. This is timed by *pl1* and *pl2* signals for alternate input vectors. The bit serial, word parallel, time skewed outputs are available on the right and are timed to start using an *lsb* signal. Meanwhile the signal, *sld*, choses outputs from the appropriate converter.

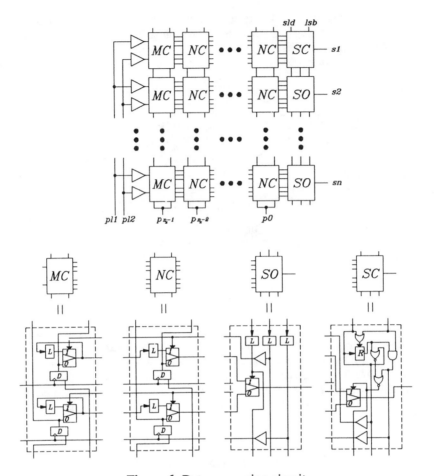

Figure 6 Data conversion circuit

7. THE VQ CHIP

A photograph of the VQ chip is shown in Fig. 7 and the layout clearly matches the floorplan in Fig. 2. All cells have been designed using full custom techniques. This produces a high transistor density. The chip dimensions are 4.8x4.6 mm² (including the padring) and it contains approximately 38,000 transistors. It has been fabricated using a 1.2 micron double level metal CMOS technology and is housed in an 84-pin PGA package. A total of 58 of these pins are actually used (including 57 connected to chip pads and 1 for package ground) with 26 unconnected.

The 58 connections and their functions are listed below.

Power and Ground There are a total of eight connections for these including

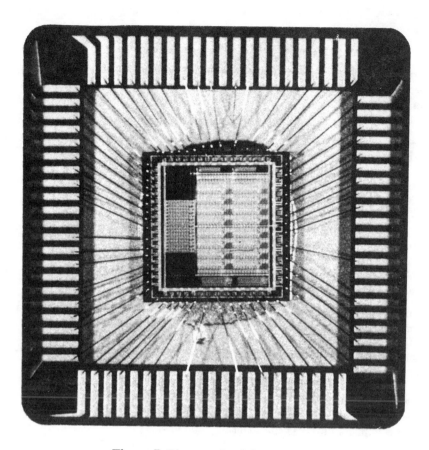

Figure 7 Photograph of the VQ chip

three for V_{dd} (power) and five for V_{ss} (ground).

Data input A total of sixteen connections are used for this purpose. Of these eight are used for input codevectors.

Data output Sixteen connections are used for both the data vectors and code vectors. These are used when two or more chips are cascaded to increase processing power.

Constant and address These functions require four pins. Two of these are for delayed output values.

Clocks and controls A total of ten connections are used for these.
- two for the two-phase non-overlapping clock signals, *ck1* and *ck2*
- one for the *lsb* and *lsbo* signals (for cascading chips)
- one for the data vector unload signal (for format conversion) and one for its delayed output.

- two for 1 data load signals
- two pins for the address output enable signals, one for each IPA.

Minimum address outputs Two connections are used for the address outputs from the two comparators. These are produced bit serially and enabled when set to HIGH.

Distortion values output Two connections are used to output inner product values. in a bit serial form computed in the two IPA's. These are generated in a bit serial form and used for chip testing purposes.

8. PERFORMANCE

The chip which has been designed is aimed mainly at VQ processing for speech applications. It can operate at a clock rate of 40 MHz. Both IPA's can accommodate vector dimensions up to 8. An initial data wordlength, n_y, of 8 bits and a system wordlength, n_s, of 12 bits are used. The performance of the VQ chip is therefore as follows.

Consider a VQ system where the code book size is 1024 entries. The data rate that the chip is capable of is *(fN/512n_s)*, where f is the clock rate. With f=40 MHz, the maximum data processing rate is therefore in the region of 52K samples per second. Alternatively, if we use the standard speech sampling rate of 8K samples per second, the maximum codebook size that the chip can process is *(fN/4000n_s)* entries ie. 6600 entries can be processed in real time. These figures are up to 500 times greater than those that are achieved by using TMS320 series fixed point DSP programmable chips. The power consumption of the chip at 40 MHz is approximately 600 mw.

9. CONCLUSIONS

In this paper, we have described a VLSI custom chip for VQ applications. It is based on highly regular, pipelined array architectures. With an overall size of only (4.8x4.6) mm^2, it is very powerful in terms of performance. This in turn demonstrates the high performance of the associated VLSI array architectures.

References

[1] R.M.Gray. Vector quantization. *IEEE ASSP Magazine*, vol 1, April 1984, pp.4-29

[2] R.Kolagotla, S.S.Yu and J.JaJa. Systolic architectures for finite state vector quantisation. *Journal of VLSI Signal Processing*, vol 5, pp 249 -259, 1993

[3] R.Kolagotla, S.S.Yu and J.JaJa. VLSI implementation of a tree searched vector quantiser. *IEEE Trans. Signal Processing*, 1993

[4] G.Davidson and A.Gersho. Application of a VLSI vector quantisation processor to real-time speech coding. *IEEE J.Selected Areas Commun.*, vol SAC-4, Jan 1986, pp112-124

[5] B.E.Nelson and C.J.Read. A bit serial VLSI vector quantiser. In *Proc Intl. Conf on Acoustics Speech and Signal Processing,* Tokyo, Japan, 1986, pp2211-2213

[6] R.Dianysian and R.Baker. A VLSI chip set for real time vector quantisation of image sequences. In *Proc IEEE Intl Symposium on Circuits and Systems,* 1987, pp221-224

[7] P.A.Ramamoorthy and Brahmaji. Bit serial systolic chip set for real-time image coding. In *Proc IEEE Intl. Conf on Acoustics Speech and Signal Processing,* Dallas, 1987, pp787-790

[8] M.Yan. VLSI architectures for speech and image coding applications. *Ph.D Thesis,* The Queen's University of Belfast, UK, 1989

[9] M.Yan, J.V.McCanny and Y.Hu. VLSI architectures for vector quantisation. accepted for publication in *Journal of VLSI Signal Processing,* 1993

[10] R.F.Lyon. Two's complement pipeline multipliers. *IEEE trans. Commun.* vol COM-24, 1979, pp.418-425

[11] Y.Hu. Computer aided design tools for high performance DSP ASICs. *Ph.D Thesis,* The Queen's University of Belfast, UK, 1992

[12] C.R.Baugh and B.A.Wooley. A two's complement parallel array multiplication algorithms. *IEEE trans. Computers,* vol C-22, Dec. 1973, pp.1045-1047

Acknowledgements

This work has been funded by the IFI Institute of Advanced Microelectronics at Queen's University of Belfast and the UK Science and Engineering Research Council.

Department of Electrical and Electronic Engineering
The Queen's University of Belfast
Belfast BT9 5AH Northern Ireland

An 8b CMOS Vector A/D Converter*

G. Tyson Tuttle, Siavash Fallahi, Asad A. Abidi

University of California, Los Angeles, CA

Almost all familiar A/D converters are scalar quantizers. There is growing interest in applications such as image and speech compression to employ block encoding, where groups of samples are quantized as vectors to the closest reference vectors in a multi-dimensional space (Figure 1) [1]. This analog circuit implementation of such a quantizer results in what might be termed a vector A/D converter. The IC stores a codebook of vectors on-chip, accepts a 16-element analog vector at the input, and calculates the Euclidean distance between the input and all codevectors (referred to as global search) to produce an 8b code indexing the codevector closest to the input prompt. At 5MHz clock rate, the chip dissipates <50mW quantizing a new 16-element analog vector every 10 clock periods, meaning that it may encode an entire 512×512 pixel gray-scale image in 4×4 blocks at 30Hz frame rate.

The vector A/D converter embodies circuits to: (a) store a codebook of arbitrary vectors on-chip. The 256 code vectors required for 8b quantization are stored as charge on capacitors; (b) compute the Euclidean distance, or metric, between a prompt vector and reference vectors; (c) search for the shortest distance, and encode the index of the closest vector.

A block diagram of the IC shows the relation between these sub-sections, and how they interface to the external support circuits (Figure 2). Circuits (a) and (b) above are embedded at every node of a 256×16 array. The array output, consisting of an unsorted set of 256 scalar distances, drives a binary tree of comparators to search for the smallest distance between the prompt and code vectors. Each circuit is as simple as possible, compact in layout, yet acceptably accurate. Simulations show that performance of the vector A/D is most sensitive to additive offsets in the distance computation, and less sensitive to input offsets or to deviations from the square law in the metric.

The storage and computation functions are merged in a self-calibrating MOS square-law cell, based on dynamic current mirrors and uncalibrated MOS square law cells (Figure 3) [2,3]. (Figure 3) The cell takes a balanced input signal, but processes it as two unipolar signals to obtain the rectification implicit in the squaring operation. During self-calibration, the inputs are connected to the code vector Y to be stored. The two pMOSFETs in all cells are diode connected, and the nMOSFET is biased by a globally-applied reference gate voltage. The input is then switched to the prompt vector X, and the nMOSFET is diode connected. $\Delta V_{gs}(j)=X(j)-Y(j)$ modulates the drain current of each pMOSFET in the 16-wide array following the MOS quadratic I-V characteristics. As a result, the diode connected nMOS voltage moves away from the global reference by an amount approximately proportional to $(\Sigma I_j)^2$, computing the Euclidean distance between X and Y. To the first order, self-calibration makes the output independent of FET threshold voltages. The entire codebook is stored and refreshed on-chip by sequentially applying each of the 256 codevectors, and addressing one column at a time into self-calibrate mode. This operation is time-interleaved with the prompt vector input. The codebook may be refreshed as infrequently as once every 10ms without degrading performance due to capacitor leakage.

A successive approximation search is carried out using a binary tree of comparators to find the least of the 256 unsorted scalar distances emerging from the previous blocks (Figure 4). Two-input latching comparators carry out pairwise comparisons down the 8-level tree. Comparisons at each level require one clock cycle. The smaller of the two comparator inputs is transmitted to the next level through an analog switch. A series-parallel layout of the feedforward switch tree gives the shortest settling time through the cascaded RC sections, where R is the switch resistance and C is dominated by the wiring capacitance. Following the strobe to the last comparator in the decision tree, asynchronous logic blocks are enabled to produce a binary encoded word indexing the column of the codevector closest to the prompt.

A simple but accurate low-power, high-speed comparator uses capacitors to sample the input and to calibrate itself against the threshold voltage mismatches in the capacitively-coupled CMOS latch (Figure 5) [4]. Spurious IR drops in the ground line running across the chip produce systematic errors in the comparator inputs. To avoid this, the ground metal is a connected mesh surrounding every node in the metric computation array, lowering its resistance to that of a perforated metal strap covering the entire chip surface. A two-segment folded layout is used for the metric computation array (Figure 6). The 6.8×6.9mm² chip is fabricated in a 2μm double-poly, double-metal CMOS process. Power supply current, and therefore the chip-power dissipation, depends on the input and stored codebook vectors. The dissipation numbers cited are average values for image encoding at the rated clock frequency. After off-chip decoupling, no measurable spikes are observed on the power supply pin.

A previously-developed codebook for quantizing 4×4 blocks of a gray-scale image was loaded on the chip, and an image outside the codebook training set was then block-wise quantized. By retrieving the codebook entries corresponding to the encoded indices, the quantized image was reconstructed. The results prove acceptable compared with the image reconstructed using the same codebook and floating point metric calculations on a SUN workstation (Figure 7). Although quantization using the chip introduces some foreground noise, key features of the original image are preserved. The vector quantized image represents a data compression by 13 relative to the original raw image.

Most of the codebook vectors happened to be clustered around zero, and offsets remaining in the circuits after self-calibration made it difficult to resolve among them. The quality of the image encoded by the chip would substantially improve with a redeveloped codebook whose elements are more evenly spread out in vector space.

References

[1] Gersho, A., R. M. Gray, *Vector Quantization and Signal Compression*, Kluwer Academic Publishers, Boston, 1992.

[2] Grouneveld, D., et al., "A Self-Calibration Technique for Monolithic High Resolution D/A Converters", IEEE J. of Solid State Circuits, vol. SC-24, pp. 1517-1522, Dec. 1989.

[3] Seevinck, E., R. Wassenaar, "A Versatile CMOS Linear Transconductor/Square Law Function Circuit", IEEE J. of Solid State Circuits, vol. SC-22, pp. 366-377, June 1987.

[4] Wu, J., B. Wooley, "A 100MHz Pipelined CMOS Comparator", IEEE J. of Solid State Circuits, vol. SC-23, pp. 1379-1385, Dec. 1988.

*Research supported by Office of Naval Research Contract N00014-89-J-1282.

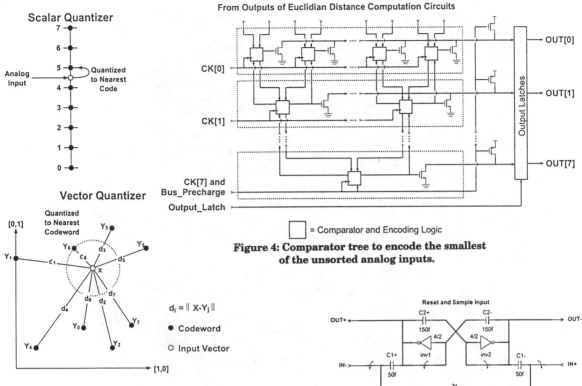

Scalar Quantizer

Figure 1: Quantization of a scalar vs. a vector.

Vector Quantizer

$d_j = \| X - Y_j \|$

● Codeword
○ Input Vector

From Outputs of Euclidian Distance Computation Circuits

□ = Comparator and Encoding Logic

Figure 4: Comparator tree to encode the smallest of the unsorted analog inputs.

Reset and Sample Input

Regenerate

Figure 5: Self-calibrating comparator.

Figures 6 and 7: See page 259.

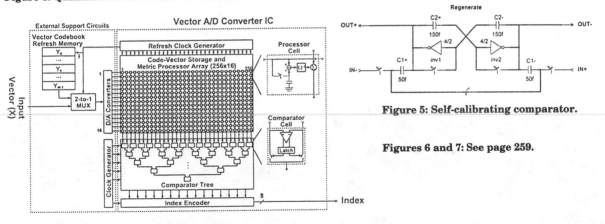

Figure 2: Block diagram of vector quantizer IC.
Required support circuits shown separately.

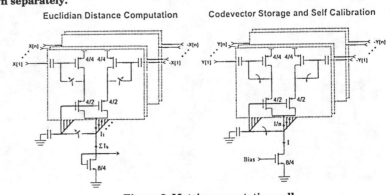

Euclidian Distance Computation

Codevector Storage and Self Calibration

Figure 3: Metric computation cell.
Calibrate, compute phases shown separately.

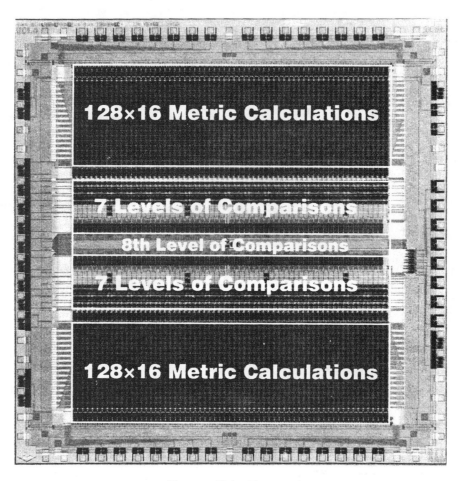

Figure 6: Chip micrograph.

Figure 7: Image reconstructed after vector quantization using chip (left), computer workstation (right).

Architectures for Wavelet Transforms

Chaitali Chakrabarti, Mohan Vishwanath and Robert M. Owens

A wide range of architectures for computing 1-D and 2-D DWT, and 1-D and 2-D CWT are presented. These architectures range from systolic arrays and parallel filters to SIMD arrays. The systolic array and the parallel filter architectures require an area that is independent of the length of the input sequence, and support single chip implementation. The SIMD architectures, on the other hand, are optimized for time, and have an area that is proportional to the size of the input.

1. INTRODUCTION

In the last few years there has been a great amount of interest in wavelet transforms, and the interest has been highly inter-disciplinary. The wavelet transforms have proved themselves to be invaluable tools for signal analysis, compression, numerical analysis etc. Research on the applications of the wavelet transforms has especially benefitted from the discovery of the Discrete Wavelet Transform by Mallat[1].

In this paper we present a wide variety of algorithms and architectures for computing both the Discrete Wavelet Transform (DWT) and the Continuous Wavelet Transform (CWT). We introduce a new 'running' (on line) algorithm for the DWT, which is a variation of the Recursive Pyramid Algorithm (RPA) [3]. The architectures range from systolic arrays and parallel filters to SIMD arrays. The systolic array and parallel filter architectures implement the RPA and the modified RPA algorithms, and compute the DWT and CWT in a word-serial manner. These architectures have an area that is independent of the input length, and can be implemented on a single chip. They are optimal w.r.t latency and period (under the word-serial model). The SIMD architectures implement the classical pyramid algorithm [2]. These architectures process the signal very fast, but have an area that is proportional to the size of the input.

2. PRELIMINARIES

The Wavelet Transform of a sequence $x(i)$ (sampled version of the continuous signal $\hat{x}(t)$), discretized on a grid whose samples are arbitrarily spaced both in time (b) and scale (a)[2], is given by,

$$W(b,a) = \frac{1}{|a|^{1/2}} \sum_{i=b}^{i=aL+b-1} x(i) h\left(\frac{i-b}{a}\right)$$

where L is the size of the support of the basic wavelet h, and h is obtained by sampling $\hat{h}(t)$. Also, a is of the form $a = ca_0^m$, $a_0 > 1$, c is a constant and $m \in Z$ The number of distinct m considered is J, in other words, J is the total number of scales. At each scale k, ie., $a = ca_0^k$, the number of samples in the time dimension is B_k, where $B_k \leq N$ (N is the number of input samples). Since the properties of the wavelet transform are heavily dependent on the properties of the basic wavelet, all the architectures that we have developed in this paper are independent of the wavelet function. Two special cases of the Wavelet Transform, the DWT and the CWT are considered in this paper. The DWT can be viewed as the multiresolution decomposition of a sequence[1]. It takes a length N sequence, $x(n)$, and generates an output sequence of length N. The output is the multiresolution representation of $x(n)$. It has $N/2$ values at the highest resolution, $N/4$ values at the next resolution, and so on. Let $N = 2^P$ and let the number of frequencies or resolutions, be J ($J \leq P$). The structure of the DWT is due to the dyadic nature of its time-scale grid, ie., the points on the grid that we are concerned with are such that $B_k = \frac{N}{a_0^{k+1}}$, $a_0 = 2$ and $a = 2^i$, $i, k \in 0, 1, \ldots, J - 1$. The CWT is usually defined as a WT with no decimation at any scale (ie., $B_k = N$ at all the J scales) and any desired frequency resolution. Thus it takes N words of input and produces N words of output at each scale. The CWT is mostly used in signal analysis while the DWT is more popular with compression and numerical analysis applications. Lower bounds on computing the Wavelet Transforms have been derived in [4]. The systolic array and the parallel filter architectures (for the DWT) described in this paper satisfy the word-serial model and are optimal, both in terms of area and time.

3. ALGORITHMS

3.1. The Recursive Pyramid Algorithm

The Recursive Pyramid Algorithm (RPA) is a reformulation of the classical pyramid algorithm [2] for the DWT, which allows computation of the N-point DWT in real-time using just $L \log N - L$ cells of storage, $L \ll N$. It consists of re-arranging the order of the N outputs such that an output is scheduled at the 'earliest' instance that it can be scheduled. The earliest 'instance' is based on the following precedence relation; if the earliest 'instance' of the ith octave clashes with that of the $(i + 1)$th octave, then the ith octave gets scheduled first. The output schedule generated by the RPA for $N = 16$ and $J = 4$ is as follows. Here $y_i(n)$ is the nth lowpass output of the ith octave.
$y_1(0), y_2(0), y_1(1), y_3(0), y_1(2), y_2(1), y_1(3), y_4(0), y_1(4), y_2(2), y_1(5), y_3(1), y_1(6)$
$, y_2(3), y_1(7), \ldots$

3.2. Modified Recursive Pyramid Algorithm

The modified Recursive Pyramid Algorithm (MRPA) is a version of the RPA that is suitable for parallel filter implementation. As in the case of RPA, the

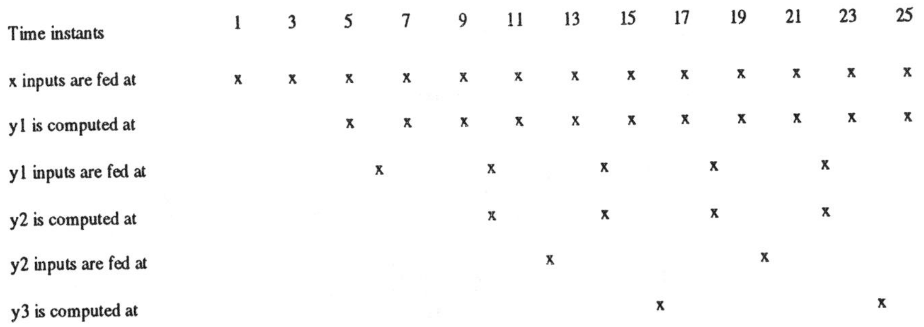

Time instants	1	3	5	7	9	11	13	15	17	19	21	23	25
x inputs are fed at	x	x	x	x	x	x	x	x	x	x	x	x	x
y1 is computed at			x	x	x	x	x	x	x	x	x	x	x
y1 inputs are fed at				x		x		x		x		x	
y2 is computed at					x		x		x		x		
y2 inputs are fed at						x				x			
y3 is computed at								x					x

Figure 1. Timing diagram illustrating the time instants at which the 1st, 2nd, and 3rd octaves are scheduled. The latency of the computation unit is $T = 4$.

first octave computations take place every other sample period, and the higher octave computations are scheduled between the first octave computations. This scheme is also based on scheduling an octave output at the 'earliest' possible instance. The earliest instance now depends on the latency of the computation unit, as well as the scheduling times of the lower octaves.

In the MRPA scheduling, the first output of any octave is scheduled such that there is no conflict with *any* of the higher octave outputs. This guarantees that there will be no conflicts for all the outputs of that octave. Figure 1 gives the timing diagram when the latency of the computation unit $T = 4$. The resulting schedule is $y_1(0), -, y_1(1), -, y_1(2), y_2(0), y_1(3), -, y_1(4), y_2(1), y_1(5), y_3(0), y_1(6), y_2(2), y_1(7), -, y_1(8), y_2(3), y_1(9), y_3(1), y_1(10), y_2(4) \ldots$

4. 1-D DISCRETE WAVELET TRANSFORM

4.1. Systolic Architecture

Systolic architectures for 1-D DWT which implement the RPA have been presented in [4] [5]. These architectures require area proportional to LJk and compute the DWT with a delay (and period) of $2N$ cycles. The first output appears only 1 cycle after the first input. These architectures are optimal both with respect to area and time under the word-serial model.

4.2. Parallel Filter Architecture

The parallel filter architecture for 1-D DWT implements the MRPA algorithm; it computes the first octave outputs every other sample period, and computes all higher octave outputs between two first octave output computations. The main components of this architecture are two parallel filters to compute the low pass and the high pass outputs, and a storage unit of size LJ to store the inputs that are required for the computation of J octave outputs. Figure 2 gives the block diagram of an architecture for computing J octaves. Each **parallel filter** consists of L fixed multipliers and a tree of

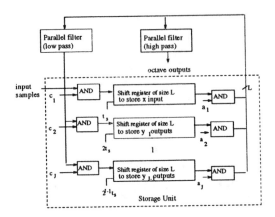

Figure 2. Block diagram of the parallel filter architecture for 1-D DWT

$(L-1)$ adders to add the products. The latency of this unit is $T_m + T_a \log L$, where T_m is the time taken to do a multiplication, and T_a is the time taken to do an addition. The **storage unit** consists of J shift registers, each of length L. The ith shift register stores the inputs required for the ith octave computation, $1 \le i \le J$. An output is written into the ith shift register every 2^{i-1} sample periods, and all L elements of the ith shift register are fed to the parallel filters every 2^i sample periods. Control signals $\{c_i\}$ and $\{a_i\}$ are generated at appropriate times to AND the signals into or out of the shift register. The hardware components of this architecture include $2L$ multipliers, $2(L-1)$ adders, JL storage units and a control unit to generate the appropriate control signals. The computation time for this architecture is $O(N)$, and the pipeline period is also $O(N)$.

4.3. SIMD Linear Array

The proposed SIMD architecture consists of a linear array of N processors with reconfigurable interconnection. Each processor contains a reconfigurable switch, which if set to 1, allows the data to pass through it without any delay. For the mth octave computation, $1 \le m \le J$, the reconfigurable switches can be used to reconfigure the N processor array to an array of size $\frac{N}{2^{m-1}}$, with processors $P(2^{m-1}j)$ being active, $0 \le j \le \lfloor \frac{N}{2^{m-1}} \rfloor$. Figure 3a describes the interconnection for the computation of 3 octaves on a 8 processor array.

In this scheme, all the outputs of any particular octave are computed at the same time. The high pass and the low pass filter coefficients are broadcast to each processor. Each processor computes the products of the pixel value and the filter coefficients, and updates the partial results of the low pass and the high pass outputs. The pixel value is sent to its active right neighbor. The computation time for computing J octaves in this architecture is LJ and the pipeline period is also LJ. The area of each processor is $O(k)$ and the overall area complexity is $O(Nk)$.

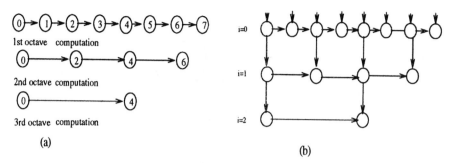

Figure 3. Interconnection among processors in (a) SIMD linear array, (b) Multigrid architecture for 1-D DWT computation, $N = 8$

Architectures	Area	Pipeline period
Aware's WTP	$O(Nk)$	$O(N \log N)$
Systolic array	$O(Lk \log N)$	$2N$
Parallel filter	$O(Lk \log N)$	N
SIMD linear array	$O(Nk)$	$L \log N$
SIMD multigrid	$O(N \log Nk)$	L

Table 1. Comparison of the area and pipeline period of the architectures for 1-D DWT

4.4. SIMD multigrid

The 1-dimensional DWT of N points can be efficiently mapped into a multigrid architecture of size N. The multigrid architecture of size N consists of $(\log N + 1)$ levels, with $\frac{N}{2^i}$ processors in level i, $0 \leq i \leq \log N$. Figure 3b describes the multigrid architecture for $N = 8$. The mapping is as follows: the ith level of the multigrid computes the $(i+1)$th octave outputs (high pass and low pass) and sends the low pass outputs to level $(i + 1)$, $0 \leq i < \log N$. The procedure to compute the ith octave outputs is the same as that of the SIMD linear array. The computation time for computing J octaves in this architecture is LJ, the pipeline period is L, and the area is $O(N \log Nk)$. The pipeline period of the multigrid architecture can be reduced to $O(1)$ by increasing the number of processors in each level by a factor of L.

5. 2-D DISCRETE WAVELET TRANSFORM

5.1. Systolic-Parallel Architecture

A systolic-parallel architecture for computing the 2-D separable DWT has been presented in [4]. It requires area proportional to NLk and computes the 2-D DWT with a delay (and period) of $N^2 + N$ cycles, the first output appearing 1 cycle after the first input.

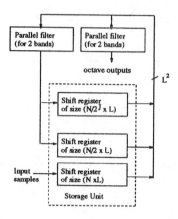

Figure 4. Block diagram of parallel filter architecture for 2-D DWT

5.2. Parallel Filter Architecture

The proposed parallel filter architecture for 2-D non-separable DWT implements the 2-D version of the MRPA algorithm. The windows of the first octave computations are centered in even-numbered rows, while the windows for higher octave computations are centered in odd-numbered rows. Along odd-numbered rows, the higher octave computations are interspersed between the 1st octave computations. The architecture consists of 2 programmable parallel filters (one filter per two bands) and a storage unit to store the lowest band outputs required for the higher octave computations. Figure 4 describes the block diagram of this architecture.

Each **parallel filter** consists of L^2 programmable multipliers and a tree of $(L^2 - 1)$ adders to add the products. The **storage unit** consists of J shift register units, the ith unit storing the inputs required for ith octave computation, $1 \leq i \leq J$. The ith unit consists of a 2-D array of storage cells of size $\frac{N}{2^{i-1}} \times L$. New data are shifted into the last row of each unit, and when the last row is filled, the data shift up by one row in the array. Each shift register unit has two clocks associated with it: a row clock and a column clock. The ith unit row clock has a period, $t_r(i)$ that is 2^{i-1} times the sample period t_s, $t_r(i) = 2^{i-1}t_s$, and has a duty cycle that is 2^{1-i} times that of the sample clock. The ith unit column clock has a period $t_c(i)$, that is N times $t_r(i)$, $t_c(i) = Nt_r(i)$, and a duty cycle of 2^{1-i}. Data is written into a shift register unit when both the row clock and the column clock are high. L^2 data are read out of the unit at time instants when every second row clock and every second column clock are high. The hardware components of this architecture include $2L^2$ multipliers, $2(L^2 - 1)$ adders, NL storage cells, and a control unit. The computation time is $O(N^2)$, and the pipeline period is also $O(N^2)$.

Architectures	Area	Pipeline period
Systolic-Parallel	$O(NLk)$	$N^2 + N$
Parallel filter	$O(NLk)$	N^2
SIMD 2-D array	$O(N^2k)$	$O(L^2 \log N)$

Table 2. Comparison of the area and time complexities of the architectures for 2-D DWT

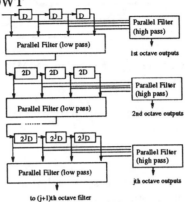

Figure 5. Parallel filter architecture for 1-D CWT. The filter size $L = 4$

5.3. SIMD 2-D array

The proposed SIMD architecture consists of a 2-D array of $(N \times N)$ processors which is reconfigured to form smaller arrays for higher octave computations. Specifically, the array is reconfigured to form an array of size $(\frac{N}{2^{m-1}} \times \frac{N}{2^{m-1}})$ for mth octave computation. All the outputs of a particular octave are computed at the same time. The area of this architecture is $O(N^2k)$. The computation time as well as the pipeline period is $O(L^2J)$.

6. 1-D CONTINUOUS WAVELET TRANSFORM

6.1. Parallel Filter Architecture

The proposed parallel filter architecture for 1-D CWT is based on the filter bank implementation of [2]. The architecture for computing J octave outputs consists of J filter units; each unit has one low pass and one high pass parallel filter. The output of the low pass filter of the jth unit is sent to the $(j + 1)$th unit. Since the inputs to the parallel filter in the jth unit have to be subsampled by a factor of 2^{j-1}, the data is stored in a delay line of size $2^{j-1}L$, and the delay line is tapped every 2^{j-1} delay units. Figure 5 describes the architecture when the filter size is $L = 4$.

The hardware components of the jth filter unit includes $2L$ multipliers, and $2(L - 1)$ adders, and $2^{J-1}(L - 1)$ storage units. Thus the total number of multipliers, adders and storage units in this design are $2JL$, $2J(L - 1)$ and

301

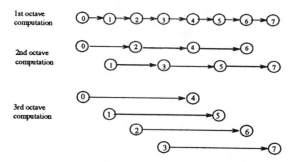

Figure 6. Interconnection among processors in a SIMD linear array for computing 1-D CWT

$(2^J-1)(L-1)$ respectively. The computation time for this design is $O(N+LJ)$ and the pipeline period is $O(N)$.

6.2. SIMD linear array

The proposed SIMD architecture consists of a linear array of N processors which can be reconfigured to form arrays of sizes $N/2$, $N/4$, and so on. For $(k+1)$th octave computation, the N processor array is reconfigured to form 2^k arrays, each of size $N/2^k$. The mth processor array is formed with processors $P(2^k j + m - 1)$, where $0 \leq k < logN$, $0 \leq j < \lfloor N/2^k \rfloor$, and $0 \leq m < 2^k$. Figure 6 describes the reconfiguration pattern when $N = 8$. The total computation time of this design is $L(1 + 2 + \ldots + 2^{J-1}) = L(2^J - 1)$. The pipeline period of this architecture is also $L(2^J - 1)$. The area is $O(Nk)$.

7. 2-D CONTINUOUS WAVELET TRANSFORM

7.1. Parallel Filter Architecture

The parallel filter architecture for 2-D CWT is an extension of the architecture for 1-D CWT. This architecture also consists of J filter units for computing J output octaves. The outputs of the low pass filter of the jth unit is sent to the $(j + 1)$th unit, $1 \leq j \leq J - 1$. The inputs to the jth filter unit have to be subsampled along the rows as well as along the columns by a factor of 2^{j-1}. Subsampling along the rows is achieved by storing the data in a delay line of size $2^{j-1}L$, and tapping the delay line every 2^{j-1} delay units. Since the inputs are fed in the line scan mode, subsampling along the columns is obtained by storing the data in a delay line of size $2^{j-1}LN$, and tapping the delay line every $2^{j-1}N$ delay units. Figure 7 describes the architecture when the filter size is (4×4).

The hardware components of the jth filter unit include $2L^2$ multipliers, $2(L^2 - 1)$ adders, and $2^{j-1}(N + L)(L - 1)$ storage units. Thus the total number of multipliers, adders and storage units are $2JL^2$, $2J(L^2 - 1)$ and $(2^J - 1)(L - 1)(N + L)$ respectively. The pipeline period for this design is $O(N^2)$.

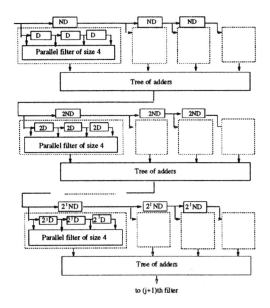

Figure 7. Parallel filter architecture for 2-D CWT when the filter is of size (4 × 4). The high pass filters have not been shown.

References

[1] S. Mallat. Multifrequency channel decompositions of images and wavelet models. *IEEE Trans. Acoustics Speech and Sig. Proc*, 37(12):2091–2110, Dec 1989.

[2] O. Rioul and P. Duhamel. Fast algorithms for wavelet transforms. *IEEE Trans. on Information Theory*, 38(2):569–586, Mar 1992.

[3] M. Vishwanath. A new algorithm for the discrete wavelet transform. To appear in the IEEE Trans. on Signal Proc., March 1994.

[4] M. Vishwanath. *Time-Frequency Distributions: Complexity, Algorithms and Architectures*. PhD thesis, Pennsylvania State University, May 1993.

[5] M. Vishwanath, R.M. Owens, and M.J. Irwin. An efficient systolic architecture for qmf filter bank trees. *Proc. of the 1992 IEEE Workshop on VLSI Signal Processing*, pages 175–184, Oct 1992.

[6] K. Parhi, and T. Nishitani. VLSI Architectures for Discrete Wavelet Transforms. To appear in IEEE Trans. on VLSI systems, June 1993.

Chaitali Chakrabarti
Arizona State Univ.
Department of EE
Tempe, AZ 85287

Mohan Vishwanath
CS Lab., Xerox
Palo Alto Res. Cntr.
3333 Coyote Hill Rd.,
Palo Alto, CA 94304

Robert M. Owens
Penn. State Univ.
Department of CSE
University Park, PA

MEMORY-I/O TRADEOFF AND VLSI IMPLEMENTATION OF LAPPED TRANSFORMS FOR IMAGE PROCESSING

Kamal NOURJI and Nicolas DEMASSIEUX

Telecom Paris, E.N.S.T, 46 rue Barrault, 75634 PARIS Cedex 13, FRANCE

ABSTRACT

This paper describes a new system architecture for real time two-dimensional image block processing by separable Lapped Transforms (LT). The main advantage of this image block 2D LT is the reduction of the transposition memory size and its on-chip integration. We shall discuss the tradeoff between Input/Output (I/O) throughput and on-chip memory size. Finally, the design of the transposition memory and its VLSI complexity performance in terms of area and speed is presented.

1-INTRODUCTION

Among existing video coding techniques, subband coding or generally Lapped Transform (LT) have received a great attention in recent years [1]. This can be attributed to the fact that this transforms reduce the blocking effects, compared to block coding methods like Discrete Cosine Transform (DCT) used for MPEG 2 normalization.

With a monodimensional LT (1D LT) [1], as with a Perfect Reconstruction Modulated Filter (PRMF) or a Pseudo Quadrature Modulated Filter (PQMF) used for subband coding [3,4,5], a N_F-sample input packet is mapped into a N_{SB} transform coefficient packet (subbands), with $N_F > N_{SB}$. N_F is the length of the LT basis functions, it is also the length of the prototype filter of the PRMF filter bank. When the packet of N_{SB} subbands is generated, the N_{SB} samples of the new packet can be shifted into the N_F-sample input buffer (figure 1). In this way, there will be an overlap of $(N_F - N_{SB})$ samples in the computation of consecutive LT packets.

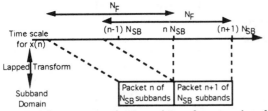

Figure 1: Signal processing with a Lapped Transform. A packet of N_F samples is mapped to compute N_{SB} transform coefficients packet.

The 1D LT implementation proposed by Malvar in [1,2] allows the computation of a finite-length signal in one step taking into account its boundaries effects. The drawbacks of this architecture are the hardware redundancy and its dependance on the signal length. Furthermore, for a 2D LT implementation, a transposition memory has to store all the image produced by the first 1D LT (figure 2). Other techniques like the processing by a 2D LT of picture stripes and the storage of the first 1D LT of the overlap between consecutive stripes allows the reduction of the transposition memory size. In spite of this optimization the memory size is still too large to allow on-chip integration for conventional picture size.

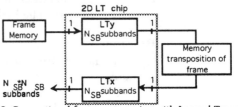

Figure 2: Conventional frame processing with Lapped Transform.

In order to reduce the size of the transposition memory, we propose to decompose the original picture into blocks (figure 4). These blocks overlap their right and down neighbors. We fetch one of these blocks and apply the 2D LT without any further access to the external frame memory. Furthermore, if the block size is not too large, the transposition memory can be put on the chip, thus reducing the I/O cost and power dissipation of the system (figure 3).

With the horizontal picture scanning (figure 4), we can take advantage of the overlapping of the input blocks by avoiding to fetch twice the right overlap zones. This is at the price of increasing the internal memory size, on storing the right overlap zone of each block into the transposition memory.

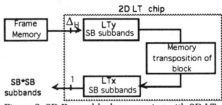

Figure 3: 2D Frame block processing with 2D LT. The transposition memory is on-chip.

The following section describes the image decomposition technique and shows how the on-chip integration of the reduced transposition memory reduces the I/O rate. A register-based and a

0-7803-0946-4/93 $3.00 © 1993 IEEE

RAM-based real-time architecture of the transposition memory, and their VLSI comparison in terms of speed and area for a 0,8μm technology, are presented in section 3.

2-IMAGE DECOMPOSITION

In the system architecture of figure 2 and figure 3, the input frame memory is connected with a 2D LT chip. The general expression of the I/O rate is defined as:

$$\Delta_{I/O} = \sum_{I/O} (\text{Number of DATA per clock cycle}) \quad (1)$$

The I/O rate allows the estimation of the redundancy of access to external memory. The limitation of the I/O rate of our system architecture will reduce power consumption and increase the overall system speed. With a conventional architecture [1,2] for a 2D LT applied to a picture of size (Nx.Ny pixels), a transposition memory size is (Nx.Ny.p bits), where p is the internal precision of coefficients. This memory size is too large to allow on chip integration. In this

case, the I/O rate will be:
$$\Delta_{I/O} = 4$$

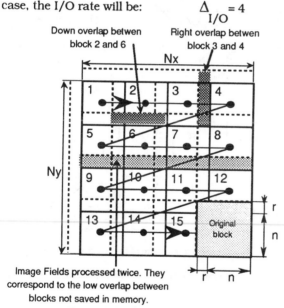

Down overlap betwen block 2 and 6

Right overlap betwen block 3 and 4

Image Fields processed twice. They correspond to the low overlap between blocks not saved in memory.

Figure 4: Horizontal image scanning.

To reduce the transposition memory size and I/O rate, we decompose the original picture into blocks of size (n+r)(n+r), with r=N_F-N_{SB} (figure 4). Each block is transformed by a separable 2D LT. The block scanning of the original picture can be done horizontally or vertically.

In the horizontal picture scanning scheme, the 1D LT is first applied to the block along the y axis and then, after transposition, along the x axis. After the first 1D LT, the transposition memory will store the right overlap. In this case, the transposition memory size will be (n.(n+r).p bits). But the down overlap must be read a second time in the external video memory. So,

for all the picture, we will access twice to $\left(\dfrac{N_y}{n} - 1\right)$

horizontal stripes of size (r.Nx) pixels. It follow this expression of the I/O rate:

$$\Delta_{I/O} = 1 + \Delta_H \quad (2)$$

with:

$$\Delta_H = 1 + \left(\frac{N_y}{n} - 1\right) \times \frac{r \cdot N_x}{N_x \cdot N_y} = 1 + \frac{r}{n} - \frac{r}{N_y} \quad (3)$$

In fact, (3) implies that (2) can be written as:

$$\Delta_{I/O} = 2 + \frac{r}{n} - \frac{r}{N_y} \quad (4)$$

In the vertical picture scanning scheme, the 1D LT along x axis is first applied. Finally, after the transposition, the 1D LT along y axis is applied. The transposition memory size will be the same as in the horizontal picture scanning scheme. The I/O rate is :

$$\Delta_{I/O} = 2 + \frac{r}{n} - \frac{r}{N_x} \quad (5)$$

The difference between the horizontal and vertical

picture scanning rates is: $\left(\dfrac{r}{N_x} - \dfrac{r}{N_y}\right)$.

In TV and HDTV applications where Nx > Ny, the horizontal picture scanning scheme provides lower I/O rate than the vertical one. The transposition memory size is the same for both. Furthermore, these two rates are lower than the rate of the existing LT system architectures which is 4 .

3-TRANSPOSITION MEMORY ARCHITECTURE

When considering two adjacent blocks (i) and (i+1) as presented in figure 5, the transposition memory has simultaneously to transpose the first block (i) composed of the A, B, and C zones, to save the right overlap C for the transposition of the next block and finally to store the D and E zones of the block (i+1). When the transposition of block (i) is completed, block (i+1) is completely stored in the transposition memory. During the block (i) transposition, the zone C will change its location. Thus, the transposition of the C, D, and E zones will be done in the correct orientation.

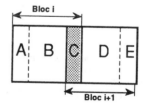

Figure 5: Conserved overlap between two adjacent blocks

We propose two architectures for the transposition memory, one is register-based solution and the other is RAM-based one.

The area and speed estimations, parametrized by the internal coefficients precision p and the block size n are done for these two architectures.

3-1 Register-based architecture.

The register-based architecture of figure 6 uses two shifting directions: vertical (top to bottom) and horizontal (left to right). To transpose sequentially two adjacent blocks (i) and (i+1) of figure 5, the first multiplexer will select the vertical input buffer which loads the A, B, and C zones using the horizontal shifting mode. The main part of the transposition memory will contain the B and C zones. The A zone will be shifted from the main part to the secondary part. The transposition of the block (i) will be done using the vertical shifting mode. Sequentially the D and E zones are loaded, through the horizontal input buffer, into the main part. The C zone changes its place in direction of auxiliary part of the transposition memory. After that, the block (i+1) will be transposed, with its zones C,D and E in the correct orientation, through the vertical output buffer. The basic cell of the main part, the secondary part and the output buffers are composed of 2->1 multiplexer and register. The input buffers have just one register in their basic cell. This regular architecture allows a real time processing of the picture blocks by the 2D separable LT.

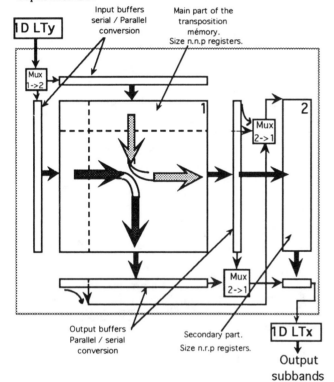

Figure 6: Transposition memory architecture register solution.

a) Area estimation:
The general area formula can be written as:

$$A_{register} = p\left(2nS_{reg1} + S_{reg2}\left[n^2 + n(2+r) + r\right]\right) \quad (6)$$

With: $r = N_F - N_{SB}$

N_{SB}: subbands number per dimension

N_F: length of the LT basis function

p: internal precision of coefficients

n: block size

S_{reg1}: 1 bit simple register area

S_{reg2}: 1 bit (2->1 multiplexer + register) area

Note that this area function is a second order polynomial of n.

b) Speed estimation:
The speed of this transposition memory depends only on the delay time introduced by the basic register cell. It is a constant depending on technology and design.

$$T_{register} = constant \quad (7)$$

3-2 RAM-based architecture.
The RAM-based transposition memory (figure 7) is implemented with 5 RAMs. Three of them (RAM1, RAM3, RAM5) have (n.r.p) bits size, and two (RAM2, RAM4) have ((n-r).n.p) bits size. The read/write cycle of these RAMs is described in figure 8. This architecture respects the general operating principle described at the beginning of this section.
After the 1D LT along y axis, each transformed block must be transposed. So, the A,B and C zones are written respectively in RAM1,RAM2 and RAM3. The reading of these RAMs is done along the x axis. During the block (i) transposition, the D and E zones are written into RAM4 and RAM5. Thus, when the block (i+1), with its C,D and E zones, is transposed, we can write the F and G zones into RAM2 and RAM1. In this way, this architecture provides a real time transposition process.

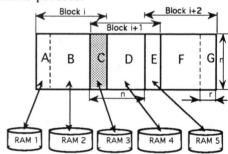

Figure 7: 5 RAM-based architecture of the transposition memory.

The global memory cost of these five RAMs is about n(2n+r)p bits.

Write (RAMs number)	Read (RAMs number)	Operations R=Read;W=Write
1/2/3	-	Write block 1
4/5	1/2/3	R (1) ; W (2)
2/1	3/4/5	R (2) ; W (3)
4/3	5/2/1	R (3) ; W (4)
2/5	1/4/3	R (4) ; W (5)
4/1	3/2/5	R (5) ; W (6)
2/3	5/4/1	R (6) ; W (7)
4/5	1/2/3	R (7) ; W (8)

Figure 8: Read/Write cycle of the 5 RAMs of the transposition memory.

For the area and speed estimation, the static RAM model proposed here suppose that the RAM has three principal parts: storage part, row address decoder and column address decoder. We also assume that a RAM has a square shape.

a) Area estimation:

The area of one RAM storing N words of p bits can be written as [6]: $S_{RAM} = S_0 Np + S_1 \sqrt{(Np)} \log_2(N)$ (8)
S_0 is the Area of the 1 bit basic cell of the storage part and S_1 the Area of the equivalent basic cell of the decoders and all RAM's control. The S_0 and S_1 coefficients can be defined with a least square error method to fit the model with actual RAM sizes provided by a particular RAM compiler.

The RAM-based transposition memory area is the addition of the areas of the 3 RAMs of size nr words of p bits and of the 2 one of size (n-r)n words of p bits.

$A_{RAM} = 3 A_{RAM1} + 2 A_{RAM2}$ (9)
$A_{RAM1} = [S_0 n r p + S_1 \sqrt{(n r p)} \log_2(n r)]$ (10)
$A_{RAM2} = [S_0 (n-r)np + S_1 \sqrt{(nrp)} \log_2(n r)]$ (11)

with:

A_{RAM1}: the area of the RAM1, RAM3 and RAM5 each one of size (nrp bits).

A_{RAM2}: the area of the RAM2 and RAM4 each one of size ((n-r)np bits).

b) Speed estimation:

The speed of one RAM storing N words of p bits can be estimated by the following formula [6]:

$T = T_{RD} + T_L + T_C + T_{CD}$ (12)

The row decoder delay time T_{RD} is proportional to the number of row address. The load time T_L of the word-line is proportional to the number of columns. The load time T_C of the column is proportional to the logarithm of the row number. Finally, T_{CD} is the column decoder delay time. It is proportional to the number of column address. Thus, after some simplifications, we can write (12) with β_1, β_2 and β_3 constants as:

$T = \beta_1 \sqrt{(Np)} + \beta_2 \log_2(N) + \beta_3 \log_2(p)$ (13)

The constant parameters β_1, β_2 and β_3 can be defined with a least square error method by using RAM speeds provided by a particular RAM compiler.

3-3 Comparison.

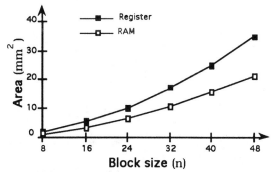

Figure 9: Area estimation of the transposition memory. Comparison of the RAM and Register solutions.

For the particular case of r=7 and p=16 bits, the area and speed comparison has been performed. The $S_0=210\mu m^2$, $S_1=1800\mu m^2$, $\beta_1=0,012ns$, $\beta_2=2,9ns$ and

$\beta_3 = -0,27ns$ parameters have been defined with the data provided by the VCC4RS2 CMOS RAM compiler of VLSI Technology for a 0.8 micron CMOS technology. The design of full custom registers with the symbolic design tool PREFORME [7] led to $S_{reg1}=660\mu m^2$, $S_{reg2}=788\mu m^2$ and $T_{register}=2,6ns$.

The plots of figures 9,10 show that the register-based architecture requires 38% more area than the RAM-based architecture. But its short critical path allows to reach clock rate of several hundreds of Mhz, thus making it possible to design HDTV decoders.

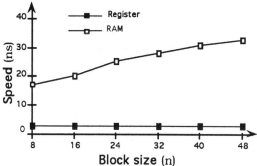

Figure 10: Speed estimation

4-CONCLUSION

In this paper we have presented an optimized 2D LT system architecture in terms of:
- independence inside the picture size
- real time processing for moving pictures
- very low internal memory size
- low I/O rate
- implementation in a single chip of the 2D LT with its transposition memory.

We have proposed two memory transposition architectures which reduce the I/O rate. We also described a method for their area and speed comparison. The results we obtained show an important gain, in term of system cost, compared to previously reported solutions [1,2].

We are now designing a VLSI subband decoder which implements the architectural ideas proposed in this paper.

5-References

[1] H.S. MALVAR, D.H. STAELIN, "The LOT: Transform Coding Without Blocking Effects", IEEE Trans on ASSP, Vol 37, No 4, pp. 553-559, (April 1989).

[2] H.S. MALVAR, "Signal Processing with Lapped Transforms", Publisher: Artech House, (1992)

[3] J.MAU, "Computationally efficient Pseudo QMF filter bank for a multi-compatible HDTV codec", ICASSP 91, (1991).

[4] M.VETTERLI, "Multi-dimentional Sub-band Coding: Some Theory and algorithms", IEEE Signal Processing, Vol 6, pp. 97-112, (April 1984).

[5] J.W. WOODS, S.D. O'NEIL, "Subband Coding of Images", IEEE Trans on ASSP, vol 34, no 5, pp. 1278-1288, (October 1986).

[6] N.DEMASSIEUX, "Architecture VLSI pour le traitement d'images: une contribution à l'étude du traitement materiel de l'information", PHD Thesis of Telecom Paris, (March 1991).

[7] J.C.DUFOURD, J.F.NAVINER, F.JUTAND, "PREFORME: A Process independent symbolic Layout System", ICCAD 90, Santa Clara, USA, (November 1990).

7.5MFLIPS Fuzzy Microprocessor
Using SIMD and Logic-in-Memory Structure

Mamoru SASAKI   Fumio UENO   Takahiro INOUE

Dept. of Electrical Engineering and Computer Science

Kumamoto University

Kumamoto, Kurokami 2-39-1, 860 Japan

(Tel.096-344-2111 Ext.3842)

Abstract

A fuzzy microprocessor, having SIMD and Logic-in-Memory structure, is developed using 1.2µm CMOS process. It can execute fuzzy inference for if-then fuzzy rules. The speed of inference including deffuzzification is 7.5MFLIPS, and it can process 960 rules and 16 input and output variables.

1 Introduction

One of the most successful applications of the fuzzy theory[1],[2] is in the field of process control[3],[4]. In the applications, the models of the processes under control are often too complicated to be formulated mathematically. While, the fuzzy-logic rule-based system allow natural translations of domain knowledge into fuzzy rules. The inference procedure derives effective control actions using the fuzzy rules.

However, the computations required in the fuzzy inference procedure are usually too sophisticated to meet the real-time requirements of process controllers. The pioneering work of Togai and Watanabe marks the beginning of the hardware approach of implementing fuzzy-logic inference engine using digital circuits [5], [6]. Yamakawa adopts a different approach in that he used analog circuits [7].

In this paper, considering that the primary aim of hardware implementation is realization of high speed processing, we introduce effective control mechanism on parallel architecture, which can realize high speed fuzzy-logic inference. First, the fuzzy rules are reconstructed and the inference sequence for the reconstructed fuzzy rules is divided between if-part and then-part. Next, we introduce different two parallel architectures suitable for the if-part and then-part. In if-part, SIMD is introduced. In then-part, in order to effectively combine the results of all rules, bit serial max/min operations with multi-input are pro-

posed and logic-in-memory suitable for implementing the bit serial max/min operations is introduced. Furthermore, in order to convert a fuzzy set being a final result to a certain value, a defuzzification operation (Center of Area method) is implemented with multi-operand parallel adders using Wallace tree.

2 Reconstruction of fuzzy rules

In the fuzzy inference sequence, the process of every fuzzy rule is independent each other. Considering parallel processing, we choose naturally one rule process as a parallel-processing unit. In fact, the previous fuzzy logic processor executes every rule in parallel [5], [6]. However, in case of many rules, it is impossible for restriction of hardware resources. So, we have to make the granularity of parallel process more large than one rule process.

The number of membership functions on one axis usually less than ten. Making use of this fact, we will reconstruct the fuzzy rules. The rules with the same membership function in then-part are gathered and a new rule is reconstructed by combining these rules. The process of one reconstructed rule is chosen as a parallel processing unit. In this case, the number of the processing elements can be equal to the number of the membership functions in then-part. Furthermore, the number of the operands of the max operation to combine the results of all rules can be also equal to the number. For example, the following rules are reconstructed as follows :

if x_1 is A_{11} and x_2 is A_{12} then y is B_1
if x_1 is A_{21} and x_2 is A_{22} then y is B_2
if x_1 is A_{31} and x_2 is A_{32} then y is B_2
if x_1 is A_{41} and x_2 is A_{42} then y is B_1

$$\downarrow$$

if (x_1 is A_{11} and x_2 is A_{12}) *or* (x_1 is A_{41} and x_2 is A_{42}) then y is B_1

if (x_1 is A_{21} and x_2 is A_{22}) or (x_1 is A_{31} and x_2 is A_{32}) then y is B_2

Where $x_i(i = 1, 2)$ and y are inputs and output, respectively. $A_{ji}(j = 1, \cdots, 4)$ and $B_k(k = 1, 2)$ are fuzzy subsets (membership functions).

Using max/min inference method, we can calculate firing grades of the rules, for example, in case that the inputs $x_1 = x_1'$ and $x_2 = x_2'$, a firing grade w_1 of the rule with B_1 is :

$$w_1 = \{A_{11}(x_1') \wedge A_{12}(x_2')\} \vee \{A_{41}(x_1')A_{42}(x_2')\} \quad (1)$$

Where $\wedge$ and $\vee$ are min and max operation, respectively. $A_{ji}(\cdot)$ is a grade of the membership function. Next, the inference result B_k' of the rule with B_k can be derived as follows.

$$B_k'(y) = w_k \wedge B_k(y) \quad (2)$$

Then, the final inference result B' can be obtained by combining the result B_k' of every rule.

$$
\begin{aligned}
B'(y) &= \bigvee_k B_k'(y) \\
&= \bigvee_k \{w_k \wedge B_k(y)\} \quad (3)
\end{aligned}
$$

When the certain value y' is required, y' can be calculated from the final result $B'(y)$ using the Center of Area method.

$$y' = \frac{\sum_y B'(y) \cdot y}{\sum_y B'(y)} \quad (4)$$

The inference sequence can be divided between if-part process (1), and then-part process (2), (3) and (4). Considering the if-part process, we understand that all firing grades can be calculated in parallel by using multiple processing elements. While, multi-input max operations are executed in then-part as mentioned (3) and it requires parallel processing on operation level. From the considerations, we introduce two parallel architectures suitable for if-part and then-part. Since same operations are executed to different data in if-part, SIMD is introduced for the if-part processor. While, logic-in-memory suitable for implementing the multi-input max operation is introduced for the then-part. The logic-in-memory can implement effectively multi-input operation by locating some logic circuits between memory elements.

3 An if-part processor

A structure of the if-part processor is shown in Fig.1. The function and feature of each block in Fig.1 are explained. A data memory can store the membership functions used in the if-part. The format of the membership function is that the number of elements is 64 (6 bits) and the number of grade levels is 16 (4 bits). Thus, one membership function can be represented with 256 bits. A structure of the data memory is shown in Fig.2. The data memory consist of four memory banks. Each memory bank can store 14 kinds of the membership functions. The labels in Fig.2 represent the kinds of the membership functions. The grade are read by 7-bit address, which consist of 1 bit being an identification of a bank and 6 bits being an identification of an element. All grade data of 14 membership functions are read on 56bits data bus. As shown in Fig.2, high-speed access time can be realized by using interlace method.

A rule memory is a local memory prepared for each processing element (refer Fig.1). The memory stores the labels representing the kinds of the membership functions. The labels correspond to A_{ji} in the rule expressions and all labels in a rule are stored in a rule memory. In the processing element, the labels become select signals for obtaining the needed grade data. The labels are read sequentially by the instruction counter. A label consists of 4 bits and can specifis 14 membership functions and 2 NOPs for the max and min operations. Each rule memory is also divided between L-block and H-block, and high-speed access-time is realized by using the interlace method.

The instruction memory stores the instruction sequence for the processing elements, the input-interface block and the variable counter. The contents in the memories are read sequentially by the instruction counter. The instruction for the processing elements consists of min-end-flag and max-end-flag. The flags correspond to the timing signals to control the processing elements. The instruction for the input-interface block consists of an address of the input registers in INPUT I/F and a identification of the data memory bank storing the membership functions of the input. In a processing element (PE), first, the selector chooses the needed grade data from the 14 kinds grade data sent from the data memory. The grade data sets of every input is sequentially sent from the data memory and the selecting signal is sent from the rule memory. Next, the min circuit calculates sequentially each min-term in a reconstructed rule. Then, the max circuit executes sequentially max operation among all min-terms. A pipeline process is implemented by preparing the register between the min circuit and the max circuit. Finally, since bit serial operations are used in the then-part processor, the shift register converts the firing grades from parallel to serial and each single bit is outputted se-

follows.

$$c_i = (Qa_i + a_i) \cdot (Qb_i + b_i) \qquad (8)$$
$$Qa_{i-1} = Qa_i + (\overline{c_i} \cdot a_i) \qquad (9)$$
$$Qb_{i-1} = Qb_i + (\overline{c_i} \cdot b_i) \qquad (10)$$

Where initial value $Qa_{n-1} = Qb_{n-1} = 0$.

The iterative equations for the max/min operations can be easily expanded to multi-input case. For example, $C = \max(A1, A2, \cdots, Aj, \cdots, Am)$ is calculated, where $A1, \cdots, Aj, \cdots, Am$, and C are the n-bit binary numbers, aj_i and c_i represent the ith bits of Aj and C, respectively. $C = \max(A1, \cdots, Aj, \cdots, Am)$ can be calculated by iterating the following equations from $i = n - 1$ to 0 with initial value $Qaj_{n-1} = 1$ $(j = 1, 2, \cdots, m)$.

$$c_i = (Qa1_i \cdot a1_i) + \cdots + (Qam_i \cdot am_i)$$
$$= c1_i + c2_i + \cdots + cm_i \qquad (11)$$
$$cj_i = Qaj_i \cdot aj_i \qquad (12)$$
$$Qaj_{i-1} = Qaj_i \cdot (\overline{c_i} + aj_i) \qquad (13)$$

A 2-input min circuit for the bit-serial operation can consists of some logic circuits and two flip-flops storing the status variables. Next, we consider a multiple inputs max circuit for bit-serial operation. The circuit can be implemented by referring the iterative equations (11), (12) and (13). Here, let us note that (12) and (13) can be locally calculated every element. Hence, the equation executed among multiple elements is only (11) and the operation is just multi-input OR operation of binary logic. The multi-input max circuit can be constructed with pre-process circuits operating locally (12) and (13), and one multi-input OR gate. The preprocess circuit consists of some logic circuits and one flip-flop storing the status variable.

A structure of the logic-in-memory is shown in Fig.4. A processing element (PE) consists of 4bit data Latch, selecter, 2-input min unit and the preprocess unit for multi-input max operation. The data Latch stores a grade data of a element. The grade data is read sequentially from MSB by the selector. In Fig.4, the outputs of the processing elements corresponding to cj_i in (11) are connected with multi-input OR gates.

4 A then-part processor

The then-part processor gets the firing grades from the if-part processor and executes (2) and (3), and calculates the center of area shown in (4). The processor consists of a logic-in-memory executing (2) and (3), a defuzzifier circuit calculating the center of area, and a sequencer controlling the two blocks. First, let us consider parallel processing of (2) and (3).

The grade $B'_k(y)$ representing a result of a rule can be calculated by min operation between the grade $B_k(y)$ and the firing grade w_k. So, the process can be executed in parallel every element in the membership function. The grade $B(y)$ representing a final result can be calculated by max operation among all grade $B'_k(y)$. Here, a structure, in which min units are put in the form of matrix as shown in Fig.3, are introduced. In Fig.3, a row corresponds to one membership function in then-part. To broadcast the firing grade w_k, inputs of all min units on a row are connected with an input-bus. The firing grades of all rules are given to all input-buses at the same time and all min units execute (2) in parallel. Then, every result $B'_k(y)$ is outputted from every min unit. The outputs of the min units are connected to the multi-input max unit every column. Since there are the grades of same element on a column, the max unit can execute (3). Although the structure can realize massive scale parallel processing, many units are required. So, the units have to be constructed as simply as possible for restriction of hardware resources.

To overcome the problem, bit-serial operation is introduced. Bit serial max and min operations are explained. For example, $C = \max(A, B)$ is calculated, where A, B, C are n-bit binary numbers, a_i, b_i, c_i represent ith bits of A, B, C, respectively. c_i can be calculated using the following iterative equations.

$$c_i = (Qa_i \cdot a_i) + (Qb_i \cdot b_i) \qquad (5)$$
$$Qa_{i-1} = Qa_i \cdot (\overline{c_i} + a_i) \qquad (6)$$
$$Qb_{i-1} = Qb_i \cdot (\overline{c_i} + b_i) \qquad (7)$$

Where $\cdot$ and $+$ are AND and OR operators of binary logic, respectively. $\bar{}$ is NOT operation. Qa_i and Qb_i are status variables keeping comparison results. $C = \max(A, B)$ can be calculated by iterating the equations from $i = n - 1$ to 0 with initial value $Qa_{n-1} = Qb_{n-1} = 1$. Iterative equations for the min operation are dual equations of the max operation as

5 Defuzzification circuit

In this section, a defuzzifier circuit operating center of area method is explained. The feature of the center of area method is that there are additions with many operands in both the numerator and the denominator as shown in (4). Although multiplications between B'(y) and element order y are executed in the numerator, the multiplication can be implemented by adding the partial products generated by bit shifting. Because element order y is a constant, amounts of shifting and number of partial products for every y can be predetected and the shifting can be implemented by wiring. So, the calculation of both the numerator and the denominator can be implemented with multi-operand adders. In this case, however, the number of operands becomes large. One of the effective implementation of the adders with many operands is using Wallace-tree, which is applied to high-speed multiplier.

A numerator calculation unit and a denominator calculation unit are implemented using the Wallace-tree. Here, we exchange calculation order in numerator. To understand easily, a case that $0 \leq y \leq 7$ is considered. As shown in the following equations, operands are classified among some groups with same shifting amount.

$$\sum_y B'(y) \cdot y =$$
$$\{B'(1) + B'(3) + B'(5) + B'(7)\} \cdot 1 +$$
$$\{B'(2) + B'(3) + B'(6) + B'(7)\} \cdot 2 +$$
$$\{B'(4) + B'(5) + B'(6) + B'(7)\} \cdot 4 \quad (14)$$

The operands are reduced every group using Wallace-tree and the reduced two operands are shifted. Then, the shifted operands of all groups are reduced to two operands using Wallace-tree again and the two operands are added by a typical adder. A defuzzification circuit consists of the numerator calculation unit, the denominator calculation unit and one divider.

6 VLSI fabrication and system configurations

The explained two processors have been fabricated as VLSI chips. $1.2\mu m$ n-well 2 metal CMOS process and standard cell design method were used. The performances are shown in Table 1.

A system configuration using the chips is shown in Fig.5. In Fig.5, a if-part chip works as master and some then-part chips work as slaves. The then-part chips get the trigger from the if-part chip and start the consequent inference sequence. Identification numbers are given to the then-part chips, and the then-part chips can detect their data and the trigger by comparing their identification numbers with a

identification outputted from the if-part chip. The identification is the content of the output variable counter in the if-part chip.

Since a rule memory is constructed with 128 wards and the number of the rule memories is 15, $128 \times 15 = 1920$ antecedents (membership functions) can be specified by using all rule memories. Hence, the number of rules (typical rules), which can be executed by the processor, is :

$$Number\ of\ rules = [1920/(i \cdot j)] \quad (15)$$

where i and j are the number of inputs and outputs, respectively. $[\cdot]$ is gaussian operation.

Since the then-part chip can execute the consequent inference including defuzzification every 4 clock cycles and the clock frequency is 30MHz, the inference speed of the then-part chip is 7.5 MFLIPS(Fuzzy Logic Inference Per second). While, the inference speed of the if-part chip is dependent on the numbers of inputs, outputs and rules. First, let us consider the dependence on the numbers of inputs and outputs. When the number of inputs and outputs are i and j respectively, $i \cdot j$ clock cycles are needed at least for the process. So, the inference speed is $30/(i \cdot j)$ MFLIPS in case of 30 MHz clock frequency. Next, let us consider the dependence on the number of rules. Since each processing element executes one antecedent per one clock cycle, 30M (clock freaxency) $\times$ 15 (the number of processing elements) = 450M antecedents can be executed per second. So, since $i \cdot j \cdot k$ antecedents are processed in case of i-input, j-output and k-rule, the inference speed is $450/(i \cdot j \cdot k)$ MFLIPS. Therefore, the inference speed of the system shown in Fig.5 can be expressed as follows:

$$Inference\ speed =$$
$$min\{7.5, 30/(i \cdot j), 450/(i \cdot j \cdot k)\} MFLIPS (16)$$

In case of one output, a relationship between the number of rules and the inference speed is described in Fig.6 with parameters being the number of inputs. A bold line in Fig.6 presents the performance of the previous fuzzy logic processor executing parallel processing every rule [6]. Under the number of rules which can be executed by the previous processor, 12.9 times higher performance can be realized in vertue of the high performance of the then-part chip and the flexiblity of the rule format in the if-part chip.

Next, an other system configuration, that is multiple if-part-chip configuration, is shown in Fig.7. The bit-serial max unit shown in Fig.7 can be easily implemented using PLA (Programable Logic Array). The inference speed can be independent of the number of rules using this configuration as shown dotted lines in Fig.6. The number of if-part chips and the inference speed is expressed as follows:

$$Number\ of\ if\ part\ processors = [k/15] + 1 \quad (17)$$

$$Inference\ speed = min\{7.5, 30/(i \cdot j)\}\ MFLIPS \quad (18)$$

where i, j and k are the numbers of inputs, outputs and rules, respectively.

A relationship between the number of if-part chips and the number of rules processed per second is described in Fig.8 with parameters being the number of inputs. As shown in Fig.8, the performances increase in proportion to the number of if-part chips without the overhead for the comunications.

7 Conclusions

Two fuzzy microprocessors have been developed as VLSI chips, one is the if-part processor with SIMD architecture and the other is the then-part processor with logic-in-memory. The features of the processors are ;

1. Maximum inference speed including defuzzification is 7.5MFLIPS (Mega Fuzzy Logic Inference Per Second). This inference speed is 12.9 times higher than the previous VLSI implementations.

2. Many fuzzy rules (960 rules), input variables (16 variables) and output variables (16 variables) can be processed. Furthermore, the rule format can be easily changed by rewriting the instructions stored in the memory.

3. The processors require no external memory since the knowledge-base (If-Then rules) can be stored in the internal memories.

Acknowledgements

The authors wish to thank the electronic devices group of the Oki electric Industry corporation, especially T. Katashiro, for fabricating the circuits.

References

[1] L.A.Zadeh, "Fuzzy sets," *Inform. Contr.*, vol.8, pp.338-358, 1965.

[2] L.A.Zadeh, "Outline of new approach to the analysis of complex systems and decision processes," *IEEE Trans. Syst., Man., Cybern.*, vol.SMC-3, pp.28-44, Jan. 1973.

[3] L.P.Holmbalad and J.J.Ostergaard, "Control of a cement kiln by fuzzy logic," in *Fuzzy Information and Decision Processes.* Amsterdam, The Netherlands: North-Holland, 1982, pp.389-399.

[4] S.Yasunobu, S.Miyamoto and H.Ihara, "A Predictive fuzzy control for automatic train operation," *Syst.Contr.Japan*, vol.28, pp.605-613, Oct. 1984

[5] M.Togai and S.Chiu, "A fuzzy chip and a fuzzy inference accelerator for real-time approximate reasoning," in *Proc. 17th IEEE Int. Symp. Multiple-Valued Logic*, May 1987, pp.25-29.

[6] H.Watanabe, W.D.Dettloff and K.E.Yount, "A VLSI fuzzy logic controller with reconfigurable, cascadable architecture," *IEEE J.Solid-State Circuits*, vol.SC-25, pp.376-382, Apr. 1990.

[7] T.Yamakawa, "Fuzzy microprocessors rule chip and defuzzifier chip," in *Proc. Int. Workshop on Fuzzy Syst. Appl.*, Aug. 1988, pp.51-52.

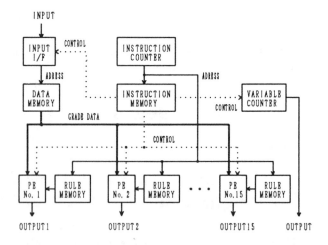

Fig.1. A structure of the if-part processor.

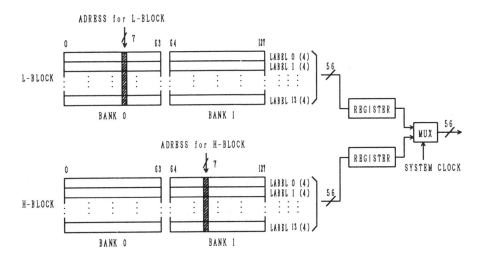

Fig.2. A structure of the data memory.

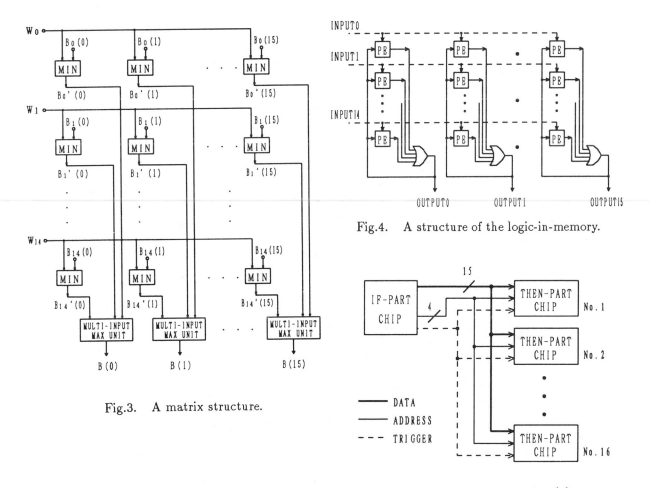

Fig.3. A matrix structure.

Fig.4. A structure of the logic-in-memory.

Fig.5. A system configuration (1).

313

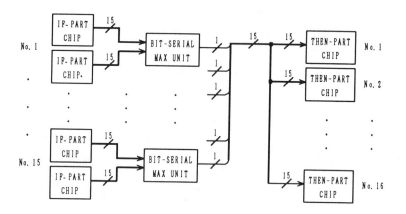

Fig.7.   A system configuration (2).

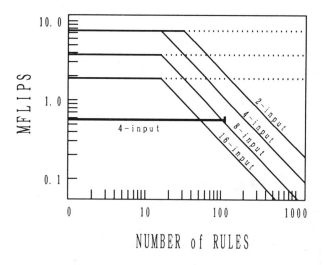

Fig.6.   Number of rules vs inference speed.

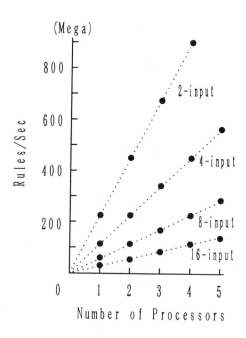

Fig.8.   Number of processors vs rules/sec.

TABLE 1   Performances.

	IF-PART Chip	THEN-PART Chip	
		MODE 1	MODE 2
Clock Frequency	30MHz		
Inference Speed	7.5M FLIPS (max)		
Number of Rules	960 (max)		
Power Supply	4.5 - 5.5 v		
Power	560 mW		
Interface	TTL compatible		
Package Type	132 pins Ceramic PGA		
Resolutions of Grade	4 bits		
Resolutions of Variables	6 bits (input)	8 bits (output)	
Number of Gates	13 0k. gates	29 2k. gates	
Number of RAMs	18.7k bits	———	
Die Size (mm)	13.79 × 13.72	13.65 × 13.65	
Number of Pins	93	95	
Number of Variables	16 (input)	1 (output)	2 (output)
Number of Labels	14 × 4	15	15 × 2
Number of Elements	64	32	16

Implementation of Several RLS Nonlinear Adaptive Algorithms Using a Commercial Floating Point Digital Signal Processor

Darioush M. Samani, Joshua Ellinger, Edward J. Powers, and Earl E. Swartzlander, Jr.

Department of Electrical and Computer Engineering
The University of Texas at Austin, Austin, Texas 78712

Abstract

We examine the potential use of the new floating point DSP chips for adaptive filtering. We have implemented several nonlinear adaptive filtering algorithms based on Least Squares estimation criterion using a floating point digital signal processor. To verify stability of an algorithm prior to implementation, we emulate single-extended precision floating point hardware using a C++ class. This elegant and transparent solution indicates an algorithm's susceptibility to errors produced by finite-precision arithmetic and the minimum precision necessary for stability. Our experience with the DSP chip indicates that the chip provides sufficient numerical accuracy and stability, and is a viable platform for such algorithms. In addition, our implementations are quick enough for some real-time applications, particularly since several processors can be used in parallel.

1: Introduction

Historically, adaptive filter algorithms have either been implemented in specialized hardware or evaluated 'off-line' on supercomputers. One possible alternative is to implement adaptive filters on digital signal processors (DSP). They offer real-time performance, can handle a wide range of system delays, and are inexpensive enough to use in embedded applications. We have implemented several adaptive filtering algorithms on Motorola's floating point DSP chip, the DSP96002. This chip supports the IEEE-754 single-extended floating point format [3].

The primary source of instability in the adaptive filters is that internal matrices become ill-conditioned or singular. Often, a system will perform well initially only to diverge and collapse after many iterations. A primary concern in implementation of adaptive filters is how errors due to finite precision will affect stability.

The issue of most concern in DSP implementations is whether the floating-point hardware has enough precision to allow the algorithms to function properly. The predominant DSP chips that support floating point (e.g., Motorola's DSP96K and Texas Instruments' TMS320C40) only implement single-extended floating point arithmetic. Only the Motorola's DSP96K fully supports IEEE arithmetic. As we will demonstrate, use of single-extended floating point does not compromise the performance of a large class of adaptive algorithms for realistic noise levels.

2: Numerical considerations

The study of **quantization** and **round-off errors** is crucial to adaptive digital filtering. The adaptive filter theory is based on the idealized framework of infinite precision. The arithmetic operations of digital computers, having limited precision, deviate from theoretical values [2]. The extent of deviation is inversely related to the number of bits used in the design of an ALU, and to the number of arithmetic operations that constitute a given algorithm. Because the precise deviation depends on the particular input parameters, an exact formulation of error produced by an algorithm is very complicated for most problems.

In real-world filtering applications, there are two major sources of errors: analog-to-digital conversions (quantization) and finite-precision arithmetic (round-off error). In this work, the emphasis is on the latter problem. We need to differentiate between **numerical instability** and **numerical inaccuracy** [2]. The numerical accuracy is a relative term and some inaccuracy is inherent to all digital filters. Numerical stability receives more attention because the onset of instability is hard to predict and difficult to correct.

The numerical stability of an algorithm is a complex function of numerical representation. Both **fixed point** and **floating point** formats suffer from **rounding error**. Floating point representation has an additional source of error, **masking error**, which occurs during addition of numbers having a wide range of magnitudes [1].

3: Limits of theoretical error calculations

One approach to determining the stability of a specific implementation is to model the data representation and derive the expected performance based on some simplifying assumptions (e.g., the input is Gaussian). A more sophisticated model uses the fact that instability is caused by internal matrices (which contains the past history of the system) becoming singular or losing positive semi-definiteness. By examining internal variables, one can estimate the condition of the matrix and thus stability. These theoretical approaches to error analysis have provided some interesting results. For example, instability in the conventional correlation-based RLS algorithm was found to originate from the lose of symmetry in the inverse autocorrelation matrix.

Unfortunately, most of the theoretical results are not applicable to our work because the simplifications required for tractable mathematical analysis. Most of the previous works on error analysis have been done for fixed point implementation. As indicated above, fixed point implementation need only consider error from multiplication/division while floating point must consider error from all operations. In the field of adaptive theory, there is very little error analysis for floating point implementations.

4: Stability modeling using a C++ class

Our interest in limited-precision floating point simulation stems from our work on DSP96K chip. Before implementing an algorithm on this processor, we use a *C++class*, that we call *Real*, to emulate the floating precision of the desired processor on a general purpose computer. If an algorithm appears unstable during simulation, we avoid the laborious process of assembly language coding on a DSP processor. [4-5]

The *Real* class is an elegant way to implement variable-precision IEEE floating point arithmetic. The class offers several advantages: 1) it is easy to install, 2) one does not have to maintain a special version of the code for simulation, 3) mixed mode simulation is possible, 4) measurement of the relationship between stability and precision is simple albeit time-consuming., 5) the Real class retains the central advantage of the IEEE standard: numerical results are reproducible independent of computer platform, and 6) since IEEE also supports an *extended double* (80 bit) format, it can be extended the Real class to support precision ranging from 1 to 80 bits.

5: Overview of a DSP processor

The Motorola DSP96K processor is the IEEE-754 floating point extension of the fixed point DSP56K. Like most DSP processors, it is optimized for vector manipulation. Its high performance stems from an inherently parallel instruction set, a wide memory bus, a special internal representation of floating point numbers, a pipelined address unit, and support for zero-overhead branching. Unlike RISC processors (which have large caches and complex pipelines), the DSP chip has very predictable timing. Most instructions take 1-2 clock cycles and there is no pipeline latency for arithmetic operations. The arithmetic and data access is inherently parallel. Perhaps the most elegant feature in the chip is the hardware *do*-loop. In addition, one can nest *do*-loops without having to save any special registers.

The DSP chip has several additional advantages. First, it supports an addressing mode that allows for row access of a matrix without performance penalties. Second, it has a wrap-around addressing mode. Third, it is optimized for fast Fourier transforms (FFT). Finally, two single-precision numbers can be loaded in the same clock cycle.

6: Division and square root on DSP chips

The DSP chip, like most digital signal processors, does not have division or square root hardware. Instead, the existence of fast addition and multiplication hardware allows one to approximate division and square root quicker in software than conventional hardware would allow. We use Newton's method to calculate both division and square root in our DSP96K implementations of adaptive filters. The power of Newton's method is its rate of convergence -- it <u>doubles</u> the number of correct bits on every iteration. The Newton's method approach is particularly useful in the square-root based adaptive filter algorithms that often use $1/\sqrt{x}$, x being a Euclidean normal of a vector. Since calculation of square root can produce $1/\sqrt{x}$ (at no extra cost), the division and square root calculation can be combined into one step.

The primary drawback of Newton's method is that it is not compliant with the IEEE floating point standard. The standard requires that all basic operations be rounded correctly in every case. <u>Newton's method occasionally rounds the last bit incorrectly</u>. As a result, we compromise one of the primary advantages of the standard -- we may not assume that if our implementation is numerically stable on a machine that supports IEEE floating point, it will definitely be stable on the DSP chip. However, the Newton's method approximation of square root and division operations on the chip are an order of magnitude faster than equivalent operations on the i486. Motorola provides IEEE division and square root routines for when exact compliance with IEEE is required. As we will show, <u>using Newton's method does not jeopardize performance at the presence of realistic noise levels</u>.

7: Algorithm timing derivation

Table 1 Operation Count and Timing of the CRLS-AMS on DSP96K

Time Initialization	$\mp$	$\times$	Mem	DSP96K
$\delta=10^{-4}\sigma_u^2; P_{MM}[0]=\delta^{-1}I_{MM}$				
Recursive Algorithm				
for$(n=1; n\leq N; n++)\{$				
$g_M^T[n]=x_M^T[n]P_{MM}[n-1]$	M^2+3M	M^2	$2M^2+4M$	$2M^2+4M$
$\kappa[n]=\lambda+g_M^T[n]x_M[n]$	$M+4$	$M+3$	$2M$	$2M$
$k_M[n]=g_M[n]/\kappa[n]$		M	$2M+3$	$2M$
$\alpha[n]=d[n]-\hat{w}_M^T[n-1]x_M[n]$	$M+1$	M	$2M$	$2M$
$\hat{w}_M[n]=\hat{w}_M[n-1]+k_M[n]\alpha[n]$	$M+1$	$M+3$	$3M$	$3M$
$P_{MM}[n]=\lambda^{-1}(P_{MM}[n-1]-k_M[n]g_M^T[n])$	M^2	$2M^2+3M$	$3M^2+4M$	$3M^2+6M$
Total Operation count	$2M^2+9M$	$3M^2+10M$	$5M^2+17$	$5M^2+19$

We have adapted several formulations of the linear Recursive-Least-Squares algorithms to model the third-order Volterra series. Two classes of algorithms are considered. The first class comprises the correlation matrix based algorithms (CRLS) [2, 4]. We consider the conventional RLS algorithm with the assumption that the inverse autocorrelation matrix is symmetric (CRLS-AMS). The second class is the data matrix based RLS (DRLS) algorithms. Orthogonal triangularization is used to achieve QR-Decomposition of the underlying data matrix. In this category, we have implemented several algorithms -- one with serial weight extraction (DRLS-SWE) [2, 4], another uses inverse QRD (DRLS-INVQR) [4], and the third uses the recursive modified Gram-Schmidt with error feedback (DRLS-MGSEF) [4]. We employ Givens rotation to do orthogonal triangularization for the DRLS-INVQR algorithm. For the DRLS-SWE algorithm, we use Givens rotation (DRLS-SWE-GR), fast Givens rotations (DRLS-SWE-GF), and Householder transformation (DRLS-SWE-HH) to implement triangularization [4]. Fast Givens rotation is known to have serious stability problems.

Since part of our motivation for using the DSP chip is that it can potentially execute algorithms in real-time environments, we consider timing issues for an implementation of the CRLS-AMS algorithm despite the known stability problems associated with this approach [2, 4]. This algorithm is less stable and requires more fraction bits for proper operation compared to the other algorithms in the RLS class that we have implemented. If the single-extended floating point of the DSP chip is sufficient for this algorithm, it should be adequate for the rest of this class.

Traditionally, the performance of an algorithm was calculated from the number of arithmetic operations performed while ignoring other overheads. Because of the increased speed of arithmetic operations, other overheads have become relatively more important, which complicates estimation of the timing of an algorithm. Arithmetic operations occur in parallel; this generates faster results but it makes timing derivation more complicated. The speed of floating-point operations has reached the point where we can no longer ignore other elements of timing if we wish to have accurate estimates. In particular, memory access is much more important that it used to be.

Table 1 contains the CRLS-AMS algorithm. For the DSP is important to realize that we can perform a multiplication, addition, and memory access in one clock cycle. As a result, our timing for each line of the algorithm is the maximum of $\times$, $\mp$ operations, and memory access. We can account for loop overhead quite easily. The DSP96K supports *hardware do-loops* which have a 3 clock cycle startup cost and no cost for additional iterations. The memory access time is the dominant term for each line. The total expected timing is shown in the Mem column . The actual timing is shown in the last column of Table 1. The discrepancy is due to extra instructions required to reset pointers.

8: Simulation Results

In this section, we present simulation results for the following algorithms: 1) DRLS-SWE, 2) DRLS-INVQR, 3) and DRLS-MGSEF. This work differs from previous published work in three significant ways. First, to our knowledge, we are the first group to extensively compare most common solutions to the LS estimation problem. Second, using *Real* class, we provide a unique insight into the stability of each algorithm by examining performance with <u>variable</u>

precision floating point. Third, we consider large nonlinear systems. Our simulation uses the third-order Volterra model. Since we have quadratic and cubic terms, our system has a much greater dynamic range than typical simulations. As such, we expect our nonlinear system to increase stability problems.

We define a standard third order Volterra model as

$$y[n] = \sum_{i=0}^{L-1} w_{l,i} u[n-i] + \sum_{i=0}^{Q-1} \sum_{j=i}^{Q-1} w_{q,ij} u[n-i] u[n-j] +$$

$$\sum_{i=0}^{C-1} \sum_{j=i}^{C-1} \sum_{k=j}^{C-1} w_{c,ijk} u[n-i] u[n-j] u[n-k] + v[n]$$

where $u[n]$ is the input, $y[n]$ is the estimated or computed output and $v[n]$ is the measurement noise at time n. The subscripts l, q, and c of the weight vector stand represent linear, quadratic, and cubic Volterra terms respectively. The upper limits on summation symbols define delay or memory span associated with that summation. In our results, we abbreviate the Volterra model as $MA(L, Q, C)=M$ where M is the total length of W. The linear(L), quadratic(Q), and cubic(C) delays generate L, $Q(Q+1)/2!$, and $C(C+1)(C+2)/3!$ terms, respectively [4].

For all simulations, desired coefficients and input data are random variables with zero-mean Gaussian distributions. The measurement noise also has Gaussian distribution with zero mean and appropriate variance (depending upon the SNR). The desired output is first contaminated with the measurement noise and then we apply an adaptive algorithm and measure how well we can recover our desired coefficients. Following standard practice, we provide the Normalized Mean Squared Error (NMSE) of the coefficients in dB which is defined as

$$NMSE_{Coefficients}(dB) = 10 \times \log_{10} \left(\frac{\sum_{i=1}^{M} (w_i - \hat{w}_i)^2}{\sum_{i=1}^{M} (w_i - \mu_w)^2} \right)$$

where $w_i, \hat{w}_i$ are the i-th desired and estimated coefficients (weights) respectively.

Because we use Newton's method to calculate division and square root, we cannot expect an exact match between simulation and implementation. However, if the algorithms remain stable under simulation for significantly fewer bits than the DSP chip has, we can be cautiously confident that they will be stable when implemented on the chip.

Our experiments indicate that we need at least 20 bits of mantissa (except for the DRLS-SWE-GF that fails with even double precision), which is far less than the 31 bits available on the DSP chip. In the following figures, the forgetting factor λ is set to 1 (infinite memory) and the number of data points considered is

represented by N. The (12,52) format, corresponding to standard IEEE double precision, provided as a reference level. The (7,23) format corresponds to standard IEEE single-precision. For the test cases, we set the exponent field to be 7 bits (which provides adequate dynamic range) and gradually reduce the number of mantissa (fraction) bits. Note that the algorithms perform well using the (7,30) format which corresponds to the DSP chip's single-extended format.

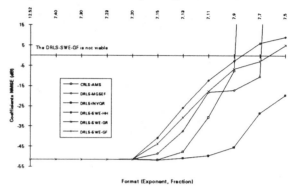

Figure 1 Variable precision error analysis using MA(9,6,4)=50, λ=1, N=10K, & SNR=30dB

Figure 1 shows a test case with realistic noise levels added to the desired signal (signal to noise ratio of the output to the added measurement noise is 30 dB).

9: DSP96K Performance

We have implemented the algorithms simulated in the previous section on the DSP96K. The algorithm with best speed was DRLS-MGSEF (which takes 8 seconds for this particular test). RLS takes about 14 seconds. The performance of DRLS-MGSEF is exaggerated because we only generate new coefficients once (after the last data point). Thus, one can expect 8 second performance from DRLS-MGSEF if one only occasionally needs the coefficients. When we include the time required to generate the coefficients, we expect its performance to deteriorate significantly. The DRLS-SWE-HH algorithm is surprisingly slow because the current implementation can only evaluate one element at a time. Normally, it is the fastest algorithm of this class when it evaluates multiple points.

We can draw two interesting conclusions from our timing results. The first is that there is no speed advantage for using fast Givens rotation over standard Givens rotation on the DSP96K. Our algorithms are limited by memory access and thus fewer multiplications in the fast Givens routine do not reduce execution time. The second point of interest is that it appears feasible to use our adaptive filter implementations in real-time applications. On the basis of our measurements, it appears that one can process (at worst) one data point every 5 milliseconds for a total filter length of 150 (RLS takes 14 seconds for 3000 data points or about 5 milliseconds per point). This implies a maximum sample rate of 200 Hz.

Assuming that filter execution time is proportional to M^2 (M is the total filter length), we can calculate expected maximum filter length for operation at 20 Hz and at 2 Hz.. For f=20 Hz, one can estimate M to be about 475, and for f=2 Hz, M is about 1500. This level of performance makes it possible to use adaptive filters for certain real-time applications.

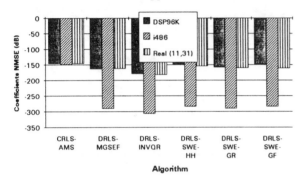

Figure 2 Coefficient error by platform using MA(5,4,3)=25, λ=1, N=25K, and SNR=Infinity

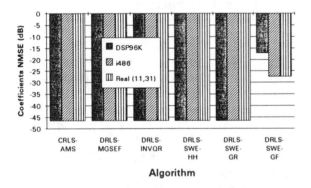

Figure 3 Coefficient error by platform using MA(5,4,3)=25, λ=1, N=25K, and SNR=30 dB

Turning to error measurements, we find several interesting results. First, we note that fast Givens rotation performs well only under no noise conditions. Second, it appears that the *Real* class simulation of the DSP96K floating point generates results that are closer to the actual results obtained by using the chip with noise than without noise. Presence of noise masks the minute errors introduced by using the Newton's method and this explains the results of Fig. 3. Third, the GF

fails when filter length is increased. Fourth, the algorithms appear stable on DSP96K.

It is also important to note that the i486 and DSP96K implementations generate identical numerical results under realistic noise levels. Without noise (Figure 2), the data matrix based algorithm generate -300 dB error because they are, in effect, solving a system of linear equations. When SNR=30 dB, (Figure 3), all the algorithms produce a NMSE of the coefficients of about -45 dB. Because the DSP and the i486 have identical accuracy for realistic noise levels, we expect that the difference in floating-point formats will not reduce the accuracy for standard applications.

10: Conclusion

We have simulated and implemented several formulations of the RLS adaptive filtering algorithm. Simulations using variable precision arithmetic indicate that the single-extended format of the DSP96K has significantly more bits than the minimum required for our algorithms. Thus, we do not expect the DSP96K to have more stability problems than comparable **C** implementations on a general purpose processor. Our functioning DSP96K implementations reveal that the degradation in numerical accuracy due to using single-extended precision is only noticeable without any noise. For realistic noise levels, IEEE double precision and single-extended precision produce equivalent error.

This work is supported in part by the Joint Services Electronics Program AFOSR Contract F-49620-92-C-0027, Department of Navy Grant N00014-92-J-1046, NSF Grant CDR-8721512, and the Texas Advanced Technology Program Project 003658-392.

References

[1] Goldberg, David, "What Every Computer Scientist Should Know About Floating-Point Arithmetic," ACM Computing Surveys, Vol. 23, No. 1, March 1991.

[2] Haykin, Simon, *Adaptive Filter Theory*, Prentice-Hall, Englewood Cliffs, NJ, 1991.

[3] IEEE, "IEEE standard for binary floating-point arithmetic," SIGPLAN Notices 22:2, 9-25, 1985.

[4] Samani, Darioush M., *Time Domain Analysis Of Adaptive Polynomial Filters With General Random Input With Applications To Nonlinear Physical Systems*, Ph.D. Dissertation, University of Texas at Austin, May 1993.

[5] Samani, Darioush, M., Joshua Ellinger, E. J. Powers, and Earl. E. Swartzlander, "Simulation of Variable Precision IEEE Floating Point Using C++ and its Application in Digital Signal Processor Design," *36th Midwest Symposium on Circuits and Systems*, Detroit, Michigan, Aug. 16-18, 1993.

Part II: *Applications*

Chapter 8: Speech Recognition

This chapter concentrates upon the most challenging form of speech recognition, the recognition of speech from realistically large vocabularies. "A Comparative Study of Signal Representations and Classification Techniques for Speech Recognition" (paper 8.1) recognizes the importance of choosing a feature analysis front end that preserves the discriminatory information available in the speech signal. The codepedence of feature extraction and pattern classification is recognized by including three classifiers with the ten feature extraction techniques that were compared, and a standard, readily-available database of speech was used for both full-bandwidth and telephone-bandwidth speech.

Examining the interaction between feature extraction and pattern classification is also the topic of "Discriminative Feature Extraction for Speech Recognition" (8.2). Here, minimization of classification errors is used to optimize the parameters of both the feature extractor and the classifier. A similar classification-based optimization for images, called relational template matching, is discussed in paper 10.5.

Through years of careful optimization, the use of continuous observation densities has emerged as a higher-performance replacement of vector quantization in the application of hidden Markov models for speech recognition. In "Vector Quantization for the Efficient Computation of Continuous Density Likelihoods" (8.3), vector quantization re-emerges. It is used here as a means to reduce computations by vector quantizing the input feature vector to identify a subset of Gaussian neighbors instead of computing full state likelihoods.

Time-synchronous hidden Markov models are used in the first of several end-to-end recognizers of this chapter, as described by "Large Vocabulary Continuous Speech Recognition of *Wall Street Journal* Data" (8.4). As in paper 8.2, the feature extractor improves recognition results by selecting parameters to minimize misclassification. "Phoneme-in-context" models are used on connected speech to achieve a speaker-independent word error rate of 13.6% on a standardized 5000-word vocabulary evaluation task using *Wall Street Journal* data. The *Wall Street Journal* task is also used in paper 8.5, "Large-Vocabulary Dictation Using SRI's DECIPHER™ Speech Recognition System: Progressive Search Techniques." Progressive search techniques are used to constrain the search space of a more accurate but slower-running technique, and a new method called the Forward-Backward Word Life Algorithm is introduced.

Field use of speech recognizers exposes them to noisy conditions and spontaneous speech patterns from talkers unfamiliar with automatic speech recognition. "Noisy Spontaneous Speech Understanding Using Noise Immunity Keyword Spotting with Adaptive Speech Cancellation" (8.6) places a speaker-independent recognizer within the context of ordering fast food. The recognizer combines those techniques listed in the paper's title with graphic indicators (open or closed lips) to cue speech, and uses noise-immunity learning to improve performance in a dialog task that combines speech input and speech output.

Large-vocabulary recognition approaches rely upon subword units from which the large number of words are built. Reducing this complex task to a single-chip implementation is described in "Golden Mandarin(II) - An Improved Single-Chip Real-Time Mandarin Dictation Machine for Chinese Vocabulary with Very Large Vocabulary" (8.7). This work uses a Motorola 96002D programmable digital signal processor to recognize the 408 base syllables of Chinese and the five tonal pitch contours, used to expand those base syllables to the 1302 different syllables that make up Mandarin speech. A speed of three characters per second, with a 5% error rate in a speaker-dependent mode, is reported.

Suggested Additional Readings:

[1]   L. R. Rabiner, "Applications of Voice Processing to Telecommunications," *Proc. IEEE,* vol. 82, no. 2, pp. 197-228, Feb. 1994.

[2]   L. R. Rabiner and B. H. Juang, "An Introduction to Hidden Markov Models," *IEEE ASSP Magazine,* vol. 3, no. 1, pp. 4-16, Jan. 1986.

[3]   H. F. Silverman and D. P. Morgan, "The Application of Dynamic Programming to Connected Speech Recognition," *IEEE ASSP Magazine,* vol 7, no. 3, pp. 6-25, July 1990.

[4]   J. Picone, "Continuous Speech Recognition Using Hidden Markov Models," *IEEE ASSP Magazine,* vol 7, no. 3, pp. 26-41.

[5]   A. Biem & S. Katagiri, "Feature Extraction Based on Minimum Classification Error/Generalized Probabilistic Descent Method," *Proc. IEEE International Conference on Acoustics, Speech, and Signal Processing 93* (Minneapolis, MN, USA) April 1993, vol. 2 pp. 275-278.

[6]   P. Kenny, P. Labute, Z. Li, R. Hollan, M. Lennig, and D. O'Shaughnessy, "A New Fast Match for Very Large Vocabulary Continuous Speech Recognition," *Proc. IEEE International Conference on Acoustics, Speech, and Signal Processing 93* (Minneapolis, MN, USA) April 1993, vol. 2 pp. 656-658.

[7]   D. M. Weber and J. A. du Preez, "A Comparison Between Hidden Markov Models and Vector Quantization for Speech Independent Speaker Recognition," *Proc. 1993 IEEE South African Symposium on Communications and Signal Processing COMSIG '93* (South Africa) August 1993, pp. 139-144.

A COMPARATIVE STUDY OF SIGNAL REPRESENTATIONS AND CLASSIFICATION TECHNIQUES FOR SPEECH RECOGNITION

Hong C. Leung and Benjamin Chigier*

Speech Technology Group
Artificial Intelligence Laboratory
NYNEX Science and Technology, Inc.
White Plains, New York 10604 USA

James R. Glass

Spoken Language Systems Group
Laboratory for Computer Science
Massachusetts Institute of Technology
Cambridge, Massachusetts 02139 USA

ABSTRACT

In this paper, we investigate the interactions of two important sets of techniques in speech recognition: signal representation and classification. In addition, in order to quantify the effect of the telephone network, we perform our experiments on both wide-band and telephone-quality speech. The spectral and cepstral signal processing techniques we study fall into a few major categories that are based on: Fourier analyses, linear prediction, and auditory processing. The classification techniques that we examine are Gaussian, mixture Gaussians, and the multi-layer perceptron (MLP). Our results indicate that the MLP consistently produces lower error rates than the other two classifiers. When averaged across all three classifiers, the Bark auditory spectral coefficients (BASC) produce the lowest phonetic classification error rates. When evaluated in our stochastic segment framework using the MLP, BASC also produces the lowest word error rate.

INTRODUCTION

Various *signal processing* techniques have been proposed for speech recognition over the past few decades. Some of these techniques are more linguistically or psychoacoustically motivated, while others are more computationally driven. Various *classification* techniques have also been applied to speech recognition. Some assume specific distance metrics or probability distributions, while others use corrective training procedures. In order to test for the effectiveness of a signal processing or classification technique, carefully controlled experiments are usually performed, with other modules of the speech recognition system held constant. For example, to test for the applicability of mixture Gaussians for classification, the signal processing and all other conditions of the entire system are often kept constant.

However, among other factors, the signal processing and classification techniques may interact with each other to affect phonetic classification and overall speech recognition. First, the result of an experiment may depend on whether the signal processing technique is well-matched with the underlying assumption of the classification technique. Second, the feature extraction procedure can affect both the dimensionality of the classifier and its accuracy. Finally, the effectiveness of a signal processing or classification technique may depend on the quality of the speech signal.

To investigate these issues, we simultaneously experimented with various signal processing and classification techniques. We chose two major categories of signal processing techniques, resulting in a total of 10 representations. We selected three classifiers, representing different classification paradigms. We also constrained our feature extraction procedures so that the number of input dimensions for all the classifiers was approximately the same. Finally, we compared results on wide-band and telephone-quality versions of the same speech database, thus enabling us to quantify the effects of the telephone network on phonetic classification.

TASKS AND CORPORA

All our experiments were performed within the stochastic explicit segment modeling (SESM) framework [8,9]. This framework can be characterized as a doubly stochastic process: broad-class segmentation and phonetic classification. Several stages of the system were evaluated, including phonetic classification, boundary detection and classification, and overall speech recognition.

Table 1 summarizes the speech databases for our study. The evaluations on phonetic classification, boundary detection, and boundary classification were based on phonetically-rich databases: TIMIT and NTIMIT [3,7]. These two databases are identical, except that the latter has been transmitted over the telephone network.[1] To facilitate our study, the same set of training and test utterances were used for both databases. Both "sx" and "si" sentences were used in our experiments. The training set consisted of 610 speakers, resulting in 4880 sentences. The test set consisted of 20 speakers, resulting in 160 sentences. There were a total of 184,492 phonetic tokens in the training set, and 6,016 phonetic tokens in the test set.

Overall evaluation of our speech recognition system was based on the recognition of isolated words drawn from a vocabulary of 25 city names in the metropolitan Boston area. The task is quite difficult in that the data were obtained from real users over dial-up telephone lines during interactions with an automated system. The speech signal is often contaminated with background noise, speech, and music, in addition to hesitation and channel distortion. The speakers can be female, male, or child. There are 4,768 utterances in the training set and 1,178 utterances in the test set. This corpus is only part of the city name database, CITRON, collected at NYNEX [17].

---

[1] The NTIMIT database is now publicly available through the LDC and NIST.

* Currently at Integrated Speech Solutions, 233 Freeman Street, Suite 3, Brookline, MA 02146 USA.

	Training		Testing	
Database	Utterances	Tokens	Utterances	Tokens
TIMIT	4880	184,492	160	6,016
NTIMIT	4880	184,492	160	6,016
CITRON	4768	-	1178	-

Table 1: Speech databases for evaluation. The TIMIT and NTIMIT databases were used to evaluate phonetic classification, boundary detection, and boundary classification. The CITRON database was used to evaluate overall recognition. It has not been phonetically transcribed.

EXPERIMENTS

Signal Processing

In our experiments, the input speech signal was first processed by one of 10 signal processing techniques. The resulting representation of the speech signal was then used for subsequent processing, including boundary detection, boundary classification, phonetic classification, and overall recognition. Whenever possible, the speech signal was analyzed once every 5 ms, with an analysis window of 28.5 ms.

Two major categories of signal representations are considered. The spectrally-based representations are the Bark auditory spectral coefficients (BASC) [1], the mel-frequency spectral coefficients (MFSC) [12], the mean-rate response (MR), the synchrony response (SR), and a synchrony auditory model (SAM) [15]. Both BASC and MFSC are DFT-based spectral representations, whereas MR, SR, and SAM are representations based on Seneff's auditory model [15].

The cepstrally-based representations are the Bark auditory cepstral coefficients (BACC) [1], mel-frequency cepstral coefficients (MFCC) [13], weighted linear predictive coding (WLPC) [18], bilinear transform LPC cepstral analysis (BLT) [16], and perceptually based linear prediction (PLP) [5]. Both BACC and MFCC are DFT-based, whereas the other three cepstral representations, WLPC, BLT, and PLP, are all LPC-based.

Classifiers

Three types of classifiers were used for phonetic classification: single full-covariance Gaussian (SFG), diagonal mixture Gaussian (DMG), and the multi-layer perceptron (MLP). Each SFG had a separate full-covariance matrix. The DMGs were trained using maximum likelihood estimates [2], and their mean vectors were initialized using k-means clustering. The MLP had 1 hidden layer of 128 units. Input normalization and center initialization were also used [10]. The *a-priori* statistics were explicitly used in the SFGs and DMGs. In the MLPs, they were embedded in the connections.

Only the MLP was tested for boundary classification. Each MLP had 1 hidden layer of 64 units. Previous experiments showed that classification accuracy is quite stable over a large range of number of hidden units [11].

Phonetic Classification

The same feature extraction procedure was used for both the spectral and cepstral representations. Each segment was first divided into three parts of equal duration. Let N be the number of spectral/cepstral coefficients averaged across each third, resulting in $3N$ average coefficients for each phonetic segment. Difference spectrum/cepstrum was computed using a procedure similar to Rabiner *et. al* [1,14]. The differences were calculated with the center at the beginning and the end of a phonetic segment, resulting in $2N$ difference coefficients. Thus, we had altogether $5N$ average and difference coefficients. In addition, duration and maximum zero-crossing count were added, resulting in a $(5N + 2)$-dimensional input vector.

For the spectral representations, N was equal to 40, resulting in a 202-dimensional raw input vector. Principal component analysis was then used to reduce the dimensionality of the raw input vector. We chose to use 70 dimensions, since this accounted for over 99% of the total variance of the original 202-dimensional vector. For the cepstral representations, N was set to 14, resulting in a 72-dimensional raw input vector for each phonetic segment.

Boundary Detection

The boundary detector that we use in our system is very similar to the one proposed by Glass [4]. Ideally, the boundaries of a phonetic segment can be hypothesized at any time in the speech signal. However, as discussed in previous work [9], this can result in a very large number of possible phonetic segments. In order to prune away unlikely segments, a boundary detector is used to look for spectral changes in the speech signal. Segments are then constrained to begin and end at these boundaries. Thus the boundary detector must be robust in that it has a very low deletion error rate. It must also make the rest of the system computationally efficient in that it does not propose boundaries too frequently.

Boundary Classification

Boundary classification can be viewed as a stochastic context-dependent broad-class segmentation process. The classifier examines each hypothesized boundary and assigns to it a probability that it is a valid boundary between 2 phones. Assuming adjacent boundaries are statistically independent, segment probabilities can be obtained directly from the boundary probabilities [8].

In our current implementation, each boundary can belong to one of 36 possible broad-class labels. Inputs to the boundary classifier are obtained from 8 abutting seed regions. From each seed region, 12 average spectral/cepstral coefficients are obtained, resulting in an input vector of 96 dimensions. For the spectral representations, the 12 average coefficients are obtained by using principal component analyses. However, for the cepstral representations, the first 12 coefficients are used.

RESULTS

Phonetic Classification

Figure 1 summarizes the phonetic classification error rates for the 10 spectral and cepstral representations on the TIMIT database, using SFG, DMG, and MLP. First, the MLP *consistently* produces error rates lower than SFG and DMG for all 10 representations. Second, the error rates on the 10 representations using the MLP fluctuate less than those using SFG or DMG. For example, the error rate using the MLP ranges from 22.0% to 27.4%, a fluctuation of 5.4%. However, the error rate using SFG fluctuates by as much as 10.7%. Third, the rank orders of the 10

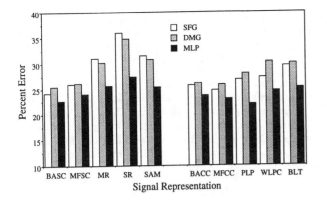

Figure 1: Error rates for phonetic classification on TIMIT. The five representations on the left are spectrally-based, whereas the five on the right are cepstrally-based.

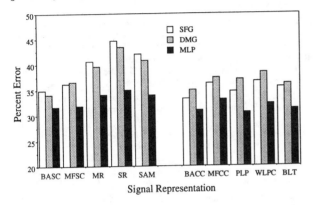

Figure 2: Errors rates for phonetic classification on NTIMIT.

signal representations are different for all three classifiers. However, when averaged across all three classifiers, BASC produces the lowest error rates. Finally, the lowest error rate of 22.0% was achieved by simultaneously using MLP and PLP.

The effectiveness of a signal processing or classification technique may depend on the quality of the speech signal. We have thus far experimented with speech transmitted over the telephone network. Figure 2 summarizes the corresponding phonetic classification error rates for the NTIMIT database. First, some of the characteristics observed for TIMIT in Figure 1 can also be seen in this figure. For example, the MLP still consistently produces lower error rates than the SFG and DMG. The fluctuation of the MLP error rates are also smaller than those of the other two classifiers. Second, averaged across all three classifiers, BACC now produces the lowest error rate. Third, the lowest error rate of 30.6% was achieved by simultaneously using MLP and PLP. Finally, the degradation due to the telephone network is quite consistent across the classifiers and signal representations. We have found that for any particular classifier and signal representation, the telephone network inflates the phonetic classification error rate by a factor of approximately *1.3*.

Boundary Detection

Boundary detection was evaluated using two measures. The detected boundaries were first compared with those boundaries provided in the phonetic transcription. *Deletion error rate* was

then defined as the percentage of the transcription boundaries not found by the boundary detector. The computational load resulted from the boundary detector was also measured by computing the average duration between adjacent detected boundaries. Figures 3 and 4 summarize the results on 6 signal representations for both the TIMIT and NTIMIT databases, respectively. First, the Bark auditory representations, BASC and BACC, have the lowest deletion error rates of less than 1% for both databases. Second, MR and MFSC are the most efficient in that they have the highest averaged detection period. Third, the telephone network, in general, causes the boundary detector to produce more boundaries. Finally, the results based on BASC are not much affected by the telephone network, in contrast to those based on other representations.

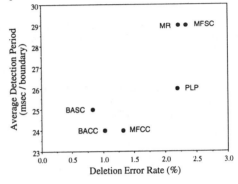

Figure 3: Boundary detection results for 6 different signal representations on TIMIT.

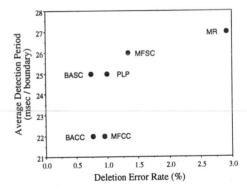

Figure 4: Boundary detection results for 6 different signal representations on NTIMIT.

Boundary Classification

Figure 5 summarizes the boundary classification results for the NTIMIT database on 5 representations using the MLP. Similar to our observation in phonetic classification when the MLP was used, PLP produced the lowest error rate. However, the error rates were quite uniform across all 5 representations.

Word Recognition

Figure 6 summarizes our overall recognition result on the task of recognizing 25 city names. We can see some correlation with other results discussed above. For example, BASC produces the lowest boundary deletion error rate on NTIMIT and it has the lowest word error rate of 3.8% on CITRON. PLP produces the

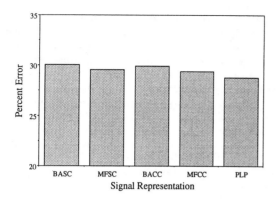

Figure 5: Percent error rates for boundary classification using 5 different signal representations on NTIMIT. Each boundary belongs to one of 36 possible classes.

rates in our experiments seem to be correlated with the results on phonetic and boundary classifications, as well as boundary detection. Future work should include investigation on the relative importance of these classification and detection techniques.

As has been previously reported, representations based on auditory processing are more insensitive to noise than others [6,12]. In the future, we plan to investigate the effect of noise on phonetic classification, in conjunction with exploring different noise-canceling techniques. Furthermore, we will study how overall recognition can be affected by using different classifiers. We will also extend our work to continuous speech.

lowest phonetic and boundary classification error rates and it has the second lowest word error rate. Finally, MFSC produces the highest boundary deletion error rate and it also has the highest word error rate.

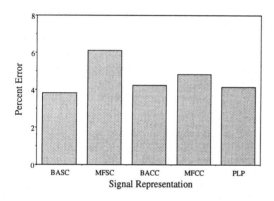

Figure 6: Overall word recognition error rates using 5 different signal representations on CITRON.

DISCUSSION

Within the bounds of our experimental conditions, our results indicate that the signal processing and classification techniques interact with each other to affect phonetic classification. Thus it is difficult to evaluate the effectiveness of a signal processing technique alone. When one signal processing technique is compared to another, we must also specify the classification technique.

However, for the task of phonetic classification, the MLP consistently produces lower error rates than the other two classifiers that we have experimented. The fluctuation of the error rates due to the MLP is also consistently lower than those due to the other two classifiers. These observations suggest that the MLP is a more flexible classifier, presumably due to the facts that the MLP does not make as strong an assumption about the underlying distribution of the data, and that it uses corrective training.

The fact that MR resulted in the highest average boundary detection period is in agreement with the earlier observation that forward masking in the auditory model attenuate many low-amplitude sound changes [4]. Furthermore, overall word error

REFERENCES

[1] Chigier, B., "Phonetic classification on wide-band and telephone quality speech," *Fifth DARPA Workshop on Speech & Natural Language*, Arden House, February 1992.

[2] Duda R. and Hart, P., *Pattern Classification and Scene Analysis*, John Wiley & Sons.

[3] Fisher, W., Doddington, G., and Goudie-Marshall, K., "The DARPA speech recognition database: specifications and status," *Proc. DARPA Workshop on Speech Recognition*, February 1986.

[4] Glass, J., *Finding Acoustic Regularities in Speech: Applications to Phonetic Recognition*, Ph.D. Thesis, MIT Department of Electrical Engineering and Computer Science, 1988.

[5] Hermansky, H. and Claude, J., "Optimization of perceptually based ASR front-ends," *Proc. ICASSP*, New York, 1988.

[6] Jankowski, C., "A comparison of auditory models for automatic speech recognition," S.M. thesis, *MIT Dept. of Electrical Engineering and Computer Science*, Cambridge, MA,1992

[7] Jankowski, C., Ashok K., Basson, S., and Spitz, J., "NTIMIT: A phonetically balanced, continuous speech, telephone bandwidth speech database", *Proc. ICASSP-90*, Albuquerque, NM, 1990.

[8] Leung, H., Hetherington, I., and Zue, V., "Speech recognition using stochastic segment neural networks," *Proc. ICASSP*, San Francisco, 1992.

[9] Leung, H., Hetherington, I., and Zue, V., "Speech recognition using stochastic explicit-segment modeling," *Proc. European Conference on Speech Communication and Technology*, Italy, 1991.

[10] Leung, H., and V. Zue, "Phonetic Classification Using Multi-Layer Perceptrons," *ICASSP*, Albuquerque, NM, 1990.

[11] Leung, H.C., "The use of artificial neural networks for phonetic recognition," Ph.D. thesis, MIT Dept. of Electrical Engineering and Computer Science, Cambridge, MA, 1989.

[12] Meng, H. and Zue, V., "Signal representation comparison for phonetic classification," *Proc. ICASSP*, Toronto, Canada, 1991.

[13] Mermelstein, P. and Davis, S., "Comparison of parametric representations for monosyllabic word recognition in continuously spoken sentences," *IEEE Transactions on Acoustics, Speech and Signal Processing*, August 1980.

[14] Rabiner, L. Wilpon, J., and Soong, F., "High performance connected digit recognition using hidden Markov models," *ICASSP*, New York, 1988.

[15] Seneff, S., "A joint synchrony/mean rate model of auditory speech processing,", *Journal of Phonetics*, Vol.16, No.1, 1988.

[16] Shikano, K., "Evaluation of LPC spectral matching measures for phonetic unit recognition," *Technical Report CMU-CS-86-102 Computer Science Dept., Carnegie Mellon University.*

[17] Spitz, J. and the Artificial Intelligence Speech Technology Group, "Collection and Analysis of Data from Real Users: Implications for Speech Recognition/Understanding Systems," *Proc. Fourth DARPA Workshop on Speech and Natural Language Workshop*, Pacific Grove, CA, February 1991.

[18] Wilpon, J., Rabiner, L., Lee, C. and Goldman, E., "Automatic recognition of keywords in unconstrained speech using hidden Markov models," *IEEE Transactions on Acoustics, Speech, and Signal Processing*, November 1990.

DISCRIMINATIVE FEATURE EXTRACTION FOR SPEECH RECOGNITION

Alain Biem, Shigeru Katagiri & Biing-Hwang Juang†
ATR Human Information Precessing Laboratories
2-2 Hikaridai, Seika-cho, Soraku-gun, Kyoto, 619-02, Japan
e-mail:biem@hip.atr.co.jp
†AT&T Bell Laboratories

Abstract- Pattern recognition consists of feature extraction and classification over the extracted features. Usually, these two processes are designed separately, entailing that a resulting recognizer is not necessarily optimal in terms of classification accuracy. To overcome this gap in recognizer design, we introduce in this paper a new design concept, named *Discriminative Feature Extraction* (DFE). DFE is based on a recent discriminative learning theory, Minimum Classification Error formalization/Generalized Probabilistic Descent method, and provides an innovative way to design the entire process of recognition. A front-end feature extractor as well as a post-end classifier is consistently optimized under a single criterion of minimizing classification errors. The concept is quite general and can be applied to a wide range of pattern recognition tasks. This paper is devoted to the application of DFE to speech recognition. Experiments on a Japanese vowel recognition task show the advantages of the proposed approach.

INTRODUCTION

Pattern *recognition* consists of both a *feature extraction* and a *classification* process. Features are often produced based on scientific expertise, such as psychological findings or statistics of pattern samples. In most cases, the post-end classifier is designed over this pre-chosen feature space. Optimality of the overall recognizer is therefore not guaranteed, due to such an inconsistency underlying the recognizer design.

In this paper, we propose a novel solution, named *Discriminative Feature Extraction* (DFE), to remedy the above gap. DFE is an extended application of a recent discriminative learning theory, called the Minimum Classification Error formalization (MCE)/ the Generalized Probabilistic Descent method (GPD) [1] [2], which has been proved useful over several classification tasks [3-6]. The idea of DFE was preliminarily reported in [7]. In this DFE framework, the MCE/GPD concept is applied to the entire recognition process: The front-end feature extractor as well as the back-end classifier is designed consistently in order to minimize misclassifications.

DFE can fundamentally handle any kind of patterns as well as be applied to various trainable recognizer structures. In this paper, we focus on applications of DFE to speech recognition. Details of both formalization and implementation are introduced. Evaluation results over an exemplar task based on cepstrum liftering and frequency scaling demonstrate the promise of the proposed approach.

DISCRIMINATIVE FEATURE EXTRACTION

As mentioned in the introduction, DFE is a method to design the overall recognizer, consisting of the feature extractor and the classifier, in a manner consistent with the minimization of classification errors. Contrary to conventional methods, features in our DFE framework are generated while focusing on the minimum misclassification properties of the classifier, assuming an interaction between classification and feature extraction. This ensures that the designed classifier will itself map an input space to a more suitable one for its proper classification.

From the viewpoint of achieving the minimization of misclassifications, the MCE concept is applied. Obviously, DFE can be formalized by using alternatives to GPD such as MCE/Steepest Descent Method or MCE/Simulated Annealing. However, taking into account the nature of the GPD adaptation scheme and previously-reported performances [3-7], we use in this paper GPD as our optimization algorithm.

The difference between previous applications of MCE/GPD and DFE is that a set of trainable system parameters, denoted by Λ, include both the feature extractor parameters and the classifier parameters. We introduce the DFE formalization in the following.

Let us consider a sample space Ω, a set of training samples $S = \{x_t\}_{t \in \{1,...,T\}}$ with $x_t \in \Omega$, where t is a training time index, and a set of classes $\{C_j\}_{j \in \{1,...,M\}}$. Also consider the following simple classification decision rule:

$$x \in C_i \quad \text{if} \quad i = \arg\max_j g_j(x, \Lambda) \tag{1}$$

where $x \in \Omega$ and $g_j(x, \Lambda)$ is a discriminant function which indicates the degree to which x belongs to C_j. The aim is to design a recognizer so as to minimize an expected loss $\mathcal{L}(\Lambda)$ which is a reflection of the misclassifications and defined as $\mathcal{L}(\Lambda) = E_x(\ell_i(x, \Lambda))$ where $\ell_i(x, \Lambda)$ is a loss occurring for an arbitrary sample x ($\in \Omega, \in C_i$) given the set of parameters Λ. The loss is a function of a misclassification measure $d_i(x)$ which is defined over the set of discriminant functions $\{g_j(x, \Lambda)\}_{j \in \{1,...,M\}}$:

$$\ell_i(x, \Lambda) = \sigma(d_i(x)) \tag{2}$$

where σ is a *smooth* (at least first differentiable) 0-1 step function such as

$$\sigma(d_i) = \frac{1}{1 + e^{-\alpha d_i + \beta}} \tag{3}$$

where α (> 0) controls the smoothness and β is a constant.

The misclassification measure emulates the classification decision in scalar values: A positive value corresponds to a misclassification and a negative value implies correct classification.

To emulate the above rule, the misclassification measure for class C_i is defined as follows:

$$d_i(x, \Lambda) = -g_i(x, \Lambda) + \left[\frac{1}{M-1} \sum_{j \neq i} \{g_j(x, \Lambda)\}^\eta \right]^{\frac{1}{\eta}} \tag{4}$$

with η being a positive number which controls the relative contribution of the classes considered.

$\mathcal{L}(\Lambda)$ is reduced by using the parameter adjustment, performed for a given training sample x_t ($\in C_i$),

$$\Lambda_{t+1} = \Lambda_t - \epsilon_t \nabla \ell_i(x_t, \Lambda_t) \tag{5}$$

where Λ_t denotes the status of Λ at the index t and ϵ_t is a small positive number. Based on the probabilistic descent theorem [2], the repetition of this adjustment reduces $\mathcal{L}(\Lambda)$ in a probabilistic sense. Moreover, if ϵ_t is selected to satisfy the stochastic approximation requirements, an infinite run of the adjustment will lead to at least a local minimum of $\mathcal{L}(\Lambda)$, which corresponds to a local optimal status of Λ.

In practice, infinite training is obviously impossible. As in other MCE/GPD applications, DFE thus tries to reduce the classification errors over the given sample set S through a finite repetition of the adjustment, while keeping the smoothness (generalization for unknown samples) embedded in the formalization.

APPLICATIONS TO SPEECH RECOGNITION

Since DFE is not tied to a particular classifier structure, nor restricted by the choice of features, it can be applied to various tasks. In particular, this paper discusses applications to speech recognition in two important feature domains, time and frequency, focusing on the following issues: 1) a quefrency (time)-domain feature extraction consisting of a lifter design, and 2) a power spectrum-domain feature extraction consisting of a nonlinear frequency scaling.

Lifter Design

Quefrency-domain feature extraction using the DFE approach is presented here. It consists of designing an optimal lifter applied to speech recognition. Research on lifter design has been carried out for several speech recognition tasks [8] [9]. Nevertheless, conventional methods are limited due to the lack of a theoretical basis for reflecting various mixed sources of information such as phoneme identity and speaker identity over cepstrum coefficients for designing a lifter properly. Designing the lifter while ignoring the back-end classifier may be inadequate. A desirable lifter should be designed in a classification-oriented framework.

The typical structure of the recognizer is illustrated in Fig. 1 (a). It is composed of two modules: the first is composed of lifter weights, one for each cepstrum coefficient and is intended to extract accurate features while the second, the classifier, is aimed at classifying speech patterns represented by the extracted features.

The two modules are trained together, without consideration of the modularity during the learning process. A chain rule might be used to adjust the liftering weights as in the error back-propagation algorithm. In the proposed-framework, this initial lifter is adjusted by the DFE training, finally producing an optimal one, in the sense of minimizing the classification errors of the back-end classifier attached to the lifter.

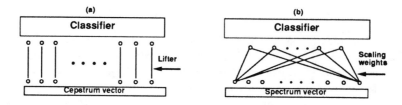

Figure 1: a) Lifter designing Recognizer and b) Frequency scaling Recognizer.

Frequency Scaling

Filter bank modeling of speech signals has been widely used in speech recognition tasks. Moreover, physiological studies have confirmed the importance of the critical bandwidths in spectral representation [10]. Although these approaches lead to a new dimensionality-reduced input space for a classifier, they use various criteria for producing center frequencies or bandwidths. There is no assumption over the optimality of these features from the point of view of the back-end classifier. An alternative would be to design the frequency scaling process together with classifier.

Within the scaling framework, the spectral sample X interacts with $Y = \mathcal{F}(X, w)$ where $Y = (Y[i])$ represents a new, N-dimensional feature produced given the filtering parameters w and the speech sample $X = (X[j])$ according to the a function $\mathcal{F}$ as illustrated in Fig. 1(b). In this regard, the optimal scaling process, in the sense of minimum error rate, corresponds to the optimal w.

In this paper, the method is intented to convert a frequency band b centered at frequency f_b into an output feature $Y[i]$ representing the weighted sum of the adjacent samples around the center frequency within the frequency band, i.e

$$Y[i] = \mathcal{F}\left(\sum_{f \in B} w_b(f)X(f)\right) \tag{6}$$

where $w_b(f)$ represents the weighting function value at frequency f and B the frequency range. The contribution of the spectral samples $X(f)$ in producing $Y[i]$ is modulated by $w_b(f)$. Therefore, the set of $w_b(f)$ for all b and all f labeled as w determines the center frequency and bandwidth for each feature in Y which can be viewed as a linear transformation of X passed through a transfer function $\mathcal{F}$. The parameters of the overall process has to be adjusted according to (5).

EXPERIMENTAL RESULTS

There are obviously many ways to implement the ideas of DFE liftering or frequency scaling: Any reasonable classifier can be used as the post-end system or other methods can be applied to determine the parameters $w_b(f)$. The recognizers used in the following are just examples among many possibilities.

As first evaluation, we present in this paper experimental results of the proposed approaches applied to the 5-class Japanese vowels recognition task. Vowel samples were extracted from 100 phonetically-balanced sentences, spoken by 5 speakers (3 males and 2 females) and digitized at a 12 kHz sample rate. The center fragment of each vowel segment was used to generate 256 FFT-based spectral coefficients through a 42ms Hamming window as the recognizer input. Note that our recognition was done just on this single frame cepstrum vector. Three thousand and five hundred samples were prepared in total; half for training and the second half for testing.

Liftering task

For this single frame vector task, we used a feed-forward Neural Network recognizer which comprises two modules.

The first module is comprised of two layers: the input layer receiving original cepstrum patterns and the liftering layer processing liftered cepstrum patterns. These two layers are composed of the same number of network cells, and each input cell is linked with one cell of the lifter layer. A cell is also associated with an observable quefrency position and its connection weight corresponds to a lifter coefficient. For liftering purposes with considerations of each layer's role, the layer's cells in this module are linear.

The second module is a usual classification-oriented feed-forward network whose aim is to classify the patterns occurring from the lifter layer.

For this recognizer, $g_j(x, \Lambda)$ is the output corresponding to class C_j for input x. The decision rule of (1) is used with the misclassification measure formulation of (4).

Contrary to this rather simple setting of task, we investigated characteristics of the proposed recognizer design using 128 FFT-based cepstral coefficients as input to the recognizer. The main experimental concerns were over the initialization of lifter weights, and the training criterion. Although the network recognizer was trained without any consideration of the modularity, there was only one difference in initializing weights between the modules, based on the difference in fundamental role. For the lifter module, two kinds of initial settings were specially investigated: 1) a uniform initialization which is equivalent to no liftering at the beginning of the process, and 2) a random initialization widely used in Neural Networks. Since DFE can theoretically handle any reasonable learning criterion other than Minimum Classification Error (MCE stands for this error measure in this section), we compared DFE, based on MCE, with the design using the widely-used Minimum Squared Error (MSE) criterion.

Table 1 summarizes the best recognition rates using MCE and MSE learning objective for two settings: the uniform initialization and the random initialization.

Table 1: Recognition rates of the liftering task.

Initial setting	MCE		MSE	
	train	test	train	test
uniform	96.8%	88.7%	96.8%	86.2%
random	93%	83.2%	89.8%	80.8%

MCE provides better generalization than MSE, proving the effectiveness of the proposed method.

Concerning the lifter initialization, the uniform initialization clearly worked better than the random initialization. This may be due to local optimality of gradient search. Similar to other gradient descent optimizations, a good initialization may be needed in the DFE framework.

Typical examples of lifters designed with the MCE or MSE training criterion are shown in Fig. 2. Both has suppressed the high quefrency region that corresponds to vocal cords excitation and enhanced the low-

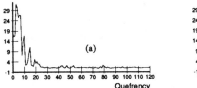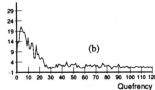

Figure 2: (a) MCE-based Lifter and (b) MSE-based Lifter

quefrency region which corresponds to vowel categorization. Moreover, the extremely low-quefrency region where speaker identity dominates is also suppressed. Comparing these two illustrations, one notes that the MCE-based lifter more effectively suppresses the range higher than around 20, resulting in a smoother feature extraction that may lead to a better generalization.

Accuracy of the proposed method could be understood through the comparison with conventional methods. Since the obtained lifters show a cutoff quefrency located between the value 20 and 40, the same architecture have been evaluated using standard rectangular lifters of cutoff quefrency ranging from 8 to 32 as summarized in Table 2. The accuracies over training as well as testing data are clearly less than those obtained by DFE. These results enable a straightforward comparison between conventional feature extraction approaches and DFE in this framework.

Table 2: Recognition rates of the Japanese vowels task using rectangular lifters.

Window size	Training	Testing
32	86.62%	84.68%
16	91.58%	85.48%
8	90.28%	85.14%

Frequency scaling task

To show the generality of DFE concept, we used a multi-template distance classifier structure described in [5]. The discriminant function for this task is

$$g_j(x, \Lambda) = \left\{ \sum_{k=1}^{V_j} |\mathcal{F}(x, w) - r_k^{(j)}|^{-\nu} \right\}^{-1/\nu} \qquad (7)$$

where $r_k^{(j)}$ represents the kth reference vector of class C_j and V_j is the number of reference vectors in class C_j. In this case, the misclassification measure is defined as:

$$d_i(x, \Lambda) = g_i(x, \Lambda) - \left[\frac{1}{M-1} \sum_{j \neq i} \{g_j(x, \Lambda)\}^{-\eta} \right]^{-\frac{1}{\eta}} \qquad (8)$$

In contrast with the previous example, a classification decision corresponds to the class with the smallest discriminant function.

For simplicity, we let $\nu \to \infty$ in (7) and $\eta \to \infty$ in (8) then (7) is simply a distance measure to the closest reference vector and for a sample x belonging to class C_α, the misclassification measure corresponding to class C_α, is approximated by

$$d_\alpha(x, \Lambda) = g_\alpha(x, \Lambda) - g_\beta(x, \Lambda) \qquad (9)$$

where $g_\beta(x, \Lambda)$ is the discriminant function corresponding to the best match over the incorrect classes.

As preliminary set up of the scaling task, 128 FFT-based log power spectral coefficients which cover the perceptual auditory range frequencies were used as input to the recognizer.

Constraining weights As first experimental set up, we simply use a piece wise transformation as transfer function. Although, each weighting parameter in (6) could be trained independently, they were constrained through the use of Gaussian weighting functions for each frequency band b such that

$$w_b(f) = e^{-\alpha_b(f_b - f)^2} \qquad (10)$$

where the trainable parameters α_b and f_b determine bandwidth and center frequency. This selection was due to our concern over avoiding statistical unstability of training results. In the following task, only the center frequencies were adjusted while the bandwidths remained constant.

Initializations. A good initialization of reference vectors and scaling weights is important to avoid "bad" local attractors that may occur due to the use of gradient descent.

A K-means clustering algorithm was used to select a set of reference vectors composed of 16 Mel-scaled log power spectra. The scaling weights were then initialized so as to select approximately the Mel-based frequency band regions, thus transforming a 128-dimensional vector into a 16-dimensional vector to be classified.

After initializations, training of the overall recognizer were performed. DFE achieved 82.45% while classical methods where only the reference vectors were trained realized 80.91% on the same testing data using only one reference vector per class. Using three reference vectors per class increased the results up to 84.57% for DFE and 83.08% for classical methods and five reference vectors realized 84.74% for DFE compared to 84.51% for classical methods. DFE provides a better accuracy than classical methods when using fewer reference vectors.

Fig. 3 presents a typical scaling curve after DFE application using one reference vector per class. The curve is obtained by extrapolation of resulting center frequencies according to their index. The channel indexes have not been reordered into a monotonic nondecreasing function of the frequencies for clarity of the adjustements within each band. It could be noticed that while lower frequencies remain unchanged, the frequency range has got wider and typical gaps are located around certain frequencies (0.5kHz, 1.5kHz, 3.5kHz) which are the regions where formants are located on average. This may illustrate that they were not directly taken into consideration in the Mel representation.

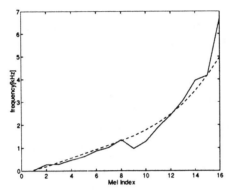

Figure 3: Center frequencies adjustment. The dashed line represents original center frequencies and the solid line represents extrapolated DFE-based center frequencies.

All the above results could be refined by use of appropriate selection of learning parameters according to the task. However, they show the feasibility of the proposed method for handling input data and interactively *adjust* them to the classifier.

CONCLUSION

A novel approach to pattern recognition, called *Discriminative Feature Extraction* has been introduced as a way to interactively handle the input data with a given classifier. The entire recognizer, consisting of the feature extractor as well as the classifier, was trained with the MCE/GPD learning algorithm. Both the philosophy and implementation examples of this approach were described. DFE realizes a significant departure from conventional approaches, providing a comprehensive base for the entire system design. By way of example, an automatic scaling process was described and experimental results for designing a cepstrum representation for vowel recognition were presented.

ACKNOWLEDGEMENTS

The authors would like to thank E. McDermott and D. Rainton for fruitful discussion concerning this research.

References

[1] S. Katagiri, C.-H. Lee & B.-H. Juang, *A generalized Probabilistic Descent Method*, Proc. Acoust. Soc. of Japan, 2-p-6. pp. 141-142 (1990.9)

[2] B. H. Juang & S. Katagiri, *Discriminative Learning for Minimum Error Classification*, IEEE trans. on ASSP, vol 40, No 12, pp. 3043-3054 (1992.12).

[3] D. Rainton & S. Sagayama, *Word Level Minimum Error Training of Phoneme HMM*, Proc. of spring conf. of ASJ, pp. 3-4 (1992.3).

[4] T. Komori & S. Katagiri, *Application of a Generalized Probabilistic Descent Method To Dynamic Time Warping-based Speech Recognition*, Proc. ICASSP 92, vol. 1, pp. 497-500 (1992.3).

[5] E. McDermott & S. Katagiri, *Prototype-Based Discriminative Training For Various Speech Units*, Proc. ICASSP 92, vol 1, pp 417-420 (1992.3)

[6] W. Chou, B.-H. Juang & C.-H. Lee, *Segmental GPD training of HMM Based Speech Recognizer*, Proc. ICASSP 92, vol. 1, pp. 473-476 (1992.3).

[7] A. Biem & S. Katagiri, *Feature Extraction Based on Minimum Classification Error/Generalized Probabilistic Descent Method*, Proc. ICASSP 93, Vol 2, pp.275-278, (1993.4)

[8] B.-H. Juang, L. R. Rabiner & J. G. Wilpon, *On the use of Bandpass Liftering in Speech Recognition*, Proc. ICASSP 86, vol. 1, pp. 765-768 (1986.4).

[9] Yoh'Ichi Tohkura, *A Weighted Cepstral Distance Measure for Speech Recognition*, IEEE Trans. on ASSP, Vol 35, No 10, pp. 301-309. (1987.11).

[10] E. Zwicker, *Subdivision of the audible frequency range into critical bands*, J. of ASA, vol. 33, pp 248 (1961).

VECTOR QUANTIZATION FOR THE EFFICIENT COMPUTATION OF CONTINUOUS DENSITY LIKELIHOODS

Enrico Bocchieri

Speech Research Dept., AT&T Bell Laboratories, MH-2C 568,
Murray Hill, N.J. 07974

ABSTRACT

In speech recognition systems based on Continuous Observation Density Hidden Markov Models, the computation of the state likelihoods is an intensive task. This paper presents an efficient method for the computation of the likelihoods defined by weighted sums (mixtures) of Gaussians. The proposed method uses vector quantization of the input feature vector to identify a subset of Gaussian *neighbors*. It is here shown that, under certain conditions, instead of computing the likelihoods of all the Gaussians, one needs to compute the likelihoods of *only* the Gaussian neighbors. Significant (up to a factor of nine) likelihood computation reductions have been obtained on various data bases, with only a small loss of recognition accuracy.

1. MOTIVATION AND ALGORITHM OVERVIEW

In speech recognition algorithms based on Hidden Markov Models (HMM), the state likelihoods of the input observation vectors are typically represented by either discrete or parametric (continuous) distributions. In the continuous case, the state likelihood computation is an intensive task, that requires up to 96% of the total computation in a typical small vocabulary application (connected digit recognition), and more than 80% of the total computation in a large vocabulary system based on Beam-Search of the state path (DARPA Resource Management). The likelihood computation is intensive also for tied-mixture systems, as discussed in [1]. Therefore, an efficient likelihood computation method can significantly reduce the computational requirements of a recognition system.

In particular, we have studied an efficient algorithm

for the computation of the state likelihoods represented by Gaussian mixtures. The generic Gaussian component likelihood of frame feature vector $\mathbf{f}_t$ is denoted by:

$$G(\mathbf{f}_t, \boldsymbol{\mu}_m, \mathbf{U}_m), \quad m = 1, M \qquad (1)$$

where $\boldsymbol{\mu}_m$ and $\mathbf{U}_m$ are the mean vector and diagonal covariance of the mth Gaussian component, respectively, and M is the total number of the mixture components of all the states of all the words (speech units) in the vocabulary. The likelihood $l_s(\mathbf{f}_t)$ of observation $\mathbf{f}_t$, given a certain state s, is computed as:

$$l_s(\mathbf{f}_t) = \sum_{m \in M_s} \varepsilon_m \, G(\mathbf{f}_t, \boldsymbol{\mu}_m, \mathbf{U}_m),$$

$$\sum_{m \in M_s} \varepsilon_m = 1 \qquad (2)$$

where ε_m is the mixture weight and M_s is the subset of mixture components (1) that belong to state s.

It is well known that the Gaussian models (1) are statistically accurate only if the input feature vector is *near* to the Gaussian means. The Gaussian model provides at best a poor approximation of the likelihood (and a small likelihood contribution) when the feature vector falls on its distribution tail (outlier feature vector). The proposed method computes only the likelihoods of those Gaussians for which the input vector is not an outlier. During system training, all the mixture components (1) are clustered into *neighborhoods*. A vector quantizer, consisting of one codeword for each neighborhood of Gaussians, is also defined. During recognition, vector quantization of the input frame vector allows one to select:

a) a small subset (neighborhood) of Gaussians whose likelihoods must be exactly computed, and

b) a complementary subset of Gaussians, such that the input vector falls on the tails of these Gaussians.

The traditional technique computes the likelihood of all the Gaussians. The proposed method *only* computes the likelihoods for the Gaussians in the set a) above. The likelihoods of the Gaussians in the set b) are quickly approximated either by table look-up or by a small constant.

2. NEIGHBORHOODS OF GAUSSIAN MIXTURE COMPONENTS

The first step for the definition of the neighborhoods, is the clustering of all the Gaussians (1). As a measure of the distance between the ith and the jth Gaussians, we use a weighted (Mahalanobis-like) Euclidean distance between the means:

$$\delta(\mu_i, \mu_j) = \frac{1}{d} \sum_{k=1}^{d} \left\{ w(k)(\mu_i(k) - \mu_j(k)) \right\}^2 \quad (3)$$

where d is the dimension of the feature vectors and $\mu_i(k)$ is the kth component of vector μ_i. The weight $w(k)$ are equal to the inverse square root of the kth diagonal element of the average of the covariances of (1). The center (codeword) of the ϕth cluster χ_ϕ is:

$$c_\phi = \frac{1}{size(\chi_\phi)} \sum_{m \in \chi_\phi} \mu_m , \quad \phi = 1, \Phi \quad (4)$$

where Φ is the specified number of clusters. In particular, the adopted clustering procedure is based on the classical k-means iteration that minimizes the average (per mixture component) distortion:

$$\delta_{avg} = \frac{1}{M} \sum_{m=1}^{M} \left\{ \min_{\phi=1}^{\Phi} \delta(\mu_m, c_\phi) \right\} \quad (5)$$

The main idea of the proposed likelihood computation algorithm is to quantize the input feature vector, and to compute the input vector likelihoods only for the Gaussian neighbors of the quantized codeword. It is assumed that the input vector falls on the tails of the Gaussians outside the neighborhood. Therefore, the (disjoint) clusters provided by the clustering algorithm are not suitable neighborhoods. In fact, if an input vector is close to the boundary between two clusters, it may be "near to" (as opposed to "on the tail of") some of the Gaussians outside the cluster of the quantized codeword.

To introduce a better definition of Gaussian neighborhood, we note that, given a threshold $\Theta \gg 1$, f is said to fall on the tail of the mth Gaussian (1) if

$$\frac{1}{d} \sum_{i=1}^{d} \frac{(f(i) - \mu_m(i))^2}{U_m(i)} > \Theta , \quad \Theta \gg 1.0 \quad (6)$$

where $U_m(i)$ is the ith diagonal element of $\mathbf{U}_m$. Therefore, for our purpose, a good definition of neighborhood ν_ϕ of codeword $\mathbf{c}_\phi$ consists of the Gaussians (1) such that:

$$G(., \mu_m, \mathbf{U}_m) \in \nu_\phi$$
$$\textit{iff} \quad \frac{1}{d} \sum_{i=1}^{d} \frac{(c_\phi(i) - \mu_m(i))^2}{U_m(i)} \leq \Theta \quad (7.a)$$

Since variance estimates are often noisy, we have also used the criterion:

$$G(., \mu_m, \mathbf{U}_m) \in \nu_\phi$$
$$\textit{iff} \quad \frac{1}{d} \sum_{i=1}^{d} \frac{(c_\phi(i) - \mu_m(i))^2}{U_{avg}(i)} \leq \Theta \quad (7.b)$$

where $U_{avg}(i)$ is the ith diagonal element of the average (or "smoothed") covariance matrices in (1). (7.b) has been more effective than (7.a) in various recognition experiments.

3. LIKELIHOOD COMPUTATION DURING RECOGNITION

The Gaussian neighborhood definition (7) is a setup operation that needs to be performed only once for a given set of HMM's. To compute the likelihood during recognition the input observation vector f_t is quantized to the nearest codeword using the metric (3). Then *only* the input vector likelihoods for the Gaussians of the codeword neighborhood are exactly computed and added into the state likelihoods (2).

Assuming that the means μ_m of the Gaussians (1) provide an accurate sampling of the acoustic space, (5) can be considered an estimate of the average quantization distortion of the input feature vector during testing. Hence, if Θ is chosen such that:

$$\Theta \gg \delta_{avg} \quad (8)$$

we are reasonably assured that the input vector falls on the tails of the Gaussians outside the neighborhood. The chosen value of Θ also controls the average size of the Gaussian neighborhoods (7), that are overlapping sets whose size is a non-decreasing function of Θ. Decreasing values of Θ will therefore improve efficiency because fewer Gaussian likelihoods will be computed during recognition. However, too small values of Θ may degrade the recognition accuracy, if the constraints in (8) and in the right-hand side of (6) do not hold.

4. ALGORITHM EVALUATION

The proposed likelihood computation method has been tested in a connected digit recognition system based on Full-Search of the optimum state path [2]. The recognizer uses eleven Hidden Markov Models (one for each vocabulary digit), each model consisting of 10 states with 64 Gaussian components, for a grand total of $M=7040$ Gaussian components. The observation vectors consist of 38 features of 12 cepstrum coefficients with their 1st and 2nd derivatives and of the 1st and 2nd derivative of the frame energy in dB.

To compare the computational load of the new technique to the traditional method, we define a "computation fraction":

$$ C = \frac{N_{new} + \Phi}{N_{tr}} \qquad (9) $$

Φ is the number of distances computed to quantize every input feature vector, assuming a full search of the codebook. N_{new} represents the average number of Gaussian likelihoods computed for each input vector by the new method. With the Full-Search of the optimum state path, N_{new} is essentially the average Gaussian neighborhood size. The denominator of (9) denotes the average number of mixture component likelihoods computed by the traditional method for each input vector ($N_{tr} = M$, with the Full-Search). (9) is therefore an estimate of the requirements of the new likelihood computation method relative to the traditional method.

Several recognition experiments have been performed with different codebook sizes. In addition, as explained in the previous Section, selecting different values of Θ allowed us to change the size of the neighborhoods and the computation fraction (9).

5. DISCUSSION

The string recognition accuracy for the Texas Instruments (TI) connected digit data base, is shown in Figure 1 as a function of the computation fraction C. For a given codebook size Φ, the computation fraction (9) values (abscissa of Figure 1) were changed by selecting different threshold Θ of equations (7).

Typically, Θ values range from 1.5 to 4.0. The traditional computation method provides a string accuracy of 98.6% [2], that is plotted in Figure 1 for the abscissa value C equal to 1.0. The comparison between Figures 1.a and 1.b shows that criterion 7.b is more effective than 7.a for computation reduction. In particular, Figure 1.b also shows that the new likelihood computation method provides computation savings of a factor ranging from 5 to 9 (C ranging from 0.20 to 0.11) without a significant accuracy loss. The new technique is also robust to the choice of the codebook size Φ.

We review additional findings, to include experiments performed on different types of data bases and to summarize some preliminary results obtained with the Beam-Search (instead of the Full-Search).

It is interesting to note that the vector quantization of the input feature vector implements a two-level search (tree-structured search) of the Gaussians, that require exact likelihood computation. The detailed structure of this search is implicitly defined by the separation of the Gaussian means relative to their covariances, as defined by distances (3) and (7). For this reason, the proposed technique automatically adapts to different numbers of mixture components and to different types of feature vectors. In fact, experiments performed with reduced feature sets (24 instead of 38 dimensions [3]) and with 16 mixture components per state (instead of 64) have shown computational savings similar to those shown in Figure 1. Our results have been further confirmed by recognition experiments performed with telephone speech data (instead of the TI high quality data), which have shown computation savings by a factor ranging from three to five, without significant loss of accuracy. The proposed technique is also suitable for DSP implementations, because it requires only a small amount of additional memory and DSP bus bandwidth, in comparison to the traditional computation methods.

Finally, we note that the proposed method is quite general, and it can be applied also to phone-based recognition systems using either Full-Search or Beam-Search of the state path, and to tied-mixture systems as well. Beam-Search systems use state likelihood computation "on demand" that, reduce both N_{new} and N_{tr} in (9). Computation on demand does not change Φ in (9). However, since Φ is a relatively small value ($\Phi = 64$ gives very good results in Figure 1), the computation fraction (9) remains significantly less than one, also with the Beam-Search. Indeed, preliminary experiments with the Beam-Search on the DARPA Resource Management task (using both context-independent and context-dependent phones), have shown computation savings by a factor ranging from two to three, without significant loss of recognition accuracy.

REFERENCES

[1] D.B. Paul, "The Lincoln Tied-Mixture HMM Continuous Speech Recognizer," Proc. 1991 Int. Conf. Acoustics Speech and Signal Processing, pp. 329-332.

[2] J.G. Wilpon, C.H. Lee, and L.R. Rabiner, "Improvements in Connected Digit Recognition Using Higher Order Spectral and Energy Features," Proc. 1991 Int. Conf. Acoustics Speech and Signal Processing, pp. 349, 352.

[3] E.L. Bocchieri and J.G. Wilpon, "Discriminative Feature Selection for Speech Recognition", to appear in Computer Speech and Language.

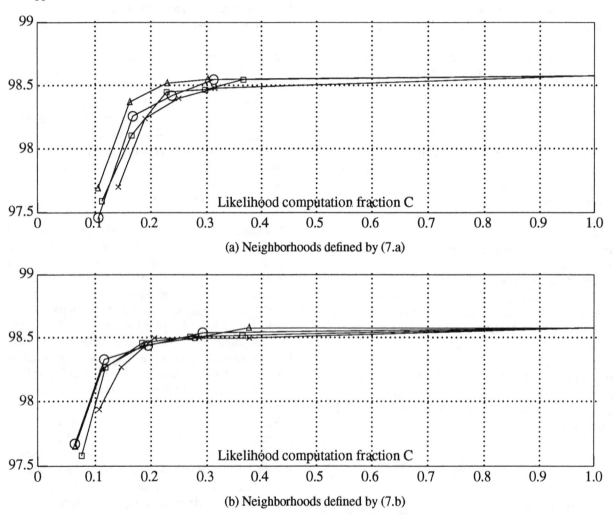

(a) Neighborhoods defined by (7.a)

(b) Neighborhoods defined by (7.b)

Figure 1. Digit string accuracy (%) versus computation fraction C for different numbers of codewords (Φ):

×: Φ=512   □: Φ=256   Δ: Φ=128   ○: Φ=64

LARGE VOCABULARY CONTINUOUS SPEECH RECOGNITION OF WALL STREET JOURNAL DATA

Robert Roth, James Baker, Janet Baker, Larry Gillick, Melvyn Hunt, Yoshiko Ito, Stephen Lowe, Jeremy Orloff, Barbara Peskin, Francesco Scattone

Dragon Systems, Inc., 320 Nevada Street, Newton, MA 02160
bob@dragonsys.com

ABSTRACT

In this summary, we report on the progress that has been made at Dragon Systems in speaker independent large vocabulary speech recognition using speech from DARPA's Wall Street Journal (WSJ) corpus. First, we present an overview of our recognition and training algorithms. Then, we describe experiments involving two recent improvements to these algorithms, moving to higher dimensional streams and using an IMELDA transformation, and present some results showing the reduction in error rates.

1. SPEECH RECOGNIZER

We will report in this paper some results of running Dragon System's continuous speech recognizer on the Wall Street Journal speech corpus [11]. This recognizer is a time synchronous Hidden Markov Model based system [1–4].

On the 16 kHz data provided, our signal processing produces 32 parameters consisting of 1 overall amplitude term, 7 spectral parameters (based on a 5.3 millisecond window and at frequencies equally spaced between 375 and 2625 Hz), 12 mel-cepstral and 12 mel-cepstral difference parameters (based on a 20 millisecond window). Training is typically done on data with a frame rate of 10 milliseconds, but recognition is done on 20 millisecond data for computational speed.

Our word models are simply a concatenation of phoneme models called PICs (phonemes in context). Each PIC model depends on the identity of preceding and succeeding phonemes as well as the current phoneme. Our PIC models are currently based on 78 phonemes, which include 3 stress levels for each vowel. Each PIC model is a linear sequence of states, each

This work was sponsored by the Defense Advanced Research Projects Agency under contract number J-FBI-92-060. The views expressed are those of the authors and do not reflect the official policy or position of the U.S. Government.

of which consists of a distribution on the frame parameters called a PEL (phonetic element) and a duration model. The duration models are Laplacian (or double exponential) with a state dependent mean and a global MAD (mean absolute deviation). Our PEL models originally were unimodal, but we currently obtain our best results by modeling them as tied mixture models ([2], [5], [9]), which we will describe in more detail.

In a tied mixtures framework, the acoustic parameters are partitioned into streams. For each stream, the probability distribution in a given state is modeled as a linear combination of a global set of basis distributions for that stream. It is assumed that, for a given PEL, the probability of the data in one stream is independent of data from other streams, so that the total probability is just the product of the individual stream probabilities.

The advantages of tied mixture models over unimodal models are twofold. First of all, they enable modeling of much more general distributions than unimodal models. With unimodal models, every model is approximated by a Laplacian (or Gaussian) with a specified mean and MAD. With tied mixtures, we can approximate much more general shapes, including nonsymmetric and multimodal distributions. Secondly, tied mixtures allows the modeling of dependence among the parameters within a stream, while in the unimodal approach, all the parameters are assumed to be independent.

When the recognizer hypothesizes that one of the active words is ending in the current frame, it asks the rapid match module to supply a relatively short list of words that may be starting at that time. Thus the task of the rapid match algorithm is to quickly pare down the number of reasonable candidates from the total vocabulary size to a more manageable number. Originally our rapid match models were trained from actual speech data, but we currently obtain the best performance by estimating them via Monte Carlo simulation from the acoustic models. The details of our rapid match algorithm can be found in references [7] and [8].

All of the results reported below use a standard digram language model constructed for this task from Wall Steet Journal text by Doug Paul of MIT Lincoln Laboratories.

2. TRAINING

The models we used for Wall Street Journal testing have been trained on data from 12 speakers, each of whom has spoken approximately 600 sentences (about 40 minutes per speaker).

A pass of training consists of the following steps. First we segment the data using our best speaker independent models. We use this segmentation to construct an IMELDA transformation that selects directions in parameter space that are useful in distinguishing speech classes (see next section). Next we use the same segmentation to create a file for each phoneme that contains the transformed speech parameters for each frame assigned to that phoneme. Finally, from each of the transformed frame files we construct a set of PIC models for which we have more than a certain number of examples (typically around 10). Each PIC model is obtained by iterating the Baum-Welch algorithm (up to 8 times) on the examples of that PIC. We smooth the data for each PIC with data from examples of the same phoneme in related contexts. In making the PIC models, we use a set of speaker independent basis distributions that have been constructed from the transformed speech data (see next section). This step is not necessary in the case of 1-parameter streams.

The training pass described above can be iterated multiple times. As a matter of practice, Dragon runs a single pass of this algorithm every time we have an improvement. Thus our current best models have been iterated on the same data many times, with the feature that in each pass, we not only obtain better segmentations, but we use a somewhat different algorithm.

3. ALGORITHM IMPROVEMENTS

We report in this section on some performance improvements we have made in speaker independent recognition. Results quoted in this section (unless stated otherwise) are on the development test data of the 5000-word closed-vocabulary speaker-independent non-verbalized punctuation version of the Wall Street Journal task.

	IMELDA	NON-IMELDA
1-parameter streams	21.0 %	26.2 %
2-parameter streams	19.3 %	22.7 %
4-parameter streams	16.5 %	—

Table I: Word error rates for baseline experiment with and without IMELDA.

We have obtained improved recognition results by transforming the parameters using the IMELDA [10] strategy. The idea behind this procedure is to select directions in parameter space that are most useful in distinguishing between speech classes. This is accomplished by first transforming the data so that the average within-class covariance matrix is the identity matrix. One then uses principal components analysis on the transformed between-class covariance matrix to select the directions of greatest variation. A subset of these principal components is kept as the final choice of parameters. These parameters have the property that they maximize the average variation of phonemes between classes relative to the average within class variation. A consequence of this for speaker independent recognition is that directions in which parameters vary greatly from one speaker to the next relative to the variation from one phoneme to the next will tend to be excluded.

Table I illustrates the improvement obtained by using the IMELDA transformation. In these experiments, we trained all PICs (a total of 12320) for which we had at least 10 examples in the training data. PICs consisted of 2 PELs, with a small number of shorter duration PIC models having 1 PEL. This resulted in about 21,000 PELs, as we did no sharing of PELs in different PICs. These PELs were the classes used to compute the IMELDA transformation, with two further restrictions. PELs were not used for the computation of the IMELDA transform if we had no examples of them from at least half the speakers or if more than 40% of the data came from a single speaker. About 18,000 PELs remained, which were weighted in proportion to the number of examples we had of each in computing the two covarience matrices, with a weight cap of 50 even if we had more than 50 examples. For the experiments in Table I, we compare the results using 16 IMELDA parameters with the corresponding experiment using all 32 original parameters. As can be seen from the table, a 15%-20% reduction in the error rate is obtained.

Our earlier Wall Street Journal results were obtained using 32 1-parameter streams. This allowed us to model more general distributions, but not de-

	Top 8	Top 12	Top 16
1-parameter streams	34.5%	23.5%	21.0%
2-parameter streams	31.4%	20.7%	19.3%
4-parameter streams	24.9%	17.6%	16.5%

Table II: Word error rates with IMELDA, using the "best" 8, 12, and 16 parameters.

pendence among the parameters. The basis distributions for each stream were chosen to be equally spaced throughout the dynamic range of the parameters and thus covered the entire space pretty well. This expedient strategy was followed to allow us to shake out the code without having to worry about how to choose a good set of basis distributions.

Since that time, we have been moving toward modeling with multiparameter streams. This has required us to intelligently choose sets of basis distributions, since in multidimensional space we cannot possibly cover the whole space. We currently use a simple one-pass clustering algorithm of adding clusters in regions of parameter space that are poorly represented and that have more than a certain number of frames. (This works somewhat better than our earlier algorithm of splitting the cluster with the most frames in the direction of greatest variation.) Table I shows that this has produced a significant reduction in the error rate as we increase the number of parameters per stream from 1 to 4.

As we mentioned above, Table I compares performance with 16 IMELDA parameters with the 32 original parameters. We made this comparison because we, as most groups working with IMELDA transformations, have found that beyond a certain point, adding more IMELDA parameters actually reduces performance. For us, the error rate starts to turn up above about 17 parameters. Although the reduction from 32 to 16 parameters already has significantly reduced the amount of computation, we thought it would be interesting to see how much we would lose by reducing the number of IMELDA parameters still further. Table II shows that a reduction from 16 to 12 parameters produces a rather modest degradation in performance. However, there is a big jump in the error rate when we run with only 8 parameters.

Finally, we experimented with gender dependent models. In Table III, we compare the performance of three systems. In the first system, we use gender independent models, while in the second we use the better scoring of two gender dependent models. In the third

Algorithm	5K Test Configuration		
	dev	eval	eval(c)
Ind	16.5 %	17.4 %	16.8 %
M+F	15.2 %	14.4 %	13.6 %
M+F+Ind	14.9 %	14.5 %	14.1 %

Table III: Word error rates for DARPA 5K development (dev) and evaluation (eval) tests, showing effects of using gender selection algorithms. Evaluation test was run a second time (eval(c)) with more conservative settings of the parameters controlling search pruning. Model designations: Ind=Gender independent, M=Male, F=Female.

system, we use the best scoring of the three models mentioned above. For the gender dependent systems, we would in principle have to run the recognizer on every one of the test sentences with two (or three) different sets of models. In practice, one can quite reliably choose the gender based on a small number of sentences for each speaker and run the rest of the sentences for that speaker with only one set of models. As can be seen from the table, both systems with gender dependent models outperform the gender independent models. The relative performance of the two gender-based systems is inconsistent and statistically insignificant. We also show our comparative performance on the development and evaluation test sets. The fact that the error rates are comparable is an indication that we have not overly tuned our system to the development test data.

As can be seen from Table III, our best result on the evaluation test data of the 5000-word closed-vocabulary speaker-independent nonverbalized punctuation version of the Wall Street Journal task is currently an error rate of 13.6%. By contrast, our best performing system on the corresponding 20000 word open-vocabulary task obtains an error rate of 24.8%.

4. CONCLUSION

All of the speaker independent recognition experiments we have reported here were performed since the February 1992 DARPA Spoken Language Systems meeting. Prior to that time, Dragon had been concentrating almost exclusively on speaker dependent recognition. We are encouraged by the rapid reduction we have obtained in our error rate on large vocabulary speaker independent recognition of Wall Street Journal data. We intend to pursue work in a number of promising directions, in addition to further work on multiparameter streams and the IMELDA strategy. Both these methods allow the system to handle dependence among parameters in the original parameter set, the former by directly modeling it and the latter by transforming to a space with less dependence. We believe that a combination of the two will be more effective than either one alone. As a result of refinements in all these areas of development, we expect to be able to report significant further progress in the future.

5. REFERENCES

[1] J.K. Baker, J.M. Baker, P. Bamberg, L. Gillick, L. Lamel, F. Scattone, R. Roth, D. Sturtevant, O. Ba and R. Benedict, "Dragon Systems Resource Management Benchmark Results - February 1991", **Proceedings of the DARPA Speech and Natural Language Workshop**, February 1991, Pacific Grove, California.

[2] J.K. Baker, J.M. Baker, P. Bamberg, K. Bishop, L. Gillick, V. Helman, Z. Huang, Y. Ito, S. Lowe, B. Peskin, R. Roth, F. Scattone, "Large Vocabulary Recognition of Wall Street Jounal Sentences at Dragon Systems", **Proceedings of the DARPA Speech and Natural Language Workshop**, February 1992, Harriman, New York.

[3] P. Bamberg, Y.L. Chow, L. Gillick, R. Roth and D. Sturtevant, "The Dragon Continuous Speech Recognition System: A Real-Time Implementation", **Proceedings of the DARPA Speech and Natural Language Workshop**, June 1990, Hidden Valley, Pennsylvania.

[4] P. Bamberg and L. Gillick, "Phoneme-in-Context Modeling for Dragon's Continuous Speech Recognizer", **Proceedings of the DARPA Speech and Natural Language Workshop**, June 1990, Hidden Valley, Pennsylvania.

[5] J.R. Bellagarda and D.H. Nahamoo, "Tied Mixture Continuous Parameter Models for Large Vocabulary Isolated-Speech Recognition", **Proc. ICASSP**, May 1989, Glasgow.

[6] L. Gillick, J.K. Baker, J.M. Baker, J. Bridle, M. Hunt, Y. Ito, S. Lowe, J. Orloff, B. Peskin, R. Roth, F. Scattone, "Application of Large Vocabulary Continuous Speech Recognition to Topic and Speaker Identification with Conversational Telephone Speech", **Proc. ICASSP-93**.

[7] L. Gillick, B. Peskin and R. Roth, "Rapid Match Training for Large Vocabularies", **Proceedings of the DARPA Speech and Natural Language Workshop**, February 1992, Harriman, New York.

[8] L. Gillick and R. Roth, "A Rapid Match Algorithm for Continuous Speech Recognition", **Proceedings of the DARPA Speech and Natural Language Workshop**, June 1990, Hidden Valley, Pennsylvania.

[9] X.D. Huang and M.A. Jack, "Semi-continuous Hidden Markov Models for Speech Recognition", **Computer Speech and Language**, Vol. 8, 1989.

[10] M.J. Hunt, D.C. Bateman, S.M. Richardson and A. Piau, "An Investigation of PLP and IMELDA Acoustic Representations and of their Potential for Combination," **Proc. ICASSP-91**, Toronto Canada, May, 1991, PP. 881-884.

[11] D.B. Paul and J.M. Baker, "The Design of the Wall Street Journal-based CSR Corpus", **Proceedings of the DARPA Speech and Natural Language Workshop**, February 1992, Harriman, New York.

LARGE-VOCABULARY DICTATION USING SRI'S DECIPHER™ SPEECH RECOGNITION SYSTEM: PROGRESSIVE SEARCH TECHNIQUES

Hy Murveit
John Butzberger
Vassilios Digalakis
Mitch Weintraub

SRI International

ABSTRACT

We describe a technique we call *Progressive Search* which is useful for developing and implementing speech recognition systems with high computational requirements. The scheme iteratively uses more and more complex recognition schemes, where each iteration constrains the search space of the next. An algorithm, the *Forward-Backward Word-Life Algorithm,* is described. It can generate a word lattice in a progressive search that would be used as a language model embedded in a succeeding recognition pass to reduce computation requirements. We show that speed-ups of more than an order of magnitude are achievable with only minor costs in accuracy.

1. INTRODUCTION

Many advanced speech recognition techniques cannot be developed or used in practical speech recognition systems because of their extreme computational requirements. Simpler speech recognition techniques can be used to recognize speech in reasonable time, but they compromise word recognition accuracy. In this paper we aim to improve the speed/accuracy trade-off in speech recognition systems using progressive search techniques.

We define *progressive search* techniques as those which can be used to efficiently implement other, computationally burdensome techniques. They use results of a simple and fast speech recognition technique to constrain the search space of a following more accurate but slower running technique. This may be done iteratively—each progressive search pass uses a previous pass' constraints to run more efficiently, and provides more constraints for subsequent passes.

We will refer to the faster speech recognition techniques as "earlier-pass techniques", and the slower more accurate techniques as "advanced techniques." Constraining the costly advanced techniques in this way can make them run significantly faster without significant loss in accuracy.

The key notions in progressive search techniques are:

1. An early-pass speech recognition phase builds a lattice, which contains all the likely recognition unit strings (e.g. word sequences) given the techniques used in that recognition pass.

2. A subsequent pass uses this lattice as a grammar that constrains the search space of an advanced technique (e.g., only the word sequences contained in a word lattice of pass p would be considered in pass p+1).

Allowing a sufficient breadth of lattice entries should allow later passes to recover the correct word sequence, while ruling out very unlikely sequences, thus achieving high accuracy and high speed speech recognition.

2. PRIOR ART

There are three important categories of techniques that aim to solve problems similar to the ones the progressive search techniques target.

2.1. Fast-Match Techniques

Fast-match techniques[1] are similar to progressive search in that a coarse match is used to constrain a more advanced computationally burdensome algorithm. The fast match, however, simply uses the local speech signal to constrain the costly advanced technique. Since the advanced techniques may take advantage of non-local data, the accuracy of a fast-match is limited and will ultimately limit the overall technique's performance. Techniques such as progressive search can bring more global knowledge to bear when generating constraints, and, thus, more effectively speed up the costly techniques while retaining more of their accuracy.

2.2. N-Best Recognition Techniques

N-best techniques[2] are also similar to progressive search in that a coarse match is used to constrain a more computationally costly technique. In this case, the coarse matcher is a complete (simple) speech recognition system. The output of the N-best system is a list of the top N most likely sentence hypotheses, which can then be evaluated with the slower but more accurate techniques.

Progressive search is a generalization of N-best—the earlier-pass technique produces a graph, instead of a list of N-best sentences. This generalization is crucial because N-best is only computationally effective for N in the order of tens or hundreds. A progressive search word graph can effectively account for orders of magnitude more sentence hypotheses. By limiting the advanced techniques to just searching the few top N sentences, N-best is destined to limit the effectiveness of the advanced techniques and, consequently, the overall system's accuracy. Furthermore, it does not make much sense to use N-best in an iterative fashion as it does with progressive searches.

2.3. Word Lattices

This technique is the most similar to progressive search. In both approaches, an initial-pass recognition system can generate a lattice of word hypotheses. Subsequent passes can search through the lattice to find the best recognition hypothesis. It should be noted that, although we refer to lattices as word lattices, they could be used at other linguistic level, such as the phoneme, syllable, e.t.c.

In the traditional word-lattice approach, the word lattice is viewed as a scored graph of possible segmentations of the input speech. The lattice contains information such as the acoustic match between the input speech and the lattice word, as well as segmentation information.

The progressive search lattice is not viewed as a scored graph of possible segmentations of the input speech. Rather, the lattice is simply viewed as a word-transition grammar which constrains subsequent recognition passes. Temporal and scoring information is intentionally left out of the progressive search lattice.

This is a critical difference. In the traditional word-lattice approach, many segmentations of the input speech which could not be generated (or scored well) by the earlier-pass algorithms will be eliminated for consideration before the advanced algorithms are used. With progressive-search techniques, these segmentations are implicit in the grammar and can be recovered by the advanced techniques in subsequent recognition passes.

3. Building Progressive Search Lattices

The basic step of a progressive search system is using a speech recognition algorithm to make a lattice which will be used as a grammar for a more advanced speech recognition algorithm. This section discusses how these lattices may be generated. We focus on generating word lattices, though these same algorithms are easily extended to other levels.

3.1. The Word-Life Algorithm

We implemented the following algorithm to generate a word-lattice as a by-product of the beam search used in recognizing a sentence with the DECIPHER™ system[4-7].

1. For each frame, insert into the table $Active(W, t)$ all words W active for each time t. Similarly construct tables $End(W, t)$ and $Transitions(W_1, W_2, t)$ for all words ending at time t, and for all word-to-word transition at time t.

2. Create a table containing the word-lives used in the sentence, $WordLives(W, T_{start}, T_{end})$. A *word-life* for word W is defined as a maximum-length interval (frame T_{start} to T_{end}) during which some phone in word W is active. That is,
    $$W \in Active(W, t), T_{start} \le t \le T_{end}$$

3. Remove word-lives from the table if the word never ended between T_{start} and T_{end}, that is, remove

$WordLives(W, T_{start}, T_{end})$ if there is time t between T_{start} and T_{end} where $End(W, t)$ is true.

4. Create a finite-state graph whose nodes correspond to word-lives, whose arcs correspond to word-life transitions stored in the *Transitions* table. This finite state graph, augmented by language model probabilities, can be used as a grammar for a subsequent recognition pass in the progressive search.

This algorithm can be efficiently implemented, even for large vocabulary recognition systems. That is, the extra work required to build the "word-life lattice" is minimal compared to the work required to recognize the large vocabulary with a early-pass speech recognition algorithm.

This algorithm develops a grammar which contains all whole-word hypotheses the early-pass speech recognition algorithm considered. If a word hypothesis was active and the word was processed by the recognition system until the word finished (was not pruned before transitioning to another word), then this word will be generated as a lattice node. Therefore, the size of the lattice is directly controlled by the recognition search's beam width.

This algorithm, unfortunately, does not scale down well—it has the property that small lattices may not contain the best recognition hypotheses. This is because one must use small beam widths to generate small lattices. However, a small beam width will likely generate pruning errors.

Because of this deficiency, we have developed the Forward/Backward Word-Life Algorithm described below.

3.2. Extending the Word-Life Algorithm Using Forward And Backward Recognition Passes

We wish to generate word lattices that scale down gracefully. That is, they should have the property that when a lattice is reduced in size, the most likely hypotheses remain and the less likely ones are removed. As was discussed, this is not the case if lattices are scaled down by reducing the beam search width.

The forward-backward word-life algorithm achieves this scaling property. In this new scheme, described below, the size of the lattice is controlled by the *LatticeThresh* parameter.

1. A standard beam search recognition pass is done using the early-pass speech recognition algorithm. (None of the lattice building steps from Section 3.1 are taken in this forward pass).

2. During this forward pass, whenever a transition leaving word W is within the beam-search, we record that probability in $ForwardProbability(W, frame)$.

3. We store the probability of the best scoring hypothesis from the forward pass, *Pbest*, and compute a pruning value
    $$Pprune = Pbest / LatticeThresh.$$

4. We then recognize the same sentence over again using the same models, but the recognition algorithm is run backwards[1].

5. The lattice building algorithm described in Section 3.1 is used in this backward pass with the following exception. During the backward pass, whenever there is a transition between words W_i and W_j at time t, we compute the overall hypothesis probability P_{hyp} as the product of $ForwardProbability(W_j,t-1)$, the language model probability $P(W_i|W_j)$, and the Backward pass probability that W_i ended at time t (i.e. the probability of starting word W_i at time t and finishing the sentence). If $P_{hyp} < P_{prune}$, then the backward transition between W_i and W_j at time t is blocked.

Step 5 above implements a backwards pass pruning algorithm. This both greatly reduces the time required by the backwards pass, and adjusts the size of the resultant lattice.

4. Progressive Search Lattices

We have experimented with generating word lattices where the early-pass recognition technique is a simple version of the DECIPHER™ speech recognition system, a 4-feature, discrete density HMM trained to recognize a 5,000 vocabulary taken from DARPA's WSJ speech corpus. The test set is a difficult 20-sentence subset of one of the development sets.

We define the number of errors in a single path p in a lattice, $Errors(p)$, to be the number of insertions, deletions, and substitutions found when comparing the words in p to a reference string. We define the number of errors in a word lattice to be the minimum of $Errors(p)$ for all paths p in the word lattice.

The following tables show the effect adjusting the beam width and *LatticeThresh* has on the lattice error rate and on the lattice size (the number of nodes and arcs in the word lattice). The grammar used by the has approximately 10,000 nodes and 1,000,000 arcs. The simple recognition system had a 1-best word error-rate ranging from 27% (beam width 1e-52) to 30% (beam width 1e-30).

Table 1: Effect Of Pruning On Lattice Size

Beam Width 1e-30

Lattice Thresh	nodes	arcs	# errors	%word error
1e-5	60	278	43	10.57
1e-9	94	541	34	8.35
1e-14	105	1016	30	7.37
1e-18	196	1770	29	7.13
1e-32	323	5480	23	5.65
1e-45	372	8626	23	5.65
inf	380	9283	23	5.65

Beam Width 1e-34

Lattice Thresh	nodes	arcs	# errors	%word error
1e-5	64	299	28	6.88
1e-9	105	613	20	4.91
1e-14	141	1219	16	3.93
1e-18	260	2335	15	3.69
1e-23	354	3993	15	3.69
1e-32	537	9540	15	3.69

Beam Width 1e-38

Lattice Thresh	nodes	arcs	# errors	%word error
1e-14	186	1338	14	3.44
1e-18	301	2674	13	3.19
1e-23	444	4903	12	2.95

Beam Width 1e-42

Lattice Thresh	nodes	arcs	# errors	%wd error
1e-14	197	1407	13	3.19
1e-18	335	2926	11	2.70
1e-23	520	5582	10	2.46

Beam Width 1e-46

Lattice Thresh	nodes	arcs	# errors	%word error
1e-14	201	1436	13	3.19
1e-18	351	3045	10	2.46
1e-23	562	5946	10	2.46

Beam Width 1e-52

Lattice Thresh	nodes	arcs	# errors	%word error
1e-14	216	1582	12	2.95
1e-18	381	3368	9	2.21

The two order of magnitude reduction in lattice size has a significant impact on HMM decoding time. Table 2 shows the per-sentence computation time required for the above test set when computed using a Sparc2 computer, for both the original grammar, and word lattice grammars generated using a *LatticeThresh* of 1e-23.

---

1. Using backwards recognition the sentence is processed from last frame to first frame with all transitions reversed.

Table 2: Lattice Computation Reductions

Beam Width	Forward pass recognition time (secs)	Lattice recognition time (secs)
1e-30	167	10
1e-34	281	16
1e-38	450	24
1e-46	906	57
1e-52	1749	65

5. Applications of Progressive Search Schemes

Progressive search schemes can be used in the same way N-best schemes are currently used. The two primary applications we've had at SRI are:

5.1. Reducing the time required to perform speech recognition experiments

At SRI, we've been experimenting with large-vocabulary tied-mixture speech recognition systems. Using a standard decoding approach, and average decoding times for recognizing speech with a 5,000-word bigram language model were 46 times real time. Using lattices generated with beam widths of 1e-38 and a *LatticeThresh* of 1e-18 we were able to decode in 5.6 times real time). Further, there was no difference in recognition accuracy between the original and the lattice-based system.

5.2. Implementing recognition schemes that cannot be implemented with a standard approach.

We have implemented a trigram language model on our 5,000-word recognition system. This would not be feasible using standard decoding techniques. Typically, continuous-speech trigram language models are implemented either with fastmatch technology or, more recently, with N-best schemes. However, it has been observed at BBN that using an N-best scheme (N=100) to implement a trigram language model for a 20,000 word continuous speech recognition system may have significantly reduced the potential gain from the language model. That is, about half of the time, correct hypotheses that would have had better (trigram) recognition scores than the other top-100 sentences were not included in the top 100 sentences generated by a bigram-based recognition system[8].

We have implemented trigram-based language models using word-lattices, expanding the finite-state network as appropriate to unambiguously represent contexts for all trigrams. We observed that the number of lattice nodes increased by a factor of 2-3 and the number of lattice arcs increased by a factor of approximately 4 (using lattices generated with beam widths of 1e-38 and a *LatticeThresh* of 1e-18). The resulting decoding times increased approximately by 50% when using trigram lattices instead of bigram lattices.

ACKNOWLEDGEMENTS

We gratefully acknowledge support for this work from DARPA through Office of Naval Research Contract N00014-92-C-0154. The Government has certain rights in this material. Any opinions, findings, and conclusions or recommendations expressed in this material are those of the authors and do not necessarily reflect the views of the government funding agencies.

REFERENCES

1. Bahl, L.R., de Souza, P.V., Gopalakrishnan, P.S., Nahamoo, D., and M. Picheny, "A Fast Match for Continuous Speech Recognition Using Allophonic Models," *1992 IEEE ICASSP*, pp. I-17-21.

2. Schwartz, R., Austin, S., Kubala, F., Makhoul, J., Nguyen, L., Placeway, P., and G. Zavaliagkos, "New uses for the N-Best Sentence Hypotheses Within the BYBLOS Speech Recognition System", *1992 IEEE ICASSP*, pp. I-1-4.

3. Chow, Y.L., and S. Roukos, "Speech Understanding Using a Unification Grammar", *1989 IEEE ICASSP*, pp. 727-730

4. H. Murveit, J. Butzberger, and M. Weintraub, "Performance of SRI's DECIPHER Speech Recognition System on DARPA's CSR Task," 1992 DARPA Speech and Natural Language Workshop Proceedings, pp 410-414

5. Murveit, H., J. Butzberger, and M. Weintraub, "Reduced Channel Dependence for Speech Recognition," 1992 DARPA Speech and Natural Language Workshop Proceedings, pp. 280-284.

6. H. Murveit, J. Butzberger, and M. Weintraub, "Speech Recognition in SRI's Resource Management and ATIS Systems," 1991 DARPA Speech and Natural Language Workshop, pp. 94-100.

7. Cohen, M., H. Murveit, J. Bernstein, P. Price, and M. Weintraub, "The DECIPHER™ Speech Recognition System," *1990 IEEE ICASSP*, pp. 77-80.

8. Schwartz, R., BBN Systems and Technologies, Cambridge MA, Personal Communication

NOISY SPONTANEOUS SPEECH UNDERSTANDING
USING NOISE IMMUNITY KEYWORD SPOTTING
WITH ADAPTIVE SPEECH RESPONSE CANCELLATION

Yoichi Takebayashi, Yoshifumi Nagata and Hiroshi Kanazawa

Toshiba Corporation, Research & Development Center
Saiwai-ku, Kawasaki, 210 Japan

ABSTRACT

We have developed a noisy spontaneous speech understanding system using Noise Immunity keyword spotting
with adaptive speech response cancellation. We have
employed a keyword-based approach to understand spontaneous speech. To facilitate robust human-computer interaction, we have introduced a new component to the system:
adaptive cancellation of synthetic speech response. We
have also extended Noise Immunity Learning to dialogue
speech contaminated by a uncancelled component of a synthetic speech response. The cancellation allows the user to
interrupt speech responses of the system for more efficient
human-computer interaction. We have integrated a user-initiated dialogue manager and a multimodal response generator to construct a speaker-independent dialogue system for
a fast-food ordering task. The experimental results have
shown the effectiveness of adaptive speech response cancellation and the extension of Noise Immunity Learning.

1. INTRODUCTION

Several high-performance HMM-based speech recognition systems have been developed to deal with speaker-
independent, large-vocabulary continuous speech[1]. However, their performance decreases with the addition of background noise and spontaneous speech, such as ellipses and
unintentional utterances. Most systems interpret only spoken words which are grammatically correct, as a result they
do not work in the case of spontaneous speech in real-world
applications. Although spoken language systems, which are
being developed in DARPA programs[2-5] and European
projects[6,7], aim to recognize spontaneous speech in the
real world, many variations in spontaneous speech pose the
significant problem.

Taking these points into account, we previously built a
prototype of a speech dialogue system TOSBURG (Task-
Oriented Speech dialogue system Based on speech Understanding and Response Generation)[8], employing a keyword-based approach for the understanding of spontaneous
speech. The system consists of a keyword spotter using
Noise Immunity Learning[9,10], a semantic keyword lattice
parser[11], a dialogue manager and a multimodal response
generator.

In this paper, our approach to real-world speech dialogue
systems is first presented. Next, problems of TOSBURG
are mentioned. Then a further developed version of TOS-
BURG, a spontaneous speech dialogue system TOSBURG
II, which is designed to allow for interruption by the user during speech response, is introduced. New extensions in TOS-
BURG II are described; synthetic speech response cancellation and the extension of Noise Immunity Learning. Finally,
experimental results are presented.

2. SPEECH DIALOGUE SYSTEM WITH MULTIMODAL RESPONSE

2.1 Approach to spontaneous speech dialogue system

In order to apply advanced speaker-independent speech
recognition to a real-world speech dialogue system, we need
to improve the real-time performance and robustness of the
system. In addition, spontaneous speech understanding
capability, placing no restrictions on grammar and utterance
manner, is desirable. A spontaneous speech understanding
system should be able to deal with unintentional utterances,
ellipses, pauses and out-of-vocabulary words. However, the
above speech phenomena are difficult to represent in terms
of the grammatical rules that form the basis of conventional
systems.

Therefore we have employed keyword-based spontaneous speech understanding on the premise that the meaning of utterance can be extracted from a combination of keyword spotting with a keyword lattice parsing. By doing so,
we can resolve inevitable ambiguity in the speech understanding process and control dialogue. We previously constructed a prototype of a task-oriented speech dialogue system
TOSBURG that handles spontaneous speech from unspecified users[8]. It has the following features:

- speaker-independent keyword-based spontaneous speech
  understanding
- dialogue management which guides the user to natural
  human-computer interaction
- multimodal response generation

In this system, the word-spotter extracts keyword candidates from spontaneous speech and passes a keyword lattice to the syntactic and semantic keyword parser. The parser generates semantic utterance representation candidates
which are passed to the dialogue manager. The dialogue
manager interprets the candidates using dialogue history
and determines a semantic response representation. The
output response generator then uses this representation to
create a multimodal response comprising synthesized
speech and visual response.

Since the response to the user is given not only via
speech but also via text and graphical information as shown
in Fig.1, the user can understand the system's response by

visual confirmation before speech response is completed. Then the user often ignores and interrupts the computer-generated speech response. However, our prototype system cannot accept the user's speech input while the system is outputting synthetic speech. The speech response transmitted from a loud-speaker generally contaminates the user's input speech at the microphone with the result that recognition performance decreases. We adopted lip shaped icons, which tell the user either to speak or to wait. However, this restriction reduces the efficiency and friendliness of the dialogue.

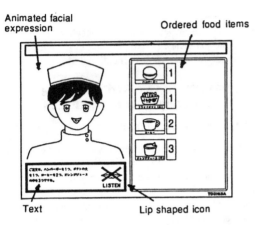

Fig.1 An example of visual response in a multimodal dialogue system

2.2 Extension to TOSBURG II

We have extended the existing system to a spontaneous speech dialogue system TOSBURG II to accept interruption by the user. This system enables spontaneous interaction, where the user can interrupt during the system's response by means of synthetic speech cancellation using active noise control technology[12]. We have also extended Noise Immunity Learning to improve keyword spotting performance using a uncancelled component of the synthetic speech response.

Our dialogue system integrates a keyword spotter, a semantic keyword lattice parser, a dialogue manager, a multimodal response generator and a speech response can-

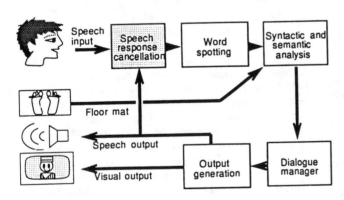

Fig.2 Configuration of speech dialogue system TOSBURG II

celler, each of which has been separately evaluated; it processes and responds to 49 Japanese keywords. The configuration is given in Fig.2. The system runs in real time using three general purpose workstations and five DSP accelerators[13]. The user initiates a dialogue with the system by standing on a weight-sensitive floor mat. The system's response is multimodal; synthesized speech is accompanied by a visual screen display of text, ordered food items and an animated cartoon face that has a range of facial expressions.

In the following sections, our newly developed components, synthetic speech response cancellation and extension of Noise Immunity Learning, are described in detail.

3. SPEECH RESPONSE CANCELLATION

To deal with the problem that the user's speech is often contaminated by the system's speech response, we have introduced an adaptive speech response cancellation to facilitate user-initiated spontaneous interaction. Figure 3 shows a block diagram of the speech dialogue system with speech response canceller. The canceller uses the LMS algorithm to adaptively estimate the impulse response between a loudspeaker and a microphone corresponding to the user's body movement during conversation as shown in Fig.4. We have also developed several techniques to improve the speech response cancellation which takes place on a digital signal containing input speech and response speech before the keyword spotting process.

In order to achieve stable and fast convergence, the system employs the Normalized Least Mean Square(NLMS) algorithm using an adaptation switch and spectral pre-

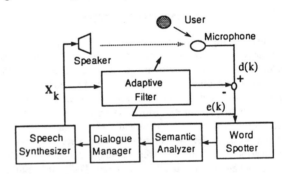

Fig.3 Speech dialogue system with speech response canceller

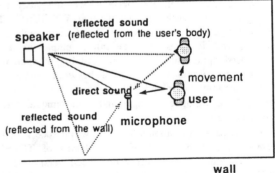

Fig.4 Examples of reflected sound

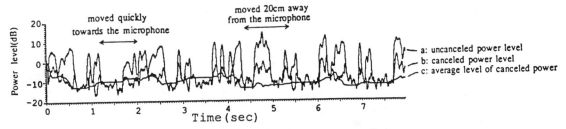

Fig.5 An example of speech response cancellation

whitening[14]. The adaptation switch, which activates the adaptation, prevents the estimate from becoming unstable due to rapid changes in signal power. Furthermore, spectral pre-whitening using a first-order difference signal accelerates convergence without increasing calculation overhead. An example of speech response cancellation is shown in Fig. 5. The speech response canceller has been implemented in a real-time speech dialogue system using a DSP accelerator and has been shown to enhance user-friendliness in human-computer interaction.

Since the user's utterances are always accepted by the speech response cancellation, lip shaped icons in Fig.1 are not needed any longer. Figure 6 shows a comparison of timing of utterance between spontaneous speech dialogue and conventional dialogue. In multimodal interaction, the meanings of utterances are understood with the aid of visual media before the end of the system's speech response. Then user's interruption enables more fluent interaction. TOS-BURG II also interrupts the user's utterance containing pauses. Mutual interruption is realized by this function.

Example of spontaneous dialogue

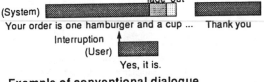

Example of conventional dialogue

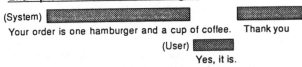

Fig.6 Comparison of timing of utterance between spontaneous dialogue and conventional dialogue

4. EXTENSION OF NOISE IMMUNITY KEYWORD SPOTTING

4.1 Keyword spotting with Noise Immunity Learning

Keyword-based speech understanding depends highly upon keyword spotting accuracy. We previously developed the Noise Immunity keyword spotter for noisy continuous speech[9,10]. The powerful features of the system are a keyword spotter based on the Multiple Similarity(MS)

method for reliable keyword detection and the use of Noise Immunity Learning for improving robustness against spontaneous speech and background noise.

The keyword spotting is based on time-frequency word patterns. The MS values are time-continuously computed for keyword spotting through pattern matching of the fixed-dimensional word pattern vectors and reference pattern vectors. During pattern matching between an input vector and reference vectors, which are generated by the learning process, an end point candidate (tj) is assumed. A series of start point candidates corresponding to tj are determined based on the maximum and minimum duration for each word class. Fixed-dimensional word pattern vectors are then extracted through uniform sampling of the time series spectrum. Time-continuous calculation of MS values is performed to detect keyword candidates.

4.2 Extension of Noise Immunity Learning using the uncancelled component of the synthetic speech response

While the speech response cancellation is effective, it is generally not perfect. The uncancelled component of the speech response contaminates the input speech as additive noise. Therefore, we extend the Noise Immunity Learning to improve keyword spotting performance for the user's interruption. Figure 7 shows a block diagram of the proposed keyword spotting method which supplements the recognition process with Noise Immunity Learning.

In the learning process, a recognition environment is simulated for noise-contaminated speech data including not only background noise and unintentional utterances but also residual response speech. The residual response speech

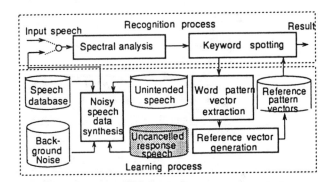

Fig.7 Block diagram of the extended Noise Immunity keyword spotting method

data is collected using real computer-generated speech response and the user's body movement in front of the microphone. Word pattern vectors are automatically extracted from the Multiple Similarity(MS) values obtained by keyword spotting. These vectors are used for covariance matrix modification in order to generate new word reference vectors by K-L expansion.

The learning process is repeated, while gradually decreasing the signal-to-noise ratio(SNR), to realize noise immunity. In addition, the learning process starts by using spontaneous speech, and then uses the uncancelled component of the speech response and background noise to achieve robustness. These noise data are used to artificially contaminate the input speech database for synthesizing noisy speech data for Noise Immunity Learning.

5. EVALUATION

To compare the new method with our previous method for noisy spontaneous speech data contaminated by system speech response, experiments were carried out with 350 sentences spoken naturally by five males for a fast-food task including 49 keywords. Table 1 shows experimental conditions. Table 2 compares the recognition scores of keyword detection rate, false alarm rate and sentence recognition rate. Without cancellation under the condition of SNR15dB, the keyword detection rate, false alarm rate and sentence recognition rate were 58.8%, 24.1 FA/H/W(False Alarms /Hour/Word) and 27.8%, respectively. While employing the new method (with cancellation and Noise Immunity Learning), rates of 89.6%, 15.4FA/H/W and 58.4% were resistered.

These results show the effectiveness of the new method for spontaneous speech contaminated by speech response in noisy environments. The results also suggest that our spontaneous speech understanding method based on keyword spotting with adaptive speech response cancellation improves the human-computer interaction in noisy environments. The newly developed real-time task-oriented dialogue system based on the speech understanding method has facilitated robust human-computer interaction. In order

Table 1 Experimental conditions

Speaker	Training: 34 males Test: 5 males
Vocabulary	350 sentences including 49 keywords
Noise	Computing room noise
Word feature vector dimension	192 (16ch x 12 frames)

Table 2 Experimental results

Condition	Without cancellation		With Cancellation			
			Without NIL		With NIL	
SNR	∞dB	15dB	∞dB	15dB	∞dB	15dB
Detection rate	64.4%	58.8%	92.5%	85.3%	94.9%	89.6%
FA/H/W	25.5	24.1	15.7	14.9	16.1	15.4
Sentence recognition rate	37.9%	27.8%	63.8%	55.6%	66.0%	58.4%

(NIL: Noise Immunity Learning)

to speed up the new method for real-time interaction, we have developed an DSP accelerator, whose peak performance is 130MFLOPS, to supplement the workstation[13].

6. CONCLUSION

We have described a real-time spontaneous speech dialogue system TOSBURG II which is an extension of the earlier system TOSBURG. This system facilitates robust spontaneous human-computer interaction for unspecified users by combining the extended Noise Immunity keyword spotting with synthetic speech response cancellation. Experimental results obtained by the real-time system have proven the effectiveness and user-friendliness of the system. We are currently improving the performance of speech understanding through evaluation of the system using real-world dialogue speech data.

ACKNOWLEDGEMENTS

We would like to thank D. Gleaves, H. Tsuboi, S. Seto, T. Nii, H. Hashimoto and H. Shinchi for their contribution to the development of the dialogue system, and M. Shimazu for her help in the preparation of this paper.

REFERENCES

[1] Lee, K. : "Automatic Speech Recognition: The Development of the SPHINX System", Kluwer Academic Publishers, Boston (1989)

[2] Goodine, D., Hirschman, L., Polifroni, J., Seneff, S. and Zue, V. : "Evaluating Interactive Spoken Language Systems", ICSLP'92, pp. 201-204 (1992)

[3] Pieraccini, R., Tzoukermann, E., Gorelov, Z., Gauvain, J., Levin, E., Lee, C. and Wilpon, J. : "A Speech Understanding System Based on Statistical Representation of Semantics", ICASSP'92, pp.I-193-I-196 (1992)

[4] Jackson, E. : "Integrating Two Complementary Approaches to Spoken Language Understanding", ICSLP'92, pp.333-336 (1992)

[5] Bates, M., Bobrow, R., Fung, P., Ingria, R., Kubala, F., Makhoul, j., Nguyen, L., Schwartz, R. and Stallard, D. : "Design and Performance of HARC, The BBN Spoken Language Understanding System", ICSLP'92, pp.241-244 (1992)

[6] Peckham, J. : "Speech Understanding and Dialogue Over The Telephone: An Overview of Progress in the SUNDIAL Project", EUROSPEECH'91, pp.1469-1472 (1991)

[7] Tuback, J. and Doignon, P. : "A System for Natural Spoken Language Queries: Design, Implementation and Assessment", EUROSPEECH'91, pp.1473-1476 (1991)

[8] Takebayashi, Y., Tsuboi, H., Sadamoto, Y., Hashimoto, H. and Shinchi, H. : "A Real-Time Speech Dialogue System Using Spontaneous Speech Understanding", ICSLP'92, pp.651-654 (1992)

[9] Takebayashi, Y., Tsuboi, H. and Kanazawa, H : "A Robust Speech Recognition System Using Word-Spotting with Noise Immunity Learning", ICASSP'91, pp.905-908 (1991)

[10] Takebayashi, Y., Tsuboi, H. and Kanazawa, H. : "Keyword-Spotting in Noisy Continuous Speech Using Word Pattern Vector Subabstraction and Noise Immunity Learning", ICASSP'92, pp.II-85-II-88 (1992)

[11] Tsuboi, H. and Takebayashi, Y.: "A Real-Time Task-Oriented Speech Understanding System Using Keyword-Spotting", ICASSP'92, pp.I-197-I-200 (1992)

[12] Gleaves, D. and Nagata, Y. : "The Cancellation of Synthetic Speech for a Man-Machine Dialogue System", Proc. Acoust. Soc. Japan, 2-5-5, pp.57-58 (1991.10)

[13] Tsuboi, H., Kanazawa, H. and Takebayashi, Y. : "An Accelerator for High-Speed Spoken Word-Spotting and Noise Immunity Learning System", ICSLP'90, pp.273-276 (1990)

[14] Chen, W : "Adaptive Pilot Filtering for LMS Algorithm", ICASSP'91, pp.1509-1512 (1991)

GOLDEN MANDARIN(II) - AN IMPROVED SINGLE-CHIP REAL-TIME MANDARIN DICTATION MACHINE FOR CHINESE LANGUAGE WITH VERY LARGE VOCABULARY

Lin-shan Lee[1,2,3], Chiu-yu Tseng[4], Keh-Jiann Chen[3], I-Jung Hung[1], Ming-Yu Lee[1], Lee-Feng Chien[2], Yumin Lee[1], Renyuan Lyu[1], Hsin-min Wang[1], Yung-Chuan Wu[1], Tung-Sheng Lin[1], Hung-yan Gu[2], Chi-ping Nee[1], Chun-Yi Liao[2], Yeng-Ju Yang[2], Yuan-Cheng Chang[2], Rung-chiung Yang[2]

National Taiwan University and Academia Sinica, Taipei, Taiwan, Republic of China *

ABSTRACT

Golden Mandarin (II) is an improved single-chip real-time Mandarin dictation machine for Chinese language with very large vocabulary for the input of unlimited Chinese sentences into computers using voice. In this dictation machine only a single chip Motorola DSP 96002D on an Ariel DSP-96 card is used, with a preliminary character correct rate around 95% in speaker dependent mode at a speed of 0.96 sec per character. This is achieved by many new techniques, primarily a segmental probability modeling technique for syllable recognition specially considering the characteristics of Mandarin syllables, and a word-lattice-based Chinese character bigram for character identification specially considering the structure of Chinese language.

I. INTRODUCTION

Today, the input of Chinese characters into computers is still a very difficult and unsolved problem. This is the basic motivation for the development of a Mandarin dictation machine. We defined the scope of this research by following limitations. The input speech is in the form of isolated syllables. The machine is speaker dependent. Reasonable errors are acceptable because they can be found on the screen and corrected from the keyboard by the user very easily. But the machine has to be able to recognize Mandarin speech with very large vocabulary and unlimited texts, because the input to computers can be arbitrary Chinese texts. Also, the machine has to work in real-time for computer input applications. A previous version of such a machine, Golden Mandarin (I), has been developed in 1990 [1][2], but the highly computation-intensive algorithms for Golden Mandarin (I) require 10 TMS 320C25 chips operating in parallel on 9 special hardware boards to meet the real time requirements. This is why the present machine is developed using completely different algorithms.

There are at least 10^5 commonly used Chinese words, each composed of one to several characters. There are at least 10^4 commonly used Chinese characters, all mono-syllabic. However, the total number of different syllables in Mandarin speech is only 1302. Based on such observation, the use of syllable as the dictation unit becomes a very natural choice. Another very special feature of Mandarin Chinese is that it is a tonal language.

Every syllable is assigned a tone in general. There are basically four lexical tones and one neutral tone in Mandarin. It has been shown that the primary difference for the tones is in the pitch contours, and the tones are essentially independent of the other acoustic properties of the syllables. If the differences in tones are disregarded, only 408 base syllables (each bearing different tones) are required for Mandarin Chinese. This means the recognition of the syllables can be divided into two parallel procedures, the recognition of the tones, and of the 408 base syllables disregarding the tones. Based on the above considerations, the overall system structure for the Golden Mandarin (II) dictation machine is shown in Fig. 1. The system is basically divided into two subsystems. The first is to recognize the Mandarin syllables, and the second is to transform the series of syllables into Chinese characters, because every syllable can be shared by many homonym characters. For the first subsystem of syllable recognition, the base syllable (disregarding the tones) and the tone are recognized independently in parallel. For the second subsystem of language model, we need to first obtain all possible word hypothesis to construct a Chinese word lattice, and then use a word-lattice-based Chinese character bigram to select the most probable concatenation of word hypotheses as the output sentence.

II. MANDARIN SYLLABLE RECOGNITION

The recognition of the Mandarin syllables includes two parts: recognition of the 408 base syllables (disregarding the tones) and recognition of the tones. The tone recognition is not too difficult. Discrete Hidden Markov Models based on feature vectors of pitch frequency, difference pitch frequency, energy and difference energy are used, and the syllable durational cues are further applied to distinguish the neutral tone from the 4 lexical tones. The recognition of the 408 base syllables (disregarding the tones), however, is very difficult, because there exist 38 confusing sets in this vocabulary. A good example is the A-set, { a, ja, cha, sha, dsa, tsa, sa, ga, ka, ha, da, ta, na, la, ba, pa, ma, fa }. Specially trained continuous density Hidden Markov Models (HMM's)[3][4] for cepstral coefficients were used in the previous version machine [1][2], which are highly computation-intensive. Considering the fact that Mandarin mono-syllables have relatively simple phonetic structure and the primary problem in base syllable recognition is to distinguish the very confusing initial consonants instead of matching the entire template, it is therefore believed that the time warping functions of the state transition probabilities of HMM's are not very important. Because state transition path searching process in HMM's is highly computation-intensive, a segmental probability model (SPM) specially for Mandarin base syllables was therefore developed, which is very similar to continuous den-

*1. Dept. of Electrical Engineering, National Taiwan University

2. Dept. of Computer Science and Information Engineering, National Taiwan University

3. Institute of Information Science, Academia Sinica

4. Institute of History and Philology, Academia Sinica

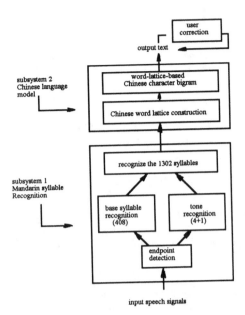

Figure 1: The overall system structure for the Golden Mandarin(II) dictation machine

sity HMM, but the state transition probabilities are deleted and the N states equally segment the syllable utterance.

In more detail, each utterance of syllable α is equally divided into N segments (or states), and each segment is modeled by M Gaussian mixtures. Each of the mixture is characterized by a mean vector $\bar{\mu}_{ij}$ and a covariance matrix $\bar{\sigma}_{ij}$, where $i = 1, 2, \ldots, N$ is the segment index, and $j = 1, 2, \ldots, M$ is the mixture index. The SPM of a syllable α is therefore represented by

$$S_{NM}(\alpha) = \{(\bar{\mu}_{ij}, \bar{\sigma}_{ij}), i = 1, 2, \ldots, N, j = 1, 2, \ldots, M\}$$

In the training phase, all training utterances for the syllable α are equally divided into N segments, and the feature vectors from the i-th segment of all training utterances are used together to train the parameters $(\bar{\mu}_{ij}, \bar{\sigma}_{ij})$, $j = 1, 2, \ldots, M$, for the i-th segment. They are first vector quantized into M clusters, and the feature vectors in the j-th cluster are used to obtain $(\bar{\mu}_{ij}, \bar{\sigma}_{ij})$. The covariance matrices $\bar{\sigma}_{ij}$ are assumed diagonal. In the recognition phase, the observation probability function $b_i(\bar{o})$ for an observed feature vector $\bar{o}$ with respect to the i-th segment of the syllable α is simply

$$b_i(\bar{o}) = \underset{j=1,2,\ldots,M}{\text{Max}} \{b_{ij}(\bar{o})\}$$

where $b_{ij}(\bar{o})$ is the Gaussian distribution function defined by ($\bar{\mu}_{ij}, \bar{\sigma}_{ij}$) . In the recognition phase, an unknown utterance U is first equally divided into N segments, assuming each with n feature vectors,

$$U = \{\bar{o}_{ik}, i = 1, 2, \ldots, N, k = 1, 2, \ldots, n\}$$

where i is the segment index and k is the vector number in a segment. The observation probability of this unknown utterance U with respect to the SPM model of a syllable $\alpha, S_{NM}(\alpha)$, is then

$$Prob(U|S_{NM}(\alpha)) = \prod_{i=1}^{N}\left[\prod_{k=1}^{n} b_i(\bar{o}_{ik})\right] \equiv Prob(U|\alpha)$$

Apparently the syllable model giving the highest observation probability for U is the recognition output. In this way, both the

training and recognition processes are simplified tremendously as compared to the continuous density HMM's.

The training data used in the experiments include 5 utterances for each of the 1302 syllables for each speaker, and the results below are the average scores obtained for two speakers. The recognition rates are listed in Table 1. The experiment (1) in the first row is for the continuous density HMM's used in the previous version machine [1][2], and experiment (2) in the second row is the initial test for SPM where N=7 and M=3, such that the number of states and mixtures are exactly the same in experiments (1) and (2) for parallel comparison. It can be seen that SPM gives an top 1 rate more than 20% lower than continuos density HMM's, apparently because SPM is a much more simplified version. However, the following series of improvements in experiments (3)~(6) in fact indicate the very high potential of SPM for Mandarin syllable recognition, if special characteristics of Mandarin syllables can be more carefully considered. Because the primary problem for Mandarin syllable recognition is to distinguish the very confusing initial consonants and some errors are often caused by the syllable ending (such as /an/ and /ang/),in experiment (3) smaller shift between adjacent speech frames was used in the first 20% and last 10% of the syllable utterances, such that finer signal characteristics can be extracted for the initial consonants and syllable ending. In the third row of Table 1, the top 1 rate was improved in this way from 69.85% to 75.49%. In experiments (4) and (5) optimal values of N and M were further found empirically and the linear prediction order P was increased from 10 to 14. Note that 3 segments (N=3) gives better results than 7 segments (N=7), probably because all the Mandarin syllables have relatively simple structure, composed of at most 3 to 4 phonemes. Without the time warping function of the state transition probabilities, too many segments (or states) may in fact cause interference among adjacent segments in the SPM. Therefore roughly one phoneme per segment turns out to be the best choice for SPM, although in HMM's 7 states gives the best results. On the other hand, because the computation load for SPM recognition is very low, increase of linear prediction order P from 10 to 14 can be easily achieved but is highly rewarding, as was indicated by the significant improvements in the top 1 rate, from 77.45% to 88.97%. Still further improvements can be achieved by a two-stage SPM approach in experiment (6) as shown in Fig. 2, in which the first stage SPM used cepstral coefficients, while the second stage SPM used regression coefficients obtained from cepstral coefficients, and the parameters M, N for the two stages can be separately optimized. The first stage selected the top L candidates $\alpha_1, \alpha_2, \ldots, \alpha_L$ and passed them to the second stage, together with the corresponding observation probabilities $P_1(U|\alpha_j), j = 1, 2, \ldots, L$. The final score of each of the L candidates is then the weighted sum of the observation probabilities obtained in the two stages. The last row of Table 1 indicates that in this way the top 1 rate can be as high as 96.57%, and the top 3 rate can be 99.75%. Note that the computation requirements in experiment (6) are still much less than those of continuous density HMM's used in experiment (1), but the performance is much more better. These results are also summarized in Fig. 3.

III. CHINESE LANGUAGE MODEL

After the base syllables and tones are recognized by the subsystem 1, the high degree of ambiguity caused by the large number of homonym characters still remain to be solved. The subsystem 2 thus acts as a linguistic decoder to identify the characters using context information. In the previous version machine [1][2], a relatively simple Chinese character bigram trained by primary

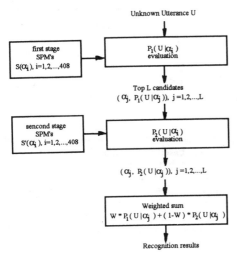

Figure 2: The two-stage approach for SPM

Experiments	top1	top 2	top 3	top 4	top 5
(1) : previous HMM	91.67	98.53	99.51	99.75	99.75
(2):initial SPM	69.85	79.66	85.04	88.73	89.71
(3): finer framing	75.49	84.80	87.01	88.24	88.48
(4) :N=3, M=4	77.45	87.50	90.44	91.91	92.16
(5): P=14	88.97	97.06	99.02	99.51	99.51
(6): Two-stage	96.57	99.26	99.75	100	100

Table 1: Base syllable recognition rates for the previous HMM and the new SPM techniques

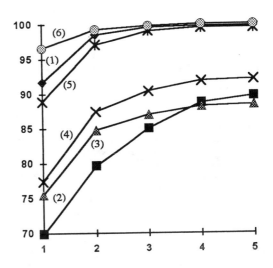

Figure 3: Recognition rates for the base syllables

school Chinese textbooks [5][6] was used, whose function was in fact limited. In Chinese language every word is composed of from one to several characters and there is no blanks between two adjacent words, thus a sentence can be considered as a sequence of words, or a sequence of characters. The 10^4 characters or 10^5 words require a character bigram of $10^4 \times 10^4$ probabilities or a word bigram of $10^5 \times 10^5$ probabilities respectively. Preliminary tests indicated that the word bigram is much more powerful than the character bigram [5], probably because the Chinese sentences are really built by words rather than by characters. But the word bigram is difficult to train and implement on a single chip because of the much larger size. A new approach considering the special structure of Chinese language using a word-lattice-based Chinese character bigram was thus developed to solve this problem. In this approach, the sequence of syllables obtained from the subsystem 1 is first matched with the words in a lexicon of 10^5 words to find all possible word hypotheses to construct a word lattice, with the help of a set of lexical rules. A word lattice is a graph of all possible paths connecting all word hypotheses, a simple example is shown in Fig. 4. The paths on the word lattice are then searched through by a word-lattice-based character bigram. The path with the highest probability is then chosen as the result, just as shown in Fig. 1.

Previous study [7][8] showed that grammatical information such as word formation are very helpful to statistical language models in grouping legal combinations of words while filtering out illegal ones. In Chinese language many compound words can be established by combining two or more words with simple rules, so they don't have to be stored in the lexicon. For example, the words " pig(豬) " and " Meat(肉) " can form a new word " Pork(豬肉) ", etc. These are the lexical rules mentioned above to help reduce the size of the lexicon. By matching the input syllable sequence with the words in the lexicon with the help of the lexical rules, all possible word hypotheses can be obtained and constructed in the word lattice. The function of the statistical language model can then be significantly reduced and simplified. For example, many noisy syllables or characters such as incorrectly recognized syllables or homonym characters which can't form word hypotheses with adjacent syllables or characters or can't be used as a mono-character word will be automatically deleted. On the other hand, if a set of adjacent word hypotheses can be grouped together earlier into a single compound word hypothesis

in the word lattice, the number of possible paths connecting the word hypothesis in the lattice can be reduced. Also, it has been observed in preliminary experiments that a longer word hypothesis is usually more reliable, and in fact very probably it is exactly the correct answer if it is really long. Therefore the establishment of such a word lattice can not only reject the interference from many noisy syllables and characters, but significantly reduce the search space of the statistical language model and improve the overall accuracy.

After a word lattice was constructed as discussed above, a specially designed word-lattice-based Chinese character bigram was used to search through the word lattice to obtain the maximum likelihood output sentence. For each word hypothesis sequence $W = W_1 W_2 ... W_m$, where W_i is the i-th component word hypothesis, let $W_i = C_{i1} C_{i2} ... C_{is_i}$, where C_{ik} is the k-th component character of W_i and s_i is the number of characters in W_i, recalling that a Chinese word is composed of several characters. Then

$$P(W) = P(W_1, W_2, ..., W_m)$$
$$= P(\underbrace{C_{11} C_{12} ... C_{1S_1}}_{W_1}, ..., \underbrace{C_{i1} ... C_{iS_i}}_{W_i}, ..., \underbrace{C_{m1} C_{m2} ... C_{mS_m}}_{W_m})$$

This probability can be approximated by

$$P(W) = P(C_{11}) P(C_{21}|C_{1S_1}) ... P(C_{m1}|C_{(m-1)S_{m-1}})$$
$$= P(C_{11}) \cdot \prod_{i=2}^{m} P(C_{i1}|C_{(i-1)S_{i-1}})$$

359

syllables: [Tzeng-1] [Jain-4] [Ji-4] [Yi-4] [Li-4]

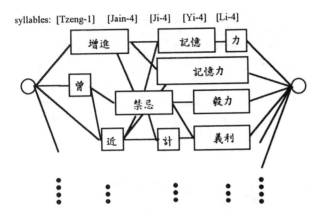

Figure 4: A partial list of an example word lattice. Each rectangle is a multi-character word, while each square is a mono-character word.

As can be found in Fig.5, this probability only considers the conditional probabilities for boundary characters in those adjacent word hypothesis, but ignores the conditional probabilities for characters within a word hypothesis, $P(C_{ik}|C_{i(k-1)}), 2 \leq k \leq S_i$ within W_i. This is because the characters within a word hypothesis are fixed and known, the conditional probability for adjacent characters can thus be assumed to be unity, when the word hypothesis is already constructed on the word lattice. In this way, the sentence hypotheses including longer word hypotheses will have higher probabilities. Therefore the longer word hypotheses will automatically have higher priority to be chosen, because they are in fact more reliable as mentioned previously. The word-lattice-based character bigram was trained in a similar way, i.e., the words in the training corpus were first segmented, and the bigram probabilities for the boundary characters were then estimated. Note that such a character bigram of $10^4 * 10^4$ probabilities is relatively easy to handle and relatively robust with respect to insufficient training corpus, but the word lattice discussed previously can effectively enhance the capabilities of the character bigram to approximate a word bigram. In Golden Mandarin (II), the character bigram was trained by a corpus of 6 million characters taken from newspapers, magazines, and so on, and the top several base syllables and tones from the subsystem 1 are included in the word lattice.

IV. REAL TIME IMPLEMENTATION AND CONCLUDING REMARKS

In the real-time implementation of the Golden Mandarin (II) all necessary computation is performed in a single chip Motorola DSP 96002D, and the complete machine is implemented on an Ariel DSP-96 card inserted into an IBM PC/AT, while a Pro-Port Model/656 acts as the front end for acoustic signals. The waveform of the input unknown syllable is filtered and sampled in ProPort and transformed into 16-bit integer format, DSP96002D then sponsors all the following processes including endpoint detection, pre-emphasis, Mandarin syllable recognition and the Chinese language model. Preliminary tests indicate that in average it takes 0.36 sec for the machine to dictate a character, which is exactly real-time, and the character correct rate is around 95%.

Golden Mandarin (II) is the second version prototype system developed in a long term project, in which the goal is to solve the difficult problem of input of arbitrary Chinese text into

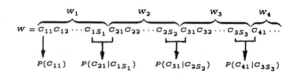

Figure 5: The probability estimation in the word-lattice-based Chinese character bigram

computers using Mandarin speech. As compared to the first version, this machine is based on a single chip with all algorithms significantly simplified, but provides improved character correct rate at higher speed. This is achieved by many new techniques, primarily a segmental probability modeling technique for syllable recognition specially considering the characteristics of Mandarin syllables, and a word-lattice-based Chinese character bigram for character identification specially considering the structure of Chinese language.

ACKNOWLEDGMENT

The work described in this paper is supported by the National Science Council of Republic of China from 1984 to 1993, and performed in the Speech Laboratory of National Taiwan University in Taipei.

References

[1] L. S. Lee, C. Y. Tseng, H. Y. Gu, K. J. Chen, F. H. Liu, C. H. Chang, S. H. Hsieh, C. H. Chen, "A Real-time Mandarin Dictation Machine for Chinese Language with Unlimited Texts and Very Large Vocabulary," *ICASSP* , Apr 1990, Albuquerque, NM, USA, pp.65-68.

[2] L. S. Lee, C. Y. Tseng, H. Y. Gu, F. H. Liu, C. H. Chang, Y. H. Lin, Y. Lee, S. L. Tu, S. H. Hsieh, C. H. Chen, "Golden Mandarin (I) - A Real-time Mandarin Speech Dictation Machine for Chinese Language with Very Large Vocabulary," to appear in *IEEE Transactions on Speech and Audio Processing*, Vol. 1, NO. 2, Apr 1993.

[3] L. S. Lee, C. Y. Tseng, F. H. Liu, C. H. Chang, H. Y. Gu, S. H. Hsieh, C. H. Chen, "Special Speech Recognition Approaches for the Highly Confusing Mandarin Syllables Based on Hidden Markov Models," *Computer Speech and Language*, Vol. 5, No. 2, Apr 1991, pp.181-201.

[4] B. H. Juang and L. R. Rabiner, "Mixture Autoregressive Hidden Markov Models for Speech Signals," *IEEE Transactions on ASSP*, pp. 1404-1413, 1985.

[5] H. Y. Gu, C. Y. Tseng and L. S. Lee, "Markov Modeling of Mandarin Chinese for Decoding the Phonetic Sequence into Chinese Characters," *Computer Speech and Language*, Vol. 5, No. 4, Oct 1991, pp.363-377.

[6] S. M. Katz, "Estimation of Probabilities from Sparse Data for the Language Model component of a Speech Recognizer," *IEEE Transactions on ASSP*, Vol. 35, pp.400- 411, 1987.

[7] L. F. Chien, K. C. Chen and L. S. Lee, "A Best-first Language Processing Model Integrating the Unification Grammar and Markov Language Model for Speech Recognition Applications, " to appear on *IEEE Transactions on Speech and Audio Processing* , Vol. 1, NO. 2, Apr 1993.

Chapter 9: Image Coding

Image Coding

Despite the growth in bandwidth available for long-distance communications, the transmission of low-bandwidth text, voice, and data remain the primary means of interaction. Compressing images is also important to integrating image storage and manipulation with personal computers and multimedia applications. The emergence of still-image and motion-video standards attest to the importance of interoperable image compression schemes.

"Rate-Constrained Picture-Adaptive Quantization for JPEG Baseline Coders" (paper 9.1) provides a brief introduction to the JPEG baseline coding algorithm, then describes an approach to generating the quantization tables, the contents of which are not specified by the JPEG standard. The quantization table calculation starts with a large step size for the first entry, then updates the table an entry at a time to maximize the ratio of decrease in distortion to increase in bit rate. "A JPEG-Based Interpolative Coding Scheme" (9.2) adjusts the scaling parameter of the quantization table to lower the bit rate, and then applies interpolative coding to address the block artifacts that result from high compression. Blocking artifacts that arise from the discrete cosine transform (DCT) that is a part of the JPEG coding standard are greatly reduced in "Iterative Projection Algorithms for Removing the Blocking Artifacts of Block-DCT Compressed Images" (9.3).

Applying image compression to real-time video signals requires the consideration of motion within a scene. "A New Approach to Decoding and Compositing Motion-Compensated DCT-Based Images" (9.4) points out that traditional techniques for combining multiple sources into a single displayed video stream require a decoding of the images prior to combining. The paper describes a technique to save computations, yet address motion compensation requirements, while remaining in the DCT domain. In "Object-Oriented Analysis-Synthesis Coding Based on Source Models of Moving 2D- and 3D-Objects" (9.5), moving objects within an image are segmented and their movement and appearance are parameterized and transmitted at a reduced data rate. The approach of extracting moving objects from an image is also used in "Motion Estimation with Wavelet Transform and the Application to Motion Compensated Interpolation" (9.6), where zero crossings of the wavelet transform provide segmentation information.

Morphological filters, which perform nonlinear signal transformations that locally modify the *geometric* features of a signal, have been applied to images and shown to preserve fine details while suppressing impulse noise. "Multiresolution Image Decomposition and Compression Using Mathematical Morphology" (9.7) applies morphology to the recursive image pyramid decomposition into low frequency and high frequency components.

An alternative to DCT-based, JPEG-standard image coding is the fractal transform, which has been described as providing higher image quality at a specified compression ratio while achieving image decoding with simpler hardware. Simplifying decoding hardware, even at the expense of more complex encoding circuitry, is useful for multimedia, personal computer, and medical picture retrieval systems. "A Hybrid Fractal Transform" (9.8) generalizes fractal coding into a framework that introduces searching for the best image regions in which to apply the tiling stage of fractal coding, in which an image is repetitively tiled with reduced copies of itself.

Suggested Additional Readings:

[1] G. W. Wornell, "Wavelet-Based Representation for the $1/f$ Family of Fractal Processes," *Proc. IEEE,* vol. 81, no. 10, pp. 1428-1450, Oct. 1993.

[2] A. E. Jacquin, "Fractal Image Coding: A Review," *Proc. IEEE,* vol. 81, no.10, pp. 1451-1465, Oct. 1993.

[3] R. B. Arps and T. K. Truong, "Comparison of International Standards for Lossless Still Image Compression," *Proc. IEEE,* vol. 82, no. 6, pp. 889-899, June 1994.

[4] P. C. Cosman, R. M. Gray, and R. A. Olshen, "Evaluating Quality of Compressed Medical Images: SNR, Subjective Rating, and Diagnostic Accuracy," *Proc. IEEE,* vol. 82, no. 6, pp. 919-932, June 1994.

[5]   B. Carpentieri and J. A. Storer, "Split-Merge Video Displacement Estimation," *Proc. IEEE,* vol. 82, no. 6, pp. 940-947, June 1994.

[6]   T. S. Huang and A. N. Netravali, "Motion and Structure from Feature Correspondences: A Review," *Proc. IEEE,* vol. 82, no. 2, pp. 252-268, Feb. 1994.

[7]   J. Serra, *Image Analysis and Mathematical Morphology.* London: Academic, 1982.

[8]   T. Denk, K. K. Parhi, and V. Cherkassky, "Combining Neural Networks and the Wavelet Transform for Image Compression," *Proc. IEEE International Conference on Acoustics, Speech, and Signal Processing 93* (Minneapolis, MN, USA) April 1993, vol. 1 pp. 637-640.

[9]   K. L. Godfrey and Y. Attikiouzel, "Self-Organized Color Image Quantization for Color Image Data Compression," *1993 IEEE International Conference on Neural Networks* (San Francisco, CA, USA) Mar. - Apr. 1993, pp. 1622-1626.

[10]   M. Gharavi-Alkhansari and T. S. Huang, "A Fractal-Based Image Block-Coding Algorithm," *Proc. IEEE International Conference on Acoustics, Speech, and Signal Processing 93* (Minneapolis, MN, USA) April 1993, vol. 5, pp. 345-348.

[11]   C. Barnes and E. S. Sjoberg, "A Comparative Study of JPEG and Residual VQ Compression of Radar Imagery," *Proc. National Telesystems Conference* (Atlanta, GA, USA) June 1993, pp. 203-207.

[12]   F. Kossentini, W. C. Chung, and M. J. T. Smith, "Low Bit Rate Coding of Earth Science Images," *Proc. Data Compression Conference '93* 1993, pp. 371-380.

RATE-CONSTRAINED PICTURE-ADAPTIVE QUANTIZATION
FOR JPEG BASELINE CODERS

Siu-Wai Wu and *Allen Gersho*

Department of Electrical and Computer Engineering
University of California, Santa Barbara, CA 93106

ABSTRACT

A recursive algorithm is presented for generating quantization tables in JPEG baseline coders from the actual statistics of the input image. Starting from a quantization table with large step sizes, corresponding to low bit rate and high distortion, we update one entry of the quantization table at a time so that at each step, the ratio of decrease in distortion to increase in bit rate is approximately maximized. This procedure is repeated until a target bit rate is reached. Simulation results demonstrate that, with picture-adaptive quantization tables designed by the proposed algorithm, the JPEG DCT coder is able to compress images with better rate-distortion performance than that achievable with conventional empirically-designed quantization tables.

1. INTRODUCTION

The JPEG baseline coding algorithm [1][2] has been established recently as an industry standard for image compression. A schematic of the JPEG baseline coding algorithm is shown in Fig 1. For each color component, the input image is first partitioned into 8×8 pixel blocks. Each block is converted by the Discrete Cosine Transform into a block of 64 transform coefficients which are quantized by a set of uniform scalar quantizers. The indices of the quantized DCT coefficients are then losslessly encoded with variable length codes. The DC indices of successive blocks are differentially coded and the AC indices within a block are runlength encoded.

In the JPEG runlength encoding scheme, the indices of the DCT coefficients are scanned in a zig-zag path as shown in Fig 2 and modeled as a sequence of runs followed by an End-of-Block symbol. Each run consists of a string of consecutive zeros terminated by a non-zero index, and the corresponding variable length codeword is determined by the runlength of zeros and the level of the non-zero terminating index. A separate codeword is also generated for the End-of-Block symbol.

The JPEG Baseline System is basically a variable bit rate coder where the quality of the coded image is determined by the matrix of quantizer step sizes called the *quantization table*, one for each color component. The quanti-

---

This work was supported by the University of California MICRO program, Rockwell International Corporation, Hughes Aircraft Company, Eastman Kodak Company, and the Center for Advanced Television Studies.

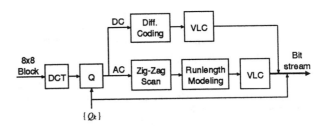

Figure 1: Schematic of JPEG baseline coding algorithm

Figure 2: Zig-Zag scan of DCT coefficients

zation tables are specified as input parameters to the system, and included in the encoded bit stream. If the default Huffman tables for the runlength encoding is used, then the quantization tables also control the compressed bit rate of the image.

Usually the quantization tables are empirically generated without consideration of the distortion-rate trade-off, hence they result in suboptimal performance. In this paper, we present an algorithm for generating a quantization table given any input image such that the image is efficiently coded by a JPEG baseline coder within a given number of bits. For simplicity, we only consider the luminance component, but the same method can be applied to the chrominance components as well.

2. PROBLEM FORMULATION

Given the standard JPEG baseline DCT coder with a default Huffman table, an input image that consist of N blocks of 8×8 pixels, and a target bit rate (B) for coding the image, we want to find a set of quantizer step sizes $\{Q_k : k = 0, \ldots, 63\}$ to minimize the overall distortion

$$D = \sum_{n=1}^{N} \sum_{k=0}^{63} D_{n,k}(Q_k) \qquad (1)$$

subject to the bit rate constraint:

$$R = \sum_{n=1}^{N} R_n(Q_0, \ldots, Q_{63}) \leq B, \qquad (2)$$

where $D_{n,k}(Q_k)$ is the distortion in the k-th DCT coefficient of the n-th block if it is quantized with step size Q_k, and $R_n(Q_0, \ldots, Q_{63})$ is the number of bits generated in coding the n-th block with the quantization table $\{Q_0, \ldots, Q_{63}\}$. For simplicity, we measure the distortion of the DCT coefficient as the squared error between the original value and the quantized value. Alternately, a perceptually weighted distortion measure could be used.

Suppose there are J_n runs in the n-th block. Let the runlengths, the levels of the terminating indices, and the lengths of the corresponding codewords be denote by r_{nj}, s_{nj}, and $L(r_{nj}, s_{nj})$ respectively, $j = 1, \ldots, J_n$. The number of bits generated for the n-th block can be calculated as

$$R_n(Q_0, \ldots, Q_{63}) = L_n^{(dc)} + \sum_{j=1}^{J_n} L(r_{nj}, s_{nj}) + L_{EOB}, \quad (3)$$

where $L_n^{(dc)}$ is the number of bits for coding the DC coefficients in the n-th block, and L_{EOB} is the length of the End-of-Block codeword.

The minimization problem of (1) is a variation of the bit allocation problem. Unfortunately, due to the runlength coding, the bit rate is not simply the sum of bits contributed by coding the individual DCT coefficients, therefore it is difficult to obtain an optimal solution with classical bit allocation techniques such as [3]. Recently, a gradient descend technique has been proposed to solve for a locally optimal solution to the quantization table design problem based on the assumption that the probability distributions of the DCT coefficients are Laplacian [4]. However, the Laplacian distribution is only an approximation of the typical statistics of the DCT coefficients, and may not hold for an arbitrary input image. In the following section, we will present a recursive algorithm to search for an approximate solution to the bit allocation problem based on the actual statistics of the input image.

3. THE QUANTIZATION TABLE DESIGN ALGORITHM

We applied a recursive technique to search for a good (although not necessarily optimal) solution to the quantization table design problem formulated above. Our approach is similar to the quantizer selection paradigm presented in [5], which is a slightly generalized version of an integer bit allocation algorithm originally proposed by Shoham and Gersho [6]. Notice that an intuitively more appealing formulation based on the framework of optimal pruning [7][8] can also be used to find the same solution to the integer bit allocation problem as obtained by [6]. In this paper, we adopt the approach of [7] and develop the quantization table design algorithm with a flavor of optimal pruning.

The main idea of the algorithm is describe as follows. Starting from an initial quantization table of large step sizes, corresponding to low bit rate and high distortion, the algorithm decreases the step size in one entry of the quantization table at a time until a target bit rate is reached. In each iteration, the objective is to obtain the best rate-distortion trade-off while the overall bit rate is decreasing. Quantitatively, we try to update the quantization table in such a way that the ratio of decrease in distortion to increase in bit rate is maximized over all possible reduced step size values for one entry of the quantization table. Thus, we seek the values of k and q that solve the maximization problem:

$$\max_k \max_q \frac{-\Delta D \mid_{Q_k \to q}}{\Delta R \mid_{Q_k \to q}}, \qquad (4)$$

where $\Delta D \mid_{Q_k \to q}$ and $\Delta R \mid_{Q_k \to q}$ are respectively the change in distortion and the change in overall bit rate when the k-th entry of the quantization table, Q_k, is replaced by q. These increments can be calculated as

$$\Delta D \mid_{Q_k \to q} = \sum_{n=1}^{N} [D_{n,k}(q) - D_{n,k}(Q_k)] \qquad (5)$$

and

$$\Delta R \mid_{Q_k \to q} = \sum_{n=1}^{N} [\ R_n(Q_0, \ldots, q, \ldots, Q_{63}) \qquad (6)$$
$$- R_n(Q_0, \ldots, Q_k, \ldots, Q_{63})].$$

To simplify the computation, we split the maximization problem in (4) into two parts, where the first part computes

$$\lambda_k = \max_q \frac{-\Delta D \mid_{Q_k \to q}}{\Delta R \mid_{Q_k \to q}} \qquad (7)$$

for all values of k and the second part solves

$$\lambda = \max_k \lambda_k. \qquad (8)$$

The algorithm maintains tables of λ_k and Q_k^* where Q_k^* is the solution of (7). In each iteration, the algorithm searches for k^* to maximize λ_{k^*} and updates the corresponding entry of the quantization table by $Q_{k^*}^*$. At the end of the iteration, the k^*-th entries in the tables $\{\lambda_k\}$ and $\{Q_k^*\}$ are updated accordingly by (7) to prepare for the next iteration.

In order to preserve the quality of the DC coefficients, we set Q_0 to the same value as that given by JPEG [2]. This step size ($Q_0 = 16$) provides virtually lossless quantization (± 1 pixel maximum error) for the DC cofficients. Thus the quantization table design algorithm operates only on the AC step sizes.

To summarize, the algorithm is outlined below:

1. Initialize the quantization table by

$$Q_k = \begin{cases} 16 & \text{for } k = 0, \\ Q_{\max} & \text{for } k = 1, \ldots, 63. \end{cases} \quad (9)$$

2. Initialize tables of λ_k and Q_k^* by searching q in $\{1, \ldots, Q_{\max} - 1\}$ to solve (7) for $k = 1, \ldots, 63$.

3. Search k in $\{1, \ldots, 63\}$ for k^* to solve (8).

4. Update the quantization table by setting $Q_{k^*} = Q_{k^*}^*$.

5. Update λ_{k^*} and $Q_{k^*}^*$ by searching q in $\{1, \ldots, Q_{k^*} - 1\}$ to solve (7) for $k = k^*$.

6. Repeat Steps 3 to 5 until $R(Q_0, \ldots, Q_{63}) \leq B$.

The maximum allowable step size in the JPEG specification [2] is 256. However we notice that for most images coded at a reasonable bit rate, initializing the quantization table with step size 128 leads to approximately the same result as that with the maximum allowable step size, but the number of iterations required to arrive at the target bit rate is greatly reduced. Therefore we use 128 instead of 256 as $Q_{\max}$.

It can be easily derived from the results of [6] or [8] that if the rate functions $R_n()$ are additively separable, i.e., if there exists a family of rate functions $\{R_n^{(k)}(q) : k = 1, \ldots, 63; \ n = 1, \ldots, N\}$ such that

$$R_n(Q_0, \ldots, Q_{63}) = \sum_{k=1}^{63} R_n^{(k)}(Q_k) \quad \text{for } n = 1, \ldots, N, \quad (10)$$

then the quantization table designed by the above algorithm is indeed optimal. In this case, the algorithm described above traces out the lower convex hull of the set of points on the rate-distortion plane corresponding to the bit rates and distortions obtained by coding the image with all possible quantization tables, and the values of λ in successive iterations are the values of the slopes of the convex hull.

However this condition is rarely satisfied in practice, thus the resulting quantization table is not optimal in general. A further complication with non-separable rate functions is that each iteration of the algorithm only approximately, rather than exactly, maximizes the ratio of decrease in distortion to the increase in bit rate. An exact, but computationally more expensive, solution would require all 63 λ_ks to be updated in every iteration instead of only updating λ_{k^*}, since the update of Q_{k^*} may affect the values of $\Delta R \mid_{Q_k \to q}$ for $k \neq k^*$. Nevertheless, as we will see in the following section, experimental results show that, compared to an empirical quantization table, non-negligible improvements in performance can be obtained with picture-adaptive quantization tables designed by the algorithm described above.

4. EXPERIMENTAL RESULTS

Computer simulations of the JPEG baseline system with black-and-white input images were used to evaluate the proposed paradigm. For each test image, we compared the performances of two different quantization tables: (1) the example quantization table given by JPEG [2], and (2) the

Figure 3: *Lenna* coded by JPEG quantization table.

Figure 4: *Lenna* coded by picture-adaptive quantization table.

367

Image	PSNR (JPEG example)	PSNR (picture-adaptive)	Improvement	Rate (bpp)
Lenna (512 × 512)	35.8 dB	36.6 dB	*0.8 dB*	0.64 bpp
Barbara (720 × 576)	32.8 dB	34.5 dB	*1.7 dB*	0.98 bpp
Baboon (512 × 512)	28.2 dB	29.7 dB	*1.5 dB*	1.39 bpp

Table 1: Performance comparison of JPEG example quantization table and picture adaptive quantization table

picture-adaptive quantization table generated by the proposed algorithm in the previous section. The target bit rate for (2) was set to the bit rate resulting from (1). Table 1 shows the coding results for the test images *Lenna*, *Barbara*, and *Baboon*. The image *Lenna* coded with the JPEG quantization table and the picture-adaptive quantization table are shown in Fig. 3 and Fig. 4 respectively. We observe that compared to the JPEG quantization table, the picture-adaptive quantization tables result in not only higher PSNR values but also better overall perceptual quality of the coded images for the same bit rate.

5. CONCLUSION

We have presented an algorithm for the design of quantization tables for JPEG baseline (DCT) coders. This algorithm efficiently trades between distortion and bit rate; the trade-off is based on the actual statistics of the input image, resulting in a quantization table that adapts to the scenes contained in the picture. As demonstated in the simulation experiments, the picture-adaptive quantization tables designed by the proposed algorithm enable the JPEG baseline algorithm to code images at the same bit rate but with lower distortion than is achieved with empirical quantization tables. Furthermore, the rate-constrained adaptive quantization technique allows the user to have precise control of the bit rate in a JPEG baseline coder. This feature is essential for some image coding applications where fixed bit rate of the entire image is required, e.g. in digital video tape recording [5].

The price paid for the enhanced performance is an increase in encoding complexity. However, for some asymmetric applications in which low decoding complexity is desired but encoding complexity is inconsequential, this picture-adaptive quantization method may be employed to effectively improve the performance of the system.

6. REFERENCES

[1] G. K. Wallace, "The JPEG still-picture compression standard," *Communications of the ACM*, vol. 34, April 1991.

[2] "JPEG Draft Technical Specification Revision 8," Aug. 1990.

[3] J.-Y. Huang and P. M. Schultheiss, "Block quantization of correlated Gaussian random variables," *IEEE Trans. Comm.*, vol. CS-11, pp. 289–296, September 1963.

[4] A. Hung and T. Meng, "Optimal quantizer step sizes for transform coders," in *Proc. Inter. Conf. Acoustics, Speech, and Signal Processing*, pp. 2621 – 2624, April 1991.

[5] S.-W. Wu and A. Gersho, "Rate-Constrained Optimal Block-Adaptive Coding for Digital Tape Recording of HDTV," *IEEE Trans. on Circuits and Systems for Video Technology*, vol. 1, pp. 100–112, March 1991.

[6] Y. Shoham and A. Gersho, "Efficient Bit Allocation for an Arbitrary Set of Quantizers," *IEEE Trans. Acoust. Speech Signal Process.*, vol. ASSP-36, pp. 1445–1453, September 1988.

[7] E. A. Riskin, "Optimal Bit Allocation via the Generalized BFOS Algorithm," *IEEE Trans. Inform. Theory*, vol. 37, pp. 400–402, March 1991.

[8] P. A. Chou, T. Lookabaugh, and R. M. Gray, "Optimal Pruning with Applications to Tree-Structured Source Coding and Modeling," *IEEE Trans. Inform. Theory*, pp. 299–315, March 1989.

A JPEG-BASED INTERPOLATIVE IMAGE CODING SCHEME

Bing Zeng

Department of Electrical and Electronic Engineering
Hong Kong University of Science and Technology
Clear Water Bay, Kowloon, Hong Kong

Anastatios N. Venetsanopoulos

Department of Electrical Engineering
University of Toronto
Toronto, Ontario M5S 1A4

ABSTRACT

JPEG has recently been recognized as the most popular and effective coding scheme for still-frame, continuous-tone images. At a moderate bit rate, JPEG can usually give a quite satisfactory solution to most of practical coding applications. However, the blocking effect produced by JPEG tends to be rather disturbing when a higher compression rate is required. In this paper, we propose a novel interpolative coding scheme which is based on the JPEG technology. As compared to the standard JPEG, the efficiency of this interpolative scheme is enhanced and, more importantly, the blocking artifacts have been reduced significantly, thus yielding a much more pleasant visual result.

I. INTRODUCTION

Development of efficient image/video coding techniques has attracted great attentions from both academic and industrial people for a long time. As a result of extensive researches during the past two decades, several standard schemes have been established recently, such as JPEG, MPEG, and H.261. JPEG (Joint Photographic Experts Group) is recommended primarily to still-frame, continuous-tone images (either monochrome or color), while MPEG (Motion Picture Experts Group) and H.261 are developed for video signals. In this paper, we focus on the JPEG technology, based on which we shall propose an interpolative coding scheme.

The organization of this paper is as follows. In Section II, we first review briefly the principle of JPEG. Then we will demonstrate by an experiment that the blocking artifacts produced by JPEG at a very low bit rate will be rather disturbing to human eyes, though the measured peak-to-peak signal-to-noise ratio (PSNR) is relatively good. Observing this phenomenon, we will propose, in Section III, an interpolative scheme to reduce the blocking artifacts. Some considerations on the design of the low-pass decimation/interpolation filters are discussed and the difference of this scheme from the sub-band method is pointed out. In Section IV, several computer simulation results are presented to compare our new scheme with the standard JPEG via measuring the PSNR as well as judging from the visual results. Some conclusions are given finally in Section V.

II. PRINCIPLE OF JPEG

The principle diagram of JPEG is shown in Fig. 1 for both encoder (the top part) and decoder (the bottom part). The encoder is composed of three main parts: forward Discrete Cosine Transform (FDCT), quantization, and entropy encoder. The decoder also consists of three main parts: entropy decoder, de-quantization, and inverse Discrete Cosine Transform (IDCT), which are all functioning oppositely to their corresponding parts in the encoder. The DCT is performed on the basis of block by block and each block has a size of 8×8. The DC component of each block is coded in a differential pulse code modulation (DPCM) format, i.e., the difference between the DC component of this block and that of the previous block is coded; while all the AC coefficients are coded in the pulse code modulation (PCM) form.

Both quantization and de-quantization are performed via a pre-determined quantization table. Currently, there are several such tables that differ only slightly from each other. The table given by JPEG is as following:

$$Q = \begin{pmatrix} 16 & 11 & 10 & 16 & 24 & 40 & 51 & 61 \\ 12 & 12 & 14 & 19 & 26 & 58 & 60 & 55 \\ 14 & 13 & 16 & 24 & 40 & 57 & 69 & 56 \\ 14 & 17 & 22 & 29 & 51 & 87 & 80 & 62 \\ 18 & 22 & 37 & 56 & 68 & 109 & 103 & 77 \\ 24 & 35 & 55 & 64 & 81 & 104 & 113 & 92 \\ 49 & 64 & 78 & 87 & 103 & 121 & 120 & 101 \\ 72 & 92 & 95 & 98 & 112 & 100 & 103 & 99 \end{pmatrix}.$$

In a practical application of JPEG, a parameter α is involved which scales the quantization table:

$$\alpha = \begin{cases} 10 - q; & \text{if } q \leq 9 \\ 1/2^{(q-9)}; & \text{otherwise.} \end{cases}$$

Before the quantization is invoked, all the 8×8 DCT coefficients of each block will be re-ordered into a zigzag format, as shown in Fig. 2. Then, each component having a non-zero value is coded as a pair in which the first entity represents its quantized value and the second one implies how many components having the zero value will follow it. In this way, all those components having

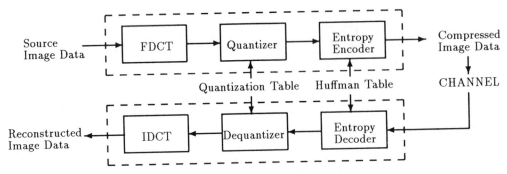

Fig. 1. The simplified block diagram of JPEG.

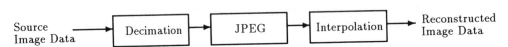

Fig. 4. The block diagram of the proposed JPEG-based interpolative scheme.

the zero value will not be coded. Finally, a special code is added to indicate the end of each block.

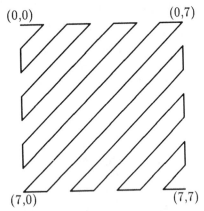

Fig. 2. The zig-zag scanning for a 8 × 8 block.

In general, the quality of a coded image by using JPEG is solely controlled by the the parameter q (or α), since all the other parts have been fixed. Experiences with working on JPEG tell us that it can usually result in a quite satisfactory solution (judged either by measuring the PSNR or by observing visually the coded image) when a moderate bit rate (around 1.0 bpp for most of natural images that are commonly used for testing) is used. When a higher compression rate (say, under 0.4 bpp) is required, however, we have also gained such an experience that JPEG will suffer seriously from the blocking effect yielded, because some (or even most, sometimes) of the AC coefficients of a block will often be quantized to be zero. To demonstrate this

result, we provide in Fig. 3 an original image (Lenna) of size 512×512 and its JPEG-coded version at 0.25 bpp ($q = 6$). It is easy to see that for such a case, the coded image is rather disturbing to human eyes due to the existence of quite many blocking artifacts, although the measured PSNR (31.86 dB) is relatively good.

III. INTERPOLATED JPEG SCHEME

In this section, we will propose a JPEG-based interpolative scheme in order to reduce the block artifacts produced by the standard JPEG at a very low bit rate.

Fig. 4 presents the principle diagram of this interpolative scheme. The main idea is as follows: First, the original image undergoes the decimation block which carries out a low-pass filtering and a down-sampling. The decimated image (with a reduced resolution) is then coded using the standard JPEG. The final image with full-resolution is obtained by feeding the coded image into an interpolation block (up-sampling and low-pass filtering). Several remarks about this scheme are in order in the following.

1. The proposed interpolative scheme differs significantly from the sub-band method. Here, we are not invoking to a filter-bank, but keep only one (low-low) band and drop all the others. Therefore, such a scheme can be judged to be much simpler than the sub-band coding method, considering in particular that JPEG is a very efficient coding method.

2. The decimated image can be coded with a high quality by adjusting the scaling factor to the JPEG's quantization table. A very low bit rate for the whole image is guaranteed by the decimation, since the

Fig. 3(a) The original Lenna of size 512 × 512.

Fig. 3(b) The JPEG-coded Lenna at 0.25 bpp
(PSNR = 31.86 dB).

compression rate resulted at coding the decimated image needs to be multiplied by N^2 when calculating the overall compression rate, where N is the decimation rate (in each direction).

3. Compared to the standard JPEG method, this interpolative scheme tries to solve the problem with blocking artifacts at a very low bit rate by a combination of decimation and interpolation. The decimation provides a contribution to the suppression of blocking artifacts through the smoothing property of the low-pass filter included. On the other hand, the major contribution comes from the interpolation. Instead of obtaining all "bad" pixels by JPEG at a very low bit rate, relatively "good" pixels could be produced via interpolation using the coded decimated image which is, in general, composed of "good" pixels.

4. It is easy to understand that one should use only low decimation rates in this interpolative coding scheme. Actually, we recommend here only the use of decimation by 2 in both directions ($N = 2$).

As we have mentioned above, the proposed interpolative scheme is significantly different from the sub-band method. This has therefore lead to a potential difference in the design of the low-pass filter for the decimator or interpolator since now we do not care the perfect reconstruction property any more. It is easy to understand that a filter with narrower pass-band will result in better correlations between pixels of the decimated image. This can amount to a larger compression rate when JPEG is used to code the decimated image. On the other hand, however, more details of the original image will be destroyed by doing so, and it is in general

not easy to recovery these details by an interpolation alone. Therefore, a comprise has to be taken into consideration in order to balance the compression rate and the quality of the coded image. In the present paper, we will not go that far to consider carefully this design problem; alternatively, we employ several very simple low-pass filters: (1) the average filter, (2) the replication filter, and (3) a Gaussian filter.

IV. COMPUTER SIMULATIONS

In order to test the proposed scheme and compare it with the standard JPEG, we have conducted extensive computer simulations. Some of the results are reported in the following.

First, we provide in Fig. 5 the PSNR's obtained for a variety of testing images (Lenna, Boat, Harbor, and Barbara) at several bit rates. In these experiments, we always use a 2 × 2 average as the low-pass filter in the decimation block. Two interpolators are employed in the interpolation block. The first is the replication filter and the second is a 5 × 5 Gaussian filter given by the following matrix:

$$\begin{pmatrix} .0025 & -.0125 & -.0300 & -.0125 & .0025 \\ -.0125 & .0625 & .1500 & .0625 & -.0125 \\ -.0300 & .1500 & .3600 & .1500 & -.0300 \\ -.0125 & .0625 & .1500 & .0625 & -.0125 \\ .0025 & -.0125 & -.0300 & -.0125 & .0025 \end{pmatrix}$$

From Fig. 5, we found that the interpolative scheme with the Gaussian filter provides a higher PSNR than the corresponding scheme with the replication interpolator. We also found that, at a relatively low com-

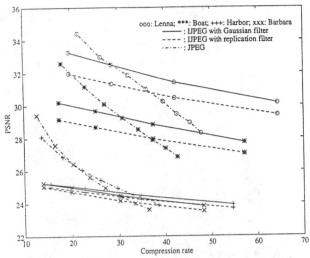

Fig. 5. PSNR versus compression rate curves of the various testing images.

pression rate, the PSNR numbers of the interpolative scheme are smaller that those of the standard JPEG. However, when the compression rate becomes larger, the PSNR yielded by the interpolative scheme surpassed the PSNR that is obtained by using the standard JPEG. Another interesting point is that the decrease in the PSNR as the compression rate is getting larger is much slower for the interpolative scheme.

In order to judge visually the proposed scheme, we present in Fig. 6 two coded Lennas (both at 0.28 bpp) using this interpolative scheme (having different interpolation filters). It is quite clear that, as compared to the result shown in Fig. 3(b), the visual effect of these two images has become much more pleasant. Although some details may have been lost, almost no obvious blocking artifacts can be percepted easily. Finally, we would like to point out that a similar visual result has also been observed for other testing images.

V. CONCLUSIONS

In this paper, we proposed an interpolative image coding scheme which has utilized the well-established JPEG technique. The primary goal was to reduce the blocking artifacts that would be produced by using the standard JPEG when the bit rate is very low. The experimental results showed clearly that such a problem has been successfully solved.

REFERENCES

[1] G. K. Wallace, "The JPEG still-picture compression standard," *Comm. of ACM*, pp. 30-44, Apr. 1991.
[2] JPEG digital compression and coding of continuous-tone still images, Draft ISO 10918, 1991.

Fig. 6(a) The coded Lenna using the interpolative scheme at 0.28 bpp. (A 2 × 2 average is used as the low-pass filter and the down-sampling by 2 in each direction is done to obtain the decimated image. The replication is used as the interpolator. PSNR = 31.33 dB.)

Fig. 6(b) The coded Lenna using the interpolative scheme at 0.28 bpp. (A 2 × 2 average is used as the low-pass filter and the down-sampling by 2 in each direction is done to obtain the decimated image. The 5 × 5 Gaussian filter is used as the interpolator. PSNR = 32.50 dB.)

ITERATIVE PROJECTION ALGORITHMS FOR REMOVING THE BLOCKING ARTIFACTS OF BLOCK-DCT COMPRESSED IMAGES

Yongyi Yang* and N. P. Galatsanos*

Department of Electrical
and Computer Engineering
Illinois Institute of Technology
Chicago, Illinois 60616

A. K. Katsaggelos

Department of Electrical
Engineering and Computer Science
Northwestern University
Evanston, Illinois 60208

ABSTRACT

In this paper, iterative projection algorithms are presented to reconstruct visually pleasing images from Block Discrete Cosine Transform (BDCT) compressed image data. Two algorithms are proposed. The first is based on the theory of Projections Onto Convex Sets(POCS). The second is motivated by the theory of POCS. Experimental results are presented which demonstrate that the proposed algorithms yield superior images to those obtained by direct reconstruction from the compressed data only.

1. INTRODUCTION

Image data compression is a very important problem for many emerging applications in the fields of visual communications. Among the many available image compression approaches, transform based methods using the Discrete Cosine Transform (DCT) are by far the most popular [1]. In these methods, the DCT is computed over a number of spatially partitioned regions (typically 16×16 or 8×8) called "blocks" [1], therefore the name Block DCT(BDCT) follows naturally. This block-by-block takes advantage of the local spatial correlation property of images and it further facilitates the VLSI hardware implementation of the DCT. However, it is well known that this independent processing of blocks induces a visually annoying *blocking artifact*, especially when high compression ratios are used. This artifact manifests itself as an artificial boundary discontinuity between adjacent blocks. It is by far the most serious obstacle in the design of BDCT based coders-decoders (codecs) and constitutes a "bottleneck" for many important present time visual communication applications.

It is clear that post-processing algorithms that can operate on images and ameliorate the blocking artifacts are extremely valuable for many practical appli-

cations. In the past such post-processing algorithms have been examined in [2], [3], and [4]. The proposed approach in [2] is just a simple low-pass filtering operation of the block border pixels which does not explicitly use all the available information about the image. The algorithms proposed in [3] are heuristic without a rigorous mathematically justification. The algorithm in [4] invokes the theory of projections onto convex sets (POCS) [5] to justify convergence, however, the smoothness set used is not rigorously defined. Thus, the convergence of the proposed algorithm cannot be justified using the classical POCS theory [5].

In this paper we have developed two iterative post-processing algorithms that treat the amelioration of blocking artifact as an image recovery problem. The first algorithms is based on the theory POCS [6]. The second algorithm is only *motivated* by POCS. The key idea behind POCS based image recovery algorithms is that known properties about the original are expressed in the form of convex constraint sets. Then, the desired image that satisfies all the imposed constraints lies in the intersection of these sets and is found by alternating projections onto these sets [6]. In order for the POCS approach to be effective, the convex sets must be as pertinent as possible to the required objective. For the de-blocking of the DCT compressed images two types of constraint sets capture most of the information pertinent to this problem. First, the set that contains the information of the transmitted DCT coefficients. Second, the sets that express the between-block smoothness properties of the desired image.

2. MATHEMATICAL FORMULATION OF THE PROBLEM

A real $N \times N$ image $\mathbf{f}$ is treated as an $N^2 \times 1$ vector in the space R^{N^2} by lexicographic ordering either by rows or columns. The BDCT is viewed as a linear transformation from R^{N^2} to R^{N^2}. Then, for an image

---

*Their work was supported by a grant from MOTOROLA

f we can write

$$\mathbf{F} = B\mathbf{f}, \quad \text{and} \quad \mathbf{f} = B^t\,\mathbf{F}, \qquad (1)$$

where **F** is the BDCT of **f** and B is the BDCT matrix.

The elements of **F** in Eq. (1) are the expression coefficients of the vector **f** using the BDCT basis in R^{N^2}. That is, **f** can be written as

$$\mathbf{f} = \sum_{n=1}^{N^2} F_n \mathbf{e}_n, \qquad (2)$$

where $\mathbf{e}_n$ denote the normalized BDCT basis vectors and F_n are the BDCT coefficients of **f**.

In image compression, only a fraction of the BDCT coefficients are coded and transmitted to the decoder. Let $\mathcal{I}$ be the set of indices of the transmitted BDCT coefficients. Then Eq. (2) can be rewritten as

$$\mathbf{f} = \sum_{n \in \mathcal{I}} F_n \mathbf{e}_n + \sum_{n \notin \mathcal{I}} F_n \mathbf{e}_n, \qquad (3)$$

Let $\mathcal{Q}$ denote the quantization operator in the coder; then in the decoder we have available the quantized BDCT coefficients $\{F_n^{'}; n \in \mathcal{I}\}$, where $F_n^{'} = \mathcal{Q}[F_n]$. In the decoder the coded image is reconstructed by assuming that the non-transmitted BDCT coefficients $\{F_n^{'}s, n \notin \mathcal{I}\}$ are equal to 0. Thus the reconstructed image is given by

$$\mathbf{f}^{'} = \sum_{n \in \mathcal{I}} F_n^{'} \mathbf{e}_n, \qquad (4)$$

which as explained earlier exhibits the visually annoying blocking artifact.

Using the previous notation, our goal in this paper is to compute estimates of both sets of coefficients $\{F_n; n \notin \mathcal{I}\}$ and $\{F_n; n \in \mathcal{I}\}$. The information used for this task is the transmitted coefficients $\{F_n^{'}; n \in \mathcal{I}\}$, knowledge of the quantization operator $\mathcal{Q}$ and prior knowledge about the smoothness properties of the original image. The set $\mathcal{I}$ of the transmitted BDCT coefficients and the quantizer $\mathcal{Q}$ depend on the *irrelevancy* reduction method used [7].

3. CONVEX SETS AND THE PROJECTIONS

As mentioned earlier, two types of constraint sets are used in this problem, namely, the constraint set(s) with the information captured by the available data and the constraint set(s) with the prior knowledge which is introduced to complement the available data. More specifically these constraint sets are: the sets that capture the knowledge about the BDCT coefficients available in the decoder and the sets that capture the smoothness properties of the desired image. Their definitions as well as the corresponding projections are given below:

The set C_1 which is based on the known BDCT coefficients is defined by

$$C_1 \triangleq \{\mathbf{f} : (\mathcal{Q}B\mathbf{f})_n = F_n^{'}, \forall n \in \mathcal{I}\}, \qquad (5)$$

where B, $\mathcal{Q}, F_n^{'}$ and $\mathcal{I}$ were defined previously. Since the quantization interval for each transmitted BDCT coefficient is assumed to be known, C_1 can be rewritten equivalently as

$$C_1 = \{\mathbf{f} : F_n^{min} \leq (B\mathbf{f})_n \leq F_n^{max}, \forall n \in \mathcal{I}\}, \qquad (6)$$

where F_n^{min} and F_n^{max} are determined by the quantizer used. By definition, C_1 is a constraint set that captures all known information about the received BDCT coefficients. It is easy to show that C_1 is closed and convex. The projection $P_1\mathbf{f}$ of an arbitrary vector **f** in R^{N^2} onto C_1 is given by [6]

$$P_1\mathbf{f} = B^t \cdot \mathbf{F}, \qquad (7)$$

where B^t is defined in Eq. (1) and **F** is determined by

$$F_n = \begin{cases} F_n^{min} & \text{if } (B\mathbf{f})_n < F_n^{min} \\ F_n^{max} & \text{if } (B\mathbf{f})_n > F_n^{max} \\ (B\mathbf{f})_n & \text{if } F_n^{min} \leq (B\mathbf{f})_n \leq F_n^{max}, \end{cases} \qquad (8)$$

for $1 \leq n \leq N^2$, $n \in \mathcal{I}$.

Constraint sets that capture the intensity continuity between block boundaries contain information that is *lost* when only the transmitted BDCT coefficients are used for reconstruction. Thus, information about smoothness in the block boundaries *complements* the information conveyed by the BDCT coefficients and is very important for this reconstruction problem. For this purpose, two new constraint set C_2 and $C_2^{'}$ are defined.

For the definition of the set C_2, the $N \times N$ image **f** is represented in column vector form as:

$$\mathbf{f} = \{\mathbf{f}_1, \mathbf{f}_2, ..., \mathbf{f}_N\}, \qquad (9)$$

where $\mathbf{f}_i$ denotes the ith column of the image. Let Q be a linear operator such that $Q\mathbf{f}$ represents the difference between the columns at the block boundaries of the image **f**. For example, for the case of $N = 512$ and 8×8 blocks,

$$Q\mathbf{f} = \begin{bmatrix} \mathbf{f}_8 - \mathbf{f}_9 \\ \mathbf{f}_{16} - \mathbf{f}_{17} \\ \cdot \\ \cdot \\ \cdot \\ \mathbf{f}_{504} - \mathbf{f}_{505} \end{bmatrix}. \qquad (10)$$

The norm of $Q\mathbf{f}$

$$\|Q\mathbf{f}\| = \left[\sum_{i=1}^{63} \|\mathbf{f}_{8\cdot i} - \mathbf{f}_{8\cdot i+1}\|^2 \right]^{\frac{1}{2}} \qquad (11)$$

is the total intensity variation between the boundary columns of adjacent blocks.

Set C_2 is defined by

$$C_2 \overset{\triangle}{=} \{\mathbf{f} : \|Q\mathbf{f}\| \le E\}, \qquad (12)$$

where E is the scalar upper bound that defines the size of this set. The value of E can be estimated by applying the operator Q to the columns within the block boundaries. It is easy to see that set C_2 is an ellipsoid which is both closed and convex.

For an image $\mathbf{f} \in R^{N^2}$ expressed in its column form $\{\mathbf{f}_1, \mathbf{f}_2, ..., \mathbf{f}_N\}$, let its projection $\tilde{\mathbf{f}} = P_2\mathbf{f}$ onto set C_2 be represented in its column form by $\{\tilde{\mathbf{f}}_1, \tilde{\mathbf{f}}_2, ..., \tilde{\mathbf{f}}_N\}$. It was shown in [8] that for a 512×512 image and 8×8 blocks

$$\begin{aligned} \tilde{\mathbf{f}}_i &= \alpha \cdot \mathbf{f}_i + (1-\alpha) \cdot \mathbf{f}_{i+1} \\ \tilde{\mathbf{f}}_{i+1} &= (1-\alpha) \cdot \mathbf{f}_i + \alpha \cdot \mathbf{f}_{i+1} \\ &\text{for } i = 8 \cdot k \text{ and } k = 1,2,...,63; \\ &\text{otherwise}, \quad \tilde{\mathbf{f}}_i = \mathbf{f}_i, \end{aligned} \qquad (13)$$

where $\alpha = \frac{1}{2}\left[\frac{E}{\|Q\cdot\mathbf{f}\|} + 1 \right]$. This result can be generalized in a straightforward manner for images of any block sizes.

In a similar fashion a set S_3 can be defined capturing the vertical smoothness between blocks. There, Qf gives the difference between the rows in the block boundaries of the image. The projection onto S_3 can be found in a same form as the projection onto S_2. However, there a row by row notation is used to represent the image and its projection.

From the above descriptions, it is clear that the same approach can be applied to define smoothness constraints between the columns and rows near the block boundaries.

For any linear operator Q, Equation (12), also defines a convex set. We denote it by S_4. The projection of an image f onto this set is given by [9], [10]

$$P_4 f = (I + \lambda Q^t Q)^{-1} f \qquad (14)$$

where I is the identity, and λ is a parameter which is determined such that $P_4 f$ satisfies the bound of Eq. (12) with equality. The choice of the operator Q dictates the nature of this projection. For example, when Q is a high-pass operator this projection in Eq. (14) results in low-pass filtering of the image f. The projections onto S_2 and S_3 are special cases of P_4. However,

P_4 cannot be found in a closed form in general because the parameter λ has to be found by numerically solving a nonlinear equation.

Other sets that can be useful for this problem are constraint sets that capture known information about the range of the values of the DCT coefficients and the intensity of pixels. These sets depend on the quantization strategies. For the rest of this paper we shall denote these sets as S_5 and the projectors onto them by P_5. These sets have been previously examined in other image recovery problems [6], [10].

4. RECONSTRUCTION ALGORITHMS

Using the previous notation, the first iterative reconstruction algorithm directly follows from the POCS theory. It is described by

$$f^{(k+1)} = P_1 P_2 P_3 P_5 f^{(k)} \qquad (15)$$

where the index k is the iteration index and P_i the projector onto set S_i. The convergence of these algorithms is guaranteed as long as the intersection of the sets S_i is not empty [5].

The second proposed iterative algorithm is motivated only from the theory of POCS. This algorithm is defined by the iteration

$$\begin{aligned} f^{(k+1)} &= B^t(Bf_0 + I_c B(I + \lambda Q^t Q)^{-1} f'^{(k)}) \\ f'^{(k)} &= P_5 f^{(k)} \end{aligned} \qquad (16)$$

where I_c is the diagonal indicator matrix of the DCT coefficients that are not known and $\mathbf{f}_0$ the blocky image. In this algorithm the parameter λ is kept unchanged from iteration to iteration. This iteration represents a projection algorithm onto the convex sets S_1, S_4 and S_5. However, since after each iteration $f^{(k)}$ changes and λ is unchanged, this algorithm projects onto a smoothness set S_4 that *changes* from iteration to iteration. Therefore, the mathematical theory of POCS cannot guarantee the convergence of this iteration [5]. However it can be shown that under some mild conditions on Q and P_5 iteration (16) is a *contractive mapping*, therefore its convergence is guaranteed [11].

5. EXPERIMENTAL RESULTS AND CONCLUSIONS

The proposed iterative reconstruction algorithms have been tested thoroughly on different DCT based compression approaches for a number of 256×256 and 512×512 images. Due to space limitations, we present in the following only a few results with the 256×256

"Lena" as the test image. For easy comparisons, the center 64×64 section of each 256×256 image is shown.

As an objective measure of the quality of the reconstructed images, the Peak Signal to Noise Ratio (PSNR) is used. For $N \times N$ images with graylevel in the range of $(0, 255)$, the PSNR is defined in DB as

$$PSNR = 10 \cdot log_{10} \left[\frac{N^2 \times 255^2}{\|\mathbf{f} - \mathbf{f_r}\|^2} \right],$$

where $\mathbf{f}$ and $\mathbf{f_r}$ are the reconstructed and the reference images respectively.

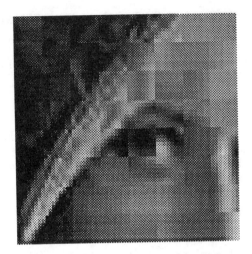

Fig. 1 the blocky "Lena", PSNR=27.9dB

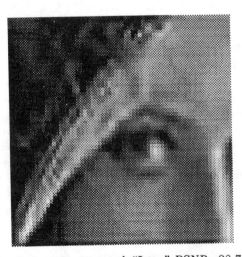

Fig. 2 the reconstructed "Lena", PSNR=28.7dB

Figure 1 shows the blocky image obtained by direct reconstruction from the BDCT coefficients from Partition Priority Coding(PPC), see [12]. The corresponding PSNR is 27.9dB. The blocking artifact is rather obvious and annoying. In Figure 2, the reconstructed

image using the first algorithm is given. The corresponding PSNR is 28.7dB. Clearly, the blocking artifact is nearly completely removed. Furthermore, it is clear that the quality of the reconstructed image is improved both subjectively and objectively using the PSNR measure.

From experiments we also observed that the convergence of the first iterative algorithm is rather quick, typically $5 \sim 15$ iterations. The image in Figure 2 is the result after 10 iterations. For the second class of iterative algorithms, our experiments show that convergence is relatively slower. However for certain choices of the operator Q, they may yield images with higher PSNR than that of the first algorithm.

REFERENCES

[1] : K. R. Rao, and P. Yip, "Discrete Cosine Transform: Algorithms, Advantages, Applications," *Academic Press,* 1990.

[2] : H. C. Reeve and J. S. Lim, "Reduction of blocking effects in image coding," *Optical Engineering,* vol. 23 no. 1, Jan/Feb 1984.

[3] : B. Ramamurthi and A. Gersho, "Nonlinear space-variant postprocessing of block coded images," *IEEE Trans. on Acoustics Speech and Signal Processing,* vol. 34, no. 5, October 1986.

[4] : A. Zakhor, "Iterative procedures for reduction of blocking effects in transform image coding", *IEEE Trans. on Circuits and Systems for Video Tech.* vol. 2, no. 1, March 1992.

[5] : L. M. Bregman, "Finding the common point of convex sets by the method of successive projection," *Dokl. Akad. Nauk SSSR,* vol. 162, pp. 487-490, 1965.

[6] : H. Stark, ed., "Image Recovery: Theory and Application," *Prentice Hall,* 1987.

[7] : H. Musmann, P. Pirsch, H. Grallert, "Advances in picture coding", *Proc. IEEE,* April,1985

[8] : Y. Yang, N. P. Galatsanos and A. K. Katsaggelos, "Regularized Reconstruction to Reduce Blocking Artifacts of Block Discrete Cosine Transform Compressed Images", *IIT-Technical Report,* IIT-ECE-TR-11-92.

[9] : A. K. Katsaggelos, "Iterative image restoration algorithms," *Optical Engineering,* Vol. 28, no. 7, pp. 735-748, July 1989.

[10] : H. Stark and T. Olsen, "Projection based image restoration," *Journal of the Optical Society of America A.,* Nov. 1992.

[11] : J. M. Ortega and W.C. Rheindoldt, "Iterative Solution of Nonlinear Equations in Several Variables", *Academic Press,* 1970.

[12] : Y. Huang, H. M. Dreizen and N. P. Galatsanos, "Prioritized DCT for Compression and Transmission of Images", *IEEE trans. on Image Processing,,* Vol. 1. No. 4, Oct. 1992.

A New Approach to Decoding and Compositing Motion-Compensated DCT-Based Images

Shih-Fu Chang and David G. Messerschmitt

Dept. of EECS, University of California, Berkeley, CA 94720

Abstract

Multi-point network video services require compositing several video sources into a single displayed video stream. Compositing video directly in the compressed domain can save computations by processing less data and avoiding the conversion process back and forth between the compressed and uncompressed data formats. Earlier work has demonstrated the computational speedup of compositing DCT-compressed video directly in the DCT domain, compared to the straightforward spatial-domain approach. Typical compositing operations include overlapping, scaling, translation, filtering, etc. In this paper, we propose a new decoding algorithm for MC-DCT compressed video which converts MC-DCT compressed video to the DCT domain and enables video compositing in the DCT compressed domain. Computational complexity is analyzed. The idea of network compositing and its impacts on multimedia network implementations are also discussed.

1. Introduction

Advanced networked video services require real-time video compression, compositing and transmission. Examples of such services are multi-point video conferencing and interactive networked video. Multiple video sources are compressed, transmitted through the network, and composited into a single displayed video stream. The compositing hardware can be located at the user site or an intermediate node (within or outside the network). In the latter case, the composited video needs to be compressed again for further transmission. It is our goal to design compositing algorithms directly in the compressed domain.

Many video compression methods, like H.261 and MPEG, include both the Discrete Cosine Transform (DCT) and the Motion Compensation (MC) algorithms, in which the former removes the spatial redundancy and the latter removes the temporal dependance between frames. It has been shown that compositing the DCT-compressed images directly in the DCT domain can save computations for many compositing operations (e.g. overlapping, scaling, translation and linear filtering), compared to the straightforward spatial-domain approach which composites images pixel by pixel [2,3]. However, obstacles exist preventing image compositing in the MC domain [1]. An example obstacle is that the reference image blocks of the background image in an overlapping scene could be replaced by the foreground image. Video sequences need to be inverse-motion-compensated before composition, and may need to be re-motion-

compensated afterwards. This may make real-time high-speed implementations difficult. In [1], we propose some heuristics for calculating new motion vectors of the composited video based on those of the original input video. Thus, the most computation-intensive operation — motion measurement, can be avoided with little performance degradation. For MC-DCT-compressed video, the DCT algorithm is applied on the prediction errors of the MC algorithm. The traditional decoder decodes DCT first and then computes inverse MC. Due to the above mentioned obstacle of compositing video in the MC domain, this decoding algorithm requires full decompression of the MC-DCT algorithm all the way back to the spatial domain and compositing in the spatial domain.

To reduce the computational complexity, we propose a new decoding algorithm for MC-DCT-based video, which performs the MC and inverse MC algorithms in the DCT domain rather than in the spatial domain. We then use this new algorithm to convert all MC-DCT-based input images to the DCT domain and use efficient algorithms proposed in [2] to composite them in the DCT domain. Another application of the proposed decoding algorithm is efficient image format conversion between different compression standards, like JPEG and MPEG. Note if the input video is further compressed with the Huffman code like that in most video compression standards, the front-stage Huffman decoder is still necessary.

2. New Decoding Algorithms for MC-DCT Compressed Video

Figure 1(a) shows the block diagram for the traditional decoder for the MC-DCT compressed video. The operations can be described as

$$P_{rec}(t, x, y) = DCT^{-1}(DCT(e(t, x, y))) + P_{rec}(t\text{-}1, x\text{-}d_x, y\text{-}d_y) \qquad (EQ\ 1)$$

where P_{rec} is reconstructed video signal, e is the prediction error, and d is the motion vector. With simple reordering, we can change it to

$$DCT(P_{rec}(t, x, y)) = DCT(e(t, x, y)) + DCT(P_{rec}(t\text{-}1, x\text{-}d_x, y\text{-}d_y)). \qquad (EQ\ 2)$$

Namely, we switch the order of the inverse DCT and the inverse MC, and perform the inverse MC algorithm in the DCT domain. The new decoder block diagram is shown in figure 1(b). However, one technical issue is that the MC algorithm can have a motion vector of arbitrary number of pixels, while in the DCT algorithm,

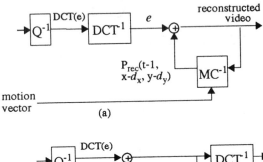

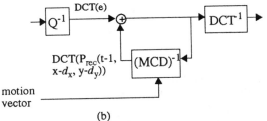

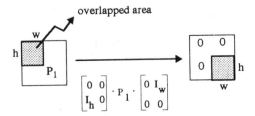

Fig. 3. Use simple matrix multiplications to perform windowing and shifting operations on image blocks.

Fig. 1. (a) Traditional decoders for MC-DCT-based videos. (b)A new decoding algorithm, in which the inverse MC algorithm is performed in the DCT domain. MCD: MC in the DCT domain.

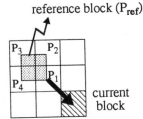

Fig. 2. The MC optimal reference block may overlap with four neighboring blocks in the DCT block structure.

all coefficients are grouped into a block-based format. The optimal reference image block may overlap with four neighboring blocks in the DCT block structure, as shown in figure 2. Kou [4] proposed an algorithm to compute DCT coefficients of a new block from two adjacent blocks, but it becomes very complicated when the overlap length is not equal to half of the block width. In [2], we have designed an algorithm to calculate the DCT coefficients of a new arbitrary-position image block directly from the DCT coefficients of four original neighboring blocks. The operations include multiplications with pre-matrices and post-matrices

$$DCT(P_{ref}) = \sum_{i=1}^{4} DCT(H_{i1})DCT(P_i)DCT(H_{i2})$$

(EQ 3)

where P_i are original neighboring image blocks and H_{ij} are special sparse matrices like

$$\begin{bmatrix} 0 & 0 \\ I_h & 0 \end{bmatrix} \text{ or } \begin{bmatrix} 0 & I_w \\ 0 & 0 \end{bmatrix}$$

whose DCT coefficients can be pre-computed and stored in the memory (h and w are overlap width). The net effects of matrix multiplication with H_{ij} are combination of *windowing* and *shift-*

ing. As shown in figure 3, the upper-left corner of P_1 is extracted (windowing) and shifted to the lower-right corner of the desired block, P_{ref}. The four summation components of equation 3 account for contributions from four original overlapped blocks, whose DCT coefficients are known. Due to the distributive property of matrix multiplication to DCT, we can perform the windowing and shifting operations directly in the DCT domain, as shown in equation 3. Therefore, the DCT coefficients of the arbitrary-position optimal reference block, i.e. $DCT(P_{rec}(t-1, x-d_x, y-d_y))$ can be calculated directly in the DCT domain. The DCT coefficients of the current frame can be obtained by adding the DCT of the prediction errors to the DCT of the reference block. The whole inverse MC in the DCT domain is represented by the MCD^{-1} block in figure 1. Fully decoded video can be obtained after applying the inverse DCT. Note, besides the numerical round-off errors, the proposed decoding approach is mathematically equivalent to the traditional decoding approach.

The required computations of equation 3 can be greatly reduced when many DCT coefficients of P_i are zero, which can be indicated by the run length code (RLC). Further, if the motion vectors are zero or integer multiples of the block width, the above block adjustment procedure can be avoided, and the inverse MC procedure requires simple additions only. Therefore, the computational complexity of the proposed DCT-domain inverse MC algorithm increases with the percentage of the non-zero motion vectors. We will discuss the complexity further in the next section. The storage overhead for storing the pre-computed DCT coefficients of H_{ij} matrices is very small. For a block size of 8✕8, a memory of 896 words (one word for each DCT coefficient) is needed.

One immediate application of this new decoding algorithm is compression format conversion, e.g. from MPEG to JPEG. The DCT coefficients of every frame can be obtained by applying the DCT-domain inverse MC directly. The conversion process back and forth between the DCT domain and the spatial domain is avoided. Another application is to composite images in the DCT domain, which will be described in the next section.

3. Compositing Networked Video in the DCT Domain

In a multi-point networked video service like video conference, multiple video sources can be composited within the network. The compositing unit takes the compressed video inputs, composites them, and then produces the compressed composited video output. The traditional straightforward approach reconstructs all images fully back to the spatial domain and composite

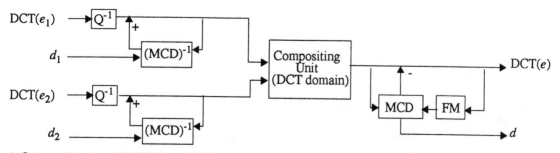

Fig. 4. Compositing two MC-DCT-based video sequences in the DCT domain. We convert input sequences to the DCT compressed domain, composite them in the DCT domain, and then convert the composited video to the original compressed format (i.e. the MC-DCT format).

them pixel by pixel. Our proposed method is to use techniques described in the previous section to convert images to the DCT domain and then composite them in the DCT domain, as shown in figure 4. The DCT and inverse DCT blocks in the traditional approach are removed, and the compositing unit needs to process less data in the DCT domain than the spatial domain. Note that in order to encode the composited video to the original full MC-DCT compressed format, the composited video still needs to be motion-compensated, as shown in the MCD block in figure 4. To prevent this block becoming the most computation-dominant in the whole compositing system, we can infer new motion vectors directly from those of original input images [1]. The required computations for motion measurement can thus be greatly reduced at the cost of little compression performance.

As mentioned above, the computational complexity for the proposed decoding algorithm (including the DCT-domain MC^{-1} only) depends on both the non-zero percentages of the motion vector and the DCT coefficients. We compare this complexity to that for the traditional decoding algorithm, which includes a fast DCT followed by an inverse MC in the spatial domain. We use the fast DCT proposed by Chen *et al* [8] in table 1. Other fast algorithms for DCT have similar or larger complexity [9]. In table 1, N is the block size, β represents the reciprocal

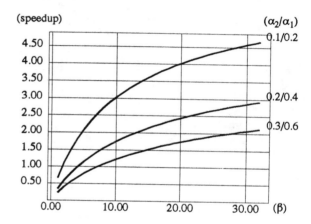

Fig. 5. Computational speedup by the proposed decoder (i.e. DCT-domain MC^{-1}), compared to the traditional full decoder for MC-DCT video.

Table 1: Comparison of computational complexity between the traditional full decoder and the proposed decoder for MC-DCT video.

	total number of real multiplications and additions
traditional decoder	$2 \cdot \log_2 N + 8/N - 3$ (mul.) $3 \cdot (\log_2 N - 1) + 4/N + 1$ (add.)
proposed decoder (DCT-domain MC^{-1} only)	$(4/\beta + 2/\sqrt{\beta}) \cdot N \cdot \alpha_2 + (2/\beta) \cdot N \cdot \alpha_1$ (mul.) $[(4/\beta + 2/\sqrt{\beta}) \cdot N + 3] \cdot \alpha_2 +$ $[(2/\beta) \cdot N + 1] \cdot \alpha_1 + 1$ (add.)

of the non-zero DCT coefficient percentage, α_2 represents the percentage of motion vectors which needs block alignment in both directions, and α_1 represents the percentage of motion vectors which requires block alignment in one direction only. We use efficient methods for calculating sparse matrix multiplication in equation 3. As we can see, the most significant computation

component for the proposed decoding algorithm is proportional to α_2, which indicates the frequency of full computation of equation 3. In figure 5, we illustrate the speedup of using the proposed decoding algorithm vs. the traditional spatial-domain decoder for some different compression parameters. For a typical β value of about 10, the speedup ranges between 1.2 and 3.0 for low-motion video sequences, e.g. head-and-shoulder video.

However, for high-motion video sequences (i.e. high α_1/α_2 values), the DCT-domain approach could become more complicated than the spatial-domain approach. This variable throughput will become an obstacle for real-time implementation which should be able to handle the worst-case situation. Fortunately, in the DCT domain, the compositing unit has the flexibility to ignore some high-order DCT coefficients during the worst case with less image quality degradation. This freedom of ignoring low-priority data is not available in the spatial domain. However, its effect on the recovered image quality needs to be investigated.

Besides the needs for compressed-domain compositing, compositing within the network has other impacts on designs of multimedia networks. The transmission cost can be reduced if multiple video sources are composited together before they are transmitted to the end users. Composited scenes can be shared among different users who subscribe the same service. Possible drawbacks for compositing video within the network are accumulation of quantization errors after repeated compression and

Fig. 6. A typical compositing scenario in video conferencing. The input images are scaled down with different ratios and translated to different positions. The image shown is reconstructed from the compressed output which is composited in the DCT domain. The input images are compressed with MC+DCT.

decompression, and increased latency for transmitting video to end users. Our experiments show that accumulation errors depends on the compositing operations performed, for example, more severe degradation for scaling, ignorable degradation for translation and overlapping.

We have simulated the DCT-domain compositing algorithms for different compositing scenarios and quality levels. For the compositing scenario shown in figure 6, our proposed DCT-domain algorithm can save the computations by 24%, compared to the straightforward spatial-domain approach. If the compression is without the MC algorithm (like JPEG), the reduction of computations is about 73%. The reconstructed images do not show clear subjective difference between our compressed-domain approach and the spatial-domain approach. (The PSNR is 42.8 dB vs. 43.2 dB.) The non-zero DCT coefficient percentage varies between 5%~22%, and the non-zero motion vector percentage ranges between 12%~90% in our simulations. The speedup can be increased if we raise the compression ratio by sacrificing some image quality.

4. Conclusions and Future Work

We design a new decoding algorithm for the MC-DCT based video, which performs inverse MC before inverse DCT. This algorithm can be applied in compositing compressed video within the network, which may take multiple compressed video sources and combine them into a single compressed output stream. The proposed algorithm converts all MC-DCT compressed video into the DCT domain and performs compositing in the DCT domain. This DCT-domain approach can reduce the required computations with a speedup factor depending on the compression ratio and the non-zero motion vector percentage. However, dropping some least-significant DCT coefficients maybe necessary for the worst case of high-motion video in real-time implementations.

Some issues of networked video compositing are also discussed. Another direct application of the proposed decoding algorithm is converting MC-DCT compressed video to the DCT compressed format directly in the DCT domain.

5. References

[1] S.-F. Chang and D.G. Messerschmitt, "Compositing Motion-Compensated Video within the Networks," IEEE 4th Workshop on Multimedia Communications, Monterey, CA, April, 1992.

[2] S.-F. Chang, W.-L. Chen, and D.G. Messerschmitt, "Video Compositing in the DCT Domain," IEEE Intern. Workshop on Visual Signal Processing and Communications, Raleigh, North Carolina, September, 1992.

[3] B.C. Smith and L. Rowe, "A New Family of Algorithms for Manipulating Compressed Images," Submitted to IEEE Computer Graphics and Applications.

[4] J.B. Lee and B.G. Lee, "Transform Domain Filtering Based on Pipelined Structure," IEEE Trans. on Signal Processing, pp.2061-4, Vol. 40, No. 8, Aug. 1992.

[5] W. Kou and T. Fjallbrant, "A Direct Computation of DCT Coefficients for a Signal Block from Two Adjacent Blocks," IEEE Trans. on Signal Processing, Vol. 39, No. 7, pp. 1692-5, July 1991.

[6] CCITT Recommendation H.261, "Video Codec for Audiovisual Services at pX64 kbits/s"

[7] Standard Draft, MPEG Video Committee Draft, MPEG 90/176 Rev. 2, Dec. 1990.

[8] W.-H. Chen, C.H. Smith, and S.C. Fralick, "A Fast Algorithm for the Discrete Cosine Transform," IEEE Trans. on Communications, Vol. COM-25, No. 9, Sep. 1977, pp. 1004-9.

[9] B.G. Lee, "FCT-A Fast Cosine Transform," IEEE ICASSP '84, San Diego, pp. 28A.3.1-4, March, 1984.

OBJECT–ORIENTED ANALYSIS–SYNTHESIS CODING BASED ON SOURCE MODELS OF MOVING 2D– AND 3D–OBJECTS

Hans Georg Musmann

Universität Hannover
Germany

ABSTRACT

For encoding moving images at very low bit rates object–oriented analysis–synthesis coding using source models of moving 2D– and 3D–objects has been investigated. According to this coding concept each moving object of an image is described and encoded by three parameter sets defining the motion, the shape and the surface colour of the moving object.

The parameters to be coded are dependent on the source model being applied. Thus, the coding efficiency of object–oriented analysis–synthesis coding can be evaluated by comparing the encoded parameter bit rates for a fixed picture quality. The results obtained with source models of moving 2D– and 3D–objects are compared to those of a standard H.261 coder at a bit rate of 64 kbit/s.

The source models of flexible 2D– and rigid 3D–objects are more efficient than source models which are based on rigid 2D–objects or on moving blocks as in the case of a H.261 coder.

1. INTRODUCTION

In order to encode moving video signals at 64 kbit/s transmission rate the standardized hybrid coding technique H.261 subdivides each image of a sequence into blocks of 16 x 16 picture elements (pels) and encodes each block by motion compensated predictive and transform coding [1]. Thus this coding technique is based on the source model of moving square blocks which can lead to visible distortions known as blocking and mosquito effects. To avoid these image distortions object–oriented analysis–synthesis coding [2] has been proposed. This coding technique is based on the source model of moving objects instead of moving square blocks. Each object is described and encoded by three parameter sets defining the motion, shape and surface colour of the object.

The parameter sets to be coded vary with the kind of object model. Three kinds of object models have been invetigated, rigid 2D–objects [3], flexible 2D–objects [4] and rigid 3D–objects [5]. In order to evaluate the coding efficiency of object–oriented analysis–synthesis coding the encoded parameter bit rates of the cited source models are compared at a fixed picture quality measured by the signal–to–noise ratio.

After a short explanation of the basic components of an object–oriented analysis–synthesis coder in section 2 the coding of the parameter sets is described in section 3. Finally the influence of the various source models on the coding efficiency is discussed in section 4.

2. GENERAL STRUCTURE OF AN OBJECT–ORIENTED ANALYSIS–SYNTHESIS CODER

The block diagram of an object oriented analysis–synthesis coder is shown in Fig.1. Instead of a frame memory as in block–oriented coding techniques object–oriented coding requires a memory to store the parameter sets $A = \{A_i\}$, $M = \{M_i\}$, $S = \{S_i\}$ defining the motion A_i, shape M_i and colour S_i of each moving object i. The memory for object parameters of the coder and decoder contains the same parameter informations and allows the coder and decoder to reconstruct a transmitted image by image synthesis. The reconstructed image I'_k is displayed at the decoder and used for image analysis of the next input image I_{k+1} at the coder.

The task of the image analysis block in Fig.1 is to estimate the parameter sets A_i, M_i. S_i for each object i of the next input image I_{k+1} by use of the reconstructed image I'_k. At the output of the image analyis block the parameter sets A_i, M_i, S_i are available in PCM representation.

Image analysis distinguishes between two types of objects. Model Compliance objects (MC–objects) can be reconstructed by transmitting only the motion A_i and shape M_i parameter sets and using the colour S_i parameter set stored in the memory. Model failure objects (MF–objects) are image areas where the motion description by the source model fails. MF–objects are encoded by the shape M_i and colour S_i parameter sets, see Table 1.

	MC–Object	MF–Object
Motion Ai	A_i	
Shape M_i	M_i	M_i
Colour S_i		S_i

Table 1. Parameter sets transmitted for MC– and MF–objects

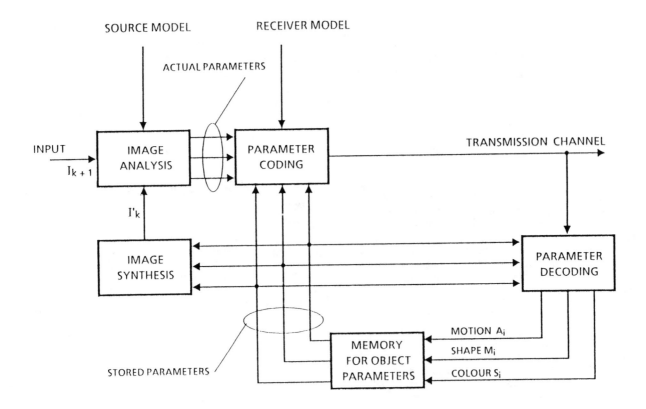

Fig. 1. Block diagram of an object–oriented analysis–synthesis coder

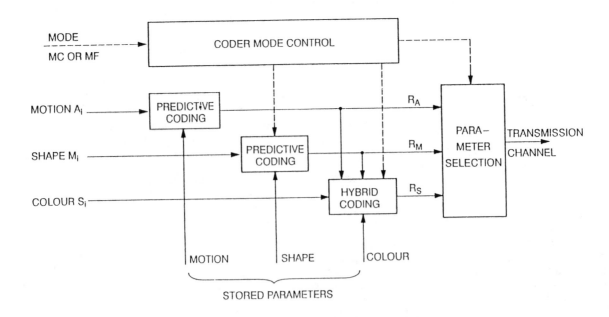

Fig. 2. Block diagram of the parameter coder

The actual parameter sets A_i, M_i, S_i of image I_{k+1} are encoded in the block parameter coding. For efficient coding known predictive and transform coding techniques are applied which make use of the stored parameter sets [4], [5], see Fig.2. Coder mode control decides with help of the MC/MF information which parameter sets of each object will be transmitted.

3. PARAMETER SETS AND PARAMETER CODING

The parameter sets to be coded depend on the source model as indicated in Table 2. In order to achieve an efficient coding of the parameter sets individual coding techniques are applied for each parameter set.

Motion parameter sets of MC–objects are predictively encoded using known DPCM techniques [3], [5]. In the case of rigid 2D–objects threedimensional motion is assumed. Therefore 8 mapping parameters $a_1,..a_8$ describing the motion have to be encoded for each MC–object [2]. In the case of flexible 2D–objects twodimensional translatory motion of each block of an MC–object is assumed [4]. The motion of 3D–objects is described by 3 translatory motion a_1, a_2, a_3 and 3 rotatory motion parameters a_4, a_5, a_6. The shape parameters of 2D–objects describe the silhouette of a MC–object by polygon and spline approximation. The vertices of this representation are predictively encoded using the actual motion and the stored shape information [3], [6]. The shape of 3D–objects is obtained only from silhouette and motion information [8]. Therefore only silhouette information has to be encoded. The accuracy of the shape approximation has to be adapted to the spatial image resolution and the camera aperture. The maximum allowable deviation of the shape approximation from the true shape contour is about 1.5 pel in order to avoid visible distortion at object boundaries.

There are less restrictions for the representation of the boundaries of MF–objects. The maximum deviation should be about 2.1 pel otherwise the bit rate for encoding the colour information of MF–objects may increase too much [4].

The colour information of MF–objects is coded by a block–oriented hybrid scheme where for each block of 8 x 8 pel a DPCM techique is used instead of a DCT whenever it allows a more efficient coding [7].

4. CODING EFFICIENCY

In order to compare the coding efficiency of object–oriented analysis–synthesis coders with different source models the bit rates R_A, R_M and R_S required for encoding the motion, shape and colour parameter sets have to be analysed. Table 3 shows the average bit rates R_A, R_M and R_S per image measured for the CCITT test sequences "Mrs. America" and "Claire" using a spatial resolution of 288 lines with 352 pel per line and 10 Hz frame frequency.

The bit rate R_M required for encoding the motion parameters of rigid objects is less than that for flexible objects, since many displacement vectors $\{a_1, a_2\}$ have to be encoded for only one MC–object.

The bit rate R_M consists of two parts, one part for MC– the other for MF–objects. R_M of rigid 2D–objects is greater than that for flexible 2D–objects since the source model of rigid 2D–objects fails more often and therefore generates more MF–objects. This can also be recognized from the number of pels N of the MF–objects to be coded. The bit rate R_S for encoding the colour parameters $\{S_1,..S_N\}$ results from the number of pels N to be coded multiplied by the bit rate r_s allocated for encoding each pel

$$R_S = N \cdot r_s \qquad (1)$$

With the choice of r_s the picture quality can be fixed. Thus N is a measure for the transmission rate required for coding the colour parameters and can be used for comparing the coding efficiency of different source models.

Table 3 shows that the two source models of 2D–objects generate almost the same bit rate $R_A + R_M$. Since N is much greater for the source model of rigid 2D–object the source model of flexible 2D–objects provides a higher coding efficiency.

Comparing the model of flexible 2D–objects with that of rigid 3D–objects the bit rate R_A is smaller for rigid 3D–objects since the number of motion parameters is less. The shape information of MC–objects is the same for flexible 2D– and rigid 3D–objects since both are encoded by silhouettes. However, the shape information required for encoding the MF–objects is greater in the case of rigid 3D–objects since the total area N of MF–objects is split into more MF–objects of smaller size. Thus R_M is higher for rigid 3D–objects although N is the same.

To achieve an acceptable picture quality r_s should be greater than 1 bit per pel. Thus the source model of flexible 2D– and rigid 3D–objects allow coding with a total bit rate of 64 kbit/s.

In a block oriented H.261 coder N is much greater in the range of 40000 pels .

5. CONCLUSION

The coding efficiency of object–oriented analysis–synthesis coders based on different source models is evaluated. At a given picture quality the bit rates

$$R = R_A + R_M + R_S \qquad (2)$$

required for encoding the parameter sets for motion R_A, shape R_M and colour R_S are compared. It has been shown that there is only a relatively low degree of freedom for encoding the motion and shape parameters so that $R_A + R_M$ is almost fixed. R_S is linearly dependent on the number N of pels in the areas of model failure at a given picture quality.

The source models of flexible 2D–objects and rigid 3D–objects are almost equivalent and superior to those of rigid 2D–objects and of moving 2D–blocks as in the case of a H.261 coder.

To improve the coding efficiency the source model of flexible 3D–objects has to be introduced [9].

	MC-object		MF-object	
	Motion A_i	Shape M_i	Colour S_i	Shape M_i
Rigid 2D–objects 3D–motion	8 mapping parameters $a_1, a_2, a_3, a_4,$ a_5, a_6, a_7, a_8	vertices of the object silhouette $\{ m_1, m_2 \}$	each pel of a MF–object $\{ s_1, ... s_N \}$	vertices of the boundary of the MF–area $\{ m_1, m_2 \}$
Flexible 2D–objects 2D–motion	2 translatory parameters for each block of an object $\{ a_1, a_2 \}$	vertices of the object silhouette $\{ m_1, m_2 \}$	each pel of a MF–object $\{ s_1, ... s_N \}$	vertices of the boundary of the MF–area $\{ m_1, m_2 \}$
Rigid 3D–objects 3D–motion	3 translatory and 3 rotary parameters $a_1, a_2, a_3, a_4, a_5, a_6$	vertices of the object silhouette $\{ m_1, m_2 \}$	each pel of a MF–object $\{ s_1, ... s_N \}$	vertices of the boundary of the MF–area $\{ m_1, m_2 \}$

Table 2. Parameter sets to be coded per object for various source models

Source model	Motion information R_A	Shape information R_M	Colour information $R_S = N \cdot r_s$
Rigid 2D–objects 3D–motion	600	1 300	$15\,000 \cdot r_s$
Flexible 2D–objects 2D–motion	1 100	900	$4\,000 \cdot r_s$
Rigid 3D–objects 3D–motion	200	1 640	$4\,000 \cdot r_s$

Table 3. Average bit rates R_A, R_M, R_S in bit per image for various source models, N is the totel number of pels of the MF–objects, r_s is the bit rate allocated per pel

REFERENCES

[1] CCITT, "Draft revision of recommendation H.261: video codec for audio visual services at p x 64 kbit/s", Study Group XV, WP/1/Q4, Specialist Group on Coding for Visual Telephony, Cambridge, USA, November 1988

[2] H. G. Musmann, M. Hötter and J. Ostermann, "Object–oriented analysis–synthesis coding of moving images", Signal Processing: Image Communications, Vol.1, No.2, pp. 117–138, October 1989

[3] M. Hötter, "Object–oriented analysis–synthesis coding based on moving two–dimensional objects", Signal Processing: Image Communications, Vol.2, No.4, pp. 409–428, December 1990

[4] M. Hötter, "Optimization and efficiency of an object–oriented analysis–synthesis coder", IEEE Trans. on Circuits and Systems for Video Technology. Accepted for publication

[5] J. Ostermann, "Object–oriented analysis–synthesis coding based on the source model of moving rigid 3D–objects", submitted for publication to Signal Processing: Image Communications

[6] M. Hötter, "Predictive contour coding for an object–oriented analysis–synthesis coder, IEEE International Symposium on Information Theory, San Diego, Californiia, USA, p. 75, January 1990

[7] H. Schiller, M. Hötter, "Colour coding in an object–oriented analysis–synthesis coder", submitted for publication in Signal Processing: Image Communication

[8] J. Ostermann, "Modelling of 3D–moving objects for an analysis–synthesis coder", SPIE/SPSE Symposium on Sensing and Reconstruction of 3D Objects and Scenes '90, B. Girod, Editor, Proc. SPIE 1260, pp.240–250, Santa Clara, California, February 1990

[9] J. Ostermann, "Analysis–synthesis coder based on moving flexible 3D–objects", accepted for publication at Picture Coding Symposium PCS 93, Lausanne, Switzerland, March 1993

MOTION ESTIMATION WITH WAVELET TRANSFORM AND THE APPLICATION TO MOTION COMPENSATED INTERPOLATION

Cha Keon Cheong[†]   Kiyoharu Aizawa[†]   Takahiro Saito[‡] and Mitsutoshi Hatori[†]

[†] Dept. of Electrical Engineering
The University of Tokyo, 7–3–1,
Hongo, Bunkyo–Ku, Tokyo, 113 JAPAN

[‡] Dept. of Electrical Engineering
The University of Kanagawa, 3–27–1,
Kanagawa–Ku, Yokohama, 221 JAPAN

Abstract

In order to estimate highly reliable motion for image sequences, a novel segment-based motion estimation is described. Segmentation is performed based on zero-crossings of the wavelet transform. Moving objects are extracted from the area of motion using the segmentation result. Multiresolution image decomposition is performed with the biorthogonal wavelet transform, and then motion is hierarchically estimated using the segmentation result. Moving objects and covered/uncovered areas are identified with the segmentation result and the estimated motion information. Finally, the application to motion compensated interpolation to the above is described.

1 INTRODUCTION

In order to accurately estimate motion, many algorithms have been proposed and tested. These algorithms can be classified into two main approaches: matching algorithms and spatiotemporal gradient-based methods [1]. Although conventional block matching methods can estimate motion using relatively simple calculations, they are inadequate in estimating real motion if objects in the same block move in different directions or if block boundaries and object boundaries do not coincide. The gradient-based method works quite well if the magnitude of the motion is not too great; however, using this method it is difficult to estimate global motion, and the algorithm tends to converge slowly and is sensitive to noise.

To solve the problem of both local and global motion estimation, hierarchical motion estimation approaches have been proposed [2–4]. It has been shown that hierarchical approaches give better estimation results than the simple block matching or gradient-based methods. However, the inter-dependency of layers is a drawback of the hierarchical motion estimation technique. If the motion estimation in a coarser layer is incorrect, it is almost impossible to overcome the mis-

leading estimation in a finer layer.

Since real images often contain many different features and these features can have very complex spatiotemporal relationships, an improvement in motion estimation can be obtained if the estimation is based on segmentation according to the real image features and/or high frequency information [5].

This paper describes a new segment-based motion estimation algorithm and its application to motion compensated interpolation as shown in Fig. 1. In the proposed method, the extraction of image features and segmentation is performed using zero-crossings of the wavelet transform, motion is hierarchically estimated based on the segmentation using a block matching method, and the identification of moving objects and covered/uncovered areas is obtained using the segmentation result and the motion information.

2 SEGMENTATION AND MOTION ESTIMATION

2.1 Zero-Crossings of a wavelet transform

If a wavelet function $\psi_s(x)$ is a second derivative of a smoothing function $\theta_s(x)$, then the zero-crossings of the wavelet transform $W_s(x)$ correspond to the local variation points of the signal $f(x)$ [6]. In an image signal, let $I(x, y)$ be the representation of the image, and let $\theta(x, y)$ be the smoothing function which is used to smooth the image. In the case of a separable smoothing function $\theta(x, y) = \theta(x)\theta(y)$, two wavelet functions are given by

$$\psi_{s,H}(x,y) = s^2 \frac{\partial^2}{\partial x^2} \theta_s(x,y) \quad (1)$$
$$= \psi_s(x)\theta_s(y),$$
$$\psi_{s,V}(x,y) = s^2 \frac{\partial^2}{\partial y^2} \theta_s(x,y) \quad (2)$$
$$= \psi_s(y)\theta_s(x),$$

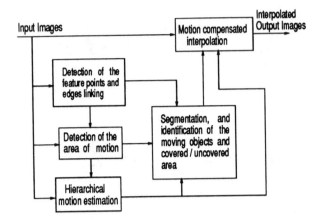

Figure 1: The proposed segment-based motion estimation and its application to MCI

and two wavelet transforms are given as follows for horizontal and vertical direction, respectively.

$$W_{s,H}(x,y) = I * \psi_{s,H}(x,y)$$
$$W_{s,V}(x,y) = I * \psi_{s,V}(x,y). \quad (3)$$

If the smoothing function is Gaussian, then detection of the zero-crossing points of the wavelet transform is equivalent to the extraction of edges with the Laplacian of Gaussian. In order to reduce the detection of false zero-crossing points caused by ripples of the wavelet filters, a short length smoothing function is derived using the B-spline function.

Although the extracted zero-crossing points represent features of the image, this set of points seldom characterize a object boundary completely because of noise and other effects that introduce spurious intensity discontinuities. Therefore, in order to assemble the zero-crossing points into meaningful object boundaries and obtain contour-based segmentation, the below edge linking and boundary following processure are applied.

1. Threshold the zero-crossing points with an appropriate window and a threshold value for horizontal and vertical, respectively.

2. Extract the edges from the resulting zero-crossings.

3. Link the disconnected edges based on minimum Hamming distance and luminance value.

4. Merge small regions into one large region based on minimum variance of the luminance value between object boundaries.

2.2 Detection of the Area of Motion

In order to estimate motion vectors only for moving objects, at first, two successive images I_{k-1} and I_k are separated into changed and unchanged regions. For this purpose, the absolute value of the frame difference is calculated using low-pass filtered images L_{k-1} and L_k of the two images. Then the resultant absolute value is compared with a threshold within a 3×3 window. The threshold value is chosen according to the variance of the noise of the images. Isolated pixels and small changed and unchanged regions are merged into their neighboring regions using median filtering and a labelling algorithm. The changed regions that are too small are neglected, and all unchanged regions lying completely inside a changed region are merged into that changed region. This detection of the area of motion is similar to the algorithm in [2].

In general, it can be assumed that edge boundaries are object boundaries. Therefore, in order to obtain the area of motion that is adapted to moving object, the changed regions are combined with the contour-based segmentation results obtained in Sec. 2.1.

2.3 Hierarchical Motion Estimation

In order to estimate both local and global motion simultaneously, a hierarchical block matching method is employed. For a given image sequence, multiresolution image decomposition is obtained by using a biorthogonal wavelet transform [8]. For low-pass filtered images, the motion vectors are estimated using a full search and the minimum mean square error (MMSE) variance with pixelwise accuracy. The motion vectors computed at a coarser layer in the multiresolution representation are used to initialize the search at a finer layer. Let $\mathbf{P}_j$ be the pixel of the image at resolution 2^{-j} where the motion vector $\mathbf{D}_j(\mathbf{P}_j)$ is to be estimated, and $\mathbf{P}_{j+1}$ the pixel of the image at resolution $2^{-(j+1)}$ where the motion vector $\mathbf{D}_{j+1}(\mathbf{P}_{j+1})$ has already been estimated. Then the estimation of the motion vector $\mathbf{D}_j(\mathbf{P}_j)$ is given by

$$\mathbf{D}_j(\mathbf{P}_j) = 2\mathbf{D}_{j+1}(\mathbf{P}_{j+1}) + \delta\mathbf{D}, \quad (4)$$

where $\delta\mathbf{D}$ is a small variation of the motion that has to be estimated at resolution 2^{-j}.

In the finest layer (original image), the motion vector is estimated only in the area of motion, and the motion vector of the unchanged regions that has already been estimated at a coarser layer is neglected. In the boundaries between the area of motion and the unchanged regions, blocksize is adjusted according to the number of pixels contained in the area of motion in the block so that local minimization of the mean square error variance due to small blocksize is prevented.

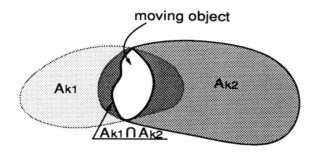

Figure 2: The identification of the moving object and uncovered/uncovered area

2.4 Identifying the Moving Objects

Identification of moving objects and covered /uncovered regions in the area of motion is performed using contour-based segmentation and estimated motion information. For this purpose, the area of motion in the image I_k is detected with three consecutive images I_{k-1}, I_k and I_{k+1} using the method of Sec. 2.2. Let A_{k1} be the area of motion for the two images I_{k-1} and I_k, and A_{k2} the area of motion for the two images I_k and I_{k+1}. The moving object is estimated using the common region between A_{k1} and A_{k2} and contour-based segmentation as shown in Fig. 2. If the error variance between the original image and the motion compensated one exceeds a given threshold, then the estimated moving object is neglected.

Secondly, a pixel ying within area A_{k1} but not lying within the moving object is assigned to the uncovered area, and a pixel lying within area A_{k2} but not lying within the moving object is identified as the covered area. Finally, a pixel lying in the common region between A_{k1} and A_{k2} that is not part of the moving object is identified as either uncovered or covered area with following procedure:

1. Calculate the average motion vector using a 3×3 window of the neighboring pixels.

2. Move the pixel in the direction of the motion vector calculated in step 1.

3. If the moving object is encountered, then the pixel is assigned to the uncovered area, otherwise the pixel is assigned to the covered area.

3 MOTION COMPENSATED INTERPOLATION (MCI)

Taking into consideration the estimated motion and segmentation results, the image $I_i(x, y)$ to be interpolated is calculated using a method to that proposed in [2] as follows:

$$
\begin{aligned}
I_i(x,y) =\ & \alpha I_k(x + (1 - t_i)dx, y + (1 - t_i)dy) \\
& + (1 - \alpha)I_{k-1}(x - t_i dx, y - t_i dy). \quad (5)
\end{aligned}
$$

where $t_i = \frac{\tau_i - t_{k-1}}{t_k - t_{k-1}}$, τ_i is time between t_{k-1} and t_k, the coefficient α is

$$
\alpha = \begin{cases}
0 & : \text{in the unchanged and covered area} \\
t_i & : \text{in the moving object} \\
1 & : \text{in the uncovered area,}
\end{cases}
$$

and dx and dy are estimated motion vectors. If the calculated pixel position is not an integer, then a bilinear interpolation filter is used.

4 SIMULATION RESULTS

Simulations were carried out on the real image sequence "table tennis" (352×240 pels). For the detection of zero-crossing points, the scale parameter $s = 2$ of the wavelet transform was used. Fig. 4 (a) and (b) show the results of contour-based segmentation using the edge linking and boundary following procedure for the original images of Fig. 3 (a) and (b), respectively. In order to prevent the detection of false edges, an authentic edge detection method [7] was applied. For each zero-crossing point, a 3×3 window and $5 \times (noise\ variance)$ of the threshold value were used.

For estimation of motion vectors, 2 multiresolution layers and ± 5 searching range for each layer were used. Blocksize for each layer was adjusted ranging from 8×8 to 16×16. Fig. 5 shows the results of the estimated motion flow.

The identified results of moving objects and covered/uncovered regions are shown in Fig. 6 (a) and (b). As an object moves from the previous to the next frame, covered and uncovered regions should exist in the present frame. In Fig. 6, these two regions and moving objects are shown, coded with the black for uncovered region, dark grey for unchanged region, light grey for covered region and white for moving object. With these results, high quality MCI images can be obtained.

5 CONCLUSION

In this paper, hierarchical estimation of the motion based on contour-based segmentation was described, and identification of covered and uncoverd regions was presented. The contour-based segmentation was performed using the zero-crossings of the wavelet transform and edge linking procedure. The main contributions of this study are that highly reliable motion estimation and successful identification of covered and uncovered regions can be obtained.

(a)   (b)

Figure 3: Original image sequences at time (a) k and (b) k+1

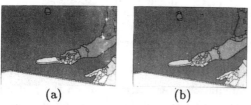

(a)   (b)

Figure 4: Contour-based segmentation for image sequences (a) k and (b) k+1

References

[1] E. Dubois and J. Konard "Review of techniques for motion estimation and motion compensation" *HDTV colloquim*, pp.3B3.1–3B3.19, Canada 1990.

[2] R. Thoma and M. Bierling "Motion compensating interpolation considering covered and uncovered background" *Signal processing: Image commun.*, vol.1, no.2, pp.192–212, 1989.

[3] C. Bergeron and E. Dubois "Gradient-based algorithm for block-oriented MAP estimation of motion compensated temporal interpolation" *IEEE Trans. on Circuits and Syst. for video technology*, vol.1, no.1, pp.72–85, March 1991.

[4] Y. Q. Zhang and S. Zafar "Motion-compensated wavelet transform coding for color video compression" *SPIE Visual commun. and Image processing'91: Visual commun.*, vol.1605, pp.301–316, Nov. 1991.

[5] N. Diehl "Object-oriented motion estimation and segmentation in image sequences" *Signal processing: Image commun.*, vol.3, no.1, pp.23–56, 1991.

[6] S. Mallat " Zero-crossings of a wavelet transform" *IEEE Trans. on Infor. Theory*, vol.37, no.4, pp.1019–1033, July 1991.

[7] J. J. Clark "Authenticating edge producing by zero-crossing algorithm" *IEEE trans. on PAMI*, vol.PAMI-11, no.1, p.43–57, Jan. 1989.

[8] C. K. Cheong, K. Aizawa, T. Saito and M. Hatori "Subband image coding with biorthogonal wavelets" *IEICE Trans. Fundament.*, vol.E75-A, no.7, pp.871–881, July 1992.

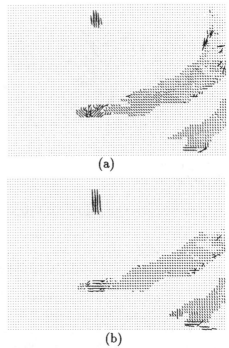

(a)

(b)

Figure 5: Estimated motion flow for image sequences (a) k and (b) k+1

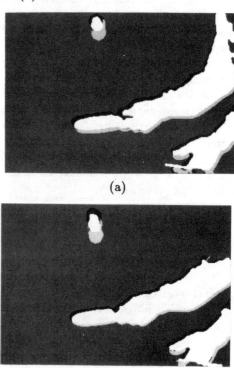

(a)

(b)

Figure 6: Identification results of moving object and covered/uncovered regions for image sequences (a) k and (b) k+1

Multiresolution Image Decomposition and Compression Using Mathematical Morphology

Ching-Han Hsu and C.-C. Jay Kuo
Signal and Image Processing Institute
Department of Electrical Engineering-Systems
University of Southern California
Los Angles, CA 90089-2546

Abstract

Multiresolution image decomposition based on linear filters has been widely used in image compression. In this research, we investigate a nonlinear multiresolution image decomposition based on mathematical morphology and obtain a morphological image pyramid. We propose to use opening and closing operations with a certain structure element to achieve image decomposition. The entropy and histogram of the difference images in the image pyramid are then examined. We use a numerical example to demonstrate that the morphological filtering approach provides some potential advantages over the conventional linear filtering approach in effective image coding.

1 Introduction

Multiresolution image decomposition and compression methods have received a lot of attention recently. One well known approach is to perform a pyramid image decomposition with linear filters such as the work of Burt and Adelson [1]. Simply speaking, the basic idea is to remove the pixel-to-pixel correlation by subtracting a lowpass filtered version of an image from the image itself and, as a result, we decompose an image into a low-frequency and a high-frequency ones. Compression is then achieved by exploiting that the low-pass filtered image can be represented at a coarser resolution while the difference image has a lower value of entropy (or variance) so that the difference image can be quantized and encoded effectively with fewer bits. The same idea can be applied recursively to the low-pass filtered images, and it leads naturally to a sequence of difference images of different sizes known as the Laplacian pyramid. They can be encoded independently. Recent research on image compression

with a pyramid structured wavelet transform is essentially based on the same idea but with different kinds of filters [7]. In the wavelet context, the decomposition is achieved by using the quadrature mirror filter (QMF) banks. The downsampling can be applied not only to the low frequency filtered subimage but also to the high frequency filtered subimage. As a consequence, the wavelet approach has one special feature. That is, the dimension of the problem remains the same when the decomposition is performed. The multiresolution approach is attractive because it can be used in progressive transmission.

However, image pyramid decomposition compression methods based on linear filtering have some disadvantages which tend to result in some unpleasant distortion in the decompressed image. For example, it has been observed that linear operators blur fine details of an image such as edges and spread out impulsive noise. Besides, only the algebraic structure of image data is taken into account in the linear filtering approach, where geometric information, e.g. the shapes of objects, is totally ignored. To overcome these shortcomings, nonlinear filters have been proposed to achieve multiresolution image decomposition and compression. One important class of nonlinear filters is the morphological filters which not only suppress impulsive noise but also preserve fine details [8]. In particular, mathematical morphology provides a powerful tool to handle the geometry of an image. In this research, we apply morphological operators with various convex structure elements to remove geometrical redundancy among image data and to achieve image compression via pyramid decomposition.

The paper is organized as follows. Basic concepts in mathematical morphology and the morphological sampling theorem are reviewed in Section 2. Some important properties of mathematical morphology are also discussed. Morphological image pyramid decomposition is studied in Section 3. Some experimental results and the potential application of the morphological pyramid decomposition in image compression are presented in Section 4.

2 Mathematical Morphology and Morphological Sampling Theorem

We review the definition of some basic gray scale morphological operators [3], [10] as follows. Let A and B be subsets in the space $\mathbf{Z}^2 \times \mathbf{Z}$, where $\mathbf{Z}^2$ denotes spatial domain and $\mathbf{Z}$ represents surface. The gray scale dilation of A by B, denoted by $A \oplus B$, is defined as

$$(A \oplus B)(s,t) = \max\{A(s - x, t - y) + B(x, y)$$
$$| (s - x, t - y) \in \mathbf{D}_A; (x, y) \in \mathbf{D}_B\}, \quad (1)$$

where $\mathbf{D}_A$ and $\mathbf{D}_B$ are the domains of A and B, respectively, and B is called the *structure element*. The gray scale erosion of A by B, denoted by $A \ominus B$, is defined as

$$(A \ominus B)(s,t) = \min\{A(s + x, t + y) - B(x, y)$$
$$| (s + x, t + y) \in \mathbf{D}_A; (x, y) \in \mathbf{D}_B\}. \quad (2)$$

The gray scale opening of A by B, denoted by $A \circ B$, is

$$A \circ B = (A \ominus B) \oplus B. \quad (3)$$

Similarly, the gray scale closing of A by B, denoted by $A \bullet B$, is

$$A \bullet B = (A \ominus B) \oplus B. \quad (4)$$

Furthermore, let $\mathbf{S} \subseteq \mathbf{Z}^2$ be a *sampling set*, and $f : \mathbf{D}_A \to \mathbf{Z}$ denote an image A. Then, the sampled version of A by S is f restricted to $\mathbf{D}_A \cap \mathbf{S}$ and denoted by $f \mid_\mathbf{S}$. The definition of $f \mid_\mathbf{S}: \mathbf{D}_A \cap \mathbf{S} \to \mathbf{Z}$ can be expressed mathematically as

$$f(x, y) \mid_\mathbf{S} = \{f(x, y) \mid (x, y) \in \mathbf{D}_A \cap \mathbf{S}\}. \quad (5)$$

We know from the Shannon's sampling theorem that a signal can be perfectly reconstructed from its samples if and only if the cutoff frequency of the signal is less than one half of sampling frequency (or Nyquist frequency) [9]. Therefore, to avoid aliasing, a lowpass filter with an appropriate cutoff frequency is usually applied to a signal before the down-sampling operation. Similarly, since we want to generate the image pyramid by using morphological operators, we are interested in some sampling conditions so that no significant shape accuracy is sacrificed during the down- and up-sampling processes. This property can be roughly stated as follows. Let us apply an opening (or closing) operation with a certain structure element B to an image A and then downsample the resulting image. The original image A can be reconstructed approximately from the downsampled version via a closing or a dilation by B. The reconstructed image $\hat{A}$ falls between an upper bound, its reconstruction by dilation, and a lower bound, its reconstruction by closing. Furthermore, the distance between these two bounds is less than or equal to the diameter of B. We refer to [4] for a more rigorous mathematical statement and its proof.

Some properties of morphological operators will be reviewed in the remainder of this section. In this work, we will focus on morphological opening and closing operators with convex structure elements. An opening operation with a convex structure element tends to smooth the contours, cut narrow isthmuses, and eliminates sharp caps, while an closing operation with a convex structure element can fuse narrow channels and slim gulfs, fill small holes and connect gaps. Both of these two operators serve as geometric lowpass filters [10]. The second important property is that both operations are idempotent which means that output will remain the same if the same operation is applied again. Thus, these two operations are complete and convergent right after a single iteration. The third interesting property is that the opening (or closing) operation with a convex structure element does not introduce extra zero-crossing so that it makes its output well ordered [2]. In addition to the convexity of structure element, we require that the center of the structure element has the value zero. This requirement is based on the consideration that we want the values of gray level in a flat region are unchanged by applying the morphological operations. This is similar to the design of linear lowpass filter design by requiring the d.c. gain to be 1. In summary, the convex structure element of our interest is of the form:

$$\begin{matrix} -a & -b & -a \\ -b & 0 & -b \\ -a & -b & -a \end{matrix},$$

where a and b are positive integers and $a \leq b$.

3 Morphological Image Pyramid

Let P_n, $n = 0, 1, 2, \ldots, N$, denote the morphological image pyramid for a given image A by using opening with a convex structure element B, and $P_0 = A$.

We can represent $P_{k-1} \circ B$ as $f_{k-1} : \mathbf{D}_{P_{k-1} \circ B} \to \mathbf{Z}$, Therefore, for resolution $k = 1, 2, \ldots, N$,

$$P_k(x, y) = f_{k-1}(x, y) \mid_{\mathbf{S}}, \qquad (6)$$

where $(x, y) \in \mathbf{D}_{P_{k-1} \circ B} \cap \mathbf{S}$ and $\mathbf{S} = \{(2x, 2y) \mid (x, y) \in \mathbf{Z}^2\}$ is the sampling set in $\mathbf{Z}^2$. Now, suppose that we have P_k and B. The reconstructed image at resolution $k - 1$ is

$$\hat{P}_{k-1} = T \bullet B, \qquad (7)$$

where T is the *up-sampled* version of P_k and can be expressed as

$$T(x, y) = \begin{cases} P_k(x, y), & \text{if } (x/2, y/2) \in \mathbf{D}_{P_k}, \\ 0, & \text{otherwise.} \end{cases} \qquad (8)$$

We can determine the *difference images* E_n, $n = 0, 1, 2, \ldots, N - 1$ from equations (6) and (7), i.e.

$$E_n(x, y) = P_n(x, y) - \hat{P}_n(x, y). \qquad (9)$$

We can see that the functionality of reconstruction is nothing but a *prediction process* from a coarser resolution to a finer one. That is, E_n indicates the *prediction error* at each resolution level. We have to use the morphological sampling theorem in order to justify how well we can achieve. To make prediction error as small as possible, different structure elements can be used. Special algorithms have been developed based on *flat* structure element [5], [6]. In our current research, only simple opening and closing operations are used in reconstruction.

4 Experiments

In this section, we use the test image *Lena* to illustrate the morphological image pyramid decomposition with the following structure element,

$$\begin{array}{ccc} -8 & -4 & -8 \\ -4 & 0 & -4 \\ -8 & -4 & -8 \end{array}.$$

The resulting morphological image pyramid with $N = 4$ are depicted in Fig. 2. To compare the performance of the image pyramids with linear and nonlinear filtering, we plot the image pyramid with linear filtering in Fig. 3, where a 2-D separable filtering kernel is used and the filtering coefficients along the vertical (or horizontal) direction are:

$$0.25 - 0.5\alpha, \ 0.25, \ \alpha, \ 0.25, \ 0.25 - 0.5\alpha.$$

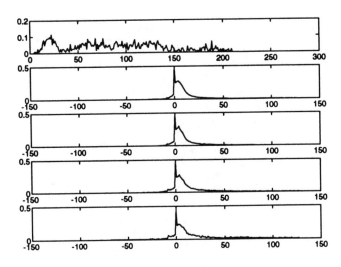

Figure 1: From top to bottom: histograms of P_4, E_0, E_1, E_2 and E_3.

The difference images at each resolution level are shown in Fig. 4. The histograms for P_4, and E_n, $n = 0, 1, 2, 3$, are given in Fig. 1

The entropy for a given digital image is defined as:

$$H = -\sum_{g_i \in G} \text{Prob}(g = g_i) \log_2[\text{Prob}(g = g_i)], \qquad (10)$$

where $\text{Prob}(g = g_i)$ is the *relative frequency* of gray level $g = g_i$ in the image and G denotes an index set which contains all possible g_i's such that $\text{Prob}(g = g_i) \neq 0$. The entropies for image pyramids and difference images are summarized in the following table, where BA denotes the Burt-Adelson image pyramid [1].

Resolution Level	H_{P_n}	H_{E_n}	BA's H_{E_n}
$n = 0$	7.594	4.43	4.48
$n = 1$	7.568	4.56	5.01
$n = 2$	7.528	4.98	5.63
$n = 3$	7.431	5.49	6.21
$n = 4$	7.192		

Based on the above experimental results, we may see some advantages with the morphological filtering approach. First, in pyramid image coding, the data to be encoded are those difference images and the final down-sampled pyramid, P_N. We have shown that the entropy of the difference images with morphological filters is less than that of the difference images generated by linear filtering such as the Laplacian pyramid. Second, the histograms for prediction error images E_n, $n = 0, 1, 2$ and 3 of the morphological image pyramid are similar, and they are all centered around 0. This

Figure 2: A 4-level morphological image pyramid. From top to bottom: P_1, P_2, P_3 and P_4.

Figure 3: A 4-level Laplacian image pyramid with $\alpha = 0.6$. From top to bottom: LAP_1, LAP_2, LAP_3 and LAP_4.

suggests that we can to use an effective quantizer for all levels. In contrast, the histogram spreads out as the level index n increases in Laplacian pyramid. This implies that different quantizers are required at different resolution levels and the resulting coding scheme will be more complicated.

5 Extension

We have shown in this research that the morphological image pyramid has several potential advantages in image coding. This will be verified in our future research. Besides, the structure element design is analogous to the selection of the parameter α associated with the laplacian pyramid. Thus, we do have the flexibility of using different filters, and it is interesting to examine the filter effect. However, due to the non-linear nature of morphological operators, performance analysis of the morphological image pyramid is a more difficult and challenging problem.

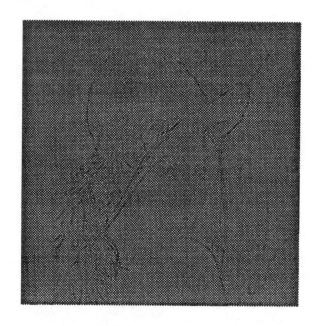

References

[1] P. J. Burt and E. H. Adelson, "The Laplacian pyramid as a compact image code," *IEEE Trans. on Communications*, Vol. COM-31, pp. 532–540, Oct. 1983.

[2] M.-H. Chen and P.-F. Yan, "A multiscaling approach based on morphological filtering," *IEEE trans. on Pattern Recognition and Machine Intelligence*, Vol. PAMI-11, pp. 694–700, July 1989.

[3] R. M. Haralick, S. R. Sternberg, and X. Zhuang, "Image analysis using mathematical morphology," *IEEE trans. on Pattern Recognition and Machine Intelligence*, Vol. PAMI-9, pp. 523–550, July 1987.

[4] R. M. Haralick, X. Zhuang, C. Lin, and J. Lee, "The digital morphological sampling theorem," *IEEE Trans. on Acoustic, Speech, and Signal Processing*, Vol. ASSP-37, pp. 2067–2090, Dec. 1989.

[5] H. J. A. M. Heijmans and A. Toet, "Morphological Sampling," *Computer Vision, Graphics, and Image Processing: Image Understanding*, Vol. 54, pp. 384–400, Nov. 1991.

[6] X. Kong and J. Goutsias, "A study of pyramid techniques for image representation and compression," tech. rep., Dept of Electrical and Computer Eng. The John Hopkins Univ., 1993.

[7] S. G. Mallat, "A Theory for multiresolution signal decomposition: the wavelet representation," *IEEE trans. on Pattern Recognition and Machine Intelligence*, Vol. PAMI-11, pp. 674–693, July 1989.

[8] P. Maragos and R. W. Schafer, "Morphological filters-Part I: Their set-theoretic analysis and relations to linear shift-invariant filters," *IEEE Trans. on Acoustic, Speech, and Signal Processing*, Vol. ASSP-35, pp. 1153–1169, Aug. 1987.

[9] A. V. Oppenheim and R. W. Schafer, *Discrete-time signal processing*, Englewood Cliffs, N.J.: Prentice Hall, 1989.

[10] J. Serra, *Image Analysis and Mathematical Morphology*, London: Academic, 1982.

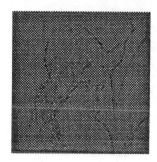

Figure 4: Difference images for 4-level morphological image pyramid. From top to bottom: E_0, E_1, E_2 and E_3.

A HYBRID FRACTAL TRANSFORM

D. M. Monro

School of Electronic and Electrical Engineering
University of Bath
Claverton Down, Bath, BA2 7AY
England

ABSTRACT

This paper presents a generalization of fractal coding of images, in which image blocks are represented by mappings derived from least squares approximations using fractal functions. Previously known matching techniques used in Fractal Transforms are subsets of this generalized method, which is called the Bath Fractal Transform (BFT). By introducing searching for the best image region for application of the BFT, a hybrid of known methods is achieved. Their fidelity is evaluated by a rms error measure for a number of polynomial instances of the BFT, over a range of searching levels using a standard test image. The method readily extends to data of higher or lower dimensions, including time as in image sequences.

1. BACKGROUND

The problem of coding images using fractals continues to attract interest, because of its intellectual appeal and the prospect of commercial applications. Recently two distinct block coding techniques have been proposed. The Iterated Transform Technique (ITT), published by Jacquin [1, 2] is similar to the system patented by Barnsley [3], but is very different from Barnsley's published work [4, 5]. In it, an image is partitioned into "range blocks" which are encoded by a simple mapping from a larger "domain block". To implement it, a search is made for the domain block which is most similar to the range block according to some measure. Beaumont [6, 7] investigated a number of implementation options, and suggested how the high cost of searching could be mitigated to some degree.

Monro and Dudbridge [8, 9] encoded a block without searching, by tiling it with reduced copies of itself using a least-squares criterion to derive an optimal mapping. The efficiency of this method suggested its feasibility for real time applications such as fractal video. This paper describes a generalization of this approach, called the Bath Fractal Transform (BFT) [10].

The matching criterion used by ITT-coding is a low order case of the BFT. By including searching with the BFT, a family of combinations is formed which are hybrids between ITT-coding and the BFT. In this communication the cost-fidelity trade is investigated over a broad class of Fractal Transforms, namely least-squares polynomial instances of the BFT with and without searching.

2. THEORY

The Bath Fractal Transform (BFT) is a general strategy for obtaining a fractal approximation to a tiling of data in any number of dimensions. To encode an image, first define an Iterated Function System (IFS) [5]

$$W = \{w_k; k = 1, ..., N\} \qquad (1$$

called the domain part of the BFT, whose attractor A is a non-overlapping tiling of the image. For each k a fractal function f_k, called the function part of the BFT, is defined on A, which approximates the greyscale of tile k. The fractal function is specified by a set of mappings v_k such that

$$f_k = v_k(x, y, f) \qquad (2$$

which have parameters $\alpha_i^{(k)}$; $k = 1, ..., N$; $i = 1, 2, ... M$. Here, N is the order of the IFS and M is the number of free parameters of the BFT.

Given the IFS W, the BFT finds f_k by minimizing $d(g, \tilde{g})$ where g is the greyscale function of the block, and $\tilde{g}$ is its mapping by the BFT, which can be evaluated at a point x through application of the function part to the greyscale at the preimage of x. This is found through the inverse of the domain part:

$$\tilde{g}(x) = v_k(w_k^{-1}(x), g(w_k^{-1}(x))) \qquad (3$$

when $x \in w_k(A)$ and d is a suitable metric, with respect to the parameters $\alpha_i^{(k)}$ of v_k. Finally solve

$$\frac{\partial \, d(g, \tilde{g})}{\partial \alpha_i^{(k)}} = 0 \qquad (4$$

for all i and k to evaluate the parameters of the BFT.

A practical case occurs when d is the mean squared difference, w_k are affine transforms and v_k are polynomials. The BFT is then a least squares approximation, which is the solution of N systems of M linear equations. Monro and Dudbridge [8, 9] tiled a block by four copies of itself ($N = 4$), using a bilinear BFT with four parameters ($M = 4$). This is described in more detail in [8] and [9], where with variable sized tiles a comparable fidelity/compression

0-7803-0946-4/93 $3.00 © **1993 IEEE**

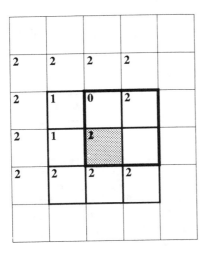

Figure 1. The shaded tile is the lower left quadrant of its level 0 parent. Other parents to level 2 are labelled in their upper left corner by the lowest search level in which they occur.

performance was demonstrated to the DCT-based JPEG standard [11], with greater computational efficiency.

In ITT-coding [1, 2, 3, 6, 7], the image is tiled by adjoint "range blocks". A set of "domain blocks" is also defined, which are necessarily larger than the range blocks. Each range block is encoded by a contraction mapping defined on one domain block, which is identified by searching the set of domain blocks. The contraction mappings may include rotation, reflection or greylevel scaling, and always shrink the domain onto the range. The ensemble of the codes for all the range blocks is an IFS of high order.

To achieve the hybrid of the BFT and ITT methods, call the range block a "tile", and a "parent" is the same as the domain block. Rather than adopt a fixed choice of parent for each tile, apply the BFT to each candidate parent block within a specified distance, and select the one which minimizes d. This gives a strategy for finding the best parent with a selected amount of searching. In this way a more general encoding strategy is applied to ITT-coding, and the scope of the BFT is increased by searching for optimum parents which may be separate from the tiles.

3. METHODS

To evaluate the hybrid transform, consider a specific instance of the procedures outlined above. An image is encoded using fixed size 4 by 4 pixel tiles. The parent block for a particular tile is chosen from among all the 8 by 8 blocks which align with tile boundaries, and are within a specified distance of the tile. In this study reflections and rotations are not considered. A level zero search consists of choosing from the subset of parent blocks which do not overlap one another. In Figure 1, the parent of the shaded tile could be the parent outlined in bold and labelled '0' in its upper left corner. In this case a tile lies within only one parent, and choosing the parents without searching reproduces the bilinear case [8,9].

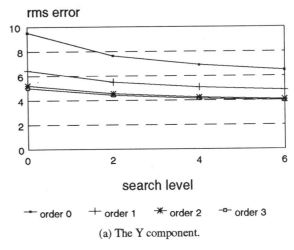

(a) The Y component.

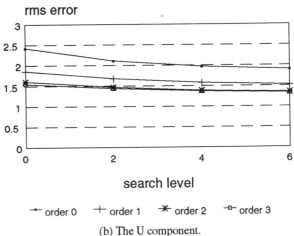

(b) The U component.

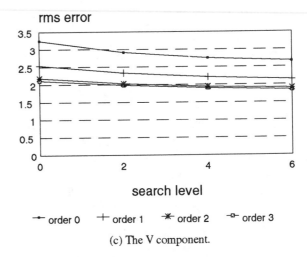

(c) The V component.

Figure 2. Rms error observed in encoding the Y-U-V color components of Gold Hill with different orders of the Bath Fractal Transform and different levels of searching. Fidelity is gained by increased searching and increased BFT complexity. (a) The Y component. (b) The U component. (c) The V component.

A level-one search also considers the three other parent blocks which overlap the tile, labelled '1' in Figure 1. Here, derive the BFT for each parent, computing the metric $d(g, \tilde{g})$ in each case, and select the best parent for each tile. A level-two search includes all the parents which overlap those in level one, labelled '2', and so on.

With the BFT, one can use polynomial fractal functions of varying complexity. In the simplest (zero-order) case, each mapping v_k has the form

$$v_k(x, y, f) = s_k f + t_k. \tag{5}$$

This reproduces ITT coding [1, 2, 3, 6, 7]. This paper also considers first, second and third-order polynomial versions of the BFT without cross products, such as the cubic

$$v_k(x, y, f) = a_3^{(k)} x^3 + a_2^{(k)} x^2 + a_1^{(k)} x$$

$$+ b_3^{(k)} y^3 + b_2^{(k)} y^2 + b_1^{(k)} y + s_k f + t_k \tag{6}$$

4. RESULTS

The hybrid fractal transform was applied to the standard test image "Gold Hill". The encoding cost increases with search level, but the accuracy of the code also improves, so that a cost-fidelity trade-off is established. The results are illustrated in Figure 2 for both monochrome and color.

Examples are given of monochrome image fragments encoded using hybrid schemes. Figure 3 shows a section of the standard image "Gold Hill", together with three combinations of searching and BFT order. The rms errors over the fragments shown are given in each case. Note that the graphs of Figure 2 include the corresponding results, but with the errors measured over the entire image.

5. CONCLUSIONS

The results show that the fidelity of the fractal transform increases with both search level and order of the polynomial approximation. Further gains in fidelity are available, by considering rotations and reflections, and by using variable sized tiles. The rms error obtained with the same Y-image using JPEG coding at 7:1 compression is 7.22. All of the results in Figure 2, except two of the zero-order ones, have lower errors than this. Significantly, the bilinear BFT without searching gives a lower error measure than ITT-coding does with level 6 searching. The quadratic case also provides a worthwhile improvement, beyond which there appears to be little reduction in error. Although further work is necessary to investigate quantization of BFT coefficients fully, the results suggest that variants of the BFT will offer better fidelity/compression tradeoffs than JPEG coding. The BFT is therefore a serious alternative as a second-generation method for still picture coding, and on fidelity grounds a quadratic form is suggested by these results.

Fractal technology offers several practical advantages. Fractal transforms are very fast to decode, and therefore in picture archiving systems where images are compressed once and decompressed often, fractal technology could be the preferred method. Low orders of the BFT with little or no searching are fast to encode also, and therefore extend the use of Fractal Transforms to new applications such as digital video. For all cases where zero-level searching is used, an optimal plotting algorithm exists [12] such that the cost of decoding is very low indeed, perhaps low enough for use in future digital television receivers. Although further work is required to specify its compression performance, this paper demonstrates the versatility and potential of the BFT with and without searching as a method of image approximation.

6. ACKNOWLEDGEMENTS

This work was supported by the UK Science and Engineering Research Council Grant No. GR/H18189, "Image Compression by Fractal Sets". The Gold Hill image was provided by the Independent Broadcasting Authority, now National Transcommunications Plc. BT Laboratories are funding the application of these methods to fast image coding and decoding on PC computers.

7. REFERENCES

[1] A. E. Jacquin, "A novel fractal block-coding technique for digital images," Proc. IEEE Int. Conf. on Acoustics, Speech and Signal Processing, pp. 2225-2228, 1990.

[2] A. E. Jacquin, "Image coding based on a fractal theory of iterated contractive image transformations," IEEE Trans. Image Processing, Vol. 1, No. 1, pp. 18-30, 1992.

[3] M. F. Barnsley, "Methods and apparatus for image compression by iterated function systems," US Patent No. 4,941,193, 10 July 1990.

[4] M. F. Barnsley, V. Ervin, D. Hardin and J. Lancaster, "Solution of an inverse problem for fractals and others sets," Proc. Natl. Acad. Sci. USA, Vol. 83, pp. 1975-1977, 1986.

[5] M. F. Barnsley, Fractals Everywhere, Academic Press, San Diego, 1988

[6] J. M. Beaumont, "Advances in block based fractal coding of still pictures," IEE Colloquium on The Application of Fractal Techniques in Image Processing, IEE Digest No. 1990/171, 1990.

[7] J. M. Beaumont, "Image data compression using fractal techniques," BT Technol. J., Vol. 9, No. 4, pp. 93-109, 1991.

[8] D. M. Monro and F. Dudbridge, "Fractal approximation of image blocks," Proc. IEEE Int. Conf. on Acoustics, Speech and Signal Processing, pp. III: 485-488, 1992.

[9] D. M. Monro and F. Dudbridge, "Fractal block coding of images," Electronics Letters, Vol. 28, no. 11, pp. 1053-1054, 21 May 1992.

[10] D. M. Monro, F, Dudbridge and J. A. Dallas, "Least squares fractal interpolation," Colloquium on Fractals in Engineering, Ecole Polytechnique, Montreal, 1992.

[11] G. K. Wallace, "The JPEG still picture compresion standard," Comm. ACM, Vol. 34, No. 4, pp. 30-34, 1991.

[12] D. M. Monro, F. Dudbridge and A. Wilson, "Deterministic rendering of self-affine fractals," IEE Colloquium on Fractal Techniques in Image Processing, IEE Digest No. 1990/171, 1990.

Figure 3. Approximations by hybrid Fractal Transforms. (a) Original. (b) BFT order 0, no search, e_{rms}= 9.53. (c) BFT order 0, search level 6, e_{rms}= 6.47. (d) BFT order 3, no search, e_{rms}= 5.02. (e) As (d), enlarged. (f) BFT order 3, search level 6, e_{rms}= 3.98.

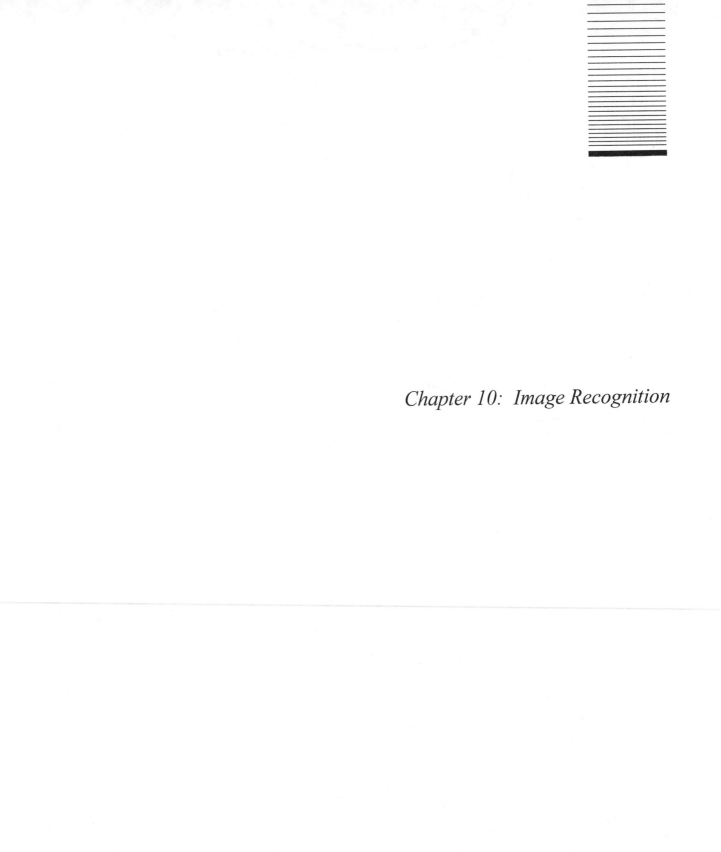

Chapter 10: Image Recognition

This chapter presents papers that address issues of extracting information from images. At the front-end analysis stage, low-level classification can be combined with the compression techniques of vector quantization, as described in "Combining Image Classification and Image Compression Using Vector Quantization" (paper 10.1). Here, the optimization criterion of squared error (for compression) and Bayes risk (for classification) are combined via a tunable weighting parameter and applied to images. An example of classification of a computerized tomography (CT) image is provided.

Hidden Markov models have emerged as one of the most successful approaches for speech recognition, and it is natural that their extension to image recognition be investigated. In "Document Image Decoding Using Markov Source Models" (10.2), a communication model of document recognition involving a source, imager, noisy channel, and decoder is used to parse columns of telephone yellow pages to recognize individual telephone listings, to recover the structural hierarchy of groups and subgroups within the column, and to recognize the reverse-video subject headings that appear. Hidden Markov model-based character recognition is applied to the very difficult recognition instances that are rejected from a commercial optical character recognition system in "Connected and Degraded Text Recognition Using Planar Hidden Markov Models" (10.3). A character error rate on a database of these rejects of less than 2% is reported on a system whose performance was improved by corrupting clean image data with Gaussian blurring and random bit flips prior to training. "Evaluation of Character Recognition Systems" (10.4) pulls together a wide variety of optical character recognition approaches from many organizations, tests them on a common set of data, and arrives at a comparison of seven different classifiers and eleven different feature sets.

Subsequent papers expand image recognition from characters to objects in scenes. "Model-Based Multi-Sensor Fusion" (10.5) develops a technique called relational template matching that, like the discriminative feature extraction described in paper 8.2, chooses features to minimize classification errors on a set of multiclass data, thereby focusing on differences between objects. The dependence of recognition upon segmenting the image into objects early in the analysis is avoided, and preliminary recognition information on a target set of 10 different objects, as viewed by second-generation forward-looking infrared (FLIR) detectors, is provided. A Baysian technique for combining FLIR imagery with millimeter-wave radar data for target recognition is described in "MUltiSensor Target Recognition System (MUSTRS)" (10.6). Both sensors are used first in a wide-area-search mode to locate targets, then are switched to a high resolution mode to place more pixels on target and achieve target classification. A fine-grained array of 92,000 processors provides the image processing. Fuzzy set theory is applied to the general problem of object recognition in "Modeling of Natural Objects Including Fuzziness and Application to Image Understanding" (10.7). Experimental examples are given in which a hierarchical model of an eye is defined and used to segment eyes from both an abstract painting and a photograph.

"Finding Similar Patterns in Large Image Databases" (10.8) addresses the problem of retrieving all images that "look like this" from a large stored collection. The Karhunen-Loéve transform is modified to achieve shift invariance, and a standard collection of textures is used as a reference set with which examples of textures can be used to retrieve the most similar ones in the database.

Suggested Additional Readings:

[1]   T. S. Huang and A. N. Netravali, "Motion and Structure from Feature Correspondences: A Review," *Proc. IEEE,* vol. 82, no. 2, pp. 252-268, Feb. 1994.

[2]   R. O. Duda and P. E. Hart, *Pattern Classification and Scene Analysis.* New York: Wiley, 1973.

[3]   J. Serra, *Image Analysis and Mathematical Morphology.* London: Academic, 1982.

[4]   R. M. Gray, K. L. Oehler, K. O. Perlmutter, R. A. Olshen, "Combining Tree-Structured Vector Quantization with Classification and Regression Trees," *The Twenty-Seventh Asilomar Conference on Signals, Systems, and Computers* (Pacific Grove, CA, USA) Nov. 1994, pp. 1494-1498.

[5]   F. R. Chen, L.D. Wilcox, and D. S. Bloomberg, "Word Spotting in Scanned Images Using Hidden Markov Models," *Proc. IEEE International Conference on Acoustics, Speech, and Signal Processing 93* (Minneapolis, MN, USA) April 1993, vol. 5 pp. 1-4.

[6]   K. S. Nathan, J. R. Bellegarda, D. Nahamoo, and E. J. Bellegarda, "On-Line Handwriting Recognition Using Continuous Parameter Hidden Markov Models," *Proc. IEEE International Conference on Acoustics, Speech, and Signal Processing 93* (Minneapolis, MN, USA) April 1993, vol. 5 pp. 121-124.

[7]   P. Salembier, J. Serra, and J. A. Bangham, "Edge Versus Contrast Estimation of Morphological Filters," *Proc. IEEE International Conference on Acoustics, Speech, and Signal Processing 93* (Minneapolis, MN, USA) April 1993, vol. 5 pp. 45-48.

[8]   T. M. English, M. P. Gomez-Gil, and W. J. B. Oldham, "A Comparison of Neural Network and Nearest-Neighbor Classifiers of Handwritten Lower-Case Letters," *1993 IEEE International Conference on Neural Networks* (San Francisco, CA, USA) Mar. - Apr. 1993, pp. 1618-1621.

[9]   J. Wang and J. Jean, "Multiresolution Neural Networks for Omnifont Character Recognition," *1993 IEEE International Conference on Neural Networks* (San Francisco, CA, USA) Mar. - Apr. 1993, pp. 1588-1593.

Combining Image Classification and Image Compression Using Vector Quantization

Karen L. Oehler   Robert M. Gray

Information Systems Lab, Department of Electrical Engineering
Stanford University, Stanford, CA 94305-4055

Abstract

We describe a technique for combining compression and low-level classification of images. The goal is to produce codes with implicit classification information, where the compressed image incorporates classification information without further signal processing. This technique can provide direct low level classification or provide an efficient front end to more sophisticated full-frame recognition algorithms. Vector quantization (VQ) is a natural choice because two of its design components, clustering and tree-structured classification methods, have obvious applications to the pure classification problem as well as to the compression problem. Here we explicitly incorporate a Bayes risk component into the distortion measure used for code design in order to permit a tradeoff of mean squared error with classification error. This method is used to analyze simulated data, identify tumors in computerized tomography (CT) lung images, and identify man-made regions in aerial images.

1 Introduction

Image compression and classification play very important roles in making digital images useful. Because lossless compression (compression without error) can typically provide a compression ratio of only 2 - 4, lossy compression techniques become necessary for further compression. VQ is a common method of lossy compression that applies statistical techniques to optimize distortion/bit rate tradeoffs [1, 2, 3].

In many scientific and medical applications, images are used by human experts to make specific decisions or inferences. The actions of human experts can sometimes be mimicked by sophisticated computer algorithms, typically requiring large amounts of computer processing on entire images in order to reach decisions. Simple low-level local classification involving only small regions of an image, however, can assist human observers by highlighting special areas of interest and can simplify future classification/analysis algorithms by incorporating pre-processing into the digital representation. The approach described here is applicable to classification tasks where the goal is to classify fairly small regions in an image, for example, to identify lung nodules in a CT lung scan, microcalcifications in mammograms, or manmade objects in an aerial photograph. Our goal is to combine such local classification with the compression process to obtain a single code that does a good job of both, with an adjustable relative importance weighting of the two aspects. VQ provides a natural means to this end because two of its design components, clustering and tree-structured classification methods, have obvious applications to the pure classification problem as well as to the compression problem.

2 Vector Quantization

Vector quantization operates on individual image subblocks or vectors (e.g. blocks of 2 pixels × 2 pixels). For each vector, the VQ encoder determines the nearest codeword and outputs the chosen codeword's index. The sequence of indices so generated can then be stored or transmitted. The VQ decoder reverses this process as it inputs each index and outputs the appropriate codeword by simple table lookup. A clear advantage of VQ is the low complexity decoding process. As the input vectors mapping into a particular codeword will generally have much in common, they are likely to belong to the same class. By attaching class significance to the codewords, we can obtain both classification and compression with a single encoding step.

The accuracy of the classifier is measured by the Bayes risk, which reduces to a simple error probability if unit costs are chosen. The more general form allowing varying costs for different error types is useful for many applications (such as classifying tissue as tumor or healthy) and our algorithm provides a convenient means of including these costs. This Bayes risk term can be incorporated into the distortion measure, and the resulting codebook is then designed to determine class membership when the image is decompressed. Our method explicitly incorporates the performance of optimal (Bayes) classifiers operating on the vector quantized data, both through the form of the distributions (which are learned by training and do not require an a priori model), and through the explicit incorporation of a Bayes risk term into the average distortion that is minimized by the design algorithm. The extra complexity is almost entirely in the code design phase; the implemented code is only slightly more complicated than an ordinary vector quantizer. The classification requires no more bits to describe than the bits required for compression alone, an important feature in low memory or low bandwidth situations. As the classification information is entirely contained in the selected codeword, it can be easily displayed with the decompressed image. For example, a physician could select either a monochrome decompressed image or one with all the suspected anomalies colored pink for highlighting.

There have been several papers devoted to the related topic of designing a classifier that uses a VQ encoder. Kohonen [4, 5, 6] proposed a variety of "likelihood vector quantizers" or LVQ to perform classification using a VQ encoder and codebook, where the encoder operates as an ordinary minimum mean squared error selection of a representative from the codebook, but the codebook is designed in a manner that attempts to reduce classification error implicitly rather than reducing mean squared error. His general goal is to imitate a Bayes classifier with less complexity than other neural network approaches, but there is no explicit minimization of Bayes risk in the code design. The Kohonen approach differs from ours in many ways. First, our approach considers mean squared error in the optimization and his does not. As we allow a weighting of relative importance, his goal can be considered as an extreme point of ours. Compression ability is not explicitly considered in LVQ. His encoder uses only minimum mean squared error, even within the learning set, while we explicitly apply an empirical Bayes classifier as part of the encoding, balancing minimum squared error with empirical Bayes risk. We explicitly incorporate the Bayes risk into the average distortion measure being minimized, while Kohonen uses a heuristic to argue that moving centroids according to nearby class membership should asymptotically have the effect of approximating a Bayes risk.

3   Approach

Given a n-dimensional vector $X = (x_1, \ldots, x_n)^T$, a full search (unconstrained) VQ maps each input vector onto a set of codewords called the codebook: $q(X) = \hat{Y}_i$, where $\hat{Y}_i \in \{\hat{Y}_1, \ldots, \hat{Y}_N\}$. Denote the partition of input vector space induced by q by $\mathcal{P} = \{\mathcal{A}_1, \ldots, \mathcal{A}_N\}$ where $X \in \mathcal{A}_i$ if $q(X) = \hat{Y}_i$. Then $\alpha(X) = i$ if $X \in \mathcal{A}_i$ is the <u>encoder</u> and $\beta(i) = \hat{Y}_i$ is the <u>decoder</u> for the VQ q. Hence $q(X) = \beta(\alpha(X))$. Because there are many more possible input vectors than codewords, the quantizer compresses the data.

Suppose we have a collection of classes $\mathcal{H} = \{H_1, H_2, \ldots, H_M\}$. To perform a classification, we associate a class label with each cell $\mathcal{A}_i$ in the partition or, equivalently, with each index i produced by the encoder α, for $i \in \{1, \ldots, N\}$. The index can then be used to decompress the vector using the codebook and to classify the vector using the decision rule. Let $\delta(i) \in \mathcal{H}$, $i = 1, \ldots, N$, denote the decision function.

Codebook construction requires a labeled learning set $\mathcal{S} = \{X_1, \ldots, X_L\}$. Each training vector X_i is assigned to a specific class in $\mathcal{H}$. The samples provide the empirical distribution of X, used to train the vector quantizer.

The average ordinary distortion corresponding to (α, β) is

$$D(\alpha, \beta) = E[\, d(X, \beta(\alpha(X)))\,] \geq E[\, \min_i d(X, \beta(i))\,] \text{ where } 1 \leq i \leq N. \tag{1}$$

The mean squared error $d(X, \beta(\alpha(X))) = \|X - \beta(\alpha(X))\|^2$ is commonly used as the distortion measure.

The classification performance can be measured by the Bayes risk [7]:

$$B(\alpha, \delta) = \sum_{i=1}^{M} \sum_{j=1}^{M} P(\delta(\alpha(X)) = H_j | X \in H_i)\, P(H_i)\, C_{ij},$$

where C_{ij} is the relative cost of error incurred when $X \in H_i$ and $\delta(\alpha(X)) = H_j$. Often, $C_{ij} = 0$ when $i = j$ (correct decisions have zero cost). Note that the reproduction values supplied by β do not affect B. Define an indicator function $I_{H_i}(H_j)$ as 1 if

$H_i = H_j$ and 0 otherwise. The Bayes risk is then given by:

$$B(\alpha, \delta) = \sum_{k=1}^{N} \sum_{j=1}^{M} I_{H_j}(\delta(k)) \sum_{i=1}^{M} P(\alpha(X) = k | X \in H_i) P(H_i) C_{ij}$$

$$\geq \sum_{k=1}^{N} \min_j \{ \sum_{i=1}^{M} P(\alpha(X) = k | X \in H_i) P(H_i) C_{ij} \} \triangleq B^*(\alpha, \delta) \qquad (2)$$

where $B^*(\alpha, \delta)$ is the minimum Bayes risk. For the simple two class case with zero cost for correct decisions (where $M = 2$ and $C_{11} = C_{22} = 0$) the above equation reduces to

$$B^*(\alpha, \delta) \triangleq \sum_{k=1}^{N} \min \{ P(\alpha(X) = k | X \in H_1) P(H_1) C_{12}, P(\alpha(X) = k | X \in H_2) P(H_2) C_{21} \}$$

4 Algorithm for Codebook Construction

In order to simultaneously consider the compression and classification abilities of the encoder, analogous to entropy-constrained VQ [8] we use a Lagrangian modified distortion expression which includes both ordinary distortion and classification error: $J_\lambda(\alpha, \beta, \delta) = D(\alpha, \beta) + \lambda B(\alpha, \delta)$. The modified distortion measure is used to determine the partitioning α of the training vectors by mapping each vector to the codeword/class label producing the minimum distortion J_λ. This measure allows flexible trade-offs between compression and classification priorities: when $\lambda \to 0$, we obtain usual VQ; when $\lambda \to \infty$, we obtain a Bayes risk classifier. Intermediate λ values allow the encoder to vary the relative emphasis of compression and classification.

We use this modified distortion measure to construct a VQ codebook by developing a descent algorithm which minimizes $J_\lambda(\alpha, \beta, \delta)$ by alternately improving α, β and δ. We start with some initial coder $(\alpha^{(0)}, \beta^{(0)}, \delta^{(0)})$ and iteratively apply an improvement transformation $(\alpha^{(t+1)}, \beta^{(t+1)}, \delta^{(t+1)}) = T(\alpha^{(t)}, \beta^{(t)}, \delta^{(t)})$ so that $J_\lambda(\alpha^{(t)}, \beta^{(t)}, \delta^{(t)})$ is nonincreasing in t. Since J_λ is bounded below by 0, we know that $J_\lambda^{(t)}$ must converge as $t \to \infty$. A simple stopping rule can be used to indicate convergence. For example, we can select a λ and iterate until the decrease in average distortion J_λ falls below a certain level.

The transformation T is implemented in three successive steps:

Step 1 Choose $\delta^{(t+1)}$ to minimize $J_\lambda(\alpha^{(t)}, \beta^{(t)}, \delta^{(t+1)})$.

Step 2 Choose $\beta^{(t+1)}$ to minimize $J_\lambda(\alpha^{(t)}, \beta^{(t+1)}, \delta^{(t+1)})$.

Step 3 Choose $\alpha^{(t+1)}$ to minimize $J_\lambda(\alpha^{(t+1)}, \beta^{(t+1)}, \delta^{(t+1)})$.

The three steps are repeated until the desired stopping conditions are met.

Given a fixed partition and codebook, Step 1 minimizes the Bayes risk associated with the codeword labels. As shown in (2), the minimization depends only on the partition represented by α; the codeword values β do not affect the minimization. For example, the minimum Bayes risk for the simple two class case is

$$\delta^{(t+1)}(i) = \begin{cases} H_1 & \text{if } \frac{P(\mathcal{A}_i|H_1)}{P(\mathcal{A}_i|H_2)} \geq \frac{P(H_2)C_{21}}{P(H_1)C_{12}} \\ H_2 & \text{otherwise.} \end{cases}$$

where the cells $\mathcal{A}_i$ are determined by $\alpha^{(t)}$.

Step 2 minimizes J_λ over the codeword values β given the partitioning α and the labeling δ. Because the codeword values do not affect the Bayes risk, J_λ is minimized when the codewords are chosen as centroids based on the distortion measure in (1).

Step 3 determines the partitioning α that minimizes J_λ given the set of codewords β and labels δ. We note that the modified distortion of the training sequence is

$$J_\lambda(\alpha,\beta,\delta) = D(\alpha,\beta) + \lambda B(\alpha,\delta)$$

$$= \sum_{X \in \mathcal{S}} P(X) \left\{ d(X,\beta(\alpha(X))) + \lambda \sum_{i=1}^{M} \sum_{j=1}^{M} I_{H_j}(\delta(\alpha(X))) \frac{P(X|H_i)}{P(X)} P(H_i)\, C_{ij} \right\}$$

$$\geq \sum_{X \in \mathcal{S}} P(X) \min_k \left\{ d(X,\beta(k)) + \lambda \sum_{i=1}^{M} \sum_{j=1}^{M} I_{H_j}(\delta(k)) P(H_i|X)\, C_{ij} \right\}.$$

The above probabilities are estimated using the vectors from the training set. Hence we chose our encoder as

$$\alpha^{(t+1)}(X) = \operatorname*{argmin}_k \left\{ d(X,\beta(k)) + \lambda \sum_{i=1}^{M} \sum_{j=1}^{M} I_{H_j}(\delta(k)) P(H_i|X)\, C_{ij} \right\} \tag{3}$$

which for the simple two class case, reduces to

$$\alpha^{(t+1)}(X) = \operatorname*{argmin}_k \left\{ d(X,\beta(k)) + \lambda \left(I_{H_2}(\delta(k)) P(H_1|X)\, C_{12} + I_{H_1}(\delta(k)) P(H_2|X)\, C_{21} \right) \right\}.$$

Because of the difficulty and complexity in estimating $P(H_i|X)$ outside the training set, in this study the optimal encoder was used only on the training set for code design and the ordinary distortion was used on test sequences. An encoder incorporating an estimate of $P(H_i|X)$ and hence extending the optimal encoder to the test sequence is currently being developed.

5 Extension to Tree-structured VQ

Full-search VQ can be computationally intensive in both codebook generation and encoding. Tree-structured VQ (TSVQ) significantly speeds up the design and encoding process with a small decrease in reproduction quality. Furthermore, full-search VQ produces a fixed-rate code. Variable-rate TSVQ [9] uses an unbalanced (usually binary) tree so that differing numbers of bits are required depending on the path that the encoder takes through the tree. The algorithm in Section 4 can be extended to TSVQ by combining it with known tree growing and pruning techniques [10]. In fact, such tree growing techniques were originally used for classification decision trees [11]. A TSVQ tree is grown by successively splitting nodes, until the desired rate is reached. By modifying the splitting criterion and codeword design methods, the TSVQ can be grown to provide both classification and compression.

6 Results

We demonstrate the compression and classification abilities of the algorithm on both simulated and real data. For the case of simulated data, the algorithm trains and encodes on separable data uniformly distributed over 2-dimensional space with a particular classification rule imposed. The simplicity of this problem provides the best means of visualizing the effects of the algorithm. The real data includes CT lung images and aerial images of the San Francisco Bay Area. The variety of problems demonstrates the flexibility of this method. Because a model is not assumed the encoder can develop its classification labeling based on the appropriate combination of intensities and vector patterns needed to distinguish between the classes.

Simulated data

We construct a training sequence of 65536 2-dimensional vectors (d_1, d_2) where d_1 and d_2 are independent identically distributed integers chosen uniformly over $[1 \ldots 256]$. The vectors in the training sequence are labeled as class 2 if $(|(d_1 - 128)| + |(d_2 - 128)|) < 128$ and as class 1 otherwise. The ideal classification is a diamond shape as shown in Figure 1; the brighter center region is class 2, and the darker corner regions are class 1. Codebooks are constructed for various values of λ using the iterative approach previously described. The partitioning and classification for a codebook with 32 codewords using $\lambda = 0$ is shown on the left in Figure 2. Again, the bright regions are labeled as class 2. This is the same quantizer as produced by the usual VQ algorithm, no classification information is used when constructing the codebook. In contrast, the partitioning and classification for the same size codebook using $\lambda = 5243$ is also shown on the right in Figure 2. The codeword boundaries are more closely aligned with the ideal classification boundaries, which improves the classification error. A plot showing the effect of λ on the classification error and on mean squared error is given in Figure 3. As expected, increasing λ generally reduces the classification error. Increasing λ generally increases the distortion, but the increases are fairly small (less than 5% over 6 decades of λ).

Medical images

We demonstrate the algorithm on CT images where we wish to both compress the images and identify pulmonary tumor nodules. Class 1 corresponds to healthy tissue and class 2 corresponds to tumor. The locations of the tumors are determined by trained radiologists. The training set consists of 10 images plus the tumor vectors from 5 additional images; the additional tumor training vectors are added because

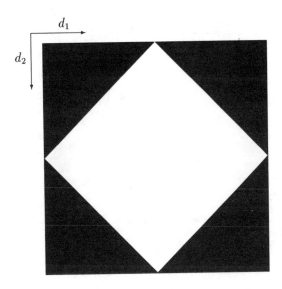

Figure 1: Ideal classification.

Figure 2: Full search codebook with 32 codewords, $\lambda = 0$ (left) and $\lambda = 5243$ (right).

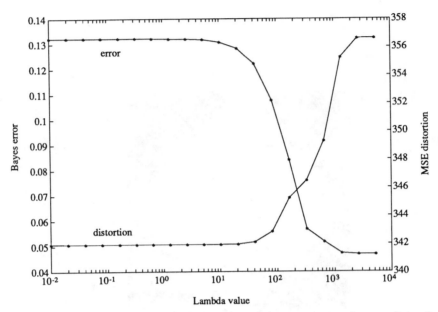

Figure 3: Effect of λ on classification error and mean squared error distortion.

of the low average percentage of tumor vectors in the data. We test on similar images outside the training set. Using classification costs of $C_{12} = 1$, $C_{21} = 100$ and $C_{11} = C_{22} = 0$ with a 2×2 vector block, full search codebooks are constructed using various values of λ. Each codebook contains 256 codewords, producing a fixed rate 2 BPP code (entropy coding would reduce this to about 1.56 BPP.) The costs are chosen to reflect the fact that classification errors have different consequences: false alarms are significantly less detrimental than missing a tumor. The algorithm provides a convenient way to incorporate such unbalanced costs.

The encoded test images were studied to determine the resulting peak signal to noise ratio (PSNR), sensitivity, specificity, predictive value positive, and overall Bayes risk. Sensitivity is the percentage of tumor vectors which are correctly classified as tumor. Specificity is the percentage of nontumor vectors which are correctly classified as nontumor. Predictive value positive (PVP) is the percentage of vectors classified as tumor that are actually tumor vectors. The algorithm was able to identify most of the tumor vectors in the test images, with up to 83.3 % sensitivity. Compression and classification results on one of the test images are shown in Figure 4 and Table 1. The images are shown after "windowing" the intensity to improve the contrast in the pixel value region of interest (the same adjustment performed by radiologists). In the image displaying the classification results, the brightest regions represent classification as tumor. This image contains three circular tumors in the left lung. The algorithm correctly identifies substantial parts of each of the three tumors.

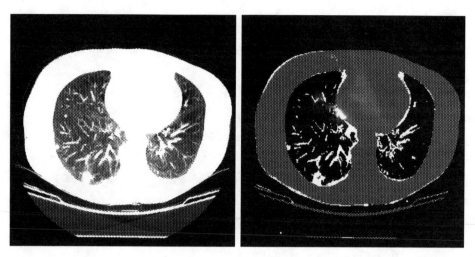

Figure 4: Compressed CT image at 2 BPP (left) and classification results (right).

Aerial images

In another application, we examined aerial images of the bay area. The classification task is to identify man-made regions rather than tumors. Again, good classification results are obtained: the test image had an overall classification error of only 25 %, comparing favorably with previous results [12].

7   Conclusions

Although better classifiers exist using more sophisticated algorithms, the classification here involves no extra decoding complexity when implemented and is based entirely on small blocks. It is significantly simpler than neural network, maximum likelihood, and other recognition algorithms and can provide a useful front end to more sophisticated algorithms yielding an overall performance improvement and complexity reduction.

Acknowledgements

This work was supported by the National Science Foundation under Grant MIP-9016974. The authors wish to thank Richard Olshen and Jack May for their many helpful comments.

PSNR	Sensitivity	Specificity	PVP	Bayes risk
39.96 dB	83.3 %	96.4 %	4.4 %	0.070

Table 1: Statistical results of algorithm on CT test image using $\lambda = 10$.

References

[1] R. M. Gray, "Vector quantization," *IEEE ASSP Magazine*, vol. 1,No. 2, pp. 4–29, April 1984.

[2] H. Abut, ed., *Vector Quantization*. IEEE Reprint Collection, Piscataway, New Jersey: IEEE Press, May 1990.

[3] A. Gersho and R. M. Gray, *Vector Quantization and Signal Compression*. Boston: Kluwer Academic Publishers, 1992.

[4] T. Kohonen, "An introduction to neural computing," *Neural Networks*, vol. 1, pp. 3–16, 1988.

[5] T. Kohonen, G. Barna, and R. Chrisley, "Statistical pattern recognition with neural networks: benchmarking studies," in *IEEE International Conference on Neural Networks*, pp. I-61–68, July 1988.

[6] T. Kohonen, *Self-organization and associative memory*. Berlin: Springer-Verlag, third ed., 1989.

[7] R. O. Duda and P. E. Hart, *Pattern Classification and Scene Analysis*. New York: John Wiley & Sons, Inc., 1973.

[8] P. A. Chou, T. Lookabaugh, and R. M. Gray, "Entropy-constrained vector quantization," *IEEE Trans. Acoust. Speech Signal Process.*, vol. 37, pp. 31–42, January 1989.

[9] J. Makhoul, S. Roucos, and H. Gish, "Vector quantization in speech coding," *Proc. IEEE*, vol. 73. No. 11, pp. 1551–1587, November 1985.

[10] E. A. Riskin and R. M. Gray, "A greedy tree growing algorithm for the design of variable rate vector quantizers," *IEEE Trans. Signal Process.*, vol. 39, pp. 2500–2507, November 1991.

[11] L. Breiman, J. H. Friedman, R. A. Olshen, and C. J. Stone, *Classification and Regression Trees*. Belmont,California: Wadsworth, 1984.

[12] K. L. Oehler, P. C. Cosman, R. M. Gray, and J. May, "Classification using vector quantization," in *Proc. Twenty-Fifth Annual Asilomar Converence on Signals, Systems, and Computers*, (Pacific Grove, Calif.), pp. 439–445, November 1991.

DOCUMENT IMAGE DECODING USING MARKOV SOURCE MODELS

Gary E. Kopec and Phil A. Chou

Xerox Palo Alto Research Center
3333 Coyote Hill Road
Palo Alto, CA 94304, USA

ABSTRACT

This paper describes a communication theory approach to document image recognition, patterned after the use of hidden Markov models in speech recognition. A document recognition problem is viewed as consisting of three elements— an image generator, a noisy channel and an image decoder. A document image generator is a Markov source which combines a message source with an imager. The message source produces a string of symbols which contains the information to be transmitted. The imager is modeled as a finite-state transducer which converts the message into an ideal bitmap. The channel transforms the ideal image into a noisy observed image. The decoder estimates the message from the observed image by finding the a posteriori most probable path through the combined source and channel models using a Viterbi-like algorithm. Application of the proposed method to decoding telephone yellow pages is described.

1. INTRODUCTION

The success of automatic speech recognition systems based on hidden Markov models (HMMs) has motivated recent attempts to apply similar methods to document image recognition problems. In particular, two approaches to bridging the gap from one-dimensional speech to two-dimensional images have been explored. The first approach is to focus on a subset of image analysis problems which may be viewed as actually one-dimensional (e.g. transcribing a single line of text) [10, 7, 2, 3]. While encourging results have been obtained, this approach is unlikely to lead to a general framework for document recognition.

The second approach is to extend the fundamental HMM grammar concepts to two dimensions [4, 9]. Previous attempts in this direction have generalized string grammar formalisms by replacing the notion of a one-dimensional phrase with that of a two-dimensional rectangular region. One disadvantage of the region-based approach is that regular (finite-state) image grammars are not particularly useful. Thus, most effort has been directed towards stochastic context-free two-dimensional grammars, which require time $O(n^3)$ to parse, where n is the number of terminal symbols (e.g. pixels) in the input. A second problem with region-based grammars is that they are based on non-overlapping regions, which makes them ill-suited for many complex graphical notations such as music.

This paper reviews an emerging alternative approach to using formal grammars for document image modeling and recognition that addresses the above problems [5]. The basis of our approach is a stochastic model of image generation that incorporates a deterministic finite state machine for sequentially imaging glyphs onto the plane. We begin with an overall formulation of the document image recognition problem in terms of classical communication theory. The three main elements in this formulation— image source, channel and decoder— are then discussed. Finally,

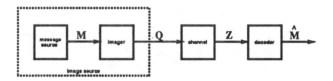

Figure 1: Communication theory view of document recognition.

we describe the application of the proposed method to the problem of decoding scanned telephone yellow pages to extract names and numbers from the listings.

2. PROBLEM FORMULATION

Fig. 1 shows a formulation of the document recognition problem based on classical communication theory, patterned after a similar formulation of speech recognition [1]. A stochastic *message source* selects a finite string M from a set of candidate strings according to a prior probability distribution $\Pr\{M\}$. A message may be a plain text string, text with embedded layout or logical structure tags, or any other encoding of information into a linear sequence of discrete symbols. The *imager* converts the message into an ideal binary image $Q = \{q_i \mid i \in \Omega\}$, drawn from a universe of binary images whose pixels are indexed by Ω. The pixel values $q_i \in \{0, 1\}$, where the foreground color (usually black) is denoted 1 and the background color (usually white) is 0. We assume that images are rectangular, so that Ω is the integer lattice $[0, W) \times [0, H)$, where W and H are the image width and height, respectively. Collectively, the message source and imager form the *image source*.

The *channel* maps the ideal image into an observed image $Z = \{z_i \mid i \in \Omega\}$ by introducing distortions due to printing and scanning, such as skew, blur and additive noise.

The *decoder* receives image Z and produces an estimate $\hat{M}$ of the original message. The decoder achieves minimum probability of error if it chooses $\hat{M}$ according to the maximum a posteriori (MAP) criterion, $\Pr\{\hat{M} \mid Z\} = \max_M \Pr\{M \mid Z\}$. It is straightforward to show that MAP decoding is equivalent to maximizing the normalized probability

$$\tilde{\Pr}\{M, Z\} = \sum_Q \tilde{\Pr}\{Z \mid Q\} \Pr\{Q, M\} \qquad (1)$$

where $\tilde{\Pr}\{Z \mid Q\} \equiv \frac{\Pr\{Z \mid Q\}}{\Pr\{Z \mid Q_0\}}$ and $\Pr\{Z \mid Q_0\}$ is the probability of observing Z given that Q is the all-white background image Q_0.

As (1) suggests, there are three main problems to be solved in the design of a document image recognition system. First, it necessary to develop a model of the image source, in order to

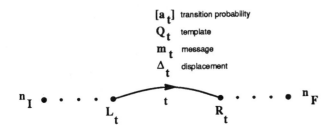

$$\begin{aligned}
[\mathbf{a}_t] & \quad \text{transition probability} \\
Q_t & \quad \text{template} \\
\mathbf{m}_t & \quad \text{message} \\
\Delta_t & \quad \text{displacement}
\end{aligned}$$

Figure 2: Markov source model for image generation.

compute $\Pr\{Q, M\}$. Second, it is necessary to understand the characteristics of the channel, in order to compute $\breve{\Pr}\{Z \mid Q\}$. Finally, a search algorithm is necessary in order to perform the maximization of (1) over M in a computationally efficient manner. The next three sections of this paper review solutions to the three document recognition system design problems.

3. IMAGE SOURCE MODEL

The structure of a set of images is captured formally by modeling image generation as a Markov source, as depicted in fig. 2. A Markov source consists of a finite set of states (nodes, vertices) $\mathcal{N}$ and a set of directed transitions (branches, edges) $\mathcal{B}$. Each transition t connects a pair of states, L_t and R_t, which are called, respectively, the *predecessor* (left) state and the *successor* (right) state of t. Two distinguished members of $\mathcal{N}$ are the initial state n_I, and the final state n_F.

With each transition t is associated a 4-tuple of attributes, $(Q_t, m_t, a_t, \vec{\Delta}_t)$, where Q_t is the *template*, m_t is the *message string*, a_t is the *transition probability* and $\vec{\Delta}_t$ is the vector *displacement* of t. A *path* π in a Markov source is a sequence of transitions $t_1 \ldots t_P$ for which $L_{t_1} = n_I$ and $R_{t_i} = L_{t_{i+1}}$ for $i = 1, \ldots, P-1$. A *complete path* is a path for which $R_{t_P} = n_F$.

Associated with each path π is a composite message M_π formed by concatenating the message strings of the transitions of the path. A Markov source defines a probability distribution on complete paths by $\Pr\{\pi\} = \prod_{i=1}^{P} a_{t_i}$ and induces a probability distribution on messages by $\Pr\{M\} = \sum_{\pi \mid M_\pi = M} \Pr\{\pi\}$, where M_π is the message associated with path π and the summation is taken over complete paths.

Also associated with each path π is a sequence of positions $\vec{x}_1 \ldots \vec{x}_{P+1}$ recursively defined by $\vec{x}_1 = \vec{0}$ and $\vec{x}_{i+1} = \vec{x}_i + \vec{\Delta}_{t_i}$, where $\vec{x}_{P+1}$ is introduced for convenience, and a composite image Q defined by $Q_\pi = \bigcup_{i=1}^{P} Q_{t_i}[\vec{x}_i]$. For images generated by a Markov source, it is not difficult to show that

$$\breve{\Pr}\{M, Z\} = \sum_{\pi \mid M_\pi = M} \breve{\Pr}\{Z \mid Q_\pi\} \Pr\{\pi\}. \tag{2}$$

4. CHANNEL MODEL

The purpose of channel modeling is to derive an expression for the normalized probability $\breve{\Pr}\{Z \mid Q\}$ for use in (1) or (2). To illustrate the role of analytic channel modeling in decoder design, we consider the simple asymmetric bit-flip noise model.

The asymmetric bit flip model assumes that each pixel of the ideal image Q is independently perturbed. The probability of a 1 (black) pixel in the ideal image Q surviving as a 1 in the observed image Z is α_1. Similarly, the probability of a 0 being observed as a 0 is α_0. The noise parameters are assumed to be constant over

the image. With these assumptions, we can show that

$$\begin{aligned}
\mathcal{L}(Z \mid Q) = & \left[(\text{\# of 1's in } Q) \cdot \log \frac{1-\alpha_1}{\alpha_0} \right] + \\
& \left[(\text{\# of 1's in } Q \wedge Z) \cdot \log \frac{\alpha_0 \alpha_1}{(1-\alpha_0)(1-\alpha_1)} \right]
\end{aligned}$$

where $\mathcal{L}(Z \mid Q) \equiv \log \breve{\Pr}\{Z \mid Q\} = \log \frac{\Pr\{Z \mid Q\}}{\Pr\{Z \mid Q_0\}}$. For images generated by a Markov source it may be further shown that

$$\mathcal{L}(Z \mid Q_\pi) = \sum_{i=1}^{P} \mathcal{L}(Z \mid Q_{t_i}[\vec{x}_i]). \tag{3}$$

5. DECODER

MAP decoding of an observed image Z with respect to a given image source involves finding a message $\hat{M}$ that maximizes $\breve{\Pr}\{M, Z\}$ in (2). One problem with direct implementation of the MAP rule is that (2) involves a summation over all complete paths π for which $M_\pi = M$. Depending on the source model, forming this sum can be computationally prohibitive. One standard approach to the analogous problem in speech recognition is to approximate the sum by its largest term, so that $\breve{\Pr}\{M, Z\} \approx \max_{\pi \mid M_\pi = M} \Pr\{\pi\} \breve{\Pr}\{Z \mid Q_\pi\}$. In that case, we can show that MAP decoding may be implemented by using a dynamic programming algorithm to compute the recursively-defined function

$$\mathcal{L}(n; \vec{x}) = \max_{t \mid R_t = n} \left\{ \mathcal{L}(L_t; \vec{x} - \vec{\Delta}_t) + \log a_t + \mathcal{L}(Z \mid Q_t[\vec{x} - \vec{\Delta}_t]) \right\} \tag{4}$$

for each $(n, \vec{x}) \in \mathcal{N} \times \Omega$. The practical significance of this result is that decoding time is $\mathbf{O}(\|\mathcal{B}\| \times \|\Omega\|)$, i.e. bilinear in the number of branches and the number of image pixels.

The decoder recursion requires a specification of the order in which the elements of $\mathcal{N} \times \Omega$ are visited. Such an order specification is called a *schedule* for the recursion. A valid schedule must satisfy the constraint that $\mathcal{L}(n; \vec{x})$ may be computed only after the computation of each $\mathcal{L}(L_t; \vec{x} - \vec{\Delta}_t)$ that appears on the right hand side of (4). Scheduling (4) is significantly more complicated than scheduling a speech HMM. We have developed an approach to this problem based on the general theory of linear schedules for regular iterative algorithms [6].

6. DECODING TELEPHONE YELLOW PAGES

Fig. 3 shows the reduced image of a sample yellow page column that illustrates most of the features found in the yellow pages published by one telephone company [8]. Columns contain two main types of material— *listings* and *display ads*. Listings fall into two main classes— individual listings and group listings. There are three main types of individual listings. The most common type is the *standard* individual listing (e.g. "Communications 2001"). In its usual form, a standard listing consists of a name line, one or more lines of address (or other information), and a telephone number. A *bold* listing (e.g. "APEX COMMUNICATIONS INC") is similar to a standard listing except that the name and phone number are in uppercase bold. A *boxed* listing (e.g. "HBC") has a bold name and a standard phone number and is enclosed by thick rulings along its top, left and bottom edges.

A *group* listing is a collection of individual listings typically used for franchises, businesses with multiple branches, brand names, etc. There are two types of group listings. A *standard group* listing (e.g. "PANASONIC AUTHORIZED SERVICENTERS") begins with a bold group name terminated by a short dash. The members of a group are divided into subgroups introduced by italic subgroup

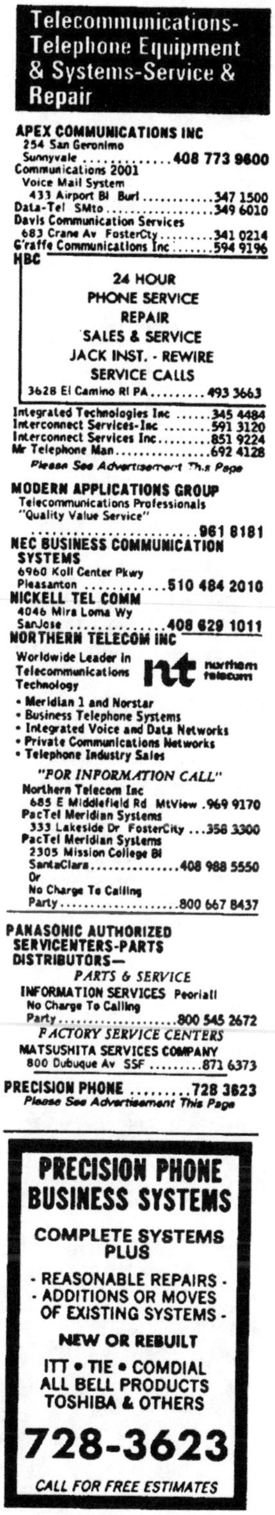

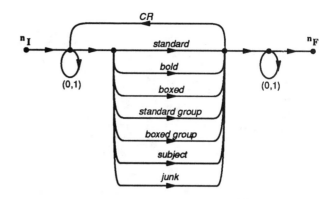

Figure 3: Example scanned yellow page column, slightly reduced.

Figure 4: Top-level source model of yellow page column. Transition probabilities omitted for simplicity.

names (e.g. "PARTS & SERVICE"). The individual listings in a standard group are called *stdgrpmem* listings and are similar to standard individual listings except that the name is uppercase. A second type of group listing is the *boxed group*, (e.g. "NORTHERN TELECOM INC") which has thin rulings along both top and bottom, similar to those of an individual boxed listing. The group members are formatted as either standard or bold individual listings. Boxed group listings frequently contain logos or other graphics.

6.1. Decoding Task and Model

We define an illustrative yellow page decoding task in terms of three objectives. The primary objective is to recognize and parse the subset of individual listings consisting of those containing a single telephone number. The desired output information for each listing is the name, telephone number and type. The second objective is to recover the hierarchical structure of groups and subgroups. The desired output for each subgroup is the subgroup name and information about its members. The output for a simple group is the group name and a description of the contained subgroups. The third objective is to recognize and transcribe the reverse-video subject headings. Collectively, the individual listings, subgroups, groups and subject headings to be decoded are called the *simple constituents* of the column.

We have developed a yellow page column model for extracting the above information which contains over 6000 branches and 1700 nodes. Fig. 4 shows the top-level column model. The basic structure of a column is a sequence of top-level constituents separated by rows of vertical whitespace. Top-level constituents comprise **standard**, **bold**, and **boxed** individual listings, **standard group** and **boxed group** group listings, **subject** headings and **junk** lines. The **CR** network returns the cursor from the right edge to the left edge of the column.

The above model has been used to decode a dataset of scanned column images, consisting of 48 columns from ten pages selected from [8] and scanned at 300 dpi. The columns contain 1208 simple constituents. Each of the 48 columns in the dataset was independently decoded, requiring 1.5 hours/column on a 28 MIPS RISC workstation. Fig. 5 shows the result of decoding the column in fig. 3.

Table 1 summarizes the performance of the decoder on the complete column database. Overall, 99.5% of the simple constituents were matched during decoding. The overall character recognition rates were 97.9% and 99.6% for the names and telephone numbers, respectively, with 98.5% of the number strings decoded exactly.

```
\subject{Telecommunications-\\Telepbone Equipment\\& Systems-Service &\\Repair}
\bold{\name{APEX COMMUNICATIONS INC} \number{408 773 9600}}
\standard{\name{Coinmuhications ZOOI} \number{347 1500}}
\standard{\name{Data-Tel 'SMto} \number{349 6010}}
\standard{\name{Davis Coinmuhication Services} \number{341 0214}}
\standard{\name{G'raffe Communicatlons Inc} \number{594 9196}}
\boxed{\name{HBC} \number{493 3663}}
\standard{\name{Integrated Tecnnologies Inc} \number{345 4484}}
\standard{\name{Interconnect Services-Inc} \number{591 3120}}
\standard{\name{Interconnect Services Inc} \number{851 9224}}
\standard{\name{Mr Telephone Man} \number{692 4128}}
\bold{\name{MOOERN APPLICATIONS GROUP} \number{961 8181}}
\bold{\name{NEC BUSINESS COMMUNICATION\\SYSTEMS} \number{510 484 2010}}
\bold{\name{NICKELL TEL COMM} \number{408 629 1011}}
\boxedgroup{\name{NORTHERN TELECOM INC}
    \subgroup{\name{''FOR INFORMATION CALL''}
        \standard{\name{Northern Telecorn Inc} \number{969 9170}}
        \standard{\name{PacTel Meridian Systems} \number{358 3300}}
        \standard{\name{PacTel Meridian Systems} \number{408 988 5550}}
        \standard{\name{' Qr} \number{0 667 8437}}}}
\standardgroup{\name{PANASONIC AUTHORIZED\\SERVICENTERS-PARTS\\DISTRIBUTDRS}
    \subgroup{\name{PARTS & SERVICE}
        \stdgrpmem{\name{INFQRMATION SERVICES 'P-J--JH} \number{545 2672}}}
    \subgroup{\name{FACTORY SERVICE CENTERS}
        \stdgrpmem{\name{MATSUSHITA SERVICES COMPANY} \number{871 6373}}}}
\bold{\name{PRECISION PHONE} \number{728 3623}}
\boxed{\name{NNNNNNNNNNNNNNNNNNNNMNNNNNNNNNNI} \number{8888881}}
```

Figure 5: Decoder output for column in fig 3.

Type	Constituent	Name		Number	
		string	character	string	char
all	99.5	62.7	97.9	98.5	99.6
individual	99.5	65.1	98.1	98.5	99.6
standard	99.6	58.9	97.7	99.2	99.8
bold	100.0	80.6	98.9	96.8	99.6
boxed	97.5	80.0	98.7	99.8	99.9
stdgrpmem	100.0	85.7	99.3	85.7	95.5
group	97.1	79.4	98.9	---	---
standard	100.0	76.9	98.9	---	---
boxed	88.9	87.5	99.2	---	---
subgroup	100.0	21.6	95.8	---	---
standard	100.0	14.3	95.3	---	---
boxed	100.0	44.4	98.5	---	---
subject	100.0	43.0	97.3	---	---

Table 1: Constituent and string recognition rates for yellow page column decoding.

REFERENCES

[1] L. Bahl, F. Jelinek, and R. Mercer, "A maximum likelihood approach to continuous speech recognition", *IEEE Trans. on Pattern Analysis and Machine Inelligence*, vol. PAMI-5, no. 2, March 1983.

[2] C. B. Bose and S. Kuo, "Connected and Degraded Text Recognition using a Hidden Markov Model", *International Conference on Pattern Recognition*, The Hague, Netherlands, 1992.

[3] F. Chen, L. Wilcox and D. Bloomberg, "Word spotting in scanned images using hidden Markov models", *Proc. 1993 IEEE International Conference on Acoustics, Speech, and Signal Processing*, Minneapolis, MN, April 27–30, 1993.

[4] P. Chou, "Recognition of equations using a two-dimensional stochastic context-free grammar", *SPIE Conf. on Visual Communications and Image Processing*, Philadelphia, PA, Nov. 1989.

[5] G. Kopec and P. Chou, "Document image decoding using Markov source models", submitted to *IEEE Trans. Pattern Analysis and Machine Intelligence*, Jan., 1992. also available as XEROX Palo Alto Research Center report P92-00075.

[6] G. Kopec, "Row-Major Scheduling of Image Decoders", submitted to *IEEE Trans. on Image Processing*, Feb. 1992. also available as XEROX Palo Alto Research Center report P92-00061.

[7] A. Kundu, Y. He, and P. Bahl, "Recognition of Handwritten Word: First and Second Order Hidden Markov Model Based Approach", *Pattern Recognition*, vol. 22, no. 3, pp. 283–297, 1989.

[8] Pacific Bell, *Smart Yellow Pages, Palo Alto, Redwood City and Menlo Park*, 1992.

[9] M. Tomita, "Parsing 2-dimensional language", *ACM International Workshop on Parsing Technologies*, 1989.

[10] J. Vlontzos and S.Y. Kung, "Hidden Markov models for character recognition", *Proceedings of the 1989 IEEE International Conference on Acoustics, Speech, and Signal Processing*, pp. 1719–1722, Glasgow, Scotland, May 23–26, 1989.

CONNECTED AND DEGRADED TEXT RECOGNITION USING PLANAR HIDDEN MARKOV MODELS

Oscar E. Agazzi[1] Shyh-shiaw Kuo[1] Esther Levin[2] Roberto Pieraccini[2]

[1]Signal Processing Research Department
[2]Speech Research Department
AT&T Bell Laboratories, Murray Hill, NJ 07974, U.S.A.

ABSTRACT

We present an algorithm for connected text recognition using enhanced Planar Hidden Markov Models (PHMMs). The algorithm we propose automatically segments text into characters (even if they are highly blurred and touching) as an integral part of the recognition process, thus jointly optimizing segmentation and recognition. Performance is enhanced by the use of state length models, transition probabilities among characters (bigrams), and grammars. Experiments are presented using: 1) A simulated database of over $24,000$ highly degraded images of city names; 2) A database of $6,000$ images rejected by a high performance commercial OCR machine with 99.5% accuracy. Measured performance on the first database is 99.65% for the most degraded images when a grammar is used, and 98.76% in the second database. Traditional OCR algorithms would fail drastically on these images.

1. INTRODUCTION

This paper describes a set of experiments for assessing the performance of the Planar Hidden Markov Model (PHMM) paradigm [1] in the recognition of connected printed text that has been severely degraded by noise, blur, and other stochastic distortions, as often happens with documents that have been repeatedly reproduced and/or transmitted by fax. Results show recognition accuracies in excess of 98% for images that have been so severely degraded that even a human observer has difficulty in recognizing them. The experiments were carried out using two large databases:

1] A simulated database of 24,600 images representing city names (Database I)

2] A database of 6126 images rejected by a high performance commercial optical character recognition (OCR) machine with 99.5% accuracy (Database II).

Database I has the advantage of allowing more control over parameters such as noise, blur, or character overlap, and was used for studying the variation of the recognition accuracy as a function of these parameters. Database II offers the advantage of testing our algorithm in a challenging "real world" application of significant commercial importance.

2. PLANAR HIDDEN MARKOV MODELS

Fig.1 shows the PHMM structure we use and the way it represents a character.

Each PHMM consists of a rectangular array of $N_X \times N_Y$ states, organized as N_X "superstates", each one of them consisting of N_Y states. In the vertical direction, transitions are allowed only among states of a given superstate. In the horizontal direction, transitions among arbitrary superstates are possible. These restrictions are necessary to ensure that the recognition algorithm runs in polynomial time [1]. In the application described in this paper, we further restrict the PHMMs, such that:

1] Superstates are strictly "top-to-bottom", i.e., from a certain state we allow transitions only to itself or to the next state.

2] The models are strictly "left-to-right", i.e., only transitions to itself or to the next superstate are allowed from a given superstate.

In an isolated character recognition application, the image is scanned from top to bottom and from left to right, and a score is computed for each PHMM representing a character. In the vertical scan, each superstate is scored using the Viterbi algorithm. These scores are then used in a second pass of the Viterbi algorithm while the image is scanned in the horizontal direction, yielding a score for the PHMM under test. All PHMMs are scored in a similar way, and the best one is chosen. The PHMMs are treated as nested one-dimensional models, rather than truly two-dimensional.

In this paper we deal with the recognition of strings, rather than isolated characters. This problem is similar to the problem of recognizing connected words in speech recognition, where the problem is solved by searching for the best path (the one with the highest likelihood) in a network that allows for all the possible combinations of phonetic HMMs [2]. Analogously, for our OCR problem, we built a network that allows for all the possible combinations of characters, each character being represented by a PHMM. The structure of this network is illustrated in Fig.2.

The network is searched with the generalized Viterbi algorithm described in [1], [3]. This network interconnects all the character-level PHMMs, allowing arbitrary transitions from character to character. The transition probabilities among PHMMs correspond to the transition probabilities among characters, or probabilities of bigrams in the English language (or any other language according to the applica-

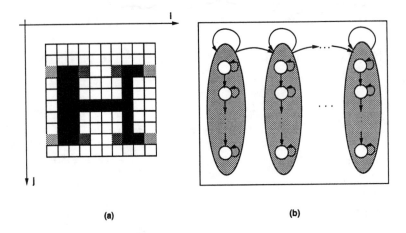

Figure 1: Structure of PHMMs. (a) The grid square located at ordinate j and abscissa i represents state j of superstate i. The grey levels represent the continuous range $[0, 1]$ of probabilities of black pixels. (b) State and superstate transition diagram corresponding to the array of (a).

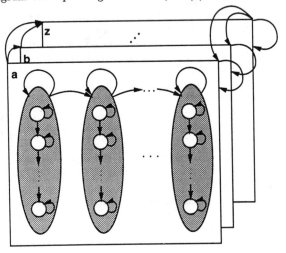

Figure 2: PHMM network used for connected character recognition

tion).

3. STATE AND SUPERSTATE LENGTH MODELING

The "dynamic planar warping" property of PHMMs [1] provides an excellent way to accommodate character shape variations, such as those encountered in multifont character recognition, or geometric distortions introduced by processes like fax transmission. However in their simplest form these models can also accommodate some unrealistic forms of distortion, like the transformation of a "p" into a "b", with very little penalty. The reason for this is that, in principle, when superstates are applied to different columns in the image, they are not forced to undergo the same or similar warping. This excessive flexibility would allow the descender of "p" to be transformed into an ascender, changing the identity of the character into a "b". Many other undesirable forms of distortion are also possible as a result of the same effect. This effect can be avoided completely while keeping all the flexibility to accommodate more real-

istic forms of distortion by modeling the lengths of the pixel sequences generated by states and superstates with a Gaussian distribution, instead of the exponential distribution inherent in the simple form of the models. The Gaussian distribution imposes a heavier penalty for extreme forms of distortion, thus preventing their occurrence.

The computational cost of introducing Gaussian state and superstate length densities can be made negligible by using the postprocessing approach described in [4].

4. SIZE NORMALIZATION

Traditionally, size normalization is applied to images *before* recognition. If text is not connected and therefore characters can be reliably segmented before recognition, it is usually not a problem to estimate the scaling factor and normalize the image. However, in the case of extremely connected and degraded text of unknown font, there is not enough information to estimate the scaling factor before recognition. As seen in section 2, the PHMM approach combines segmentation and recognition, thus avoiding one of the common causes of failure of traditional OCR systems when presented with connected text [5], namely, the inability of the recognition engine to recover from errors introduced by the segmentation engine. Another distinct advantage of PHMMs is that they also allow size normalization to be combined with recognition, therefore avoiding another possible failure mode of non-PHMM systems, i.e., the inability of the recognition engine to recover from inaccurate size normalization.

5. FEATURES

The feature vector can be simply the binary value of the pixels. However performance can be improved by the use of a more elaborate feature vector, where the pixel value is augmented by four more binary components representing the presence or absence of straight line segments at $0°$, $45°$, $90°$, and $135°$ in the neighborhood of the pixel. These segments are extracted by a Hough transform applied in a small square window around the pixel.

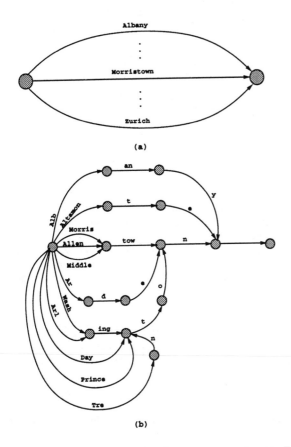

(a)

(b)

Figure 3: Grammars used in the city names experiment. For clarity, only a few nodes are shown. (a) Simple grammar. (b) Grammar obtained by minimization of (a).

6. USE OF GRAMMARS

In character recognition systems it is common to correct recognition errors by searching in a dictionary for the strings that best match the ones provided by the recognizer. Dictionary search is in these cases a postprocessing operation. When using PHMMs, it is possible to constrain the recognized strings to be members of a certain lexicon by using a grammar. Unlike the traditional postprocessing approach, this method prevents the recognizer from generating erroneous strings, instead of correcting them a posteriori, resulting in enhanced performance. Fig.3 shows a simplified diagram of a grammar used in the experiments with Database I described in the next section.

7. EXPERIMENTS

Experiments to evaluate the performance of the algorithm have been conducted on a database of 24,600 images representing blurred and noisy instances of 205 city names printed in Times-Roman size 10 characters (Database I). The original binary images represented by 250×40 pixel arrays, were corrupted by pixel-flip noise with a probability P_N, then they were blurred using a Gaussian filter with standard deviation σ_B, and finally thresholded at a level T. Values of σ_B and T were kept fixed at 1.5 pixels and 160 (in a scale where 0 represents black and 255 represents white), whereas P_N was varied between 0. and 0.25 in intervals of

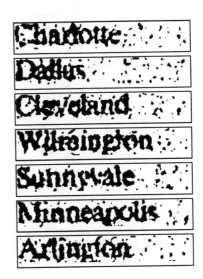

Figure 4: Images of city names Charlotte, Dallas, Cleveland, Wilmington, Sunnyvale, Minneapolis, and Arlington in the simulated database. In these images the probability of noise is $P_n = 0.25$.

P_n	Char. Accuracy	String Accuracy
≤ 0.20	100.00%	100.00%
0.25	99.65%	99.62%

Table 1: Recognition accuracy on Database I. In these experiments a grammar was used to improve performance.

0.05. These parameter values were chosen to produce realistic forms of distortion, closely resembling the effect of multiple reproductions and/or fax transmissions. The largest values of P_N generate severe levels of distortion, making some of the images difficult to recognize even for a human observer (Fig.4). For each value of P_N, a set of 4100 images was generated (20 instances of each city name). Half of them were used for training and the rest for testing.

The results of our experiments are given in Table 1.

The accuracy is extremely high even for highly degraded images such as the ones shown in Fig.4, and it is 100% for $P_N \leq 0.20$. Traditional OCR algorithms would fail drastically on these images. A major factor contributing to this excellent performance is the use of a grammar (Fig.3) that constrains the recognized string to be one of the 205 city names.

A second set of experiments was conducted on another database consisting of 6,126 strings rejected by a commercial OCR machine (Database II). This machine has a 99.5% accuracy, however in the high volume application where it is used, the cost of manually processing the 0.5% rejects is high. A robust algorithm that can recognize a large fraction of these rejects would be desirable. Fig.5 shows some typical images of this database. Although characters are not very connected, the resolution and image quality are poor, and there is a mixture of different fonts and sizes. In this case no probabilities of bigrams, grammars, or any other form of contextual information was available to improve

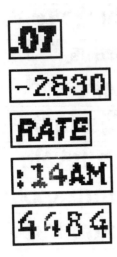

Figure 5: Images from Database II.

Character Accuracy	98.76%
String Accuracy	96.13%

Table 2: Recognition accuracy on database of rejects of commercial OCR machine.

the recognition results. Approximately half of the database was used for training and the rest for testing. The results of the experiment are summarized in Table 2.

If this level of accuracy could be maintained in the field, the compounded error rate of the combination of the commercial OCR machine with the HMM algorithm would be reduced to 0.0062%, or 1 error in 16,000 characters. This integration of the two algorithms, however, has not been done.

More than half of the residual errors reported in Table 2 are substitutions caused by highly confusable pairs, such as O and 0, which could be corrected by the use of contextual information, so that a reduction of the error rate of at least a factor two could be achieved if this information were incorporated.

Experiments were also conducted on Database II to assess the relative importance of some of the enhancements of the basic algorithm that were described in previous sections. Table 3 shows the impact on performance of removing these enhancements.

It can be seen that the use of Gaussian state duration

Enhancement Removed	Char. Acc.	Str. Acc.
State dur. post.	91.73%	75.80%
Size normaliz.	94.46%	80.96%
Vector features	91.40%	75.53%

Table 3: Effect on performance of eliminating algorithm enhancements.

Beam Width	Char. Acc.	Str. Acc.	Time/char
50	97.62%	93.87%	0.65s
100	98.76%	96.13%	0.81s
200	98.83%	96.41%	1.15s
500	98.94%	96.75%	2.17s

Table 4: Accuracy and computation time as functions of the beam width.

postprocessing and the use of vector features are the most important. The importance of size normalization depends on the size variability in the test set.

In these experiments, beam search was used to reduce computation time. Table 4 shows the computation time on an SGI Indigo workstation for several values of beam width, and the associated effect on performance (the results of Table 2 correspond to a beam width of 100).

8. CONCLUSION

We described an algorithm for the recognition of connected and degraded text based on PHMMs. A number of enhancements have been added to the basic algorithm of [1], including incorporation of Gaussian state duration densities, size normalization done jointly with recognition, and the use of a feature vector that incorporates information about the orientation of line segments. We also reported on the results of extensive experiments on two databases. In one of them, use of a PHMMs and a grammar results in recognition accuracy of 99.65% in the presence of text so degraded that it is almost unrecognizable even for a human observer. In the other, accuracy is 98.76%.

REFERENCES

[1] E.Levin and R.Pieraccini, "Dynamic planar warping for optical character recognition", *Proceedings of the 1992 International Conference on Acoustics, Speech, and Signal Processing*, San Francisco, California, March 23-26, 1992.

[2] L.R.Rabiner, "A tutorial on hidden Markov models and selected applications in speech recognition", *Proceedings of the IEEE*, Vol. 77, No. 2, pp. 257-286, Feb. 1989.

[3] S.Kuo and O.E.Agazzi, "Machine vision for keyword spotting using pseudo 2D hidden Markov models", *Proceedings of the 1993 International Conference on Acoustics, Speech, and Signal Processing* Minneapolis, April 27-30, 1993.

[4] L.R.Rabiner, B.H.Juang, S.E.Levinson, and M.M.Sondhi, "Recognition of isolated digits using hidden Markov models with continuous mixture densities", *AT&T Technical Journal*, Vol. 64, No.6, pp. 1211-1233, July-August 1985.

[5] M.Bokser, "Omnidocument technologies", *Proceedings of the IEEE*, vol.80, No.7, pp.1066-1078, July 1992.

Evaluation of Character Recognition Systems

C. L. Wilson

National Institute of Standards and Technology
Gaithersburg, MD 20899

Abstract

At the first Census Optical Character Recognition Systems (COCRS) Conference, National Institute of Standards and Technology (NIST) produced accuracy data for more than 40 character recognition systems. The recognition experiments were performed on sample sizes of 58,000 digits, and 12,000 upper and lower case alphabetic characters. The algorithms used by the 26 conference participants included rule-based methods, image-based methods, statistical methods, and neural networks. The neural network methods included Multi-Layer Perceptron's, Learned Vector Quantitization, Neocognitrons, and cascaded neural networks.

In this paper 11 different COCRS systems are evaluated using correlations between the answers of different systems, comparing the decrease in error rate as a function of confidence of recognition, and comparing the writer dependence of recognition. This comparison shows that methods that used different algorithms for feature extraction and recognition performed with very high levels of correlation. Subsquent experiments were performed by NIST to compare the OCR accuracy of various neural network and statistical classification systems. For each neural network system a statistical system of comparable accuracy was developed. These experiments tested seven different classifiers using 11 different feature sets and obtained OCR error levels between 2.5% and 5.1% for the best feature set sizes. This similarity in accuracy is true for neural network systems and statistically based systems and leads to the conclusion that neural networks have *not* yet demonstrated a clear superiority to more conventional statistical methods in either the COCRS test or in independent tests at NIST.

1 Introduction

At the first COCRS Conference a large number of systems (40 for digits) were used to recognize the same sample of characters [1]. A summary of these results is given in Table 1. Neural network systems, systems combining neural network methods with other methods (hybrid system), and systems based entirely on statistical pattern recognition methods were submitted to the COCRS conference. This provides a large test sample which can be used to detect differences between these various methods. In addition, subsequent test at NIST [2] using a seven different neural network (NN) and statistical classifiers confirmed that using a fixed set of features both types of methods have similar accuracies.

In this paper 11 different COCRS conference systems are discussed in some detail. These system are itemized by type in Table 2. These systems are broken into NN based systems, hybrid systems, and non-NN systems. The author realizes that this distinction is subject to interpretation, but it does allow some useful comparisons to be made. The COCRS conference systems were all designed to use different methods of feature extraction. In order to seperate feature extraction and classification, the image recognition group at NIST performed classification experiments using seven methods of classification and a common set of Karhunen-Loève (KL) based features [3]. The results of these tests are shown in Table 3.

In the past few years NN's have become important as a possible method for constructing computer programs that can solve problems, such as speech and character recognition, where "human-like" response or artificial intelligence is needed. The most useful characteristics of NN's are their ability to learn from examples, their ability to operate in parallel, and their ability to perform well using data that are noisy or incomplete. Many of these characteristics are shared by various statistical pattern recognition methods. These characteristics of pattern recognition systems are important for solving real problems from the field of character recognition exemplified by this paper.

It is important to understand that the accuracy of the trained OCR system produced will be strongly dependent on both the size and the quality of the training data. Many common test examples used to demonstrate the properties of pattern recognition system contain on the order of 10^2 examples. These examples show the basic characteristics of the system but provide only an approximate idea of the system accuracy.

As an example, the first version of an OCR system was built at NIST using 1024 characters for training and testing. This system has an accuracy of 94%. As the sample size was increased the accuracy initially dropped as more difficult cases were included. As the test and training sample reached 10000 characters the accuracy began to slowly improve. The poorest accuracy achieved was with sample sizes near 10^4 and was 85%. The 58,000 digit sample discussed in this paper is well below the 10^5 character sample size which we have estimated is necessary to saturate the learning process of the NIST system [3]. The best system developed by NIST uses probabalistic NNs (PNN) [13] and achieved an accuracy of 2.5% when trained on 7480 digits.

The goal of this paper is to discuss the different kinds of methods used at the COCRS Conference in a way that will illustrate why NN's and statistical methods achieved similar levels of performance. The various methods used are summarized in Figure 1 for classification and

Entered	Percentage Classification Error		
System	Digits	Uppers	Lowers
AEG	3.43 ± 0.23	3.74 ± 0.82	12.74 ± 0.75
ASOL	8.91 ± 0.39	11.16 ± 1.05	21.25 ± 1.36
ATT_1	3.16 ± 0.29	6.55 ± 0.66	13.78 ± 0.90
ATT_2	3.67 ± 0.23	5.63 ± 0.63	14.06 ± 0.95
ATT_3	4.84 ± 0.24	6.83 ± 0.86	16.34 ± 1.11
ATT_4	4.10 ± 0.16	5.00 ± 0.79	14.28 ± 0.98
COMCOM	4.56 ± 0.91	16.94 ± 0.99	48.00 ± 1.87
ELSAGB_1	5.07 ± 0.32		
ELSAGB_2	3.38 ± 0.20		
ELSAGB_3	3.35 ± 0.21		
ERIM_1	3.88 ± 0.20	5.18 ± 0.67	13.79 ± 0.80
ERIM_2	3.92 ± 0.24		
GMD_1	8.73 ± 0.35	14.04 ± 1.00	22.54 ± 1.22
GMD_2	15.45 ± 0.64	24.57 ± 0.91	28.61 ± 1.25
GMD_3	8.13 ± 0.39	14.22 ± 1.09	20.85 ± 1.25
GMD_4	10.16 ± 0.35	15.85 ± 0.95	22.54 ± 1.22
GTESS_1	6.59 ± 0.18	8.01 ± 0.59	17.53 ± 0.75
GTESS_2	6.75 ± 0.30	8.14 ± 0.59	18.42 ± 1.09
HUGHES_1	4.84 ± 0.38	6.46 ± 0.52	15.39 ± 1.10
HUGHES_2	4.86 ± 0.35	6.73 ± 0.64	15.59 ± 1.08
IBM	3.49 ± 0.12	6.41 ± 0.80	15.42 ± 0.95
IFAX	17.07 ± 0.34	19.60 ± 1.26	
KAMAN_1	11.46 ± 0.41	15.03 ± 0.79	31.11 ± 1.15
KAMAN_2	13.38 ± 0.49	20.74 ± 0.88	35.11 ± 1.09
KAMAN_3	13.13 ± 0.45	19.78 ± 0.60	33.55 ± 1.37
KAMAN_4	20.72 ± 0.44	27.28 ± 1.30	46.25 ± 1.23
KAMAN_5	15.13 ± 0.41	33.95 ± 1.22	42.20 ± 0.96
KODAK_1	4.74 ± 0.37	6.92 ± 0.78	14.49 ± 0.77
KODAK_2	4.08 ± 0.26		
MIME	8.57 ± 0.34	10.07 ± 0.81	
NESTOR	4.53 ± 0.20	5.90 ± 0.68	15.39 ± 0.90
NIST_1	7.74 ± 0.31	13.85 ± 0.83	18.58 ± 1.12
NIST_2	9.19 ± 0.32	23.10 ± 0.88	31.20 ± 1.16
NIST_3	9.73 ± 0.29	16.93 ± 0.90	20.29 ± 0.99
NIST_4	4.97 ± 0.30	10.37 ± 1.28	20.01 ± 1.06
NYNEX	4.32 ± 0.22	4.91 ± 0.79	14.03 ± 0.96
OCRSYS	1.56 ± 0.19	5.73 ± 0.63	13.70 ± 0.93
REI	4.01 ± 0.26	11.74 ± 0.90	
RISO	10.55 ± 0.43	14.14 ± 0.88	21.72 ± 0.98
SYMBUS	4.71 ± 0.38	7.29 ± 1.07	
THINK_1	4.89 ± 0.24		
THINK_2	3.85 ± 0.33		
UBOL	4.35 ± 0.20	6.24 ± 0.66	15.48 ± 0.81
UMICH_1		5.11 ± 0.94	15.08 ± 0.92
UPENN	9.08 ± 0.37		
VALEN_1	17.95 ± 0.59	24.18 ± 1.00	31.60 ± 1.33
VALEN_2	15.75 ± 0.32		

Table 1: Mean zero-rejection-rate error rates and standard deviations in percent calculated over 10 partitions of the COCRS conference test data. See [1] for details

feature extraction. Most of the systems presented at the Conference used separate methods of feature extraction and classification. In the discussion presented here any image processing which preceded the feature extraction is combined with feature extraction. The results of these comparisons are presented in sections 2 by algorithm type and in section 3 for NN and statistical algorithms.

Since the results of the COCRS conference were different than was originally expected, NIST conducted a set of pattern classification experiments using KL features sets of different sizes and using seven different classification methods. These experiments confirm the COCRS conference results. These results are discusssed in section 5.

2 Types of Algorithms Used

The discriminant function and classification sections of the systems are of two types: adaptive learning based and rule-based. The most common approach to machine learning based systems used at the Conference was NNs. The neural approach to machine learning was originally devised by Rosenblat [4] by connecting together a layer of artificial neurons [5] on a perceptron network. The observations which were present in this approach were analyzed by Minski and Papert [6]. The results of this Conference suggest that many of these weaknesses are still relevant. The advent of new methods for network construction and training during the last ten years led to rapid expansions in NN research in the late 1980s. Many of the methods referred to in Figure 1 were developed in this period. Adaptive learning is further subdivided into two types, supervised learning and self-organization. The material presented in this paper does not cover the mathematical detail of these methods, but the bibliographic references provided with many of the systems [1] discuss these methods in detail.

The principal difference between NN methods and rule-based methods is that the former attempt to simulate intelligent behavior by using adaptive learning and the latter use logical symbol manipulation. The two most common rule-based approaches at the Conference were those derived from mathematical image processing and those derived from statistics. Image based methods are usually used for feature extraction while statistical methods are usually used for classification.

Most of the OCR implementations discussed in this report combine several methods to carry out preprocessing (filtering) and feature extraction. Many of the filtering methods used are based on methods described in texts on image processing such as [7] and on methods based on KL transforms [3]. In these methods, the recognition is done using features extracted from the primary image by rule based techniques. The filtering and feature extraction processes start with an image of a character.

The features produced are then used as the input for classification.

In a self-organizing method, such as [8], data is applied directly to the NN and any filtering is learned as features are extracted. In a supervised method, the features are extracted using either rule-based or adaptive methods and classification is carried out using either type of method.

In Figure 1, rules based on mathematical image processing are distinguished from rules based on statistics. These two types of rules are similar in that they both derive features based on a model of the images. Statistical rules derive these model parameters based on the data presented. For example, typical model parameters might be sample means and variances. Mathematical rules operate on the data based on external model parameters or on the specific data being analyzed. The model parameters might be designed to detect strokes, curvature, holes, or concave or convex surfaces.

All of the methods shown in Figure 1 can also be categorized broadly into linear methods, such as LVQ [9], and nonlinear methods, such as Multi-Layer Perceptrons (MLPs) [10]. This separation into linear and non-linear algorithms also extends to mathematical and statistical methods. Many of the convolution and transform methods, such as combinations of Gabor transforms [11] are linear. Other methods start with linear operations such as correlation matrices and become non-linear by removing information with low statistical significance; KL transforms [7] and principal component analysis (PCA) [12] are examples of this.

When training data is used to adjust statistical model parameters to train MLPs, certain methods may be classified as either NN or statistical methods. The PNN [13] is an example of this type of method. In another context PNN methods can be regarded as one class of a radial basis function (RBF) method [14]. The information in Figure 1 classifies methods of this kind in an arbitrary way when statistical accumulation or NN models of a given method are equivalent.

System	Features	Classification
Neural Net		
ATT_2	receptor fields	MLP
Hughes_1	neocognitron	
Nestor	necognitron	MLP
Symbus	raw	self-Org. NN
Hybrid		
ERIM_1	morophological	MLP
Kodak_2	Gabor	MLP
NYNEX	model	MLP
NIST_4	K-L	PNN
Non Neural Net		
Think_1	template	distance maps
UBOL	rule based	KNN
Elsagb_1	shape func.	KNN

Table 2: Feature extraction and classification methods used for the 11 system discussed.

System	24	28	32	36	40	44	48	52	56	60	64
KNN:1	2.9	2.7	2.7	2.7	**2.6**	2.6	2.6	2.7	2.7	2.7	2.7
KNN:3	2.8	2.7	2.7	2.7	**2.6**	2.7	2.7	2.7	2.7	2.8	2.7
KNN:5	2.9	2.8	2.8	**2.7**	2.8	2.8	2.8	2.8	2.8	2.8	2.8
WSNN:1.1	2.8	2.7	2.6	2.6	**2.5**	2.6	2.6	2.6	2.6	2.5	2.6
PNN:3.0	2.7	2.7	2.6	2.6	**2.5**	2.6	2.6	2.6	2.6	2.5	2.5
MLP:32	5.8	5.6	5.7	5.5	5.6	5.5	**5.3**	5.4	5.4	5.3	5.4
MLP:48	5.2	5.2	5.0	4.7	4.9	5.0	4.7	**4.6**	4.9	5.0	4.9
MLP:64	4.6	4.5	4.6	4.5	4.5	4.5	4.5	4.3	4.5	4.4	4.5
RBF1:1	13.2	13.1	13.9	13.0	**12.6**	13.4	12.6	13.2	13.3	13.2	13.2
RBF1:2	8.5	8.5	8.4	8.2	8.4	8.2	8.1	8.3	8.1	**7.9**	7.9
RBF1:3	6.7	6.6	6.5	6.5	6.5	6.4	6.4	**6.2**	6.4	6.2	6.3
RBF1:4	5.7	5.5	5.5	5.5	5.4	5.5	5.4	5.4	**5.3**	5.3	5.4
RBF1:5	5.0	4.7	4.9	5.0	4.9	4.8	4.7	4.9	4.9	4.7	**4.6**
RBF1:6	4.6	4.4	4.3	4.5	4.3	4.3	**4.2**	4.2	4.4	4.3	4.4
RBF2:1	8.7	9.5	9.1	9.1	9.2	**8.6**	8.8	8.8	8.9	8.9	8.9
RBF2:2	6.7	6.4	**6.1**	6.1	6.3	6.3	6.2	6.3	6.2	6.2	6.5
RBF2:3	5.6	5.5	5.0	6.0	5.4	**4.9**	5.7	4.9	5.0	5.6	5.0
RBF2:4	4.4	5.6	5.0	**4.3**	4.5	4.6	4.6	4.5	4.4	4.8	4.7
RBF2:5	4.5	4.6	4.4	4.6	4.4	4.4	4.4	4.2	4.1	4.1	**4.0**
RBF2:6	4.3	4.5	4.0	4.0	4.2	**3.9**	4.2	4.0	3.9	4.0	4.0
EMD:1	15.2	15.1	15.0	15.0	14.9	14.9	**14.8**	14.8	14.8	14.8	14.8
EMD:2	11.0	10.8	10.7	10.7	10.7	10.7	10.7	**10.6**	10.6	10.6	10.6
EMD:3	8.8	8.8	8.7	**8.6**	8.6	8.7	8.7	8.7	8.7	8.7	8.7
EMD:4	7.3	7.3	7.4	7.3	**7.1**	7.2	7.1	7.1	7.1	7.1	7.1
EMD:5	6.7	6.6	6.6	6.5	6.3	6.7	6.6	**6.2**	6.2	6.2	6.3
EMD:6	6.1	5.9	6.1	6.0	**5.7**	6.0	5.8	5.9	6.0	5.9	6.1
EMD:7	5.6	5.3	5.5	5.3	5.2	5.4	**5.1**	5.2	5.4	5.4	5.6
QMD:1	**4.8**	4.9	5.1	5.1	5.2	5.3	5.6	5.6	5.8	5.8	5.9
QMD:2	**4.7**	4.9	4.9	5.0	5.2	5.3	5.5	5.6	5.7	5.8	5.9
QMD:3	**4.0**	4.5	4.7	4.9	5.1	5.3	5.4	5.6	5.9	6.0	6.3
QMD:4	**4.5**	4.9	5.0	5.3	5.5	6.1	6.3	6.5	6.9	7.2	7.6
NRML	**4.8**	4.9	5.0	5.0	5.2	5.3	5.5	5.6	5.5	5.5	5.6

Table 3: Dependence of Classification Error on KL Transform Feature Set Dimensionality. Given with the classifier acronym are: For k-NN the value of k, for WSNN the value of α, for PNN the value of σ, for MLP networks the number of hiddens units, for RBF networks the number of centers per class, and for EMD and QMD classifiers the number of clusters per class. Bold type indicates the dimensionality yielding minimum error for each classifier. See [2] for more detailed discussion.

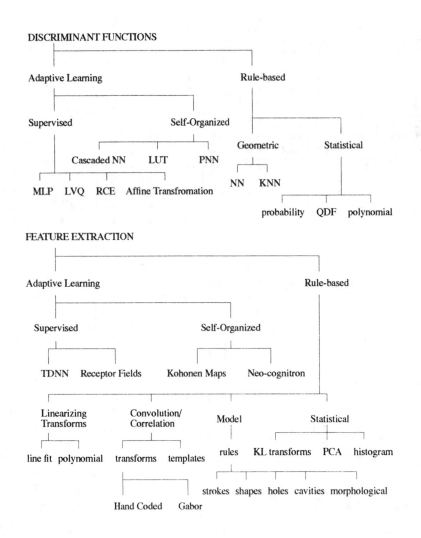

Figure 1: Types of methods used for feature extraction and classification.

Two types of data will be used to compare the neural and non-neural recognition systems. First the recognition accuracy as a function of reject rate is used and second the writer dependence as a function of reject rate is used.

Comparison of NN and statistical systems shows that with no rejection the neural and hybrid systems have errors between 3.67% (ATT_2) and 4.84% (HUGHES_1). The statistical systems have errors between 4.35% (UBOL) and 5.07% (ELSAGB_1). Since the standard deviations on these numbers is typically ±0.3% a significant overlap in performance exists. The best and worst neural systems are 4 standard deviations apart and the statistical system are about 2 standard deviations apart. Across the range of measured performance, the statistical systems can not be distinguished from each other. Across this same range of performance the neural systems can be distinguished form each other. As the fraction of characters rejected increases, the variation in accuracy increases for the NN system while the statistical systems remain tightly grouped. At 30% rejection the best NN system has an error of 0.15% (ATT_2) and the worst NN system has an error of 0.52% (SYMBUS). At the same rejection rate THINK_1 has an error of 0.27% and NIST_4 has an error rate of 0.21%. At high reject rates the statistical systems are nearing the performance of better NN systems and are significantly better than the worst NN system. For further details see [1],[15].

For the writer dependence of NN and statistical systems, the greatest writer differentiation, 50 writers, occurs at a reject rate of 5%. The best systems in terms of error have the least writer sensitivity. This is not because these systems get more writer correct at zero reject but because no system from either group gets over 80 writers correct at zero rejection. This separation of systems exists because when the worst characters from each writer are removed the best system from each group obtains a 50 writer advantage as the first 5% of the characters are rejected. Writer dependence is less significant in distinguishing systems than error performance. For further details see [1],[15].

3 NIST Classification Experiments

NIST evaluated four statistical classifiers and three NN classifiers. The statistical classifiers are Euclidean Minimum Distance (EMD), Quadratic Minimum Distance (QMD), Normal (NRML), and k-Nearest Neighbor (k-NN). The three neural classifiers included in the evaluation are the MLP, RBF, and PNN. For a given application, all the classifiers were given the same feature sets. Misclassification errors using a 23140 dataset are tabulated as a function of feature dimension and classifier parameters such as the number of prototypes. Table 3 shows for each classifier the estimated probabilities of *error*, expressed as percentages, for increasing dimensionality of the KL feature set. Note that the optimal number of features yielding lowest classification error (shown in bold) is not the same for all classifiers, the parametric classifiers, QMD and NRML, being noticeably more parsimonious in the number of features required. It is also apparent that most of the classifiers essentially attain a plateau as the number of features reaches approximately 32 thereafter only gaining several tenths of a percent. The best classifiers are the computationally expensive nearest neighbor classifiers and the related PNN. They achieve one third less errors than the NNs and parametric classifiers. The optimum value of $\alpha = 1.1$ for WSNN corresponds to a 1-NN scheme for most test patterns. Accordingly, k-NN is seen to have a higher error rate for increasing k.

Two caveats should be made about the table. First, the MLP and RBF results depend on the initial guesses for the parameters. Often a number of different random guesses are tried assess the effect of the initial guess; for this table, because of the magnitude of the calculation necessary, only one initial guess was used.

These results show that for character classification accuracy NN methods and statistical methods have comparable accuracies confirming the COCR results.

4 Conclusions

Examination of the results of 11 OCR systems using a wide variety of recognition algorithms has shown that in accuracy and writer independence NN systems have not demonstrated a clear cut superiority over statistical methods. Some neural system have higher accuracy than statistical methods; other have lower accuracy. The performance of statistical methods is more closely grouped and is approximately the same as the performance of an average NN system considered here. One area where NN's may have an advantage is in speed of implementation and recognition.

Examination of Table 3 show that on OCR classification the ranking of the methods is similar. The neighbor-based methods are the most accurate with PNN being the best of these. The comparison of MLP and RBF methods show that RBF is usually the better method. When MLP and RBF methods are compared to multicluster EMD and QMD methods the NN methods are more straightforward to implement but do not show a clear accuracy advantage. All of the experiments presented here also suggest that the training set sizes used, although large, are not sufficient to fully saturate most of the machine learning methods studied here.

Acknowledgement

The author would like to acknowledge Patrick Grother for providing tables 1 and 3, Jon Geist for assistance in interpretation of results of classifier comparisons, and Rama Chellappa for help in designing the comparison experiments.

References

[1] R. A. Wilkinson, J. Geist, S. Janet, P. J. Grother, C. J. C. Burges, R. Creecy, B. Hammond, J. J. Hull, N. J. Larsen, T. P. Vogl, and C. L. Wilson. The First Optical Character Recognition Systems Confernce. Technical Report NISTIR 4912, National Institute of Standards and Technology, August 1992.

[2] Patrick J. Grother and Gerald T. Candela. Comparison of Hand-printed Digit Classifiers. Technical Report NISTIR 51??, National Institute of Standards and Technology, June 1993.

[3] P. J. Grother. Karhunen Loève feature extraction for neural handwritten character recognition. In *Proceedings: Applications of Artificial Neural Networks III*. Orlando, SPIE, April 1992.

[4] F. Rosenblatt. The perceptron: a probabilistic model for information storage and organization in the brain. *Psychological Review*, 65:386–408, 1958.

[5] W. S. McCulloch and W. Pitts. A logical calculus of the ideas immanent in nervous activity. *Bull. Math. Biophysics*, 9:115–133, 1943.

[6] M. Minsky and S. Papert. *Perceptrons*. MIT Press, Cambridge, MA, 1969.

[7] Anil K. Jain. *Fundamentals of Digital Image Processing*, chapter 5.11, pages 163–174. Prentice Hall Inc., prentice hall international edition, 1989.

[8] K. Fukushima. Neocognitron: A self-organizing neural network model for mechanism of pattern recognition unaffected by shift in position. *Biological Cybernetics*, 36:193–202, 1980.

[9] T. Kohonen. *Self-Organization and Associative Memory*. Springer-Verlag, Berlin, second edition, 1988.

[10] D. E. Rumelhart, G. E. Hinton, and R. J. Williams. Learning internal representations by error propagation. In D. E. Rumelhart and J. L. McClelland, et al., editors, *Parallel Distributed Processing: Explorations in the Microstructure of Cognition. Volume 1: Foundations*, chapter 8, pages 318–362. MIT Press, Cambridge, MA, 1986.

[11] J. G. Daugman. Complete discrete 2-d Gabor transform by neural networks for image analysis and compression. *IEEE Trans. on
Acoustics, Speech, and Signal Processing*, 36:1169–1179, 1988.

[12] T. P. Vogl, K. L. Blackwell, S. D. Hyman, G. S. Barbour, and D. L.
Alkon. Classification of Japanese Kanji using principal component
analysis as a preprocessor to an artificial neural etwork. In *International Joint Conference on Neural Networks*, volume 1, pages
233–238. IEEE and International Neural Network Society, 7 1991.

[13] Donald F. Specht. Probabilistic neural networks. *Neural Networks*,
3(1):109–118, 1990.

[14] T. Poggio and F. Girosi. Networks for approximation and learning.
Proceedings of the IEEE, 78(9):1481–1497, 1990.

[15] Charles L. Wilson. Effectiveness of Feature and Classifier Algorithms in Character Recognition Systems. In D. P. D'Amato, editor, , volume 1906. SPIE, San Jose, 1993.

Model-Based Multi-Sensor Fusion

Mark K. Hamilton and Teresa A. Kipp

U.S. Army Research Laboratory

Abstract

A new class of ATR algorithms known in general as model-based have recently demonstrated some promising results. One major advantage of these new methods is that they are not only data-driven but they are also model-driven. The fact that they are model-driven removes the dependence of the recognition process on a low level segmentation process. These new model-based approaches accomplish segmentation and recognition simultaneously. In other words, the process entertains multiple segmentation possibilities until a sufficient amount of high level and low level information is gathered so that an intelligent interpretation of what the image contains can be formed. A second major advantage of these new methodologies is the fact that the fusion of information from multiple sensors is a continual process not a one shot deal which is typical of first generation algorithms. This paper will concentrate on the model-based ATR algorithm called relational template matching.

1: Introduction

This paper describes the U.S. Army's efforts under the ATR relational template matching (ARTM) program to develop both a single sensor flir only ATR algorithm as well as a multi-sensor flir and laser radar ATR algorithm. The program's target set is ground targets at low depression angles of less than 6 and ranges of 500 to 6000m. However, the algorithm is directly applicable to other target sets, including air targets, and ground targets at high depression angles. The ARTM program is a three phase effort. The goal of the first phase (phase I) was to provide a proof-of-principle demonstration of this new model-based methodology. This was accomplished on synthetic flir imagery for the three class problem of M60A3 tank, M113 apc , and M35 truck. A comprehensive test and evaluation was preformed at the Night Vision & Electronics Sensors Directorate. Interested parties are referred to the report by C. P. Walters [1]. The goal of the second phase (phase II) was to demonstrate this new

algorithm approach on realistic field collected second generation flir imagery with a larger target set. The target set used for the phase II effort consisted of the M60A1, M35, M113A2, M2, M1A1, HMMWV, 2S1, BMP, ZSU23, and T72. The phase II effort was just coming to a close at the time this paper was being written so only limited results on second generation flir imagery will be provided in this paper; however, overall preliminary results are once again promising. The goal of the third phase (phase III) is to demonstrate this methodology for multiple sensors, namely FLIR and Laser Radar. This paper will concentrate on describing the general methodology and it's extension into the multi-sensor realm of flir and laser radar. Preliminary results obtained during developmental testing will also be given to illustrate the potential of this approach.

2: RELATIONAL TEMPLATE MATCHING

2.1: Background

For the last several years, the results obtained from traditional approaches to aided or automatic target recognition (ATR) have strongly suggested that these approaches have reached their performance limits, i.e. the point of diminishing returns. In general, traditional approaches include statistical classifiers and some template matchers. Specifically, traditional approaches refer to approaches that are characterized by a sequence of separate functions, frequently called, detection, segmentation, feature extraction, and classifier (statistical). In the case of template matching, the separate function of feature extraction becomes essentially the template and the classifier becomes some global measure of fit, e.g. correlation, matched filter, or one of a set of infinite possibilities. Traditional approaches fail for several reasons; however, this paper will only briefly mention two of them. The first reason is that they depend on a (low level) segmentation process. Errors made in the segmentation module are then propagated down the line and are in some cases magnified. Many investigators have and still are looking for the perfect (low level) segmentor. The second reason for failure is especially evident when global measures of fit are used, e.g. correlation. These approaches uniformily emphasize the whole target template instead of weighting the various elements or details according to the model of target signature. The crux of the matter are these differences. Yet, with traditional approaches, the template for the m60 tank is designed totally independently of the template for the m1 tank and the template for the T72 tank. After coming to a fundamental understanding of the problems that plague traditional approaches which are also called largely bottom-up approaches for obvious reasons, the ATR Research Branch of the Army Research Laboratory began to investigate a radically new approach to ATR algorithm design, called relational template matching.

The relational template matching algorithm is based on principles of rigid object recognition. The approach was originally developed for optical character recognition (OCR) by Mathematical Technologies Inc. It has also been modified and implemented for laser radar (synthetic) imagery, see E. Bienenstock et al [2]. More recently, the approach has been modified to run on flir imagery both synthetic, see E. Bienenstock et al [3]; and real, see D. Geman et al [4]. This approach is model-based with a small number of parameters that control the false alarm rate, missed detections, and incorrect recognitions or confusions. The algorithm has demonstrated stable performance over a large class of variations including, contrast, range, target aspects, internal target variations (multi-modal targets), and limited occlusion across separate government training and evaluation sets.

2.2: General Approach

ATR relational template matching, although based on the principles of rigid object recognition, does not approach this problem in the typical brute force manner. In rigid object recognition, each different view of a 3D target relative to the sensor is regarded as a separate object to be recognized. The obvious brute force method is to store all of these different views or target poses and then preform some sort of global measure of fit, e.g. correlation. The first major drawback comes from the fact that as the number of targets increase the number of distinct objects that must be stored grows rapidly very large. This has the obvious potential drawback of quickly limiting the processing speed to an unacceptable level. The second and more important reason is that global measures of fit uniformily emphasize the whole target template instead of weighting the various elements or details according to the model of target signature.

In contrast to traditional or brute force methods, i.e. global measures of fit like correlation, relational template matching solves the above two problems by the following. First, instead of uniformly emphasizing the whole object, relational template matching focuses on the differences between objects. This methodology approaches the problem of recognition as the problem of separating one object from another object rather than trying to recognize objects in isolation with an absolute criteria. The speed issue is handled by a clever but well known search strategy, i.e binary search. The idea here is to always try and cut the search space in half with each question asked, e.g. the twenty questions game played by children.

In order to keep the amount of computation to an ultimate minimum in the field, the strategy of relational template matching decomposes the ATR problem into two parts. The first part is the off-line construction of an efficient search tree. This process is done once for a given set of targets. Each node of the search tree can be thought of as containing a question which is designed to try and cut the search space in as close to half as possible. Simply put, the job of the off-line algorithm is to decide a priori what the best questions are that will accomplish the recognition task. Therefore, the off-line algorithm produces an on-line or ATR algorithm with O(n log n) run time in the average case where n is the number of objects to be recognized (i.e. remember each distinct view or pose of a 3D target in the rigid object recognition problem is viewed as a separate or distinct object that must be recognized).

2.3: Signature Model

Currently, relational template matching uses only silhouette information and makes only weak assumptions regarding flir target signature. The assumption made is that the intensity changes across target boundaries are in general larger than the intensity changes found in the local background. This general assumption allows for the transition from target to background to be either white on black or black on white. It is typical to find both types of target to background transitions on a single target even against a uniform background.

2.4: Probes and Relational Templates

(U) A probe is the most basic or primitive operator used by relational template matching. It is the simple question of whether or not a boundary exists between point A and point B, e.g. pixel A and pixel B. The precise location and direction of an edge is unreliable information due to discretization and noise. Therefore, probes are used to determine the presence or absence of an edge only to within a given approximation. A relational template is a collection of probes (i.e. basic questions) and their interrelationship.

2.5: Automatic Design of the Decision Tree or ATR Algorithm

The decision tree or ATR algorithm is designed off-line once and for all for a given set of targets. If the target set changes or the addition of a new target is necessary, then the off-line process must be run again. The off-line process consists of constructing a tree (i.e. a recursive partitioning of the different aspects of the targets of interest) and a relational template or set of questions for each node of the tree that will accomplish this partition on-line. Because only silhouette information is currently used, BRL CAD models can be used to automatically generate the 2D silhouettes from distinct poses of the 3D targets of interest. Figure 1 shows a BRL CAD model of an M60 tank and the corresponding silhouette of this model for the aspect of the M60 seen by the reader.

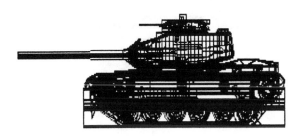

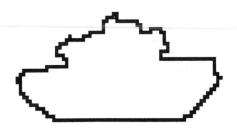

Figure 1

The BRL CAD models are very detailed models. However, this level of detail is too fine and is therefore not totally consistent with the features seen through a flir at interesting ranges. Therefore, these features were removed from the CAD models and do not appear in the silhouettes generated from them. Features removed include antenna's and usually gun barrels.

(U) Once the silhouettes have been constructed, the next step is to decide the finest level of aspect identification (i.e.

identification of vehicle type and aspect seen by viewer) that is reasonable to obtain. Silhouettes for the same target that only differ by a small amount as defined by a similarity measure are combined. The combined silhouettes represent the finest level of aspect identification possible and are called ribbons for obvious reasons. Figure 2 shows two fine level ribbons (i.e. terminal nodes of the decision tree) one for the T72 and one for the M60A1.

T72 - (85-100°)

M60A1 - (75-105°)

Figure 2

(U) Once the fine level ribbons have been constructed, there are two ways to construct the tree. One option is to recursively partition the set of silhouettes into two disjoint sets based of a measure of similarity, i.e., build the tree from the root down. The second option is to group silhouettes in a recursive pairwise manner based on a measure of similarity, i.e., build the tree from the bottom up. The second option was chosen by our team and Figure 3 illustrates the top portion of a 10-class tree showing the boundary type of relational templates. The targets for which this tree was constructed are the M60A1, M35, M113A2, M2, M1A1, HMMWV, 2S1, BMP, ZSU23, and T72.

435

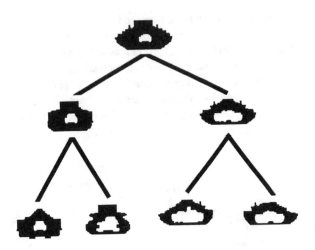

Figure 3

The relational template matching approach uses two types of relational templates. The two types are called boundary and competition. The boundary type of relational template is designed for each node of the search tree by an off-line process while the competition templates are designed on-line. The boundary relational templates use only the silhouette information contained in the particular set of objects that a particular node of the search tree represents. In other words, the boundary templates try only to establish that an object within a given approximation could be located at a particular x and y location in an image. Boundary templates basically try to distinguish objects of interest from clutter whereas competition templates only distinguish one object of interest from another object of interest. The competition relational templates are designed on-line, and their job is to make the final determination as to whether the object is; for example, an M60A1 or an T72, i.e. compete the fine level ribbons that have been confirmed by the boundary templates as being potentially the identity of the object in the image.

The competition relational templates have to be designed on-line because it is not really feasible to reliably predict off-line which objects will survive the boundary tests to become competitors. The construction of competition templates off-line was tried with the predicted competitors being determine based on a similarity measure between silhouettes. It was observed that internal thermal structure (which is not modeled/used at this point) caused two different objects that had extremely different silhouettes to become competitors. For example, a HMMWV frequently fit very nicely on the hot engine of a tank and frequently fit with a near prefect score causing it to be the winner for boundary tests alone. However, although off-line competition relational templates could not correct this, on-line competition templates have no problem in choosing

the correct answer in this case. Figure 4 illustrates the relational boundary templates for the near broadside views of the M60A1 and the T72. Figure 5 illustrates the relational competitions templates for the M60A1 being confused with the M1A1.

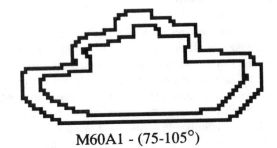

M60A1 - (75-105°)

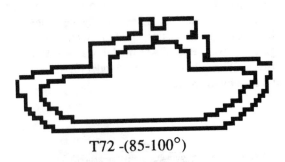

T72 -(85-100°)

Figure 4

Without Probes

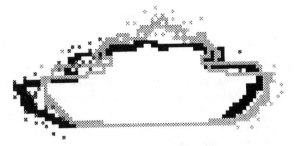

With Probes

Figure 5

2.6: Multi-Sensor Fusion with Relational Template matching

Relational template matching provides a natural framework for multi-sensor fusion. In fact, the relational template matching paradigm is a method to assure intelligent inference from the interaction between data and the model of target. Consequently, target model structure for multiple sensors is sufficient to obtain multi-sensor fusion at the pixel level by this method of local probes and since the target structures for different sensors are connected/related locally, fusion can be obtained at the pixel level or probe level.

This can be easily seen by considering the fusion of the laser radar or TV with the flir. All three sensors produce images that have obvious connections through the geometry of the targets, e.g. the silhouettes. If we limit this example to the silhouette information only then it is very easy to see the connection. The locations of probes/ questions used for the flir, laser radar, and TV could be the same. And, therefore, the coupling between the target model and all the data (i.e. single or multiple sensors) takes place at each node of the search tree. However, the specific tests at these locations although of the same nature, may be somewhat different in terms of specifics due to the nature of the information provide by each sensor. Figure 6 depicts the information used to automatically construct the multi-sensor ATR algorithm.

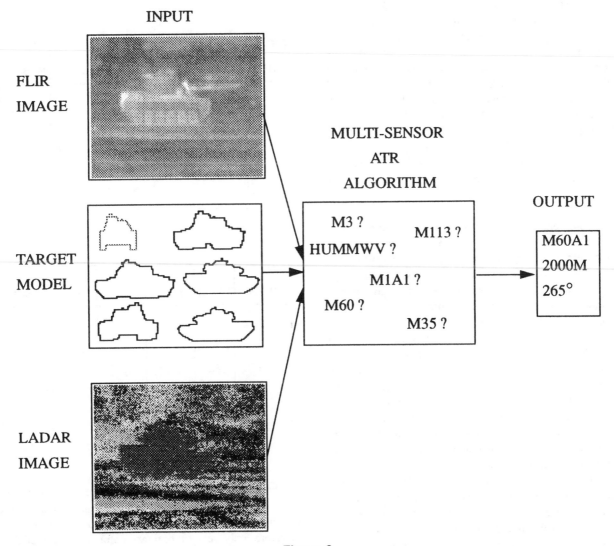

Figure 6

3.0: EXPERIMENTAL RESULTS

The ARTM algorithm developed in Phase I has undergone extensive Army testing on simulated flir imagery collected on the Night Vision & Electronic Sensors Directorate's physical terrain board. The Phase I algorithm was designed for the three class problem of M60A3 tank, M113 apc, and M35 truck, ranges of 1km to 5km, and flat (ground) plane rotations of these objects. The government established two distinct data sets of terrain board simulated flir imagery, one for training and one for independent government testing and evaluation. The first set or training set consisted of images taken of targets at four different ranges, and three different clutter levels. This data set was distributed widely to ATR algorithm developers. The other data set was not distributed to anyone. The results of the independent government evaluation are classified confidential and so the interested reader is referred to two reports by C.P. Walters ([1], [5]) describing both the sequestered data set and the performance of the ARTM algorithm on this data. The results detailed below were achieved on the training data at 1500m with the algorithm being trained only on the top 2/3 of the target models. The bottom 1/3 of the models were avoided due to the realistic potential of bottom occlusion in real world scenarios. The algorithm tested on the training data set and the sequestered data sets are exactly the same, including all parameter settings which in the case of this algorithm are few.

A comprehensive evaluation on the training data at 1500m consisted of 126 images or 504 targets, where the targets were at all relevant aspects, multiple contrast levels, and multiple clutter levels. In this data set, each image contained 4 targets of the same type, one high contrast uniform, one medium contrast uniform, one low contrast uniform, and one signature target. The low contrast uniform targets typically appeared nearly invisible or ghost like. The experimental results are summarized in the confusion matrices in tables 1-3. Table 1 shows the performance over all aspects, signature types, contrast levels, and clutter levels. The overall probability of detection was 76% with conditional identification of 92%. This is with 0 false alarms per field of view in the low clutter images and 1.5 false alarms per field of view in the high clutter images. Table 2 depicts nearly perfect performance on the uniform targets. Table 3 depicts the performance on signature targets. The performance in table 3 should more closely predict the performance on real flir imagery. Identification performance is not effected by clutter level. It is; however, effect by target contrast as was evident with the signature targets. The performance on medium-to-high contrast signature targets was better than

90% conditional identification with near perfect detection. For low contrast signature targets the identification performances dropped to approximately 50% with the probability of detection falling to 60%.

GROUND TRUTH	TOTAL	M113	M60	M35	MISSED
M113	126	90	0	0	36
M60	126	3	89	0	34
M35	126	4	0	95	27

Table 1

GROUND TRUTH	TOTAL	M113	M60	M35	MISSED
M113	168	125	2	0	21
M60	168	12	107	5	44
M35	168	8	3	121	36

Table 2

GROUND TRUTH	TOTAL	M113	M60	M35	MISSED
M113	42	35	2	0	5
M60	42	9	18	5	10
M35	42	4	3	26	9

Table 3

The ARTM algorithm developed under Phase II of this effort for real second generation flir imagery was nearing completion at the time this paper was being written; and therefore, only very preliminary numbers were available. The following confusion matrix shown in table 4 depicts the results of the 10 class problem of M60A1, M35,

M113A2, M2, M1A1, HMMWV, 2S1, BMP, ZSU23, and T72 on limited second generation flir imagery taken at 2km at Yuma Arizona. Other imagery taken at Yuma was used to set the three on-line parameters. Two hundred and forty four (244) images that represent 244 target opportunities at approximately 15° intervals of aspects were used for this test. The probability of detection for this matrix achieved was .99 with a conditional probability of identification of .62 and unconditional probability of identification of .61. The unconditional probability of identification of each target achieved was as follows: M113 60%, M35 76%, M60 60%, HMMWV 62.5%, M1A1 68%, M3 33%, 2S1 72%, ZSU 50%, T72 80%, and BMP 44%.

GROUND TRUTH	TOTAL	M113	M35	M60	HMWV	M1A1	M3	2S1	ZSU	T72	BMP	MISSED
M113	25	15	4	0	1	1	1	3	0	0	0	0
M35	23	0	19	0	0	2	0	1	1	0	0	0
M60	25	0	1	15	0	1	2	2	0	0	2	2
HMWV	24	3	0	0	15	0	0	2	0	3	1	0
M1A1	25	1	2	2	0	17	0	2	1	0	0	0
M3	24	1	0	2	1	4	8	0	1	5	2	0
2S1	25	1	1	1	0	2	1	18	0	1	0	0
ZSU	24	1	0	0	0	3	3	1	12	1	3	0
T72	24	1	0	0	0	0	0	1	0	20	2	0
BMP	25	0	1	0	0	0	2	9	1	1	11	0

Table 4

3. 0: CONCLUSION

The model-based approach of relational template matching has demonstrated and continues to demonstrate promising performance results. The goal over the next year is to implement this approach for the multiple sensor suite of FLIR and Laser Radar.

4.0: REFERENCES

[1] C. P. Walters, "Initial Performance Evaluation of Model-Based and Template Matching Automatic Target Recognition Algorithms: Volume II: Algorithm Description and Results for the Relational Template Matching Approach", Technical Report, Night Vision & Electronic Sensors Directorate, to appear 1993.

[2] E. Bienenstock, D. Geman, S. Geman, D. McClure, "Development of Laser Radar ATR Algorithms: Phase II-Military Objects, Technical Algorithm Description", Technical Report, Night Vision & Electronic Sensors Directorate, October 1990.

[3] E. Bienenstock, D. Geman, S. Geman, A. Kramer, D. McClure, "ATR Relational Template Matching - Phase I (final report)", Technical Report, Night Vision & Electronic Sensors Directorate, September 1990 - August 1991.

[4] D. Geman, S. Geman, A. Kramer, K. Manbeck, D. McClure, "ATR Relational Template Matching - Phase II (final report)", Technical Report, Night Vision & Electronic Sensors Directorate, August 1993.

MUltiSensor Target Recognition System (MUSTRS)

Arthur V. Forman, Jr., David B. Brown, James H. Hughen,
Rebecca R. Pressley, Albert R. Sanders, Daniel J. Sullivan

Martin Marietta Technologies, Inc.
Electronics & Missiles
P. O. Box 555837 MP 1160
Orlando, FL 32855-5837

Abstract

This paper describes the Advanced Research Projects Agency (ARPA) MUltiSensor Target Recognition System (MUSTRS). A smart sensor manager controls forward-looking infrared (FLIR) and millimeter wave (MMW) radar sensors to obtain multiple looks at targets on the ground. Targets in IR images are recognized using a variation on minimum average correlation energy (MACE) filtering and/or a model-based algorithm called Key Features. Radar data are processed using quadratic distance composite filtering techniques. Evidence is combined using a Bayesian method. The system has been designed to correctly classify time critical mobile targets with very low false alarm rates.

1: Introduction

The MUSTRS Program will demonstrate key capabilities to search for and recognize time critical targets (TCTs) such as mobile tactical ballistic missile launchers, resupply vehicles, and the related command posts [DePersia 92]. The recognition of hidden TCTs presents a particularly challenging problem, as evidenced by events during Operation Desert Storm. Large areas must be searched quickly using affordable sensors. Targets are masked by terrain, foliage, clutter, and camouflage. Decoys and deception techniques are frequently employed to further confound the search process. Operational requirements mandate that systems be effective during the day and at night, with minimal degradation due to adverse weather. Derived requirements for this application include a probability of correct target declaration ≥ 0.65 for a 5-type target identification problem, and a false target declaration rate ≤ 0.01 per square km, in areas where up to 4 confuser vehicles may exist per square kilometer.

2: Hardware

The MUSTRS Smart Sensor Subsystem (SSS - Figure 1) contains three major hardware subsystems: an 8-12 micron dual field-of-view second generation forward-looking infrared (FLIR) sensor; an independently-gimballed multimode millimeter wave (MMW) radar, and a multi-sensor signal processor (MSSP) that provides realtime autonomous target recognition (ATR) and sensor management.

The wide field of view (WFOV) FLIR is used for detection of target areas and for initial detection of individual targets. The narrow field of view (NFOV) FLIR provides more pixels-on-target and is used for target classification. The MMW radar is a 35 GHz dual circular polarized fully coherent system which operates in three modes—real beam, doppler beam sharpened (DBS) and synthetic aperture radar (SAR). (See Figure 2.) The real beam and DBS modes are used for broad area search. In the real beam mode, high range resolution (HRR) profiles are developed for target detection and classification. In the DBS mode, the azimuthal beamwidth is sharpened by a factor of eight. The synthetic aperture radar (SAR) mode provides imagery to refine target classification for previously detected targets. The SAR mode is cued to interesting locations by the sensor manager.

The two sensors are controlled by a dynamic sensor manager that establishes search strategies, controls sensor operations, maintains and updates track files, and controls the ATR functions. The system provides multiple looks with either or both sensors at each candidate target. The multiple sensor ATR results are combined over time in a process called evidence accumulation. The use of the multisensor approach makes deployment of effective ATR countermeasures more expensive for a potential adversary, since two sensor regimes must now be defeated simultaneously.

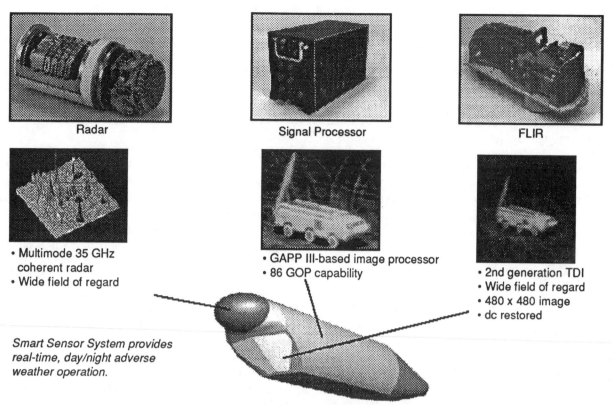

Radar

Signal Processor

FLIR

- Multimode 35 GHz coherent radar
- Wide field of regard

- GAPP III-based image processor
- 86 GOP capability

- 2nd generation TDI
- Wide field of regard
- 480 x 480 image
- dc restored

Smart Sensor System provides real-time, day/night adverse weather operation.

Figure 1. MUSTRS smart sensor elements.

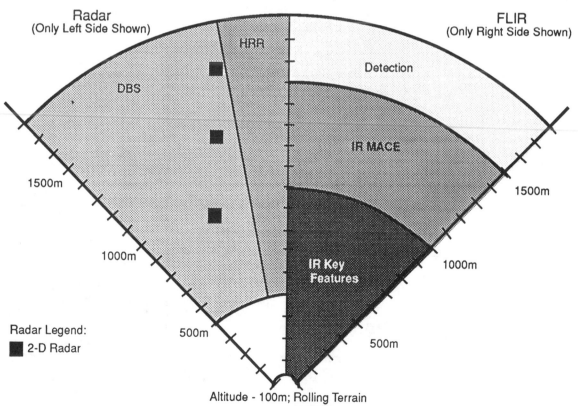

Figure 2. MUSTRS field of regard.

The MSSP has three major parts. The Image Processor consists of a single instruction, multiple data (SIMD) array of 92,000 processors utilizing third generation Geometric Arithmetic Parallel Processor (GAPP^{TM}) chips. Each of the 92,000 processors has 128 bits of local random access memory and a one bit arithmetic logic unit which runs at 25 MHz. The Image Processor is used for FLIR and 2-d radar image processing. The Radar Signal Processor utilizes a bank of four dual-i860 processors with associated Fast Fourier Transform (FFT) co-processors for all HRR and DBS detection and SAR image formation processing. Finally, the Control Data Processor utilizes three dual-SPARC processors for sensor management and multi-sensor, multi-look evidence accumulation functions. The MSSP architecture is fully programmable in C.

3: Algorithms

An overview of the ATR processing is depicted in Figure 3 and has three stages: sensor control and detection, multiple ATRs, and multiple-level fusion. The sensor manager sends appropriate commands to the dual field of view FLIR and multimode MMW radar sensors. Every image from the wide field of view IR system is processed through a regions algorithm to detect roads and treelines (probable target locations). Every image from the wide and narrow fields of view is processed through a detection algorithm to nominate

target-sized regions having high thermal contrast. Similarly, the multimode radar sensor is operated in a high range resolution profile or doppler beam sharpened mode to nominate targets based on radar cross section. The sensor/process manager examines the detections from the multiple sensors as they are tracked over time and determines which ATR algorithm(s) should be invoked to further process the most interesting detections.

After nominated detections have been centered, one of two IR ATR algorithms is invoked depending on range to target. Range to target is estimated using sensor estimates of platform altitude, sensor elevation angle, and stored digital map data. At the longer ranges in the narrow field of view, a variation on a technique called minimum average correlation energy (MACE - [Mahalanobis 88]) filters is applied to classify detected targets. This filter bank constitutes a kind of distributed associative memory (DAM), and has been demonstrated to be very effective at ranges where the pixels-on-target count is too low for the Key Features structural approach, which is the IR ATR algorithm of choice at closer ranges, where more internal structure is evident.

The IR MACE filter algorithm begins by warping the detected image to account for differences between the current estimated range and roll to target and the range and roll used when the MACE filter was trained. A simplification of the Canny edge operator is then applied to find edges on the target. Next a bank of filters is applied by loading the FFTs

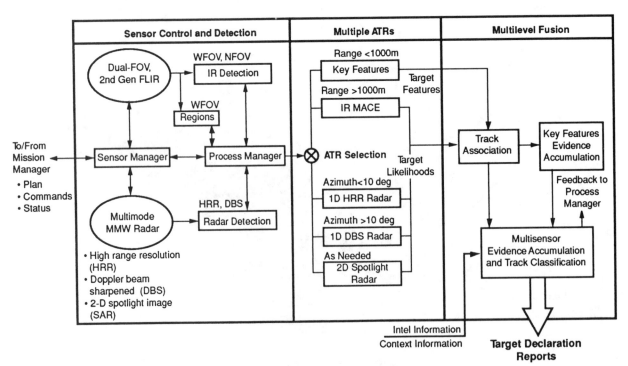

Figure 3. ATR processing for time-critical targets.

for these filters, doing a SIMD multiplication of the image FFT by the filter FFTs, and then performing a single SIMD inverse FFT. The filters are created using the MACE filter training procedure, which is described in [Sullivan 92]. The cross correlation surfaces are then scanned to determine where on each of the surfaces the peak correlation occurs. These correlations are later converted to six likelihood values—one for each of five classes of target, and one for clutter.

The Key Features ATR [Pemberton 88, Curtis 93] is a model-based approach which is specifically designed to recognize well-resolved targets even when they are obscured by as much as 50%. Indexing is achieved through an aspect estimation process. Aspect estimates are refined (and alternate hypotheses tested) via feedback from the multi-level fusion processes. The algorithm carries a wireframe model of each of the key features of each of the five targets to be recognized and manipulates the model in realtime to achieve target identification. First the image is processed to estimate the aspect of the target in the detected image. This estimate is combined with input values of range to target, sensor elevation angle, and image roll angle, to rotate and scale the five three-dimensional target models to match the current hypothesized target azimuth, elevation, and range.

The model for each target consists of a collection of up to eight of the target's key discriminating features or subparts. The rotated wireframe model for each feature becomes the r-table for a generalized Hough transform (GHT) [Ballard 81]. The GHT matches the edge directions in the image to the predicted edge directions from the rotated wireframe model. Each pixel in the image receives a detection metric value corresponding to the number of votes, or degree of match, the image at that pixel had with the feature being tested. Any location where at least a preset percentage of the feature is detected is retained as a potential feature location. After all of the features have been tested, a second "structural" GHT is run to determine which feature sets (if any) have been detected in the proper geometrical arrangement for a target at the hypothesized aspect.

There are three radar ATRs, one for each mode of the multimode MMW radar. The ATRs for the two 1-d search modes of the radar use a quadratic extension of samples from the high range resolution profile to develop a discriminant function which minimizes a measure of mean squared classification error. Variations on this approach are used for the real beam and DBS modes. The 1-d radar training procedures and preliminary test results are presented in [Hughen 91, 92].

For the 2-d synthetic aperture radar (SAR) mode of the radar ATR, an extension of MACE called a distance classifier is used to recognize detected images of nominated objects. The distance classifer design equations and the procedure for training the distance classifier are described in [Mahalanobis 93]. The application of the filters is similar to the application of MACE filters to IR imagery. The distance classifier has proven very effective for target classification and for rejection of natural clutter and confuser vehicles.

Embedded Kalman filters link together all of the ATR reports which correspond to the same location on the ground. It is this list of reports, called a "track", which is the input to the evidence accumulation process for target classification. To help resolve ambiguous tracks, a multiple-hypothesis tracking (MHT) algorithm [Blackman 86] has been implemented which determines the most likely global interpretation of all of the tracks based on the distances from each report to each feasible track.

The collection of ATR tracks corresponding to the most likely hypothesis are combined using a Bayesian evidence combination scheme to create a single cumulative likelihood vector to represent a track. Each ATR has been constrained to report its results in the form of a likelihood vector. The likelihood vector consists of a single likelihood for each of five target classes plus a single likelihood for clutter. Bayesian combination involves forming the product of all of the likelihood values from all of the looks and all of the sensors for a given track for each target separately These products are then normalized so that the sum of the target and clutter likelihoods is unity in the cumulative likelihood vector for the track.

If the target class of maximum cumulative likelihood has a sufficiently high likelihood, and if the difference between the highest and second highest likelihood exceeds a second threshold, a target is declared. Both thresholds are dynamically adjusted based on prior probabilities for each class and the constantly changing priorities of the mission manager during the flight.

4: Summary

The MUSTRS program will culminate with a series of realtime flight demonstrations against time critical targets in 1994. Technologies successfully applied to the difficult problem of targeting time critical targets are immediately applicable to a number of other problems.

Acknowledgements

The authors express their appreciation to Arthur Chang, Nathalie Day, James Curtis, Eric Grajales, and Robert Wood of Martin Marietta and Dr. Rubin Johnson of Operations Research Concepts Applied, Inc. for their contributions to this work.

References

[Ballard 82] D. Ballard and C. Brown, Computer Vision, Prentice-Hall, Englewood Cliffs, NJ, 1982

[Blackman 86] S. Blackman, Multiple Target Tracking with Radar Applications, Chapter 10, Artech House, Norwood, MA, 1986

[Curtis 93] J. Curtis, R. Pressley, and A. Sanders, "Key Features Target Recognition System", Third ATR Systems and Technology Conference, Monterey, CA, June 29, 1993

[DePersia 92] A. Trent DePersia, N. Johnson, and R. Pomeroy, "Thirsty Saber Autonomous Air Vehicle Avionics Suite", Joint Cruise Missile Association, Association of Unmanned Systems Symposium, Sept. 22, 1992

[Forman 92] A. Forman, W. Pemberton, A. Sanders, "Multialgorithm automatic target recognition using second-generation infrared sensors", Second ATR Systems and Technology Conference, Fort Belvoir, VA, Feb. 17, 1992

[Hughen 91] J. Hughen and K. Hollon, "Millimeter Wave Radar Stationary Target Classification Using a Higher Order Neural Network", SPIE Conference, Orlando, FL, SPIE 1469-43, April 1991

[Hughen 92] J. Hughen and K. Hollon, "Second order neural network for ATR in real beam radar", SPIE Conference, Los Angeles, CA, SPIE 1630, pp. 122-130, January 1992

[Mahalanobis 88] A. Mahalanobis, B. Kumar, and D. Casasent, "Minimum Average Correlation Energy Filters", Applied Optics, Vol. 26, No. 17, pp. 3633-3640, 1988

[Mahalanobis 93] A. Mahalanobis, A. Forman, M. Bower, N. Day, R. Cherry, "A quadratic distance classifier for multi-class ATR using correlation filters", SPIE 1875-13, Los Angeles, CA, Jan. 1993

[Pemberton 88] W. Pemberton, A. Sanders, and J. Sura, "Techniques for extraction of mobile missiles from infrared imagery", Conference on Pattern Recognition for Advanced Missile Systems, Huntsville, AL, Nov. 1988

[Sullivan 92] D. Sullivan, A. Forman, and A. Chang, "Realtime distortion-tolerant filters for automatic target identification", SPIE Conference, SPIE 1701, Orlando, FL, April 1992

[Forman 93] A. Forman, D. Brown, J. Hughen, A. Mahalanobis, A. Sanders, D. Sullivan, "Multisensor target recognition system", Third ATR Systems and Technology Conference, Monterey, CA, June 29, 1993

Modeling of Natural Objects including Fuzziness and Application to Image Understanding

Koji MIYAJIMA, Anca RALESCU[1,2]

Laboratory for International Fuzzy Engineering Research
Siber Hegner Bldg. 3Fl. 89-1 Yamashita-Cho, Naka-Ku, Yokohama-Shi 231 JAPAN

Abstract - In this paper, we present an object recognition method which takes into account fuzziness both in the object model and the matching operation between the model and the image processing results. Objects are described by a hierarchic model. Attributes of components such as color, shape and size are described by fuzzy sets. The (ambiguous) importance and correlation between attributes and components are described using a fuzzy measure. We examine ways of determining this fuzzy measure. Considering these, importance and correlations, the objects are recognized by integrating the outcome of the matching between the results of image processing and the attributes in the model. Finally, at the experimental level, the method is applied to recognition both of an abstract painting and real photograph.

I. INTRODUCTION

The purpose of a computer based image understanding is to realize the recognition of contents of an image (objects, scenes, etc.) at a level comparable to that of the human behavior. However, the mechanisms underlying the human image understanding system have not been completely cleared. Therefore, currently the typical approach in computer vision is to use the results of the image processing, features which can be measured, in conjunction with a probabilistic method to identify the objects in an image. By contrast it is plausible that in addition to image processing, image understanding in humans is also the result of reasoning in terms of knowledge about the world. This knowledge necessarily includes uncertainty and fuzziness to reflect the incompleteness and imprecision inherent in any human knowledge of the real world. For example, when thinking about a classification task that assigns some objects to categories, we are very aware of the fact that real world categories are by excellence fuzzy. It is thus only natural to make use of this knowledge in our judgment about what an object may be. Traditionally, in computer vision the treatment of uncertainty has been probability based. This approach limits the kind of uncertainty which can be modeled, and severely hampers the process of recognition of an object, and ultimately that of image understanding.

Alternative methods to the treatment of uncertainty in computer vision have been under investigation. Thus, in [1], the concepts necessary for the geometry of fuzzy sets were derived and applied to image segmentation. In [2] a probability based ambiguity reduction method for scene labeling was introduced. Evidential reasoning, as proposed in [5] was used in [3], and [4] for pattern classification. Several different aspects of uncertainty were modeled: evidence for and evidence against some hypothesis was gathered. The methods concentrate on solving the conflict between these evidences. In [6] an evidential approach, based on using the concepts of fuzzy measure and fuzzy integral (as a rule of evidence combination) was presented. In any case none of the methods dealing with the uncertainty aspect of object recognition took into account the aspect of the (fuzzy) correlation between pieces of evidence (such as features detected).

In this paper, we propose an alternative method, in which fuzziness plays an important role in describing an object model as well as in specifying the mutual effects (correlation and importance) of these components. Note that it is very important to be able to capture different kinds of correlation (dependence or support), and that the type of correlation varies among the items considered. This kind of information is essential in the matching operation between the model and the result of features extraction (which also includes fuzziness) from image data. The object model is a hierarchy, in terms of components of the objects, their physical characteristics, as well as relations between these components; the attributes of components are represented when appropriate (e.g. for components such as color, shape and size) as fuzzy sets. The ambiguous correlation and relative importance between attributes and components are described using the fuzzy measure. This in turn brings up the issue of determining a fuzzy measure. Considering the relative importance and these correlations, objects are recognized by integrating the outcome of the matching between the results of image processing and the attributes of the model. Finally, the method is applied to both the recognition of an abstract painting and of a real photograph as experimental results and its effects are discussed.

II. OUTLINE OF IMAGE UNDERSTANDING

Roughly speaking, the process of image understanding is divided into (1) image segmentation, which extracts image features (such as edges, regions, etc.), and (2) a matching that compares image features and models of the objects. The main difficulty is to realize a kind of soundness and completeness of this process: extracting all the meaningful features and only those. The difficulty of the match lies in, among others, the management of uncertainty, more specifically of the uncertainty which is not of probabilistic nature. The overall challenge is then to integrate the two steps of image understanding while solving, or avoiding the difficulties mentioned above.

The general process flow of image understanding is as follows: first, using some image processing primitive features

[1] On leave from the Computer Science Department, University of Cincinnati, U.S.A.
[2] This work was partially supported from the NSF Grant INT-9108632

such as edge detection, regions, textures are extracted from image data. Next primitive symbols such as size and main axis of the regions, inclination and length of the edges are calculated. Finally, these primitive symbols are matched to the primitive features of the model. The (level of) recognition is obtained based on the result of the matching (Fig.1).

A. Object Modeling.

The analysis of a complex image requires a large volume of knowledge about the objects in the image. Knowledge representation techniques, such as production rules, semantic networks, frames and so on, can be used for this purpose. Spectrum features, shape features, texture features which an object may have are thought of as knowledge sources and the recognition is based on the matching of these features to the image characteristics. However, comprehensive and expressive models are difficult to obtain, especially for natural objects. By contrast, for artificial objects (such houses, buildings, roads, used in remote sensing image recognition) the models are relatively easy. The recognition is also relatively easy. This is so because their shapes are clear and the difference of contrast between such objects and background is very large. On the other hand, the contrast between background and natural objects (such as forests, grasslands, rivers, etc.) is very ambiguous, and their shapes are highly irregular. This entails difficulties both in describing them (obtain a model) and in detecting their physical characteristics.

B. Reasoning about the Result of the Recognition.

In image understanding, reliability of information is usually represented by numerical values. Probability reasoning represents similarity on matching or reliability of image features, then obtains the interpretation of the highest reliability. Using this approach, evidential reasoning was proposed as the method of integrating information. Some pieces of information on a object are derived, then the method uses an evidence combination rule (such as Dempster-Shafer) as the method of integration. In probability reasoning, it is possible to deal with uncertainty of information and error information, but not to interpolate incomplete information. In addition, usually an underlying independence assumption exists, thus effectively ignoring, mutual effects of these sources of evidence. An effective method which can deal with correlations (eventually partially specified) has not been proposed yet.

The problem of recognition of natural objects especially those for which a model is difficult to obtain is the central goal of this paper. The recognition of such objects is achieved by

integrating partial knowledge about their components, and information on the mutual effects between these components.

III. INCLUDING FUZZINESS INTO THE OBJECT MODEL

In this paper we choose to represent an object as part-of hierarchy. A node in this hierarchy represents a component of the object. Each node stores information on the corresponding component: attributes and features of that component (Fig.2). The features can be represented by linguistic expressions (linguistic variables). This representation is more in agreement with the way a person might describe some of these attributes. Thus we can represent natural descriptions of color, such as "color is red", or shapes, such "shape is kind of round" better than by using some formulae or exact numerical values (which ultimately may not be detected). Fuzzy sets are used as values for the linguistic variables in question, such that in addition to high expressive power we achieve robustness, meaning that some variability is not only allowed, but exploited.

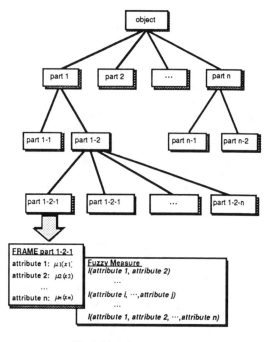

Fig.2. Modeling of an object

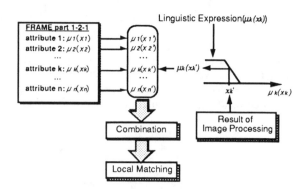

Fig.3. Local matching of primitive features

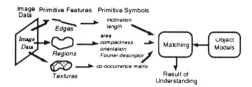

Fig.1 Outline of image understanding.

IV. MATCHING TO MODEL

A. Local Matching of Features Extracted from Image Data.

We now discuss in detail the matching of the primitive features obtained from image processing, color, shape, etc., to the model. Let $\mu_k()$ denote the membership function associated to an attribute of a component of the object, as indicated by the model. Let x_k denote the attribute value calculated from image data. Then the matching degree of the attribute to the model can be expressed by $\mu_k(x_k)$ (Fig.3). The matching degrees of each attribute associated to a node of the model are calculated in a similar way. The results of the matching are combined using an integral, with respect to a fuzzy measure which conveys information on the correlation between attributes. The matching associated to the components of one node is referred below as "local matching".

The combination of the results of attribute matching is done as follows: Let Ω denote the set of attributes in a node. For our problem Ω is finite, that is $\Omega = \{\omega_1, \cdots, \omega_n\}$. The result, $\mu_k(x_k)$, of matching of each attribute is treated as evidence. Without loss of generality we assume that Ω is presented in non increasing order of these evidences. We define, $h: \Omega \cup \{\omega_{n+1}\} \Rightarrow [0,1]$ as

$$h(\omega_i) = \begin{cases} \mu_i(x_i) & if \quad i = 1, \cdots, n \\ 0 & if \quad i = n+1 \end{cases} \quad (1)$$

We introduce the dummy attribute ω_{n+1}. To each of these attributes we associate an importance degree, I. We require for I to be a fuzzy measure. That is, $I: \wp(\Omega) \Rightarrow [0,1]$, and I satisfies:

(i) $I(\Phi) = 0, \quad I(\Omega) = 1$ (2)

(ii) If $A, B \subseteq \Omega, A \subseteq B \Rightarrow I(A) \leq I(B)$ (3)

With these definitions we can define the Choquet integral of h with respect to the fuzzy measure I. Its value M, is the result of the match (local, or global). From [7] we have:

$$M = (C)\int_0^1 I(x|h(x) > \alpha)d\alpha$$
$$= \sum_{i=1}^{i=n} \{h(\omega_i) - h(\omega_{i+1})\} I(\{\omega_1, \cdots, \omega_n\}) \quad (4)$$

The local matching is obtained by integrating the matching results of the attributes using (4). Of course, the important part in this discussion is the determination of the fuzzy measure I. This is explained later in Section V.

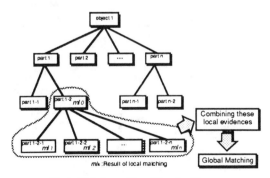

Fig.4. Global matching of local evidences

B. Global Matching based on the Results of the Local Matching.

The results of the local match are combined, in the global matching step, to achieve final object recognition. The global matching is also based on evaluating a Choquet integral as in (4), with respect to a new fuzzy measure on the space of components. The results of the local matching are further used to derive this fuzzy measure.

V. DETERMINING THE FUZZY MEASURE

In this section, we consider the issues which appear in connection with determining the fuzzy measures needed in the steps of local matching and global matching. In both steps the fuzzy measures are used to represent information on correlation between, and importance of items (features in the case of the local match, and components in the case of the global match). However, in each case, the information needed may be difficult to obtain: for example, if the number of attributes and/or components is very large, or the needed information is represented incompletely and/or by linguistic expressions, as it is very likely when this information is elicited from a human expert/user.

A method for deriving the fuzzy measure which corresponds to a user's preferential structure with respect to a collection of criteria was presented in [8]. According to this method the fuzzy measure was determined as a solution to a quadratic programming problem, namely minimizing the squared error between this measure and the uniform fuzzy measure, subject to constraints corresponding to partial and total evaluations of the measure. Reference [8] made use of a total evaluation given by the user to derive constraints for the measure to be determined. However, such an overall evaluation is not available in our case, thus we consider the modification of the method presented in [8].

According to [8] in order to determine a fuzzy measure one must determine the parameters λ and γ in equations (5) and (6) respectively. The fuzzy measure is known to model different situations of dependence or support (interaction) between pieces of evidence. The dependence case is expressed as:

$I(A \cup B) = I(A) + \lambda I(B)$ where $I(A) \geq I(B)$, $0 \leq \lambda \leq 1$ (5)

where A and B are perfectly dependent if $\lambda = 0$ and independent if $\lambda = 1$. The support case is expressed by

$I(A \cup B) = I(A) + I(B) + \gamma(1 - I(A) - I(B))$

where $I(A) + I(B) \leq 1$, $0 \leq \gamma \leq 1$ (6)

where the relation between A and B is perfect support if $\gamma = 1$ and independence if $\gamma = 0$.

Note that the independence case is a special case both of dependence and support cases. Both cases express the extent to which the measure I is not additive: more precisely it is subadditive in the case of dependence (5), and superadditive in the case of support (6). The dependence and support could be expressed linguistically as "perfect dependent/perfect support", "a little dependent/a little support", etc. These expressions are mapped into constraints intervals for λ and γ respectively.

TABLE I
QUANTITATIVE REPRESENTATION OF LINGUISTIC CORRELATION

Dependence		Support	
$I(A\cup B)=I(A)+\lambda I(B)$ where $I(A)\geq I(B)$		$I(A\cup B)=I(A)+I(B)+\gamma(1-I(A)-I(B))$ where $I(A)+I(B)\leq 1$	
Linguistic Representation	Range of λ	Linguistic Representation	Range of γ
"perfect dependent"	0.0	"perfect support"	1.0
"almost dependent"	$0<\lambda\leq 0.3$	"almost support"	$0.7\leq\gamma<1.0$
"dependent"	$0.3<\lambda\leq 0.7$	"support"	$0.3\leq\gamma<0.7$
"a little dependent"	$0.7<\lambda\leq 1.0$	"a little support"	$0<\gamma<0.3$

TABLE II
QUANTITATIVE REPRESENTATION OF LINGUISTIC RELATIVE IMPORTANCE

Linguistic representation	Value of η constraint $I(A)\geq\eta I(B)$
"A as important as B"	$0.9<\eta\leq 1.1$
"A is a little more important than B"	$1.1<\eta\leq 1.3$
"A is more important than B"	$1.3<\eta\leq 1.7$
"A is much more important than B"	$1.7<\eta\leq 2.5$

Table I shows the ranges for λ and γ as given in [8] corresponding to different linguistic expressions.

In addition, the user's subjective relative importance of pieces of evidence can be expressed pairwise as:

$$I(A) \geq \eta I(B) \quad (7)$$

where η denotes the relative importance degree corresponding to a linguistic expression. These linguistic expressions ("*same importance*", "*a little more important*", "*more important*", etc.) are translated into intervals constraining the values for η. Table II shows these intervals as given in [8].

A. Constraints on the Fuzzy Measure needed for Local Match.

At the level of the local match, the measure I is determined over the collection of features described in the node of each component according to the constraints induced by Table I and Table II. Two kinds of knowledge are used to express correlations and importance: (i) knowledge of image processing, and (ii) (subjective) knowledge about the object. However, at the local level the second type of knowledge may be limited given that it requires a very detailed knowledge of the object. Section VI illustrates the process of deriving the fuzzy measure necessary for local matching.

B. Constraints on the Fuzzy Measure needed for Global Match.

As in the case of the local match, the linguistic expressions for correlations and importance converted by Tables I and II respectively are used to determine the fuzzy measure. In addition, at this level we let the result of the local match to possibly affect the computation of the fuzzy measure needed. We implement the following heuristic regarding the importance of a component: "assign more importance to components with higher degree of recognition". To appreciate the effect of this heuristic let us consider the following situation: Suppose that for a given component the knowledge of the difficulty of image processing required to detect that component leads us to assign it a very low priority. Suppose, on the other hand that for

another component knowledge of the object assigns a high priority. Next, suppose that for a given image the result of the local match is higher for the first component than for the second. (For instance this could happen when the first component is exaggerated in the image, or the second is understated, or occluded). In this situation we would like the result of the global combination to be higher than that strictly prescribed by the model. We will illustrate this point in the abstract painting example considered in Section VI.

VI. EXPERIMENTAL RESULTS

In this section we illustrate the approach described in the previous sections on two images of a natural object, *an eye*, in a real photograph, and in an abstract painting. Fig.5 shows the model of "*eye*" described in terms of components such as: "eyelash", "eyelid", "pupil", and "white of the eye". The node associated to each component has the following attributes: "size", "shape", "color", "location" and "orientation". We assume that in both cases attributes of the eye itself are unknown.

A. Local Match of the "Eye."

Suppose that the correlations of these primitive symbols are as follows: "size" is *support* on "shape", "color" is *almost dependent* on "size", "location" is *support* for "orientation". Using Table I we obtain from

$I(\{size, shape\}) = I(\{size\}\cup\{shape\}) =$
$I(\{size\})+I(\{shape\})+\gamma(1-I(\{size\})-I(\{shape\}))$,

where $I(\{size\})+I(\{shape\}) \leq 1, 0.3\leq\gamma<0.7$ the constraints:

$$0.7(I(\{size\})+I(\{shape\})) - I(\{size\}\cup\{shape\}) \leq -0.3, \quad (C1)$$

$$I(\{size\}\cup\{shape\}) - 0.3(I(\{size\})+I(\{shape\})) < 0.7. \quad (C2)$$

for "size" and "shape". Similarly, for "size" and "color" we obtain

$$I(\{size\}\cup\{color\}) > I(\{size\}), \quad (C3)$$

$$I(\{size\})+0.3I(\{color\}) \geq I(\{size\}\cup\{color\}), \quad (C4)$$

and for "location" and "orientation" we obtain:

$$0.7(I(\{location\})+I(\{orientation\})) - I(\{location\}\cup\{orientation\}) \leq -0.3, \quad (C5)$$

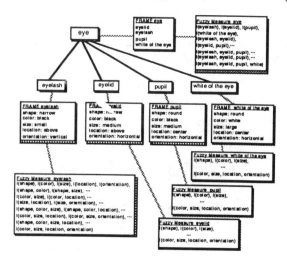

Fig. 5. Example of model for "*eye*"

TABLE III
THE FUZZY MEASURE SATISFYING THE CONSTRAINS C1-C8:
x1="size", x2="shape", x3="color", x4="location", x5="orientation".

The fuzzy measure after constraints satisfaction				
$I(\{x_1\})=0.236$	$I(\{x_1,x_3\})=0.249$	$I(\{x_3,x_4\})=0.308$	$I(\{x_1,x_3,x_4\})=0.600$	$I(\{x_3,x_4,x_5\})=0.523$
$I(\{x_2\})=0.085$	$I(\{x_1,x_4\})=0.400$	$I(\{x_3,x_5\})=0.400$	$I(\{x_1,x_3,x_5\})=0.600$	$I(\{x_1,x_2,x_3,x\})=0.800$
$I(\{x_3\})=0.045$	$I(\{x_1,x_5\})=0.462$	$I(\{x_4,x_5\})=0.350$	$I(\{x_1,x_4,x_5\})=0.600$	$I(\{x_1,x_2,x_3,x_5\})=0.721$
$I(\{x_4\})=0.017$	$I(\{x_2,x_3\})=0.400$	$I(\{x_1,x_2,x_3\})=0.525$	$I(\{x_2,x_3,x_4\})=0.630$	$I(\{x_1,x_2,x_4,x_5\})=0.866$
$I(\{x_5\})=0.054$	$I(\{x_2,x_4\})=0.630$	$I(\{x_1,x_2,x_4\})=0.630$	$I(\{x_2,x_3,x_5\})=0.600$	$I(\{x_1,x_3,x_4,x_5\})=0.800$
$I(\{x_1,x_2\})0.525$	$I(\{x_2,x_5\})=0.400$	$I(\{x_1,x_2,x_5\})=0.630$	$I(\{x_2,x_4,x_5\})=0.665$	$I(\{x_1,x_3,x_4,x_5\})=0.800$

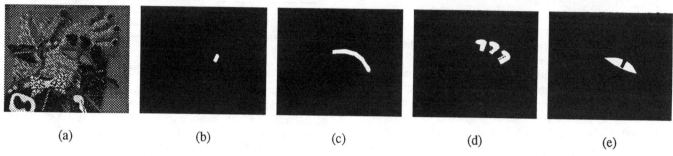

Fig. 6. Abstract painting and its segmentation. (a)Abstract Painting of "eye", (b) segmentation of "pupil", (c) segmentation of "eyelid", (d) segmentation of "eyelash", (e) segmentation of "white of the eye"

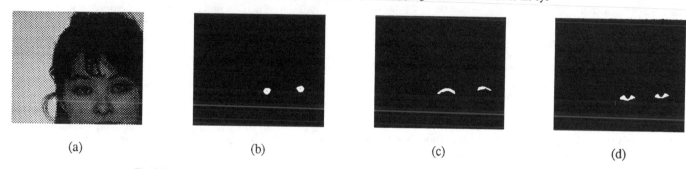

Fig. 7. Real photograph and its segmentation.(a)Real photograph of "eye", (b) segmentation of "pupil", (c) segmentation of "eyelid", (d) segmentation of "white of the eye"

$I(\{location\} \cup \{orientation\})$
$\quad - 0.3(I(\{location\}) + I(\{orientation\})) < 0.7.$ (C6)

Suppose next that the relative importance of attributes is specified as: "shape" is *more important than* "color", "orientation" is *a little important than* "color", Using Table II these importances are converted into the constraints:

$\quad I(\{shape\}) \geq 1.5I(\{color\}).$ (C7)
$\quad I(\{orientation\}) \geq 1.2I(\{color\}).$ (C8)

In these cases, η in Table II is the middle of its range. Based on C1-C8, the quadratic programming problem of Section V is solved. Table III shows the result of the solution. Fig.6 shows an abstract painting and the segmentation result for "eyelash", "eyelid", "pupil" and "white of the eye". Fig.7 shows a real photograph and its segmentation.

In the real photograph, the component "eyelash" was not extracted. In the abstract painting, the colors of components are not expressed. In fact these colors were dark purple for the "eyelash", light blue for the "eyelid", pale purple for the "pupil" and pink for the "white of the eye". The primitive symbols "shape" and "size" were obtained by calculating the number of pixels and degree of roundness (ratio of size and perimeter). These were then matched to the fuzzy sets in the model. The

primitive symbol "color" is obtained by converting RGB into HVC color specification and calculating the average of HVC. The "location" is obtained by calculating a direction between two components, the "orientation" is obtained by calculating the angle between horizon and the direction of maximum length on the region, then matched to the model. Table IV shows the result of matching the primitive symbols to the model.

Using the fuzzy measure in Table III, the local matching is obtained by the Choquet integral (4). Table V shows the results of the local matching using both the integral and the weighted sum of the attribute values detected.

TABLE IV
RESULT OF THE MATCHING OF PRIMITIVE SYMBOLS

	Abstract painting				Real picture			
	pupil	eye-lid	eye-lash	eye-white	pupil	eye-lid	eye-lash	eye-white
size	0.5	0.6	0.2	0.6	0.7	0.5	0	0.7
shape	0.5	0.8	0.6	0.8	0.9	0.7	0	0.8
color	0.5	0.5	0.5	0.2	0.9	0.7	0	0.6
location	0.8	0.8	0.8	0.7	0.8	0.9	0	0.7
orienta-tion	0.7	0.8	0.7	0.7	0.8	0.8	0	0.9

TABLE V
RESULT OF THE LOCAL MATCHING

	Abstract painting				Real picture			
	pupil	eye-lid	eye-lash	eye-white	pupil	eye-lid	eye-lash	eye-white
Fuzzy integral	0.57	0.72	0.54	0.62	0.82	0.70	0	0.73
Weighted sum	0.54	0.66	0.39	0.61	0.78	0.61	0	0.73

B. Global Match of the "Eye"

Suppose that the user specifies the following correlation information regarding components: "pupil" is *support* for "eye white", "eyelash" is *almost dependent* on "eyelid". Using Table I these are converted into constraints for the measure. As mentioned in Section V, the results of the local matching can be used to determine the fuzzy measure for the global match. For example, in the case of the abstract painting, based on Table V we have that: "eyelid" is more important than "white of eye", "white of eye" is more important than "pupil", "pupil" is more important than "eyelash". For the real photograph the results of the local match produce the following information: "pupil" is more important than "white of eye", "white of eye " is more important than "eyelid", "eye lid" is more important than "eyelash". In each case the information is converted into constraints using Table II. As in the case of the local match the fuzzy measure corresponding to these constraints is determined. The results are shown in Table VI. Using these results, the global matching is obtained by applying (4) and the results are shown in Table VII.

From the results displayed in Table V and Table VII we see that both at the local match and the global match levels, and for each item considered, the recognition result of the fuzzy measure based method is greater than or equal to that using weighted sums (equality occurred only for two features of the real photograph). It is very important to stress here that this is a qualitative result, in the sense that the improvement and not the quantity of the improvement were of primary concern for us. The amount of improvement can be increased if we are willing to refine some of our steps. For example, in using the results of the local match to derive the constraints on the measure at the global level we took into account only that the local match was better for some features than for others, and not how much better was in different cases. Consequently we converted the result of the local match into only one type of constraint: "...*more important than*...". Were we to implement a rule which would convert the results in more than one constraint, the result of the global match would be increased.

TABLE VI
THE FUZZY MEASURE SATISFYING THE CONSTRAINTS FOR THE GLOBAL MATCH
x1="pupil", x2="eyelid", x3="eyelash", x4="eye white"

(A) ABSTRACT PAINT

The value of the fuzzy measure		
$I(\{x_1\})=0.122$	$I(\{x_1x_2\})=0.400$	$I(\{x_3x_4\})=0.400$
$I(\{x_2\})=0.288$	$I(\{x_1x_3\})=0.400$	$I(\{x_1x_2x_3\})=0.600$
$I(\{x_3\})=0.082$	$I(\{x_1x_4\})=0.514$	$I(\{x_1x_3x_4\})=0.600$
$I(\{x_4\})=0.183$	$I(\{x_2x_3\})=0.312$	$I(\{x_2x_3x_4\})=0.600$
	$I(\{x_2x_4\})=0.400$	

(B) REAL PICTURE

The value of the fuzzy measure		
$I(\{x_1\})=0.144$	$I(\{x_1x_2\})=0.400$	$I(\{x_3x_4\})=0.400$
$I(\{x_2\})=0.064$	$I(\{x_1x_3\})=0.400$	$I(\{x_1x_2x_3\})=0.468$
$I(\{x_3\})=0.043$	$I(\{x_1x_4\})=0.468$	$I(\{x_1x_2x_4\})=0.937$
$I(\{x_4\})=0.096$	$I(\{x_2x_3\})=0.077$	$I(\{x_1x_3x_4\})=0.468$
	$I(\{x_2x_4\})=0.400$	$I(\{x_2x_3x_4\})=0.468$

TABLE VII
RESULT OF THE GLOBAL MATCHING

	Abstract painting	Real picture
Fuzzy integral	0.607	0.683
Weighted sum	0.592	0.639

VII. CONCLUSION

The main goal of this paper was to investigate the use of fuzzy based techniques in the following aspects of recognition of natural objects: (i)modeling of the objects which contain fuzzy information and (ii)a fuzzy measure based method of matching the results of the image processing to the model. We have shown that the use of fuzzy methods in model representation and reasoning plays an important role in the recognition of natural objects, and it can exploit both the results of image processing and the user's cognitive structure to improve recognition. An important part of our paper is the dynamic determination of the fuzzy measure underlying the matching process, allowing the use of knowledge on image processing and its results, and the model for this determination. As applications we presented two completely different images of an eye: a natural photograph, and abstract painting. Several issues remain to be discussed in connection with the method presented. Among them we mention its further possible application for recognition of complex objects in complex conditions, such as (partial) occlusion.

ACKNOWLEDGMENT

We like to thank Professor Minoru Asada, and Professor Toshiro Terano for advice during the research and implementation of the issues discussed in this paper.

REFERENCES

[1] A. Rosenfeld, "The fuzzy geometry of image subsets", *Pattern Recognition Letters 2*, pp. 311-317, 1984.
[2] A. Rosenfeld, R. A. Hummel, S. W. Zucker, "Scene labeling by relaxation operations, *IEEE Trans. System, Man., and Cybernetics*, vol. SMC-6, No.6, pp.420-433, 1976.
[3] Faugeras. O.D., "Relaxation labeling and evidence gathering", *6th ICPR*, pp.405-412, 1982.
[4] Andress. K.M., Kak. A.C., "Evidence accumulation & flow of control in a hierarchical spatial reasoning system", *AI Magazine*, vol.9, No.2, pp. 75-94, 1988.
[5] Shafer. G., A "Mathematical theory of evidence", *Princeton Univ. Press*, 1976.
[6] Keller J., Qiu H., Tahani H., "The fuzzy integral and image segmentation", *Proceedings, NAFIPS-86*, New Orleans, pp 324-338, 1986.
[7] T. Murofushi, M. Sugeno: "An interpretation of fuzzy measure and the Choquet as an integral with respect to fuzzy measure", *Fuzzy Sets and Systems*, vol.29, No.2, 201-227, 1989.
[8] M. Yoneda, S. Fukami., M. Grabisch: "Interactive determination of a utility function represented as a fuzzy integral", *Information Sciences*, in press.

FINDING SIMILAR PATTERNS IN LARGE IMAGE DATABASES

R. W. Picard and T. Kabir

Media Laboratory, E15-392
Massachusetts Institute of Technology, Cambridge, MA 02139
picard@media.mit.edu

ABSTRACT

We address a new and rapidly growing application, automated searching through large sets of images to find a pattern "similar to this one." Classical matched filtering fails at this problem since patterns, particularly textures, can differ in every pixel and still be perceptually similar. Most potential recognition methods have not been tested on large sets of imagery. This paper evaluates a key recognition method on a library of almost 1000 images, based on the entire Brodatz texture album. The features used for searching rely on a significant improvement to the traditional Karhunen-Loéve (KL) transform which makes it shift-invariant. Results are shown for a variety of false alarm rates and for different subsets of KL features.

1. INTRODUCTION

As vastly increasing amounts of image and video are stored in computers it becomes harder for humans to locate a particular scene or video clip. It is currently impossible, in the general case, to semantically describe an image to the computer and have it retrieve it. A simpler and more immediate solution might be to have the user show the computer example image data, or to speak to it keywords with which it has previously associated image features, and then have the computer search for similar patterns or features. However, no existing image processing tools are known to solve this problem for a large general set of images.

Most people know of the ease with which a computer can perform text-string matching. Similar techniques based on matched filters have been successful for signal detection in noise. These types of solutions, however, fail for the problem described above. Two uncorrupted patterns, especially two textures, can be visually similar and still differ in every pixel. Consider for example two video clips of a waterfall taken several minutes apart. Although the pixel values in these two "temporal texture" images will differ, their semantic appearances are likely equivalent. When a human asks a computer to find a particular picture, one would like the computer to understand the human's similarity criteria.

Ideally, we could define a measure of perceptual or semantic similarity and use it instead of the ubiquitous mean-squared error measure of similarity. A step toward this ideal is to transform the data so that perceptually similar things become measurably close to one another in some new space.

This work was supported in part by BT, U.K.

The mean-squared error, or a weighted version of it is then used to measure "closeness" in this new space.

Most studies of potential transformation algorithms have been run on small sets of test data, typically four to sixteen images at once from the standard Brodatz library of natural textures [1]. Moreover, selected test images have typically exhibited a lot of visual and semantic dissimilarity, as well as a lot of within-class homogeneity. The selected subsets usually do not include the less homogeneous samples from the Brodatz library or those samples which are different shots of the same material, possessing "semantic" similarity. In this study, the inclusion of non-homogeneous patterns gives a scenario closer to what one would expect searching through an image database.

The Brodatz library is limited in that most patterns do not include perspective distortions, most are uniformly illuminated, and there is not much diversity in orientation. Nevertheless, it still challenges the existing texture discrimination tools. To our knowledge, this study is the first which uses the entire available Brodatz library[1]. From the interior of each of 111 original 8 bit 512×512 Brodatz images, nine 128×128 subimages were cropped. This yields a set of $d = 999$ 128×128 images which are called the "Brodatz database."

2. FEATURE SELECTION

Optimal feature selection remains an open research problem. For texture, it has been shown that local second order statistics are important, and that statistics such as co-occurrences incorporate perceptually significant changes [2]. However, features based purely on co-occurrences have been out-performed by features based on outputs of various local filters, and emphasis has shifted to the choice of these filters [3, 4, 5]. This study begins with the "eigenfilters," or principle components of the texture covariance. Eigenfilters have been shown to provide good texture discrimination on small sets of data [3, 4].

The eigenfilter method is more commonly known in the image coding community as the Karhunen-Loéve (KL) transform or principal components analysis, and is optimal for decorrelating the features. Although there is no direct evidence that the human visual system uses an eigenvector-

---

[1] Note the borders were not used since in several cases their data was corrupted by the imaging process. Also, the original Brodatz library has 112 images, one of which appears to have been omitted from the digital library.

Figure 1: Photobook displays images in raster scan order, by their similarity to the upper left image.

based method, there are some who believe that humans construct basic "templates" of commonly occurring configurations and then use combinations of these for recognition. However, based on currently available understanding of the human visual system, it is highly unlikely anyone can prove that a given algorithm imitates the human notion of "visual similarity" on more than a trivial set of data.

3. PRINCIPAL COMPONENTS ANALYSIS

The principal components analysis is conducted as follows. Let $x_i \in \mathbf{R}^{n \times 1}$, $i = 1, \ldots, d$, $n = 128^2$ be a vector representing the DFT magnitude of one of the images in the database. The vectors x_i are formed by raster scan ordering the rows of the image into one long vector. Ten percent of the images, $p = 100$, are picked at random for training, $x_{t_1}, \ldots, x_{t_p}$, and used to form an estimate of the "pooled covariance", $\mathbf{C} = \frac{1}{p} \sum_{j=1}^{p} (x_{t_j} - \mathbf{m})(x_{t_j} - \mathbf{m})' = \mathbf{XX}'$, where $\mathbf{m}$ is the training set sample mean, $\mathbf{m} = \frac{1}{p} \sum_{j=1}^{p} x_{t_j}$. Each principal component, q_j, is an eigenvector of this covariance matrix having associated eigenvalue λ_j:

$$\mathbf{C} q_j = \mathbf{XX}' q_j = \lambda_j q_j \qquad (1)$$

Taking into account that $\mathbf{C}$ will have at most $p \ll n$ eigenvectors we can save computation by first solving for the eigenvectors of the problem:

$$\mathbf{X}' \mathbf{X} u_j = \lambda_j u_j. \qquad (2)$$

Left multiplying both sides of (2) by $\mathbf{X}$ gives

$$\mathbf{XX}'(\mathbf{X} u_j) = \lambda_j(\mathbf{X} u_j)$$

so that the desired vectors for projection can be obtained by $q_j = \mathbf{X} u_j$. With this analysis the calculations are greatly reduced. The associated eigenvalues are used to order the eigenvectors. Note that in recognition, unlike in coding, $\lambda_j > \lambda_i$ does not imply that u_j will be more useful in reducing error than u_i [6].

For a given database, these values and the projection of each x_i, $i = 1, \ldots, d$, onto the p eigenvectors are precomputed and stored. These projections (or KL transform coefficients) are the features used for comparing patterns.

Earlier texture applications of eigenfilters computed the coefficients directly from the covariance estimate of the spatial data. However, using the DFT magnitudes makes the KL features invariant to spatial translation. This one difference made an immediately noticeable improvement in the recognition ability of the algorithm. This improvement is similar to one noticed by Akamatsu, *et. al.* [7] in face recognition. Incorporating translation-invariance makes the recognition algorithm perform a little more closely like a human. Other invariants not investigated here [8] may also be similar to those used by humans. Although phase is important for structural image reconstruction, the linear part of the phase appears to be unimportant for much of pattern recognition. We suggest that selective use of phase components should be better than discarding all the phase.

It should be noted that horizontal/vertical boundary effects were mitigated by first windowing each image with a 2D isotropic Gaussian filter, $\sigma = 24$, before computing its DFT. The windowing eliminated the corner regions, in most cases reducing the information available for comparisons.

4. PHOTOBOOK TEST ENVIRONMENT

The Photobook test environment is illustrated in Fig. 1 running with features based on the phase invariant KL coefficients. For the example shown, the first twenty coefficients were used as features. The Euclidean distance was used on these features to measure similarity. Using twenty coefficients gave surprisingly good performance while maintaining the interactive speed of the system. (Actually, the re-display time was found to be the limiting factor speed wise.)

Initially, Photobook displays forty randomly selected images from the database. The user selects an image of interest, and after about 2 seconds on a DECstation 5000 the display is updated with the "nearest" patterns. In the case shown, the user selected D51-raffia and the algorithm found the eight other subimages from the original D51 image. It also found several other textures, such as D72-tree stump which has subimages visually similar to D51.

5. RECOGNITION RESULTS

First using a zero "false alarm" constraint, the very strictest, we determined the classification performance for every sample in the Brodatz database. This test required that Photobook return the eight other samples from the source image as the first eight nearest samples. This was repeated for all nine samples of each image, averaging the number recognized in the first eight. The patterns are ordered by these recognition rates in Table 1.

These results can be used in many ways. If a researcher picks only a subset of the Brodatz library, and happens to pick patterns at the high end of this table, then they can claim better than 90-100% performance. If they pick patterns located near the left, then this method will result in reduced performance. Hence, these results can be used to estimate a measure of "difficulty" for a particular subset of patterns. It is interesting to observe where the patterns people have traditionally used appear in this table. One can also use these results to typify the types of patterns the eigenvector method seems best/worst suited for.

Note that the results reported here correspond to a "worst case" scenario, or a lower bound on performance. Typically researchers evaluate performance on a smaller subset of the Brodatz patterns where the discrimination has less potential to choose a wrong pattern. These results are also conservative for another reason. As mentioned, the Brodatz database contains different images (e.g. D3 and D36, both of reptile skin) that humans may judge to look similar, but which our results declare to be a "miss" because the source images differ. This case appears to occur more frequently than the opposing case where two visually different samples from the same source image are declared to match. When a source image was very inhomogeneous, the performance was poor as one would expect for both the human and the computer algorithms.

The data presented above was for the strictest case, zero false-alarms. Fig. 2 shows the results of a study that allows up to 32 false alarms. This limit is appropriate as the screen only displays 40 images at once. It is acceptable in database search applications to declare success if all the

Avg. Recogn. Rate	Brodatz Texture	Avg. Recogn. Rate	Brodatz Texture
12.35	D88	60.49	D64
14.81	D89	60.49	D75
17.28	D31	60.49	D92
18.52	D43	60.49	D68
18.52	D58	62.96	D10
19.75	D42	64.20	D47
20.99	D107	65.43	D106
22.22	D40	65.43	D2
23.46	D72	65.43	D79
23.46	D30	66.67	D93
24.69	D23	66.67	D18
24.69	D69	66.67	D3
25.93	D60	66.67	D46
25.93	D41	66.67	D71
25.93	D91	67.90	D52
25.93	D98	69.14	D24
25.93	D13	69.14	D87
27.16	D90	70.37	D15
27.16	D73	71.60	D80
32.10	D27	74.07	D76
34.57	D100	74.07	D12
34.57	D7	75.31	D82
34.57	D108	76.54	D81
35.80	D44	76.54	D38
35.80	D97	80.25	D35
35.80	D63	81.48	D4
35.80	D112	82.72	D51
37.04	D28	82.72	D1
37.04	D54	82.72	D50
38.27	D39	85.19	D25
39.51	D62	86.42	D16
39.51	D99	86.42	D85
40.74	D59	86.42	D105
42.00	D36	87.65	D17
42.00	D111	88.89	D8
42.00	D66	90.12	D84
43.21	D61	90.12	D95
43.21	D70	91.36	D32
45.68	D48	95.06	D57
45.68	D109	96.30	D21
46.91	D5	97.53	D83
48.15	D86	97.53	D37
49.38	D74	97.53	D56
50.62	D104	98.77	D65
51.85	D29	98.77	D49
51.85	D94	98.77	D110
51.85	D33	98.77	D6
51.85	D67	98.77	D77
54.32	D19	100.00	D34
55.56	D11	100.00	D55
55.56	D78	100.00	D101
56.79	D26	100.00	D102
56.79	D22	100.00	D20
58.02	D9	100.00	D14
58.02	D96	100.00	D53
59.26	D103		

Table 1: Brodatz textures ordered by recognition rates.

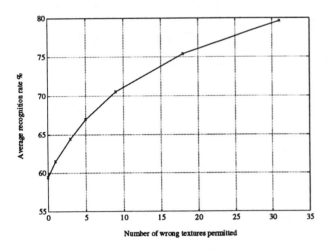

Figure 2: An approximate "operating characteristic" for the Brodatz database.

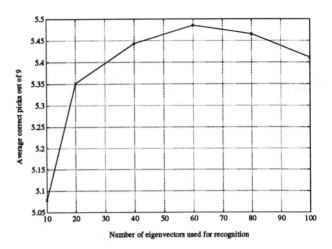

Figure 3: How average recognition rate, with zero false detects, varies while increasing the number of features.

patterns it should have found get displayed. Each point on this plot was formed from an average of the performance for all the 999 database images. Similar to the Neyman-Pearson operating characteristic [6], these curves show that the performance of the shift-invariant principal components will increase monotonically with the permitted number of false detections. In the limit of course, as the number of false detects goes to 990, all the curves will reach 100%. A method which has a curve lying above this one can be considered an improvement. Thus, this data can be used as a bench mark for subsequent methods.

As noted, the coefficients for the largest eigenvalues (except DC) may be optimal for representation, but this is not necessarily true for discrimination. To find the best optimal subset of size p requires a search over the power set of coefficients, a size 2^p problem. Clearly this is unreasonable for large p. A suboptimal alternative was taken here, picking subsets of size $s = 10, 20, 40, 60, 80,$ and 100 from the coefficients corresponding to the s largest eigenvalues. The DC coefficient was only included in the last case.

The results of this study are shown in Fig. 3. Each point here is the average over a whole table of data like that in Table 1. This study verifies that using more coefficients, although better for representation, is not necessarily better for discrimination or classification. After the first sixty coefficients the performance degrades. The greatest increase can be seen to occur at twenty coefficients, where we concluded the best price-performance was for the interactive Photobook system.

6. SUMMARY

The "similar pattern" recognition problem has been investigated in the context of a set of 999 images based on the entire Brodatz library of textures, the standard source in texture classification. The performance of a shift-invariant principal components algorithm was characterized for each pattern, for various false alarm rates, and for various subsets of features. These results provide a bench mark for

comparing alternate algorithms. Additionally, they provide one ordering of the images from "difficult" to "easy" for their recognition from within the Brodatz database.

Acknowledgments

The authors would like to thank A. P. Pentland for the idea of the Photobook interface, T. E. Starner for its initial implementation, and C. H. Perry for his help implementing some of the experiments.

7. REFERENCES

[1] P. Brodatz. *Textures: A Photographic Album for Artists and Designers*. Dover, New York, 1966.

[2] R. Haralick. Statistical and structural approaches to texture. *Proc. IEEE*, 67:786–804, May 1979.

[3] F. Ade. Characterization of textures by 'eigenfilters'. *Signal Processing*, 5:451–457, 1983.

[4] M. Unser. Local linear transforms for texture measurements. *Signal Processing*, 11:61–78, 1986.

[5] R. Vistnes. Texture models and image measures for texture discrimination. *Int. J. of Computer Vision*, 3:313–336, 1989.

[6] C. W. Therrien. *Decision Estimation and Classification*. John Wiley and Sons, Inc., New York, 1989.

[7] S. Akamatsu, T. Sasaki, H. Fukamachi, and Y. Suenaga. A robust face identification scheme – KL expansion of an invariant feature space. In *Proc. SPIE Conf. on Intell. Robots and Comp. Vis.*, volume 1607, pages 71–84, Boston, MA, Nov. 1991.

[8] L. Jacobson and H. Wechsler. Invariant analogical image representation and pattern recognition. *Patt. Rec. Lett.*, 2:289–299, 1984.

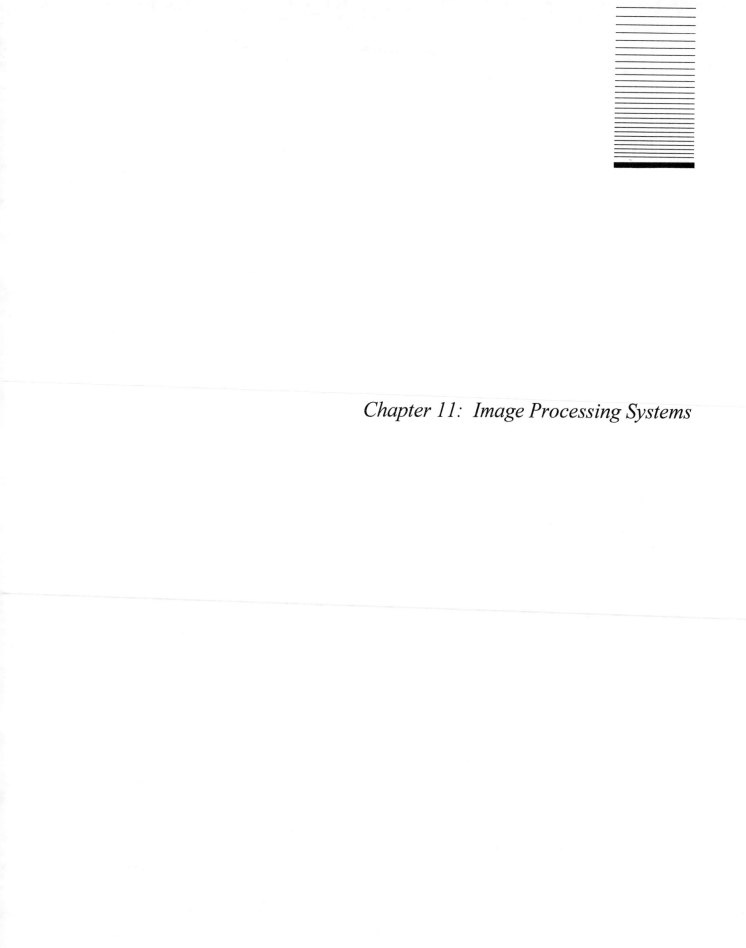

Chapter 11: Image Processing Systems

Image Processing Systems

Topics in this chapter pertain to applications of signal processing to imaging technology, but in areas outside image coding and recognition. This chapter emphasizes the pervasive topic of image quality assessment and describes several system- or application-specific image processing solutions.

Developing objective techniques that correlate with human perception is important to avoid the frequent use of panels of human subjects to assess the quality of image coders. "An Objective Technique for Assessing Video Impairments" (paper 11.1) describes a set of objective measurements that highly correlate with human quality assessment of impaired motion video. Impairments included degradation from coding at a very high compression ratio, bit errors, and problems in the analog domain. Measurements include both spatial impairments (i.e., normalized energy differences between corresponding original and impaired frames) and temporal impairments (based on the first-order frame difference sequence). Objective correlates to subjective perceptions are further discussed in "Psycho-Visual Based Distortion Measures for Monochrome Image and Video Compression" (11.2). This paper introduces a psychovisual model of perception that emphasizes errors that are uncorrelated with local image features over correlated errors, and it describes several distortion measures that are based on this model. As with the previous paper, high correlation of these measurements with image quality assessment by a panel of observers is reported.

"Recent Trends in Compression of Multispectral Imagery" (11.3), like papers in Chapter 9 on motion coding, introduces a third dimension to the two-dimensional image compression problem. Here, the third dimension is wavelength rather than time, but the techniques of interpolating across images of the same scene at different wavelengths are reminiscent of those used in motion encoding. A hierarchy of coding approaches is reviewed, with particular emphasis given to vector quantization and transform coding.

Converting a standard analog television signal to a digital component video or computer graphics format entails clock and sync acquisition, color reference acquisition, sampling and quantization, filtering, linear algebra, and quadrature demodulation. "Video to VGA and Back" (11.4) provides a description of recently-available commercial components that perform these functions.

High definition television (HDTV) is a topic of current interest, and its demand for high information transfer through limited-bandwidth channels motivates the inclusion of "Designing a Video Compression System for High Definition Television" (11.5). Requirements that the specialized nature of television place on the image coder are reviewed. These include the need for lowest-cost decoding circuitry (even when accompanied by additional encoding cost), the need for rapid coder initialization to support channel surfing, the requirements for a constant bit rate, and the difficulty of forward leakage in a coder prediction loop as interfering with reverse playback on video cassette recorders. A coding method that meets most of these requirements is proposed which has been presented for consideration as the U.S. HDTV standard.

"Programmable Facsimile Image Processor Including Fuzzy-Based Decision" (11.6) describes an application specific integrated circuit that handles the facsimile functions of encoding documents that mix text and gray-scale images, applying halftoning techniques that use patterns of binary pixels to reproduce the appearance of gray scale, and accommodating the requirements of several generations of modem standards. A fuzzy logic dithering scheme is described, and transmission times for image and text documents are reported. Digital halftoning is extended into the color regime and applied to color printers in "Printer Models and Color Halftoning" (11.7). Common commercial printers are modeled and the paper then addresses the problem of creating the illusion of a continuous-tone color image under the constraints of a limited number of colors under a variety of distorting conditions such as "dot overlap."

Suggested Additional Readings:

[1]   N. Jayant, J. Johnston, and R. Safranek, "Signal Compression Based on Models of Human Perception," *Proc. IEEE,* vol. 81, no. 10, pp. 1385-1422, Oct. 1993.

[2]   D. Anastassiou, "Digital Television," *Proc. IEEE,* vol. 82, no. 4, pp. 510-519, April 1994.

[3]   R. Hopkins, "Choosing an American Digital HDTV Terrestrial Broadcasting System," *Proc. IEEE,* vol. 82, no. 4, pp. 554-563, April 1994.

[4]   K. Aoyama and R. Ishii, "Image Magnification by using Spectrum Extrapolation," *Proc. IECON '93 International Conference on Industrial Electronics, Control and Instrumentation,* pp. 2266-2271.

[5]   C. F. Barnes and E. S. Sjoberg, "A Comparative Study of JPEG and Residual VQ Compression of Radar Imagery," *Proc. National Telesystems Conference* (Atlanta, GA, USA) June 1993, pp. 203-207.

[6]   F. Kossentini, W. C. Chung, and M. J. T. Smith, "Low Bit Rate Coding of Earth Science Images," *Proc. Data Compression Conference '93* 1993, pp. 371-380.

[7]   V. R. Algazi and N. Hiwasa, "Perceptual Criteria and Design Alternatives for Low Bit Rate Video Coding," *The Twenty-Seventh Asilomar Conference on Signals, Systems, and Computers* (Pacific Grove, CA, USA) Nov. 1994, pp. 831-834.

[8]   J. J. Nicolas and J. S. Lim, "Equalization and Interference Rejection for the Terrestrial Broadcast of Digital HDTV," *Proc. IEEE International Conference on Acoustics, Speech, and Signal Processing 93* (Minneapolis, MN, USA) April 1993, vol. 4 pp. 176-179.

AN OBJECTIVE TECHNIQUE FOR ASSESSING VIDEO IMPAIRMENTS

Stephen Voran and Stephen Wolf

Institute for Telecommunication Sciences, U.S. Department of Commerce
NTIA/ITS.N3, 325 Broadway, Boulder, Colorado 80303

ABSTRACT

The Institute for Telecommunication Sciences is deriving new techniques for assessing impairments induced by video transport and storage systems. These techniques are based on digital image processing operations performed on digitized original and impaired video sequences. Measurements that quantify perceptual video attributes in both the spatial and temporal domains are extracted from the digitized video. These measurements are then used to compute a single score that quantifies the perceptual impact of the impairments present in the video sequence. This objective impairment score is well-correlated (r=.92) with impairment assessments made by human viewers. Thus, it can be used to augment, or possibly to replace, the expensive and time-consuming subjective viewing tests that are typically used to evaluate video coding and transmission techniques.

1: INTRODUCTION

When faced with the task of assessing the video impairments in a video storage or transport system in a meaningful and repeatable way, the video engineer has relatively few practical options. First of all, carefully conducted subjective tests can provide very useful data, but these tests must be conducted in a controlled viewing environment, and they require the participation of tens of viewers. Secondly, while many digital image processing techniques for assessing the displayed quality of a *single* image have been proposed, these techniques do not easily extend to cover the *continuing sequence* of images that are generated by a video signal. Thirdly, the objective video test signals typically used in the analog television industry (e.g., multiburst, color bars) are either static or are deterministic functions of time. While these signals and the associated measurements are useful for the characterization of the electrical performance of time-invariant analog video systems, the measurements often do not correlate well with the impairment level perceived by the users of the video system. Furthermore, video signals are now commonly transmitted and stored in compressed digital form and impairments may arise from the compression process or the presence of errors in the transmitted bit stream.

Effective compression algorithms are dynamic, with the input video signal dictating the overall behavior of the algorithm through sub-algorithms that perform motion prediction, adaptive transforms, and adaptive quantization, to name only a few. The resulting systems are clearly time-varying and signal dependent. Static, deterministic test signals cannot provide an accurate characterization of their performance on real video sequences.

These observations motivated us to design a video impairment assessment technique that uses actual video signals. This approach provides a realistic measurement environment and allows measurements to be made while a video system is in use. The design process is described in Section 2. The process uses data from an extensive series of subjective video impairment tests, which is then correlated against a family of proposed objective video impairment measurements. Two effective measurements are presented in Sections 3.1 and 3.2. A measure of spatial impairment is based on normalized energy differences of Sobel-filtered video frames and a complementary temporal impairment measurement is based on the first-order frame difference sequence. A linear combination of these measurements generates a single objective impairment score that is well correlated with the subjective test data. In Section 3.3, we discuss the performance of this video impairment assessment technique and identify areas for continuing research.

2. SUBJECTIVE AND OBJECTIVE VIDEO IMPAIRMENT TESTS

We are seeking an objective assessment technique that quantifies video impairments in a way that correlates with subjective evaluations made by those who actually view the impaired video. It is clear that we must incorporate knowledge of human perception in the design of the assessment technique. Toward this end, we have conducted an extensive set of subjective video impairment tests. The tests were conducted following a methodology and viewing laboratory environment specified in the CCIR Recommendation 500-3[1]. Forty-eight viewers rated a collection of 132 impaired video sequences. The viewers, grouped 3 at a time, were shown a 9-second original video sequence followed by an impaired version of that sequence and were then asked to rate the impact of the impairment. The five possible responses were: 5=imperceptible, 4=perceptible but not annoying, 3=slightly annoying, 2=annoying, and 1=very annoying. The responses of the 48 viewers were then averaged to obtain a mean value on a 5-point scale for each of the 132 impaired sequences.

The impaired video sequences were selected from a collection of 36 video sequences that were passed through a group of 28 video systems, both analog and digital. Digital systems range from slightly compressed (45 Mbps) to highly compressed (56 Kbps). Controlled bit errors were introduced into some of the digital video systems. Analog impairments included NTSC encode/decode cycles, VHS record/play cycles, and a noisy RF channel. In order to derive the most general system possible, the video sequences contained widely varying amounts of spatial and temporal information. Examples include sports events, newscaster, still shots, and graphics. The spatial and temporal information content of test scenes are important considerations because, in digital systems, information content determines sample rates and plays a crucial role in determining the level of impairment that is suffered when the video is transmitted over a fixed-rate digital channel. Similarly, in analog video systems, the spatial information content governs the relationship between system passband and video impairment level.

In addition to the subjective impairment tests, a set of candidate measurements based on digital image processing techniques were applied to digitized versions (24 bits/pixel, 756 pixels/line, 486 lines/frame, approximately 30 frames/second) of the original and impaired video sequences. In an exact parallel to the subjective tests, the objective measurements were all differential. That is, they involved both the original and the impaired versions of each video sequence. All video impairments can be described as distortions of the amplitude or the timing of the video waveform. When displayed on a monitor, this one-dimensional voltage waveform is interpreted as a continuously evolving, multicolored, multidimensional signal. Useful impairment measures must take note of this human interpretation and mimic it to the extent possible. Thus, the candidate set of objective measures included those designed to measure temporal, luminance, chrominance, and spatial impairments to the video image sequence.

The design process hinges on a joint statistical analysis of the *objective* and *subjective* video impairment data sets. This analysis reveals which of the objective video impairment measurements are meaningful, and how they might be combined to create a video impairment assessment technique that correlates well with subjective test results. An effective pair of spatial and temporal video impairment measurements is discussed in the following section. Because color impairments in the data set were very small relative to other

impairments, the measurements discussed below are applied only to the luminance portion of the video signal. More complete descriptions of the subjective and objective video impairment experiments can be found in [2], [3], and [4].

3. RESULTS AND DISCUSSION

The scanned nature of video signals makes the true separation of their spatial and temporal dimensions a rather complicated interpolation issue. While a single frame is not truly "a set of spatial samples at a fixed time instant," we make that approximation. Accordingly, we acknowledge that for the "spatial" and "temporal" video measurements presented here, the separation is not complete and the classification is only approximate.

3.1. Spatial Video Impairment Measurement

The selected measurement of spatial impairment is based on normalized energy differences of Sobel-filtered video frames. It is given by

$$m_s = \frac{\mid \text{mean}^2_{time}(x_t) - \text{mean}^2_{time}(y_t) \mid}{\text{mean}^2_{time}(x_t)}, \quad (1)$$

$$x_t = \text{std}_{space}(\text{Sobel}(X_t)), \quad y_t = \text{std}_{space}(\text{Sobel}(Y_t)),$$

X_t: Original Video Frame t, Y_t: Impaired Video Frame t.

In Equation 1, std_{space} denotes a standard deviation calculation conducted over the visible portion of the pixel array. The temporal means are calculated over the entire 9-second sequence that was subjectively rated. Since the Sobel filter operation emphasizes edges or high-frequency content of a video frame[5], the measurement tracks changes in edge content or spatial resolution, quantities that are known to be perceptually important. Due to the absolute value function, either an increase (e.g., noise, false edges) or a decrease (e.g., reduced resolution) in these quantities causes a positive deflection of m_s. As the impairment level is reduced to zero, y_t must approach x_t, and m_s goes to zero. If one defines x_t to be the spatial information content of frame X_t, then m_s is a normalized measure of the change in spatial information content caused by the video system under test. Computations of m_s over a large set of video sequences have indicated that the Sobel filtering operation can be replaced with a "pseudo-Sobel" operation without significant impact on m_s. This substitution can provide a significant savings in computations since, for every pixel, the true Sobel filter combines the outputs of the vertical and horizontal convolutions via $[h^2+v^2]^{\frac{1}{2}}$, while the "pseudo-Sobel" approximates this with $\mid h \mid + \mid v \mid$.

3.2. Temporal Video Impairment Measurement

As mentioned earlier, we have only approximately separated the temporal and spatial dimensions of the video signal.

Because of this overlap, the temporal impairment measurement, m_t, that best complements the spatial impairment measurement m_s, may not be optimal for quantifying pure temporal impairments. Rather, the measurement was selected to correlate with the overall subjective impairment variations that are not tracked by m_s. This allows a linear combination of m_s and m_t to track the subjective data set as closely as possible in the mean-squared error sense.

The temporal impairment measurement is based on the first-order frame difference sequence. First, define

$$\Delta F_t = \text{mean}_{space}(\,|\,F_t - F_{t-1}\,|\,), \qquad (2)$$

where F_t is a video frame at time t, and the absolute value operation is applied to each pixel of the frame difference. When two successive frames are identical, ΔF will be zero. When motion is present in the video sequence, successive frames are different and this is reflected by ΔF. Thus, ΔF_t tends to track the amount of motion present in the video sequence at time t. Figure 1 demonstrates how the ΔF sequences provide a wealth of information regarding frame rate, continuity of motion, and noise levels, all of which are highly relevant to human perception. The broken line in Figure 1 shows values of ΔF for 180 frames (≈ 6 seconds) of an original video sequence. The solid lines indicate values of ΔF for two impaired versions of that same sequence. The lower line displays runs of up to 10 zeros, indicating that this digital video encoder achieves bit-rate reduction by reducing its frame rate from approximately 30 frames/second to as low as 3 frames/second. Notice that the frame rate is adaptive. The contents of the video sequence determine how many bits are required by a frame update, and hence how often updates may be encoded. This temporal impairment is perceived as a variable level of "jerkiness" in the video sequence. The upper solid line in Figure 1 shows the output of a noisy video system that always generates approximately 30 frames/second. Here the random noise adds a somewhat constant amount of "false motion" to the video sequence.

When we compute the sequence defined in Equation 2 for a pair of original and impaired video sequences, we refer to them as ΔX_t and ΔY_t, respectively. To measure the dissimilarity between the two sequences, we compute temporal statistics of their log ratio and form the temporal impairment measurement m_t

$$m_t = [\,\max_{time}(s_t) - \min_{time}(s_t)\,] + \tfrac{3}{4}\,\text{mean}_{time}(s_t), \qquad (3)$$

$$\text{where} \quad s_t = \log_{10}\!\left(\frac{\Delta Y_t}{\Delta X_t}\right).$$

Here the spread in the log ratio is augmented by a fraction of the mean of the log ratio to form the overall measurement of temporal impairment. Notice that as impairments become small, ΔX_t will match ΔY_t, the log ratio will become zero, as

will m_t. If one defines ΔF_t to be the temporal information content of frame F_t, then m_t is a measure of the "spreading" in temporal information content caused by the video system under test.

Since the "spreading" or "change in smoothness" of temporal information is a good indicator of temporal impairment, the flatness ratio should also be considered:

$$\frac{\text{flatness}(\Delta Y_t)}{\text{flatness}(\Delta X_t)} = \frac{\left[\prod_{i=0}^{N-1} \Delta Y_{t-i}\right]^{\frac{1}{N}} \left[\frac{1}{N}\sum_{i=0}^{N-1} \Delta Y_{t-i}\right]^{-1}}{\left[\prod_{i=0}^{N-1} \Delta X_{t-i}\right]^{\frac{1}{N}} \left[\frac{1}{N}\sum_{i=0}^{N-1} \Delta X_{t-i}\right]^{-1}}. \qquad (4)$$

Here flatness(ΔF_t) is simply a discrete version of the spectral flatness measure[6]. It is also the ratio between the geometric mean and the arithmetic mean of ΔF_t. The flatness measure is a positive quantity that attains its maximum value of 1 when the sequence of ΔF_t maintains a constant value over the N sample measurement window. We have found that a window size of $N=5$ allows the temporal statistics of the flatness ratio to provide similar information to m_t but at somewhat greater computational expense.

We have also investigated a more complex algorithm that performs weighted integrals of ΔX_t between the spikes in ΔY_t. This attempt to more directly characterize the amount of motion that is "lost" or "displaced" due to frame rates below 30 frames/second shows good correlation to overall subjective impairment levels, but does not serve as an effective companion to m_s.

3.3. Discussion

Recall that the subjective assessment results were averaged across viewers to form a single score on the 5-point scale. The linear combination of the spatial and temporal impairment measurements (m_s and m_t) that minimizes the mean-squared error between the objective scores and the subjective scores is

$$\text{objective score} = 4.95 - 3.41 \cdot m_s - .46 \cdot m_t. \qquad (5)$$

Since m_s and m_t are positive quantities that tend toward zero for unimpaired video sequences, the maximum objective score produced by Equation 5 is 4.95. This is consistent with our observation that when presented with identical video sequences, a small fraction of viewers will respond that the second of the two is visibly impaired relative to the first, resulting in a mean subjective impairment value that is slightly less than 5. The negative coefficients for m_s and m_t cause the objective score to decrease as spatial or temporal impairments increase. Estimates of the variance of the 3 regression coefficients in Equation 5 are .01 for the constant, .09 for the coefficient of m_s, and .01 for the coefficient of m_t. These estimates include a variance inflation factor of 1.7 to account for the correlation between m_s and m_t. Thus, one would expect only small changes in the regression coefficients when the tests are repeated on a statistically similar data set. On the other hand, different coefficients may result when tests are conducted on data sets that have more restricted ranges of impairment levels.

For each of the 132 video sequences, the objective score is plotted against the subjective score in Figure 2. The RMS error between the two data sets is .54 impairment units, and the coefficient of correlation is r=.92. Examination of the errors reveals that the error distribution is nearly Gaussian, and has a constant variance. The objective assessment technique does not systematically err on a particular video system. This random nature of the errors means that the averaging of N results will tend to decrease the error by a factor of $N^{\frac{1}{2}}$. This is demonstrated in Figure 3, where groups of 3 to 7 objective and subjective scores have been averaged to generate a single objective score and a single subjective score for each of the 28 video systems in the data set. The mean value of the averaging factor, N, is 132/28 = 4.7, so we would expect the RMS error to decrease to $.54/4.7^{\frac{1}{2}}$ = .25. In fact, Figure 3 reveals an RMS error of .25 impairment units, and a correlation of r=.98. These results are very encouraging, especially in light of the wide range of video sequences, video systems, and resulting impairment levels included in the data set.

The refinement, improvement, and implementation of this assessment technique continues. While the video digitizing and processing presented here were conducted in batch mode on a fairly powerful computer, we are currently constructing a PC-based implementation of the assessment technique. The PC is outfitted with a pair of commercially available image acquisition and processing cards that will enable it to digitize original and impaired video signals and to evaluate the perceptual impact of video impairments in real time. Additional details are available in [4].

Several additional facets of this work deserve further discussion. The first is the issue of time alignment. When testing video systems, there is always some delay between the system input and output. For the objective assessment technique to perform at its best, this delay must be accounted for when computing m_s and m_t. In compressed digital systems, the delay is often a function of the input video sequence. As a further complication, Figure 1 demonstrates that input frames might not generate a corresponding output frame. We have found that shifting smoothed versions of the ΔX_t and ΔY_t sequences to attain a minimum in the standard deviation of the difference sequence $(\Delta X_t - \Delta Y_{t-\tau})$ provides a fairly reliable and robust time alignment technique.

The measurements presented here are calculated on entire video frames. We are currently investigating their application over specific regions of interest. As an example, the video frame could be divided into a grid of 12 rectangular regions. The measurements m_s and m_t could then be applied in the region with the most motion and the region with the least motion, as determined by regional evaluations of ΔX_t. This approach is motivated by the fact that human resolution judgements and noise assessments are more critical in still areas than in moving areas, but human attention is normally drawn to moving objects. Thus, when properly combined, possibly through some dynamic weighting functions, the resulting four measurements might offer improved assessments over the existing technique. Computing a multiplicity of localized measurements has the additional advantage that individual errors may cancel as measurements are averaged or otherwise combined. Also, localized processing of very specifically shaped regions (e.g., narrow horizontal or vertical stripes) could allow the impairment assessment technique to detect and account for the different types of structure in video impairments. As an example, noise that is periodic in space or time tends to have a perceptual impact that differs from the perceptual impact of noise that is random.

When a video system has a constant non-unity gain or non-zero bias, these values must be normalized out of y_t and ΔY_t before m_s and m_t are calculated. Removing constant system gain and bias from the objective assessment technique does not destroy relevant impairment information since this is exactly what a viewer does when adjusting the contrast and brightness of a video display.

The 9-second video sequence length used in this work was selected for ease of subjective testing. In order to match the subjective tests, objective assessments were also performed over this 9-second window. Separate experiments should be conducted to determine what assessment window is most appropriate for continuous impairment assessment of extended video sequences. One possible experiment would ask viewers to note impairment changes as they perceive them throughout a 15-minute video selection with controlled impairment levels. A series of variable time-lag correlations between the subjective responses and the controlled impairment levels might then reveal what "temporal assessment windows" the viewers used. It would then be natural to use these same windows in the objective assessment technique.

4. CONCLUSION

We have presented an objective video impairment assessment technique that appears to quantify the perceptual impact of video impairments in an accurate way. The development uses 132 impaired video sequences that cover a remarkably wide range of motion, detail, impairment type, and impairment level. The objective impairment measurements show a correlation of r=.92 with carefully conducted subjective assessments. We have demonstrated that this correlation can be improved by repeating assessments and averaging the results. These results are very encouraging. Refinement and improvement of the assessment technique continues, and several directions for continuing research have been noted. In addition, we are constructing a PC-based implementation of the technique. This instrument will digitize original and impaired video signals and evaluate the perceptual impact of the impairments in real time. The video impairment assessment technique is being considered for inclusion in the video teleconferencing performance standard that is being drafted by the ANSI Accredited Standards Committee T1, Working Group T1A1.5.

The research described here is being conducted at the Institute for Telecommunication Sciences in Boulder, Colorado, under sponsorship of the U.S. Department of Commerce and National Communications System. In addition to the authors, research participants include Arthur Webster, Coleen Jones, Margaret Pinson, and Paul King, who are members of the System Performance Standards Group.

REFERENCES

[1] CCIR, Recommendation 500-3, *Method for the Subjective Assessment of the Quality of Television Pictures*, 1986.

[2] S. Wolf, et. al., "Objective Quality Assessment of Digitally Transmitted Video" and "The Development and Correlation of Objective and Subjective Video Quality Measures", *Proceedings of IEEE Pacific Rim Conference on Communication, Computers, and Signal Processing*, Victoria, BC., Canada, May 1991.

[3] S. Voran and S. Wolf, "The Development and Evaluation of an Objective Video Quality Assessment System that Emulates Human Viewing Panels", *International Broadcasting Convention Technical Papers*, Amsterdam, The Netherlands, July 1992.

[4] A. Webster, et. al., "An Objective Video Quality Assessment System Based on Human Perception", *Proceedings of IS&T/SPIE 1993 International Symposium on Electronic Imaging: Science & Technology*, San Jose, CA, US, February 1993.

[5] A.K. Jain, *Fundamentals of Digital Image Processing*, Prentice-Hall Inc., 1989.

[6] N.S. Jayant and P. Noll, *Digital Coding of Waveforms*, Prentice-Hall Inc., 1984.

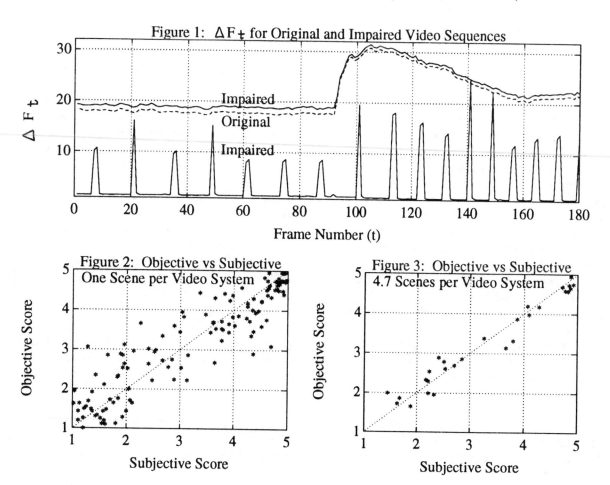

Figure 1: ΔF_t for Original and Impaired Video Sequences

Figure 2: Objective vs Subjective One Scene per Video System

Figure 3: Objective vs Subjective 4.7 Scenes per Video System

Psycho-Visual based Distortion Measures for Monochrome Image and Video Compression

Navin Chaddha and Teresa H.Y. Meng
Information Systems Laboratory
Stanford University, Stanford, CA 94305.

Abstract

In this paper we describe quantitative distortion measures for compressed monochrome image and video based on a psycho-visual model. Our model follows the human vision perception in that the distortion as perceived by a human viewer is dominated by the compression error uncorrelated with the local features of the original image and for a video sequence the distortion is perceived from two sources, the still areas and the motion areas of a video frame. We have performed subjective tests to obtain ranking results for compressed images and video sequences which were compressed using different compression algorithms and compared the results with the rankings obtained using our distortion measure and other existing mean-square error based measures. We have found that our distortion measure's ranking matches the subjective ranking accurately where as the mean-square error and its variants performed poorly in matching the subjective ranking on an average.

1. Introduction

There is an ever increasing demand for transmission and storage of vast amounts of information in image and video processing environments today. To reduce the large costs involved, image and video compression are widely used for minimizing the amount of data to be stored or transmitted. Numerous image and video compression techniques exist with the common goal of reducing the number of bits needed to store or transmit image sequences. The efficiency of the compression algorithm depends on three criteria: compression ratio, implementation complexity and quality of the compressed image or video.

The first two criteria are easy to quantify but the third criterion is difficult to evaluate for a lossy compression algorithm, as there is no reliable and consistent distortion measure for determining the magnitude of distortion resulting from a lossy compression algorithm. Such a measure is not only needed for comparing the quality of images or video produced by different compression algorithms, but also for designing image and video compression systems.

By far the most popular distortion measure of compression in use is the *pixel-squared* error between the original and the compressed image or video.

Variants of this measure include the normalized *mean-square* error, *maximum absolute deviation*, the *signal-noise ratio* and the *peak signal-noise ratio*. These error measures have been computed between bit-maps generated from the original and the compressed image or video, without taking into consideration the human visual system.

Though many attempts have been made in the past to model the human visual system, few have been successfully applied to a distortion measure for compressed images. Earlier work in [2] addressed the problem of developing a subjective model using non-linear transduction of stimulus intensity followed by spatial filtering. Later, in [3] a crude quantitative model of the human viewer was developed which consisted of a number of operations on the error signal to assess image quality. However, its performance was reported to be the same to that of the simple mean-square error measure in most cases. Recently there has been some work in perceptual and psychophysical rating of images using some image compression techniques [4,5,6]. However, most studies have reported only limited success in ranking accuracy compared with human viewers as these studies assumed random additive white noise, though the error introduced by compression is often highly correlated.

The motivation for our model [1] lies in the fact that to the human viewer, errors uncorrelated with the local features are more important than the errors correlated with the local features in judging the quality of a compressed image. Hence we give more weighting to uncorrelated errors in our image distortion measure. For a video sequence the main sources of distortion are spatial blurring, loss in edge sharpness in areas of motion and jerkiness. Therefore in our distortion measure for video compression we concentrate on measuring the distortion of edges in areas of motion and jerkiness in motion.

This paper is organized as follows. Section 2 describes the distortion measure for monochrome images. Section 3 develops the distortion measure for compressed video. Section 4 presents the subjective tests. Section 5 compares the ranking results of our distortion measure with those obtained by using the mean-square error and its variants. Section 6 gives the conclusion and future work.

2. Monochrome Image Distortion Measure

We define the error signal (e) to be the difference between the pixels of the original and the compressed image. We first split the error signal into two parts: the *orthogonal-space* error signal, the part of the error signal which is correlated with the local features of the original image and the *sub-space* error signal, the part of the error signal which is uncorrelated with the local features of the original image. The splitting operation is performed by a 2-D adaptive filter, as shown in Figure 1. using a modified 2-D LMS [8] algorithm which is described in [1].

Figure 1. Structure of adaptive filter

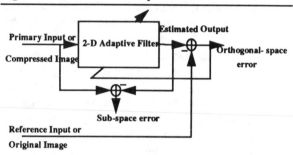

The 2-D adaptive filter works as follows. The pixel under consideration is at the centre of a 9x9 filter block and has four neighbors to the left, right, top and bottom of it. The filtering operation is shown in Figure 2, in which Figure 2.a shows the sliding filter moving to the right. When the filtering window reaches the end of a row, as shown in Figure 2.b, the filter block just shifts down by one row and continues in the reverse direction. This process is repeated until all the pixels have been at centre of the filter block once. The edges are zero padded at the end where the moving filter crosses the boundary of the image.

Figure 2. Filtering operation

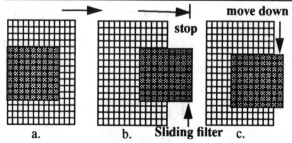

The second stage is a full wave rectifier which outputs absolute value of the errors in case they are negative, thus preparing the data for the following non-linear logarithmic operation. The presence of logarithmic sensitivity in vision has been known for a long time [7], supported by the fact that light sensitive neurons fire at rates proportional to the logarithm of the incident energy on them. The fourth stage consists of a non-linear power stage. In this stage the two components of the error signal obtained after the logarithmic stage are raised to a power 2. The fifth stage consists of a weighted summer in which the two error components obtained after the fourth stage are summed over the entire image Next a relative weighting of the two distortion measures D_o and D_s is done to obtain a final distortion measure. The following equation describes the measure

$$D_{img} = a \times D_s + b \times D_o \qquad (1)$$

We verified the model for different values of a and b for one of the images (Lena) in the USC database. To minimize the error between subjective distortion and our distortion measure we have found that *a equal to 0.80* and *b equal to 0.20*, give the best representation of distortion as perceived by human viewers [1]. We need to use a non-zero value of b as the filter does not converge and there is some leak associated with it. We verified the ranking results obtained using our distortion measure with these values of a and b for the remaining five images in the database. The entire distortion measure model is shown in Figure 3.

Figure 3. Processing stages

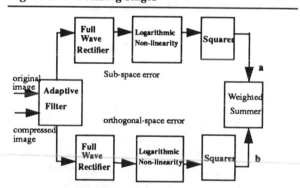

3. Distortion measure for video

The ability of a human viewer to resolve the details in a video sequence is dependent on how much motion is present and the ability of the eye to track it. The stationary portions of a video sequence can be resolved in greater detail whereas the moving portions are usually resolved in less detail by the human viewer. The time averaged information content of a still video scene is much less than the time-averaged information content of a moving scene. Thus, it is desirable to separate the moving portions from the still portions of a video sequence. Section 3.1. describes an algorithm for motion-still segmentation. Section 3.2 develops a distortion measure for the still portion of the video scene. Section 3.3. discusses a distortion measure for the moving portion of the video scene. Section 3.4. combines the two to form a total video distortion measure.

3.1. Motion-still segmentation algorithm

The motion-still segmentation algorithm is used for separating the motion pixels from the still pixels of a frame in a video sequence. The partioning of the *kth* frame is done using the *(k-1)th* and the *(k+1)th* frame in the video sequence. The first step involves taking an absolute difference of the two frames on a pixel-by-pixel basis. The second step involves comparing the difference at each pixel with a *motion detection threshold* (T). The pixels which have absolute difference greater than T are considered as motion pixels where-as the other pixels are considered as still pixels. The motion pixels are assigned a value of 1 while the still pixels are assigned a value 0. Thus a binary motion mask is obtained. The algorithm is summarized in Figure. 4.

Figure 4. Motion-still segmentation algorithm

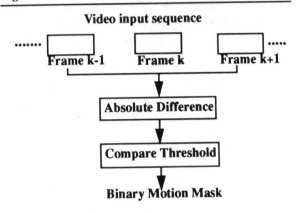

The determination of T is not a trivial task. Ideally if the video sequence had no camera and system noise then the value of T would be zero. However this is not the case because of the system and camera noise. To determine the value of T we took 100 video sequences of 10 frames with a size of 352x240 pixels each and eight bit luminance range. In each sequence we indicated which portion appears as motion and which appears as still to us and determined the corresponding pixels corresponding to the two portions. Then we tried out different values of T and chose the value that gave the minimum error in determination of the still pixels and the motion pixels. We found the value of T=30 gives the minimum error in detection of the still and the motion pixels.

3.2. Distortion measure for still pixels

The purpose of this measure is to capture the effects of the distortion in the still portion of a video sequence. We use the same distortion measure as was used for monochrome images described in Section 2, and call it still-area distortion measure ($D_{stl,k}$).

3.3. Distortion measure for motion pixels

The most noticeable degradations produced in a video sequence are blurring, distortion of edges in areas of motion and jerkiness. The edge blurring is caused due to the loss of high frequency components. The distortion of edges can be due to either presence of false edges in output video or loss in sharpness of the edges. Jerkiness is caused because of missing frames in output video.

The effect of jerkiness is reflected in D_{stl} because if a frame is missing, it is either repeated based on a previous frame or interpolated between adjacent frames. The error between the original frame and the compressed frame is usually large and is reflected in the sub-space distortion measure in the still area of the frame.

The other two artifacts present in the motion areas of a video sequence are extracted using a spatial Sobel filter for edge detection [9]. We have used an eight-neighbor Sobel filter as it has the advantage of smoothing and reducing the sensitivity of the derivative operations to noise. The Sobel filter uses two masks, one for horizontal edges and the other for vertical edges as shown below.

-1	-2	-1
0	0	0
1	2	1

Horizontal Sobel Filter Mask

-1	0	1
-2	0	2
-1	0	1

Vertical Sobel Filter Mask

The component of the gradient vector in the *x* direction is given by:

$$G_x = (x_7 + 2x_8 + x_9) - (x_1 + 2x_2 + x_3) \quad (2)$$

and the component of the gradient vector in the y direction is given by:

$$G_y = (x_3 + 2x_6 + x_9) - (x_1 + 2x_4 + x_7) \quad (3)$$

where x_i correspond to the following 3x3 image region:

x_1	x_2	x_3
x_4	x_5	x_6
x_7	x_8	x_9

The values of G_x and G_y are combined to give the vector magnitude (G) for edge detection using:

$$G = \sqrt{(G_x^2 + G_y^2)} \quad (4)$$

The filtering operation is repeated over the entire frame until all the motion pixels have been at the centre of the 3x3 image region and the corresponding value of G is obtained for each pixel. The value of G is compared with an edge threshold E_t. If G is greater than E_t then the pixel under consideration corresponds to an edge pixel, otherwise it corresponds to a non-edge pixel. Each pixel is assigned a value (e_i) depending on if it's an edge pixel or not according to the following equation:

$$e_i = \begin{pmatrix} 1, & G \geq E_t \\ 0, & G < E_t \end{pmatrix} \tag{5}$$

The edge threshold (E_t) is determined for each frame individually. For each frame a measure, M_k, of edge energy is calculated using:

$$M_k = \left(\sum_i e_i^2 \right)^2 \tag{6}$$

This measure M_k of edge energy for a frame is calculated both for the original and the compressed video frame. The distortion measure, $D_{mot,k}$, for motion areas of a frame is an absolute measure of the decrease or increase in the compressed edge energy, $M_{out,k}$, relative to the original edge energy, $M_{inp,k}$, and is given by:

$$D_{mot,k} = \frac{\left| M_{inp,k} - M_{out,k} \right|}{M_{inp,k}} \tag{7}$$

3.4. Total video distortion measure

The total video distortion measures for a video sequence is obtained by relative weighting of the normalized distortion measure for still and motion pixels. The normalization is calculated according to their maximum values $D_{stl,max}$ and $D_{mot,max}$ in the video sequence. The relative contribution of the still and motion distortion measures in a frame is obtained by weighting the normalized distortion measures by the ratio of still pixels and motion pixels in the frame respectively. The video distortion measure for a particular frame is given by:

$$D_{vid,k} = s_k \frac{D_{stl,k}}{D_{stl,max}} + (1 - s_k) \frac{D_{mot,k}}{D_{mot,max}} \tag{8}$$

where s_k is the ratio of still pixels in a frame and $(1-s_k)$ is the ratio of motion pixels in a frame.

The total video distortion measure, D_{vid}, is the combined distortion from all frames and is expressed as:

$$D_{vid} = \left(\frac{1}{M} \right) \sum_k D_{vid,k} \tag{9}$$

where M is the total number of frames in the video sequence.

4. Subjective tests

An experiment was conducted in which sixty human viewers were asked to rank the compressed images of six different scenes on a scale of 0-10 in accordance with the 10-point impairment scale. A high score indicated that the distortion was imperceptible while a low score indicated that the distortion was very annoying. The original image was displayed always while each compressed image was displayed until the viewers had made their decision. The experiment was repeated many times with a different order of compressed images each time.

For each image a set of eight compressed images displaying various compression artifacts were generated using eight different compression schemes: JPEG compression, sub-band compression, vector quantization, block averaging, decimation by a factor of 4 followed by JPEG compression followed by interpolation, block averaging followed by JPEG compression, decimation by a factor of 4 followed by sub-band compression followed by interpolation, block averaging followed by sub-band compression. There were various kinds of distortion developed in the compressed images. Some had blockiness, some had ringing, some had blurring artifacts while some had a combination of all of the above.

After collection of the scores obtained from human viewers, the average scores for each compressed image were computed. Next a ranking scheme was developed with the compressed image having the highest score being ranked first and the compressed image with the lowest score being ranked eight. Similar ranking schemes were calculated using our distortion measure and mean-square error based measures where the compressed image having the lowest distortion was ranked first and the compressed image having highest distortion was ranked eight.

A similar experiment was performed for compressed video in which fifty human viewers were asked to rank the compressed video of eight different sequences on a 5-point impairment scale. The five scales of quality are: excellent, good, fair, poor and unacceptable. The original video sequence was shown first and then the different compressed video sequences were shown again and again in a random manner.

For each video sequence a set of seven compressed video sequences were generated using different compression techniques. There were various kinds of distortion introduced in the output video sequence like spatial blurring, temporal blurring, blocking artifacts, jerkiness, ringing, edge busyness and image persistence.

A similar ranking scheme as for image compression is developed with subjective scores from (0-5).

5. Results

We have compared our distortion measure with the mean-square error based measures for compressed image/video. Among the six still images experimented with we describe here the results for one of the images, Couple, in the database. Table 1 shows the average scores obtained from human viewers for the eight compressed images, our distortion measure, mean-square error, mean absolute error and maximum absolute deviation. It can be seen that the ranking obtained using our distortion measure is exactly the same as the subjective ranking given by human viewers where-as the ranking obtained using mean-square error makes four errors compare to the subjective ranking.

We have found that on the set of six images that the rankings obtained using our distortion measure are 100% correct with a correlation of 0.95 compared to the subjective ranking whereas the rankings obtained using the mean-square error and its variants were only 60% correct on an average with a correlation of 0.53 compared to the subjective ranking.

Table 1. Comparison of image distortion measures

Image Number	Subjective Scores	Our distortion measure	Mean square error	Mean abs. error	Max. deviation
1.	8.75	1.95	29.74	4.01	52
2.	6.0	2.73	91.1	6.12	100
3.	1.7	3.78	172.39	9.53	165
4.	5.4	2.81	101.70	6.55	107
5.	1.9	3.61	209.81	9.85	176
6.	7.43	2.54	71.97	5.84	90
7.	7.01	2.65	94.63	6.46	101
8.	3.3	3.26	137.29	7.92	126

A similar comparison was done for eight different video sequences. Table 2 summarizes the results for one of the video sequence, bike, a scene from terminator-II.

Table 2. Comparison of video distortion measures

Image Number	Subjective Scores	Our distortion measure	Mean square error	Mean abs. error	Max. deviation
1.	3.29	0.54	34.681	4.01	59
2.	1.67	0.77	51.096	6.045	135
3.	2.87	0.61	20.213	2.968	82
4.	4.54	0.36	8.635	2.110	23
5.	0.45	0.91	41.933	4.789	102
6.	3.83	0.47	14.289	2.567	41
7.	2.41	0.69	29.066	3.456	47

We have found that on the set of eight video sequences the rankings obtained using our distortion measure were (96%correct) with a correlation of 0.92 compared to the subjective ranking where-as the rank-

ings obtained using the mean-square error and its variants were only (57% correct) on an average with a correlation of 0.47 with the subjective scores.

6. Conclusion

We have developed a universal distortion measure of quality for monochrome image compression. This distortion measure is based on uncorrelated local errors (sub-space error) to quantify various artifacts in compressed images. This measure of distortion is found to match very well with the quality degradation often experienced by humans in viewing compressed images.

We have also developed a distortion measure for compressed video, which separates compression distortion to still portion effects and motion effects. We verified that this video compression distortion measure corresponds well with the subjective results. More work has to be done on determining the relative importance of the motion and still pixels in a frame and also the relative importance of a particular frame in an entire video sequence.

For future work, we plan to use these measures to derive a realistic rate-distortion curve, based on which an optimal scalable compression algorithm will be designed to accommodate the varying bandwidth available in a dynamic wireless network.

7. References

[1] Navin Chaddha and Teresa H.Y. Meng, "Psycho-visual based distortion measure for Monochrome Image compression", SPIE Proc., Visual Communications and Image Processing, Nov. 8-11, 1993.

[2] J.L. Mannos & D.J. Sakrison, "The effects of visual fidelity criterion on the encoding of images", IEEE Trans. Info. Theory, Vol. IT-20, No. 4, pp. 525-536, July 1974.

[3] J.O. Limb, "Distortion Criteria of the Human Viewer", IEEE Trans. on Sys., Man, and Cybernetics, Vol. SMC-9, No. 12, pp. 778-793, Dec. 1979.

[4] C.S. Stein, A.B. Watson & E.H. Lewis, "Psychophysical rating of image compression techniques", SPIE Vol. 1077: Human Vision, Visual Processing and Digital Display, pp. 198-208.

[5] J.E. Farrell & A.E. Fitzhugh, "Discriminability metric based on human contrast sensitivity", J. of Optical Society of America A, Vol. 7, No. 10, pp. 1976-1984, Oct. 1990.

[6] J. Farell, H. Trontelj, C. Rosenberg, J. Wiseman, "Perceptual metrics for monochrome image compression", SID 91 Digest, pp. 631-634.

[7] T.N. Cornsweet, Visual Perception, academic Press, New York, 1970.

[8] S.J. Orafanidis, Optimum Signal Processing, McGraw-Hill, New York, 1988.

[9] W.K. Pratt, Digital Image Processing, John. Wiley & Sons, New York, 1991.

Recent Trends in Compression of Multispectral Imagery *

S. Gupta, A. Gersho
Dept. of ECE
University of California
Santa Barbara, CA 93106

A. G. Tescher
Lockheed R&D
3251 Hanover Street
Palo Alto, CA 94304-1191

Abstract

In this work, we provide a comprehensive trade-off analysis of major compression algorithms as applied to multispectral imagery. Our goal is to identify a real-time, low cost codec for multispectral imagery. Hence the major criteria used for performance and ranking of various algorithms emphasize implementation complexity in addition to compression performance and robustness. Multispectral imagery codecs are designed by suitably extending and modifying the state-of-art still image codecs so that the resultant codec simultaneously benefits from the *spatial* and *spectral* correlations inherent in multispectral imagery. In particular, two different approaches of interest which emerge are *vector quantization (VQ)* and *transform coding (TC)*. This paper describes and analyzes some recently developed multispectral imagery codecs utilizing the above two approaches. For completeness, a brief overview of alternate approaches is also included.

1 Introduction

Recent years have seen greatly expanded interest in the compression of remote-sensed data. The motivation for this interest stems from the massive increase anticipated in the volume of remote-sensed data coupled with existing limitations on channel bandwidth and storage capacity. The Earth Observing System (EOS) scheduled for launch in the late 1990s will carry a High Resolution Imaging Spectrometer (HIRIS) with 192 spectral channels compared with 4 channels on the Multi-Spectral (MSS) scanner and 7 bands on the Thematic Mapper (TM) scanner. Effective information transfer from satellites with high-resolution

*This work was supported by the University of California MICRO program, Rockwell International Corporation, Hughes Aircraft Company, Qualcomm Inc., and the Center for Advanced Television Studies.

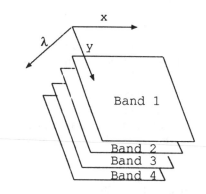

Figure 1: A Multispectral Image Set.

scanners, commercial utilization of multispectral images and development of GIS (Geographic Information Systems) databases with fast image retrieval capabilities are making it imperative to address the need for compression of remote-sensed imagery in general and multispectral imagery in particular.

This paper describes and analyzes applications of major compression algorithms to multispectral imagery. Multispectral imagery consists of sets of images where each image or band is generated by sensing the radiation from a scene in different parts of the electro-magnetic spectrum [1]. The spectral signature (the set of pixel amplitudes across different bands for a fixed spatial location) is determined by the geography of the terrain. Thus images within a multispectral data set are characterized by strong spectral (inter-image) correlation in addition to the spatial (intra-image) correlation present within each individual image [1].

2 Intraband vs. Interband Coding

A coder for multispectral imagery has to compress multispectral image sets. A multispectral image set with four bands is shown in Figure 1. Since each multispectral image set is essentially a collection of bands or still images, it may be compressed by applying still image coding techniques to each band individually and independently. Such an approach termed *intraband coding* utilizes the intra-image or spatial correlation but ignores the inter-image or spectral correlation. For optimum compression performance, multispectral imagery coders should simultaneously utilize both spectral and spatial correlation. Such an approach is termed *interband coding*. This paper primarily concentrates on interband approaches to multispectral image compression.

Since the exploitation of spatial redundancies is critical for effective multispectral image compression and can provide a starting point for the design interband multispectral image codecs, an overview of single-frame image coding approaches can provide an insight into feasible coding approaches for multispectral imagery. Next section provides this overview and identifies the coding approaches of most interest in the context of multispectral image compression.

3 Coding Approaches

3.1 Lossless vs. Lossy Compression

Depending on the fidelity of the reconstructed or decompressed data from the compressed bit stream, compression can be categorized as either *lossy* or *lossless*. Lossless or reversible coding uses a probabilistic model to eliminate statistical redundancies and permits exact reconstruction of the original data from the compressed format. Examples of lossless data compression are Huffman codes, Arithmetic codes and even the Morse code [2]. If fixed codes are used performance degradation occurs for sources with varying statistics. Dynamic Huffman codes, adaptive Arithmetic coding and in particular the Rice algorithm [3] can be used for generating adaptive codes. In fact, Rice algorithm is widely used for lossless compression of remote sensing data.

The reduction in data volume achievable through lossless compression is limited and rarely exceeds a factor of two-to-four. Lossy or irreversible compression can provide significantly higher compression ratios than lossless compression but introduces some irreversible error or distortion in the reconstructed data so that original data cannot be exactly recovered. However with proper optimization the error can be made sufficiently small so as to render lossy compression attractive for applications needing higher compression.

3.2 Scalar vs. Block Quantization

Lossy compression can be achieved by operating on scalars or blocks of scalars (vectors). Traditional predictive coding approaches such as differential pulse code modulation (DPCM) employ scalar quantization but utilize the previous inputs in the encoding process. The performance objective in the design of a lossy source coder is to minimize a desired *distortion* which is a measure of dissimilarity between original and reconstructed data for a given reduction in data volume or minimize data volume for a given maximum allowable distortion.

Shannon's rate distortion theory [4] establishes that better performance can be achieved by coding *vectors* rather than *scalars*. For images, this implies that a maximum possible number of pixels should be grouped together and coded as an entity to maximize performance. In the block coding approach to image coding, the image plane is tessellated with non-overlapping square blocks. Pixels within each block are grouped together to form the input vector.

Since we are interested in comparative performance analysis of state-of-the-art codecs capable of providing good performance at sufficiently high compression ratios, this paper describes coders which focus primarily on the major block coding approaches. As described below, these are structured *vector quantization (VQ)* and *transform coding (TC)*.

3.3 Vector Quantization vs. Transform Coding

Vector Quantization (VQ) [5] is essentially a pattern matching technique which operates on vectors formed by grouping pixels in the spatial domain and codes by searching for the minimum distortion match from a set of templates commonly referred to as a codebook. Although VQ is an optimal coding technique theoretically capable of attaining theoretical rate-distortion bounds for a large enough input vector dimension, its storage and encoding complexity grows exponentially with increase in rate and dimension. Thus complexity constraints limit the dimensionality of the input block and preclude the attainment of rate-distortion bounds.

Alternate sub-optimal approaches with lower complexity can exploit correlation over a larger area than unstructured VQ for a fixed complexity constraint. These can be divided into *transform* based methods where quantization is done in the transform domain and *structured* vector quantization based methods where the input waveforms are directly quantized

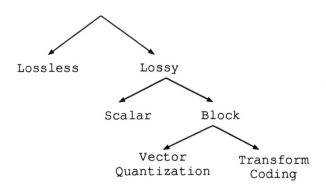

Figure 2: Image Coding Approaches.

in the spatial domain.

Transform coders model images as stationary random processes and apply either a data-dependent transform such as the KLT or a data-independent transform such as the DCT to decorrelate the input [2]. The decorrelated coefficients are independently scalar quantized. Structured VQ works by imposing structural constraints on the codebook. Thus the code vectors cannot have arbitrary locations but are distributed in a restricted manner on the k-dimensional Euclidean space which allows for a lower complexity search at the encoder. In this paper, we will examine structured VQ as well as transform coding approaches to multispectral image compression. Figure 2 provides a schematic hierarchy of the coding approaches mentioned so far.

4 Multispectral Imagery Codecs

Earlier work on the compression of multispectral imagery includes block coders which utilized *only* the spectral redundancy in corresponding pixels of different bands [6], [7]. On the other hand, the application of any existing still-image codec to multispectral image compression exploits *only* the spatial redundancy within a band but ignores the spectral redundancy across bands. Recently *interband* multispectral image codecs are being designed which simultaneously exploit spectral and spatial redundancies [8], [9], [10], [11]. These interband codecs utilize either transform coding or structured VQ. This paper will provide examples of codecs based on both of the above approaches and analyzes their relative performance to provide an insight into the performance of these approaches for multispectral image compression.

In order to simultaneously benefit from both spectral and spatial correlation in the image data set, we will focus primarily on *interband* codecs. Therefore, the entity to be quantized is a 3-dimensional data block formed by concatenating the pixels in blocks at the same spatial location in different spectral bands (see Figure 1).

The specific examples of interband multispectral image codecs described include a feature predictive vector quantizer, a multistage vector quantizer and a three-dimensional transform coder. Their relative performance is analyzed to derive conclusions regarding their application to multispectral image compression. (A detailed description of these codecs can be found in [12].)

Due to the popularity of the JPEG still image coding standard, an intraband coder which individually compresses each image in the data set through a JPEG baseline coder [13] is included as a benchmark.

4.1 Feature Predictive VQ (FPVQ)

In FPVQ [8], initially a small subset of images is vector quantized thus effectively exploiting the correlation within each image in the subset. The remaining images in the data set are predicted from the quantized images. The prediction process exploits the inter-image redundancy between these remaining images and the previously quantized images. The residual error images are encoded at a second stage based on the magnitude of the energy in the prediction residual. These residual quantizers are designed to benefit from the intra-image correlation in the predicted images. Since the inter-image and the intra-image correlations exhibited by each image in a data set vary according to its spectral location, a different predictor and vector quantizer are designed for each image. The resulting codec thus simultaneously exploits spectral and spatial redundancies by employing vector prediction for complexity reduction.

4.2 Multistage VQ (MSVQ)

Another structured VQ approach to multispectral image compression is based on variable rate multistage vector quantization [11]. In addition to the simultaneous exploitation in the spectral and spatial correlation, the wide variation in spatial/spectral entropy across an image data set is efficiently exploited by an adaptive bit allocation algorithm based on a *greedy* approach where the rate-distortion trade-off is locally optimized for each successive encoding stage.

Variable bit-allocation or multi-resolution quantization within an MSVQ framework is readily performed by quantizing each input vector with a different number of stages. Performance is thereby approximately optimized for a given average bit-rate and an increase in performance over the FPVQ is observed [11].

Next, we describe a three-dimensional transform coder for multispectral image compression which is tailored to multispectral image compression.

4.3 Three-dimensional Transform Coding

In the three-dimensional transform coding approach described here, signal dependent KLT is applied along the spectral direction for maximum compaction along the spectral dimension [12]. Following an approach similar to a JPEG baseline coder, the resultant spectrally transformed image set is next spatially transformed using a 2-dimensional DCT for further energy compaction.

These transform coefficients are selectively quantized based on their magnitude. Uniform scalar quantization followed by entropy coding is employed for maximum compression performance. Laplacian modeling of the transform coefficients enables tailoring of the entropy codes to the input data. Due to the low complexity of the method, correlation exploitation over 8 × 8 spatial area is possible along with exploitation of spectral correlation over all bands for a 4 band data set. The structured VQ approaches described earlier used a 4×4 block size due to complexity constraints.

5 Relative Performance

This section describes the relative performance of the proposed multispectral imagery codecs. Since the JPEG encoder codes each band individually, it cannot exploit the spectral correlation in multispectral image sequences. However compared to the interband VQ codecs employing MSVQ and FPVQ where feasibility constraints limit the the block dimension to 4 × 4 along the spatial directions, the JPEG baseline coder uses a larger block size of 8 × 8. It remains to be seen whether the larger input block dimension in JPEG encoding can offset the gain in interband VQ codecs due to exploiting of interband correlation. It should be obvious though that with proper design the performance of the intraband JPEG coder and perhaps even that of interband VQ codecs can be exceeded by a suitable

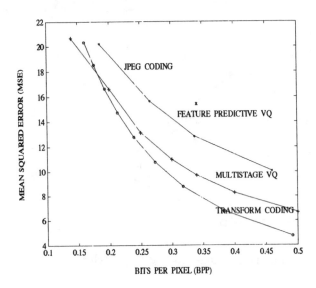

Figure 3: Rate-Distortion Performance.

three-dimensional transform coder which can simultaneously exploit spectral and spatial correlation over a large area in a multispectral image data set.

Figure 3 provides the relative performance curves. Actual simulations were performed for the interband coders and for an intraband JPEG coder using the first four bands of data sets acquired through the Thematic Mapper multispectral scanner for a test image set of an agricultural area near river Adige in Italy. The resulting mean-squared error (MSE) is obtained by averaging over the four bands. A block size of 4 × 4 was used for MSVQ and FPVQ simulations while a block size of 8 × 8 was used for the transform coders.

Let us first examine the performance of the intraband JPEG coder compared to the interband codecs. Among the interband codecs, FPVQ cannot exceed the JPEG coder in *average* performance. Further analysis shows that the quality of the predicted bands in FPVQ exceeds that of the corresponding bands coded through JPEG. However since the feature band is quantized intraband in the FPVQ, the corresponding band coded through JPEG utilizes correlation over a larger area and has a better rate-distortion performance. This lowers the overall performance of FPVQ. However the interband MSVQ and the three-dimensional TC consistently and substantially exceed the intraband JPEG codec in performance.

Compared to FPVQ, MSVQ and three-dimensional TC utilize spectral and spatial correlation *simultaneously over all bands*. Also MSVQ employs a superior bit allocation strategy. The three-dimensional transform coder provides the best performance except at very high compression ratios. The high performance of three-dimensional transform coder can be attributed to the simultaneous exploitation of spectral and spatial redundancies over a large area and a flexible robust bit allocation by selective quantization of the transformed coefficients.

The cross-over of the performance curves for MSVQ and three-dimensional TC occurs at very low bit rates where the MSVQ encoder emulates an optimal VQ more closely than the TC. Additionally, our MSVQ scheme employs a global bit allocation strategy unlike the localized bit allocation in TC. The improvement in distribution of bits through a global strategy is more at lower bit rates and helps the MSVQ exceed the TC in performance at low bit rates. However, three-dimensional TC can provide high performance at a variety of bit rates with reasonable complexity.

We would like to caution the reader that the performance curves should be used to gain an insight into the relative rather than the absolute performance since the absolute performance will vary with the test data. However we believe that our results are indicative of the relative performance of the different coding approaches and lead us to conclude that three-dimensional transform coding provides the best choice for multispectral imagery compression.

6 Conclusion

We have described major coding approaches within the context of their application to multispectral image coding. After identifying suitable coding approaches, specific multispectral image codecs were described. A comparative analysis shows that a three-dimensional transform coder capable of simultaneously exploiting spectral and spatial correlation in the data set can be implemented at a relatively low cost and provides the best performance except at very high compression ratios.

It should however be mentioned that on most occasions, the end use of multispectral imagery is not for human viewing. Rather it is used for geographical classification of terrain. In order to minimize the performance loss for such applications, it is imperative that suitable distortion measure which accurately reflect the loss in classification accuracy due to lossy compression are developed. Such a formulation requires that the scientists who analyze multispectral data join forces with the image compression community to ensure that the codecs developed satisfy the needs of the end-users.

References

[1] J. G. Moik, *Digital Processing of Remotely Sensed Images.* SP-431: NASA, 1980.

[2] M. Rabbani and P. Jones, *Digital Image Compression Techniques,* pp. 116 128. SPIE Press, 1991.

[3] R. F. Rice, "Some Practical Universal Noiseless Coding Techniques," *Technical Report: JPL Publication,* vol. NASA, 79-22 1979.

[4] C. E. Shannon, "Coding theorems for a discrete source with a fidelity criterion," in *IRE National Convention Record, Part 4,* pp. 142 163, 1959.

[5] A. Gersho and R. Gray, *Vector Quantization and Signal Compression.* Kluwer Academic Publishers, 1991.

[6] R. L. Baker and Y. T. Lee, "Compression of High Spectral Resolution Imagery," in *Proc. SPIE on Applications of Digital Image Processing XI,* August 1988.

[7] E. E. Hilbert, "Joint Pattern Recognition/Data Compression Concept for ERTS Multispectral Imaging," in *Proc. SPIE Vol. 66,* pp. 122 137, 1975.

[8] S. Gupta and A. Gersho, "Feature Predictive Vector Quantization of Multispectral Images," *IEEE Trans. on Geoscience and Remote Sensing,* vol. 30, pp. 491 501, May 1992.

[9] J. A. Saghri and A. G. Tescher, "Near-Lossless Bandwidth Compression for Radiometric Data," *Journal of Optical Engineering,* pp. 934 939, July 1991.

[10] D. Tretter and C. Bouman, "A Model-Based Approach to Multispectral Image Coding," in *Proc. ICASSP,* pp. V 361 V 364, April 1993.

[11] S. Gupta and A. Gersho, "Variable Rate Multistage Vector Quantization of Multispectral Imagery with Greedy Bit Alocation," in *Proc. SPIE Visual Commun. and Image Proc.,* Nov. 1993.

[12] S. Gupta, *Interframe Processing and Coding of Image Sequences.* PhD thesis, University of California, Santa Barbara, 1993.

[13] "JPEG Draft Technical Specification Revision 8," Aug. 1990.

VIDEO TO VGA AND BACK

John A. Eldon

Raytheon Semiconductor
La Jolla CA 92038 USA

Converting a standard analog television signal into a digital component video or computer graphics format entails: clock and sync acquisition, color reference acquisition, sampling and quantization, filtering, linear algebra, and quadrature demodulation. This paper traces a high performance video decoding algorithm and maps it to the architecture of a cost-effective new mixed-signal integrated circuit set from Raytheon Semiconductor.

1. Background and Problem Statement

Many applications require that computer graphics and/or text be mixed with or overlaid onto a standard NTSC or PAL television signal for highlighting, captioning, special effects, design rendering, etc. To obtain visually acceptable results, the system must synchronize the computer graphics data with the video signal and modulate its chroma components onto a subcarrier whose phase and frequency are locked to those of the video signal. Time synchronization errors can yield spatial distortion and wavering of the graphical overlay, whereas subcarrier phase errors will distort its colors. Even without the graphic overlays, many personal computer users would like to be able to archive television images or sequences digitally and/or to display them on the computer's monitor. These functions require full translation of the analog television signal into a computer graphics format, such as RGB or VGA.

Converting a standard analog television signal into a digital computer graphics format entails:clock and sync acquisition, color reference acquisition, two-dimensional digital filtering, linear algebra, quadrature demodulation, two-dimensional resampling, and nonlinear data mapping.

If the incoming video signal is analog, then the genlock subsystem must derive its own pixel-rate sampling clock, keyed to the incoming horizontal sync pulses. It then uses this clock to sample the actual video signal, including its reference chroma burst signal. From the digitized samples of the chroma burst, the genlock determines the frequency and phase of the incoming video signal's color subcarrier.

Once the clock and color reference have been extracted, the video decoder must separate luma (brightness) from chroma information, using simple low-pass and band-pass filters. However, since the luma and chroma spectra overlap, we sacrifice spatial resolution and induce visual artifacts by nominally identifying all low-frequency signal as luma and high-frequency signals as chroma. Because video data are transmitted in a line-scanned raster format, and because there a half-integer (NTSC) or quarter-integer (PAL) number of color subcarrier cycles per line, the decoder can employ comb filtering to enhance the separation between the luma and chroma information streams. In a uniform video field, the pixel-by-pixel sum of two consecutive NTSC lines will have enhanced luma and suppressed chroma, whereas the difference will comprise enhanced chroma and suppressed luma.

Once the phase and frequency of the color burst reference are known, the video decoder demodulates the chroma portion of the data into its two quadrature components, corresponding vaguely to red and blue axes on a color chart, in which angle correlates with hue and amplitude indicates degree of saturation.

The video encoding portion of an overlay system accepts horizontal sync, pixel clock, digitized video, and subcarrier phase and frequency information from the genlocking digitizer, plus any desired graphical overlay data from the host computer system.

2. Algorithm

2.1 Sync acquisition and clock extraction

The chip's clock generator and horizontal sync phase-locked loop incorporate a programmable direct digital synthesizer, an analog frequency multiplier, a phase error detector, and an off-chip analog low pass filter. Starting at the center of each horizontal sync falling edge in the incoming video signal, this part of the chip generates and counts sampling clock pulses. Upon receiving the next horizontal sync falling edge, the chip generates a frequency error signal by comparing the number of counts between sync edges to a preprogrammed nominal value. It uses this error signal to increase or decrease the frequency of its clock synthesizer. The chip uses the sync level at the sample nearest to the center of the sync falling edge for sub-pixel timing resolution.

2.2 Color reference burst acqusition

The burst phase-locked loop comprises a gated Hilbert transformer, a direct digital synthesizer, and a low-pass filter. When each new video line's chroma burst signal arrives, the sampling gate opens, and the Hilbert transformer estimates its phase. If the phase of the internal direct digital synthesizer is retarded compared to that of the incoming burst, then its frequency is increased slightly to enable it to "catch up." Likewise, if the internal synthesizer gets ahead of the incoming burst in phase, its frequency is decreased. A low-pass filter enhances loop stability, allowing the loop to track frequency drifts in the incoming data without being confounded by short-term random noise effects.

2.3 Sync strip

Whereas for historical reasons some 30% of the dynamic range of a composite television signal is reserved for horizontal and vertical sync tips, the VGA computer graphics standard uses almost the full range for picture information alone. Thus, part of a television decoder's job is to strip away the sync signals and remap the remainder of the incoming signal onto the full dynamic range of the display drivers. This is accomplished by comparing each incoming composite sample against the video blank level (also extracted from the incoming video signal), and hard-limiting negative excursions to zero.

2.4 Chroma - luma separation

To first order, we can assume that the lower frequency components (say, 0 to 2 MHz) of a baseband composite video signal correspond to brightness (luma) information, whereas the higher frequencies (above 2 MHz) are chroma information, modulated by the 3.58 (NTSC) or 4.43 (PAL) MHz chroma subcarrier. The decoder's band-splitting filters keep most of the luma energy out of the chroma subchannel and most of the chroma information out of the luma subchannel. However, any high frequency luma information present in sharp-edged, highly detailed pictures can still find its way into the chroma subband, where it causes color distortions known as "cross-luminance."

2.5 Comb filtering

To improve the chroma-luma separation, high-quality decoders employ various sorts of comb filtering, which takes advantage of the spectral line nature of the luma and chroma signals. When NTSC television was defined in 1952, the committee wisely decided that there should be a half-integer (actually, 227.5) number of color subcarrier cycles per horizontal line. Because of vertical raster scanning, the luma spectrum comprises a series of spectral lines, spaced by the horizontal scan frequency. The result of the half-integer subcarrier frequency is that the chroma spectrum comprises another set of spectral lines, also spaced by the scanning frequency, but offset halfway between the luma lines. A comb filter, comprising a one-line delay and an adder, can exploit this relationship and enhance the separation of chroma and luma signals. This technique, used extensively in "HQ" VCRs, noticeably improves the performance of the encoder, particularly where highly detailed color information (e.g. plaids or stripes) is received.

In PAL, a quarter-integer subcarrier was used (1135/4 cycles per line) to avoid artifacts which would have occurred had the NTSC half-integer relationship been employed. (In PAL, or phase-alternating-line, format, one of the two color components is inverted on alternate lines. A half integer subcarrier would make this component appear without line-to-line cancellation, causing some colors to have visibly and objectionably smaller apparent vertical resolution than others.) A special comb filter with two line delays addresses this quarter-line-frequency offset between PAL luma and PAL chroma information by filtering three adjacent lines instead of two.

2.6 Conversion to VGA

Finally, some linear algebra permits the decoder to translate the luma and demodulated chroma components into computerese "true color" red, green, and blue color components. Although true color permits the system to display any of 16 million colors accurately on a computer, most computers use either standard VGA, with 8 bits per pixel, or high-color VGA, with 16 bits per pixel, instead. In these systems, the host system loads a RAM to select a 256 (or 64K for high-color) color palette subset of the 16-million-color universe. A video-to-VGA decoder can employ a programmable window-addressable memory (WAM) to match each decoded 24-bit true color value to its nearest stored palette equivalent.

3. Architecture

A video decoder comprises an analog front end to set blank level and gain, an 8-bit A/D converter, phase-locked loops for pixel clock and chroma subcarrier extraction, a direct digital synthesizer, band-separation, and a bypassable matrix to compute red, green, and blue components from luma and chroma data. Figures 1 and 2 depict the architectures of a video genlock and a digital video decoder, respectively. This two-chip set permits the user to blend television and computer graphics images and display or store them on the computer. The companion video encoder chip [4] converts the output back into television format, if desired. The genlocking digitizer chip samples and quantizes the analog video signal, extracts its pixel clock and color burst, and sends this information to the decoder. The decoder chip accepts a digital television signal (or the digitized output of the genlocking digitizer chip), separates its luma and chroma portions, and demodulates the chroma information into the two orthogonal color components. The user may select either RGB (computer graphics) or YUV (digital television) output format.

4. Summary and Conclusions

This paper has traced a high-performance video decoding algorithm and mapped it to the architecture of a cost-effective new mixed-signal integrated circuit from Raytheon Semiconductor. The problems of clock acqusition, color demodulation, filtering, and resampling, which formerly required a large board, have been solved in an inexpensive integrated circuit set.

References

1. G. Hutson, P. Shepherd, J. Brice, Colour Television, 2nded, McGraw-Hill, London, 1990

2. K. Benson, Television Engineering Handbook, McGraw-Hill, New York

3. J. Eldon, "Two Video Encoder Integrated Circuits," SPIE 1992 Symposium on Optical Applied Science and Engineering, San Diego CA

4. J. Eldon, "Computer Graphics -- As Seen on TV," 1992 Asilomar Conference on Signals, Systems, and Computers

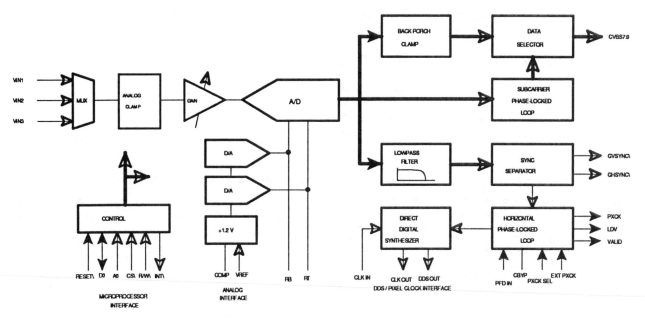

Figure 1. TMC22071 Functional Block Diagram

24381B

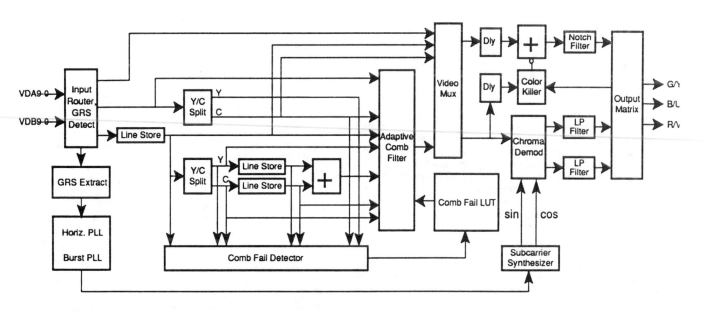

Figure 2. TMC22153 Functional Block Diagram

477

DESIGNING A VIDEO COMPRESSION SYSTEM FOR HIGH DEFINITION TELEVISION

John G. Apostolopoulos   Peter A. Monta   Julien J. Nicolas   Jae S. Lim

Advanced Television and Signal Processing Group
Research Laboratory of Electronics
Massachusetts Institute of Technology
Cambridge, MA

ABSTRACT

There has been significant interest in the design of a digital television system for terrestrial HDTV broadcasting. To design such a system, a number of issues must be addressed. This paper will examine the design of a video compression system for the efficient delivery of digital television signals. A number of considerations that arise when designing the video compression system will be discussed, and a sample system which meets most of these requirements will be presented.

1. INTRODUCTION

When television was commercially introduced, spectrum was readily available and signal processing was very expensive. Therefore, the video signal was simply scanned and used to modulate a radio frequency carrier. Little signal processing or compression was used, resulting in a very inefficient use of the spectrum. The advent of high-definition television (HDTV), with a larger amount of information to be delivered, coupled with a scarcity of available bandwidth (and a simulcast scenario with NTSC), demands a much more efficient method of delivering television signals to the home. In 1990 the Federal Communication Commission (FCC) ruled that in order to apportion the limited RF spectrum among the services that may need it, without realigning the current services, any HDTV system used for terrestrial broadcast in the U.S. must use the same 6 MHz channel bandwidth as the current television system. Comparing the raw data rate for a typical HDTV video signal, about 1.3 Gb/s, with the transmission capacity of today's state of the art digital transmission systems, about 20 Mb/s within a terrestrial 6 MHz channel, it is evident that a large amount of compression is required.

1.1. Current television and HDTV

An appreciation of the important features of HDTV, and the amount of information it contains, can be obtained by a comparison with a conventional television system. The current color television system in the U.S. is an interlaced 525 scan line, 60 fields/sec or 30 frames/sec, 4:3 aspect ratio (width to height) system. The perceptual effects of interlace result in a spatial resolution of about 340 lines with about 420 resolvable elements per line. This is rather poor resolution which is especially evident on large screen displays. The interlaced scanning format can result in a number of degrading artifacts such as interline flicker and improper motion rendition. Also, the chrominance resolution is severely

This research has been sponsored principally by the Center for Advanced Television Studies (CATS). Current Consortium members are Ampex, Capital Cities/ABC, Eastman Kodak, General Instrument, Motorola, PBS, and Tektronix.

restricted, and the system is highly susceptible to transmission impairments such as multipath and interference.

An HDTV system may process a multitude of different video formats. In some HDTV systems, the baseline video format is a progressively scanned, 720×1280 square pixel, 60 frames/sec video signal, providing six times the spatial resolution and improved motion rendition over the current system. This increased spatio-temporal resolution significantly enhances the feeling of realism for the viewer. Also enhancing the realism is the 16:9 aspect ratio of the video signal which better emulates the greater horizontal than vertical human field of view. The improved chrominance reproduction and high-fidelity (CD-quality) digital audio add to the crispness of the entire experience. Furthermore, the displayed video signal will not be degraded by transmission impairments such as multipath and random noise. Through the choice of progressive scanning and square pixels, the HDTV system is also easily interoperable with computer systems, facilitating the merging of HDTV and computer-related tasks and services.

A digital HDTV system uses both digital source coding and digital channel coding, affording great flexibility for the representation and communication of the video signal. The digital channel coding ensures that, under predefined channel conditions, the compressed video signal can be recovered with a small probability of error. Under this assumption the encoder will know exactly what the decoder will receive, facilitating the use of highly aggressive forms of video compression.

1.2. Application Profile

A number of application issues directly affect the design of an HDTV video compression system. These include the fact that the system (1) is to be employed in a broadcast environment, (2) requires real-time processing, and (3) couples with a constant bit rate channel. In a broadcast environment there will be few encoders and many decoders. The encoders may be expensive, but the decoders should be realizable at low cost. Therefore, most of the complexity should be localized at the encoder rather than at the decoder. Within a broadcast environment, consumers randomly turn on their receivers and change the channel, without the encoder having any knowledge of these changes. This necessitates fast receiver initialization and channel acquisition. The real-time application requires that the encoder/decoder time delay is significantly small so as not to hinder any "live" applications. Coupling a possibly variable bit rate video encoder to a constant bit rate channel requires a buffering mechanism. To achieve the necessary buffering without using large amounts of memory and while still attaining continuous high-quality video, a sophisticated buffer control algorithm is required.

Perhaps the most controversial issue today regarding

HDTV is about the specific choice for the spatio-temporal sampling structure to be used. When the current television system was originally designed, *interlaced scanning* was an excellent solution for the problems of the day. With the advent of HDTV those problems no longer exist, but there still is a powerful "tradition" to still use interlace. Interlaced scanning results in a number of degrading artifacts including improper motion rendition, interline flicker, line crawl, and vertical aliasing. Also, interlaced scanning complicates video processing and general interoperability with computers. A simpler approach is *progressive scanning*, where consecutive scan lines within each frame are read sequentially. Progressive scanning with square pixels lends itself easily to interoperability with computers, which use the same format. Similarly, progressive scanning and square pixels are extremely useful for format conversion, frame capture, computer graphics, and general video processing, since they allow processing to be performed without any added complexity from the sampling structure.

Implementation considerations such as computation and memory requirements, delay, algorithm complexity, and amenability to parallel processing are also important, although to a lesser extent, as VLSI continues to increase in performance and drop in price.

2. THE VIDEO COMPRESSION SYSTEM

Many excellent books and papers have discussed the principles of video compression [1, 2, 3]. The goal of video compression is the reduction of the *redundancy* and the *irrelevancy* inherent in the video signal. Redundancy exists along the temporal, spatial, and color space dimensions. Irrelevancy corresponds to those qualities that are perceptually unimportant to the human visual system (HVS).

Any digital compression system can be expressed as the combination of three distinct, though interrelated, operations. *Representation*, where the signal is expressed in a manner that facilitates the process of compression. *Quantization*, where typically most of the actual compression is achieved. And *Symbol Encoding*, which reduces the average bit rate necessary to describe the data.

2.1. Representation

The primary approaches taken toward creating an efficient and compact representation may be classified broadly into the categories of *predictive processing, transform/subband filtering,* and *model-based processing*. Of these three categories, only the first two have currently reached a state of maturity and quality appropriate for an HDTV application. Approaches of this form may be intermixed and applied along the temporal, spatial, and color space dimensions of the video signal. We will briefly examine their application along the different dimensions.

Color Space Processing. The color space redundancy is typically reduced by performing a linear transformation from red, green, and blue (RGB) to the YIQ or YUV color spaces. This transformation reduces the correlation among the three color components and also enables processing to exploit the differing HVS response to the luminance/chrominance components.

Temporal Processing. A transform/subband filtering scheme may be applied along the temporal dimension, but any nonuniform motion would result in the signal energy being distributed throughout the 3-D frequency space. This inefficiency hinders the performance of this approach. Also, a temporal transform/subband filtering scheme requires the storage of a large number of frames. On the other hand, predictive coding schemes typically perform very well while using only 2 frame stores. By accounting for the motion

between frames, a motion-compensated (MC) prediction scheme can perform exceedingly well. The error in the MC-prediction, or MC-residual, is typically much less than the original frame, and will be further compressed in the spatial domain. The motion estimation (ME) algorithm typically used is block matching, since it yields high performance while having a simple, periodic structure, allowing relatively easy VLSI implementation.

MC-processing may fail, either locally (with the appearance of new imagery) or globally (with a scene change), and in these cases the MC-residual may be more difficult to encode than the original image. To achieve high-quality video, the compression algorithm must identify these situations, suppress the MC-processing, and process in a more efficient manner. Two possible approaches are spatially-adaptive inter/intra processing and spatially-adaptive leakage. In the first approach the MC-processing can be turned on or off in a spatially adaptive manner, while in the second a portion of the original frame is allowed to "leak" into the residual signal. Note that the latter is a more general version of the former. On the other hand, the former approach may be more convenient (less sophistication/computation).

Inter/intra processing and leakage also supply simple methods to solve the problems of receiver initialization and channel acquisition, where the decoder prediction loop must begin tracking the encoder loop. For example, by periodically intra encoding an entire frame the prediction loops at both the encoder and decoder can be periodically reinitialized. This also restricts the time duration of effect of uncorrected channel errors, and also provides a mechanism such that if the decoder loses synchronization for any reason, it can rapidly reacquire.

Spatial Processing. Predictive schemes are not generally viewed as viable approaches for compression along the spatial dimension. However, transform/subband filtering schemes perform very well, by concentrating most of the energy and information into only a few coefficients for subsequent quantization and symbol encoding. In order to exploit the nonstationary characteristics within a frame of video, each MC-residual is typically partitioned into nonoverlapping blocks which are then independently transformed and adaptively coded based on their individual characteristics. The Block Discrete Cosine Transform (Block DCT) is the most well-known example of this. Subband filtering and Wavelet schemes also have been successfully applied toward encoding the MC-residual. These schemes have overlapping basis functions which may be of the same length (uniform-band filterbank) or multiples of a given length. Wavelet schemes are especially interesting, as they include basis functions which achieve precise spatial localization required to efficiently represent high-frequency features, such as edges or other transients, and basis functions which yield precise frequency localization for the important low-frequency components within a video signal.

When examining the different spatial processing approaches in the context of spatially-adaptive inter/intra and spatially-adaptive leakage, a number of issues become evident. Since adjacent blocks are processed independently with the Block DCT, alternating the amount of MC-processing among adjacent blocks is easy to perform. However, for typical subband filtering and Wavelet schemes, adjacent regions are processed together. These schemes require a more complex process for efficiently encoding adjacent regions (e.g., one region in an inter manner and an adjacent region in an intra manner). More sophisticated processing is required to implement spatially-adaptive inter/intra or spatially-adaptive leakage for these schemes as opposed to the Block DCT [4].

2.2. Quantization and Symbol Encoding

The representation for the video signal is therefore formed by a color space conversion, followed by MC-prediction along the temporal dimension, and a spatial transform/subband filtering of the error signal. The computed transform/subband coefficients must be adaptively quantized to exploit the characteristics of the HVS. This is a crucial step to produce high-quality video. The quantized coefficients, motion vectors, and other information must then be entropy coded for transmission. Huffman and arithmetic coding are the two most promising techniques. Arithmetic coding has a major advantage over Huffman coding in that it enables sophisticated adaptive models to be applied to the data. However, there are currently no high-speed arithmetic coding VLSI chips on the market, and Huffman coding is therefore used nearly universally. By using different Huffman tables tuned to the statistics of the data to be transmitted, significant increases in performance can be achieved. A buffer and buffer control mechanism must also be used to couple the variable bit rate output of the video encoder to the constant bit rate channel.

3. IMPORTANT FEATURES AND ISSUES

Important features essential for a successful HDTV system include *Interoperability*, *Extensibility*, and *Scope of Services and Features*. Interoperability will be especially important in the future because there will be many different means of information exchange. An HDTV system may have to process video from a number of sources which have different formats, e.g., spatial resolution, frame rate, aspect ratio, color/monochrome, etc. Whereas the current television system converts the various video sources to a single format for processing and display, an HDTV system should process and display different source formats so as to exploit their individual characteristics. The HDTV system should facilitate interaction with computers and communications over the growing digital networks. Simple scalability may allow a range of price/performance receivers.

Useful extensibility of an HDTV system may come in many forms, including the capability to process at higher bit rates for studio-quality processing or for even higher resolution television. Also, there must be provisions to add new features to the system or enhancements to the compression algorithms. This is partly facilitated in that an HDTV standard will specify the core encoder and decoder algorithms and the bit stream syntax, but not how the information is generated at the encoder or used at the decoder. For example, improved methods for estimating the motion field or concealing channel errors may be developed. These schemes, as well as optional pre- and post-processing modules, may be added in a compatible manner.

The scope of services and features that an HDTV system can facilitate will partly determine the overall usefulness of the system. An HDTV system must be able to dynamically allocate its data capacity among the video and other services to be transmitted. For example, the audio may range from one channel to five-channel surround sound to multilingual stereo. Similarly, there should be provisions for an ancillary data channel, text, closed captioning, and encryption and addressing for pay-TV. Amenability to typical VCR functions, such as random access, fast forward/reverse search, slow motion, freeze frame, etc., is important. The capability to splice or edit the video at the bit stream level may also be important. Sustained high quality even after multiple concatenated encode/decode operations facilitates other applications.

A crucial system issue concerns *bit stream integrity*. The compressed video bit stream is highly susceptible to channel errors, with greater susceptibility at higher compression. An effective and robust delivery system is required. Error control coding as well as appropriate error concealment techniques at the decoder can guard against errors. Catastrophic errors may occur for any entropy coding scheme, since whenever an uncorrected channel error occurs, the decoding process becomes unsynchronized with the encoding process. This effect must be minimized.

4. DIGITAL HDTV SYSTEM EXAMPLE

We will briefly illustrate how the video compression principles and system issues can be integrated into a framework for designing a high-performance digital HDTV system. The Channel-Compatible DigiCipher (CCDC) digital HDTV system was developed by the Massachusetts Institute of Technology and General Instrument Corporation for possible adoption as the U.S. HDTV standard [5]. The system can operate at 32-QAM (preferred mode) or 16-QAM, depending on the desired system threshold and coverage considerations. This corresponds to available video data rates of approximately 18.88 and 13.60 Mb/s.

A high-level diagram of the video encoder is shown in Figure 1. The analog video signal undergoes anti-aliasing filtering and A/D conversion in the pre-processor (not shown). An RGB to YUV matrix conversion is performed (approximately the SMPTE 240M standard). The chrominance signals (U and V) are filtered and decimated by a factor of two along both the horizontal and the vertical dimensions, producing chrominance signals with one-fourth the sampling density of the luminance signal.

A full-search block matching scheme is employed to perform the motion estimation. The ME is performed on the luminance frames only, and the computed motion vectors are applied to both the luminance and chrominance components. The criterion for choosing a match is minimum absolute error (MAE) between blocks. The system is designed to exploit both the higher prediction efficiency provided by a dense motion field estimate and the lower amount of information to be transmitted by a sparser motion field estimate. Over each local region of the video, the resolution of the motion field is chosen based upon the local characteristics. The motion field resolution (blocksize) is chosen between 16×16 and 8×8 blocksize, where the criterion for selection will be discussed shortly. The search range for the motion vectors is +15/-16 pixels horizontally and +7/-8 pixels vertically. The larger horizontal search range corresponds to the increased likelihood of rapid motion along the horizontal direction. The search range will allow the encoder to track objects moving at up to 0.75 frame width and 0.67 frame height per second. This is especially useful for encoding video with large frame-to-frame movements, as in sporting events. The motion vectors are estimated with $\frac{1}{2}$ pixel accuracy.

MC-prediction typically performs very well. Nevertheless, the CCDC system compares for each 8×8 pixel block the *efficiency* when using MC-prediction and when using intraframe encoding. A comparison is made among three possibilities: (1) MC-prediction with 16×16 blocksize, (2) MC-prediction with 8×8 blocksize, and (3) purely intraframe encoding of the block. Each block is processed independently with each of the three approaches, and the approach requiring the minimum number of bits to achieve the same reconstructed video quality is selected. Through these processing steps, the CCDC system can exploit the advantages of both coarse and fine motion field estimates, as well as suppress the MC-processing and use purely intraframe processing if that leads to improved performance.

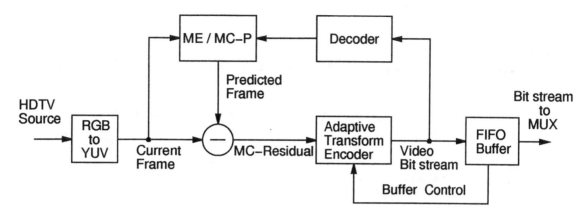

Figure 1: The CCDC HDTV video encoder.

The capability of spatially-adaptive inter/intra encoding of different blocks lends itself very easily to an elegant solution for a number of problems. When using any temporal predictive scheme like MC-prediction, the decoder must be able to accurately track the encoder, especially true in the scenario of a broadcast environment. Two previously discussed approaches to achieve this are periodically intra-encoding an entire frame, and applying leakage within the prediction loop. Both of these approaches have drawbacks. Periodically intra-encoding an entire frame may result in artifacts in the reconstructed frame because of the higher bit rate required for intra-encoding and the limited distribution capabilities of the buffer. Leakage is designed for one-way processing (forward only) and hinders simple execution of VCR functions such as reverse playback.

The CCDC system uses spatially-adaptive inter/intra processing to solve these problems in a very simple and elegant manner. For each frame, successive columns of the video are encoded in an intra manner. In each predetermined length of time the entire video frame is encoded using intraframe encoding, and the corresponding increase in bit rate is uniformly distributed over time. For example, with the prototype CCDC system, the baseline video signal has 1/20th of each frame intra-encoded, producing a "refresh" of the video 3 times per second. The increase in required bit rate is therefore spread over 20 frames, as opposed to a single frame receiving the full hit in performance. Through this approach of partial refreshing of each frame, the video acquisition time is .33 sec, excluding synchronization of the communication systems, and the maximum propagation of uncorrected channel errors is also limited to .33 sec. Inter/intra decision-making therefore provides highly adaptive and efficient video compression, robustness against uncorrected channel errors, and also enables simple VCR functionality.

Of the many approaches for transform/subband filtering the residual, the Block DCT is currently the most mature and well understood. This, coupled with its high performance, simple spatially-adaptive processing, and hardware availability, led to its choice for processing the residual. The MC-residual is partitioned into 8×8 pixel blocks, each of which is independently transformed using the 2-D DCT.

The key step toward reducing the bit rate is the quantization and entropy coding of the DCT coefficients. The quantization is performed in an adaptive manner for each local region in order to exploit the human visual perception. Each 8×8 block of DCT coefficients is weighted or normalized by a matrix of 8×8 weighting factors. The DCT coefficients are then quantized with a uniform quantizer. The matrix of weighting factors is chosen based upon the inter/intra and luminance/chrominance characteristics of each block, as well as measures of the local scene complexity (to exploit the spatial masking that may exist) and buffer fullness. The weighting factors effectively scale each individual DCT coefficient quantizer, making it more coarse (increasing the stepsize) or fine (decreasing the stepsize) in order to take advantage of the differing HVS sensitivity. A feedback mechanism from the buffer regulates the coarseness/fineness of the quantization to ensure high video quality while preventing buffer overflow or underflow.

The location of the nonzero quantized DCT coefficients is encoded through a Vector Coding (VC) approach. In this scheme, each 8×8 block of coefficients is divided into 4 regions each containing 16 coefficients. The division is such that the first region contains the low-frequency coefficients which are most likely to be nonzero, and the last region contains the high-frequency coefficients which are least likely to be nonzero. A 16 bit pattern identifies the location of the nonzero coefficients within each group, and these are Huffman coded for transmission. The DC coefficients are differentially Huffman encoded and the nonzero AC coefficients are Huffman coded with a codebook chosen based upon the position of the coefficient. Improved performance for encoding the location and amplitude information is achieved through the use of different codebooks based upon the inter/intra and luminance/chrominance characteristics of each block.

In this paper, we have discussed various issues related to the design of an HDTV system. Additional details can be found in [1, 2, 3, 4, 5].

REFERENCES

[1] J. Apostolopoulos and J. Lim, "Video compression for digital ATV systems," in *Motion Analysis and Image Sequence Processing* (M. Sezan and R. Lagendijk, eds.), ch. 15, Hingham, MA: Kluwer Academic Publishers, 1993.

[2] J. S. Lim, *Two-Dimensional Signal and Image Processing*. Englewood Cliffs, N.J.: Prentice Hall, Inc., 1990.

[3] A. Netravali and B. Haskell, *Digital Pictures, Representation and Compression*. New York: Plenum Press, 1988.

[4] P. Monta, *Signal Processing for High Definition Television*. PhD thesis, MIT, To be submitted 1993.

[5] American Television Alliance, Massachusetts Institute of Technology and General Instrument Corp., *Channel-Compatible DigiCipher HDTV System*, April 1992.

Programmable Facsimile Image Processor including Fuzzy-Based Decision

Bang W. Lee, Jin W. Lee, Si H. Bae, Suh K. Kim, In H. Hwang, and Jae H. Kim*

Samsung Electronic Co.,
82-3, DoDang-Dong, Chung-Gu, Buchun, KyungGi-Do 421- 130, Korea
* currently with Dept. of Electronic Engineering in Pusan University
Tel : 2-740-6743, Fax : 2-740-6708, E-mail : buchun@saitgw.sait.samsung.co.kr

ABSTRACT

In this paper, we present an application-specific DSP that is optimally partitioned for functional flexibility into a programmable 16-bit integer type DSP core and special hardwares. This LSI can handle up to 7 bi-level image processing schemes including three novel features. For mixed text and picture documents, a fuzzy algorithm is proposed to discriminate text regions and picture regions. An 8 x 8 dithering matrix for higher data compression of these picture regions and a smoothing scheme for doubling vertical resolution at normal mode receiving are also proposed. With these proposed schemes, the facsimile transmission time can be reduced by more than half. This proposed chip fabricated in 80mm² die area with a 1.2 μm CMOS technology consumes 0.35 watt at 5 volt operating voltage.

1. INTRODUCTION

A facsimile (fax) is widely used to exchange information in more natural way. Compared with an electronic mail system, fax is inferior in transmission speed and data reliability. But, many business documents are still delivered with fax because of an unique feature that original shapes such as signature, pictures, and drawings can be retained as they are. Besides, a fax delivers more mixed-type documents containing both texts and pictures in one page. The text-only document generates one binary data - black or white - per one pixel. Conversely, a picture document generates usually more than 4 gray-level bits per one pixel. In order to express the gray-level data into black-and-white printing devices such as laser printer and fax, halftoning techniques [1] are used. Since frequency of black dots occurrences expresses the gray level in a fax, a picture document generates a huge amount of data in the scanning process. This becomes a key cause to increase fax transmission time.

There are three main blocks to decide fax transmission speed, which are modem, entropy codec, and document image processor (DIP) [2]. In addition to old modem standards such as V.27ter and V.29, new standard V.33 and V.17 based on a Trellis coded modulation start to use for 14,400 bps data speed. Data compression and decompression are conduct by an entropy codec [2-4], where the compression table are made by statistical occurrences of binary data patterns. By assigning small number of bits to high probable patterns, data compression of binary data from a text document is usually by a factor of 5. Basic features in conventional DIPs are pre-processing for compensation of scanning data and post-processing for a printing device. Extended features for picture documents are mainly focus to reduce data amount of the gray level expression.

In this paper, we will present an application specific LSI and image processing schemes for document image processing. The proposed DIP can conduct pre/ post processing for nonlinear compensations as well as up to 7 image enhancement algorithms. This proposed schemes are smoothing at receiving mode, fuzzy-based discrimination of picture/text regions, and an 8 x 8 dithering matrix.

2. CHIP ARCHITECTURE

Figure 1 shows block diagram of proposed programmable document image processor. This DIP communicates with other devices with two buses (image memory bus and CPU bus) and two I/O ports (sensor input and recorder output). The shading memory stores shading correction data, while the line memory stores one previous line pixel data for several image processing schemes. All control signals and data I/O with an external processor are done with the CPU interface unit, which consists of several control registers, direct memory access (DMA) controller, and interface with CPU bus and internal main bus. This DMA interface block operates a direct data transfer between line memory and external processor memory.

Analog interface unit (AIU) receives analog signals from a scanning device and converts to 6-bit digital signals, as shown in Fig. 2. The 6-bit programmable counter provides sampling signal of the 6-bit ADC, which depends upon characteristics of a scanning device. Voltage reference of this flash ADC is provided from the 8-bit DAC. This 64 gray level digital signal is latched and transferred to the DSP core for further image processing. Feedback loop consisting of ADC, Vpeak counter, and DAC allows an automatic background control (ABC) for various brightness of documents.

Figure 3 shows pixel processing unit (PPU), which is a specialized hardware engine working with the DSP core. By implementing heavy computing algorithms with this customized block, computing loading of the DSP core can be minimized. From data in the pixel window, minimum/ maximum/ average values are generated. Key features of the DSP core [5]

0-7803-0826-3/93 $3.00 © 1993 IEEE

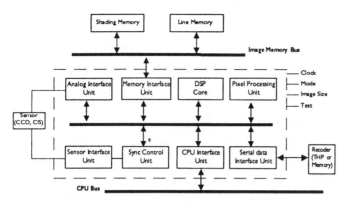

Fig. 1. Block diagram of the proposed DSP

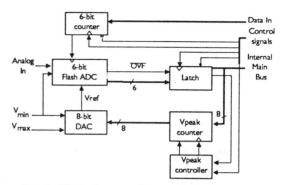

Fig. 2. Block diagram of analog interface unit

are 100 nSec/ a MAC cycle with 16-bit fixed point arithmetic, a 16 × 16-bit multiplier with a 24-bit accumulator, and two 256-word internal data RAMs

Memory interface unit provides data transfer between internal main bus and image memory bus according to DSP core instruction. Sync control unit (SCU) generates all necessary timing signals such as pixel sync, line sync, sensor clocks and the ABC enable signal for AIU, interrupt signals for DSP Core, and pixel clocks for PPU. Serial data interface unit provides serial output for print or storage. Interaction between a host processor and the proposed chip for a normal operation can be done by setting parameters of a control register in this DSP. Programming parameters for sampling time and number of pixels per line allow to use either charge-coupled device or contact image sensor. Programming gamma correction table also allows to use any printing device such as laser printer, ink-jet, and thermal head printer. After initialization, the proposed chip operates as a co-processor of external host processor.

3. OPERATION WITH CONVENTIONAL ALGORITHMS

Fax image algorithms can be classified into distortion correction, image quality enhancement, and image compression. Image distortion is mainly caused by scanning and printing devices. Inevitable non-uniformities of light intensity can be compensated by comparing reference levels from white panel under light bulb. In this proposed LSI, shading function is stored at external shading memory and shading correction is done with the DSP core in a software format. Here, limited lamp size, non uniform mechanical distance between lamp and line sensor, difference in light reflection, uncleanness by dusts, and electronic parts aging can cause the shading distortions.

Once the shading correction procedure is done, 6-bit shading corrected data is transformed into 4-bit gamma corrected data. In this calculation, RAM in the DSP core can be used to map with a tabulated method that 6-bit data is used for RAM address and 4-bit gamma corrected data is stored as RAM content. Gamma correction compensates differences in human visual sensitivity and tone scale intensity of black dots. Depending on dot size, paper material, and printer grid shape, same dot density may give different feeling to human. The gamma correction table is downloaded into SRAM in DSP Core during the initialization procedure.

The simplest scheme for text document is to compare current pixel level with a fixed threshold value. This fixed threshold scheme can be realize with very low cost hardware, but can not distinguish darker patterns in dark background. Image enhancement is to emphasize edges of text documents and to adjust the threshold value according to adjacent pixel intensity. With this adaptive threshold method, edge emphasis for text boundary and patterns inside dark background can be obtained to improve text readability. In this proposed chip, local adaptive threshold is used [4,8].

For the picture document transmission, continuous gray scaled image should be converted into a binary format. Printing device using in fax is a binary system, while a picture image is continuous gray pattern. Halftoning techniques [6-8] make an illusion of continuous gray image from an artificial treatment on binary image. Ordered dithering techniques and error diffusion technique [3] are widely used. In this dithering technique, dithering matrix is used as threshold value of each pixel. Since a document is divided into a certain matrix size and the threshold value is not random, the ordered dithering technique can make regular grainy patterns. Since low frequency components generated by ordered dithering matrix make this undesirable pattern, blue noise that has only high frequency noise components is used in the error diffusion technique [6,8]. In the error diffusion technique, gray level of current pixel that is quantized into black or white is diffused into neighbor pixels. Special weights are forced to each diffused gray level of neighbor pixels and summed for binary quantization. In spite of the beauty of blue noise characteristics, the error diffusion technique also gives usually correlated artifacts in continuous gray image, direc-

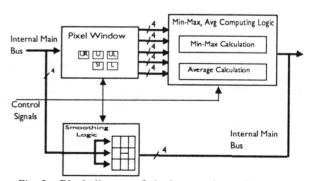

Fig. 3. Block diagram of pixel processing unit

tional hysteresis in very white and black images, and poor transient performance. Thus, many modified methods are still investigating [7].

In addition to the large amount of gray scaled data, the other obstacle to let transmission time increase must be entropy codec standards (MH : Modified Huffman and MR : Modified Read) of current Group-3 fax. Since this entropy codec was made at late 70's, coding table was optimized to minimize scanning data of a mainly text-only document [2,3]. Thus, image data of picture documents is not adequate for the MH or MR entropy codec. Data amount of an picture document can be conversely increased after the entropy encoding process

4. PROPOSED ALGORITHMS

At document receiving, smoothing mode can double vertical resolution without increasing transmission time. Similar vertical resolution can be also obtained with fine mode operation at the price of twice longer transmission time. Notice that horizontal resolution of Group-3 fax is fixed to 200 dot per inch (dpi), while vertical resolution is 100 dpi in normal mode and 200 dpi in fine mode. Smoothing logic operation in the PPU is shown in Fig. 4. Pixel Y5 is divided into two pixels according to patterns of surrounding pixels.

Figure 5 shows a novel 8 x 8 dithering matrix for the picture documents, which helps transmission time three times less than conventional dithering matrices. This 8 x 8 dithering matrix consists of four identical 4 x 4 matrices, which is made by reflecting one 4 x 4 matrix in vertical and horizontal directions. This significant data reduction after either MH or MR encoding is caused by artificial wave-like patterns.

In the case of mixed text/ picture documents, an optimal processing strategy must be to separately process text regions and picture regions. However, discrimination of text and picture regions is usually crucial in many cases. In this proposed DIP chip, a fuzzy rule based discrimination can be used. Each three membership functions for pixel intensity SI { $M_d(.), M_g(.), M_b(.)$ } and edge intensity EG_{SI} { $M_s(.), M_m(.), M_L(.)$ } are used to determine probabilities, which are $P_d, P_g, P_b, P_s, P_m,$ and $P_L,$ respectively. Here, EG_{SI} is given as difference of maximum value and minimum value of the pixel window. Based on

3	2	1	0	4	5	6	7
11	10	9	8	12	13	14	15
15	14	13	12	8	9	10	11
7	6	5	4	0	1	2	3
0	1	2	3	7	6	5	4
8	9	10	11	15	14	13	12
12	13	14	15	11	10	9	8
4	5	6	7	3	2	1	0

Fig. 5. Proposed dithering matrix for fuzzy mode

probability values of each membership function, the discrimination is done. Following pseudo C-code implies these operation.

$$P_{MAX}= max\{P_dP_s, P_dP_m, P_dP_L, P_gP_s, P_gP_m, P_gP_L, P_bP_s, P_bP_m, P_bP_L\}$$

If $P_{MAX} = P_bP_s$ or P_bP_L,
    then { treat SI as a white pixel.}
If $P_{MAX} = P_dP_L$ or P_dP_m,
    then { treat SI as a black pixel in text. }
If $P_{MAX} = P_gP_L$,
    then { use *local adaptive threshold method.* }
If $P_{MAX} = P_dP_m, P_gP_s, P_gP_m$ or P_bP_m
    then { use *wave dither method.* }

For faster transmission with reasonable picture quality, we choose the wave dithering scheme for picture regions and the local adaptive threshold for text regions.

5. EXPERIMENTAL RESULTS

Figure 7 shows experimental results using a fixed threshold algorithm and an error diffusion algorithm. Since the fixed threshold algorithm is mainly for text only documents, processed image is just too simple for the picture only document. On the other hand, the image processed with error diffusion technique shows a naturally smoothing tone. Transmission time of various algorithms are listed in Table 1. Transmission time of the fuzzy mode operation is slight longer than that of adaptive threshold, but almost 2 times shorter than that of error diffusion.

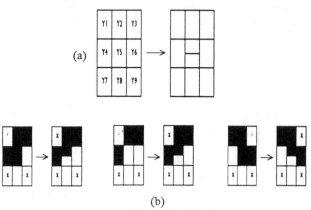

(b)

Fig. 4. Operation of smoothing logic
    (a) Pixel windows format
    (b) Examples

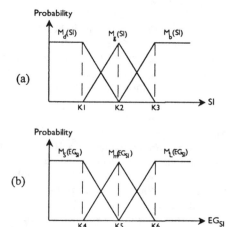

Fig. 6. Fuzzy mode membership function
    (a) Pixel intensity
    (b) Edge intensity

In the simple mixed mode, difference of maximum value and minium value among all pixels in the pixel window in PPU is used to discriminate text and picture regions. Figure 8 shows experimental results in the smoothing mode. Since vertical resolution of a G3 fax is half of horizontal resolution in normal mode, many discontinuities can be observed. On the other hand, continuous patterns can be obtained with the proposed smoothing mode. Here, transmitting fax uses a conventional DIP and receiving fax uses this proposed DIP.

The image processor has been fabricated with an 1.2 μm CMOS technology in about 80 mm² silicon area as shown in Fig. 9. The chip includes a 16-bit integer type DSP core, 15000 standard cells, an 6-bit flash ADC, and 8-bit DAC. Operating frequency is 10 MHz and power dissipation is 0.35 Watt at 5 Volt power supply. Since this LSI uses a general-purpose DSP as a supercell, adding functions such as voice codec can be easily done. On the contrast, conventional fax image processors with fully hardwired style do not have any flexibility.

6. DISCUSSION

Heavy computing parts are implemented with special hardware engines, while soft-decision part is implemented with the

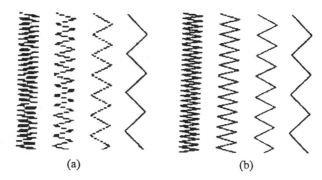

(a)   (b)

Fig. 8. Receiving mode example
(a) Without smoothing, (b) With smoothing

programmable 16-bit DSP core. Compared with conventional micro-controller approach, the application-specific DSP approach gives more computing horsepower and great flexibility. In addition to error diffusion algorithm, fuzzy mode operation also can be handled for mixed text and picture document. In addition, the proposed wave dithering matrix can help to send picture document with very short time duration. These proposed techniques can be also used for many printing devices and display devices that can handle black-and-white images.

REFERENCE

[1] Robert Ulichney, *Digital Halftoning,* The MIT Press, 1987.
[2] R. Hunter and A. H. Robinson, "International Digital Facsimile Coding Standards," *Proc. of the IEEE*, vol. 68, no. 7, pp. 854-867, July 1980.
[3] K. R. McConnell, D. Bodson, and R. Schaphorst, *Digital Facsimile Technology & Applications,* Arrech House Inc., 1989.
[4] M. Rabbani and P. W. Jones, *Digital Compression Techniques*, vol. TT7, SPIE Press, 1991.
[5] Samsung Electronics, *SSP1600 User's Manual,* Mar. 1991.
[6] H. Ochi and N. Tetsutani, "A new half-tone reproduction and transmission method using standard black and white facsimile code", *IEEE Trans. on Communications,* vol. COM-35, no. 4, Apr. 1987.
[7] R. A. Ulichney, "Dithering with blue noise," *Proc. of IEEE*, vol. 76, no. 1, Jan. 1988.
[8] B. W. Lee, J. H. Kim, S. H. Bae, S. K. Kim, I. H. Hwang, and J. W. Lee, "An Application-Specific Digital Signal Processor for New Generation FAX Image Processing," *Inter. Conf. on DSP Applications and Technologies*, pp. 494 - 502, vol. 1, Nov. 2-5, 1992, Boston, MA.

(a)

(b)

Fig. 7. Example of a picture document (Real size)
(a) Fixed threshold, (b) Error Diffusion

Table 1. Transmission Time of Various Algorithms

Document	Resolution	Binary	Local Adaptive	Wave Dithering	Error Diffusion	Simple Mixed	Fuzzy Based
Document # 1 (40% Text/ 60% Picture)	Normal	0'51"	1'18"	1'10"	2'20"	1'19"	1'15"
	Fine	1'27"	2'03"	2'20"	4'57"	2'48"	2'26"
Document # 2 (100% Picture)	Normal	0'51"	3'39"	1'49"	5'51"	2'11"	2'09"
	Fine	1'18"	7'46"	4'13"	13'17"	5'22"	5'20"

Fig. 9. Die photo of the proposed chip

485

PRINTER MODELS AND COLOR HALFTONING

Thrasyvoulos N. Pappas

Signal Processing Research Department
AT&T Bell Laboratories
Murray Hill, NJ

ABSTRACT

We present new models for color laser printers. They form the basis of "model-based" techniques that exploit printer distortions to increase both the spatial and color-scale resolution of printed images. We consider two model-based techniques, the modified error diffusion algorithm and the least-squares model-based algorithm. The models account for distortions caused by "dot-overlap" and imperfect inks. We show that when the inks are assumed to have perfect light absorption properties, both algorithms are separable. When the inks are not assumed to be perfect, the algorithms are not separable and, moreover, the modified error diffusion becomes unstable. The separable printer models offer computational simplicity and robustness to errors in color registration, while the non-separable models produce images with better colors.

1. INTRODUCTION

Digital halftoning is the process of generating a pattern of pixels with a limited number of colors (typically 2-8) that create the illusion of a continuous-tone image. Digital halftoning is necessary for display of continuous-tone images in media (most commonly paper) in which the direct rendition of the tones is impossible.

Many halftoning techniques can be found in the literature [1]. Most of these techniques assume perfect printing. In [2, 3, 4] we presented "model-based" halftoning techniques that exploit printer models to produce high quality gray-scale images using standard black and white (BW) laser printers. Recently, several different color printers have become available in the market. A variety of technologies exist: electrographic, ink jet, thermal, etc. Most of the color printers produce halftone images. The quality of the images that these printers produce can be improved significantly by the use of model-based halftoning techniques.

Color printers present many of the problems that we encounter in the BW case. Thus, they also generate "distortions" such as "dot overlap." However, they also present some problems that are unique to color printing. Color printers use ink of three (or four) different colors, typically cyan, magenta, yellow, (and black), while BW printers use only black ink. Thus, we can have overlap between dots of the same or different colors. Also, the light absorption properties of the inks are usually not perfect, thus resulting in color distortions which can be significant for some printers. Finally, inexact registration of the different colors can also distort the printed images.

Conventional halftoning techniques resist printer distortions by printing colored dots in clusters or *macrodots*. One such technique is the "classical" clustered-dot ordered dither. The image intensity is represented by the size of the macrodots, while their spacing is fixed. In color printing, macrodots of the three (or four) different colors are printed. Since the macrodots of different colors typically overlap, the

printer distortions affect both the brightness and the color of the image. Thus the color "classical" screen is not as robust to printer distortions as the BW screen. In contrast, the model-based techniques can exploit both dot overlap and interactions between colors to increase the spatial and color scale resolution of the printed images. We consider two model-based techniques, the modified error diffusion algorithm and the least-squares model-based (LSMB) algorithm. The basis of model-based techniques is an accurate printer model.

In [5] we assumed perfect inks and considered only "dot-overlap" distortions. We showed that (separable) model-based techniques produce images that are sharper, less noisy, and with richer and better color tones than the commonly used "classical" screening techniques. In this paper, we present printer models that account for both distortions caused by overlap between neighboring dots of the same and different colors and imperfect inks. We show that when the inks are assumed to have perfect light absorption properties, both algorithms are separable. When the inks are not perfect, the algorithms are not separable. Moreover, the modified error diffusion algorithm becomes unstable.

We used a simple 300 dpi thermal printer and a more sophisticated 400 dpi electrographic printer as our test vehicles. We experimented with both separable and non-separable printer models corresponding to the two printers. We used a flatbed scanner and a reflection densitometer to measure the absorption densities of the inks of the two printers. We also used the absorption densities of typical inks provided in [8, p. 572].

Our results indicate that the separable printer models offer computational simplicity, robustness to errors in color-plane registration, and, in the case of error diffusion, stability. The non-separable models produce images with better colors, assuming good color registrations and an accurate device for estimating the ink parameters. Unfortunately, the error diffusion algorithm with a non-separable printer model diverges for several images.

2. PRINTER MODELS

In this section we present models for color laser printers. The models are independent of the characteristics of the human visual system. We used a CANON CLC300 printer as the principal test vehicle. It is an electrographic 400 dpi printer. We also used a QMS ColorScript 100 printer, which is a thermal 300 dpi printer.

To a first approximation, such printers are capable of producing colored spots (usually called dots) on a piece of paper, at any and all sites of a Cartesian grid with horizontal and vertical spacing of T inches. The reciprocal of T is the "printer resolution" in dots per inch (dpi). Color printers use cyan (C), magenta (M), and yellow (Y) inks to produce color dots. These colors form the basis for the subtractive system of colors. The relationship to the additive

0-7803-0946-4/93 $3.00 © 1993 IEEE

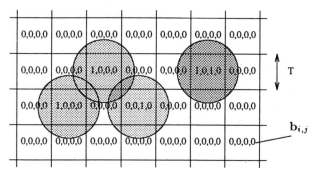

Figure 1. Actual color-dot overlap.

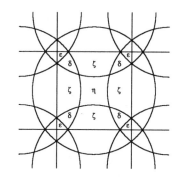

Figure 2. Overlapping segments of color

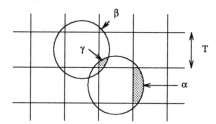

Figure 3. Definition of α, β, and γ for printer model.

colors (RGB) is very simple. Cyan ink absorbs red light, magenta absorbs green, and yellow absorbs blue. Different dots of ink can be printed on top of each other to produce red, green, blue, and black dots. In practice, however, the inks are not perfect, i.e. there are unwanted absorptions. Thus, many printers use a separate black ink (K) to produce better black dots. In the remainder of this paper we will assume that the printer uses all four types of ink.

The printer is controlled by an $N_W \times N_H$ array of four-dimensional vectors with binary components

$$\mathbf{b}_{i,j} = \left(b_{i,j}^C, b_{i,j}^M, b_{i,j}^Y, b_{i,j}^K \right) \qquad (1)$$

where $b_{i,j}^C = 1$ indicates that a cyan dot is to be placed at site (i,j) and $b_{i,j}^C = 0$ indicates that no cyan dot is to be placed at the site. The magenta $b_{i,j}^M$, yellow $b_{i,j}^Y$, and black $b_{i,j}^K$ components are defined similarly. When all components are zero, the site is to remain white. We'll refer to the latter as a "white" dot. When more than one component is equal to 1, different inks are printed on top of each other to produce red, green, blue, or black dots. In principle, we can specify $2^4 = 16$ different colors for each dot, but 9 of these colors are variations of black. Usually, the black ink is used only in combination with the other three inks to produce solid black dots, thus reducing the number of colors that each dot can take to $2^3 = 8$.

As we saw in [2], printers produce circular rather than square black dots. Fig. 1 illustrates the most elementary distortion introduced by most printers: their dots are larger than the minimal covering size, as if "ink spreading" occurred. A variety of other distortions are present in actual printers, caused by the heat finishing, reflections of light within the paper, and other phenomena. Also, the inks are not perfect. As a result, the color level produced by the printer in the vicinity of site (i,j) depends in some complicated way on $\mathbf{b}_{i,j}$ and neighboring dots. However, due to the close spacing of dots and the limited spatial resolution of the eye, the color level can be modeled as having a constant value $\mathbf{p}_{i,j}$ within the area of the square pixel at site (i,j), equal to the average color level of the pixel. The color printer model takes the form [5]

$$\mathbf{p}_{i,j} = \left(p_{i,j}^R, p_{i,j}^G, p_{i,j}^B \right) = \mathbf{P}(\mathbf{W}_{i,j}) \qquad (2)$$

where $\mathbf{W}_{i,j}$ consists of $\mathbf{b}_{i,j}$ and its neighbors, and $\mathbf{P}(.)$ is some function thereof.

For the methods we study here, it is essential that $\mathbf{p}_{i,j}$ be entirely determined by the dots in a finite window around $\mathbf{b}_{i,j}$. The possible values of $\mathbf{P}$ can be listed in a table. The number of elements in the table, however, is 8^9 for a 3×3 window. In principle, the individual elements of this table can be derived from measurements of the color that results when various dot patterns are printed. In practice, this is

very difficult, so a simpler model must be derived from a physical understanding of the printing mechanism.

2.1. Circular-Dot-Overlap Color Printer Model

We now develop a specific color printer model that accounts for the "dot-overlap" distortions illustrated in Fig. 1. In the BW case, all the overlapping segments have the same color as the dots. In the color case, the overlapping segments can take different colors depending on the color of the neighboring dots. Fig. 2 shows the segments of different colors that we can get within a pixel.

The color of each segment is specified in additive (RGB) coordinates and depends on the absorption properties of the overlapping dots of ink. Assuming that the paper saturates with each type of ink, the color of each segment depends on which types of ink overlap the segment. A table provides the color for each of the 16 different ink combinations. The elements of the table are trivial to obtain in the case of *perfect inks*, that is, when cyan ink blocks 100% of the red component of the light and no green or blue, magenta ink blocks 100% of the green component and no red or blue, and yellow ink blocks 100% of the blue component and no red or green. For *imperfect inks*, the elements of the table can be obtained by measuring the reflectance of different patches of printed color.

Once the color of each segment is determined, the printer model $\mathbf{P}$ specifies the average color $\mathbf{p}_{ij}$ of the pixel (i,j) in the RGB (additive) domain as a weighted sum of the colors of the different segments with the weight being the area of the segment. The area of each segment is easy to calculate in terms of the parameters α, β, and γ shown in Fig. 3.

$$\epsilon = \beta, \quad \delta = \gamma - \beta, \quad \zeta = \alpha - 2\gamma, \quad \eta = 1 - 4\alpha + 4\gamma \qquad (3)$$

The parameters α, β, and γ are the ratios of the areas of the shaded regions shown in Fig. 3 to T^2, and can be expressed in terms of the ratio ρ of the actual dot radius to the ideal dot radius $T/\sqrt{2}$ [2]. For both the CANON and the QMS printer, we found that the dot size is close to the ideal, i.e. $\rho = 1$, and $\alpha = .143$, $\beta = 0$, and $\gamma = 0$.

If we assume that the inks are perfect, the color of each segment can be determined independently for each of the color components. This is because cyan ink affects only

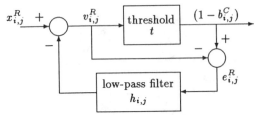

Figure 4. Standard error diffusion

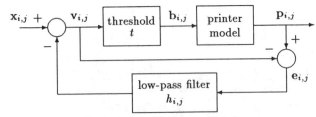

Figure 5. Modified error diffusion

the red component, magenta affects only the green, and yellow affects only the blue. The black is unnecessary in this case, since a combination of the other three inks produces a perfect black. Thus, the average color $\mathbf{p}_{i,j}$ for the site (i,j) can be determined independently for each color component, and the printer model of Eq. (2) becomes separable:

$$\mathbf{p}_{i,j} = \left(p_{i,j}^R, p_{i,j}^G, p_{i,j}^B\right) = \left(P^R(W_{i,j}^C), P^G(W_{i,j}^M), P^B(W_{i,j}^Y)\right) \tag{4}$$

where each of the components is specified by a circular-dot-overlap model which is identical to the gray-scale model [2].

3. ERROR DIFFUSION

Error diffusion [7] is a popular method for generating sharp halftone images for displays, such as some CRT's that do not suffer from substantial dot overlap or other distortions. The standard color error diffusion algorithm is applied to each color component independently. Let $[\mathbf{x}_{i,j}]$ be a color image, where

$$\mathbf{x}_{i,j} = (x_{i,j}^R, x_{i,j}^G, x_{i,j}^B) \qquad 1 \le i \le N_W, \; 1 \le j \le N_H \tag{5}$$

denotes the pixel located at the i-th column and the j-th row. The intensity of each color component is assumed to vary between 0 and 1. We assume that the image has been sampled so there is one pixel per dot to be generated. Let $[\mathbf{b}_{i,j}]$ be the halftone image produced by error diffusion with $\mathbf{b}_{i,j}$ as defined in Eq. (1). Without loss of generality, we assume that the image is scanned left to right top to bottom. The cyan component of the halftoned image is obtained by the following set of equations

$$v_{i,j}^R = x_{i,j}^R - \sum_{m,n} h_{m,n} e_{i-m,j-n}^R \tag{6}$$

$$b_{i,j}^C = \begin{cases} 0, & \text{if } v_{i,j}^R > t \\ 1, & \text{otherwise} \end{cases} \tag{7}$$

$$e_{i,j}^R = (1 - b_{i,j}^C) - v_{i,j}^R \tag{8}$$

Here $v_{i,j}^R$ is the "corrected" value of the red component of the continuous-tone image. The error $e_{i,j}^R$ at any "instant" (i,j) is defined as the difference between the "corrected" continuous-tone image component and the halftone image component (converted to RGB domain). The "past" errors are low-pass filtered and subtracted from the current image value $x_{i,j}^R$ before it is thresholded to obtain $b_{i,j}^C$. $[h_{i,j}]$ is the impulse response of the low-pass filter. The other two color components are defined similarly. Assuming that the halftone color components are perfect, the black component $b_{i,j}^K$ is not useful and can be set to zero.

The diagram of the algorithm is shown in Fig. 4. We fixed the threshold t at .5, the middle of the range of each color. The low-pass filter $h_{i,j}$ has non-symmetric half-plane support, and the coefficients are positive and sum to one (to guarantee stability). We used the filter proposed by Jarvis, Judice and Ninke [1, p. 241].

In the presence of dot overlap, error diffusion produces dark color images. This problem can be corrected by using the color printer models we developed in Section 2. The printer models can also account for the distortions introduced by nonperfect inks. The modified error diffusion algorithm that compensates for printer distortions is shown in Fig. 5. The error is now defined as the difference between the "corrected" color image $\mathbf{v}_{i,j}$ and the output of the printer model $\mathbf{p}_{i,j}$, rather than the halftone image. Thus, it accounts for printer distortions as well as quantization effects. The modified error diffusion equations are

$$\mathbf{v}_{i,j} = \mathbf{x}_{i,j} - \sum_{m,n} h_{m,n} \mathbf{e}_{i-m,j-n}^{i,j} \tag{9}$$

$$b_{i,j}^C = \begin{cases} 0, & \text{if } v_{i,j}^R > t \\ 1, & \text{otherwise} \end{cases} \tag{10}$$

$$b_{i,j}^M = \begin{cases} 0, & \text{if } v_{i,j}^G > t \\ 1, & \text{otherwise} \end{cases} \tag{11}$$

$$b_{i,j}^Y = \begin{cases} 0, & \text{if } v_{i,j}^B > t \\ 1, & \text{otherwise} \end{cases} \tag{12}$$

$$\mathbf{e}_{m,n}^{i,j} = \mathbf{p}_{m,n}^{i,j} - \mathbf{v}_{m,n} \qquad \text{for } (m,n) \prec (i,j) \tag{13}$$

where $(m,n) \prec (i,j)$ means (m,n) precedes (i,j) in the scanning order and

$$\mathbf{p}_{m,n}^{i,j} = \mathbf{P}(\mathbf{W}_{m,n}^{i,j}) \qquad \text{for } (m,n) \prec (i,j) \tag{14}$$

where $\mathbf{W}_{m,n}^{i,j}$ consists of $\mathbf{b}_{m,n}$ and its neighbors as in (2), but here the neighbors $\mathbf{b}_{k,l}$ have been determined only for $(k,l) \prec (i,j)$; they are assumed to be zero (white) for $(k,l) \succeq (i,j)$. Since only the dot-overlap contributions of the "past" pixels can be used in (14), the "past" errors keep getting updated as more pixel values are computed. Hence the dependence of the error and the printer model output on the "instant" (i,j). In the modified error diffusion equations above, we assume that the black component $b_{i,j}^K$ cannot be specified independently, and is equal to 1 if and only if all the other three components are equal to 1.

Note that if the printer model is *separable* (e.g. circular-dot-overlap model with perfect inks), there is no coupling between the three color components, and the modified error diffusion algorithm can be applied independently to each color component. When the printer model is *non-separable* (e.g. imperfect inks), the modified error diffusion equations are coupled. This affects not only the ease of implementation, but also the stability of the algorithm. It can be shown that if the inks are not perfect, the algorithm is unstable.

We tested the modified error diffusion algorithm on several images and compared it to clustered ordered dither ("classical") and the least-squares model-based technique of the next section. We used both separable and non-separable printer models corresponding to the 400 dpi CANON printer and the 300 dpi QMS printer. We used both a HOWTEK flatbed scanner and a MACBETH RD922 Reflection Densitometer to measure the ink absorption densities. We also used the set of absorption densities of typical inks provided in [8, p. 572].

The separable printer models offer the advantages of computational simplicity, robustness to errors in color-plane

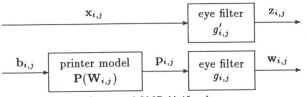

Figure 6. LSMB Halftoning

registration, and, most importantly, the stability of the modified error diffusion algorithm. As discussed in [5], the separable modified error diffusion algorithm produces images that are sharper, less noisy, and with richer and better color tones than the commonly used "classical" screening techniques.

However, the inks used by most printers are not perfect and for many printers there are significant unwanted absorptions. Unfortunately, the non-separable algorithm diverges for several images. When it does converge, it produces images with better colors and all the properties of error diffusion mentioned above. Convergence of the algorithm depends on both the ink parameters and the image.

4. LEAST-SQUARES MODEL-BASED HALFTONING

The least-squares model-based (LSMB) halftoning approach minimizes the squared error between the output of the cascade of the printer and visual models in response to the halftone image and the output of the visual model in response to the original gray-scale image [3, 4]. The color LSMB halftoning algorithm is shown in Fig. 6. The original continuous-tone color image is denoted by $[\mathbf{x}_{i,j}]$. We assume that the image has been sampled so there is one pixel per dot to be generated. We assume a printer model with the form of Eq. (2), and an eye model which has the form of an FIR filter with impulse response $[g_{i,j}]$. We seek the halftone image $[\mathbf{b}_{i,j}]$ that minimizes the squared error

$$E = \sum_{i,j} ||\mathbf{z}_{i,j} - \mathbf{w}_{i,j}||^2 \qquad (15)$$

where

$$\mathbf{z}_{i,j} = \mathbf{x}_{i,j} * g'_{i,j} \quad \text{and} \quad \mathbf{w}_{i,j} = \mathbf{p}_{i,j} * g_{i,j} \qquad (16)$$

and $*$ indicates convolution. The boundary conditions assume that no ink is placed outside the image borders. Note that we have allowed different impulse responses $g_{i,j}$, $g'_{i,j}$ for the eye filters corresponding to the continuous-tone and halftone images. In fact, we found in [3] that when we do not filter the continuous-tone image, the resulting halftone images are sharper. Thus, we will drop the filter $g'_{i,j}$ completely for the remainder of this paper.

We used a simple eye model that consists of a two-dimensional finite impulse response (FIR) filter. We used the same impulse response that was used in [3, 5]. It is based on estimates of the spatial frequency sensitivity of the eye, often called the modulation transfer function (MTF). It was obtained as a separable combination of one-dimensional approximations to the eye MTF and corresponds to a viewing distance of 30 inches at 300 dpi, or 22.5 inches at 400 dpi. The impulse and frequency responses of the filter can be found in [3, 5]. We used the same filter for all the color components. No attempt was made to here to exploit the different sensitivity of the human visual system to the luminance and the chromatic dimensions, as suggested in [9].

When the printer model is *separable* as in Eq. (4), the least-squares problem can be solved independently for each color component. When the printer model is *non-separable*, the least-squares problem is also not separable. In the non-separable case, the selection of the black component $b_{i,j}^K$ is

not constrained, and thus we can use it to produce an even richer variety of colors (the black ink produces a different color than the combination of the other three or all four inks).

The least-squares solution is obtained by iterative techniques [3, 5]. Such techniques assume that an initial estimate of the halftone image $[b_{i,j}]$ is given. This could be a trivial image, e.g. all white or all black, or the output of any halftoning algorithm like the modified error diffusion of the previous section. The iterative technique we consider in this paper, updates the value of one pixel at a time, in the following way. Given an initial estimate of $[b_{i,j}]$, for every image site (i, j) (in some fixed or random order, usually a raster scan), we find the value $\mathbf{b}_{i,j}$ that minimizes the squared error

$$E_{i,j} = ||\mathbf{z}_{i,j} - \mathbf{w}_{i,j}||^2 \qquad (17)$$

The error computation involves only pixels in the neighborhood of (i, j) since the value of each pixel affects only the model outputs $\mathbf{w}_{k,l}$ in its neighborhood. An iteration is complete when the minimization is performed once at each image site. A few iterations are required for convergence, typically 5-10 when the starting point is an all black or all white image. When the starting point is the modified error diffusion result, then even fewer iterations are required. The images produced using the above method are only local minima of the least-squares problem. There could in fact be several local minima for a given image, depending on the starting point, as we showed in [3].

We tested the LSMB algorithm on several images using both separable and non-separable printer models. The separable printer models are more robust to errors in color-plane registration, and require significantly less computation because the minimization is done independently for each color component. The non-separable models produce images with better colors than those corresponding to the separable case. Note that, in contrast to the error diffusion case, there are no stability problems. For both printer models, the LSMB algorithm produces images that are sharper than those produced by the modified error diffusion algorithm. When the multi-pass [6] modified error diffusion is used as the starting point, it preserves the texture of error diffusion in the smooth areas of the image. Unfortunately though, as we saw in the previous section, the non-separable modified error diffusion algorithm may diverge. In such cases the LSMB algorithm may converge to a local optimum that does not possess the visually pleasant texture of error diffusion.

REFERENCES

[1] R. Ulichney, *Digital Halftoning.* The MIT Press, 1987.

[2] T.N. Pappas, D.L. Neuhoff, "Model-Based Halftoning," *SPIE,* San Jose, CA, Feb. 1991.

[3] T.N. Pappas, D.L. Neuhoff, "Least-Squares Model-Based Halftoning," *SPIE,* San Jose, CA, Feb. 1992.

[4] D.L. Neuhoff, T.N. Pappas, and N. Seshadri, "One-Dimensional LSMB Halftoning," *ICASSP-92.*

[5] T. N. Pappas, "Model-Based Halftoning of Color Images," *IS&T's 8th Int. Cong. Adv. Non-Impact Printing Techn.,* Williamsburg, VA, Oct. 25-30, 1992.

[6] Chen-Koung Dong, "Perceptual Printing of Gray Scale Images," M.S. Thesis, MIT, May 1992.

[7] R.W. Floyd, L. Steinberg, "An Adaptive Algrthm for Spatial Grey Scale," *Pr. SID,* vol. 17/2, p. 75–77, 1976.

[8] R.W.G. Hunt, *The Reproduction of Colour in Photography, Printing & Television.* Fountain Press, 1987.

[9] J.B. Mulligan, A.J. Ahumada, Jr., "Principled Methods for Color Dithering Based on Models of the Human Visual System," *SID'92 Dig. of Tech. Pap.,* p. 194–197.

Chapter 12: Acoustic and Biomedical Applications

Acoustic and Biomedical Applications

The field of acoustics is where many of the techniques of signal processing began, and the increasing use of signal analysis in biomedicine points the way to the future growth in signal processing applications. It is fitting to end *Signal Processing: Technology and Applications* with a chapter on these two areas.

Architectural acoustics requires measuring and modeling the impulse response of a room. In "Parametric Approximation of Room Impulse Responses by Multirate Systems" (paper 12.1), subband analysis and wavelets are used to approximate the long time impulse response and perform real-time simulation. Acoustic beamforming, used in phase-based steerable microphone arrays, suffers from the fact that the width of the beam depends upon the frequency of the signal. This presents difficulties for broadband applications such as the microphone and loudspeaker arrays for teleconferencing. "Constant Bandwidth Beamforming" (12.2) describes several methods of preserving a constant-bandwidth beam over a broad frequency range. "Acoustic Feedback Cancellation in Hearing Aids" (12.3) develops a maximum phase scheme that cancels the open-loop phase delay within the primary audio frequency region and simultaneously suppresses the new open-loop transfer function outside the primary frequency region.

A means of locating electrically-active muscle tissue following myocardial infarction is described in "A Filtering Approach to Electrocardiography Volume Conductor Inverse Source Problems" (12.4). The paper addresses the reconstruction of the electrical potential on an active layer of cells using two-dimensional filtering, the location estimation of a single active layer by maximum likelihood techniques, and identifying locations of still-active layers via a multiple-hypothesis generalized likelihood ratio test.

Image compression in the biomedical domain has been required to be fully reversible and lossless, since the effects of compression artifacts from lossy compression techniques upon diagnostic interpretation is unknown. "Reversible Compression of Medical Images Using Decomposition and Correlation Methods" (12.5) splits an image into most-, intermediate-, and least-significant bit images to achieve lossless compression ratios ranging from 3.2:1 to 4:1 for magnetic resonance, computerized tomography, and chest X-ray images.

Reconstruction of biomedical images is explored in "A Baysian Segmentation Approach to 3-D Tomographics Reconstruction from Few Radiographs" (12.6). In this paper, the development of a three-dimensional model from a minimal number of noisy two-dimensional slices is discussed using the maximum *a posteriori* probability basis. It provides a method for segmenting a three-dimensional object from digitized radiographs. Finally, "An Approach Toward Automatic Diagnosis of Breast Cancer from Mammography" (12.7) approaches one of the most difficult tasks in radiograph interpretation. The method begins with high-intensity regions of the radiograph as potential tumor sites, then applies scale-space filtering to trace edge elements through successive views, from low resolution to high. Finally, Delaunay arcs and Voronoi boundaries are determined to connect the edge elements into closed contours that bound tumors. An example is given that shows the results of tumor identification.

Suggested Additional Readings:

[1] S. J. Elliot and P. A. Nelson, "Active Noise Control," *IEEE Signal Processing Magazine,* vol. 10, no. 4, pp. 12-35, October 1993.

[2] B. X. Li and S. Haykin, "Chaotic Detection of Small Target in Sea Clutter," *Proc. IEEE International Conference on Acoustics, Speech, and Signal Processing 93* (Minneapolis, MN, USA) April 1993, vol. 1 pp. 237-240.

[3] P. Chu and J. M. Mendel, "First Break Refraction Event Picking Using Fuzzy Logic Systems," *Second IEEE International Conference on Fuzzy Systems* (San Francisco, CA, USA) Mar. - Apr. 1993, pp. 889-894.

[4] D. M. Honea and S. D. Stearns, "Lossless Waveform Compression: A Case Study," *The Twenty-Seventh Asilomar Conference on Signals, Systems, and Computers* (Pacific Grove, CA, USA) Nov. 1994, pp. 1514-1518.

[5]   P. C. Cosman, R. M. Gray, and R. A. Olshen, "Evaluating Quality of Compressed Medical Images: SNR, Subjective Rating, and Diagnostic Accuracy," *Proc. IEEE,* vol. 82, no. 6, pp. 919-932, June 1994.

[6]   L. Marple, T. Brotherton, R. Barton, K. Lugo, and D. Jones, "Travels Through the Time-Frequency Zone: Advanced Doppler Ultrasound Processing Techniques," *The Twenty-Seventh Asilomar Conference on Signals, Systems, and Computers* (Pacific Grove, CA, USA) Nov. 1994, pp. 1469-1473.

[7]   M. Sun, M. S. Scher, R. E. Dahl, N. D. Ryan, S. Iyengar, B. Kosanovic, and R. J. Sclabassi, "Analysis of Aliasing and Quantization Problems in EEG Data Quantization," *Proc. Twelfth Southern Biomedical Engineering Conference* (New Orleans, LA, USA) Apr. 1993, pp. 280-282.

[8]   T. G. Xydis and A. E. Yagle, "Performance Bound on Depth Estimation of Bioelectrical Source," *Proc. IEEE International Conference on Acoustics, Speech, and Signal Processing 93* (Minneapolis, MN, USA) April 1993, vol. 4 pp. 396-399.

PARAMETRIC APPROXIMATION OF ROOM IMPULSE RESPONSES BY MULTIRATE SYSTEMS

M. Schoenle, N. Fliege , U. Zoelzer

Telecommunications Group, Technical University of Hamburg
Eissendorfer Str. 40, D-2100 Hamburg 90, Germany

ABSTRACT

In this paper, we propose a new approach to the approximation and real-time simulation of room impulse responses. A time-frequency representation of impulse response data is obtained by wavelet decomposition using a multirate analysis filter bank. In a second step the parameters of cascaded moving average comb filter structures in the frequency subbands are calculated. Combining the output signals of the subband models by a synthesis filter bank with perfect reconstruction properties gives an approximation of the broadband impulse reponse. Due to the multirate signal decomposition with a high frequency resolution at low frequencies very long room impulse responses can be modeled accurately and simulated in real-time.

1 INTRODUCTION

The subject of simulating impulse responses of real rooms by means of digital reverberation systems has been addressed by several authors during the last 30 years [1-4]. Until now it is impossible to perform the real-time convolution of a long impulse response and some input signal by state of the art signal processors. Therefore many structures for approximation have been proposed. It is common practice to simulate the so-called early reflections (Fig. 1) by FIR filters and the exponentially decaying reverberant part by recursive structures like comb filters and allpass filters. All approaches involve the problem of finding appropriate model parameters and most of them are describing the properties of real and approximated impulse responses merely from a time-domain view.

It is known that each room impulse response has different decay properties at different frequencies. Therefore a decomposition of the impulse response data, which gives insight into the time-frequency behaviour of the system within its main subbands, is essential. This can be performed by tree-structured analysis filter banks offering a logarithmic spacing of the frequency range and decimated signals in the subbands. If there exists a synthesis filter bank with perfect reconstruction properties the subband signals can be interpreted as coefficients of the discrete-time wavelet transform (DTWT) of the input signal [5]. This decomposition gives a high frequency resolution and a corresponding low time resolution for low frequencies and vice versa

for high frequencies. There is a close relation between this analysis method and the frequency analysis performed by the cochlea of the human ear [6]. Some analysis results will be presented in section 2. Section 3 is dealing with the approximation of the subband signals by parametric models. Based on the wavelet decomposition the models in the low-frequency subbands can be designed accurately. According to the reduced sampling rates they can still be processed in real-time. Since stationary systems are considered, the parameters of the subband models are not adjusted using an adaptive algorithm like the LMS. Due to the arising large data arrays another approach, which consists of two steps, is used: First subband reference impulse responses are calculated by cross-correlation, which can be performed by effective FFT algorithms. The Prony-Method is then applied to obtain the parameters of a moving average comb filter structure. A synthesis filter bank, that guarantees perfect reconstruction, combines the model output signals and gives an approximation of the broadband impulse response.

2 MULTIRATE SIGNAL DECOMPOSITION

Fig. 1 shows the measured impulse response signal of a concert hall. From this, only the arrival time of the first reflections and the global decay characteristic can be determined, whereas a decomposition into frequency subbands gives insight into the decay properties of the subband signals. A transformation ideally adapted to this purpose is the discrete-time wavelet transformation (DTWT). The DTWT provides a splitting of the frequency range into octaves, resulting in a dense pattern of subbands at low frequencies. This conforms with a high frequency resolution and, according to the reduced bandwidth, allows the re-

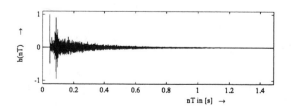

Figure 1: Measured Impulse Response

0-7803-0946-4/93 $3.00 © 1993 IEEE

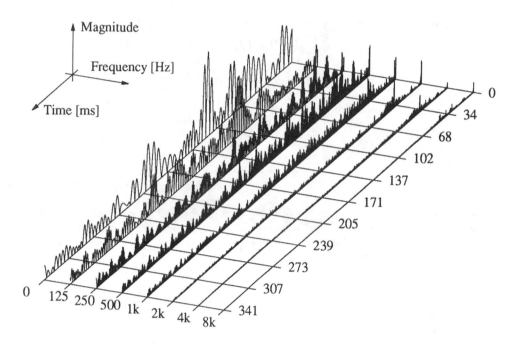

Figure 2: Wavelet Decomposition

duction of the sampling rate in the subbands. These properties are essential, because systems like real rooms contain most of their energy at low frequencies. Especially for large rooms the impulse response consists mainly of the reverberant part, in which the high frequencies usually have decayed.

Fig. 2 shows the DTWT of the impulse response in Fig. 1. The decomposition has been realized by a tree-structured multi-complementary filter bank, which is non-critically subsampled and therefore avoids aliasing [8]. For a smooth graphical representation the wavelet coefficients in the subbands have been interpolated to the original sampling frequency of 48 kHz and only the magnitude of the interpolated signals is shown. On the first glance, the different decay characteristics in the frequency subbands can be recognized. Furthermore it can be seen, that the first reflections of the room impulse response are distributed over the whole frequency range, while the reverberant part is dominated by low frequencies.

3 SUBBAND APPROXIMATION

The structure used for approximation is known from adaptive filtering in acoustic systems [7] and is depicted in Fig. 3. The input sequence $x(n)$ to the system with the transfer function $H(z)$ and the output sequence $y(n)$ are both decomposed by an analysis filter bank giving the subband signals $x_1(n_1) \ldots x_p(n_p)$ and $y_1(n_1) \ldots y_p(n_p)$. The signals $y_1(n_1) \ldots y_p(n_p)$ are approximated by parametric models $H_i(z) = B_i(z)/A_i(z)$, producing the output signals $\hat{y}_1(n_1) \ldots \hat{y}_p(n_p)$. Combining these signals by a synthesis

filter bank with perfect reconstruction properties gives an approximation $\hat{y}(n)$ of the broadband signal $y(n)$. If the input sequence is an unit pulse sequence $x(n) = \delta(n)$, a decomposition of the broadband impulse response signal $y(n) = h(n)$ into octaves is obtained.

A variant of the Prony-Method [9] is used to obtain the parameters of the subband models from a subband reference impulse response $h_{ref,i}(n_i)$.

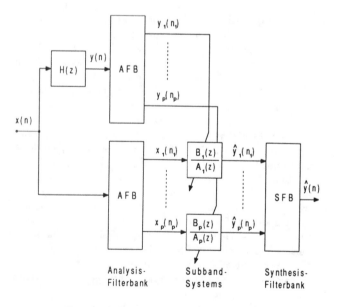

Figure 3: Structure for Approximation

As we are dealing with stationary systems, the coefficients $h_{ref,i}(n_i)$ need not be adjusted by an adaptive algorithm like the LMS. The optimal vector $h_i^* = [h_{ref,i}(0), \ldots, h_{ref,i}(K_i)]^T$ in a mean-square sense is obtained by solving the Wiener-Hopf equation [10]:

$$h_i^* = R_{x_i x_i}^{-1} \, r_{x_i y_i} \,, \qquad (1)$$

where $r_{x_i y_i}$ is the cross-correlation vector of the subband signals $x_i(n_i)$ and $y_i(n_i)$ and $R_{x_i x_i}$ is the autocorrelation matrix of the $x_i(n_i)$. This approach gives good results under the assumption that aliasing effects produced by the analysis filter banks are kept small or are even avoided as in noncritically subsampled filter bank systems. While the cross-correlation can be computed be effective FFT routines, the inversion of the autocorrelation matrix $R_{x_i x_i}$ is difficult, due to the large data arrays. If the input signal $x(n)$ is a pseudorandom sequence, the signals $x_i(n_i)$ may be considered to be approximately white within the subbands. The multiplation with the inverse of the autocorrelation matrix can thus be avoided:

$$h_i \approx r_{x_i y_i} \,. \qquad (2)$$

Since the subband signals $x_i(n_i)$ and $y_i(n_i)$ are the outputs of corresponding branches of the analysis filter banks in Fig. 3, the same resulting branch transfer function has been applied to them: $X_i(z^{k_i}) = A_i(z, .., z^{k_i}) X(z)$ and $Y_i(z^{k_i}) = A_i(z, .., z^{k_i}) Y(z)$. The k_i are powers of two and express the decimation of the signals in the subbands (in relation to the original sampling rate). It can be seen that, while calculating the cross-correlation, these branch transfer functions are multiplied, resulting in a cascaded transfer function $\tilde{A}_i(z^2, .., z^{2k_i}) = A_i(z, .., z^{k_i}) A_i(z, .., z^{k_i})$ with a steeper transition slope. The reduced transition bandwidth is matched by the parametric models. After synthesis this leads to spectral gaps in the transfer function of the estimated impulse response. Further investigations, especially on the acoustical consequences of these spectral gaps, still need to be done.

As the reverberant part of a room impulse response behaves like white noise, damped by an exponentially decaying envelope, this behaviour can be approximated by a comb filter having uniformly distributed poles on a circle concentric to the unit circle. The transfer function of the comb filter is given by $H_{C,i}(z) = 1/(1 - g_i z^{-N_i})$, where N_i is the order of the comb filter. The first reflections are approximated by the moving average (MA) transfer function $H_{M,i}(z) = b_0 + b_1 z^{-1} + \ldots + b_{M_i} z^{-M_i}$. The overall transfer function $H_i(z) = H_{C,i}(z) H_{M,i}(z)$ is set equal to the Z-transform of $h_{ref,i}(n_i)$:

$$H_i(z) = \frac{b_0 + \ldots + b_{M_i} z^{-M_i}}{1 - g_i z^{-N_i}} = \sum_{n_i=0}^{\infty} h_{ref,i}(n_i) z^{-n_i} \qquad (3)$$

Truncating $h_{ref,i}(n_i)$ after K_i samples, multiplying both sides of (3) with the denominator $1 - g_i z^{-N_i}$ and comparing the coefficients of the powers of z gives the matrix equation

$$b = Ha \qquad (4)$$

with

$$b = \begin{bmatrix} b_0 & b_1 & \ldots & b_M & 0 & \ldots & 0 \end{bmatrix}^T$$

$$a = \begin{bmatrix} 1 & 0 & \ldots & 0 & 0 & \ldots & -g \end{bmatrix}^T$$

and

$$H = \begin{bmatrix} h_0 & 0 & 0 & \ldots & \ldots & 0 \\ h_1 & h_0 & 0 & \ldots & \ldots & 0 \\ h_2 & h_1 & h_0 & \ldots & \ldots & 0 \\ \vdots & \vdots & \vdots & & & \vdots \\ h_M & h_{M-1} & h_{M-2} & \ldots & \ldots & h_{M-N} \\ \hline h_{M+1} & h_M & h_{M-1} & \ldots & \ldots & h_{M-N+1} \\ \vdots & \vdots & \vdots & & & \vdots \\ h_K & h_{K-1} & h_{K-2} & \ldots & \ldots & h_{K-N} \end{bmatrix}$$

For means of simplicity the coefficients $h_{ref,i}(n_i)$ have been denoted by h_n and the subband index has been omitted. From (4) it can be seen, that the equations for the numerator and the denominator coefficients can be separated. First the comb filter feedback coefficient g is determined and the solution is then used in the first M equations of (4) to obtain the coefficients b_k of the MA filter, which are equal to the coefficients h_k for $k = 0 \ldots M - 1$ and $b_M = h_M - g h_0$ for $M = N$. As a modification of the Prony-Method the comb filter coefficient g is not determined from the last $K - M$ equations of (4) by a least-squares approach. Because of many zero coefficients in the denominator polynomial the data used to solve the normal equations is reduced to only two vectors. This gives poor results for the estimation of g. A better way is to determine g from the exponentially decaying envelope $h_e(l)$ of the subband impulse response, which is obtained by averaging in time direction. Calculating the natural logarithm of $h_e(l)$ gives a straight line. The impulse response of a comb filter with the transfer function $H_C(z) = 1/(1 - g z^{-N})$ is given by $h_c(l = Nn) = g^l$, which provides exponentially decaying values. So the coefficient g can be determined from the slope of the straight line:

$$\log g = \frac{\log h_e(l_1) - \log h_e(l_2)}{l_1 - l_2}, \qquad l_1 < l_2 \qquad (5)$$

This approach avoids the shortcomings of pure comb filter based reverberant structures. The output signal of the comb filter, which is nonzero only every N_{th} sample, due to the inherent delay of this structure, is convolved with the impulse response of the MA filter to obtain a dense pattern of reflections. If $M = N$ the maximum density is achieved. Furthermore the equally spaced resonance peaks of the comb filter in the frequency domain, which have the same magnitude and lead to unnatural sounds, are compensated by the zeros of the MA filter. From a time domain view these zeros provide the first reflections.

4 SIMULATION RESULTS

The simulation results have been obtained using measured impulse response data $h(n)$, representing the system

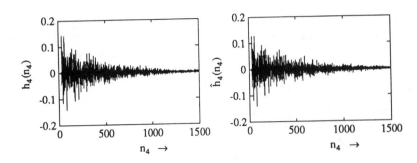

Figure 4: Subband 4 - Impulse Response Approximation

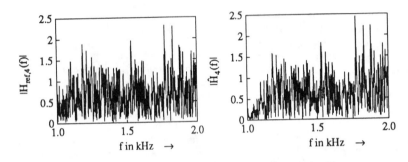

Figure 5: Subband 4 - Transfer Function Approximation

with transfer function $H(z)$ in Fig. 3. For the estimation of the subband parameters the system was excited by a binary maximum-length sequence. For comparison Fig. 4 shows the first 1500 samples of a subband impulse response $h_4(n_4)$ in the subband from 1 kHz to 2 kHz, as it is obtained by the DTWT of $h(n)$, and the estimated one $\hat{h}_4(n_4)$. The number K of approximated samples was 2048, the order of the MA comb filter model was $M = N = 512$. According to the used method the first M samples are matched nearly exactly, while for the remaining samples only the exponentially decaying envelope is approximated. Fig. 5 shows the magnitude of the transfer function in the corresponding subband. The attenuation effect at the band edges is produced by cascaded analysis filter transfer functions, as it was explained in section 3.

5 CONCLUSION

We have proposed a new approach to the approximation and real-time simulation of long room impulse responses. Based on a wavelet analysis using logarithmic multirate filter banks parametric models in the frequency subbands are calculated. In this way the time-frequency behaviour of real rooms can be approximated realistically. Finally the multirate technique allows an efficient implementation on existing signal processors.

REFERENCES

[1] M.R. Schroeder, "Natural Sounding Artificial Reverberation", *J. Audio Eng. Soc.*, Vol. 10, 1962, pp. 219-223.

[2] J.A. Moorer, "About This Reverberation Business", *Comp.Music J.*, Vol. 3/2, 1978, pp.13-28.

[3] J. Stautner, M. Puckette, "Designing Multichannel Reverberators", *Comp. Mus. J.*, Vol. 6/1, 1982, pp.52-62.

[4] J. Jot, "Digital Delay Networks For Designing Artificial Reverberators", *Proc. 90th AES Convention 1991, Paris.*

[5] P. P. Vaidyanathan, *Multirate Systems and Filter Banks*, Prentice Hall, 1993.

[6] E. Zwicker, *Psychoacoustics*, Springer, 1982.

[7] A. Gilloire, M. Vetterli, "Adaptive Filtering in Subbands" *Proc. ICASSP 1988, New York*, pp. 1572-1575.

[8] N. J. Fliege, U. Zoelzer, "Multi-Complementary Filter Bank: A new Concept with Aliasing-Free Subband Signal Processing and Perfect Reconstruction", *Proc. EUSIPCO-92, Brussels*, pp. 207-210.

[9] J.H. McClellan, "Parametric Signal Modeling" in Lim, J.S.; Oppenheim, A.V. (editors), *Advanced Topics in Signal Processing*, Prentice Hall, 1988.

[10] B. Widrow, S. D. Stearns, *Adaptive Signal Processing*, Prentice Hall, 1985.

CONSTANT BEAMWIDTH BEAMFORMING

Michael M. Goodwin and Gary W. Elko

Acoustics Research Department
AT&T Bell Laboratories
600 Mountain Avenue P.O. Box 636
Murray Hill, NJ 07974-0636

ABSTRACT

The beamwidth of a linear array decreases as frequency increases. For wideband beamformers such as microphone arrays intended for teleconferencing, this frequency dependence implies that signals incident on the outer portions of the main beam are subject to the undesirable effects of lowpass filtering. In this paper we discuss several ways of attaining beamwidth constancy and present a novel method based on superimposing several marginally steered beams to form a constant beamwidth multi-beam. This method provides an analytically tractable framework for designing constant beamwidth beamformers.

1. INTRODUCTION

The beamwidth of a linear array decreases as frequency increases. In narrowband beamformers, this inverse proportionality of beamwidth with respect to frequency is generally insignificant. On the other hand, in applications such as microphone and loudspeaker arrays for audio teleconferencing, where broadband operation is desirable and in fact necessary, this beamwidth variation proves detrimental in that signals incident on the outer portions of the main beam are subject to lowpass filtering.

In this paper we discuss several methods of constant beamwidth beamforming and propose a novel approach in which a constant beamwidth *multi-beam* is synthesized by superimposing several marginally steered single beams. We show that this technique provides an analytical framework for a previously proposed method which has been neglected due to computational difficulties. First, however, we present relevant results from the theory of beamforming.

2. BEAMFORMING THEORY

For a weighted linear array of $2N + 1$ equally spaced omnidirectional elements, the far field response for a time-harmonic plane wave inclined at an angle θ with respect to broadside is given by

$$H(k, \theta) = \sum_{n=-N}^{N} a_n e^{-jn\omega\tau_0 \sin\theta} \qquad (1)$$

where the a_n are the tap weights and $\tau_0 = d/c$ is the interelement spacing divided by the speed of sound. The mapping $z = \exp(j\omega\tau_0 \sin\theta)$ indicates that a linear array is analogous to a discrete-time FIR filter, implying that standard FIR filter synthesis techniques can be readily applied to array design. For instance, an array can be weighted with Chebyshev coefficients to establish equi-level sidelobes.

The inter-null beamwidth of a uniformly excited array is given by

$$\theta_{BW} = 2\sin^{-1}\left(\frac{2\pi}{M\omega\tau_0}\right) \approx \frac{4\pi}{M\omega\tau_0} \qquad (2)$$

where $M = 2N + 1$. This expression clearly indicates the inverse proportionality of beamwidth and frequency; it also implies that an increase in either the number of elements or the interelement spacing results in a decrease in the beamwidth as well. If the array is shaded with Chebyshev weights to keep the sidelobes below a specified threshold, the beamwidth is given by

$$\theta_{BW_{cheb}} = 2\sin^{-1}\left\{\frac{2}{\omega\tau_0}\cos^{-1}\left(\frac{1}{x_0}\cos\frac{\pi}{4N}\right)\right\} \qquad (3)$$

where x_0 determines the sidelobe threshold [1]. For increased x_0, corresponding to a lower sidelobe level, the array beamwidth is likewise greater. This expression exhibits the same functional dependencies on frequency, spacing, and number of elements as the expression for a uniform array.

Arrays intended for use in dynamic environments are typically capable of beam steering. If progressive delays $n\tau$ are imposed on the array elements, the main beam rotates to an angle $\phi = \sin^{-1}(\tau/\tau_0)$ from broadside. Then, the array response is:

$$A(\omega, \theta) = \sum_{n=-N}^{N} a_n e^{-jn\omega\tau_0(\sin\theta - \sin\phi)} \qquad (4)$$

For a uniformly excited array, the inter-null beamwidth of such a steered beam is given by

$$\theta_{BW} = \theta_{null1} - \theta_{null2} \qquad (5)$$

where the terms

$$\theta_{null1} = \sin^{-1}\left(\frac{2\pi}{M\omega\tau_0} + \sin\phi\right) \qquad (6)$$

$$\theta_{null2} = \sin^{-1}\left(\frac{-2\pi}{M\omega\tau_0} + \sin\phi\right) \qquad (7)$$

represent the locations of the nulls flanking the main beam [2].

An analysis of Eq. 4 indicates that peaks occur in the angular response where $\tau_0(\sin\theta - \sin\phi) = 2\pi m/\omega$ for integer values of m. The nominal case ($m = 0$) corresponds to the main beam at $\theta = \phi$; other values of m correspond to extraneous peaks known as *grating lobes*. This *spatial aliasing*, analogous to frequency-domain aliasing in a sampled data system, does not occur if the interelement spacing obeys the inequality

$$d < \frac{\lambda}{1 + |\sin\phi|} \qquad (8)$$

which is essentially a Nyquist criterion for the spatial sampling rate [1, 2]. For the extreme case of endfire beamforming, this upper bound corresponds to half-wavelength spacing. This can be expressed in terms of frequency as the requirement $f < 1/2\tau_0$. A lower bound on the spacing is enforced by the physical size of the array elements.

3. CONSTANT BEAMWIDTH BEAMFORMING

3.1. Harmonic Nesting

A broadband constant beamwidth can be achieved by combining the outputs of two colinear geometrically scaled arrays. This is referred to as the SHA technique, named after Smith [3] and Hixson and Au [4]. Specifically, the output of an array with the desired beamwidth at $\omega = \omega_0$ is combined in a frequency-dependent fashion with the output of an array C times smaller so as to maintain a constant beamwidth over a frequency range extending from ω_0 to $C\omega_0$. Given the anti-aliasing constraint, the SHA technique with $C = 2$ conforms to a *harmonic nesting* approach, wherein progressively higher frequency octaves are processed by

progressively smaller arrays [2]. The scale factor of two permits an efficient nesting of the various single-octave *subarrays*; some elements are shared between multiple subarrays. Smaller (*i.e.* non-integral) scale factors are physically impractical; thus, if aliasing is to be avoided, the SHA technique simply reduces the extent of beamwidth variation to that which occurs within a single octave. We therefore consider other methods with the insight that the beamwidth need only be constant over one octave.

3.2. Elemental Lowpass Filtering

A constant beamwidth can be attained by placing appropriate lowpass filters in the signal paths of the respective array elements. The cutoff frequency of such a filter depends on the position of its corresponding element with respect to the array center; the farther the element is from the center, the lower the cutoff of its filter. In this approach, then, the array appears long at low frequencies since all of the elements are used; for increasing frequency, the array shortens in length since the signals corresponding to the outer elements are progressively attenuated. The design of the elemental lowpass filters has been neglected in the literature due to the apparent computational difficulty. In a later section we show that these filters can be accurately approximated by reasonably simple expressions.

3.3. Multi-Beamforming

In the multi-beamforming approach proposed by Tucker [5], a broadside beam and several marginally steered beams are simultaneously formed; the steered beams are formed symmetrically with respect to broadside at fixed steering angles. The inter-null beamwidth of the resultant *multi-beam* is roughly the angular distance between the outermost nulls of the outermost beams; this an approximation due to the effects of interfering sidelobes from the constituent beams. Since the outermost beams narrow as frequency increases in accordance with Eq. 5, the beamwidth of the multi-beam likewise decreases by roughly the same amount as that of a single beam. Thus, this technique does not exhibit beamwidth constancy; its improvement over the single-beam case is simply a lower percentage decrease of beamwidth with frequency. In addition to this marginal improvement, the beamwidth of the multi-beam is too large for highly directional applications. We therefore propose an alternative method.

4. A NOVEL MULTI-BEAMFORMER

The aforementioned similarity between the far field array response (Eqs. 1,4) and the Z-transform suggests that a beamformer can be modeled as a tapped delay line where the delays determine the beam steering angle. In a standard delay line, these delays are constant as a function of frequency, resulting in a frequency-independent steering angle. If instead the beamformer is modeled as a tapped allpass filter line, the beam-steering delays incurred by the array signals depend on frequency; they are equivalent to the phase delays of the constituent allpass filters. If these allpass filters have linear phase with zero phase at DC, the delays are constant, implying that the fixed delay line is a specific case of this revised beamformer model.

Constant beamwidth can be attained by incorporating the capability of frequency-dependent beam steering into the multi-beamforming approach discussed in Section 3.3. In that scheme, the multi-beam beamwidth decreases as frequency increases due to the narrowing of the outermost beams. If the steering angles are made frequency-dependent as in the revised beamformer model, that narrowing can be counteracted by steering the outermost beams to larger angles as frequency increases. In this way the angular position of the outermost nulls can be held constant, resulting in a constant beamwidth.

In the revised beamformer, a constant beamwidth can be achieved over an extended frequency range. At the low end of the frequency band, $2r + 1$ beams are formed simultaneously at broadside. As frequency increases, the outermost beams are steered so as to keep the outermost nulls stationary. The inner beams are steered proportionally to maintain the integrity of the beam shape. For excessively high frequencies, the constituent beams begin to visibly separate, resulting in dips in the angular magnitude of the multi-beam; for an octave band, however, this beam separation is not visibly manifest.

For a uniformly excited array, the approximate inter-null beamwidth of the multi-beam is twice the angle at which the outermost null is located. Using Eq. 6 and expressing the frequency-dependent steering angle of the outermost beam as $\phi_r = \sin^{-1}(\tau(\omega)/\tau_0)$ where $\tau(\omega)$ denotes the phase delay of the allpass filters,

$$\theta_{BW} = 2\sin^{-1}\left(\frac{2\pi}{M\omega\tau_0} + \frac{\tau(\omega)}{\tau_0}\right) \qquad (9)$$

If the phase delay $\tau(\omega)$ is chosen to keep the arcsine argument constant as a function of frequency, a constant beamwidth results. The appropriate phase delay corresponds to the phase function

$$f(\omega) = \frac{2\pi}{M}\left(1 - \frac{\omega}{\omega_0}\right) \qquad (10)$$

where ω_0 denotes the low frequency limit, namely the frequency at which all of the constituent beams are formed at broadside. A similar derivation for a Chebyshev-shaded array yields the phase function

$$f(\omega) = 2\left[\cos^{-1}\left(\frac{1}{x_0}\cos\frac{\pi}{4N}\right)\right]\left(1 - \frac{\omega}{\omega_0}\right) \qquad (11)$$

These phase functions are both *affine*, meaning that they are linear in ω but have a DC offset. Figure 1 shows the affine phase for the outermost multi-beam of an 11-element -25 dB Chebyshev shaded array for one-half of the array elements.

Since $2r+1$ beams are being formed simultaneously, the far field response of the multi-beamformer is

$$A_m(\omega,\theta) = \sum_{i=-r}^{r}\sum_{n=-N}^{N} a_n e^{-jn\omega\tau_0\sin\theta}e^{jnf(\omega)\frac{i}{r}} \qquad (12)$$

where $f(\omega)$ denotes the proper phase function to achieve constant beamwidth given the taps a_n. Interchanging the order of summation and invoking the symmetry of the multi-beam yields

$$A_m(\omega,\theta) = \sum_{n=-N}^{N} B_n(\omega)e^{-jn\omega\tau_0\sin\theta} \qquad (13)$$

where

$$B_n(\omega) = a_n\left[1 + \sum_{i=1}^{r} 2\cos\left(nf(\omega)\frac{i}{r}\right)\right] \qquad (14)$$

The zero-phase lowpass filters $B_n(\omega)$ are the appropriate elemental lowpass filters discussed in Section 3.2. Figure 2 shows the resulting lowpass filters for the same configuration of Figure 1. Of course, the filters need only take this form in the octave passband; it is assumed that the signal is bandpass-filtered elsewhere.

If the $B_n(\omega)$ derived above are used as frequency-dependent shading weights for an equi-spaced linear array, a nearly constant inter-null beamwidth results. The beamwidth deviates from constancy only when the outermost nulls of the outermost beams do not correspond closely to the nulls of the multi-beam; this is the case when high-level sidelobes from other constituent beams coincide with the angular region in which the outermost null is located. For Chebyshev shading, the equi-level sidelobes can be attenuated, thus nullifying the effect of the interfering sidelobes. Since sidelobe attenuation is accompanied by increased beamwidth, this discussion suggests that a greater nominal beamwidth results in improved beamwidth constancy as motivated briefly in Section 3.3. Figures 3 and 4 show the directional responses with the configuration used in Figures 1 and 2 for the cases of single beamforming and multi-beamforming to provide constant directivity, respectively. The constant inter-null beamwidth can clearly be seen in Figure 4.

In this scheme, the use of the elemental lowpass filters results in a lowpass broadside response, which is undesirable since it reduces the fidelity of signals corresponding to the main beam. This broadside nonuniformity can be removed by the compensation filter:

$$G(\omega) = \frac{1}{\sum_{n=-N}^{N} B_n(\omega)}.$$

5. CONCLUSIONS

In this paper we have discussed several techniques of constant beamwidth beamforming, and have proposed a novel method. We analyzed this improved method from a multi-beamforming point of view for the cases of uniform array excitation and Chebyshev shading. The symmetry of the beamformer enabled a simplification of the array response which indicated the equivalence of the novel technique to the elemental lowpass filtering method. The primary result of our analysis is a simple expression for the appropriate lowpass filters to be affixed to the respective array elements.

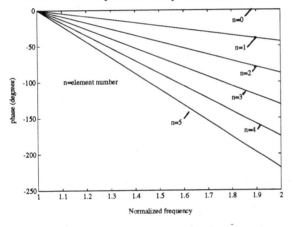

Figure 1: Phase of the first 6 elements for the outermost beam of a -25 dB Chebyshev multi-beamformer.

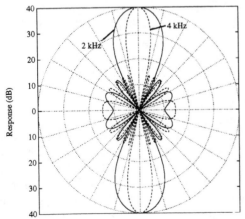

Figure 3: Single beam -25 dB shaded 11-element Chebyshev array with 4.3 cm spacing directional response at 2 kHz and 4 kHz.

6. REFERENCES

[1] M. Goodwin and G. Elko. Beam dithering. In *Proc. of the Audio Engineering Society*, Oct 1992.

[2] M. Goodwin. Implementation and applications of electroacoustic array beamformers. Master's thesis, M.I.T., May 1992.

[3] R. Smith. Constant beamwidth receiving arrays for broad band sonar systems. *Acustica*, 23:21–26, 1970.

[4] E. Hixson and K. Au. Broadband constant beamwidth acoustical arrays. Technical Report 19, Acoustics Research Lab, U.T.Austin, 1970.

[5] D. Tucker. Arrays with constant beam-width over a wide frequency range. *Nature*, 180:496–497, Sept 1957.

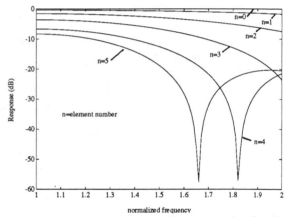

Figure 2: Elemental lowpass filters for the first 6 elements of a constant beamwidth 11-element -25 dB Chebyshev-shaded array.

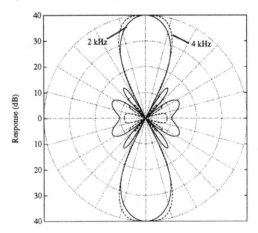

Figure 4: Multi-beam -25 dB shaded 11-element Chebyshev array with 4.3 cm spacing at 2 kHz and 4 kHz.

ACOUSTIC FEEDBACK CANCELLATION IN HEARING AIDS

Rongtai Wang and Ramesh Harjani

Department of Electrical Engineering, University of Minnesota

200 Union Street S.E., Minneapolis, MN 55455

USA

ABSTRACT

We describe a maximum phase cancellation scheme for hearing aids that prevents acoustic oscillations. In this scheme, the open-loop phase delay in the primary audio frequency region is canceled to the largest extent possible. At the same time, the magnitude response of the new open-loop transfer function outside the primary audio frequency region is suppressed by using negative feedback. The use of both these techniques increase the stability and the maximum usable gain of the overall hearing aid system. Computer simulations, using real measured data, are used to confirm our results.

1. INTRODUCTION

Hearing aid research has made tremendous strides both in miniaturization and in limiting power consumption in the last few years. However, a fundamental problem that places severe limitations on the maximum usable gain is acoustic feedback. The hearing aid can be modeled as a positive feedback system where acoustic oscillations are initiated when the open-loop gain of this feedback system is unity and the open-loop phase is a multiple of 2π radian. A number of techniques have been tried to alleviate the acoustic feedback problem by either altering the gain response or the phase response of the overall loop; these include time-delay, inverse filtering and feedback cancellation.

Figure 1 shows a signal flow graph of the hearing aid modeled as a feedback system. In this figure T_M represents the transfer function of the microphone, T_{HA} represents the transfer function of the hearing aid including the preamplifier and filters. T_R represents the transfer function for the receiver and the output power amplifier, and T_F represents the transfer function of the mechanical and acoustic feedback that initiates acoustic oscillations. In figure 1, the filled in blocks represent electrical signals while the clear blocks represent acoustic signals.

The transfer function of the overall hearing aid including acoustic feedback is given by

$$T_{system} = \frac{T_M T_{HA} T_R}{1 - T_M T_{HA} T_R T_F} \qquad (1)$$

By defining the open-loop transfer function as

$$T_{open} = T_M T_{HA} T_R T_F \qquad (2)$$

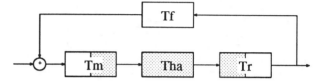

Figure 1: Control system model of a hearing-aid system

it is possible to see that when the magnitude of the open-loop transfer function is equal to unity and the phase is a multiple of 2π then the system transfer function is undefined and the hearing aid becomes unstable. The effects of acoustic feedback can be reduced by altering either the magnitude or the phase relationships of the feedback-loop of the hearing aid. Phase altering approaches include frequency shift [1], [2] where the input spectrum entering at the microphone is shifted by a few Hertz prior to being fed it to the receiver. Though successful with public address systems for years this approach has not had much success in hearing aids due to the large percentage variation of the feedback path. The phase information can also be altered by providing a time-varying delay in the signal path. This approach provides a maximum of 1-2dB of extra gain and also suffers from a audible warbling sound [3]. Other phase altering techniques [4] have not enjoyed much success either. In gain altering approaches the primary aim is to reduce the gain of the system at the frequencies where oscillations are likely to occur. The usual method that is employed to accomplish this is the use of a narrowband notch filter [5], [3] or a number of narrowband notch filters [6]. Unfortunately, even adaptive notch filtering techniques [3], [6] have provided only 3-5dB of additional usable gain and is not sufficient for high gain hearing aids. In feedback cancellation an attempt is made to cancel the entire effect of the feedback signal by providing an additional feedback path that is equal to but 180 degrees out of phase from the normal problematic feedback path. Adaptive versions of the feedback cancellation scheme has provided the maximum increase in usable gain ([7], [3], [8], [9]). Though feedback cancellation has enjoyed greater success it too is limited by some inherent problems. During normal use the acoustic feedback path changes quite dramatically and if the internal feedback path does not adapt to this change then the overall hearing aid system is likely to become unstable primarily due to the effects of the internal feedback path itself.

The idea of a phase cancellation scheme is inherently more immune to both gain and phase variations. Theoretically it is capable of providing 180 degrees of phase margin and is totally immune to gain variations if both poles and zeros of the open-loop transfer function lie in the left-half-plane. But, unfortunately, the open-loop transfer functions of real hearing aids do contain right-half-plane zeros. Because of these right-half-plane zeros complete phase cancellation is not possible. We introduce maximum phase cancellation scheme where the phase delay of the original open-loop is canceled in the primary audio frequency region (100Hz-5kHz) to the largest extent possible. At the same time the zero-phase frequencies in the resulting open-loop transfer function are moved to a region outside the primary audio frequency range. The magnitude outside the primary audio frequency range is further suppressed to be less than unity by using negative feedback. As mentioned earlier the feedback path transfer function changes during normal operation. However, for this paper, we we shall assume it to be fixed at a value close to the maximum seen in normal use. Any reduction in the value only makes the system more stable. In the future, we plan to publish techniques that are capable adapting to variations in the feedback path.

In section 2, we describe the method of maximum phase cancellation in the primary audio frequency range for a hearing aid system. In section 3, we illustrate the use of negative feedback in suppressing the magnitude response of hearing aid outside the primary audio frequency range. And finally, we provide a summary of the results discussed in this paper.

2. MAXIMUM PHASE CACELLATION (MPC)

In general, we can express the open-loop transfer function of an ordinary hearing aid system as

$$T_{open} = A(s-z_1)...(s-z_m)\frac{(s+z_{m+1})...(s+z_n)}{(s+p_1)...(s+p_p)} \quad (3)$$

where A is a constant, m is the number of left-half-plane zeros, $n-m$ is the number of right-half-plane zeros, and p is the number of left-half-plane poles. These zeros and poles can either be real or complex (complex ones appear in pairs). The value for the poles and zeros are extracted from curve fitting measured values [10]. Our studies show that there always exists some right-half-plane zeros in the transfer functions of the receiver and the acoustic feedback path. As a result, it is impossible to provide complete phase cancellation while maintaining the stability of the entire system. Nevertheless, the open-loop phase delay can still be canceled to a large extent. Theoretically, the maximum phase cancellation, under the constraint of system stability, can be reached by implementing an equalization block T_{EQ} inserted between the amplifier and the receiver as shown in figure 2. The transfer

function of T_{EQ} is equal to the inverse of the fractional part of equation (2), i.e.

$$T_{EQ} = \frac{(s+p_1)...(s+p_p)}{(s+z_{m+1})...(s+z_n)} \quad (4)$$

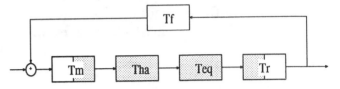

Figure 2: Control system model of a hearing-aid system with a maximum phase equalization block

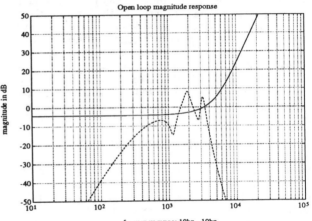

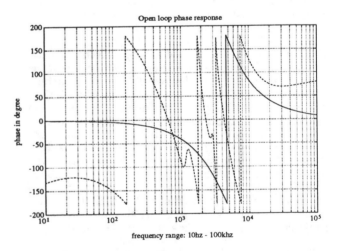

Figure 3: Open-loop transfer function

As an example, in figure 3, the response curves of the original open-loop and the open-loop with T_{EQ} inserted in between the amplifier and the receiver are plotted in dashed and solid lines respectively. From this figure, we see that the phase delay of the new open-loop with T_{EQ} inserted is much smaller than the original one.

In fact, in the original open-loop phase response curve there are three zero-phase frequencies (i.e. 0.6khz, 2.2khz, 4.2khz). However the magnitude of the open-loop transfer function exceeds unity at 2.2kHz. As a result of this, oscillations occur, and is shown in figure 6. The dashed curve in figure 6 shows the closed-loop magnitude response of the original system, where the peak located at 2.2kHz represents acoustic oscillations. Since in normal use of the hearing aid the transfer function of acoustic feedback path is fixed or varies unpredictably, the only way to avoid acoustic oscillation is to reduce the gain of the forward-path. Therefore the usable gain is limited by onset of acoustic oscillations. If the forward gain is reduced sufficiently, then no oscillations occur. An example of such a closed-loop magnitude response is plotted as a dotted curve in figure 6.

The advantage of inserting T_{EQ} into the loop is that the phase is maximally canceled, and the zero-phase frequencies become zero and infinity. No zero-phase point appears in the primary audio frequency region. Unfortunately, the magnitude of the open-loop including T_{EQ} increases rapidly with increased frequencies. Additionally, it is also desirable to reduce the gain at lower frequencies to increase intelligibility [11]. But this problem can be solved if we consider the following: first, we notice that, there are some zeros in equation (2) located at the origin. Second, there are a number of poles in the receiver transfer function that lie beyond 10kHz. A simple explanation for this is that the impedance of receivers increases rapidly with increased frequencies. These poles do not contribute much phase delay in the primary audio frequency region. Therefore, T_{EQ} can be chosen as the inverse of the partial original open-loop transfer function in which only the left-half-plane poles and zeros located in the primary frequency region are included. We call this approach as the primary audio frequency maximum phase cancellation scheme(PAFMPC). In this scheme, the magnitude response decreases at lower frequencies and does not increase as rapidly as shown in figure 2 in the high frequency region.

3. NEGATIVE FEEDBACK

Further magnitude reduction in the high frequency region can be achieved by implementing a negative feedback-loop using blocks T_{C1} and T_{C2} as shown in figure 4.

If we set the transfer function T_{C2} to be equal to $1/T_{EQ}$, the transfer function of the inner-loop can then be simplified as

$$T_{inner} = \frac{T_{EQ}}{1 + T_{C1}} \qquad (5)$$

Obviously, in order to maintain the stability of the system and suppress any acoustic oscillation, T_{C1} should be chosen to meet the following conditions:

- T_{C1} must be a polynomial in s of order m, i.e. the number of right-half-plane zeros.

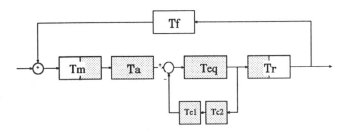

Figure 4: Control system model of a hearing-aid system with a maximum phase cancellation block and a negative feedback loop

- The coefficients of this polynomial should be chosen such that the transfer function of the resulting inner-loop $T_{EQ}/(1 + T_{C1})$ will have only left-half-plane poles.

- The zero-phase crossing points introduced by additional phase delay of $1/(1 + T_{C1})$ should lie outside the primary audio frequency region.

In our example, m is equal to 4, where one pair of right-half-plane zeros $z_{1,2} = (3.5026 \pm i1.5156)10^3$ come from the receiver transfer function, and another pair $z_{3,4} = (3.4009 \pm i4.0998)10^3$ comes from the acoustic feedback path. It is easy to show that the following simple choice of T_{C1}

$$T_{c1} = (s/3500 + 1)^4 \qquad (6)$$

will satisfy the above three conditions if the gain of forward-path is adjusted appropriately. Computer simulation results of the new open-loop transfer function with T_{EQ} and the negative feedback-loop discussed above is plotted in figure 5. From the curves in this figure, we see that the zero-phase points of the new-open loop are located below 800Hz or above 5khz. However, the magnitude of the open-loop response at these zero-phase frequencies is less than unity. Therefore acoustic oscillations will not occur.

Computer simulation results of the closed-loop transfer function for the PAFMPC scheme is plotted in solid in figure 6. In comparison to the original closed-loop response curves which are plotted in dashed and dotted lines, we see that the new closed-loop transfer function not only prevents acoustic oscillations but also improves the maximum usable gain.

4. SUMMARY

In summary, we have presented a maximum phase cancellation scheme (MPC) for a hearing aid system which can prevent acoustic oscillations and increase the usable gain. Two techniques have been introduced, one, the phase equalization block T_{EQ} which cancels the phase delay in the primary audio frequency range; and two, the blocks T_{C1} and T_{C2} that are used to reduce the magnitude of the open-loop outside the primary audio frequency region. A number of additional techniques exist to increase the maximum usable gain and are to appear in another paper in the near future [12].

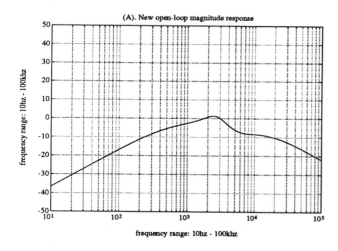

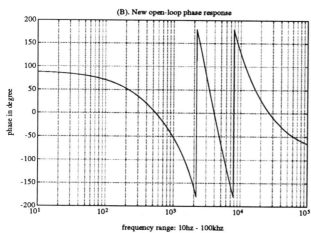

Figure 5: New open-loop frequency response

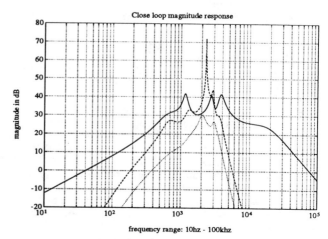

Figure 6: Closed-loop magnitude response

ACKNOWLEDGEMENT

The authors would like to thank Doctor D. A. Preves, J. Newton and B. Woodruff of Argosy Electronics, Inc. for their help with this project.

References

[1] S. M.J.Bennet and L.M.H.Browne, "A controlled feedback hearing aid," *Hearing Aid Journal*, May 1980.

[2] M. Schroeder, "Improvement of acoustic feedback stability in public address systems," in *Third International Congress on Acoustics, Stuttgart*, 1959.

[3] D.K.Bustamante, T.L.Worrail, and M.J.Williamson, "Measurement and adaptive suppression of acoustic feedback in hearing aids," in *1989 International Conference on Acoustics, Speech and Signal Processing*, 1989.

[4] D. Preves, *Evaluation of Phase Compensation for Enhancing the Signal Processing Capabilities of Hearing Aids in situ*. PhD thesis, University of Minnesota, 1985.

[5] D. Egolf, "Review of acoustic feedback literature from a control systems point of view," in *Amplification for Hearing Impaired: Research Needs*, Monographs in Contemporary Audiology, 1982.

[6] D. Egolf, "Acoustic feedback suppression in hearing aids," tech. rep., V.A. Medical Center, 1984.

[7] D. Egolf, V. Larson, C. Ahlstrom, H. Rainbolt, and T. McConnell, "The development and evaluation of a prototype acoustic feedback suppressor," *ASHA Conference*, 1987.

[8] A. Engebretson, M. O'Connell, and F. Gong, "An adaptive feedback equalization algorithm for the cid digital hearing aid," *Annual International Conference of IEEE Engineering in Medicine and Biology Society*, 1990.

[9] O.Dyrlund and N.Bisgaard, "Acoustic feedback margin improvement in hearing instruments using protype dfs (digital feedback suppression) system," *Scand Audiology*, 1991.

[10] D. Preves and J. Newton, "Modelling a vented push-pull hearing aid fitting in situ as a feedback control system." *

[11] D. Preves, "Approaches to noise reduction in analog, digital, and hybrid hearing aids," *Seminars in Hearing*, 1991.

[12] R. Wang and R. Harjani, "Suppression of acoustic oscillations in hearing aids using minimum phase techniques," *International Conference on Circuits and Systems*, 1993.

A FILTERING APPROACH TO ELECTROCARDIOGRAPHY VOLUME CONDUCTOR INVERSE SOURCE PROBLEMS

Thomas G. Xydis and Andrew E. Yagle

Dept. of Electrical Engineering and Computer Science, University of Michigan, Ann Arbor, MI 48109

ABSTRACT

We summarize the application of several signal processing techniques to several volume conductor inverse source problems. These include: (1) reconstructing the electrical potential on an active layer of cells from surface measurements using 2-D filtering; (2) identifying locations of still-active layers following myocardial infarction using a multiple-hypothesis generalized likelihood ratio test; and (3) estimating the location of a single still-active layer by maximum likelihood estimation, and determining the Cramer-Rao bound for the mean-square estimator error. These problems all have applications in electrocardiography.

1 INTRODUCTION

1.1 Inverse Planar Source Problem

In electrocardiology electrical measurements are made on the pericardium (essentially at the surface of the heart). The heart muscle consists of layers of electrically-active muscle tissue. The electrical potential distribution in these layers indicates current muscle function. Features of interest include the point of activation of the depolarization wave associated with muscle activity, and cell-to-cell propagation irregularities (unobtainable using invasive techniques). A considerable amount of current cardiac research focuses on the correlation of activation propagation through the heart with the occurrence of cardiac arrhythmias. Using a quasistatic approximation, the propagating action potential will be laid out spatially, rather than temporally.

The problem of reconstructing the potential distribution in a layer from surface measurements can be formulated as a volume conductor inverse problem. The 2-D voltage source is an electrically-active layer of cells, and the volume conductor is interstitial fluid or electrically-inert tissue. Since cardiac muscle tissue is layered, the layered volume conductor model is appropriate. Since the quasistatic assumption is reasonable, the image of the distributed voltage source shows the form of the depolarization wave (associated with muscle activity).

We solve this problem using a *medium filter*. This is a 2-D filter which relates the 2-D measured potential at the surface of the heart to the 2-D source potential. This approach allows fast 2-D filtering and regularization techniques to be applied to this problem. The 1-D medium filter was proposed in [1]; the extension to 2-D was made in [2]. The source is modelled by an extracellular potential; current sources or transmembrane potentials can also be used [3].

1.2 Locations of Multiple Active Layers

Following myocardial infarction, many of the layers of electrically-active tissue are dead. Determining the number and locations of still-active layers is useful in diagnosis. This location estimation problem for planar bioelectric sources can be formulated as follows: Given some knowledge of the functional form of planar 2-D bioelectric sources at unknown locations in a horizontally layered volume conductor with known structure, and measurements of the resulting extracellular field in a plane parallel to the sources, determine the number and locations of any sources present.

Here we use the medium filter noted above to transform the location estimation problem into one of estimating the parameters of the filter. The problem is formulated as a composite multiple hypothesis testing problem, solved using a generalized likelihood ratio test. The hypotheses are that any layer or combination of layers, at pre-specified depths, are still active. The unknown parameters for each hypothesis are the unknown lateral locations (within a layer) of action potentials; for details see [3].

1.3 Cramer-Rao Bound for One Layer

Again using the 2-D medium filter, we may easily compute the maximum-likelihood estimate of the location of a single still-active layer from noisy surface measurements of electrical potential (additive white Gaussian noise field). We may also compute the Cramer-Rao lower bound on the mean square error of any unbiased depth estimator. Here the source depth need not be restricted to any pre-specified locations.

Although this problem itself is of limited interest, it furnishes valuable insight into the performance of the multiple-layer detection algorithm. In particular, it turns out that the maximum likelihood estimator is asymptotically efficient (attains the bound) at signal-to-noise ratios of interest. Also, the wavenumber characteristics of the 2-D signal that are most valuable in detecting its presence are present in cardiac action potentials.

2 ASSUMPTIONS

We make the following assumptions, all of which are physiologically reasonable [4]:

1. The extracellular medium consists of isotropic, horizontal, and electrically passive layers with known conductivities and thicknesses;

2. The problem is quasistatic, implying that the medium has negligible reactance and propagation delay. Thus all potentials are computed at an instant in time;

3. For the single-layer and inverse planar source problems, there is only a single active layer at depth $z = z'$ parallel to the plane $z = 0$ of measurements. For the inverse planar source problem z' is known.

Although the surface of the heart is obviously not a plane, the brush electrode used to perform electric field measurements and the scale of the problem itself are both so small (on the order of millimeters) that regarding the heart as locally planar is reasonable. The planar heart approximation has been used in the past for the volume conductor forward problem at small distances, as well as in the interpretation and simulation of tissue bath experiments in which the planar assumption is rigorously met (see reference list in [2] and [3]).

Due to lack of space we simply summarize results. Details are available in [2], [3], and elsewhere in these proceedings.

3 INVERSE PLANAR SOURCE PROBLEM

3.1 Problem Statement

The basic problem is to solve Laplace's equation $\nabla^2 \phi(x, y, z) = 0$ with the following boundary conditions (σ_n is conductivity in the nth layer):

1. $\phi(x, y, 0) = s(x, y) =$ source; $\phi(x, y, \infty) = 0$;

2. $\phi(x, y, z)$ and $\sigma_n \frac{\partial \phi}{\partial z}$ are continuous in z across a layer interface.

The medium is assumed to have infinite extent in x and y; since cardiac muscle comes in flat sheets on the scale of interest, this is quite reasonable. Taking Fourier transforms $x \rightarrow k_x$ and $y \rightarrow k_y$ shows that potential on a horizontal plane $z = z_o$ is related to the source potential by

$$\Phi(k_x, k_y, z_o) = S(k_x, k_y) H(k_x, k_y, z_o) \qquad (1)$$

where $H(k_x, k_y, z)$ is the medium filter.

3.2 Medium Filter Computation

Here we summarize from [2]. The medium filter in the nth layer has form

$$H(k_x, k_y, z) = A_n e^{-\sqrt{k_x^2 + k_y^2}\,z} + B_n e^{\sqrt{k_x^2 + k_y^2}\,z} \qquad (2)$$

and the coefficients A_n and B_n can be recursively computed using the relation

$$\begin{bmatrix} A_{n-1} \\ B_{n-1} \end{bmatrix} = \sqrt{\frac{\sigma_n}{\sigma_{n-1}}} \frac{1}{\sqrt{1 - r_n^2}} \begin{bmatrix} 1 & a_n r_n \\ r_n/a_n & 1 \end{bmatrix} \begin{bmatrix} A_n \\ B_n \end{bmatrix}$$

$$(3)$$

$$a_n = e^{2\sqrt{k_x^2 + k_y^2}\,z_n}; \quad r_n = \frac{\sigma_n - \sigma_{n-1}}{\sigma_n + \sigma_{n-1}} \qquad (4)$$

in which σ_n is the conductivity in the nth layer and z_n is the thickness of the nth layer.

The recursion (3) is *initialized* as follows. Assume the top (N^{th}) layer $z_N < z < \infty$ has finite non-zero conductivity σ_N. Clearly $B_N = 0$, but A_N is unknown. However, we also know $A_0 + B_0 = 1$. Hence by linearity we may proceed as follows:

1. Initialize (3) using $A_N = 1$ and $B_N = 0$;

2. Propagate the recursion (3) in decreasing n;

3. Multiply each A_n and B_n by $1/(A_0 + B_0)$.

If the top layer is a perfect conductor or perfect insulator, a slight variation is necessary [2].

3.3 Implementation

The forward problem of determining measured potential $\phi(x, y, z_o)$ from source potential $s(x, y)$ is solved using $H(k_x, k_y, z_o)$. The inverse problem of determining source potential from measured potential is solved using the inverse filter $1/H(k_x, k_y, z_o)$.

For an infinite homogeneous volume conductor the medium filter simplifies to

$$H(k_x, k_y, z) = e^{-\sqrt{k_x^2 + k_y^2}\, z}. \tag{5}$$

Note the effect of the volume conductor is to smooth and attenuate the 2-D source potential. The inverse filter $1/H(k_x, k_y, z)$ grows exponentially with wavenumbers k_x and k_y and distance z. Hence high-wavenumber noise is amplified, and the problem is ill-conditioned, so regularization is needed; e.g., spatially lowpass filter the data $\phi(x, y, z_o)$ [2].

4 LOCATIONS OF MULTIPLE ACTIVE LAYERS

4.1 Assumptions

To formulate the problem, we assume:

1. Sources can only exist at certain known depths $z_i, i = 1 \ldots N$, with unknown lateral location (x_i', y_i') at each depth. If a source exists at depth z_i, the ith layer is said to be "active";

2. If a source exists at a particular depth, its orientation is known (the method can be extended to sources with unknown orientations);

3. The *a priori* probability that the ith layer is active is 0.5, and independent of the jth layer being active. Then the number of active layers is binomially distributed with mean $N/2$;

4. The goal is to minimize the probability of error in choosing the active layers, while computing the maximum likelihood estimates of the locations of the source in each active layer.

Now the problem can be formulated as a composite multiple hypothesis minimum error probability test, with each hypothesis corresponding to a particular combination of active depths. This means that for each hypothesis, the locations of the sources are estimated, and then the most likely hypothesis given the estimates is chosen. If there are N possible active layers, then there are 2^N hypotheses.

Our data are noisy measurements $r(x, y)$ of the surface potential $\phi(x, y)$

$$r(x, y) = \phi(x, y) + n(x, y); \tag{6}$$

$n(x, y)$ is a zero-mean white Gaussian noise field.

4.2 Likelihood Function

As formulated, the problem can be solved using a generalized likelihood ratio test. The total potential due to multiple sources is the superposition of the potential due to each source separately. Hence the observed potential $\phi(x, y)$ at the surface of the volume conductor due to N active layers (i.e., N sources $s_i(x - x_i', y - y_i'), i = 1 \ldots N$} at depths $\{z_i, i = 1 \ldots N\}$) is given by

$$\Phi(k_x, k_y) = \sum_{i=1}^{N} H(k_x, k_y, z_i) e^{-j(k_x x_i' + k_y y_i')} S_i(k_x, k_y) \tag{7}$$

where N is the number of sources and (x_i', y_i', z_i) is the location of the ith source within the ith layer.

Let $H_{i_1, i_2, \ldots i_N}$, $i_j = 0$ or 1, represent the hypothesis that the layers corresponding to $i_j = 1$ are active. These hypotheses are equally likely *a priori*. Let $\{(x_j', y_j', z_j), i_j = 1\}$ be the locations of the active sources, and let $R(k_x, k_y)$ be the Fourier transform of noisy observations of $\phi(x, y)$. Then the log-likelihood function for minimum error probability detection with equal *a priori* probabilities is

$$\log \Lambda[R(k_x, k_y)] = \frac{2}{2^N N_o} \int_{-\infty}^{\infty} \int_{-\infty}^{\infty} R(k_x, k_y) \times$$

$$\sum_{n:i_n=1} S_n(k_x, k_y) H(k_x, k_y, z_n) e^{-j(k_x x'_n + k_y y'_n)} dk_x \, dk_y$$

$$-\frac{1}{2^N N_o} \int_{-\infty}^{\infty} \int_{-\infty}^{\infty} | \sum_{n:i_n=1} S_n(k_x, k_y) H(k_x, k_y, z_n) \times$$

$$e^{-j(k_x x'_n + k_y y'_n)}|^2 dk_x \, dk_y. \qquad (8)$$

For each of the 2^N sets of $\{i_1, \ldots i_N\}$, this is maximized over $\{(x'_j, y'_j), i_j = 1\}$ (recall that the depths z_i are all known a priori). Then the maximum of the resulting 2^N numbers is found; the corresponding hypothesis $H_{i_1, i_2, \ldots i_N}$ denotes which layers are active. The previously-computed locations $\{(x'_j, y'_j), i_j = 1\}$ for that hypothesis then denote the location of the source in each active layer.

5   CRAMER-RAO BOUND FOR ONE LAYER

Here our data is the same as (6). However, there is now only a single known source $s(x, y)$ at an unknown depth $z = z'$. Our goal is to estimate z'.

The likelihood function can be found by setting $N = 1$ in (8), and the maximum likelihood estimate $\hat{z}$ of depth z' computed by maximizing this.

The Cramer-Rao bound is a lower bound on the variance of any unbiased estimator. The Cramer-Rao bound for estimating z' is

$$E[(z' - \hat{z})^2] \geq$$

$$4\pi^2 N \left(\int_0^{\infty} \int_0^{2\pi} |\frac{\partial H(k, \phi, z')}{\partial z'} S(k, \phi)|^2 k \, d\phi \, dk \right)^{-1}$$
$$(9)$$

where $\hat{z}$ is the estimate of depth z', $k = (k_x^2 + k_y^2)^{1/2}$ is radial wavenumber, and $S(k, \phi)$ is the Fourier transform of source potential in polar coordinates.

For an infinite homogeneous volume conductor,

$$E[(z' - \hat{z})^2] \geq$$

$$4\pi^2 N \left(\int_0^{\infty} \int_0^{2\pi} k^3 e^{-2kz'} |S(k, \phi)|^2 d\phi \, dk \right)^{-1}. \quad (10)$$

The bound (10) illustrates two interesting points:

1. Low-wavenumber components of the source potential are important since the factor $e^{-2kz'}$ attenuates high-wavenumber components of the source. This is reasonable since the latter are attenuated by the medium;

2. Very low wavenumber components in the source potential are not helpful, since they are attenuated by the factor k^3. This is reasonable: the medium has little effect on these components, so they are less useful in estimating z'.

Numerical results presented elsewhere in these proceedings show that the maximum likelihood estimator is efficient (attains the Cramer-Rao bound) at SNRs as low as -15.

6   ACKNOWLEDGMENTS

The first author would like to thank Professor C.B. Sharpe and Dr. Alan H. Kadish for their helpful consultation. The work of the second author was supported in part by the National Science Foundation under grant MIP-8858082.

References

[1] N. Ganapathy, J. W. Clark, O. B. Wilson, and W. Giles, "Forward and inverse potential field solutions for cardiac strands of cylindrical geometry," *IEEE Transactions on Biomedical Engineering*, vol. 32, no. 8, pp. 566–577, 1985.

[2] T. G. Xydis and A. Yagle, "A filtering approach to the two-dimensional volume conductor forward and inverse problems," vol. 7, pp. 307–327, April 1991.

[3] T. G. Xydis, A. Yagle, and A. Kadish, "Estimation of locations of bioelectric sources using an equivalent filter," *Inverse Problems*, vol. 6, pp. 696–708, August 1990.

[4] R. Plonsey, *Bioelectric Phenomena*. New York: McGraw-Hill, 1969.

Reversible Compression of Medical Images Using Decomposition and Decorrelation Methods

Y. S. Fleming Lure[s], C.S. Joe Liou[s#], and John J. Metzner[#]

[s] Caelum Research Corporation, Silver Spring, MD 20901
[#] Department of Electrical Engineering
The Pennsylvania State University
University Park, PA 16802

Abstract

A new reversible image compression technique is presented for compressing and reconstructing various types of medical images. This technique involves a decomposition method to split the image into the most, intermediate, and the least significant bit images in order to reduce the noise effect, followed by decorrelation methods which involve differential pulse coded modulation (DPCM) and novel interpolative technique. Three types of medical images including 60 MR, 40 CT, and 5 chest X-ray images are considered in our experiment. Depending upon the image quality and characteristics, this error-free compression technique has improved the compression efficiency by approximate 30-50% over the conventional DPCM/Huffman coding. The practical compression ratios vary from 3.2:1 to 4:1 and the true storage saving ratio reduces up to 5.35:1.

1. Introduction

The transmission and storage of digital images is becoming increasingly important in clinical environment through the use of picture archiving and communications system (PACS). Therefore, the capacity to compress the size of these images plays a pivotal role in the health care system. Compression techniques can be divided into reversible (or lossless) and irreversible (or lossy) methods. Due to the sensitivity of medical diagnosis images, the loss of image fidelity introduced in the lossy compression method is often not acceptable by medical community, even lossy techniques compress the size by a factor of ten or more. Hence, a reversible compression technique is necessary. Because of the entropy limitation, reversible techniques reduce amount of data by a factor ranging from only 1.5 to 2.5 [1,2].

In this research work, based on the original image quality and characteristics (e.g. noise level, detailing of the image, etc.), we develop and apply image decomposition method as well as interpolative decorrelation method to remove noise-bit effect and reduce data redundancy, respectively. In the experiments, we tested the impact of this new technique on image compression by employing it with arithmetic [3,4] and L-Z [5,6] codings on a set of clinically used images. The results show that the new technique provides higher compression ratio than conventional methods by approximately 50% depending on the noise level of the image. Specifically, through examining the quality and characteristics of images in terms of noise levels and entropy level, we derive the operational parameters and results based upon the following procedures: (1) a decomposition method to split the image into the most, intermediate, and the least significant bit images, (2) decorrelation techniques which involve DPCM and novel interpolative decorrelation, (3) selection of the kernel size in interpolative decorrelation method to obtain the best compression; (4) coding techniques including Lempel-Ziv coding, and arithmetic coding with various orders.

2. Medical image

Three most widely used clinical images are considered, i.e., magnetic resonance (MR), computerized tomography (CT), and chest X-ray radiograph. These films are digitized by using several scanners (e.g., Konica Laser Film Scanner and TruVel's TruScan Scanner) into 256x256x12, 512x512x12 and 2000x2048x12 bits of computer data, for MR, CT and chest X-ray images, respectively. Those images are acquired for the compression study: (1) 20 sets of sagittal view MRI, (2) 20 sets of transverse view MRI, and (3) 20 sets of coronal view MRI, (4) 40 sets of CT image including abdomen and head images, (5) 5 sets of chest X-ray radiographs. Although dynamic range over the whole image is 12 bits, the image has low contrast locally. Therefore, coding techniques developed for source coding may perform differently at coding medical image files.

3. Image decomposition

Since noise associated with physical processes is not compressible, investigation of noise level by using correlation method is necessary before decomposition is applied. A correlation analysis of each sub-image, which is generated from original image by stripping higher order bits off, is performed at various bit plans to obtain series correlation coefficients as function of bit levels and to distinguish signal from noise at such bit level. A remapping algorithm [7] was applied to the split image to avoid any discontinuity during correlation analysis in the split LSB image. It is found that noise bits for 12-bit MR, CT and X-ray images are 4, 2 and 4 bits, respectively, which are also confirmed by the medical devices.

Because of the noise impact on the image, a decomposition technique is employed to split the image into three parts: (1) the signal, (2) the mixed, and (3) the noise regions which are usually characterized as the most significant bits (MSB), intermediate significant bits (ISB) and the least significant bits (LSB) areas, respectively. Both the boundaries between signal and mixed regions and between mixed and pure noise regions are digital device dependent. If the noise range contained in the data sequence is minimized, the model can be reduced to categories (1) and (2). The splitting of the L least significant bits (LSB), I intermediate significant bits (ISB), and the M most significant bits (MSB) from image f(x,y) is performed at bit plan such that

$$f(x,y) = f_M(x,y) \cdot 2^{(I+L)} + f_I(x,y) \cdot 2^L + f_L(x,y)$$

where subscripts **M**, **I**, and **L** denote split MSB, ISB, and LSB images, respectively. The summation of **M**, **I**, and **L** is equal to 12, the total number of significant bits. It can be shown that with M(MSB)-L(LSB) splitting decomposition, the total entropy for the summation of the split images is smaller than that from original image. Arrangement of M, I, and L such as 3(MSB)-6(ISB)-3(LSB) split and 3-5-4 split in the splitting decomposition is further explored in the experiment.

4. Image decorrelation

The image decorrelation technique refers to the process of removing redundancy due to pixel-to-pixel dependency within image. In this study, the popular DPCM method and the specifically-designed interpolative method are investigated. The DPCM method computes the predicted values from the weighted sum over the pixels closest to the position of the pixel to be predicted and the difference between original and predicted values is coded.

A specially designed fast interpolative technique was developed and applied to the medical images, based upon bilinear interpolation method for lossless predictive compression. Bilinear interpolation method is used instead of spline methods [7-9], since they produce similar type of probability density function of the differential signals. Through manipulation and derivation of its matrix form, this technique is able to perform extremely fast compression/decompression at better compression ratio [10]. The image is first sub-sampled (under-sampled) at every n pixels horizontally and vertically. The sub-sampled values (considered as header) are coded either directly or through the DPCM method. The bilinear interpolation method is used to interpolate those values at un-sampled pixels within (n+1) x (n+1) cell (or called image block) from the four sub-sampled values (y_1, y_2, y_3, y_4) at the corners of the cell. The interpolated values $[y']$ within the image block are obtained by multiplying four (n+1)x(n+1) zero-doubly-bordered symmetric matrices $([M_1], [M_2], [M_3], [M_4])$, e.g.,

$$n^2 \cdot [M_1] = \begin{bmatrix} 0 & 0 & ..... & 0 & 0 \\ n \cdot 1 & (n-1) \cdot 1 & ..... & 1 \cdot 1 & 0 \\ ... & ... & ..... & ... & ... \\ n(n-1) & (n-1)(n-1) & ..... & 1 \cdot (n-1) & 0 \\ n \cdot n & (n-1)n & ..... & 1 \cdot n & 0 \end{bmatrix}$$

,with four sub-sampled values, i.e.,

$$[y'] = [M_1] \cdot y_1 + [M_2] \cdot y_2 + [M_3] \cdot y_3 + [M_4] \cdot y_4$$

where $[y']$ is a (n+1)x(n+1) matrix. The four matrices $([M_1], [M_2], [M_3], [M_4])$ are inter-related through operations including matrix transpose and matrix similarity transformation, i.e., $M_2 = M_1 C$, $M_3 = M_1{}^T$, $M_4 = CM_1$ where

$$C = \begin{bmatrix} 0 & 0 & ..... & 0 & 1 \\ 0 & 0 & ..... & 1 & 0 \\ ... & ... & 1 & ... & ... \\ 0 & 1 & ..... & 0 & 0 \\ 1 & 0 & ..... & 0 & 0 \end{bmatrix}$$

This algorithm is then implemented in a fast

processing procedure by taking advantage of the properties of the zero-doubly-bordered symmetric matrices. By applying this method in every (n+1)x(n+1) image block without overlapping of the image block, the differential image between interpolative and original images is generated over the whole image. Because of sub-sampling involved in every n+1 pixels, the histogram of the differential image becomes narrower than those from other predictive DPCM methods since more zeros (which is the mode of the histogram) are produced. Therefore, the entropy of the signal from interpolative prediction technique is lower than that from both original image and from the DPCM method. In addition, these zeros at sub-sampled location are removed from the compressed data file, hence higher compression efficiency can be further achieved. Reconstruction of the sub-sampled points can be easily obtained due to the knowledge of their locations. Size n in interpolative (n+1)x(n+1) cell is investigated, since it has an effect on the decorrelation performance involving both DPCM and interpolative methods. Application of any coding technique to the LSB images, primarily formed by the noise, gains little compression effect since no correlation or redundancy exists within noise. A bit packing is employed by joining all the low bits from the noise portion and storing them as a unit of character (8 bits).

5. Experimental results

The experiments performed involve the investigation of the above mentioned procedures (decomposition, decorrelation, coding, and size of interpolative cell). Systematic experiment in a tree structure is developed to perform the experiment which involves over 14 different experimental procedures to perform compression for single image. An optimal procedure is then determined based on the compression performance. Since it is found that the noise bits normally occupy 4 LSB, the decomposition techniques are primarily applied on this bit level. Decorrelation methods include DPCM, 3x3 and 4x4 window interpolative techniques. Direct LZ coding (without preprocessing of image) and DPCM/LZ coding (apply LZ coding after DPCM) are used as references for the comparison with our results.

In the coding methods, arithmetic coding outperforms LZ coding. Interpolation with 3x3 window achieves better compression result than interpolation with 4x4 (and larger) window. Because the interpolation method removes most spatial redundancy, arithmetic coding with models of order higher than 2 demonstrate no advantage over models of lower order. The best performance of coding for MR and CT in experiment is arithmetic with order 1 or 2.

In the following experiments, 8-4/3Int/8MSB represents the compression procedure: Decomposition using 8(MSB)-4(LSB) splitting decomposition, application of fast interpolative decorrelation with 3x3 window to the 8 MSB split image to generate difference image, then application of arithmetic coding applied to the differential images. 'Seg.' (segmentation) represents the process that the compression is applied to the anatomic region of those medical images, since only this region is meaningful during diagnostic procedure.

After processing 60 sets of MR images which consists of 3 types of MR images: sagittal view, transverse view, and coronal view, the average compression ratio for procedure excluding segmentation is 5.95 bits/pixel (as shown in table 1), which is much better than conventional LZ or DPCM/LZ techniques by an average of 36% and 28%, respectively.

Table 1 Compression and storage ratio of MR images

A(B) represents that A is in the unit of bits/pixel and B is the storage tatio, obtained from B=16/A

Methods \ Images	Sagittal	Coronal	Transverse	Average
LZ	10.14(1.58)	10.34(1.55)	7.38(2.17)	9.29(1.72)
DPCM/LZ	8.79(1.82)	8.84(1.81)	7.06(2.11)	8.23(1.94)
DPCM/Arith	6.57(2.44)	7.26(2.20)	5.36(2.99)	6.39(2.50)
8-4/4Int/8MSB	6.13(2.61)	6.70(2.39)	5.11(3.13)	5.98(2.68)
8-4/3Int/8MSB	6.08(2.63)	6.69(2.39)	5.10(3.14)	5.95(2.69)
8-4/3INt/8MSB/Seg.	3.90(4.10)	4.01(3.99)	1.05(15.2)	2.99(5.35)

Compression results for computerized tomography (CT) images are acquired after examining 40 images (as shown in table 2). Since CT images are little contaminated by noise, compression with decomposition

Table 2 Compression and storage ratio of CT images

A(B) represents that A is in the unit of bits/pixel and B is the storage tatio, obtained from B=16/A

Methods \ Images	Abdomen	Head	Average
LZ	7.91(2.02)	9.16(1.75)	8.54(1.87)
DPCM/LZ	6.39(2.50)	8.41(1.90)	7.40(2.16)
DPCM/Arith	4.92(3.25)	6.38(2.51)	5.67(2.82)
8-4/3Int/8MSB	5.56(2.88)	6.70(2.39)	6.13(2.61)
3Int/Coding/12b	4.98(3.21)	6.32(2.53)	5.65(2.83)
3Int/Coding/12b/Seg.	2.85(5.61)	3.89(4.11)	3.37(4.75)

does not produce the best result. The average compression ratio is 5.65 bits/pixel for interpolation on the whole image, which is 24% better than DPCM/LZ method.

Compression ratio for chest X-ray images is 3.9 bits/pixel (as shown in table 3) or 4.1:1 in true storage ratio after examining 5 images. An average compression ratio of 2.69:1, 2.83:1, and 4.1:1 is obtained for MR, CT, and chest X-ray images, respectively which are 36%, 34%, and 48% better than the standard LZ technique used in UNIX system.

Table 3 Compression and storage ratio of X-ray images

A(B) represents that A is in the unit of bits/pixel and B is the storage tatio, obtained from B=16/A

Methods	Average of Chest X-ray Images
LZ	7.56(2.12)
DPCM/LZ	5.12(3.13)
DPCM/Arith	4.01(3.99)
3Int/Coding/12b	3.90(4.10)
3Int/Coding/12b/Seg.	3.82(4.19)

When consideration of anatomic region is involved, further improvement of 50%, 40% and 2% on compression ratio over the procedure without segmentation can be achieved for MR, CT and chest X-ray respectively. The storage ratio can be reduced by a factor of 5.35, 4.75 and 4.19 for MR, CT and chest X-ray respectively. The improvement of chest X-ray is the minimum because the anatomic region occupies almost all the film. Since compression results obtained from various compression techniques give the best compression ratio for different images, adaptive selection of the techniques based on individual image content is expected to provide an optimal compression performance.

Acknowledgement

This work was supported by the National Institute of Health under Grant R43 RR07735-01 and by the National Science Foundation under Grant NCR-9104750. Its contents are solely the responsibility of the authors and do not necessarily represent the official view of NIH or NSF.

References

[1] T.V. Ramabadran and K. Chen, "The Use of Contextual Information in the Reversible Compression of Medical Images,"IEEE Transactions on Medical Imaging, vol. 11, No.2, pp. 185-195, June 1992.

[2] N. Tavakoli, "Lossless Compression of Medical Images," Proceedings of the 4th Annual IEEE symposium, Baltimore, MD, pp.200-207, May 1991.

[3] T.C. Bell, J.G. Cleary and I.H. Witten, Text Compression, Prentice Hall, New Jersey, 1990.

[4] I.H. Witten, R.M. Neal and J.G. Cleary, "Arithmetic Coding for Data Compression," Communications of the ACM, Vol. 30, pp. 520-540, June 1987.

[5] J. Ziv and A. Lempel, "Compression of Individual Sequences via Variable-Rate Coding," IEEE Transactions on Information Theory, Vol. IT-24, pp. 530-536, Sept. 1978.

[6] A. Lempel and J. Ziv, "Compression of Two-Dimensional Data," IEEE Transactions on Information Theory, vol. IT-32, No. 1, pp.2-8, 1986.

[7] S.C. Lo, E. Shen, S.K. Mun and J. Chen, "A Method for Splitting Digital Value in Radiological Image Compression," Medical Physics Journal, August 1991.

[8] P. Roos, M.A. Viergerver, et al.,"Reversible Interframe Compression of Medical Images," IEEE Transactions on Medical Imaging, MI-3, pp.179-186, 1988.

[9] P. Roos and M.A. Viergerver, "Reversible Image Data Compression Based on HINT Decorrelation and Arithmetic Coding," SPIE vol. 1444, Image Capture, Formatting and Display, pp. 283-290, 1991.

[10] C.S. Joe Liou, Y.M. Fleming Lure and John J. Metzner, "Lossless Interpolative Compression of Radiological Images," Porc. Eighth Workshop on Image and Multidimensional Signal Processing, IEEE Signal Processing Society, Cannes, France, Sept. 8-10, 1993.

A BAYESIAN SEGMENTATION APPROACH TO 3-D TOMOGRAPHIC RECONSTRUCTION FROM FEW RADIOGRAPHS

Catherine Klifa and Ken Sauer

Laboratory for Image and Signal Analysis
Department of Electrical Engineering
University of Notre Dame
Notre Dame, IN 46556

ABSTRACT

Three-dimensional tomographic reconstruction is typically achieved by combining the results of planar reconstructions from repeated 2-D scans. This approach is infeasible when only very sparse three-dimensional cone-beam data is available. We have developed a MAP 3-D reconstruction technique to estimate a solid object directly from sparse cone-beam data. In this paper, we focus on radiographic flaw detection in solid materials, which can be viewed as a segmentation of the object into a binary-valued reconstruction. Optimization is performed by iterated conditional modes, with deterministic convergence, but a solution dependent on the initial condition. To speed convergence and improve the estimate, we use an initial condition based on maximum-likelihood estimation of flaw location.

1. INTRODUCTION

Many problems in medical imaging and nondestructive evaluation can be cast in the framework of three-dimensional reconstruction from integral projections. Typical transmission tomographic algorithms achieve a 3-D reconstruction by "stacking" and interpolating slices from closely spaced 2-D scans, or other deterministically based methods[1]. A more difficult problem arises when 3-D features must be estimated from only a very few projections onto 2-D planes, e.g. radiographic films. These are a type of cone-beam projection data, frequently beset by restrictions in angles of view, high noise in radiographs, uncertainty in placement of the radioactive source, and scarcity of data for 3-D reconstructions of the object under test. In many situations, this problem involves finding exceptions to otherwise homogeneous material, which invites a binary model for detecting the presence of "objects." A

solution to this type of problem for 2-D objects, and parallel projections was presented in [2]. But this object detection/estimation approach requires parametric modelling of the object(s) to be detected, and becomes very difficult computationally when model geometry in generalized.

We formulate the problem of 3-D tomographic reconstruction from very sparse data in the maximum *a posteriori* probability (MAP) estimation framework, with a binary Markov random field model[3]. The estimate is in the form of a segmentation of the 3-D volume, with voxel values as the parameters in question. The formulation of the radiographic system includes Compton scattering compensation derived from a Taylor series approximation of the log likelihood as a function of voxel values in the estimate. Finding the true MAP estimate is possible with algorithms such as simulated annealing, but is computationally infeasible for problems of the size we consider. We therefore use iterated conditional modes (ICM)[4] to deterministically converge to an approximate MAP segmentation. In most cases including poor quality data and significant weighting of the *a priori* distribution of the object, the solution arrived at by ICM represents a local maximum in *a posteriori* likelihood, and is dependent on the initial condition. To improve the choice of this local maximum, we use an initial segmentation derived from a likelihood function for parametric object estimation[2].

2. STATISTICAL MODELS

We use the notation $p = \mathbf{A}f$ for the discrete approximation of the projection operator of the volume vector f into the projection vector (radiograph) p. The entry A_{ij} represents the interaction between the j-th 3-D cell (voxel) of the object (f_j)and the i-th datum, measuring a photon count λ_i at pixel i among a set of radiographs. The count λ_i is transformed into an integral projection estimate p_i. The radiographic problem

This research supported by Electricité de France under Grant no. P21L03/2K3208/EP542.

has a cone-beam (divergent) geometry, which can be solved in the complete data case by deterministic cone-beam methods[1]. The statistical methods we employ are more general, and are not essentially dependent on scanning geometry. In the classical straight-line model for transmission tomography, a source rate of λ_T photons per ray gives the Poisson-distributed random photon count λ_i an expected value of $\lambda_T \exp\{-(\mathbf{A}f)_i\}$. This implicitly assumes attenuated photons are simply absorbed. But Compton scattering is the major mode of attenuation, and a major source of distortion for common gamma-ray sources[5], and has a decided forward orientation for both the common Cobalt and Iridium sources. For specimens with large inhomogeneities, forward scatter causes substantial artifacts.

The angular distribution and energy loss of Compton-scattered photons is expressed in the Klein-Nishina formula[6]. Completely accurate modeling of the scattering phenomenon is very complex, due largely to the presence of multiply-scattered photons, many of which remain in the range to which radiographic film is sensitive. We have taken a point of view more accurate than typical approximations, but within the realm of computational feasibility. We base our model on the simple dependence of the Poisson distribution at pixel i on the densities of rays directed toward other pixels:

$$E[\lambda_i] = \lambda_T [e^{-[\mathbf{A}f]_i} + \sum_k (1 - e^{-[\mathbf{A}f]_k} b(i,k)] \quad (1)$$

The function $b(i,k)$ expresses the probability that a photon scattered from the k-th ray will be received by the i-th pixel. The form of this kernel depends on the energy of the source, thickness and composition of material, and distance between the object and each radiograph. In spite of this inter-relation among the expected values of separate pixels, the counts at different pixels, *conditioned on the voxel values*, are independent random variables. To simplify the estimation process, we approximate the log-likelihood function of the photon counts with its Taylor series expansion. Let $\theta_i = E[\lambda_i]$. Because θ_i is deterministically related to the vector f, we now express the likelihood function in terms of θ for the sake of simplicity, rather than f. The log likelihood of the entire set of photon counts is

$$\begin{aligned} L(\lambda|\theta) &= \log\left(\prod_i \frac{e^{-\theta_i}\theta_i^{\lambda_i}}{\lambda_i!}\right) \\ &= \sum_i -\theta_i + \lambda_i \log \theta_i + c(\lambda) \quad (2) \end{aligned}$$

The function $c(\lambda)$ will be dropped, since it is fixed by the observations.

We perform a Taylor series expansion in the domain of integral projections, where parameters are the integral densities $\{[\mathbf{A}f]_i\}$, which we will abbreviate g_i for the derivation. The first derivative is

$$\begin{aligned} \frac{\partial L(\lambda|\theta)}{\partial g_i} &= \sum_k \left[-1 + \frac{\lambda_k}{\theta_k}\right] \frac{\partial \theta_k}{\partial g_i} \quad (3) \\ &= \sum_k \left[\frac{\lambda_k}{\theta_k} - 1\right] \lambda_T e^{-g_i}[b(k,i) - \delta(i-k)] \end{aligned}$$

To place the expansion at a point $\tilde{g}$ of zero gradient, we set

$$\theta_i(g)|_{g=\tilde{g}} = \lambda_i, \quad (4)$$

with the dependence of θ on g made explicit here, and the form of dependence given in (1). The vector $\tilde{g}$ will be a function of modified photon counts:

$$\tilde{g}_j = \ln\left(\frac{\lambda_T}{q(\lambda)_j}\right). \quad (5)$$

Combining (1) and (5), $q(\lambda)$ is a transformation of the data

$$\lambda_i = q(\lambda)_i + \sum_k (\lambda_T - q(\lambda)_k)b(i,k), \quad (6)$$

or, in vector notation:

$$\lambda = q(\lambda) + \mathbf{B}(\lambda_T \mathbf{1} - q(\lambda)). \quad (7)$$

The matrix $\mathbf{B}$ has entries $b(i,k)$, and $\mathbf{1}$ is the vector all of whose entries are unity. A simple manipulation gives us

$$q(\lambda) = (\mathbf{I} - \mathbf{B})^{-1}(\lambda - \lambda_T \mathbf{B1}). \quad (8)$$

Thus the desired point of expansion requires a preprocessing of the vector of photon counts, which has a linear form related to the scatter pattern. We note here that the transformation of (5) requires that all entries of $q(\lambda)$ be positive. While this is a potential problem in the case of very wide variation of integral density, we do not expect it to occur often in our case of flaw reconstruction in solids, and it is easily treated as a special case.

The Hessian matrix has a relatively simple form, with

$$\mathbf{Q} = diag\{q(\lambda)_1, \ldots, q(\lambda)_M\}$$

and

$$\tilde{\mathbf{D}} = diag\{\lambda_1^{-1}, \ldots, \lambda_M^{-1}\},$$

$$\nabla_g^2 L(\lambda|\theta)|_{g=\tilde{g}} = -\frac{1}{2}\mathbf{Q}(\mathbf{I} - \mathbf{B})^T \tilde{\mathbf{D}}(\mathbf{I} - \mathbf{B})\mathbf{Q}. \quad (9)$$

The approximated log likelihood in terms of the reconstruction f is

$$L(\lambda|f) \approx \tag{10}$$
$$-\frac{1}{2}(\tilde{g} - \mathbf{A}f)^T \mathbf{Q}(\mathbf{I} - \mathbf{B})^T \tilde{\mathbf{D}}(\mathbf{I} - \mathbf{B})\mathbf{Q}(\tilde{g} - \mathbf{A}f).$$

This has the same basic form as the straight-line quadratic approximation in [7], with

$$L(\lambda|f) \approx -\frac{1}{2}(\tilde{p} - \tilde{\mathbf{A}}f)^T \tilde{\mathbf{D}}(\tilde{p} - \tilde{\mathbf{A}}f), \tag{11}$$

using the obvious substitutions from (10).

The feasibility of computing statistical estimates with this scatter compensation will depend on a few further approximations. Most importantly, we estimate the operator $\mathbf{B}$ as convolution with a stationary kernel composed of the superposition of a constant background scatter process, and a windowed local element for effects of strongly forward oriented scatter. This allows the storage of a set of coefficients which can be simply accessed at each voxel update.

The *a priori* distribution is a binary MRF, with neighborhood size of up to 26 voxels. Although there is limited gain in using this large a neighborhood, the computational cost associated with the prior term is quite small compared with that of the likelihood term.

3. ICM AND INITIAL ESTIMATE

For any state of the estimate of the object f, and voxel f_j with neighborhood N_j, ICM makes a greedy choice between the two feasible densities $\{\alpha, \beta\}$ according to

$$\hat{f}_j = \tag{12}$$
$$\arg \min_{f_j \in \{\alpha, \beta\}} \sum_i \tilde{d}_i(\tilde{p}_i - [\tilde{\mathbf{A}}f]_i)^2 + \sum_{k \in N_j} \gamma b_{jk}|f_j - f_k|.$$

The summation over i need only include those pixels for which $\tilde{\mathbf{A}}_{ij} \neq 0$, hence the efficiency of localizing the non-constant portion of the scatter kernel approximation.

Using the default initial condition of all f_j being the same value can lead to ICM's being trapped in a local minimum. Our initial condition is therefore based on the principle of object detection in [2]. When the entities sought in a reconstruction can be modelled with only location parameters, a likelihood function for detecting and estimating them can be form similarly to convolution backprojection. In the multi-object case[8], a bit more computation is necessary to choose location estimates, but this is optional in our case, since the initial estimate does not require these decisions. We simply use the thresholded likelihood function, with a

spherical object model of 2 pixels radius, as the initial estimate of f. This requires no object-based *decision*, but takes advantage of the binary nature of the information we seek, and the agglomeration of at least several voxels assumed to constitute a void in the solid under study.

4. SIMULATIONS

The synthetic test object for our experiments is shown in Fig. 2, and represents a block of steel approximately 3 cm on each of its longer sides. The simulated 3-D object is displayed as differentially transparent, with the lighter regions having density 0, representing voids. The radiographs of Fig. 1 are simulated with isotropic Compton scatter under small gamma-ray dosages. The fraction of radiation reaching each radiograph due to scatter was 1/2, which is typical of this source and material choice[9].

The $\mathbf{B}$ used in preprocessing for the trials shown here was constant, corresponding to isotropic scatter. Thus preprocessing of radiographs consists of only scatter subtraction, similar to that in [5]. The estimate shown in Fig. 3 resulted from 4 iterations of ICM, after which there were essentially no changes. Though the presence and locations of the flaws are ambiguous at best in the radiographs, their locations in the segmented reconstruction are clear.

5. CONCLUSION

This MAP tomographic estimation algorithm provides a simple, robust method for segmenting a 3-D object from digitized radiographs. Its output is appropriate for either automated decision-making, or visual inspection, but avoids the necessity of making decisions independently on separate radiographs, as is currently typical in application. Although the estimation process is computationally costly in terms of cost per pixel at each iteration, its convergence is very rapid in terms of iteration counts.

6. REFERENCES

[1] L.A. Feldkamp, L.C. Davis, and J.W. Kress, "Practical Cone-Beam Algorithm," *J. Opt. Soc. Amer. A,* 612, 1984.

[2] D.J. Rossi and A.S. Willsky, "Reconstruction from Projections Based on Detection and Estimation of Objects-Parts I and II: Performance Analysis and Robustness Analysis," *IEEE Trans. Acoust. Speech and Sig. Proc.,* vol. ASSP-32, no. 4, pp. 886-906, Aug. 1984.

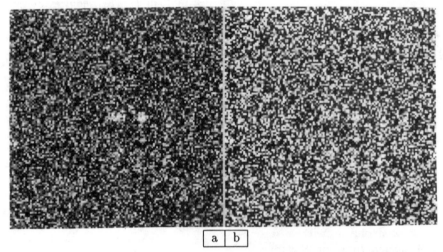

Figure 1: (a)Simulated radiographs, with image of voids perceptible near the center. (b) Identical view as in (a), but approximately 4 times larger noise variance due to lower dosage.

[3] R. Kindermann and J. L. Snell, *Markov Random Fields and their Applications*. Providence: American Mathematical Society, 1980.

[4] J. Besag, "On the Statistical Analysis of Dirty Pictures," *J. Royal Stat. Soc.*, vol. 48, no. 3, pp. 259-279, 1986.

[5] G.H. Glover, "Compton Scatter Effects in CT Reconstructions," *Med. Phys.*, vol. 9, no. 6, pp. 860-867, Nov./Dec. 1982.

[6] E. Segre, *Nuclei and Particles*, W.A. Benjamin, Inc., 1977.

[7] K. Sauer and C. Bouman, "A Local Update Strategy for Iterative Reconstruction from Projections," to appear Feb. 1993, *IEEE Trans. on Sig. Proc.*

[8] K. Sauer and B. Liu, "Image Reconstruction from a Limited Number of Projections Using Multiple Object Detection and Estimation," *Proc. of IEEE Int'l Conf. on Acoustics, Speech, and Signal Processing*, Albuquerque, NM, April 3-6, 1990, pp. 1861-1864.

[9] R. Halmshaw, *Industrial Radiography: Theory and Practice*, Applied Science Publishers, London, 1982.

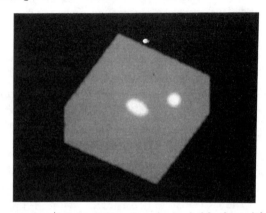

Figure 2: Original simulated steel block, with two voids. Approximate dimensions are $3 \times 3 \times 1.5cm$.

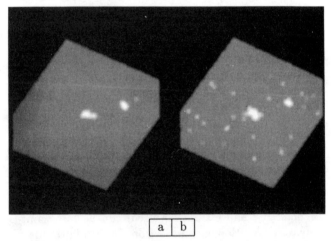

Figure 3: MAP segmentations from sets of radiographs as in Fig. 1. (a) Photon dosage/pixel (λ_T) of 3000, 3 radiographs; (b) $\lambda_T \approx 800$, with 7 radiographs. Each estimate resulted from 4 iterations of ICM.

An Approach toward Automatic Diagnosis of Breast Cancer from Mammography

M. Terauchi, Y. Takeshita

Department of Circuits and Systems, Hiroshima University
1-4-1 Kagamiyama, Higashi-Hiroshima, 724 Japan

ABSTRACT

In this paper, we propose an approach to extract an exact shape of breast cancer tumor from X-ray mammograms. The tumors usually have a lot of variety in their shape and size. Therefore, it is difficult to utilize exact object models which are often used in industrial applications. However, as tumors in the grown stage have comparably a large structure, it is considered to be possible to recognize it even in a blurred image. Therefore we first obtain high intensity pixel clusters as initial candidate regions on the input image by using the image restoration technique. Then we adopt the scale-space filtering, in which we can trace edge elements from the lowest resolution to the highest one. Furthermore, we utilize Delaunay arcs in order to get a connected contour of the tumor. Finally, we show some experimental results to confirm its performance.

1 Introduction

The breast cancer has been examined by the physical examination now in Japan. This means the present examination of the breast cancer depends on human resources, i.e. experienced medical doctors, because it is difficult to detect the breast cancer if the surgeons are not skillful. In the present group examination, as the number of candidates who are detected to have tumor in the examination is very small, the surgeons usually waste a lot of time. Furthermore, the earlier breast cancer cannot be detected by the physical examination, because the tumor is too small to be picked up. From these reasons, the study of the automatic image processing of X-ray mammograms is strongly desired. The mammogram is the X-ray slide of vertically or horizontally pressed human breast. In this research area, the study of the improvement of the quality of X-ray mammograms has been widely made, because of low quality of the X-ray image due to low S/N (signal-noise ratio). For example, blurring noise processing in the real X-ray image and enhancement of the tumor shape in X-ray mammograms were studied [1]. For example, Jin et al studied the decision rule if the tumor given as the model pattern is benign or malignant from the shape of the contour [2]. But there is few study on automatic recognition of tumors from the real mammography. Using the technique in computational geometry, Aisaka et al. proposed the method [3]. In their study the border line of the stomach was detected

by using the Delaunay net, a method in computational geometry. In this paper, we extract a contour of the tumor that is considered as candidates of breast cancer from a digitized image of a film-mammogram by using the features of breast cancer.

In general, as X-ray images contain many kinds of tissues (dense breast, ducts, blood vessels etc.), other than objective tumors, and noise, it is impossible to extract a tumor contour just using simple edge extraction technique. So we adopt the technique of scale-space filtering, in which we can trace edges continuously from lower resolution images to higher ones. Furthermore we utilize the Delaunay net, a planer graph representing "neighboring" in order to connect edge segments on the higher resolution images.

2 Pictorial characteristics of tumors

The surgeons have used the features of the breast cancer, when they detect the breast cancer in the X-ray mammography. The information, which could be available for pattern recognition, is divided into two groups according to the type of lesions. One of the feature is the characteristics of tumors, where tumor shadow can be detected as high density region. If the shape of the tumor shadow is not regular, the tumor is considered malignant. In this case tumor usually has spicules. The other feature is the existence of calcification, where the calcium particles with high contrast can be seen. If there is a cluster of calcification in the tumor, it could form a malignant one. Therefore the existence of calcification can be utilized for the detection of breast cancer in early stage, which can not be inspected in a physical examination because of their too small size. As the tumor does not, however, have a particular shape or size, it is difficult to detect tumors by using the specified characteristics of their shapes. Therefore we should adopt simple and general features of tumors as follows:

· the contrast of the tumor shadow is comparatively high,
· the size of one is usually larger than other tissues.

In general smaller patterns can be eliminated by applying low-pass filter in terms of spatial frequency, but the contour of the tumor would be also smoothed. Therefore, in order to extract exact contours of tumors, we introduce the technique of scale-space filtering. It isn't, however, guaranteed that the extracted contour would become closed. For the purpose we use Delaunay net a planar graph studied in computational geometry. The details of these two techniques are discussed in the following section.

3 Recognition of tumor

3.1 Initial estimation for tumor region

Boundary detection is usually performed by computing zero-crossing in Laplacian-Gaussian filtered image. In this study we also adopt such a method to obtain a contour of shape. However it is not including the utilization of high level intensity in tumor contours, which is considered as one of the most dominant characteristics for tumors. Then we adopt Bayesian resotration method, one of the stochastic relaxations, which is robust, and was applied to the image restoration [6]. An example of restored image is shown in Fig.1 a). The tumor region is initially decided along the boundary shown in Fig.2 b).

 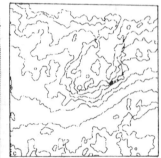

a) restorated image   b) boundaries
Fig.1 Tumor region estimation and its boundaries.

3.2 Coarse-to-fine strategy

When we see a noisy image, we first tend to pay attention to the global structure of the image, and gradually try to find the details. To realize such hierarchical viewpoint, the scale-space filtering has been studied [4]. We can see the image in the different resolution as such shown in Fig.2 by degradating the image by convoluting with two-dimensional Gaussian as follows.

$$F(\mathbf{x}, \sigma) = f(\mathbf{x}) \cdot g(\mathbf{x}, \sigma), \ \mathbf{x} \in R^2 \qquad (1)$$

The Fig.2 shows examples of traced zero-crossings in the scale-space for one-dimensional signal. In this method, it is considered that we can extract a fine contour of a tumor suppressing the influence of noise by tracing the zero-crossings starting from ones in the lower resolution image as shown as the dashed lines in Fig.2.

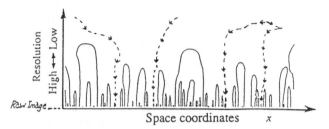

Fig.2 1D example of edge tracing in scale space.

3.3 Delaunay arc enclosure

The edge segments, which are considered as candidates for elements of the tumor contour, are generally disconnected due to noise etc. However, closed contours of tumors are desired as a result for further medical analysis. Therefore it is necessary to connect these edge segments. So we introduce a planar graph called the Delaunay net. The graph represents "neighboring" between points in terms of distance and their placement. It was applied to contour detection of stomach in X-ray image [3]. Let $P = \{p_1, p_2, \cdots, p_n\}$ denote a set of points in the plane,

$$V(p_i) = \{R^2 | d(p, p_i) < d(p, p_j) \text{ for } j = 1, \cdots, n; j \neq i\},$$

where $d(p, p_i)$ denotes Euclidean distance between the points p and p_i, and $V(p_i)$: Voronoi region. The Voronoi regions $V(p_1), V(p_2), \cdots, V(p_n)$ is called Voronoi diagram. The points in the set P is called the sites of Voronoi diagram. The Voronoi diagram is the regions, points in which are closer to the site than any other sites. Two sites of the Voronoi regions, which have a boundary edge in common, are regarded as "neighboring" ones. The graph drawn by connecting the sites corresponding to the neighboring Voronoi regions is called the Delaunay net.

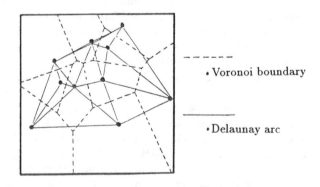

- - - - -   •Voronoi boundary

—————   •Delaunay arc

Fig.3 An example of Voronoi graph and Delaunay net.

3.4 Tumor contour exrtaction

Our purpose is to extract the exact contours of the tumors from an image of a film-mammogram. The procedures for the extraction are described as follows. First, the edge points are extracted in the tumor region on each resolution image obtained by scale-space filtering. Then the edge points are traced between different resolution images. Furthermore the divided contours are connected by using arcs of a Delaunay net in each stage of resolution. The details of each procedure are explained in the following sections.

3.4.1 Extracting edge points

As the first, the original image is filtered by Humming window function for anti-aliasing. Next, to get the image in the different resolution, we use the low-pass filters with the different cut-off frequency. Then the zero-crossings are extracted as edge points from each filtered image.

3.4.2 Tracing edge points

In order to relate the edge points on neighboring resolution images, we introduce two constraints: 1) the distance between two points is sufficiently short, 2) the directions of the gradient vectors at edge points are similar. In other words, edges are traced in scale space by using their similarity. The constraints are shown as follows.

(a) Distance between edge points
   Let (x_i, y_i) and (x_j, y_j) denote the coordinates of edge points in neighboring resolution images.

$$(x_j - x_i)^2 + (y_j - y_i)^2 \leq d \qquad (2)$$

(b) Difference between directions of gradient vectors
   For each edge, the direction of the gradient vector is computed. Let θ_i and θ_j denote the directions in neighboring resolution images.

$$|\theta_j - \theta_i| \leq \phi \qquad (3)$$

The edge points which hold the constraints (2) and (3) are regarded as candidates of corresponding ones.

3.4.3 Connecting edge segments

The Delaunay net is used to connect the divided contours as explained before. The net is getting more popular to be applied to image processing [5]. We first make a Voronoi diagram from a set of end points of the edge segments, then draw a Delaunay net, a dual graph, from it. We select the paths, which connect the end points each other, from a set of arcs in this net. The process is as follows:

> First, the end points of edge segments, which should be connected, are extracted by eliminating connecting edge points between them based on eight neighborhood connectivity of pixels. We select pairs of end points whose directions face each other within a certain angle. If the candidates of opponents are two or more, we choose the shortest arc between two points(Fig.4).

Thus we select the Delaunay arc between the end points. Above operation is performed in each resolution image.

In general, we should consider all the combination of points, i.e. arcs of a complete graph, as a connecting arc, however the operation enables us to reduce the number of arcs to be involved using the concept of neighborhood.

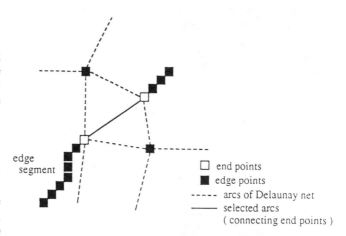

Fig.4 Edge segments and Delaunay arcs.

4 Experimental results

The real film-mammogram is used in the experiment. The input image is scanned from a copy of a simple film-mammogram, and consists of 256x256 pixels with 256 levels of intensity. Several images filtered by using Gaussian convolution with different variances are shown in Fig.6. The extracted contour in each resolution is shown in Fig.6. In this results each extracted contour is connected, however it does not have a closed form. This is considered why the rules for tracing and connecting edges work so locally that the edge segments involve noise, or miss edges to be included. In order to get a complete contour, further investigation on diagnostic heuristics by surgeons is required.

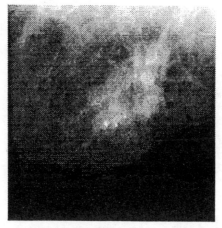

Fig.5 Input mammogram image.

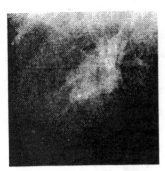

raw image (original input)

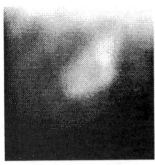

$f_c = 6.5$ Hz

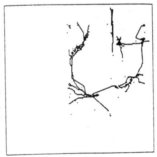

$f_c = 2.5$ Hz

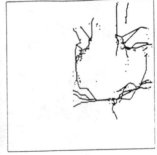

$f_c = 1.8$ Hz

a) Filtered Images　　b) Extracted tumor contours
Fig.6 Experimental results.

5　Concluding remarks

In this paper, we proposed a method to extract contours of the tumors from X-ray mammography by using scale-space filtering, and Delaunay net. The edge points are traced from the lowest resolution image to highest, where edge segments are connected by adding arcs out of Delaunay net. However as these processes are performed locally, more global constraints should be introduced in order to obtain closed contours. Moreover, in earlier stage of cancer, there often exists microcalcification only. For further study it is necessary to consider the utilization of analysis of micro-calcification.

Acknowledgement

The approach presented in this paper is studied along the project with Dr.Ohsaki and Dr.Tanaka of Department of Surgery, and Department of Biostatistics, Research Institute for Nuclear Medicine and Biology, Hiroshima University respectively. The authors wish to express their appreciations for their suggestions and encouragements.

References

[1] A.P.Dhawan et al, "Enhancement of Mammographic Features by Optimal Adaptive Neighborhood Image Processing," IEEE Transactionson Medical Imaging, Vol.MI-5, No.1 (1986).

[2] Jin H.R., and H.Kobatake, "Feature Extraction of Cancer by Morphology Analysis," Proc. of SICE'90, Tokyo, pp.179-180 (1990) in Japanese.

[3] K.Aisaka et al, "Finding Stomach in X-ray Film Using Computational Geometry," Proc. of IAPR Workshop on Machine Vision Application, pp.85-88 (1990).

[4] A.P.Witkin, "Scale-space Filtering," Proc. of IJCAI'83, pp.1019-1022 (1983).

[5] K.Sugihara, "Computational-Geometry Method and Image Analysis —Concentrated on Applications of Voronoi Diagrams—," Trans. of Information Processing Society of Japan, Vol.30, No.9, pp.44-52 (1989).

[6] S.Geman and D.Geman, "Stochastic Relaxation, Gibbs Distributions, and the Bayesian Restoration of Images," IEEE Trans. on Pattern Anal. and Machine Intell., Vol.PAMI-6, pp.721-741 (1984).

[7] Y.Takeshita, M.Terauchi, K.Onaga, "A Method for Tumor Shape Extraction from X-ray Mammography," Proc. of IAPR workshop on Machine Vision App., pp.133-136 (1992).

Author Index

John G. Ackenhusen

John G. Ackenhusen (BS '75, BSE '75, MS '76, MSE '76, PhD '77, all from the University of Michigan) has performed and directed research and development in real-time signal processing since 1978, both designing computer architectures to accommodate signal processing algorithms and developing algorithms to fit existing computer architectures. He is Manager of the Image and Signal Processing Laboratory at the Environmental Research Institute of Michigan (ERIM), Ann Arbor, MI, USA.

At AT&T Bell Laboratories, he was Head of the Signal Processing Systems Design Department (1988-91), Supervisor of the Speech Recognition Group (1981-88), and Member of Technical Staff (1978-81). He developed and patented an early special-purpose computer for speech recognition and instigated its evolution into an AT&T product. More recently at AT&T Bell Labs, he led the systems engineering department that defined and directed the design of the U.S. Navy's Enhanced Modular Signal Processor (EMSP).

Dr. Ackenhusen joined the Environmental Research Institute of Michigan (ERIM) in 1991 to lead its Image and Signal Processing Laboratory, which is devoted to computers and algorithms for real-time image processing, with capabilities in architecture, hardware, software, and multichip module electronic packaging. He served as manager of the Advanced Target Cueing and Recognition Engine (ATCURE), a program that delivered to the U.S. Government a heterogeneous-architecture computer for image recognition that included a wafer-scale multichip module image processing pipeline.

Dr. Ackenhusen was elected a Fellow of the Institute of Electrical and Electronics Engineers (IEEE) for his contributions in real-time signal processing. He served as Technical Program Chair in 1988 for the IEEE International Conference on Acoustics, Speech, and Signal Processing (ICASSP-88), chaired the IEEE Signal Processing Society Conference Board (1986-89), and serves as Vice Chair of ICASSP-95. Dr. Ackenhusen served as President of the IEEE Signal Processing Society (1990-91) and received the 1992 IEEE Signal Processing Society Meritorious Service Award for service as Society President and sustained contributions to the conference operations of the Society. He was elected to the IEEE Board of Directors for 1994-95, where he leads its Signals and Applications Division.